DESIGN OF MACHINERY

AN INTRODUCTION TO THE SYNTHESIS AND
ANALYSIS OF MECHANISMS
AND MACHINES

Fifth Edition

McGraw-Hill Series in Mechanical Engineering

DESIGN OF MACHINERY

AN INTRODUCTION TO THE SYNTHESIS AND ANALYSIS OF MECHANISMS AND MACHINES

Fifth Edition

Robert L. Norton

Worcester Polytechnic Institute

Worcester, Massachusetts

Connect
Learn
Succeed™

DESIGN OF MACHINERY: An Introduction to the Synthesis and Analysis of
Mechanisms and Machines, Fifth Edition

Published by McGraw-Hill, a business unit of The McGraw-Hill Companies, Inc., 1221 Avenue of the Americas, New York, NY, 10020. Copyright © 2012 by The McGraw-Hill Companies, Inc. All rights reserved. Previous editions © 2008, 2004, 2001, 1999, and 1992. No part of this publication may be reproduced or distributed in any form or by any means, or stored in a database or retrieval system, without the prior written consent of The McGraw-Hill Companies, Inc., including, but not limited to, in any network or other electronic storage or transmission, or broadcast for distance learning. Some ancillaries, including electronic and print components, may not be available to customers outside the United States.

This book is printed on acid-free paper containing 10% postconsumer waste.

8 9 LCR 21 20 19 18 17

ISBN 978-0-07-352935-6
MHID 0-07-352935-4

Vice President and Editor-in-Chief: *Marty Lange*
Vice President EDO/Central Publishing Services: *Kimberly Meriwether David*
Publisher: *Raghotham Srinivasan*
Sponsoring Editor: *Bill Stenquist*
Marketing Manager: *Curt Reynolds*
Developmental Editor: *Darlene Schueller*
Senior Project Manager: *Lisa Bruflodt*
Design Coordinator: *Brenda A. Rolwes*
Cover Designer: *Studio Montage, St. Louis, MO*
Cover Image: *Chrysler Group LLC*
Buyer: *Susan K. Culbertson*
Media Project Manager: *Balaji Sundararaman*
Printer: *LSC Communications*

All credits appearing on page or at the end of the book are considered to be an extension of the copyright page.

All text, drawings, and equations in this book were prepared and typeset electronically, by the author, on a *Macintosh*® computer using *Freehand*®, *MathType*®, and *InDesign*® desktop publishing software. The body text is set in Times Roman, and headings set in Avant Garde. Color-separated printing plates were made on a direct-to-plate laser typesetter from the author's files. All *clip art* illustrations are courtesy of *Dubl-Click Software Inc.*, 22521 Styles St., Woodland Hills CA 91367, reprinted from their *Industrial Revolution* and *Old Earth Almanac* series with their permission (and with the author's thanks). The cover photo, courtesy of Chrysler Group LLC, is a cut-away of their 1.4-liter, in-line four-cylinder engine that uses Fiat's Multiair technology to significantly reduce emissions while improving fuel economy and power delivery.

Library of Congress Cataloging-in-Publication Data

Norton, Robert L.
 Design of machinery : an introduction to the synthesis and analysis of mechanisms and machines / Robert L. Norton. -- 5th ed.
    p. cm. -- (Mcgraw-Hill series in mechanical engineering)
 Includes bibliographical references and index.
 1. Machine design. 2. Machinery, Kinematics of. 3. Machinery, Dynamics of. I. Title.
 TJ175.N58 2012
 621.8'15--dc22   2010044542

Book's web page is http://www.mhhe.com/norton.   Author's web page is http://www.designofmachinery.com

ABOUT THE AUTHOR

Robert L. Norton earned undergraduate degrees in both mechanical engineering and industrial technology at Northeastern University and an MS in engineering design at Tufts University. He is a registered professional engineer in Massachusetts. He has extensive industrial experience in engineering design and manufacturing and many years' experience teaching mechanical engineering, engineering design, computer science, and related subjects at Northeastern University, Tufts University, and Worcester Polytechnic Institute.

At Polaroid Corporation for 10 years, he designed cameras, related mechanisms, and high-speed automated machinery. He spent three years at Jet Spray Cooler Inc., designing food-handling machinery and products. For five years he helped develop artificial-heart and noninvasive assisted-circulation (counterpulsation) devices at the Tufts New England Medical Center and Boston City Hospital. Since leaving industry to join academia, he has continued as an independent consultant on engineering projects ranging from disposable medical products to high-speed production machinery. He holds 13 U.S. patents.

Norton has been on the faculty of Worcester Polytechnic Institute since 1981 and is currently the Milton Prince Higgins II Distinguished Professor in the mechanical engineering department, head of the design group in that department, and the director of the Gillette Project Center at WPI. He teaches undergraduate and graduate courses in mechanical engineering with emphasis on design, kinematics, vibrations, and dynamics of machinery.

He is the author of numerous technical papers and journal articles covering kinematics, dynamics of machinery, cam design and manufacturing, computers in education, and engineering education and of the texts *Machine Design: An Integrated Approach, 4ed* and the *Cam Design and Manufacturing Handbook, 2ed*. He is a Fellow and life member of the American Society of Mechanical Engineers and a member of the Society of Automotive Engineers. In 2007, he was chosen as *U. S. Professor of the Year* for the State of Massachusetts by the *Council for the Advancement and Support of Education (CASE)* and the *Carnegie Foundation for the Advancement of Teaching*, who jointly present the only national awards for teaching excellence given in the United States of America.

This book is dedicated to the memory of my father,

Harry J. Norton, Sr.

who sparked a young boy's interest in engineering;

to the memory of my mother,

Kathryn W. Norton

who made it all possible;

to my wife,

Nancy Norton

who provides unflagging patience and support;

and to my children,

Robert, Mary, and Thomas,

who make it all worthwhile.

CONTENTS

PREFACE
to the Fifth Edition

How many times do you have to do this to get it right, Norton?
THE EDITORS ASKED, RE. THE FOURTH EDITION

I plead the fifth.
THE AUTHOR

The fifth edition is an evolutionary improvement over the fourth and earlier editions. See the updated *Preface to the First Edition* (overleaf) for more detailed information on the book's purpose and organization. The principal changes in this edition are:

- A new chapter (16) has been added on cam-driven and servo-driven linkages based on the author's recent consulting experience that indicates increased use of servomotors to drive mechanisms in machinery.

- Over 50 new problem assignments have been added. All of the problem figures are included as PDF files on the DVD so that students can easily print hard copies on which to work the solutions.

- Several of the author-written programs that come with the book have been completely rewritten to improve their interface and usability, and they are now compatible with the latest operating systems and computers. The programs FOURBAR, FIVEBAR, SIXBAR, and SLIDER have been combined in a new program called LINKAGES that does everything those programs collectively did and has some new features added. Program DYNACAM also has been completely rewritten and is much improved. Programs ENGINE and MATRIX are unchanged. These computer programs undergo frequent revision to add features and enhancements. Professors who adopt the book for a course may register to download the latest student versions of these programs from: http://www.designofmachinery.com/registered/professor.html.

- The Working Model program that was provided with earler editions was not a full-featured version of the program. It did not allow the user to save or print a model. The version of Working Model available with this text is a **fully capable version** but is provided only by means of a one-time download from a website and will run for one year from the time of installation. A password is provided with this text to enable its download.

- Many small improvements have been made to the discussion of a variety of topics in many chapters, based largely on user feedback, and all known errors have been corrected.

The extensive DVD content that was introduced in the Fourth Edition is included with the Fifth Edition as well. These include:

- The entire Hrones and Nelson Atlas of Coupler Curves and the Zhang Atlas of Geared Fivebar Coupler Curves are provided as PDFs on the DVD.

- Eighteen instructional videos that provide several lecture/demonstration tutorials on subjects such as position synthesis, coupler curves and cognates, the Grashof condition, etc. Others show real machinery that uses linkages, cams and gears for such tasks as spring manufacture, bottle printing, and pick-and-place mechanisms for automated assembly.

- Wang's Mechanism Simulation in a Multimedia Environment that contains over 100 Working Model (WM) files based on the book's figures and 19 Matlab® models for kinematic analysis and animation is included.

- Videos of two "virtual laboratories" that replicate labs created by the author at WPI are provided. These include demonstrations of the machines and spreadsheet files of the acceleration and force data taken during the experiments. The intent is to allow students at other schools to do these exercises as virtual laboratories.

ACKNOWLEDGMENTS   The sources of photographs and other nonoriginal art used in the text are acknowledged in the captions and opposite the title page, but the author would also like to express his thanks for the cooperation of all those individuals and companies who generously made these items available. The author is indebted to, and would like to thank, a number of users who kindly notified him of errors or suggested improvements in all editions since the first. These include: Professors *Chad O'Neal* of Louisiana Tech, *Bram Demeulenaere* of Leuven University, *Eben Cobb* of WPI, *Diego Galuzzi* of University of Buenos Aires, *John R. Hall* of WPI, *Shafik Iskander* of U. Tennessee, *Richard Jakubek* of RPI, *Cheong Gill-Jeong* of Wonkwang University, Korea, *Swami Karunamoorthy* of St. Louis University, *Pierre Larochelle* of Florida Tech, *Scott Openshaw* of Iowa State, *Francis H. Raven* of Notre Dame, *Arnold E. Sikkema* of Dordt College, and *Donald A. Smith* of U. Wyoming.

Professors *M. R. Corley* of Louisiana Tech, *R. Devashier* of U. Evansville, *K. Gupta* of U. Illinois-Chicago, *M. Keefe* of U. Delaware, *J. Steffen* of Valparaiso University, *D. Walcerz* of York College, and *L. Wells* of U. Texas at Tyler provided useful suggestions or corrections. Professors *L. L. Howell* of BYU, and *G. K. Ananthasuresh* of U. Penn supplied photographs of compliant mechanisms. Professor *C. Furlong* of WPI generously supplied MEMS photos and information. Special thanks to *James Cormier* and *David Taranto* of WPI's Academic Technology Center for their help in creating the videos on the DVD. Reviewers of this edition were: Professors *J. Dave* of University of Cincinnati, *K. Fisher* of Montana State, *S. Gellin* of Buffalo State College, *D. Halsmer* of Oral Roberts University, *L. Howell* of Brigham Young University, *R. A. Lucas* of Lehigh University, *J. Mirth* of U. Wisconsin-Platteville, *C. Rubin* of Vanderbilt University, and *T. J. Thompson* of Cedarville University.

The author would like to express his appreciation to Professor *Sid Wang* of NCAT and his students for their efforts in creating the *Working Model* and *Matlab* files on the DVD. Professor *Thomas A. Cook*, Mercer University (Emeritus), provided most of the new problem sets as well as their solutions in his impressive and voluminous solutions manual and its accompanying *Mathcad*® solution files. The author is most grateful for Dr. Cook's valuable contributions.

Robert L. Norton
Mattapoisett, Mass.
August, 2010

If you find any errors or have comments or suggestions for improvement, please email the author at norton@wpi.edu. Errata as discovered, and other book information, will be posted on the author's web site at http://www.designofmachinery.com.

PREFACE
to the First Edition

When I hear, I forget
When I see, I remember
When I do, I understand
ANCIENT CHINESE PROVERB

This text is intended for the kinematics and dynamics of machinery topics which are often given as a single course, or two-course sequence, in the junior year of most mechanical engineering programs. The usual prerequisites are first courses in statics, dynamics, and calculus. Usually, the first semester, or portion, is devoted to kinematics and the second to dynamics of machinery. These courses are ideal vehicles for introducing the mechanical engineering student to the process of design, since mechanisms tend to be intuitive for the typical mechanical engineering student to visualize and create.

While this text attempts to be thorough and complete on the topics of analysis, it also emphasizes the synthesis and design aspects of the subject to a greater degree than most texts in print on these subjects. Also, it emphasizes the use of computer-aided engineering as an approach to the design and analysis of this class of problems by providing software that can enhance student understanding. While the mathematical level of this text is aimed at second- or third-year university students, it is presented *de novo* and should be understandable to the technical school student as well.

Part I of this text is suitable for a one-semester or one-term course in kinematics. Part II is suitable for a one-semester or one-term course in dynamics of machinery. Alternatively, both topic areas can be covered in one semester with less emphasis on some of the topics covered in the text.

The writing and style of presentation in the text are designed to be clear, informal, and easy to read. Many example problems and solution techniques are presented and spelled out in detail, both verbally and graphically. All the illustrations are done with computer-drawing or drafting programs. Some scanned photographic images are also included. The entire text, including equations and artwork, is printed directly from the author's computer disk by laser typesetting for maximum clarity and quality. Many suggested readings are provided in the bibliography. Short problems and, where appropriate, many longer, unstructured design project assignments are provided at the ends of chapters. These projects provide an opportunity for the students *to do and understand.*

The author's approach to these courses and this text is based on over 40 years' experience in mechanical engineering design, both in industry and as a consultant. He has taught these subjects since 1967, both in evening school to practicing engineers and in day school to younger students. His approach to the course has evolved a great deal in that time, from a traditional approach, emphasizing graphical analysis of many structured problems, through emphasis on algebraic methods as computers became available, through requiring students to write their own computer programs, to the current state described above.

The one constant throughout has been the attempt to convey the art of the design process to the students in order to prepare them to cope with *real* engineering problems in practice. Thus, the author has always promoted design within these courses. Only recently, however, has technology provided a means to more effectively accomplish this goal, in the form of the graphics microcomputer. This text attempts to be an improvement over those currently available by providing up-to-date methods and techniques for analysis and synthesis that take full advantage of the graphics microcomputer, and by emphasizing design as well as analysis. The text also provides a more complete, modern, and thorough treatment of cam design than any existing texts in print on the subject.

The author has written four interactive, student-friendly computer programs for the design and analysis of mechanisms and machines.[*] These programs are designed to enhance the student's understanding of the basic concepts in these courses while simultaneously allowing more comprehensive and realistic problem and project assignments to be done in the limited time available than could ever be done with manual solution techniques, whether graphical or algebraic. Unstructured, realistic design problems which have many valid solutions are assigned. Synthesis and analysis are equally emphasized. The analysis methods presented are up to date, using vector equations and matrix techniques wherever applicable. Manual graphical analysis methods are deemphasized. The graphics output from the computer programs allows the student to see the results of variation of parameters rapidly and accurately and reinforces learning.

These computer programs are distributed on DVD with this book, which also contains instructions for their use on any Windows NT/2000/XP/Vista/Windows7 capable computer. Program LINKAGES analyzes the kinematics and dynamics of fourbar, geared fivebar, sixbar, and fourbar slider linkages. It also will synthesize fourbar linkages for two and three positions. Program DYNACAM allows the design and dynamic analysis of cam-follower systems. Program ENGINE analyzes the slider-crank linkage as used in the internal combustion engine and provides a complete dynamic analysis of single- and multicylinder engine inline, V, and W configurations, allowing the mechanical dynamic design of engines to be done. Program MATRIX is a general-purpose linear equation system solver. These are student editions of professional programs that are written by the author and that he provides to companies the world over.

All these programs, except MATRIX, provide dynamic, graphical animation of the designed devices. The reader is strongly urged to make use of these programs in order to investigate the results of variation of parameters in these kinematic devices. The programs are designed to enhance and augment the text rather than be a substitute for it. The converse is also true. Many solutions to the book's examples and to the problem sets are provided on the DVD as files to be opened in these programs. Most of these solutions can be animated on the computer screen for a better demonstration of the concept than is

* One of these four programs replaces four of the original set of seven programs that came with earlier editions of the book. These were called FOURBAR, FIVEBAR, SIXBAR, and SLIDER. All four have been combined into the new program LINKAGES in the 5th edition.

possible on the printed page. The instructor and students are both encouraged to take advantage of the computer programs provided. Instructions for their use are in Appendix A.

The author's intention is that synthesis topics be introduced first to allow the students to work on some simple design tasks early in the term while still mastering the analysis topics. Though this is not the "traditional" approach to the teaching of this material, the author believes that it is a superior method to that of initial concentration on detailed analysis of mechanisms for which the student has no concept of origin or purpose.

Chapters 1 and 2 are introductory. Those instructors wishing to teach analysis before synthesis can leave Chapters 3 and 5 on linkage synthesis for later consumption. Chapters 4, 6, and 7 on position, velocity, and acceleration analysis are sequential and build upon each other. In fact, some of the problem sets are common among these three chapters so that students can use their position solutions to find velocities and then later use both to find the accelerations in the same linkages. Chapter 8 on cams is more extensive and complete than that of other kinematics texts and takes a design approach. Chapter 9 on gear trains is introductory. The dynamic force treatment in Part II uses matrix methods for the solution of the system simultaneous equations. Graphical force analysis is not emphasized. Chapter 10 presents an introduction to dynamic systems modeling. Chapter 11 deals with force analysis of linkages. Balancing of rotating machinery and linkages is covered in Chapter 12. Chapters 13 and 14 use the internal combustion engine as an example to pull together many dynamic concepts in a design context. Chapter 15 presents an introduction to dynamic systems modeling and uses the cam-follower system as the example. Chapters 3, 8, 11, 13, and 14 provide open-ended project problems as well as structured problem sets. The assignment and execution of unstructured project problems can greatly enhance the student's understanding of the concepts as described by the proverb in the epigraph to this preface.

ACKNOWLEDGMENTS   The sources of photographs and other nonoriginal art used in the text are acknowledged in the captions and opposite the title page, but the author would also like to express his thanks for the cooperation of all those individuals and companies who generously made these items available. The author would also like to thank those who reviewed various sections of the first edition of the text and who made many useful suggestions for improvement. Mr. John Titus of the University of Minnesota reviewed Chapter 5 on analytical synthesis and Mr. Dennis Klipp of Klipp Engineering, Waterville, Maine, reviewed Chapter 8 on cam design. Professor William J. Crochetiere and Mr. Homer Eckhardt of Tufts University, Medford, MA., reviewed Chapter 15. Mr. Eckhardt and Professor Crochetiere of Tufts, and Professor Charles Warren of the University of Alabama taught from and reviewed Part I. Professor Holly K. Ault of Worcester Polytechnic Institute thoroughly reviewed the entire text while teaching from the prepublication, class-test versions of the complete book. Professor Michael Keefe of the University of Delaware provided many helpful comments.   Sincere thanks also go to the large number of undergraduate students and graduate teaching assistants who caught many typos and errors in the text and in the programs while using prepublication versions. Since the book's first printing, Profs. D. Cronin, K. Gupta and P. Jensen and Mr. R. Jantz have written to point out errors or make suggestions that I have incorporated and for which I thank them. The author takes full responsibility for any errors that may remain and invites from all readers their criticisms, suggestions for improvement, and identification of errors in the text or programs, so that both can be improved in future versions. Contact norton@wpi.edu.

Robert L. Norton
Mattapoisett, Mass.
August, 1991

PART

KINEMATICS OF
MECHANISMS

Chapter **1**

INTRODUCTION

Inspiration most often strikes
those who are hard at work
ANONYMOUS

1.0 PURPOSE

In this text we will explore the topics of **kinematics** and **dynamics of machinery** in respect to the **synthesis of mechanisms** in order to accomplish desired motions or tasks, and also the **analysis of mechanisms** in order to determine their rigid-body dynamic behavior. These topics are fundamental to the broader subject of **machine design**. On the premise that we cannot analyze anything until it has been synthesized into existence, we will first explore the topic of **synthesis of mechanisms**. Then we will investigate techniques of **analysis of mechanisms**. All this will be directed toward developing your ability to design viable mechanism solutions to real, unstructured engineering problems by using a **design process**. We will begin with careful definitions of the terms used in these topics.

1.1 KINEMATICS AND KINETICS

KINEMATICS   *The study of motion without regard to forces.*

KINETICS   *The study of forces on systems in motion.*

These two concepts are really *not* physically separable. We arbitrarily separate them for instructional reasons in engineering education. It is also valid in engineering design practice to first consider the desired kinematic motions and their consequences, and then subsequently investigate the kinetic forces associated with those motions. The student should realize that the division between **kinematics** and **kinetics** is quite arbitrary and is done largely for convenience. One cannot design most dynamic mechanical systems without taking both topics into thorough consideration. It is quite logical to consider them in the order listed since, from Newton's second law, $\mathbf{F} = m\mathbf{a}$, one typically needs to know the **accelerations** ($\mathbf{a}$) in order to compute the dynamic **forces** ($\mathbf{F}$) due to the motion of the system's **mass** (*m*). There are also many situations in which the applied forces are known and the resultant accelerations are to be found.

One principal aim of **kinematics** is to create (design) the desired motions of the subject mechanical parts and then mathematically compute the positions, velocities, and accelerations that those motions will create on the parts. Since, for most earthbound mechanical systems, the mass remains essentially constant with time, defining the accelerations as a function of time then also defines the dynamic forces as a function of time. **Stresses**, in turn, will be a function of both applied and inertial (*m***a**) forces. Since engineering design is charged with creating systems that will not fail during their expected service life, the goal is to keep stresses within acceptable limits for the materials chosen and the environmental conditions encountered. This obviously requires that all system forces be defined and kept within desired limits. In machinery that moves (the only interesting kind), the largest forces encountered are often those due to the dynamics of the machine itself. These dynamic forces are proportional to acceleration, which brings us back to kinematics, the foundation of mechanical design. Very basic and early decisions in the design process involving kinematic principles can be crucial to the success of any mechanical design. A design that has poor kinematics will prove troublesome and perform badly.

1.2   MECHANISMS AND MACHINES

A mechanism

A **mechanism** is a device that transforms motion to some desirable pattern and typically develops very low forces and transmits little power. Hunt[13] defines a mechanism as "a means of *transmitting, controlling, or constraining relative movement*." A **machine** typically contains mechanisms that are designed to provide significant forces and transmit significant power.[1] Some examples of common mechanisms are a pencil sharpener, a camera shutter, an analog clock, a folding chair, an adjustable desk lamp, and an umbrella. Some examples of machines that possess motions similar to the mechanisms listed above are a food blender, a bank vault door, an automobile transmission, a bulldozer, a robot, and an amusement park ride. There is no clear-cut dividing line between mechanisms and machines. They differ in degree rather than in kind. If the forces or energy levels within the device are significant, it is considered a machine; if not, it is considered a mechanism. A useful working **definition of a mechanism** is *A system of elements arranged to transmit* **motion** *in a predetermined fashion*. This can be converted to a definition of a **machine** by adding the words **and energy** after **motion**.

A machine

Mechanisms, if lightly loaded and run at slow speeds, can sometimes be treated strictly as kinematic devices; that is, they can be analyzed kinematically without regard to forces. Machines (and mechanisms running at higher speeds), on the other hand, must first be treated as mechanisms, a kinematic analysis of their velocities and accelerations must be done, and then they must be subsequently analyzed as dynamic systems in which their static and dynamic forces due to those accelerations are analyzed using the principles of kinetics. **Part I** of this text deals with **Kinematics of Mechanisms**, and **Part II** with **Dynamics of Machinery**. The techniques of mechanism synthesis presented in Part I are applicable to the design of both mechanisms and machines, since in each case some collection of movable members must be created to provide and control the desired motions and geometry.

1.3   A BRIEF HISTORY OF KINEMATICS

Machines and mechanisms have been devised by people since the dawn of history. The ancient Egyptians devised primitive machines to accomplish the building of the pyramids and other monuments. Though the wheel and pulley (on an axle) were not known to the Old Kingdom Egyptians, they made use of the lever, the inclined plane (or wedge), and probably the log roller. The origin of the wheel and axle is not definitively known. Its first appearance seems to have been in Mesopotamia about 3000 to 4000 B.C.

A great deal of design effort was spent from early times on the problem of time-keeping as more sophisticated clockworks were devised. Much early machine design was directed toward military applications (catapults, wall scaling apparatus, etc.). The term **civil engineering** was later coined to differentiate civilian from military applications of technology. **Mechanical engineering** had its beginnings in machine design as the inventions of the industrial revolution required more complicated and sophisticated solutions to motion control problems. **James Watt** (1736-1819) probably deserves the title of first kinematician for his synthesis of a straight-line linkage (see Figure 3-29a on p. 140) to guide the very long stroke pistons in the then new steam engines. Since the planer was yet to be invented (in 1817), no means then existed to machine a long, straight guide to serve as a crosshead in the steam engine. Watt was certainly the first on record to recognize the value of the motions of the coupler link in the fourbar linkage. **Oliver Evans** (1755-1819), an early American inventor, also designed a straight-line linkage for a steam engine. **Euler** (1707-1783) was a contemporary of Watt, though they apparently never met. Euler presented an analytical treatment of mechanisms in his *Mechanica Sive Motus Scienta Analytice Exposita* (1736-1742), which included the concept that planar motion is composed of two independent components, namely, translation of a point and rotation of the body about that point. Euler also suggested the separation of the problem of dynamic analysis into the "geometrical" and the "mechanical" in order to simplify the determination of the system's dynamics. Two of his contemporaries, **d'Alembert** and **Kant**, also proposed similar ideas. This is the origin of our division of the topic into kinematics and kinetics as described on p. 3.

In the early 1800s, L'Ecole Polytechnic in Paris, France, was the repository of engineering expertise. **Lagrange** and **Fourier** were among its faculty. One of its founders was **Gaspard Monge** (1746-1818), inventor of descriptive geometry (which incidentally was kept as a military secret by the French government for 30 years because of its value in planning fortifications). Monge created a course in elements of machines and set about the task of classifying all mechanisms and machines known to mankind! His colleague, **Hachette**, completed the work in 1806 and published it as what was probably the first mechanism text in 1811. **Andre Marie Ampere** (1775-1836), also a professor at L'Ecole Polytechnic, set about the formidable task of classifying "all human knowledge." In his *Essai sur la Philosophie des Sciences*, he was the first to use the term **cinematique**, from the Greek word for motion,[*] to describe the study of motion without regard to forces, and suggested that "this science ought to include all that can be said with respect to motion in its different kinds, independently of the forces by which it is produced." His term was later anglicized to *kinematics* and germanized to *kinematik*.

Robert Willis (1800-1875) wrote the text *Principles of Mechanism* in 1841 while a professor of natural philosophy at the University of Cambridge, England. He attempted to systematize the task of mechanism synthesis. He counted five ways of obtaining

[*] Ampere is quoted as writing "(The science of mechanisms) must therefore not define a machine, as has usually been done, as an instrument by the help of which the direction and intensity of a given *force* can be altered, but as an instrument by the help of which the direction and *velocity* of a given motion can be altered. To this science . . . I have given the name Kinematics from μ —motion." in Maunder, L. (1979). "Theory and Practice." *Proc. 5th World Cong. on Theory of Mechanisms and Machines*, Montreal, p. 1.

relative motion between input and output links: rolling contact, sliding contact, linkages, wrapping connectors (belts, chains), and tackle (rope or chain hoists). **Franz Reuleaux** (1829-1905) published *Theoretische Kinematik* in 1875. Many of his ideas are still current and useful. **Alexander Kennedy** (1847-1928) translated Reuleaux into English in 1876. This text became the foundation of modern kinematics and is still in print! (See bibliography at end of chapter.) He provided us with the concept of a kinematic pair (joint), whose shape and interaction define the type of motion transmitted between elements in the mechanism. Reuleaux defined six basic mechanical components: the link, the wheel, the cam, the screw, the ratchet, and the belt. He also defined "higher" and "lower" pairs, higher having line or point contact (as in a roller or ball bearing) and lower having surface contact (as in pin joints). Reuleaux is generally considered the father of modern kinematics and is responsible for the symbolic notation of skeletal, generic linkages used in all modern kinematics texts.

In the 20th century, prior to World War II, most theoretical work in kinematics was done in Europe, especially in Germany. Few research results were available in English. In the United States, kinematics was largely ignored until the 1940s when **A. E. R. de Jonge** wrote *What Is Wrong with 'Kinematics' and 'Mechanisms'?*[2] which called upon the U.S. mechanical engineering education establishment to pay attention to the European accomplishments in this field. Since then, much new work has been done, especially in kinematic synthesis, by American and European engineers and researchers such as **J. Denavit, A. Erdman, F. Freudenstein, A. S. Hall, R. Hartenberg, R. Kaufman, B. Roth, G. Sandor**, and **A. Soni** (all of the United States) and **K. Hain** (of Germany). Since the fall of the "iron curtain" much original work done by Soviet Russian kinematicians has become available in the United States, such as that by **Artobolevsky**[3] Many U.S. researchers have applied the computer to solve previously intractable problems, of both analysis and synthesis, making practical use of many of the theories of their predecessors[4] This text will make much use of the availability of computers to allow more efficient analysis and synthesis of solutions to machine design problems. Several computer programs are included with this book for your use.

1.4   APPLICATIONS OF KINEMATICS

One of the first tasks in solving any machine design problem is to determine the kinematic configuration(s) needed to provide the desired motions. Force and stress analyses typically cannot be done until the kinematic issues have been resolved. This text addresses the design of kinematic devices such as linkages, cams, and gears. Each of these terms will be fully defined in succeeding chapters, but it may be useful to show some examples of kinematic applications in this introductory chapter. You probably have used many of these systems without giving any thought to their kinematics.

Virtually any machine or device that moves contains one or more kinematic elements such as links, cams, gears, belts, chains. Your bicycle is a simple example of a kinematic system that contains a chain drive to provide torque multiplication and simple cable-operated linkages for braking. An automobile contains many more examples of kinematic devices. Its steering system, wheel suspensions, and piston engine all contain linkages; the engine's valves are opened by cams; and the transmission is full of gears. Even the windshield wipers are linkage-driven. Figure 1-1a shows a spatial linkage used to control the rear wheel movement over bumps of a modern automobile.

(a) Spatial linkage rear suspension
Courtesy of Daimler Benz Co.

(b) Utility tractor with backhoe
Courtesy of John Deere Co.

(c) Linkage-driven exercise mechanism
Courtesy of ICON Health & Fitness, Inc.

FIGURE 1-1

Examples of kinematic devices in general use

Construction equipment such as tractors, cranes, and backhoes all use linkages extensively in their design. Figure 1-1b shows a small backhoe that is a linkage driven by hydraulic cylinders. Another application using linkages is that of exercise equipment as shown in Figure 1-1c. The examples in Figure 1-1 are all of consumer goods that you may encounter in your daily travels. Many other kinematic examples occur in the realm of producer goods—machines used to make the many consumer products that we use. You are less likely to encounter these outside of a factory environment. Once you become familiar with the terms and principles of kinematics, you will no longer be able to look at any machine or product without seeing its kinematic aspects.

1.5 THE DESIGN PROCESS

Design, Invention, Creativity

These are all familiar terms but may mean different things to different people. These terms can encompass a wide range of activities from styling the newest look in clothing, to creating impressive architecture, to engineering a machine for the manufacture of facial tissues. **Engineering design**, which we are concerned with here, embodies all three of these activities as well as many others. The word **design** is derived from the Latin **designare**, which means *"to designate, or mark out."* Webster's gives several definitions, the most applicable being *"to outline, plot, or plan, as action or work. . . . to conceive, invent – contrive."* **Engineering design** has been defined as ". . . *the process of applying the various techniques and scientific principles for the purpose of defining a device, a process or a system in sufficient detail to permit its realization . . . Design may be simple or enormously complex, easy or difficult, mathematical or nonmathematical; it may involve a trivial problem or one of great importance."* **Design** is a universal constituent of engineering practice. But the complexity of engineering subjects usually

1

TABLE 1-1
A Design Process

1 Identification of
   Need

2 Background Re-
   search

3 Goal Statement

4 Performance
   Specifications

5 Ideation and Inven-
   tion

6 Analysis

7 Selection

8 Detailed Design

9 Prototyping and
   Testing

10 Production

Blank paper syndrome

requires that the student be served with a collection of **structured, set-piece problems** designed to elucidate a particular concept or concepts related to the particular topic. These textbook problems typically take the form of "*given A, B, C, and D, find E.*" Unfortunately, real-life engineering problems are almost never so structured. Real design problems more often take the form of "*What we need is a framus to stuff this widget into that hole within the time allocated to the transfer of this other gizmo.*" The new engineering graduate will search in vain among his or her textbooks for much guidance to solve such a problem. This **unstructured problem** statement usually leads to what is commonly called "**blank paper syndrome.**" Engineers often find themselves staring at a blank sheet of paper pondering how to begin solving such an ill-defined problem.

Much of engineering education deals with topics of **analysis**, which means *to decompose, to take apart, to resolve into its constituent parts.* This is quite necessary. The engineer must know how to analyze systems of various types, mechanical, electrical, thermal, or fluid. Analysis requires a thorough understanding of both the appropriate mathematical techniques and the fundamental physics of the system's function. But, before any system can be analyzed, it must exist, and a blank sheet of paper provides little substance for analysis. Thus the first step in any engineering design exercise is that of **synthesis**, which means *putting together.*

The design engineer, in practice, regardless of discipline, continuously faces the challenge of *structuring the unstructured problem.* Inevitably, the problem as posed to the engineer is ill-defined and incomplete. Before any attempt can be made to *analyze the situation,* he or she must first carefully define the problem, using an engineering approach, to ensure that any proposed solution will solve the right problem. Many examples exist of excellent engineering solutions that were ultimately rejected because they solved the wrong problem, i.e., a different one than the client really had.

Much research has been devoted to the definition of various "design processes" intended to provide means to structure the unstructured problem and lead to a viable solution. Some of these processes present dozens of steps, others only a few. The one presented in Table 1-1 contains 10 steps and has, in the author's experience, proved successful in over 40 years of practice in engineering design.

ITERATION   Before we discuss each of these steps in detail, it is necessary to point out that this is not a process in which one proceeds from step one through ten in a linear fashion. Rather it is, by its nature, an iterative process in which progress is made haltingly, two steps forward and one step back. It is inherently *circular*. To **iterate** means *to repeat, to return to a previous state.* If, for example, your apparently great idea, upon analysis, turns out to violate the second law of thermodynamics, you can return to the ideation step and get a better idea! Or, if necessary, you can return to an earlier step in the process, perhaps the background research, and learn more about the problem. With the understanding that the actual execution of the process involves iteration, for simplicity, we will now discuss each step in the order listed in Table 1-1.

Identification of Need

This first step is often done for you by someone, boss or client, saying, "What we need is . . .." Typically this statement will be brief and lacking in detail. It will fall far short of providing you with a structured problem statement. For example, the problem statement might be "We need a better lawn mower."

Background Research

This is the most important phase in the process, and is unfortunately often the most neglected. The term **research**, used in this context, should *not* conjure up visions of white-coated scientists mixing concoctions in test tubes. Rather this is research of a more mundane sort, gathering background information on the relevant physics, chemistry, or other aspects of the problem. Also it is desirable to find out if this, or a similar problem, has been solved before. There is no point in reinventing the wheel. If you are lucky enough to find a ready-made solution on the market, it will no doubt be more economical to purchase it than to build your own. Most likely this will not be the case, but you may learn a great deal about the problem to be solved by investigating the existing "art" associated with similar technologies and products. Many companies purchase, disassemble, and analyze their competitors' products, a process sometimes referred to as "**benchmarking**."

Identifying the need

The **patent** literature and technical publications in the subject area are obvious sources of information and are accessible via the World Wide Web. The U.S. Patent and Trademark Office operates a web site at www.uspto.gov where you can search patents by keyword, inventor, title, patent number, or other data. You can print a copy of the patent from the site. A commercial site at www.delphion.com also provides copies of extant patents including those issued in European countries. The "disclosure" or "specification" section of a patent is required to describe the invention in such detail that anyone "skilled in the art" could make the invention. In return for this full disclosure the government grants the inventor a 20-year monopoly on the claimed invention. After that term expires, anyone can use it. Clearly, if you find that the solution exists and is covered by a patent still in force, you have only a few ethical choices: buy the patentee's existing solution, design something that does not conflict with the patent, or drop the project.

Reinventing the wheel

Technical publications in engineering are numerous and varied and are provided by a large number of professional organizations. For the subject matter of this text, the *American Society of Mechanical Engineers* (ASME), which offers inexpensive student memberships, and the *International Federation for the Theory of Machines and Mechanisms* (IFToMM) both publish relevant journals, the ASME *Journal of Mechanical Design* and *Mechanism and Machine Theory*, respectively. Your school library may subscribe to these, and you can purchase copies of articles from their web sites at www.asme.org/pubs/journals/ and www.elsevier.com/inca/publications, respectively.

The World Wide Web provides an incredibly useful resource for the engineer or student looking for information on any subject. The many search engines available will deliver a wealth of information in response to selected keywords. The web makes it easy to find sources for purchased hardware such as gears, bearings, and motors, for your machine designs. In addition, much machine design information is available from the web. A number of useful web sites are catalogued in the bibliography of this chapter.

Grass shorteners

It is very important that sufficient energy and time be expended on this research and preparation phase of the process in order to avoid the embarrassment of concocting a great solution to the wrong problem. Most inexperienced (and some experienced) engineers give too little attention to this phase and jump too quickly into the ideation and invention stage of the process. *This must be avoided!* You must discipline yourself to *not* try to solve the problem before thoroughly preparing yourself to do so.

1

1   Device to have self-contained power supply.

2   Device to be corrosion resistant.

3   Device to cost less than $100.00.

4   Device to emit < 80 dB sound intensity at 10 m.

5   Device to shorten 1/4 acre of grass per hour.

6   etc. . . . etc.

Performance Specifications

Lorem
Ipsum
Dolor amet
Euismod
Volutpat
Laoreet
Adipiscing

Goal Statement

Once the background of the problem area as originally stated is fully understood, you will be ready to recast that problem into a more coherent goal statement. This new problem statement should have three characteristics. It should be concise, be general, and be uncolored by any terms that predict a solution. It should be couched in terms of **functional visualization**, *meaning to visualize its function*, rather than any particular embodiment. For example, if the original statement of need was *"Design a Better Lawn Mower,"* after research into the myriad of ways to cut grass that have been devised over the ages, the wise designer might restate the goal as **"Design a Means to Shorten Grass."** The original problem statement has a built-in trap in the form of the *colored* words "lawn mower." For most people, this phrase will conjure up a vision of something with whirring blades and a noisy engine. For the **ideation** phase to be most successful, it is necessary to avoid such images and to state the problem generally, clearly, and concisely. As an exercise, list 10 ways to shorten grass. Most of them would not occur to you had you been asked for 10 better lawn mower designs. You should use **functional visualization** to avoid unnecessarily limiting your creativity!

Performance Specifications [*]

When the background is understood, and the goal clearly stated, you are ready to formulate a set of *performance specifications* (also called *task specifications*). These should **not** be design specifications. The difference is that **performance specifications** define **what** *the system must do*, while **design specifications** define **how** *it must do it*. At this stage of the design process it is unwise to attempt to specify *how* the goal is to be accomplished. That is left for the **ideation** phase. The purpose of the performance specifications is to carefully define and constrain the problem so that it both *can be solved* and *can be shown to have been solved* after the fact. A sample set of performance specifications for our "grass shortener" is shown in Table 1-2.

Note that these specifications constrain the design without overly restricting the engineer's design freedom. It would be inappropriate to require a gasoline engine for specification 1, because other possibilities exist that will provide the desired mobility. Likewise, to demand stainless steel for all components in specification 2 would be unwise, since corrosion resistance can be obtained by other, less-expensive means. In short, the performance specifications serve to define the problem in as complete and as general a manner as possible, and they serve as a contractual definition of what is to be accomplished. The finished design can be tested for compliance with the specifications.

Ideation and Invention

This step is full of both fun and frustration. This phase is potentially the most satisfying to most designers, but it is also the most difficult. A great deal of research has been done to explore the phenomenon of **creativity**. It is, most agree, a common human trait. It is certainly exhibited to a very high degree by all young children. The rate and degree of development that occurs in the human from birth through the first few years of life certainly requires some innate creativity. Some have claimed that our methods of Western education tend to stifle children's natural creativity by encouraging conformity and restricting individuality. From "coloring within the lines" in kindergarten to imitating the

[*] Orson Welles, famous author and filmmaker, once said, *"The enemy of art is the absence of limitations."* We can paraphrase that as *The enemy of design is the absence of specifications.*

textbook's writing patterns in later grades, individuality is suppressed in favor of a socializing conformity. This is perhaps necessary to avoid anarchy but probably does have the effect of reducing the individual's ability to think creatively. Some claim that creativity can be taught, others that it is only inherited. No hard evidence exists for either theory. It is probably true that one's lost or suppressed creativity can be rekindled. Other studies suggest that most everyone underutilizes his or her potential creative abilities. You can enhance your creativity through various techniques.

CREATIVE PROCESS   Many techniques have been developed to enhance or inspire creative problem solving. In fact, just as design processes have been defined, so has the *creative process* shown in Table 1-3. This creative process can be thought of as a subset of the design process and to exist within it. The ideation and invention step can thus be broken down into these four substeps.

IDEA GENERATION   is the most difficult of these steps. Even very creative people have difficulty inventing "on demand." Many techniques have been suggested to improve the yield of ideas. The most important technique is that of *deferred judgment*, which means that your criticality should be temporarily suspended. Do not try to judge the quality of your ideas at this stage. That will be taken care of later, in the **analysis** phase. The goal here is to obtain as large a *quantity* of potential designs as possible. Even superficially ridiculous suggestions should be welcomed, as they may trigger new insights and suggest other more realistic and practical solutions.

BRAINSTORMING   is a technique for which some claim great success in generating creative solutions. This technique requires a group, preferably 6 to 15 people, and attempts to circumvent the largest barrier to creativity, which is *fear of ridicule*. Most people, when in a group, will not suggest their real thoughts on a subject, for fear of being laughed at. Brainstorming's rules require that no one be allowed to make fun of or criticize anyone's suggestions, no matter how ridiculous. One participant acts as "scribe" and is duty bound to record all suggestions, no matter how apparently silly. When done properly, this technique can be fun and can sometimes result in a "feeding frenzy" of ideas that build upon each other. Large quantities of ideas can be generated in a short time. Judgment on their quality is deferred to a later time.

When you are working alone, other techniques are necessary. **Analogies** and **inversion** are often useful. Attempt to draw analogies between the problem at hand and other physical contexts. If it is a mechanical problem, convert it by analogy to a fluid or electrical one. Inversion turns the problem inside out. For example, consider what you want moved to be stationary and vice versa. Insights often follow. Another useful aid to creativity is the use of **synonyms**. Define the action verb in the problem statement, and then list as many synonyms for that verb as possible. For example:

Problem statement: *Move this object from point A to point B.*

The action verb is "move." Some synonyms are push, pull, slip, slide, shove, throw, eject, jump, spill.

By whatever means, the aim in this **ideation** step is to generate a large number of ideas without particular regard to quality. But, at some point, your "mental well" will go dry. You will have then reached the step in the creative process called **frustration**. It is time to leave the problem and do something else for a time. While your conscious mind is occupied with other concerns, your subconscious mind will still be hard at work on the

---

TABLE 1-3
The Creative Process

5a Idea Generation

5b Frustration

5c Incubation

5d Eureka!

Brainstorming

Frustration

Eureka!

1

problem. This is the step called **incubation.** Suddenly, at a quite unexpected time and place, an idea will pop into your consciousness, and it will seem to be the obvious and "right" solution to the problem . . . **Eureka!** Most likely, later analysis will discover some flaw in this solution. If so, back up and **iterate!** More ideation, perhaps more research, and possibly even a redefinition of the problem may be necessary.

In "Unlocking Human Creativity"[5] Wallen describes three requirements for creative insight:

- Fascination with a problem.

- Saturation with the facts, technical ideas, data, and the background of the problem.

- A period of reorganization.

The first of these provides the motivation to solve the problem. The second is the background research step described on p. 9. The period of reorganization refers to the frustration phase when your subconscious works on the problem. Wallen[5] reports that testimony from creative people tells us that in this period of reorganization they have no conscious concern with the particular problem and that the moment of insight frequently appears in the midst of relaxation or sleep. So to enhance your creativity, saturate yourself in the problem and related background material. Then relax and let your subconscious do the hard work!

Analysis

Once you are at this stage, you have structured the problem, at least temporarily, and can now apply more sophisticated analysis techniques to examine the performance of the design in the **analysis phase** of the design process. (Some of these analysis methods will be discussed in detail in the following chapters.) Further iteration will be required as problems are discovered from the analysis. Repetition of as many earlier steps in the design process as necessary must be done to ensure the success of the design.

	Cost	*Safety*	*Performance*	*Reliability*	*RANK*
Weighting Factor	.35	.30	.15	.20	1.0
Design 1	3 / 1.05	6 / 1.80	4 / .60	9 / 1.80	5.3
Design 2	4 / 1.40	2 / .60	7 / 1.05	2 / .40	3.5
Design 3	1 / .35	9 / 2.70	4 / .60	5 / 1.00	4.7
Design 4	9 / 3.15	1 / .30	6 / .90	7 / 1.40	5.8
Design 5	7 / 2.45	4 / 1.20	2 / .30	6 / 1.20	5.2

FIGURE 1-2

A decision matrix

Selection

When the technical analysis indicates that you have some potentially viable designs, the best one available must be **selected** for **detailed design**, **prototyping,** and **testing.** The selection process usually involves a comparative analysis of the available design solutions. A **decision matrix** sometimes helps to identify the best solution by forcing you to consider a variety of factors in a systematic way. A decision matrix for our better grass shortener is shown in Figure 1-2. Each design occupies a row in the matrix. The columns are assigned categories in which the designs are to be judged, such as cost, ease of use, efficiency, performance, reliability, and any others you deem appropriate to the particular problem. Each category is then assigned a **weighting factor**, which measures its relative importance. For example, reliability may be a more important criterion to the user than cost, or vice versa. You as the design engineer have to exercise your judgment as to the selection and weighting of these categories. The body of the matrix is then filled with numbers that rank each design on a convenient scale, such as 1 to 10, in each of the categories. Note that this is ultimately a *subjective ranking* on your part. You must examine the designs and decide on a score for each. The scores are then multiplied by the weighting factors (which are usually chosen so as to sum to a convenient number such as 1) and the products are summed for each design. The weighted scores then give a ranking of designs. Be cautious in applying these results. Remember the source and subjectivity of your scores and the weighting factors! There is a temptation to put more faith in these results than is justified. After all, they look impressive! They can even be taken out to several decimal places! (But they shouldn't be.) The real value of a decision matrix is that it breaks the problem into more tractable pieces and forces you to think about the relative value of each design in many categories. You can then make a more informed decision as to the "best" design.

Detailed Design

This step usually includes the creation of a complete set of assembly and detail drawings or **computer-aided design** (CAD) part files, for *each and every part* used in the design. Each detail drawing must specify all the dimensions and the material specifications necessary to make that part. From these drawings (or CAD files) a prototype test model (or models) must be constructed for physical testing. Most likely the tests will discover more flaws, requiring further **iteration**.

Prototyping and Testing

MODELS   Ultimately, one cannot be sure of the correctness or viability of any design until it is built and tested. This usually involves the construction of a prototype physical model. A mathematical model, while very useful, can never be as complete and accurate a representation of the actual physical system as a physical model, due to the need to make simplifying assumptions. Prototypes are often very expensive but may be the most economical way to prove a design, short of building the actual, full-scale device. Prototypes can take many forms, from working scale models to full-size, but simplified, representations of the concept. Scale models introduce their own complications in regard to proper scaling of the physical parameters. For example, volume of material varies as the cube of linear dimensions, but surface area varies as the square. Heat trans-

1

fer to the environment may be proportional to surface area, while heat generation may be proportional to volume. So linear scaling of a system, either up or down, may lead to behavior different from that of the full-scale system. One must exercise caution in scaling physical models. You will find as you begin to design linkage mechanisms that a **simple cardboard model** of your chosen link lengths, joined together with thumbtacks for pivots, will tell you a great deal about the quality and character of the mechanism's motions. You should get into the habit of making such simple articulated models for all your linkage designs.

TESTING of the model or prototype may range from simply actuating it and observing its function to attaching extensive instrumentation to accurately measure displacements, velocities, accelerations, forces, temperatures, and other parameters. Tests may need to be done under controlled environmental conditions such as high or low temperature or humidity. The microcomputer has made it possible to measure many phenomena more accurately and inexpensively than could be done before.

Production

Finally, with enough time, money, and perseverance, the design will be ready for production. This might consist of the manufacture of a single final version of the design, but more likely will mean making thousands or even millions of your widget. The danger, expense, and embarrassment of finding flaws in your design after making large quantities of defective devices should inspire you to use the greatest care in the earlier steps of the design process to ensure that it is properly engineered.

The **design process** is widely used in engineering. Engineering is usually defined in terms of what an engineer does, but engineering can also be defined in terms of *how* the engineer does what he or she does. **Engineering** is *as much a method, an approach, a process, a state of mind for problem solving, as it is an activity.* The engineering approach is that of thoroughness, attention to detail, and consideration of all the possibilities. While it may seem a contradiction in terms to emphasize "attention to detail" while extolling the virtues of open-minded, freewheeling, creative thinking, it is not. The two activities are not only compatible, they are also symbiotic. It ultimately does no good to have creative, original ideas if you do not, or cannot, carry out the execution of those ideas and "reduce them to practice." To do this you must discipline yourself to suffer the nitty-gritty, nettlesome, tiresome details that are so necessary to the completion of any one phase of the creative design process. For example, to do a creditable job in the design of anything, you must *completely* define the problem. If you leave out some detail of the problem definition, you will end up solving the wrong problem. Likewise, you must *thoroughly* research the background information relevant to the problem. You must *exhaustively* pursue conceptual potential solutions to your problem. You must then *extensively* analyze these concepts for validity. And, finally, you must *detail* your chosen design down to the last nut and bolt to be confident it will work. If you wish to be a good designer and engineer, you must discipline yourself to do things thoroughly and in a logical, orderly manner, even while thinking great creative thoughts and iterating to a solution. Both attributes, creativity and attention to detail, are necessary for success in engineering design.

1.6   OTHER APPROACHES TO DESIGN

In recent years, an increased effort has been directed toward a better understanding of design methodology and the design process. Design methodology is the study of the process of designing. One goal of this research is to define the design process in sufficient detail to allow it to be encoded in a form amenable to execution in a computer, using "artificial intelligence" (AI).

Dixon[6] defines a design as a *state of information* which may be in any of several forms:

> . . . words, graphics, electronic data, and/or others. It may be partial or complete. It ranges from a small amount of highly abstract information early in the design process to a very large amount of detailed information later in the process sufficient to perform manufacturing. It may include, but is not limited to, information about size and shape, function, materials, marketing, simulated performance, manufacturing processes, tolerances, and more. Indeed, any and all information relevant to the physical or economic life of a designed object is part of its design.

He goes on to describe several generalized states of information such as the *requirements* state that is analogous to our **performance specifications**. Information about the physical concept is referred to as the *conceptual* state of information and is analogous to our **ideation** phase. His *feature configuration* and *parametric* states of information are similar in concept to our **detailed design** phase. Dixon then defines a design process as

> The series of activities by which the information about the designed object is changed from one information state to another.

Axiomatic Design

N. P. Suh[7] suggests an *axiomatic approach* to design in which there are four domains: **customer** domain, **functional** domain, **physical** domain, and **process** domain. These represent a range from "what" to "how," i.e., from a state of defining what the customer wants through determining the functions required and the needed physical embodiment, to how a process will achieve the desired end. He defines two axioms that need to be satisfied to accomplish this:

1   Maintain the independence of the functional requirements.

2   Minimize the information content.

The first of these refers to the need to create a complete and nondependent set of performance specifications. The second indicates that the best design solution will have the lowest information content (i.e., the least complexity). Others have earlier referred to this second idea as *KISS,* which stands, somewhat crudely, for "*keep it simple, stupid.*"

The implementation of both Dixon's and Suh's approaches to the design process is somewhat complicated. The interested reader is referred to the literature cited in the bibliography to this chapter for more complete information.

1.7   MULTIPLE SOLUTIONS

Note that by the nature of the design process, there is **not** any **one** correct answer or solution to any design problem. Unlike the structured "engineering textbook" problems, which most students are used to, there is no right answer "in the back of the book" for any real design problem.* There are as many potential solutions as there are designers willing to attempt them. Some solutions will be better than others, but many will work. Others will not! There is no "one right answer" in design engineering, which is what makes it interesting. The only way to determine the relative merits of various potential design solutions is by thorough analysis, which usually will include physical testing of constructed prototypes. Because this is a very expensive process, it is desirable to do as much analysis on paper, or in the computer, as possible before actually building the device. Where feasible, mathematical models of the design, or parts of the design, should be created. These may take many forms, depending on the type of physical system involved. In the design of mechanisms and machines it is usually possible to write the equations for the rigid-body dynamics of the system, and solve them in "closed form" with (or without) a computer. Accounting for the elastic deformations of the members of the mechanism or machine usually requires more complicated approaches using **finite difference** techniques or the **finite element method** (FEM).

1.8   HUMAN FACTORS ENGINEERING

With few exceptions, all machines are designed to be used by humans. Even robots must be programmed by a human. **Human factors engineering** is the study of the human-machine interaction and is defined as *an applied science that coordinates the design of devices, systems, and physical working conditions with the capacities and requirements of the worker*. The machine designer must be aware of this subject and design devices to "fit the man" rather than expect the man to adapt to fit the machine. The term **ergonomics** is synonymous with *human factors engineering*. We often see reference to the good or bad ergonomics of an automobile interior or a household appliance. A machine designed with poor ergonomics will be uncomfortable and tiring to use and may even be dangerous. (Have you programmed your VCR lately, or set its clock?)

Make the machine
fit the man

There is a wealth of human factors data available in the literature. Some references are noted in the bibliography. The type of information that might be needed for a machine design problem ranges from dimensions of the human body and their distribution among the population by age and gender, to the ability of the human body to withstand accelerations in various directions, to typical strengths and force-generating ability in various positions. Obviously, if you are designing a device that will be controlled by a human (a grass shortener, perhaps), you need to know how much force the user can exert with hands held in various positions, what the user's reach is, and how much noise the ears can stand without damage. If your device will carry the user on it, you need data on the limits of acceleration that the body can tolerate. Data on all these topics exist. Much of it was developed by the government which regularly tests the ability of military personnel to withstand extreme environmental conditions. Part of the background research of any machine design problem should include some investigation of human factors.

---

* A student once commented that "*Life is an odd-numbered problem.*" This (slow) author had to ask for an explanation, which was, "*The answer is not in the back of the book.*"

1.9   THE ENGINEERING REPORT

Communication of your ideas and results is a very important aspect of engineering. Many engineering students picture themselves in professional practice spending most of their time doing calculations of a nature similar to those they have done as students. Fortunately, this is seldom the case, as it would be very boring. Actually, engineers spend the largest percentage of their time communicating with others, either orally or in writing. Engineers write proposals and technical reports, give presentations, and interact with support personnel and managers. When your design is done, it is usually necessary to present the results to your client, peers, or employer. The usual form of presentation is a formal engineering report. Thus, it is very important for the engineering student to develop his or her communication skills. *You may be the cleverest person in the world, but no one will know that if you cannot communicate your ideas clearly and concisely.* In fact, if you cannot explain what you have done, you probably don't understand it yourself. To give you some experience in this important skill, the design project assignments in later chapters are intended to be written up in formal engineering reports. Information on the writing of engineering reports can be found in the suggested readings in the bibliography at the end of this chapter.

1.10   UNITS

There are several systems of units used in engineering. The most common in the United States are the **U.S. foot-pound-second (fps) system**, the **U.S. inch-pound-second (ips) system**, and the **Systeme International (SI)**. All systems are created from the choice of three of the quantities in the general expression of Newton's second law

$$F = \frac{m\,l}{t^2} \tag{1.1a}$$

where F is force, m is mass, l is length, and t is time. The units for any three of these variables can be chosen and the other is then derived in terms of the chosen units. The three chosen units are called *base units*, and the remaining one is then a *derived unit*.

Most of the confusion that surrounds the conversion of computations between either one of the U.S. systems and the SI system is due to the fact that the SI system uses a different set of base units than the U.S. systems. Both U.S. systems choose *force*, *length,* and *time* as the base units. Mass is then a derived unit in the U.S. systems, and they are referred to as *gravitational systems* because the value of mass is dependent on the local gravitational constant. The SI system chooses *mass*, *length,* and *time* as the base units and force is the derived unit. SI is then referred to as an *absolute system* since the mass is a base unit whose value is not dependent on local gravity.

The **U.S. foot-pound-second (fps)** system requires that all lengths be measured in feet (ft), forces in pounds (lb), and time in seconds (sec). Mass is then derived from Newton's law as

$$m = \frac{F t^2}{l} \tag{1.1b}$$

and the units are pound seconds squared per **foot** (lb-sec^2/ft) = **slugs.**

1

The **U.S. inch-pound-second (ips)** system requires that all lengths be measured in inches (in), forces in pounds (lb), and time in seconds (sec). Mass is still derived from Newton's law, equation 1.1b, but the units are now:

Pound-seconds squared per **inch** (lb-sec²/in) = **blobs**

This mass unit is not slugs! It is worth twelve slugs or one blob.*

Weight is defined as the force exerted on an object by gravity. Probably the most common units error that students make is to mix up these two unit systems (**fps** and **ips**) when converting weight units (which are pounds force) to mass units. Note that the gravitational acceleration constant (g) on earth at sea level is approximately 32.2 **feet** per second squared, which is equivalent to 386 **inches** per second squared. The relationship between mass and weight is:

Mass = weight / gravitational acceleration

$$m = \frac{W}{g} \tag{1.2}$$

It should be obvious that, if you measure all your lengths in **inches** and then use g = 32.2 **feet/sec²** to compute mass, you will have an error of a *factor of twelve* in your results. This is a serious error, large enough to crash the airplane you designed. Even worse off is the student who neglects to convert weight to mass *at all* in his calculations. He will have an error of either 32.2 or 386 in his results. This is enough to sink the ship!†

To even further add to the student's confusion about units is the common use of the unit of **pounds mass** (lb_m). This unit is often used in fluid dynamics and thermodynamics and comes about through the use of a slightly different form of Newton's equation:

$$F = \frac{ma}{g_c} \tag{1.3}$$

where m = mass in lb_m, a = acceleration and g_c = the gravitational constant.

The value of the **mass** of an object measured in **pounds mass** (lb_m) is *numerically equal* to its **weight** in **pounds force** (lb_f). However the student *must remember to divide* the value of m in lb_m by g_c when substituting into this form of Newton's equation. Thus the lb_m will be divided either by 32.2 or by 386 when calculating the dynamic force. The result will be the same as when the mass is expressed in either slugs or blobs in the $F = ma$ form of the equation. Remember that in round numbers at sea level on earth:

$1\ lb_m = 1\ lb_f$　　　　　　$1\ slug = 32.2\ lb_f$　　　　　　$1\ blob = 386\ lb_f$

The **SI** system requires that lengths be measured in meters (m), mass in kilograms (kg), and time in seconds (sec). This is sometimes also referred to as the **mks** system. Force is derived from Newton's law, equation 1.1b, and the units are:

kilogram-meters per second² (kg-m/s²) = newtons

Thus in the SI system there are distinct names for mass and force which helps alleviate confusion. When converting between SI and U.S. systems, be alert to the fact that mass converts from kilograms (kg) to either slugs (sl) or blobs (bl), and force converts from newtons (N) to pounds (lb). The gravitational constant (g) in the SI system is approximately 9.81 m/s².

---

* It is unfortunate that the mass unit in the **ips** system has never officially been given a name such as the term *slug* used for mass in the **fps** system. The author boldly suggests (with tongue only slightly in cheek) that this unit of mass in the **ips** system be called a *blob* (bl) to distinguish it more clearly from the *slug* (sl), and to help the student avoid some of the common units errors listed above.

Twelve slugs = one blob

Blob does not sound any sillier than slug, is easy to remember, implies mass, and has a convenient abbreviation (bl) which is an anagram for the abbreviation for pound (lb). Besides, if you have ever seen a garden slug, you know it looks just like a "*little blob.*"

---

† A 125-million-dollar space probe was lost because NASA failed to convert data that had been supplied in *ips* units by its contractor, Lockheed Aerospace, into the metric units used in the NASA computer programs that controlled the spacecraft. It was supposed to orbit the planet Mars, but instead either burned up in the Martian atmosphere or crashed into the planet because of this units error. *Source: The Boston Globe, October 1, 1999, p. 1.*

Table 1-4   Variables and Units
Base Units in Boldface – Abbreviations in ( )

Variable	Symbol	ips unit	fps unit	SI unit
Force	F	**pounds (lb)**	**pounds (lb)**	newtons (N)
Length	l	**inches (in)**	**feet (ft)**	**meters (m)**
Time	t	**seconds (sec)**	**seconds (sec)**	**seconds (sec)**
Mass	m	lb–sec^2/in = bl	lb–sec^2/ft = sl	**kilograms (kg)**
Weight	W	pounds (lb)	pounds (lb)	newtons (N)
Velocity	v	in/sec	ft/sec	m/s
Acceleration	a	in/sec^2	ft/sec^2	m/s^2
Jerk	j	in/sec^3	ft/sec^3	m/s^3
Angle	θ	degrees (deg)	degrees (deg)	degrees (deg)
Angle	θ	radians (rad)	radians (rad)	radians (rad)
Angular velocity	ω	rad/sec	rad/sec	rad/s
Angular acceleration	α	rad/sec^2	rad/sec^2	rad/s^2
Angular jerk	φ	rad/sec^3	rad/sec^3	rad/s^3
Torque	T	lb–in	lb–ft	N–m
Mass moment of inertia	I	lb–in–sec^2	lb–ft–sec^2	N–m–s^2
Energy	E	in–lb	ft–lb	joules (J)
Power	P	in–lb/sec	ft–lb/sec	watts (W)
Volume	V	in^3	ft^3	m^3
Weight density	γ	lb/in^3	lb/ft^3	N/m^3
Mass density	ρ	bl/in^3	sl/ft^3	kg/m^3

The principal system of units used in this textbook will be the U.S. **ips** system. Most machine design in the United States is still done in this system. Table 1-4 shows some of the variables used in this text and their units. The inside front cover contains a table of conversion factors between the U.S. and SI systems.

The student is cautioned to always check the units in any equation written for a problem solution, whether in school or in professional practice after graduation. If properly written, an equation should cancel all units across the equal sign. If it does not, then you can be *absolutely sure it is **incorrect.*** Unfortunately, a unit balance in an equation does not guarantee that it is correct, as many other errors are possible. Always double-check your results. You might save a life.

1.11   A DESIGN CASE STUDY

Of all the myriad activities that the practicing engineer engages in, the one that is at once the most challenging and potentially the most satisfying is design. Doing calculations to analyze a clearly defined and structured problem, no matter how complex, may be difficult, but the exercise of creating something from scratch, to solve a problem that is often poorly defined, is *very* difficult. The sheer pleasure and joy at conceiving a viable solution to such a design problem is one of life's great satisfactions for anyone, engineer or not.

Some years ago, a very creative engineer of the author's acquaintance, George A. Wood Jr., heard a presentation by another creative engineer of the author's acquaintance, Keivan Towfigh, about one of his designs. Years later, Mr. Wood himself wrote a short paper about creative engineering design in which he reconstructed Mr. Towfigh's presumed creative process when designing the original invention. Both Mr. Wood and Mr. Towfigh have kindly consented to the reproduction of that paper here. It serves, in this author's opinion, as an excellent example and model for the student of engineering design to consider when pursuing his or her own design career.

Educating for Creativity in Engineering[9]

by GEORGE A. WOOD JR.

One facet of engineering, as it is practiced in industry, is the creative process. Let us define creativity as Rollo May does in his book, The Courage to Create.[10] It is "the process of bringing something new into being." Much of engineering has little to do with creativity in its fullest sense. Many engineers choose not to enter into creative enterprise, but prefer the realms of analysis, testing and product or process refinement. Many others find their satisfaction in management or business roles and are thus removed from engineering creativity as we shall discuss it here.

From the outset, I wish to note that the less creative endeavors are no less important or satisfying to many engineers than is the creative experience to those of us with the will to create. It would be a false goal for all engineering schools to assume that their purpose was to make all would-be engineers creative and that their success should be measured by the "creative quotient" of their graduates.

On the other hand, for the student who has a creative nature, a life of high adventure awaits if he can find himself in an academic environment which recognizes his needs, enhances his abilities and prepares him for a place in industry where his potential can be realized.

In this talk I will review the creative process as I have known it personally and witnessed it in others. Then I shall attempt to indicate those aspects of my training that seemed to prepare me best for a creative role and how this knowledge and these attitudes toward a career in engineering might be reinforced in today's schools and colleges.

During a career of almost thirty years as a machine designer, I have seen and been a part of a number of creative moments. These stand as the high points of my working life. When I have been the creator I have felt great elation and immense satisfaction. When I have been with others at their creative moments I have felt and been buoyed up by their delight. To me, the creative moment is the greatest reward that the profession of engineering gives.

Let me recount an experience of eight years ago when I heard a paper given by a creative man about an immensely creative moment. At the First Applied Mechanisms Conference in Tulsa, Oklahoma, was a paper entitled The Four-Bar Linkage as an Adjustment Mechanism.[11] It was nestled between two "how to do it" academic papers with graphs and equations of interest to engineers in the analysis of their mechanism problems. This paper contained only one very elementary equation and five simple illustrative figures; yet, I remember it now more clearly than any other paper I have ever heard at mechanism conferences. The author was Keivan Towfigh and he described the application of the geometric characteristics of the instant center of the coupler of a four bar mechanism.

His problem had been to provide a simple rotational adjustment for the oscillating mirror of an optical galvanometer. To accomplish this, he was required to rotate the entire galvanometer assembly about an axis through the center of the mirror and perpendicular to the pivot axis of the mirror. High rigidity of the system after adjustment was essential with very limited space available and low cost required, since up to sixteen of these galvanometer units were used in the complete instrument.

His solution was to mount the galvanometer elements on the coupler link of a one-piece, flexure hinged, plastic four bar mechanism so designed that the mirror center was at the instant center* of the linkage at the midpoint of its adjustment. (See Fig 4.) It is about this particular geometric point (see Fig 1.) that pure rotation occurs and with proper selection of linkage dimensions this condition of rotation without translation could be made to hold sufficiently accurately for the adjustment angles required.

Unfortunately, this paper was not given the top prize by the judges of the conference. Yet, it was, indirectly, a description of an outstandingly creative moment in the life of a creative man.

Let us look at this paper together and build the steps through which the author probably progressed in the achievement of his goal. I have never seen Mr. Towfigh since, and I shall therefore describe a generalized creative process which may be incorrect in some details but which, I am sure, is surprisingly close to the actual story he would tell.

The galvanometer problem was presented to Mr. Towfigh by his management. It was, no doubt, phrased something like this: "In our new model, we must improve the stability of the adjustment of the equipment but keep the cost down. Space is critical and low weight is too. The overall design must be cleaned up, since customers like modern, slim-styled equipment and we'll lose sales to others if we don't keep ahead of them on all points. Our industrial designer has this sketch that all of us in sales like and within which you should be able to make the mechanism fit."

Then followed a list of specifications the mechanism must meet, a time when the new model should be in production and, of course, the request for some new feature that would result in a strong competitive edge in the marketplace.

I wish to point out that the galvanometer adjustment was probably only one hoped-for improvement among many others. The budget and time allowed were little more than enough needed for conventional redesign, since this cost must be covered by the expected sales of the resulting instrument. For every thousand dollars spent in engineering, an equivalent increase in sales or reduction in manufacturing cost must be realized at a greater level than the money will bring if invested somewhere else.

In approaching this project, Mr. Towfigh had to have a complete knowledge of the equipment he was designing. He had to have run the earlier models himself. He must have adjusted the mirrors of existing machines many times. He had to be able to visualize the function of each element in the equipment in its most basic form.

(research)

Secondly, he had to ask himself (as if he were the customer) what operational and maintenance requirements would frustrate him most. He had to determine which of these might be improved within the design time available. In this case he focused on the mirror adjustment. He considered the requirement of rotation without translation. He determined the maximum angles that would be

---

* The theory of instant centers will be thoroughly explained in Chapter 6.

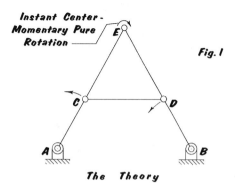

Instant Center - Momentary Pure Rotation

Fig. 1

The Theory

Fig. 2

The Development

Fig. 3

The Mechanism

Screen   Mirror   Fig.4

Lamp   Adj. screw

The Final Product of Keivan Towfigh

(ideation)

(frustration)

(incubation)

(Eureka!)

necessary and the allowable translation that would not affect the practical accuracy of the equipment. He recognized the desirability of a one screw adjustment. He spent a few hours thinking of all the ways he had seen of rotating an assembly about an arbitrary point. He kept rejecting each solution as it came to him as he felt, in each case, that there was a better way. His ideas had too many parts, involved slides, pivots, too many screws, were too vibration sensitive or too large.

He thought about the problem that evening and at other times while he proceeded with the design of other aspects of the machine. He came back to the problem several times during the next few days. His design time was running out. He was a mechanism specialist and visualized a host of cranks and bars moving the mirrors. Then one day, probably after a period when he had turned his attention elsewhere, on rethinking of the adjustment device, an image of the system based on one of the elementary characteristics of a four bar mechanism came to him.

I feel certain that this was a visual image, as clear as a drawing on paper. It was probably not complete but involved two inspirations. First was the characteristics of the instant center.* (See Figs 1, 2, 3.) Second was the use of flexure hinge joints which led to a one-piece plastic molding. (See Fig 4.) I am sure that at this moment he had a feeling that this solution was right. He knew it with certainty. The whole of his engineering background told him. He was elated. He was filled with

---

* Defined in Chapter 6.

joy. His pleasure was not because of the knowledge that his superiors would be impressed or that his security in the company would be enhanced. It was the joy of personal victory, the awareness that he had conquered.

The creative process has been documented before by many others far more qualified to analyze the working of the human mind than I. Yet I would like to address, for the remaining minutes, how education can enhance this process and help more engineers, designers and draftsmen extend their creative potential.

The key elements I see in creativity that have greatest bearing on the quality that results from the creative effort are visualization and basic knowledge that gives strength to the feeling that the right solution has been achieved. There is no doubt in my mind that the fundamental mechanical principles that apply in the area in which the creative effort is being made must be vivid in the mind of the creator. The words that he was given in school must describe real elements that have physical, visual significance. $F = ma$ must bring a picture to his mind vivid enough to touch.

If a person decides to be a designer, his training should instill in him a continuing curiosity to know how each machine he sees works. He must note its elements and mentally see them function together even when they are not moving. I feel that this kind of solid, basic knowledge couples with physical experience to build ever more critical levels at which one accepts a tentative solution as "right."

It should be noted that there have been times for all of us when the inspired "right" solution has proven wrong in the long run. That this happens does not detract from the process but indicates that creativity is based on learning and that failures build toward a firmer judgment base as the engineer matures. These failure periods are only negative, in the growth of a young engineer, when they result in the fear to accept a new challenge and beget excessive caution which then stifles the repetition of the creative process.

What would seem the most significant aspects of an engineering curriculum to help the potentially creative student develop into a truly creative engineer?

First is a solid, basic knowledge in physics, mathematics, chemistry and those subjects relating to his area of interest. These fundamentals should have physical meaning to the student and a vividness that permits him to explain his thoughts to the untrained layman. All too often technical words are used to cover cloudy concepts. They serve the ego of the user instead of the education of the listener.

(analysis)

Second is the growth of the student's ability to visualize. The creative designer must be able to develop a mental image of that which he is inventing. The editor of the book Seeing with the Mind's Eye,[12] by Samuels and Samuels, says in the preface:

". . . visualization is the way we think. Before words, images were. Visualization is the heart of the bio-computer. The human brain programs and self-programs through its images. Riding a bicycle, driving a car, learning to read, baking a cake, playing golf - all skills are acquired through the image making process. Visualization is the ultimate consciousness tool."

Obviously, the creator of new machines or products must excel in this area.

To me, a course in Descriptive Geometry is one part of an engineer's training that enhances one's ability to visualize theoretical concepts and graphically reproduce the result. This ability is essential when one sets out to design a piece of new equipment. First, he visualizes a series of

1

complete machines with gaps where the problem or unknown areas are. During this time, a number of directions the development could take begin to form. The best of these images are recorded on paper and then are reviewed with those around him until, finally, a basic concept emerges.

The third element is the building of the student's knowledge of what can be or has been done by others with different specialized knowledge than he has. This is the area to which experience will add throughout his career as long as he maintains an enthusiastic curiosity. Creative engineering is a building process. No one can develop a new concept involving principles about which he has no knowledge. The creative engineer looks at problems in the light of what he has seen, learned and experienced and sees new ways for combining these to fill a new need.

Fourth is the development of the ability of the student to communicate his knowledge to others. This communication must involve not only skills with the techniques used by technical people but must also include the ability to share engineering concepts with untrained shop workers, business people and the general public. The engineer will seldom gain the opportunity to develop a concept if he cannot pass on to those around him his enthusiasm and confidence in the idea. Frequently, truly ingenious ideas are lost because the creator cannot transfer his vivid image to those who might finance or market it.

Fifth is the development of a student's knowledge of the physical result of engineering. The more he can see real machines doing real work, the more creative he can be as a designer. The engineering student should be required to run tools, make products, adjust machinery and visit factories. It is through this type of experience that judgement grows as to what makes a good machine, when approximation will suffice and where optimization should halt.

It is often said that there has been so much theoretical development in engineering during the past few decades that the colleges and universities do not have time for the basics I have outlined above. It is suggested that industry should fill in the practice areas that colleges have no time for, so that the student can be exposed to the latest technology. To some degree I understand and sympathize with this approach, but I feel that there is a negative side that needs to be recognized. If a potentially creative engineer leaves college without the means to achieve some creative success as he enters his first job, his enthusiasm for creative effort is frustrated and his interest sapped long before the most enlightened company can fill in the basics. Therefore, a result of the "basics later" approach often is to remove from the gifted engineering student the means to express himself visually and physically. Machine design tasks therefore become the domain of the graduates of technical and trade schools and the creative contribution by many a brilliant university student to products that could make all our lives richer is lost.

As I said at the start, not all engineering students have the desire, drive and enthusiasm that are essential to creative effort. Yet I feel deeply the need for the enhancement of the potential of those who do. That expanding technology makes course decisions difficult for both student and professor is certainly true. The forefront of academic thought has a compelling attraction for both the teacher and the learner. Yet I feel that the development of strong basic knowledge, the abilities to visualize, to communicate, to respect what has been done, to see and feel real machinery, need not exclude or be excluded by the excitement of the new. I believe that there is a curriculum balance that can be achieved which will enhance the latent creativity in all engineering and science students. It can give a firm basis for those who look towards a career of mechanical invention and still include the excitement of new technology.

I hope that this discussion may help in generating thought and providing some constructive suggestions that may lead more engineering students to find the immense satisfaction of the creative moment in the industrial environment. In writing this paper I have spent considerable time reflecting on my years in engineering and I would close with the following thought. For those of us who have known such times during our careers, the successful culminations of creative efforts stand among our most joyous hours.

Mr. Wood's description of his creative experiences in engineering design and the educational factors which influenced them closely parallels this author's experience as well. The student is well advised to follow his prescription for a thorough grounding in the fundamentals of engineering and communication skills. A most satisfying career in the design of machinery can result.

1.12 WHAT'S TO COME

In this text we will explore the **design of machinery** in respect to the **synthesis of mechanisms** in order to accomplish desired motions or tasks, and also the **analysis of mechanisms** in order to determine their rigid-body dynamic behavior. On the premise that we cannot analyze anything until it has been synthesized into existence, we will first explore the synthesis of mechanisms. Then we will investigate the analysis of those and other mechanisms for their kinematic behavior. Finally, in Part II we will deal with the **dynamic analysis** of the forces and torques generated by these moving machines. These topics cover the essence of the early stages of a design project. Once the kinematics and kinetics of a design have been determined, most of the conceptual design will have been accomplished. What then remains is **detailed design**—sizing the parts against failure. The topic of *detailed design* is discussed in other texts such as reference [8].

1.13 RESOURCES WITH THIS TEXT

There are a number of useful resources included on the DVD with this text that include computer programs, sample files for those programs, PDF files with all problem figures for use in solving them, two linkage atlases (the Hrones and Nelson fourbar atlas, and the Zhang, Norton, Hammond geared fivebar atlas), and digital videos with tutorial information on various topics in the book, program use, and views of actual machines in operation to show applications of the theory. An index of the DVD contents is in the Appendices.

Programs

The commercial program Working Model (WM) is included in a "textbook edition" that has some limitations (see the Preface for more details). It will run all the WM files of book figures and examples that are on the DVD. Seven programs written by the author for the design and analysis of linkages and cams are provided: FOURBAR, FIVEBAR, SIXBAR, SLIDER, DYNACAM, ENGINE, and MATRIX. User manuals, sample files, and tutorial videos for some of these programs are provided on the DVD and are also accessible from within those programs.

Videos

The videos provided are in four categories: tutorials on topics in the text, tutorials on program use, virtual laboratories, and depictions of actual mechanisms and machines.

1

TOPICAL TUTORIALS　　The tutorials on topics in the text (e.g., Position Synthesis in Chapter 3) typically provide much more information on the topic than can be presented on the page and also provide a "show and tell" advantage. These are all footnoted in the sections of the text where the topics are addressed.

PROGRAM TUTORIALS　　The tutorials on program use serve as an introduction to the programs. Additional short video clips on specific topics within each program are also provided. All of these videos can be viewed from within the programs provided that the computer has an Internet connection.

VIRTUAL LABORATORIES　　There are two virtual laboratory videos provided, one on linkages and one on cams. These show and describe laboratory machines used by the author at WPI to introduce students to the measurement and analysis of kinematic and dynamic parameters on real machines. It is instructive to see the differences between theoretical predictions of a machine's behavior and actual measured data. All the data taken in a typical lab session from these machines is provided on disk along with descriptions of the lab assignment so that any reader can do a virtual laboratory exercise similar to that done at WPI.

MACHINES IN ACTION　　These range from commercially produced videos about a company's products or manufacturing processes to student-produced videos about their projects that involved mechanisms. Most students have not had an opportunity to visit a manufacturing plant or see the inner workings of machinery, and the hope is that these videos will give some insight into applications of the theories presented in the text.

1.14　REFERENCES

1　　**Rosenauer, N., and A. H. Willis**. (1967). *Kinematics of Mechanisms*. Dover Publications: New York, p. 275 ff.

2　　**de Jonge, A. E. R.** (1942). "What Is Wrong with 'Kinematics' and 'Mechanisms'?" *Mechanical Engineering*, **64** (April), pp. 273-278.

3　　**Artobolevsky, I. I.** (1975). *Mechanisms in Modern Engineering Design*. N. Weinstein, translator. Vols. 1 - 5. MIR Publishers: Moscow.

4　　**Erdman, A. E.**, ed. (1993). *Modern Kinematics: Developments in the Last Forty Years*. Wiley Series in Design Engineering, John Wiley & Sons: New York.

5　　**Wallen, R. W.** (1957). "Unlocking Human Creativity." *Proc. of Fourth Conference on Mechanisms*, Purdue University, pp. 2-8.

6　　**Dixon, J. R.** (1995). "Knowledge Based Systems for Design." *Journal of Mechanical Design*, **117b**(2), p. 11.

7　　**Suh, N. P.** (1995). "Axiomatic Design of Mechanical Systems." *Journal of Mechanical Design*, **117b**(2), p. 2.

8　　**Norton, R. L.** (2006). *Machine Design: An Integrated Approach*. 4ed, Prentice-Hall: Upper Saddle River, NJ.

9　　**Wood, G. A. Jr.** (1977). "Educating for Creativity in Engineering," Presented at the 85th Annual Conf. ASEE, Univ. of No. Dakota.

10　　**May, R**. (1976). *The Courage to Create*, Bantam Books: New York.

11　　**Towfigh, K**. (1969). "The Four-Bar Linkage as an Adjustment Mechanism," 1st Applied Mechanism Conf., Oklahoma State Univ., Tulsa, OK, pp. 27-1– 27-4.

12   **Samuels. M.** (1975). *Seeing with the Mind's Eye: The History, Techniques and Uses of Visualization*, Random House: New York.

13   **Hunt, K. H.** (1978). *Kinematic Geometry of Mechanisms*, Oxford University Press: Oxford, p. 1.

1.15   BIBLIOGRAPHY

For additional information on the **history of kinematics**, *the following are recommended:*

Artobolevsky, I. I. (1976). "Past Present and Future of the Theory of Machines and Mechanisms." *Mechanism and Machine Theory*, **11**, pp. 353-361.

Brown, H. T. (1869). *Five Hundred and Seven Mechanical Movements*. Brown, Coombs & Co.: New York, republished by USM Corporation, Beverly, MA, 1970.

de Jonge, A. E. R. (1942). "What Is Wrong with 'Kinematics' and 'Mechanisms'?" *Mechanical Engineering*, **64** (April), pp. 273-278.

de Jonge, A. (1943). "A Brief Account of Modern Kinematics." *Transactions of the ASME*, pp. 663-683.

Erdman, A. E., ed. (1993). *Modern Kinematics: Developments in the Last Forty Years*. Wiley Series in Design Engineering, John Wiley & Sons: New York.

Ferguson, E. S. (1962). "Kinematics of Mechanisms from the Time of Watt." *United States National Museum Bulletin*, **228**(27), pp. 185-230.

Freudenstein, F. (1959). "Trends in the Kinematics of Mechanisms." *Applied Mechanics Reviews*, **12**(9), September, pp. 587-590.

Hartenberg, R. S., and J. Denavit. (1964). *Kinematic Synthesis of Linkages*. McGraw-Hill: New York, pp. 1-27.

Nolle, H. (1974). "Linkage Coupler Curve Synthesis: A Historical Review - I. Developments up to 1875." *Mechanism and Machine Theory*, 9, pp. 147-168.

Nolle, H. (1974). "Linkage Coupler Curve Synthesis: A Historical Review - II. Developments after 1875." *Mechanism and Machine Theory*, 9, pp. 325-348.

Nolle, H. (1975). "Linkage Coupler Curve Synthesis: A Historical Review - III. Spatial Synthesis and Optimization." *Mechanism and Machine Theory*, 10, pp. 41-55.

Reuleaux, F. (1963). *The Kinematics of Machinery*, A. B. W. Kennedy, translator. Dover Publications: New York, pp. 29-55.

Strandh, S. (1979). *A History of the Machine*. A&W Publishers: New York.

For additional information on **creativity and the design process**, *the following are recommended:*

Alger, J., and C. V. Hays. (1964). *Creative Synthesis in Design*. Prentice-Hall: Upper Saddle River, NJ.

Allen, M. S. (1962). *Morphological Creativity*. Prentice-Hall: Upper Saddle River, NJ.

Altschuller, G. (1984). *Creativity as an Exact Science*. Gordon and Breach: New York.

Buhl, H. R. (1960). *Creative Engineering Design*. Iowa State University Press: Ames, IA.

Dixon, J. R., and C. Poli. (1995). *Engineering Design and Design for Manufacturing—A Structured Approach*. Field Stone Publishers: Conway, MA.

Fey, V., et al. (1994). "Application of the Theory of Inventive Problem Solving to Design and Manufacturing Systems." *CIRP Annals*, **43**(1), pp. 107-110.

Fuller, R. B. (1975). *Synergetics: Explorations in the Geometry of Thinking*. Macmillan.

Fuller, R. B. (1979). *Synergetics 2*. Macmillan.

Glegg, G. C. (1969). *The Design of Design.* Cambridge University Press: Cambridge, UK.

Glegg, G. C. (2009). *The Science of Design.* Cambridge University Press: Cambridge, UK.

Glegg, G. C. (2009). *The Selection of Design.* Cambridge University Press: Cambridge, UK.

Gordon, W. J. J. (1962). *Synectics.* Harper & Row: New York.

Grillo, P. J. (1960) *Form Function and Design.* Dover Publications: New York.

Haefele, W. J. (1962). *Creativity and Innovation.* Van Nostrand Reinhold: New York.

Harrisberger, L. (1982). *Engineersmanship.* Brooks/Cole: Monterey, CA.

Johnson, C. L. (1985). *Kelly: More than My Share of it All.* Smithsonian Inst. Press: Washington, DC.

Kim, S. (1981). *Inversions.* Byte Books, McGraw-Hill: New York.

Moore, A. D. (1969). *Invention, Discovery, and Creativity.* Doubleday Anchor Books: New York.

Norman, D. A. (1990). *The Design of Everyday Things.* Doubleday: New York.

Norman, D. A. (1992). *Turn Signals are the Facial Expressions of Automobiles.* Addison-Wesley: Reading, MA.

Osborn, A. F. (1963). *Applied Imagination.* Scribners: New York.

Pleuthner, W. (1956). "Brainstorming." *Machine Design*, January 12, 1956.

Suh, N. P. (1990). *The Principles of Design.* Oxford University Press: New York.

Taylor, C. W. (1964). *Widening Horizons in Creativity.* John Wiley & Sons: New York.

Unknown. (1980). *Ed Heinemann: Combat Aircraft Designer.* Naval Institute Press, 1980.

Von Fange, E. K. (1959). *Professional Creativity.* Prentice-Hall: Upper Saddle River, NJ.

*For additional information on **human factors**, the following are recommended:*

Bailey, R. W. (1982). *Human Performance Engineering: A Guide for System Designers.* Prentice-Hall: Upper Saddle River, NJ.

Burgess, W. R. (1986). *Designing for Humans: The Human Factor in Engineering.* Petrocelli Books.

Clark, T., and E. Corlett. (1984). *The Ergonomics of Workspaces and Machines.* Taylor and Francis.

Huchinson, R. D. (1981). *New Horizons for Human Factors in Design.* McGraw-Hill: New York.

McCormick, D. J. (1964). *Human Factors Engineering.* McGraw-Hill: New York.

Osborne, D. J. (1987). *Ergonomics at Work.* John Wiley & Sons: New York.

Pheasant, S. (1986). *Bodyspace: Anthropometry, Ergonomics & Design.* Taylor and Francis.

Salvendy, G. (1987). *Handbook of Human Factors.* John Wiley & Sons: New York.

Sanders, M. S. (1987). *Human Factors in Engineering and Design.* McGraw-Hill: New York.

Woodson, W. E. (1981). *Human Factors Design Handbook.* McGraw-Hill: New York.

*For additional information on **writing engineering reports**, the following are recommended:*

Barrass, R. (1978). *Scientists Must Write.* John Wiley & Sons: New York.

Crouch, W. G., and R. L. Zetler. (1964). *A Guide to Technical Writing.* The Ronald Press: New York.

Davis, D. S. (1963). *Elements of Engineering Reports.* Chemical Publishing Co.: New York.

Gray, D. E. (1963). *So You Have to Write a Technical Report.* Information Resources Press: Washington, DC.

Michaelson, H. B. (1982). *How to Write and Publish Engineering Papers and Reports.* ISI Press: Philadelphia, PA.

Nelson, J. R. (1952). *Writing the Technical Report.* McGraw-Hill: New York.

*Some useful **web sites** for design, product, and manufacturing information:*

http://www.machinedesign.com

> *Machine Design* magazine's site with articles and reference information for design (searchable).

http://www.motionsystemdesign.com

> *Motion System Design* magazine's site with articles and reference information for design and data on motors, bearings, etc. (searchable).

http://www.thomasregister.com

> Thomas Register is essentially a national listing of companies by product or service offered (searchable).

http://www.howstuffworks.com

> Much useful information on a variety of engineering devices (searchable).

http://www.manufacturing.net/dn/index.asp

> *Design News* magazine's site with articles and information for design (searchable).

http://iel.ucdavis.edu/design/

> University of California Davis Integration Engineering Laboratory site with applets that animate various mechanisms.

http://kmoddl.library.cornell.edu/

> A collection of mechanical models and related resources for teaching the principles of kinematics including the Reuleaux Collection of Mechanisms and Machines, an important collection of 19th-century machine elements held by Cornell's Sibley School of Mechanical and Aerospace Engineering.

http://www.mech.uwa.edu.au/DANotes/design/home.html

> A good description of the design process from Australia.

*Suggested **keywords** for searching the web for more information:*

machine design,   mechanism,   linkages,   linkage design,   kinematics,   cam design

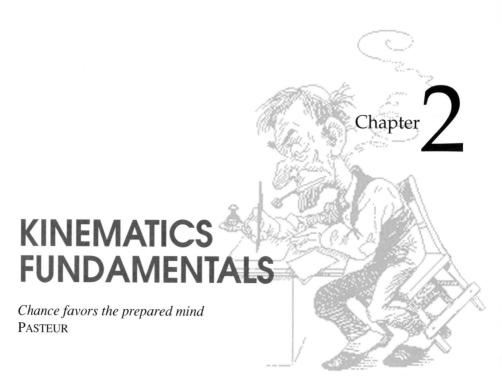

Chapter 2

KINEMATICS FUNDAMENTALS

Chance favors the prepared mind
PASTEUR

2.0 INTRODUCTION

This chapter will present definitions of a number of terms and concepts fundamental to the synthesis and analysis of mechanisms. It will also present some very simple but powerful analysis tools that are useful in the synthesis of mechanisms.

2.1 DEGREES OF FREEDOM (DOF) OR MOBILITY

A mechanical system's **mobility** (M) can be classified according to the number of **degrees of freedom** (*DOF*) that it possesses. The system's *DOF* is equal to *the number of independent parameters (measurements) that are needed to uniquely define its position in space at any instant of time*. Note that *DOF* is defined with respect to a selected frame of reference. Figure 2-1 shows a pencil lying on a flat piece of paper with an x, y coordinate system added. If we constrain this pencil to always remain in the plane of the paper, three parameters (*DOF*) are required to completely define the position of the pencil on the paper, two linear coordinates (x, y) to define the position of any one point on the pencil and one angular coordinate (θ) to define the angle of the pencil with respect to the axes. The minimum number of measurements needed to define its position is shown in the figure as $x, y,$ and θ. This system of the pencil in a plane then has **three** *DOF*. Note that the particular parameters chosen to define its position are not unique. Any alternate set of three parameters could be used. There is an infinity of sets of parameters possible, but in this case there must be three parameters per set, **such as two lengths and an angle**, to define the system's position because *a rigid body in plane motion has three DOF*.

2

FIGURE 2-1

A rigid body in a plane has three *DOF*

Now allow the pencil to exist in a three-dimensional world. Hold it above your desktop and move it about. You now will need six parameters to define its **six** *DOF*. One possible set of parameters that could be used is **three lengths**, (x, y, z), plus **three angles** (θ, ϕ, ρ). *Any rigid body in three-space has six degrees of freedom.* Try to identify these six *DOF* by moving your pencil or pen with respect to your desktop.

The pencil in these examples represents a **rigid body**, or **link**, which for purposes of kinematic analysis we will assume to be incapable of deformation. This is merely a convenient fiction to allow us to more easily define the gross motions of the body. We can later superpose any deformations due to external or inertial loads onto our kinematic motions to obtain a more complete and accurate picture of the body's behavior. But remember, we are typically facing a *blank sheet of paper* at the beginning stage of the design process. We cannot determine deformations of a body until we define its size, shape, material properties, and loadings. Thus, at this stage we will assume, for purposes of initial kinematic synthesis and analysis, that *our kinematic bodies are rigid and massless.*

2.2   TYPES OF MOTION

A rigid body free to move within a reference frame will, in the general case, have **complex motion**, which is a simultaneous combination of **rotation** and **translation**. In three-dimensional space, there may be rotation about any axis (any skew axis or one of the three principal axes) and also simultaneous translation that can be resolved into components along three axes. In a plane, or two-dimensional space, complex motion becomes a combination of simultaneous rotation about one axis (perpendicular to the plane) and also translation resolved into components along two axes in the plane. For simplicity, we will limit our present discussions to the case of **planar (2-D) kinematic systems**. We will define these terms as follows for our purposes, in planar motion:

Pure rotation

The body possesses one point (center of rotation) that has no motion with respect to the "stationary" frame of reference. All other points on the body describe arcs about that center. A reference line drawn on the body through the center changes only its angular orientation.

Pure translation

All points on the body describe parallel (curvilinear or rectilinear) paths. A reference line drawn on the body changes its linear position but does not change its angular orientation.

Complex motion

A simultaneous combination of rotation and translation. Any reference line drawn on the body will change both its linear position and its angular orientation. Points on the body will travel nonparallel paths, and there will be, at every instant, a center of rotation, which will continuously change location.

Translation and **rotation** represent independent motions of the body. Each can exist without the other. If we define a 2-D coordinate system as shown in Figure 2-1 (p. 31), the x and y terms represent the translation components of motion, and the θ term represents the rotation component.

2.3   LINKS, JOINTS, AND KINEMATIC CHAINS

We will begin our exploration of the kinematics of mechanisms with an investigation of the subject of **linkage design**. Linkages are the basic building blocks of all mechanisms. We will show in later chapters that all common forms of mechanisms (cams, gears, belts, chains) are in fact variations on a common theme of linkages. Linkages are made up of links and joints.

A **link**, as shown in Figure 2-2, is an (assumed) rigid body that possesses at least two **nodes** that are *points for attachment to other links.*

Binary link   *- one with two nodes.*

Ternary link   *- one with three nodes.*

Quaternary link   *- one with four nodes.*

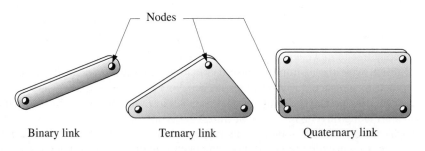

Binary link   Ternary link   Quaternary link

FIGURE 2-2

Links of different order

A **joint** is *a connection between two or more links (at their nodes), which allows some motion, or potential motion, between the connected links.* **Joints** (also called **kinematic pairs**) can be classified in several ways:

1 By the type of contact between the elements, line, point, or surface.

2 By the number of degrees of freedom allowed at the joint.

3 By the type of physical closure of the joint: either **force** or **form** closed.

4 By the number of links joined (order of the joint).

Reuleaux[1] coined the term **lower pair** to describe joints with surface contact (as with a pin surrounded by a hole) and the term **higher pair** to describe joints with point or line contact. However, if there is any clearance between pin and hole (as there must be for motion), so-called surface contact in the pin joint actually becomes line contact, as the pin contacts only one "side" of the hole. Likewise, at a microscopic level, a block sliding on a flat surface actually has contact only at discrete points, which are the tops of the surfaces' asperities. The main practical advantage of lower pairs over higher pairs is their better ability to trap lubricant between their enveloping surfaces. This is especially true for the rotating pin joint. The lubricant is more easily squeezed out of a higher pair, nonenveloping joint. As a result, the pin joint is preferred for low wear and long life, even over its lower pair cousin, the prismatic or slider joint.

Figure 2-3a (p. 34) shows the six possible lower pairs, their degrees of freedom, and their one-letter symbols. The revolute (R) and the prismatic (P) pairs are the only lower pairs usable in a planar mechanism. The screw (H), cylindric (C), spherical (S), and flat (F) lower pairs are all combinations of the revolute and/or prismatic pairs and are used in spatial (3-D) mechanisms. The R and P pairs are the basic building blocks of all other pairs that are combinations of those two as shown in Table 2-1.

A more useful means to classify joints (pairs) is by the number of degrees of freedom that they allow between the two elements joined. Figure 2-3 (p. 34) also shows examples of both one- and two-freedom joints commonly found in planar mechanisms. Figure 2-3b (p. 34) shows two forms of a planar, **one-freedom** joint (or pair), namely, a rotating (revolute) pin joint (R) and a translating (prismatic) slider joint (P). These are also referred to as **full joints** (i.e., full = 1 *DOF*) and are **lower pairs**. The pin joint allows one rotational *DOF*, and the slider joint allows one translational *DOF* between the joined links. These are both contained within (and each is a limiting case of) another common, one-freedom joint, the screw and nut (Figure 2-3a). Motion of either the nut or the screw with respect to the other results in helical motion. If the helix angle is made zero, the nut rotates without advancing and it becomes the pin joint. If the helix angle is made 90 degrees, the nut will translate along the axis of the screw, and it becomes the slider joint.

Figure 2-3c shows examples of two-freedom joints (higher pairs) that simultaneously allow two independent, relative motions, namely translation and rotation, between the joined links. Paradoxically, this **two-freedom joint** is sometimes referred to as a "**half joint**," with its two freedoms placed in the denominator. The **half joint** is also called a **roll-slide joint** because it allows both rolling and sliding. A spherical, or ball-and-socket joint, (Figure 2-3a) is an example of a three-freedom joint, which allows three independent angular motions between the two links joined. This *joystick* or *ball joint* is typically used in a three-dimensional mechanism, one example being the ball joints in an automotive suspension system.

TABLE 2-1
The Six Lower Pairs

Name (Symbol)	DOF	Contains
Revolute (R)	1	R
Prismatic (P)	1	P
Helical (H)	1	RP
Cylindric (C)	2	RP
Spherical (S)	3	RRR
Planar (F)	3	RPP

2

Revolute (R) joint—1 *DOF*

square X-section

Prismatic (P) joint—1 *DOF*

Helical (H) joint—1 *DOF*

Cylindric (C) joint—2 *DOF*

Spherical (S) joint—3 *DOF*

Planar (F) joint—3 *DOF*

(*a*) The six lower pairs

Rotating full pin (R) joint (form closed)

Translating full slider (P) joint (form closed)

(*b*) Full joints - 1 *DOF* (lower pairs)

Link against plane (force closed)

Pin in slot (form closed)

(*c*) Roll-slide (half or RP) joints - 2 *DOF* (higher pairs)

First order pin joint - 1 *DOF*
(two links joined)

Second order pin joint - 2 *DOF*
(three links joined)

(*d*) The order of a joint is one less than the number of links joined

May roll, slide, or roll-slide, depending on friction

(*e*) Planar pure-roll (R), pure-slide (P), or roll-slide (RP) joint - 1 or 2 *DOF* (higher pair)

FIGURE 2-3

Joints (pairs) of various types

A joint with more than one freedom may also be a **higher pair** as shown in Figure 2-3c. Full joints (lower pairs) and half joints (higher pairs) are both used in planar (2-D), and in spatial (3-D) mechanisms. Note that if you do not allow the two links in Figure 2-3c connected by a roll-slide joint to slide, perhaps by providing a high friction coefficient between them, you can "lock out" the translating (Δx) freedom and make it behave as a full joint. This is then called a **pure rolling joint** and has rotational freedom ($\Delta \theta$) only. A common example of this type of joint is your automobile tire rolling against the road, as shown in Figure 2-3e. In normal use there is pure rolling and no sliding at this joint, unless, of course, you encounter an icy road or become too enthusiastic about accelerating or cornering. If you lock your brakes on ice, this joint converts to a pure sliding one like the slider block in Figure 2-3b. Friction determines the actual number of freedoms at this kind of joint. It can be **pure roll**, **pure slide**, or **roll-slide**.

To visualize the degree of freedom of a joint in a mechanism, it is helpful to "mentally disconnect" the two links that create the joint from the rest of the mechanism. You can then more easily see how many freedoms the two joined links have with respect to one another.

Figure 2-3c also shows examples of both **form-closed** and **force-closed** joints. A **form-closed** joint is kept together or *closed by its geometry*. A pin in a hole or a slider in a two-sided slot is form closed. In contrast, a **force-closed** joint, such as a pin in a half-bearing or a slider on a surface, *requires some external force to keep it together or closed*. This force could be supplied by gravity, a spring, or any external means. There can be substantial differences in the behavior of a mechanism due to the choice of force or form closure, as we shall see. The choice should be carefully considered. In linkages, form closure is usually preferred, and it is easy to accomplish. But for cam-follower systems, force closure is often preferred. This topic will be explored further in later chapters.

Figure 2-3d shows examples of joints of various orders, where **joint order** is defined as *the number of links joined minus one*. It takes two links to make a single joint; thus the simplest joint combination of two links has joint order one. As additional links are placed on the same joint, the joint order is increased on a one-for-one basis. Joint order has significance in the proper determination of overall degree of freedom for the assembly. We gave definitions for a **mechanism** and a **machine** in Chapter 1. With the kinematic elements of links and joints now defined, we can define those devices more carefully based on Reuleaux's classifications of the kinematic chain, mechanism, and machine. [1]

A kinematic chain is defined as:

An assemblage of links and joints, interconnected in a way to provide a controlled output motion in response to a supplied input motion.

A mechanism is defined as:

A kinematic chain in which at least one link has been "grounded," or attached, to the frame of reference (which itself may be in motion).

A machine is defined as:

A combination of resistant bodies arranged to compel the mechanical forces of nature to do work accompanied by determinate motions.

2

By Reuleaux's[*] definition[1] a machine is *a collection of mechanisms arranged to transmit forces and do work.* He viewed all energy or force transmitting devices as machines that utilize mechanisms as their building blocks to provide the necessary motion constraints.

We will now define a **crank** as *a link that makes a complete revolution and is pivoted to ground*, a **rocker** as *a link that has oscillatory (back and forth) rotation and is pivoted to ground*, and a **coupler** (or connecting rod) *as a link that has complex motion and is not pivoted to ground.* **Ground** is defined as *any link or links that are fixed* (nonmoving) with respect to the reference frame. Note that the reference frame may in fact itself be in motion.

2.4   DRAWING KINEMATIC DIAGRAMS

Analyzing the kinematics of mechanisms requires that we draw clear, simple, schematic kinematic diagrams of the links and joints of which they are made. Sometimes it can be difficult to identify the kinematic links and joints in a complicated mechanism. Beginning students of this topic often have this difficulty. This section defines one approach to the creation of simplified kinematic diagrams.

Real links can be of any shape, but a "kinematic" link, or link edge, is defined as a line between joints that allow relative motion between adjacent links. Joints can allow rotation, translation, or both between the links joined. The possible joint motions must be clear and obvious from the kinematic diagram. Figure 2-4 shows recommended schematic notations for binary, ternary, and higher-order links, and for movable and grounded joints of rotational and translational freedoms plus an example of their combination. Many other notations are possible, but whatever notation is used, it is critical that your diagram indicate which links or joints are grounded and which can move. Otherwise nobody will be able to interpret your design's kinematics. Shading or crosshatching should be used to indicate that a link is solid.

Figure 2-5a shows a photograph of a simple mechanism used for weight training called a leg press machine. It has six pin-jointed links labeled L_1 through L_6 and seven pin joints. The moving pivots are labeled A through D; O_2, O_4 and O_6 denote the grounded pivots of their respective link numbers. Even though its links are in parallel planes sepa-

---

[*] Reuleaux created a set of 220 models of mechanisms in the 19th century to demonstrate machine motions. Cornell University acquired the collection in 1892 and has now put images and descriptions of them on the web at: http://kmoddl. library.cornell.edu. The same site also has depictions of three other collections of machines and gear trains.

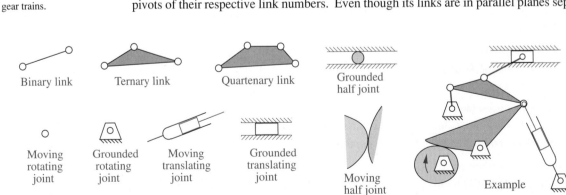

Binary link   Ternary link   Quaternary link   Grounded half joint

Moving rotating joint   Grounded rotating joint   Moving translating joint   Grounded translating joint   Moving half joint   Example

FIGURE 2-4

Schematic notation for kinematic diagrams

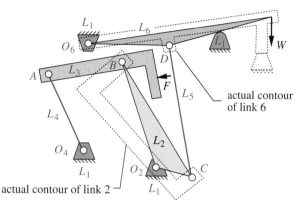

(a) Weight-training mechanism   (b) Kinematic diagram

FIGURE 2-5

A mechanism and its kinematic diagram

rated by some distance in the *z*-direction, it can still be analyzed kinematically as if all links were in a common plane.

To use the leg press machine, the user loads some weights on link 6 at top right, sits in the seat at lower right, places both feet against the flat surface of link 3 (a coupler) and pushes with the legs to lift the weights through the linkage. The linkage geometry is designed to give a variable mechanical advantage that matches the human ability to provide force over the range of leg motion. Figure 2-5b shows a kinematic diagram of its basic mechanism. Note that here all the links have been brought to a common plane. Link 1 is the ground. Links 2, 4, and 6 are rockers. Links 3 and 5 are couplers. The input force *F* is applied to link 3. The "output" resistance weight *W* acts on link 6. Note the difference between the actual and kinematic contours of links 2 and 6.

The next section discusses techniques for determining the mobility of a mechanism. That exercise depends on an accurate count of the number of links and joints in the mechanism. Without a proper, clear, and complete kinematic diagram of the mechanism, it will be impossible to get the count, and thus the mobility, correct.

2.5   DETERMINING DEGREE OF FREEDOM OR MOBILITY

The concept of **degree of freedom** (*DOF*) is fundamental to both the synthesis and analysis of mechanisms. We need to be able to quickly determine the *DOF* of any collection of links and joints that may be suggested as a solution to a problem. Degree of freedom (also called the **mobility *M*)** of a system can be defined as:

Degree of Freedom

the number of inputs that need to be provided in order to create a predictable output;

also:

the number of independent coordinates required to define its position.

2

At the outset of the design process, some general definition of the desired output motion is usually available. The number of inputs needed to obtain that output may or may not be specified. Cost is the principal constraint here. Each required input will need some type of actuator, either a human operator or a "slave" in the form of a motor, solenoid, air cylinder, or other energy conversion device. (These devices are discussed in Section 2.19 on p. 74.) These multiple-input devices will have to have their actions coordinated by a "controller," which must have some intelligence. This control is now often provided by a computer but can also be mechanically programmed into the mechanism design. There is no requirement that a mechanism have only one *DOF*, although that is often desirable for simplicity. Some machines have many *DOF*. For example, picture the number of control levers or actuating cylinders on a bulldozer or crane. See Figure 1-1b (p. 7).

Kinematic chains or mechanisms may be either **open** or **closed**. Figure 2-6 shows both open and closed mechanisms. A closed mechanism will have no open attachment points or **nodes** and may have one or more degrees of freedom. An open mechanism of more than one link will always have more than one degree of freedom, thus requiring as many actuators (motors) as it has *DOF*. A common example of an open mechanism is an industrial robot. *An open kinematic chain of two binary links and one joint* is called a **dyad**. The sets of links shown in Figure 2-3b and c (p. 34) are **dyads**.

Reuleaux limited his definitions to closed kinematic chains and to mechanisms having only one *DOF*, which he called *constrained*.[1] The somewhat broader definitions above are perhaps better suited to current-day applications. A multi-*DOF* mechanism, such as a robot, will be constrained in its motions as long as the necessary number of inputs is supplied to control all its *DOF*.

Degree of Freedom (Mobility) in Planar Mechanisms

To determine the overall *DOF* of any mechanism, we must account for the number of links and joints, and for the interactions among them. The *DOF* of any assembly of links can be predicted from an investigation of the **Gruebler condition**.[2] Any link in a plane has 3 *DOF*. Therefore, a system of *L* unconnected links in the same plane will have 3*L*

(*a*) Open mechanism chain   (*b*) Closed mechanism chain

FIGURE 2-6

Mechanism chains

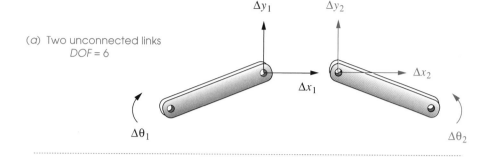

(a) Two unconnected links
     DOF = 6

(b) Connected by a full joint
     DOF = 4

(c) Connected by a roll-slide (half) joint
     DOF = 5

FIGURE 2-7

Joints remove degrees of freedom

DOF, as shown in Figure 2-7a where the two unconnected links have a total of six *DOF.*
When these links are connected by a **full joint** in Figure 2-7b, Δy_1 and Δy_2 are combined
as Δy, and Δx_1 and Δx_2 are combined as Δx. This removes two *DOF*, leaving four *DOF.*
In Figure 2-7c the half joint removes only one *DOF* from the system (because a half joint
has two *DOF*), leaving the system of two links connected by a half joint with a total of five
DOF. In addition, when any link is grounded or attached to the reference frame, all three
of its *DOF* will be removed. This reasoning leads to **Gruebler's equation:**

$$M = 3L - 2J - 3G \qquad (2.1a)$$

where: M = *degree of freedom or mobility*
   L = *number of links*
   J = *number of joints*
   G = *number of grounded links*

2

Note that in any real mechanism, even if more than one link of the kinematic chain is grounded, the net effect will be to create one larger, higher-order ground link, as there can be only one ground plane. Thus G is always one, and Gruebler's equation becomes:

$$M = 3(L-1) - 2J \qquad (2.1b)$$

The value of J in equations 2.1a and 2.1b must reflect the value of all joints in the mechanism. That is, half joints count as 1/2 because they only remove one *DOF*. It is less confusing if we use **Kutzbach's** modification of Gruebler's equation in this form:

$$M = 3(L-1) - 2J_1 - J_2 \qquad (2.1c)$$

where: M = *degree of freedom or mobility*
         L = *number of links*
         J_1 = *number of 1 DOF (full) joints*
         J_2 = *number of 2 DOF (half) joints*

The value of J_1 and J_2 in these equations must still be carefully determined to account for all full, half, and multiple joints in any linkage. Multiple joints count as one less than the number of links joined at that joint and add to the "full" (J_1) category. The *DOF* of any proposed mechanism can be quickly ascertained from this expression before investing any time in more detailed design. It is interesting to note that this equation has no information in it about link sizes or shapes, only their quantity. Figure 2-8a shows a mechanism with one *DOF* and only full joints in it.

Figure 2-8b shows a structure with zero *DOF* that contains both half and multiple joints. Note the schematic notation used to show the ground link. The ground link need not be drawn in outline as long as all the grounded joints are identified. Note also the joints labeled **multiple** and **half** in Figure 2-8a and b. As an exercise, compute the *DOF* of these examples with **Kutzbach's** equation.

Degree of Freedom (Mobility) in Spatial Mechanisms

The approach used to determine the mobility of a planar mechanism can be easily extended to three dimensions. Each unconnected link in three-space has 6 *DOF*, and any one of the six lower pairs can be used to connect them, as can higher pairs with more freedom. A one-freedom joint removes 5 *DOF*, a two-freedom joint removes 4 *DOF*, etc. Grounding a link removes 6 *DOF*. This leads to the Kutzbach mobility equation for spatial linkages:

$$M = 6(L-1) - 5J_1 - 4J_2 - 3J_3 - 2J_4 - J_5 \qquad (2.2)$$

where the subscript refers to the number of freedoms of the joint. We will limit our study to 2-D mechanisms in this text.

2.6   MECHANISMS AND STRUCTURES

The degree of freedom of an assembly of links completely predicts its character. There are only three possibilities. *If the DOF is positive, it will be a **mechanism**, and the links will have relative motion. If the DOF is exactly zero, then it will be a **structure**, and no motion is possible. If the DOF is negative, then it is a **preloaded structure**,* which means that no motion is possible and some stresses may also be present at the time of assembly. Figure 2-9 (p. 42) shows examples of these three cases. One link is grounded in each case.

2

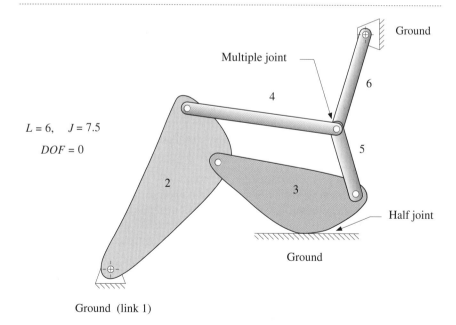

Note:
There are no
roll-slide
(half) joints
in this
linkage

Ground

Sliding full joint

8

7

Ground

5

6

$L = 8,\quad J = 10$

$DOF = 1$

3

ω

2

4

Multiple joint

Ground (link 1)

Ground

(a) Linkage with full and multiple joints

Ground

Multiple joint

4

6

$L = 6,\quad J = 7.5$

$DOF = 0$

2

3

5

Half joint

Ground

Ground (link 1)

(b) Linkage with full, half, and multiple joints

FIGURE 2-8

Linkages containing joints of various types

2

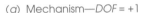

(a) Mechanism—DOF = +1   (b) Structure—DOF = 0   (c) Preloaded structure—DOF = −1

FIGURE 2-9

Mechanisms, structures, and preloaded structures

---

* If the sum of the lengths of any two links is less than the length of the third link, then their interconnection is impossible.

---

† The concept of *exact constraint* also applies to mechanisms with positive *DOF*. It is possible to provide redundant constraints to a mechanism (e.g., making its theoretical *DOF* = 0 when 1 is desired) yet still have it move because of particular geometry (see Section 2.8 *Paradoxes*). Non-exact constraint should be avoided in general as it can lead to unexpected mechanical behavior. For an excellent and thorough discussion of this issue see Blanding, D. L., *Exact Constraint: Machine Design Using Kinematic Principles*, ASME Press, 1999.

---

§ Not to be confused with "joint order" as defined earlier, which refers to the number of *DOF* that a joint possesses.

Figure 2-9a shows four links joined by four full joints which, from the Gruebler equation, gives one *DOF*. It will move, and only one input is needed to give predictable results.

Figure 2-9b shows three links joined by three full joints. It has zero *DOF* and is thus a **structure**. Note that if the link lengths will allow connection,* all three pins can be inserted into their respective pairs of link holes (nodes) without straining the structure, as a position can always be found to allow assembly. This is called *exact constraint.*†

Figure 2-9c shows two links joined by two full joints. It has a *DOF* of minus one, making it a **preloaded structure**. In order to insert the two pins without straining the links, the center distances of the holes in both links must be exactly the same. Practically speaking, it is impossible to make two parts exactly the same. There will always be some manufacturing error, even if very small. Thus you may have to force the second pin into place, creating some stress in the links. The structure will then be preloaded. You have probably met a similar situation in a course in applied mechanics in the form of an indeterminate beam, one in which there were too many supports or constraints for the equations available. An indeterminate beam also has negative *DOF*, while a *simply supported* beam has zero *DOF*.

Both structures and preloaded structures are commonly encountered in engineering. In fact the true structure of zero *DOF* is rare in civil engineering practice. Most buildings, bridges, and machine frames are preloaded structures, due to the use of welded and riveted joints rather than pin joints. Even simple structures like the chair you are sitting in are often preloaded. Since our concern here is with mechanisms, we will concentrate on devices with positive *DOF* only.

2.7   NUMBER SYNTHESIS

The term **number synthesis** has been coined to mean *the determination of the number and order of links and joints necessary to produce motion of a particular DOF*. **Link order** in this context refers to the number of nodes per link,§ i.e., **binary**, **ternary**, **quaternary**, etc. The value of number synthesis is to allow the exhaustive determination of all possible combinations of links that will yield any chosen *DOF*. This then equips the designer with a definitive catalog of potential linkages to solve a variety of motion control problems.

As an example we will now derive all the possible link combinations for one *DOF*, including sets of up to eight links, and link orders up to and including hexagonal links. For simplicity we will assume that the links will be connected with only single, full rotating joints (i.e., a pin connecting two links). We can later introduce half joints, multiple joints, and sliding joints through linkage transformation. First let's look at some interesting attributes of linkages as defined by the above assumption regarding full joints.

Hypothesis: If all joints are full joints, an odd number of *DOF* requires an even number of links and vice versa.

Proof: **Given:** All even integers can be denoted by $2m$ or by $2n$, and all odd integers can be denoted by $2m - 1$ or by $2n - 1$, where n and m are any positive integers. The number of joints must be a positive integer.

Let : L = number of links, J = number of joints, and $M = DOF = 2m$ (i.e., all even numbers)

Then: Rewriting Gruebler's equation 2.1b to solve for J,

$$J = \frac{3}{2}(L-1) - \frac{M}{2} \qquad (2.3a)$$

Try: Substituting $M = 2m$, and $L = 2n$ (i.e., both any even numbers):

$$J = 3n - m - \frac{3}{2} \qquad (2.3b)$$

This cannot result in J being a positive integer as required.

Try: $M = 2m - 1$ and $L = 2n - 1$ (i.e., both any odd numbers):

$$J = 3n - m - \frac{5}{2} \qquad (2.3c)$$

This also cannot result in J being a positive integer as required.

Try: $M = 2m - 1$, and $L = 2n$ (i.e., odd-even):

$$J = 3n - m - 2 \qquad (2.3d)$$

This is a positive integer for $m \geq 1$ and $n \geq 2$.

Try: $M = 2m$ and $L = 2n - 1$ (i.e., even-odd):

$$J = 3n - m - 3 \qquad (2.3e)$$

This is a positive integer for $m \geq 1$ and $n \geq 2$.

So, for our example of one-*DOF* mechanisms, we can only consider combinations of 2, 4, 6, 8, . . . links. Letting the order of the links be represented by:

B = *number of binary links*
T = *number of ternary links*
Q = *number of quaternaries*
P = *number of pentagonals*
H = *number of hexagonals*

the total number of links in any mechanism will be:

$$L = B + T + Q + P + H + \cdots \qquad (2.4a)$$

Since *two link nodes* are needed to make *one joint*:

$$J = \frac{nodes}{2} \qquad (2.4b)$$

and

$$nodes = order\ of\ link \times no.\ of\ links\ of\ that\ order \qquad (2.4c)$$

then

$$J = \frac{(2B + 3T + 4Q + 5P + 6H + \cdots)}{2} \qquad (2.4d)$$

Substitute equations 2.4a and 2.4d into Gruebler's equation (2.1b, on p. 40)

$$M = 3(B + T + Q + P + H - 1) - 2\left(\frac{2B + 3T + 4Q + 5P + 6H}{2}\right) \qquad (2.4e)$$

$$M = B - Q - 2P - 3H - 3$$

Note what is missing from this equation! The ternary links have dropped out. The *DOF* is independent of the number of ternary links in the mechanism. But because each ternary link has three nodes, it can only create or remove 3/2 joints. So we must add or subtract ternary links in pairs to maintain an integer number of joints. *The addition or subtraction of ternary links in pairs will not affect the DOF of the mechanism.*

In order to determine all possible combinations of links for a particular *DOF*, we must combine equations 2.3a (p. 43) and 2.4d:[*]

$$\frac{3}{2}(L - 1) - \frac{M}{2} = \frac{2B + 3T + 4Q + 5P + 6H}{2} \qquad (2.5)$$

$$3L - 3 - M = 2B + 3T + 4Q + 5P + 6H$$

Now combine equation 2.5 with equation 2.4a to eliminate *B*:

$$L - 3 - M = T + 2Q + 3P + 4H \qquad (2.6)$$

We will now solve equations 2.4a and 2.6 simultaneously (by progressive substitution) to determine all compatible combinations of links for *DOF* = 1, up to eight links. The strategy will be to start with the smallest number of links, and the highest-order link possible with that number, eliminating impossible combinations.

(Note: *L must be even for odd DOF.*)

CASE 1.   **L = 2**

$$L - 4 = T + 2Q + 3P + 4H = -2 \qquad (2.7a)$$

This requires a negative number of links, so $L = 2$ is impossible.

* Karunamoorthy [17] defines some useful rules for determining the number of possible combinations for any number of links with a given degree of freedom.

CASE 2. $L = 4$

$$L - 4 = T + 2Q + 3P + 4H = 0 \qquad\qquad \text{so: } T = Q = P = H = 0$$

(2.7b)

$$L = B + 0 = 4 \qquad\qquad\qquad B = 4$$

The simplest one-*DOF* linkage is four binary links—the **fourbar linkage**.

CASE 3. $L = 6$

$$L - 4 = T + 2Q + 3P + 4H = 2 \qquad \text{so: } \qquad P = H = 0 \qquad (2.7c)$$

T may only be 0, 1, or 2; Q may only be 0 or 1

If $Q = 0$ then T must be 2 and:

$$L = B + 2T + 0Q = 6 \qquad\qquad B = 4 \qquad\qquad T = 2 \qquad (2.7d)$$

If $Q = 1$, then T must be 0 and:

$$L = B + 0T + 1Q = 6 \qquad\qquad B = 5 \qquad\qquad Q = 1 \qquad (2.7e)$$

There are then two possibilities for $L = 6$. Note that one of them is in fact the simpler fourbar with two ternaries added as was predicted above.

CASE 4. $L = 8$

A tabular approach is needed with this many links:

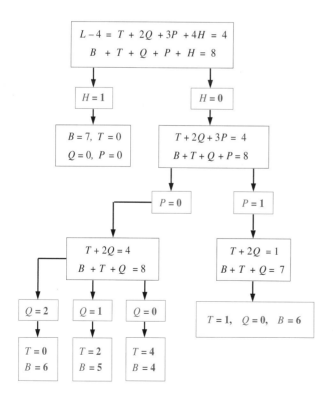

(2.7f)

TABLE 2-2   1-*DOF* Planar Mechanisms with Revolute Joints and Up to 8 Links

Total Links	Link Sets				
	Binary	Ternary	Quaternary	Pentagonal	Hexagonal
4	4	0	0	0	0
6	4	2	0	0	0
6	5	0	1	0	0
8	7	0	0	0	1
8	4	4	0	0	0
8	5	2	1	0	0
8	6	0	2	0	0
8	6	1	0	1	0

From this analysis we can see that, for one *DOF,* there is only one possible four-link configuration, two six-link configurations, and five possibilities for eight links using binary through hexagonal links. Table 2-2 shows the so-called "link sets" for all the possible linkages for one *DOF* up to 8 links and hexagonal order.

2.8   PARADOXES

Because the Gruebler criterion pays no attention to link sizes or shapes, it *can give misleading results* in the face of unique geometric configurations. For example, Figure 2-10a shows a structure (*DOF* = 0) with the ternary links of arbitrary shape. This link arrangement is sometimes called the "**E-quintet**," because of its resemblance to a capital **E** and the fact that it has five links, including the ground.[*] It is the next simplest **structural** building block to the "**delta triplet**."

Figure 2-10b shows the same E-quintet with the ternary links straight and parallel and with equispaced nodes. The three binaries are also equal in length. With this very unique geometry, you can see that it will move despite Gruebler's prediction to the contrary.

Figure 2-10c shows a very common mechanism that also disobeys Gruebler's criterion. The joint between the two wheels can be postulated to allow no slip, provided that sufficient friction is available. If no slip occurs, then this is a one-freedom, or full, joint that allows only relative angular motion ($\Delta\theta$) between the wheels. With that assumption, there are 3 links and 3 full joints, from which Gruebler's equation predicts zero *DOF*. However, this linkage does move (actual *DOF* = 1), because the center distance, or length of link 1, is exactly equal to the sum of the radii of the two wheels.

There are other examples of paradoxes that disobey the Gruebler criterion due to their unique geometry. The designer needs to be alert to these possible inconsistencies. Gogu[†] has shown that none of the simple mobility equations so far discovered (Gruebler, Kutzbach, etc.) are capable of resolving the many paradoxes that exist. A complete analysis of the linkage motions (as described in Chapter 4) is necessary to guarantee mobility.

[*] It is also called an Assur chain.

[†] Gogu, G. (2005), Mobility of Mechanisms: A Critical Review." *Mechanism and Machine Theory* (40) pp. 1068-1097.

(a) The E-quintet with $DOF = 0$
    —agrees with Gruebler equation

(b) The E-quintet with $DOF = 1$
    —disagrees with Gruebler equation
    due to unique geometry

Full joint -
pure rolling
no slip

(c) Rolling cylinders with $DOF = 1$
    —disagrees with Gruebler equation
    which predicts $DOF = 0$

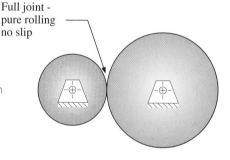

FIGURE 2-10

Gruebler paradoxes—linkages that do not behave as predicted by the Gruebler equation

2.9 ISOMERS

The word **isomer** is from the Greek and means *having equal parts.* Isomers in chemistry are compounds that have the same number and type of atoms but which are interconnected differently and thus have different physical properties. Figure 2-11a shows two hydrocarbon isomers, n-butane and isobutane. Note that each has the same number of carbon and hydrogen atoms (C_4H_{10}), but they are differently interconnected and have different properties.

Linkage isomers are analogous to these chemical compounds in that the **links** (like atoms) have various **nodes** (electrons) available to connect to other links' nodes. The assembled linkage is analogous to the chemical compound. Depending on the particular connections of available links, the assembly will have different motion properties. The number of isomers possible from a given collection of links (as in any row of Table 2-2) is far from obvious. In fact mathematically prediction of the number of isomers of all link combinations has been a long-unsolved problem. Many researchers have spent much

TABLE 2-3

Number of Valid Isomers

Links	Valid Isomers
4	1
6	2
8	16
10	230
12	6856

effort on this problem with some recent success. See references [3] through [7] for more information. Dhararipragada [6] presents a good historical summary of isomer research to 1994. Table 2-3 shows the number of valid isomers found for one-*DOF* mechanisms with revolute pairs, up to 12 links.

Figure 2-11b shows all the isomers for the simple cases of one *DOF* with 4 and 6 links. Note that there is only one isomer for the case of 4 links. An isomer is only unique if the interconnections between its types of links are different. That is, all binary links are considered equal, just as all hydrogen atoms are equal in the chemical analog. Link lengths and shapes do not figure into the Gruebler criterion or the condition of isomerism. The 6-link case of 4 binaries and 2 ternaries has only two valid isomers. These are known as **Watt's chain** and **Stephenson's chain** after their discoverers. Note the different interconnections of the ternaries to the binaries in these two examples. Watt's chain has the two ternaries directly connected, but Stephenson's chain does not.

There is also a third potential isomer for this case of six links, as shown in Figure 2-11c, but it fails the test of **distribution of degree of freedom**, which requires that the overall *DOF* (here 1) be uniformly distributed throughout the linkage and not concentrated in a subchain. Note that this arrangement (Figure 2-11c) has a **structural subchain** of *DOF* = 0 in the triangular formation of the two ternaries and the single binary connecting them. This creates a truss, or **delta triplet**. The remaining three binaries in series form a fourbar chain (*DOF* = 1) with the structural subchain of the two ternaries and the single binary effectively reduced to a structure that acts like a single link. Thus this arrangement has been reduced to the simpler case of the fourbar linkage despite its six bars. This is an **invalid isomer** and is rejected.

Franke's "Condensed Notation for Structural Synthesis" method can be used to help find the isomers of any collection of links that includes some links of higher order than binary. Each higher-order link is shown as a circle with its number of nodes (its valence) written in it as shown in Figure 2-11. These circles are connected with a number of lines emanating from each circle equal to its valence. A number is placed on each line to represent the quantity of binary links in that connection. This gives a "molecular" representation of the linkage and allows exhaustive determination of all the possible binary link interconnections among the higher links. Note the correspondence in Figure 2-11b between the linkages and their respective Franke molecules. The only combinations of 3 integers (including zero) that add to 4 are: (1, 1, 2), (2, 0, 2), (0, 1, 3), and (0, 0, 4). The first two are, respectively, Stephenson's and Watt's linkages; the third is the invalid isomer of Figure 2-11c. The fourth combination is also invalid as it results in a 2-DOF chain of 5 binaries in series with the 5th "binary" comprised of the two ternaries locked together at two nodes in a preloaded structure with a subchain *DOF* of –1. Figure 2-11d shows all 16 valid isomers of the eightbar 1-*DOF* linkage.

2.10 LINKAGE TRANSFORMATION

The number synthesis techniques described above give the designer a tool kit of basic linkages of particular *DOF*. If we now relax the arbitrary constraint that restricted us to only revolute joints, we can transform these basic linkages to a wider variety of mechanisms with even greater usefulness. There are several transformation techniques or rules that we can apply to planar kinematic chains.

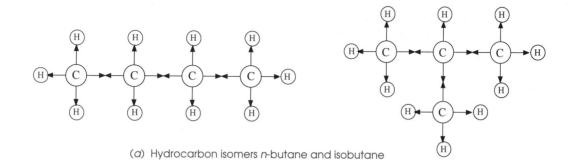

(a) Hydrocarbon isomers *n*-butane and isobutane

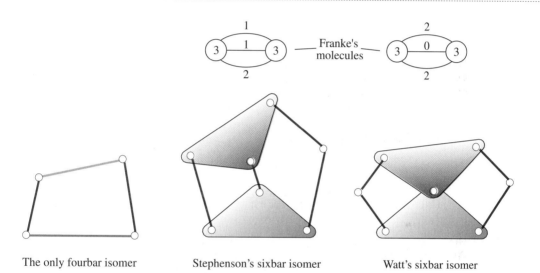

The only fourbar isomer   Stephenson's sixbar isomer   Watt's sixbar isomer

(b) All valid isomers of the fourbar and sixbar linkages

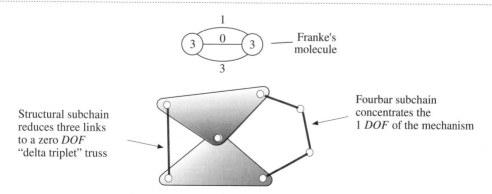

(c) An invalid sixbar isomer which reduces to the simpler fourbar

FIGURE 2-11 Part 1

Isomers of kinematic chains

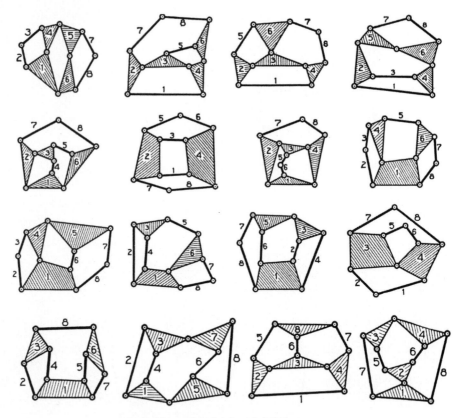

(*d*) All the valid eightbar 1-*DOF* isomers

FIGURE 2-11 Part 2

Isomers of kinematic chains (*Source: Klein, A. W., 1917. Kinematics of Machinery, McGraw-Hill, NY.*)

1 Revolute joints in any loop can be replaced by prismatic joints with no change in *DOF* of the mechanism, provided that at least two revolute joints remain in the loop.*

2 Any full joint can be replaced by a half joint, but this will increase the *DOF* by one.

3 Removal of a link will reduce the *DOF* by one.

4 The combination of rules 2 and 3 above will keep the original *DOF* unchanged.

5 Any ternary or higher-order link can be partially "shrunk" to a lower-order link by coalescing nodes. This will create a multiple joint but will not change the *DOF* of the mechanism.

6 Complete shrinkage of a higher-order link is equivalent to its removal. A multiple joint will be created, and the *DOF* will be reduced.

Figure 2-12a† shows a fourbar crank-rocker linkage transformed into the fourbar slider-crank by the application of rule #1. It is still a fourbar linkage. Link 4 has become a sliding block. Gruebler's equation is unchanged at one *DOF* because the slider block

Grashof crank-rocker

Grashof slider-crank

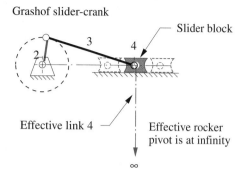

(a) Transforming a fourbar crank-rocker to a slider-crank

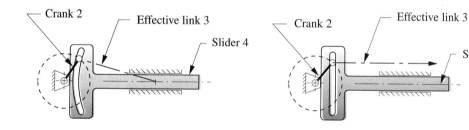

(b) Transforming the slider-crank to the Scotch yoke

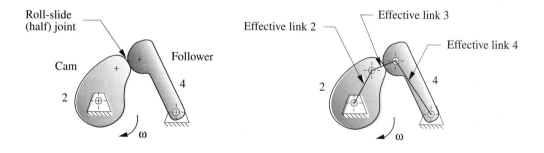

(c) The cam-follower mechanism has an effective fourbar equivalent

FIGURE 2-12

Linkage transformation

provides a full joint against link 1, as did the pin joint it replaces. Note that this transformation from a rocking output link to a slider output link is equivalent to increasing the length (radius) of rocker link 4 until its arc motion at the joint between links 3 and 4 becomes a straight line. Thus the slider block is equivalent to an infinitely long rocker link 4, which is pivoted at infinity along a line perpendicular to the slider axis as shown in Figure 2-12a (p. 51).[*]

Figure 2-12b[*] shows a fourbar slider-crank transformed via rule #4 by the substitution of a half joint for the coupler. The first version shown retains the same motion of the slider as the original linkage by use of a curved slot in link 4. The effective coupler is always perpendicular to the tangent of the slot and falls on the line of the original coupler. The second version shown has the slot made straight and perpendicular to the slider axis. The effective coupler now is "pivoted" at infinity. This is called a **Scotch yoke** and gives exact *simple harmonic motion* of the slider in response to a constant speed input to the crank.

Figure 2-12c shows a fourbar linkage transformed into a **cam-follower** linkage by the application of rule #4. Link 3 has been removed and a half joint substituted for a full joint between links 2 and 4. This still has one *DOF*, and the cam-follower is in fact a fourbar linkage in another disguise, in which the coupler (link 3) has become an effective link of *variable length*. We will investigate the fourbar linkage and these variants of it in greater detail in later chapters.

Figure 2-13a shows **Stephenson's sixbar chain** from Figure 2-11b (p. 49) transformed by *partial shrinkage* of a ternary link (rule #5) to create a multiple joint. It is still a one-*DOF* Stephenson sixbar. Figure 2-13b shows **Watt's sixbar chain** from Figure 2-11b (p. 49) with one ternary link *completely shrunk* to create a multiple joint. This is now a structure with *DOF* = 0. The two triangular subchains are obvious. Just as the fourbar chain is the basic building block of one-*DOF* mechanisms, this threebar triangle **delta triplet** is the *basic building block* of zero-*DOF* structures (trusses).

[*] This figure is provided as animated AVI and Working Model files on the DVD. Its filename is the same as the figure number.

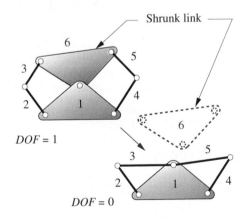

(a) Partial shrinkage of higher link retains original *DOF*

(b) Complete shrinkage of higher link reduces *DOF* by one

FIGURE 2-13

Link shrinkage

2.11 INTERMITTENT MOTION

Intermittent motion is *a sequence of motions and dwells*. A **dwell** is *a period in which the output link remains stationary while the input link continues to move*. There are many applications in machinery that require intermittent motion. The **cam-follower** variation on the fourbar linkage as shown in Figure 2-12c (p. 51) is often used in these situations. The design of that device for both intermittent and continuous output will be addressed in detail in Chapter 8. Other pure linkage **dwell mechanisms** are discussed in Chapter 3.

GENEVA MECHANISM A common form of intermittent motion device is the **Geneva mechanism** shown in Figure 2-14a (p. 54).[*] This is also a transformed fourbar linkage in which the coupler has been replaced by a half joint. The input crank (link 2) is typically motor driven at a constant speed. The **Geneva wheel** is fitted with at least three equispaced, radial slots. The crank has a pin that enters a radial slot and causes the Geneva wheel to turn through a portion of a revolution. When the pin leaves that slot, the Geneva wheel remains stationary until the pin enters the next slot. The result is intermittent rotation of the Geneva wheel.

The crank is also fitted with an arc segment, which engages a matching cutout on the periphery of the Geneva wheel when the pin is out of the slot. This keeps the Geneva wheel stationary and in the proper location for the next entry of the pin. The number of slots determines the number of "stops" of the mechanism, where *stop* is synonymous with *dwell*. A Geneva wheel needs a minimum of three stops to work. The maximum number of stops is limited only by the size of the wheel.

RATCHET AND PAWL Figure 2-14b[*] shows a ratchet and pawl mechanism. The **arm** pivots about the center of the toothed **ratchet wheel** and is moved back and forth to index the wheel. The **driving pawl** rotates the ratchet wheel (or **ratchet**) in the counterclockwise direction and does no work on the return (clockwise) trip. The **locking pawl** prevents the ratchet from reversing direction while the driving pawl returns. Both pawls are usually spring-loaded against the ratchet. This mechanism is widely used in devices such as "ratchet" wrenches, winches, etc.

LINEAR GENEVA MECHANISM There is also a variation of the Geneva mechanism that has linear translational output, as shown in Figure 2-14c.[*] This mechanism is analogous to an open Scotch yoke device with multiple yokes. It can be used as an intermittent conveyor drive with the slots arranged along the conveyor chain or belt. It also can be used with a reversing motor to get linear, reversing oscillation of a single slotted output slider.

2.12 INVERSION

It should now be apparent that there are many possible linkages for any situation. Even with the limitations imposed in the number synthesis example (one *DOF*, eight links, up to hexagonal order), there are eight linkage combinations shown in Table 2-2 (p. 46), and these together yield 19 valid isomers as shown in Table 2-3 (p. 48). In addition, we can introduce another factor, namely mechanism inversion. An **inversion** is *created by grounding a different link in the kinematic chain*. Thus there are as many inversions of a given linkage as it has links.

_____

[*] This figure is provided as animated AVI and Working Model files on the DVD. Its filename is the same as the figure number.

2

Geneva wheel

ω_{out}

3

Crank

2

ω_{in}

Arc

(a) Four-stop Geneva mechanism

Driving pawl

Arm

Locking pawl

Spring

ω_{in}

ω_{out}

Ratchet wheel

(b) Ratchet and pawl mechanism

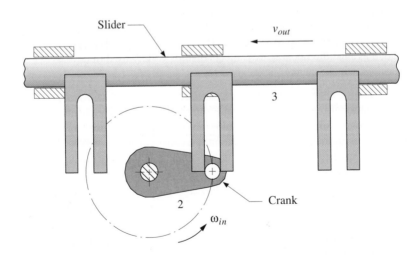

Slider

v_{out}

3

Crank

2

ω_{in}

(c) Linear intermittent motion "Geneva" mechanism

See also Figures P3-7 (p. 160) and P4-6 (p. 220) for other examples of linear intermittent motion mechanisms

FIGURE 2-14

Rotary and linear intermittent motion mechanisms

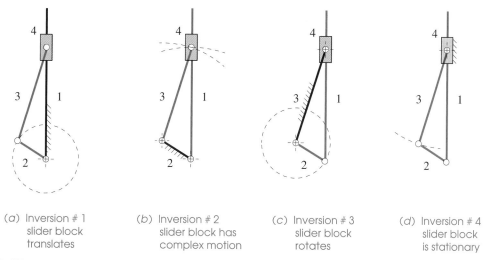

(a) Inversion # 1
slider block
translates

(b) Inversion # 2
slider block has
complex motion

(c) Inversion # 3
slider block
rotates

(d) Inversion # 4
slider block
is stationary

FIGURE 2-15

Four distinct inversions of the fourbar slider-crank mechanism (each black link is stationary—all red links move)

The motions resulting from each inversion can be quite different, but some inversions of a linkage may yield motions similar to other inversions of the same linkage. In these cases only some of the inversions may have distinctly different motions. We will denote the *inversions that have distinctly different motions* as **distinct inversions.**

Figure 2-15[*] shows the four inversions of the fourbar slider-crank linkage, all of which have distinct motions. Inversion #1, with link 1 as ground and its slider block in pure translation, is the most commonly seen and is used for **piston engines** and **piston pumps**. Inversion #2 is obtained by grounding link 2 and gives the **Whitworth** or **crank-shaper** quick-return mechanism, in which the slider block has complex motion. (Quick-return mechanisms will be investigated further in the Chapter 3.) Inversion #3 is obtained by grounding link 3 and gives the slider block pure rotation. Inversion #4 is obtained by grounding the slider link 4 and is used in hand-operated, **well pump** mechanisms, in which the handle is link 2 (extended) and link 1 passes down the well pipe to mount a piston on its bottom. (It is upside down in the figure.)

Watt's sixbar chain has two distinct inversions, and **Stephenson's sixbar** has three distinct inversions, as shown in Figure 2-16.[†] The pin-jointed fourbar has four distinct inversions: the crank-rocker, double-crank, double-rocker, and triple-rocker which are shown in Figures 2-17[*] (p. 57) and 2-18 (p. 58).[*]

2.13   THE GRASHOF CONDITION[§]

The **fourbar linkage** has been shown above to be the *simplest possible pin-jointed mechanism* for single-degree-of-freedom controlled motion. It also appears in various disguises such as the **slider-crank** and the **cam-follower**. It is in fact the most common and ubiquitous device used in machinery. It is also extremely versatile in terms of the types of motion that it can generate.

[*] These figures are provided as animated AVI and Working Model files on the DVD. Its filename is the same as the figure number.

[§] The Watt I is the only sixbar that has a floating binary link separated from ground by two links at each node, so it is good for long-reach applications and as a parallel motion generator. The Watt II is good for amplifying force or motion and is often used for function generation. The Stephenson III is often used to improve transmission angles by connecting a driven dyad to its coupler. It is also stable due to its three fixed pivots (as is the Watt II). The other two Stephenson inversions are not as often used.

[§] A video on *The Grashof Condition* is included on the book's DVD.

2

(a) Stephenson s sixbar inversion I   (b) Stephenson s sixbar inversion II   (c) Stephenson s sixbar inversion III

(d) Watt s sixbar inversion I   (e) Watt s sixbar inversion II

FIGURE 2-16

All distinct inversions of the sixbar linkage

Simplicity is one mark of good design. The fewest parts that can do the job will usually give the least expensive and most reliable solution. Thus the **fourbar linkage** should be among the first solutions to motion control problems to be investigated. The **Grashof condition**[8] is a very simple relationship that predicts the *rotation behavior* or **rotatability** of a fourbar linkage's inversions based only on the link lengths.

Let: S = length of shortest link

            L = length of longest link

            P = length of one remaining link

            Q = length of other remaining link

Then if: $S + L \le P + Q$ (2.8)

* According to Hunt[18] (p. 84), Waldron proved that in a Grashof linkage, no two of the links other than the crank can rotate more than 180° with respect to one another, but in a non-Grashof linkage (which has no crank) links can have more than 180° of relative rotation.

the linkage is **Grashof** and at least one link will be capable of making a full revolution with respect to the ground plane. This is called a **Class I** kinematic chain. If the inequality is not true, then the linkage is **non-Grashof** and *no* link will be capable of a complete revolution relative to any other link.*[†] This is a **Class II** kinematic chain.

Note that the above statements apply regardless of the order of assembly of the links. That is, the determination of the Grashof condition can be made on a set of unassembled links. Whether they are later assembled into a kinematic chain in S, L, P, Q or S, P, L, Q, or any other order, will *not* change the Grashof condition.

† See footnote on opposite page.

(*a*) Two non-distinct crank-rocker inversions (GCRR)

(*b*) Double-crank inversion (GCCC)
(drag link mechanism)

(*c*) Double-rocker inversion (GRCR)
(coupler rotates)

FIGURE 2-17

All inversions of the Grashof fourbar linkage

The motions possible from a fourbar linkage will depend on both the Grashof condition and the **inversion** chosen. The inversions will be defined with respect to the shortest link. The motions are:

For the Class I case, $S + L < P + Q$:

Ground either link adjacent to the shortest and you get a **crank-rocker**, in which the shortest link will fully rotate and the other link pivoted to ground will oscillate.

Ground the shortest link and you will get a **double-crank**, in which both links pivoted to ground make complete revolutions as does the coupler.

Ground the link opposite the shortest and you will get a **Grashof double-rocker**, in which both links pivoted to ground oscillate and only the coupler makes a full revolution.

For the Class II case, $S + L > P + Q$:

All inversions will be **triple-rockers** [9] in which no link can fully rotate.

For the Class III case, $S + L = P + Q$:

† The fourbar slider is a special case. Because two of its links are effectively infinite in length (the effective slider and the effective ground link are parallel and "meet" at infinity), the Grashof condition for a fourbar slider is always true, provided that the link lengths are such that they can physically connect. If so, $S + \infty$ is always $<= P + \infty$.

2

(a) Triple-rocker #1 (RRR1)   (b) Triple-rocker #2 (RRR2)

(c) Triple-rocker #3 (RRR3)   (d) Triple-rocker #4 (RRR4)

FIGURE 2-18

All inversions of the non-Grashof fourbar linkage are triple rockers

Referred to as **special-case Grashof** and also as a **Class III** kinematic chain, all inversions will be either **double-cranks** or **crank-rockers** but will have "**change points**" twice per revolution of the input crank when the links all become colinear. At these change points the output behavior will become indeterminate. Hunt[18] calls these "**uncertainty configurations**." At these colinear positions, the linkage behavior is unpredictable as it may assume either of two configurations. Its motion must be limited to avoid reaching the change points or an additional, out-of-phase link must be provided to guarantee a "carry through" of the change points. (See Figure 2-19c.)

Figure 2-17* (p. 57) shows the four possible inversions of the **Grashof case**: two crank-rockers, a double-crank (also called a drag link), and a double-rocker with rotating coupler. The two crank-rockers give similar motions and so are not distinct from one another. Figure 2-18* shows four nondistinct inversions, all triple-rockers, of a **non-Grashof linkage**.

Figure 2-19a and b shows the **parallelogram** and **antiparallelogram** configurations of the **special-case Grashof** linkage. The parallelogram linkage is quite useful as it exactly duplicates the rotary motion of the driver crank at the driven crank. One common use

---

* This figure is provided as animated AVI and Working Model files on the DVD. Its filename is the same as the figure number.

(a) Parallelogram form

(b) Antiparallelogram form

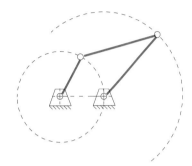

(c) Double-parallelogram linkage gives parallel
motion (pure curvilinear translation) to coupler
and also carries through the change points

(d) Deltoid or kite form

FIGURE 2-19

Some forms of the special-case Grashof linkage

is to couple the two windshield wiper output rockers across the width of the windshield on an automobile. The coupler of the parallelogram linkage is in curvilinear translation, remaining at the same angle while all points on it describe identical circular paths. It is often used for this parallel motion, as in truck tailgate lifts and industrial robots.

The antiparallelogram linkage (also called "butterfly" or "bow-tie") is also a double-crank, but the output crank has an angular velocity different from the input crank. Note that the change points allow the linkage to switch unpredictably between the parallelogram and antiparallelogram forms every 180 degrees unless some additional links are provided to carry it through those positions. This can be achieved by adding an out-of-phase companion linkage coupled to the same crank, as shown in Figure 2-19c. A common application of this double parallelogram linkage was on steam locomotives, used to connect the drive wheels together. The change points were handled by providing the duplicate linkage, 90 degrees out of phase, on the other side of the locomotive's axle shaft. When one side was at a change point, the other side would drive it through.

The **double-parallelogram** arrangement shown in Figure 2-19c is quite useful as it gives a translating coupler that remains horizontal in all positions. The two parallelogram stages of the linkage are out of phase so each carries the other through its change points. Figure 2-19d shows the **deltoid** or **kite** configuration that is a double-crank in which the shorter crank makes two revolutions for each one made by the long crank. This is also called an **isoceles** linkage or a **Galloway** mechanism after its discoverer.

There is nothing either bad or good about the Grashof condition. Linkages of all three persuasions are equally useful in their place. If, for example, your need is for a motor driven windshield wiper linkage, you may want a non-special-case Grashof crank-rocker linkage in order to have a rotating link for the motor's input, plus a special-case parallelogram stage to couple the two sides together as described above. If your need is to control the wheel motions of a car over bumps, you may want a non-Grashof triple-rocker linkage for short stroke oscillatory motion. If you want to exactly duplicate some input motion at a remote location, you may want a special-case Grashof parallelogram linkage, as used in a drafting machine. In any case, this simply determined condition tells volumes about the behavior to be expected from a proposed fourbar linkage design prior to any construction of models or prototypes.[*]

Classification of the Fourbar Linkage

Barker[10] has developed a classification scheme that allows prediction of the type of motion that can be expected from a fourbar linkage based on the values of its link ratios. A linkage's angular motion characteristics are independent of the absolute values of its link lengths. This allows the link lengths to be normalized by dividing three of them by the fourth to create three dimensionless ratios that define its geometry.

Let the link lengths be designated r_1, r_2, r_3, and r_4 (all positive and nonzero), with the subscript 1 indicating the ground link, 2 the driving link, 3 the coupler, and 4 the remaining (output) link. The link ratios are then formed by dividing each link length by r_2 giving: $\lambda_1 = r_1 / r_2$, $\lambda_3 = r_3 / r_2$, $\lambda_4 = r_4 / r_2$.

Each link will also be given a letter designation based on its type of motion when connected to the other links. If a link can make a full revolution with respect to the other links, it is called a crank (C), and if not, a rocker (R). The motion of the assembled linkage based on its Grashof condition and inversion can then be given a letter code such as GCRR for a Grashof crank-rocker or GCCC for a Grashof double-crank (drag link) mechanism. The motion designators C and R are always listed in the order of input link, coupler, output link. The prefix G indicates a Grashof linkage, S a special-case Grashof (change point), and no prefix a non-Grashof linkage.

Table 2-4 shows Barker's 14 types of fourbar linkage based on this naming scheme. The first four rows are the Grashof inversions, the next four are the non-Grashof triple-rockers, and the last six are the special-case Grashof linkages. He gives unique names to each type based on a combination of their Grashof condition and inversion. The traditional names for the same inversions are also shown for comparison and are less specific than Barker's nomenclature. Note his differentiation between the Grashof crank-rocker (subclass -2) and rocker-crank (subclass -4). To drive a GRRC linkage from the rocker requires adding a flywheel to the crank as is done with the internal combustion engine's slider-crank mechanism (which is a GPRC linkage). See Figure 2-12a (p. 51).

Barker also defines a *solution space* whose axes are the link ratios λ_1, λ_3, λ_4 as shown in Figure 2-20. These ratios' values theoretically extend to infinity, but for any practical linkages the ratios can be limited to a reasonable value.

In order for the four links to be assembled, the longest link must be shorter than the sum of the other three links,

$$L < (S + P + Q) \tag{2.9}$$

[*] See the video "The Grashof Condition" on the book's DVD for a more detailed and complete exposition of this topic.

TABLE 2-4   Barker's Complete Classification of Planar Fourbar Mechanisms

Adapted from ref. (10). s = shortest link, l = longest link, Gxxx = Grashof, RRRx = non-Grashof, Sxx = Special case

Type	$s + l$ vs. $p + q$	Inversion	Class	Barker's Designation	Code	Also Known As
1	<	$L_1 = s$ = ground	I-1	Grashof crank-crank-crank	GCCC	Double-crank
2	<	$L_2 = s$ = input	I-2	Grashof crank-rocker-rocker	GCRR	Crank-rocker
3	<	$L_3 = s$ = coupler	I-3	Grashof rocker-crank-rocker	GRCR	Double-rocker
4	<	$L_4 = s$ = output	I-4	Grashof rocker-rocker-crank	GRRC	Rocker-crank
5	>	$L_1 = l$ = ground	II-1	Class 1 rocker-rocker-rocker	RRR1	Triple-rocker
6	>	$L_2 = l$ = input	II-2	Class 2 rocker-rocker-rocker	RRR2	Triple-rocker
7	>	$L_3 = l$ = coupler	II-3	Class 3 rocker-rocker-rocker	RRR3	Triple-rocker
8	>	$L_4 = l$ = output	II-4	Class 4 rocker-rocker-rocker	RRR4	Triple-rocker
9	=	$L_1 = s$ = ground	III-1	Change-point crank-crank-crank	SCCC	SC* double-crank
10	=	$L_2 = s$ = input	III-2	Change-point crank-rocker-rocker	SCRR	SC crank-rocker
11	=	$L_3 = s$ = coupler	III-3	Change-point rocker-crank-rocker	SRCR	SC double-rocker
12	=	$L_4 = s$ = output	III-4	Change-point rocker-rocker-crank	SRRC	SC rocker-crank
13	=	Two equal pairs	III-5	Double change point	S2X	Parallelogram or deltoid
14	=	$L_1 = L_2 = L_3 = L_4$	III-6	Triple change point	S3X	Square

* SC = special case

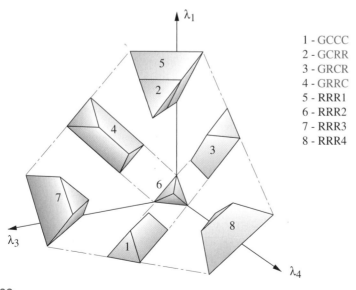

1 - GCCC
2 - GCRR
3 - GRCR
4 - GRRC
5 - RRR1
6 - RRR2
7 - RRR3
8 - RRR4

FIGURE 2-20

Barker's solution space for the fourbar linkage *Adapted from reference (10).*

If $L = S + P + Q$, then the links can be assembled but will not move, so this condition provides a criterion to separate regions of no mobility from regions that allow mobility within the solution space. Applying this criterion in terms of the three link ratios defines four planes of zero mobility that provide limits to the solution space.

$$1 = \lambda_1 + \lambda_3 + \lambda_4$$
$$\lambda_3 = \lambda_1 + 1 + \lambda_4$$
$$\lambda_4 = \lambda_1 + 1 + \lambda_3 \qquad (2.10)$$
$$\lambda_1 = 1 + \lambda_3 + \lambda_4$$

Applying the $S + L = P + Q$ Grashof condition (in terms of the link ratios) defines three additional planes on which the change-point mechanisms all lie.

$$1 + \lambda_1 = \lambda_3 + \lambda_4$$
$$1 + \lambda_3 = \lambda_1 + \lambda_4 \qquad (2.11)$$
$$1 + \lambda_4 = \lambda_1 + \lambda_3$$

The positive octant of this space, bounded by the λ_1–λ_3, λ_1–λ_4, λ_3–λ_4 planes and the four zero-mobility planes (equation 2.10), contains eight volumes that are separated by the change-point planes (equation 2.11). Each volume contains mechanisms unique to one of the first eight classifications in Table 2-4 (p. 61). These eight volumes are in contact with one another in the solution space, but to show their shapes, they have been "exploded" apart in Figure 2-20 (p. 61). The remaining six change-point mechanisms of Table 2-4 (p. 61) exist only in the change-point planes that are the interfaces between the eight volumes. For more details on this solution space and Barker's classification system than space permits here, see reference [10].

2.14   LINKAGES OF MORE THAN FOUR BARS

Geared Fivebar Linkages

We have seen that the simplest one-*DOF* linkage is the fourbar mechanism. It is an extremely versatile and useful device. Many quite complex motion control problems can be solved with just four links and four pins. Thus in the interest of simplicity, designers should always first try to solve their problems with a fourbar linkage. However, there will be cases when a more complicated solution is necessary. Adding one link and one joint to form a fivebar (Figure 2-21a) will increase the *DOF* by one, to two. By adding a pair of gears to tie two links together with a new half joint, the *DOF* is reduced again to one, and the **geared fivebar mechanism (GFBM)** of Figure 2-21b[*] is created.

The geared fivebar mechanism provides more complex motions than the fourbar mechanism at the expense of the added link and gearset as can be seen in Appendix E. The reader may also observe the dynamic behavior of the linkage shown in Figure 2-21b by running the program FIVEBAR provided with this text and opening the data file F02-21b.5br. See Appendix A for instructions on running the program. Accept all the default values, and animate the linkage.

---

[*] This figure is provided as animated AVI and Working Model files on the DVD. Its filename is the same as the figure number.

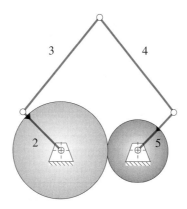

(a) Fivebar linkage—2 DOF   (b) Geared fivebar linkage—1 DOF

FIGURE 2-21

Two forms of the fivebar linkage

Sixbar Linkages

We already met Watt's and Stephenson's sixbar mechanisms. See Figure 2-16 (p. 56). **Watt's sixbar** can be thought of as *two fourbar linkages connected in series* and sharing two links in common. **Stephenson's sixbar** can be thought of as *two fourbar linkages connected in parallel* and sharing two links in common. Many linkages can be designed by the technique of combining multiple fourbar chains as *basic building blocks* into more complex assemblages. Many real design problems will require solutions consisting of more than four bars. Some Watt's and Stephenson's linkages are provided as built-in examples to the program SIXBAR supplied with this text. You may run that program to observe these linkages dynamically. Select any example from the menu, accept all default responses, and animate the linkages.

Grashof-Type Rotatability Criteria for Higher-Order Linkages

Rotatability is defined as *the ability of at least one link in a kinematic chain to make a full revolution with respect to the other links* and defines the chain as Class I, II, or III. **Revolvability** refers to *a specific link in a chain and indicates that it is one of the links that can rotate.*

ROTATABILITY OF GEARED FIVEBAR LINKAGES Ting[11] has derived an expression for rotatability of the geared fivebar linkage that is similar to the fourbar's Grashof criterion. Let the link lengths be designated L_1 through L_5 in order of increasing length,

then if:
$$L_1 + L_2 + L_5 < L_3 + L_4 \qquad (2.12)$$

the two shortest links can revolve fully with respect to the others and the linkage is designated a **Class I** kinematic chain. If this inequality is *not* true, then it is a **Class II** chain and may or may not allow any links to fully rotate depending on the gear ratio and phase angle between the gears. If the inequality of equation 2.12 is replaced with an equal sign, the linkage will be a **Class III** chain in which the two shortest links can fully revolve but it will have change points like the special-case Grashof fourbar.

Reference [11] describes the conditions under which a Class II geared fivebar linkage will and will not be rotatable. In practical design terms, it makes sense to obey equation 2.12 in order to guarantee a Grashof condition. It also makes sense to avoid the Class III change-point condition. Note that if one of the short links (say L_2) is made zero, equation 2.12 reduces to the Grashof formula of equation 2.8 (p. 56).

In addition to the linkage's rotatability, we would like to know about the kinds of motions that are possible from each of the five inversions of a fivebar chain. Ting[11] describes these in detail. But if we want to apply a gearset between two links of the fivebar chain (to reduce its *DOF* to 1), we really need it to be a double-crank linkage, with the gears attached to the two cranks. A Class I fivebar chain will be a **double-crank** mechanism if *the two shortest links are among the set of three links that comprise the mechanism's ground link and the two cranks pivoted to ground.* [11]

ROTATABILITY OF *N*-BAR LINKAGES   Ting et al. [12], [13] have extended rotatability criteria to all single-loop linkages of *N*-bars connected with revolute joints and have developed general theorems for **linkage rotatability** and the **revolvability** of individual links based on link lengths. Let the links of an *N*-bar linkage be denoted by L_i ($i = 1, 2, \ldots, N$), with $L_1 \leq L_2 \leq \cdots \leq L_N$. The links need not be connected in any particular order as rotatability criteria are independent of that factor.

A single-loop, revolute-jointed linkage of *N* links will have ($N - 3$) *DOF*. The necessary and sufficient condition for the **assemblability** of an *N*-bar linkage is:

$$L_N \leq \sum_{k=1}^{N-1} L_k \tag{2.13}$$

A link *K* will be a so-called *short* link if

$$\{K\}_{k=1}^{N-3} \tag{2.14a}$$

and a so-called *long* link if

$$\{K\}_{k=N-2}^{N} \tag{2.14b}$$

There will be three long links and ($N - 3$) short links in any linkage of this type.

A single-loop *N*-bar kinematic chain containing only first-order revolute joints will be a Class I, Class II, or Class III linkage depending on whether the sum of the lengths of its longest link and its ($N - 3$) shortest links is, respectively, less than, greater than, or equal to the sum of the lengths of the remaining two long links:

Class I: $L_N + (L_1 + L_2 + \cdots + L_{N-3}) < L_{N-2} + L_{N-1}$

Class II: $L_N + (L_1 + L_2 + \cdots + L_{N-3}) > L_{N-2} + L_{N-1}$ (2.15)

Class III: $L_N + (L_1 + L_2 + \cdots + L_{N-3}) = L_{N-2} + L_{N-1}$

and, for a Class I linkage, there must be one and only one long link between two noninput angles. These conditions are necessary and sufficient to define the rotatability.

The **revolvability** of any link L_i is defined as its ability to rotate fully with respect to the other links in the chain and can be determined from:

$$L_i + L_N \leq \sum_{k=1,\,k \neq i}^{N-1} L_k \qquad\qquad (2.16)$$

Also, if L_i is a revolvable link, any link that is not longer than L_i will also be revolvable.

Additional theorems and corollaries regarding limits on link motions can be found in references [12] and [13]. Space does not permit their complete exposition here. Note that the rules regarding the behavior of geared fivebar linkages and fourbar linkages (the Grashof law) stated above are consistent with, and contained within, these general rotatability theorems.

2.15 SPRINGS AS LINKS

We have so far been dealing only with rigid links. In many mechanisms and machines, it is necessary to counterbalance the static loads applied to the device. A common example is the hood hinge mechanism on your automobile. Unless you have the (cheap) model with the strut that you place in a hole to hold up the hood, it will probably have either a fourbar or sixbar linkage connecting the hood to the body on each side. The hood may be the coupler of a non-Grashof linkage whose two rockers are pivoted to the body. A spring is fitted between two of the links to provide a force to hold the hood in the open position. The spring in this case is an additional link of variable length. As long as it can provide the right amount of force, it acts to reduce the *DOF* of the mechanism to zero, and holds the system in static equilibrium. However, you can force it to again be a one-*DOF* system by overcoming the spring force when you pull the hood shut.

Another example, which may now be right next to you, is the ubiquitous adjustable arm desk lamp, shown in Figure 2-22.* This device has two springs that counterbalance the weight of the links and lamp head. If well designed and made, it will remain stable over a fairly wide range of positions despite variation in the overturning moment due to the lamp head's changing moment arm. This is accomplished by careful design of the geometry of the spring-link relationships so that, as the spring force changes with increasing length, its moment arm also changes in a way that continually balances the changing moment of the lamp head.

A linear spring can be characterized by its spring constant, $k = F / x$, where F is force and x is spring displacement. Doubling its deflection will double the force. Most coil springs of the type used in these examples are linear.

2.16 COMPLIANT MECHANISMS

The mechanisms so far described in this chapter all consist of discrete elements in the form of rigid links or springs connected by joints of various types. Compliant mechanisms can provide similar motions with fewer parts and fewer (even zero) physical joints. Compliance is the opposite of stiffness. A member or "link" that is compliant is capable of significant deflection in response to load. An ancient example of a compliant mechanism is the bow and arrow, in which the bow's deflection in response to the archer pulling back

FIGURE 2-22

A spring-balanced linkage mechanism

---

* This figure is provided as animated AVI and Working Model files on the DVD. Its filename is the same as the figure number.

2

FIGURE 2-23

A tackle box with "living hinge" *Courtesy of Penn Plastics Inc, Bridgeport, CT*

the bowstring stores elastic strain energy in the flexible (compliant) bow, and that energy launches the arrow.

The bow and bowstring comprise two parts, but in its purest form a compliant mechanism consists of a single link whose shape is carefully designed to provide areas of flexibility that serve as pseudo joints. Probably the most commonly available example of a simple compliant mechanism is the ubiquitous plastic tackle box or toolbox made with a "living hinge" as shown in Figure 2-23. This is a dyad or two-link mechanism (box and cover) with a thin section of material connecting the two. Certain thermoplastics, such as polypropylene, allow thin sections to be flexed repeatedly without failure. When the part is removed from the mold, and is still warm, the hinge must be flexed once to align the material's molecules. Once cooled, it can withstand millions of open-close cycles without failure. Figure 2-24 shows a prototype of a fourbar linkage toggle switch made in one piece of plastic as a compliant mechanism. It moves between the on and off positions by flexure of the thin hinge sections that serve as pseudo joints between the "links." The case study discussed in Chapter 1 describes the design of a compliant mechanism that is also shown in Figure 6-13 (p. 307).

Figure 2-25a shows a forceps designed as a one-piece compliant mechanism. Instead of the conventional two pieces connected by a pin joint, this forceps has small cross sections designed to serve as pseudo joints. It is injection molded of polypropylene thermoplastic with "living hinges." Note that there is a fourbar linkage 1, 2, 3, 4 at the center whose "joints" are the compliant sections of small dimension at A, B, C, and D. The compliance of the material in these small sections provides a built-in spring effect to hold it open in the rest condition. The other portions of the device such as the handles and jaws are designed with stiffer geometry to minimize their deflections. When the user closes the jaws, the hooks on the handles latch it closed, clamping the gripped item. Figure 2-25b shows a two-piece snap hook that uses the compliance of the spring closure that results from either ear of the wire spring being pivoted at different locations A_1 and A_2.

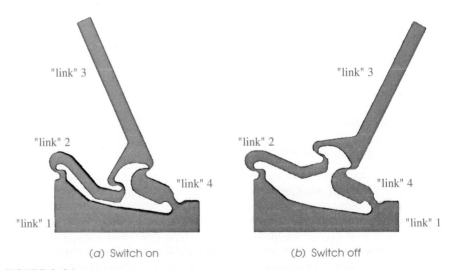

(a) Switch on   (b) Switch off

FIGURE 2-24

One-piece compliant switch *Courtesy of Professor Larry L. Howell, Brigham Young University*

(a) One-piece molded polypropylene compliant forceps
(Nalge Nunc International, Sweden)

(b) Compliant snap hook ™
(Wichard USA, Portsmouth, RI)

FIGURE 2-25

Compliant mechanisms

These examples show some of the advantages of compliant mechanisms over conventional ones. No assembly operations are needed, as there is only one part. The needed spring effect is built in by control of geometry in local areas. The finished part is ready to use as it comes out of the mold. These features all reduce cost.

Compliant mechanisms have been in use for a long time (e.g., the bow and arrow, fingernail clipper, paper clips), but found new applications in the late 20th century due in part to the availability of new materials and modern manufacturing processes. Some of their advantages over conventional mechanisms are the reduction of number of parts, elimination of joint clearances, inherent spring loading, and potential reductions in cost, weight, wear, and maintenance compared to conventional mechanisms. They are, however, more difficult to design and analyze because of their relatively large deflections that preclude the use of conventional small-deflection theory. This text will consider only the design and analysis of noncompliant (i.e., assumed rigid) links and mechanisms with physical joints. For information on the design and analysis of compliant mechanisms see reference [16].

2.17 MICRO ELECTRO-MECHANICAL SYSTEMS (MEMS)*

Recent advances in the manufacture of microcircuitry such as computer chips have led to a new form of mechanism known as micro electro-mechanical systems or MEMS. These devices have features measured in micrometers, and micromachines range in size from a few micrometers to a few millimeters. They are made from the same silicon wafer material that is used for integrated circuits or microchips. The shape or pattern of the desired device (mechanism, gear, etc.) is computer generated at large scale and then photographically reduced and projected onto the wafer. An etching process then removes the silicon material where the image either did or did not alter the photosensitive coating on the silicon (the process can be set to do either). What remains is a tiny reproduction

* More information on MEMS can be found at: http://www.sandia.gov/ and http://www.memsnet.org/mems/

2

of the original geometric pattern in silicon. Figure 2-26a shows silicon microgears made by this method. They are only a few micrometers in diameter.

Compliant mechanisms are very adaptable to this manufacturing technique. Figure 2-26b shows a micromotor that uses the gears of Figure 2-26a and is smaller than a few millimeters overall. The motor drive mechanism is a series of compliant linkages that are oscillated by an electrostatic field to drive the crank shown in the enlarged view of Figure 2-26b. Two of these electrostatic actuators operate on the same crank, 90° out of phase to carry it through the dead center positions. This motor is capable of continuous speeds of 360 000 rpm and short bursts to a million rpm before overheating from friction at that high speed.

Figure 2-27 shows "a compliant bistable mechanism (known as the Young mechanism) in its two stable positions. Thermal actuators amplify thermal expansion to snap the device between its two positions. It can be used as a microswitch or a microrelay. Because it is so small, it can be actuated in a few hundred microseconds."[†]

Applications for these micro devices are just beginning to be found. Microsensors made with this technology are currently used in automobile airbag assemblies to detect sudden deceleration and fire the airbag inflator. MEMS blood pressure monitors that can be placed in a blood vessel have been made. MEMS pressure sensors are being fitted to automobile tires to continuously monitor tire pressure. Many other applications are being and will be developed to utilize this technology in the future.

† Professor Larry L. Howell (2002), personal communication.

compliant electrostatic actuators

300 μm

50 μm

motor drive

(a) Microgears   (b) Micromotor and gear train

FIGURE 2-26

MEMS of etched silicon (a) microgears *Courtesy of Sandia National Laboratories* (b) micromotor by Sandia Labs
SEM photos courtesy of Professor Cosme Furlong, Worcester Polytechnic Institute

(a)   (b)

FIGURE 2-27

Compliant bistable silicon micromechanism in two positions *Courtesy of Professor Larry L. Howell, Brigham Young University*

2.18 PRACTICAL CONSIDERATIONS

There are many factors that need to be considered to create good-quality designs. Not all of them are contained within the applicable theories. A great deal of art based on experience is involved in design as well. This section attempts to describe a few such practical considerations in machine design.

Pin Joints versus Sliders and Half Joints

Proper material selection and good lubrication are the key to long life in any situation, such as a joint, where two materials rub together. Such an interface is called a **bearing**. Assuming the proper materials have been chosen, the choice of joint type can have a significant effect on the ability to provide good, clean lubrication over the lifetime of the machine.

REVOLUTE (PIN) JOINTS   The simple revolute or pin joint (Figure 2-28a) is the clear winner here for several reasons. It is relatively easy and inexpensive to design and build a good-quality pin joint. In its pure form—a so-called **sleeve** or **journal** bearing— the geometry of pin-in-hole traps a lubricant film within its annular interface by capillary action and promotes a condition called *hydrodynamic lubrication* in which the parts are separated by a thin film of lubricant as shown in Figure 2-29 (p. 70). Seals can easily be provided at the ends of the hole, wrapped around the pin, to prevent loss of the lubricant. Replacement lubricant can be introduced through radial holes into the bearing interface, either continuously or periodically, without disassembly.

A convenient form of bearing for linkage pivots is the commercially available **spherical rod end** shown in Figure 2-30 (p. 70). This has a spherical, sleeve-type bearing that *self-aligns* to a shaft that may be out of parallel. Its body threads onto the link, allowing links to be conveniently made from round stock with threaded ends that allow adjustment of link length.

Relatively inexpensive **ball** and **roller bearings** are commercially available in a large variety of sizes for revolute joints as shown in Figure 2-31 (p. 70). Some of these

(a) Pin joint

(b) Slider joint

(c) Half joint

FIGURE 2-28

Joints of various types

FIGURE 2-29

Hydrodynamic lubrication in a sleeve bearing—clearance and motions exaggerated

(a) Ball bearing

(b) Roller bearing

(c) Needle bearing

FIGURE 2-31

Ball, roller, and needle bearings for revolute joints *Courtesy of NTN Corporation, Japan*

bearings (principally ball type) can be obtained prelubricated and with end seals. Their rolling elements provide low-friction operation and good dimensional control. Note that *rolling-element bearings* actually contain higher-joint interfaces (half joints) at each ball or roller, which is potentially a problem as noted below. However, the ability to trap lubricant within the roll cage (by end seals) combined with the relatively high rolling speed of the balls or rollers promotes elastohydrodynamic lubrication and long life. For more detailed information on bearings and lubrication, see reference [15].

For revolute joints pivoted to ground, several commercially available bearing types make the packaging easier. **Pillow blocks** and **flange-mount bearings** (Figure 2-32, p. 71) are available fitted with either rolling-element (ball, roller) bearings or sleeve-type journal bearings. The pillow block allows convenient mounting to a surface parallel to the pin axis, and flange mounts fasten to surfaces perpendicular to the pin axis.

PRISMATIC (SLIDER) JOINTS   require a carefully machined and straight slot or rod (Figure 2-28b, p. 69). The bearings often must be custom made, though linear ball bearings (Figure 2-33, p. 71) are commercially available but must be run over hardened and ground shafts. Lubrication is difficult to maintain in any sliding joint. The lubricant is not geometrically captured, and it must be resupplied either by running the joint in an oil bath or by periodic manual regreasing. An open slot or shaft tends to accumulate airborne dirt particles that can act as a grinding compound when trapped in the lubricant. This will accelerate wear.

FIGURE 2-30

Spherical rod end *Courtesy of Emerson Power Transmission, Ithaca, NY*

(a) Pillow-block bearing   (b) Flange-mount bearing

FIGURE 2-32

Pillow block and flange-mount bearing units. *Courtesy of Emerson Power Transmission, Ithaca, NY.*

FIGURE 2-33

Linear ball bushing
Courtesy of Thomson Industries, Port Washington, NY

HIGHER (HALF) JOINTS   such as a round pin in a slot (Figure 2-28c, p. 69) or a cam-follower joint (Figure 2-12c, p. 51) suffer even more acutely from the slider's lubrication problems, because they typically have two oppositely curved surfaces in line contact, which tend to squeeze any lubricant out of the joint. This type of joint needs to be run in an oil bath for long life. This requires that the assembly be housed in an expensive, oil-tight box with seals on all protruding shafts.

These joint types are all used extensively in machinery with great success. As long as the proper *attention to engineering detail* is paid, the design can be successful. Some common examples of all three joint types can be found in an automobile. The windshield wiper mechanism is a pure pin-jointed linkage. The pistons in the engine cylinders are true sliders and are bathed in engine oil. The valves in the engine are opened and closed by cam-follower (half) joints that are drowned in engine oil. You probably change your engine oil fairly frequently. When was the last time you lubricated your windshield wiper linkage? Has this linkage (not the motor) ever failed?

Cantilever or Straddle Mount?

Any joint must be supported against the joint loads. Two basic approaches are possible as shown in Figure 2-34. A cantilevered joint has the pin (journal) supported only, as a cantilever beam. This is sometimes necessary as with a crank that must pass over the coupler and cannot have anything on the other side of the coupler. However, a cantilever beam is inherently weaker (for the same cross section and load) than a straddle-mounted (simply supported) beam. The straddle mounting can avoid applying a bending moment to the links by keeping the forces in the same plane. The pin will feel a bending moment in both cases, but the straddle-mounted pin is in double shear—two cross sections are sharing the load. A cantilevered pin is in single shear. It is good practice to use straddle-mounted joints (whether revolute, prismatic, or higher) wherever possible. If a cantilevered pin must be used, then a commercial shoulder screw that has a hardened and ground shank as shown in Figure 2-35 (p. 72) can sometimes serve as a pivot pin.

(a) Cantilever mount —single shear

(b) Straddle mount —double shear

FIGURE 2-34

Cantilever, and straddle-mounted pin joints

Short Links

It sometimes happens that the required length of a crank is so short that it is not possible to provide suitably sized pins or bearings at each of its pivots. The solution is to design the link as an **eccentric crank**, as shown in Figure 2-36. One pivot pin is enlarged to the point that it, in effect, contains the link. The outside diameter of the circular crank becomes the journal for the moving pivot. The fixed pivot is placed a distance *e* from the center of this circle equal to the required crank length. The distance *e* is the crank's eccentricity (the crank length). This arrangement has the advantage of a large surface area within the bearing to reduce wear, though keeping the large-diameter journal lubricated can be difficult.

FIGURE 2-35

Shoulder screw
*Courtesy of Cordova Bolt
Inc., Buena Park, CA*

Bearing Ratio

The need for straight-line motion in machinery requires extensive use of linear translating slider joints. There is a very basic geometrical relationship called *bearing ratio*, which if ignored or violated will invariably lead to problems.

The **bearing ratio** (BR) is defined as *the effective length of the slider over the effective diameter of the bearing*: $BR = L / D$. For smooth operation **this ratio should be greater than 1.5, and never less than 1**. The larger it is, the better. **Effective length** is defined as *the distance over which the moving slider contacts the stationary guide*. There need not be continuous contact over that distance. That is, two short collars, spaced far apart, are effectively as long as their overall separation plus their own lengths and are kinematically equivalent to a long tube. **Effective diameter** is *the largest distance across the stationary guides*, in any plane perpendicular to the sliding motion.

If the slider joint is simply a rod in a bushing, as shown in Figure 2-37a, the effective diameter and length are identical to the actual dimensions of the rod diameter and bushing length. If the slider is a platform riding on two rods and multiple bushings, as shown in Figure 2-37b, then the effective diameter and length are the overall width and length, respectively, of the platform assembly. It is this case that often leads to poor bearing ratios.

A common example of a device with a poor bearing ratio is a drawer in an inexpensive piece of furniture. If the only guides for the drawer's sliding motion are its sides running against the frame, it will have a bearing ratio less than 1, since it is wider than it

(*a*) Eccentric crank-rocker   (*b*) Eccentric slider-crank

FIGURE 2-36

Eccentric cranks

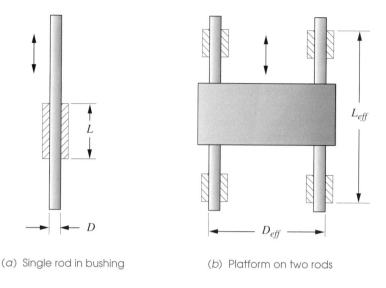

(a) Single rod in bushing   (b) Platform on two rods

FIGURE 2-37

Bearing ratio

is deep. You have probably experienced the sticking and jamming that occurs with such a drawer. A better-quality chest of drawers will have a center guide with a large L/D ratio under the bottom of the drawer and will slide smoothly.

Commercial Slides

Many companies provide standard linear slides that can be used for slider crank linkages and cam-follower systems with translating followers. These are available with linear ball bearings that ride on hardened steel ways giving very low friction. Some are preloaded to eliminate clearance and backlash error. Others are available with plain bearings. Figure 2-38 shows an example of a ball-bearing linear slide with two cars riding on a single rail. Mounting holes are provided for attaching the rail to the ground plane and in the cars for attaching the elements to be guided.

Linkages versus Cams

The pin-jointed linkage has all the advantages of revolute joints listed above. The cam-follower mechanism (Figure 2-12c, p. 51) has all the problems associated with the half joint listed above. But both are widely used in machine design, often in the same machine and in combination (cams driving linkages). So why choose one over the other?

The "pure" pin-jointed linkage with good bearings at the joints is a potentially superior design, all else equal, and it should be the first possibility to be explored in any machine design problem. However, there will be many problems in which the need for a straight, sliding motion or the exact dwells of a cam-follower are required. Then the practical limitations of cam and slider joints will have to be dealt with accordingly.

FIGURE 2-38

Ball bearing linear slide
*Courtesy of THK America
Inc., Schaumberg, IL*

Linkages have the disadvantage of relatively large size compared to the output displacement of the working portion; thus they can be somewhat difficult to package. Cams tend to be compact in size compared to the follower displacement. Linkages are relatively difficult to synthesize, and cams are relatively easy to design (as long as a computer program such as DYNACAM is available). But linkages are much easier and cheaper to manufacture to high precision than cams. Dwells are easy to get with cams, and difficult with linkages. Linkages can survive very hostile environments, with poor lubrication, whereas cams cannot, unless sealed from environmental contaminants. Linkages have better high-speed dynamic behavior than cams, are less sensitive to manufacturing errors, and can handle very high loads, but cams can match specified motions better.

So the answer is far from clear-cut. It is another *design trade-off situation* in which you must weigh all the factors and make the best compromise. Because of the potential advantages of the pure linkage it is important to consider a linkage design before choosing a potentially easier design task but an ultimately more expensive solution.

2.19 MOTORS AND DRIVERS

Unless manually operated, a mechanism will require some type of driver device to provide the input motion and energy. There are many possibilities. If the design requires a continuous rotary input motion, such as for a Grashof linkage, a slider-crank, or a cam-follower, then a motor or engine* is the logical choice. Motors come in a wide variety of types. The most common energy source for a motor is electricity, but compressed air and pressurized hydraulic fluid are also used to power air and hydraulic motors. Gasoline or diesel engines are another possibility. If the input motion is translation, as is common in earth-moving equipment, then a hydraulic or pneumatic cylinder is usually needed.

Electric Motors

Electric motors are classified both by their function or application and by their electrical configuration. Some functional classifications (described below) are **gearmotors**, **servomotors**, and **stepping motors**. Many different electrical configurations as shown in Figure 2-39 are also available, independent of their functional classifications. The main electrical configuration division is between **AC** and **DC** motors, though one type, the **universal motor**, is designed to run on either AC or DC.

AC and **DC** refer to *alternating current* and *direct current* respectively. AC is typically supplied by the power companies and, in the United States, will be alternating sinusoidally at 60 hertz (Hz), at about ±120, ±240, or ±480 volts (V) peak. Many other countries supply AC at 50 Hz. Single-phase AC provides a single sinusoid varying with time, and 3-phase AC provides three sinusoids at 120° phase angles. DC is constant with time, supplied from generators or battery sources and is most often used in vehicles, such as ships, automobiles, aircraft, etc. Batteries are made in multiples of 1.5 V, with 6, 12, and 24 V being the most common. Electric motors are also classed by their rated power as shown in Table 2-5. Both AC and DC motors are designed to provide continuous rotary output. While they can be stalled momentarily against a load, they cannot tolerate a full-current, zero-velocity stall for more than a few minutes without overheating.

* The terms *motor* and *engine* are often used interchangeably, but they do not mean the same thing. Their difference is largely semantic, but the "purist" reserves the term *motor* for electrical, hydraulic and pneumatic motors and the term *engine* for thermodynamic devices such as external combustion (steam, stirling) engines and internal combustion (gasoline, diesel) engines. Thus, a conventional automobile is powered by a gasoline or diesel engine, but its windshield wipers and window lifts are run by electric motors. The newest hybrid automobiles have one or more electric motors to drive the wheels plus an engine to charge the battery and supply auxiliary power directly to the wheels. Diesel-electric locomotives are hybrids also, using electric motors at the wheels for direct drive and diesel engines running generators to supply the electricity. Modern commercial ships use a similar arrangement with diesel engines driving generators and electric motors turning the propellers.

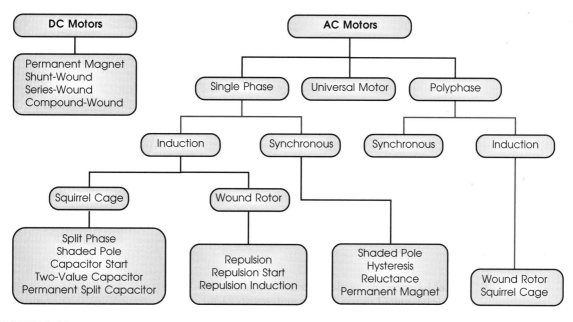

FIGURE 2-39

Types of electric motors *Source: Reference (14)*

DC MOTORS   are made in different electrical configurations, such as *permanent magnet (PM), shunt-wound, series-wound, and compound-wound.* The names refer to the manner in which the rotating armature coils are electrically connected to the stationary field coils—in parallel (shunt), in series, or in combined series-parallel (compound). Permanent magnets replace the field coils in a PM motor. Each configuration provides different *torque-speed* characteristics. The *torque-speed* curve of a motor describes how it will respond to an applied load and is of great interest to the mechanical designer as it predicts how the mechanical-electrical system will behave when the load varies dynamically with time.

PERMANENT MAGNET DC MOTORS   Figure 2-40a (p. 76) shows a torque-speed curve for a permanent magnet (PM) motor. Note that its torque varies greatly with speed, ranging from a maximum (stall) torque at zero speed to zero torque at maximum (no-load) speed. This relationship comes from the fact that *power = torque × angular velocity.* Since the power available from the motor is limited to some finite value, an increase in torque requires a decrease in angular velocity and vice versa. Its torque is maximum at stall (zero velocity), which is typical of all electric motors. This is an advantage when starting heavy loads: e.g., an electric-motor-powered vehicle needs no clutch, unlike one powered by an internal combustion engine that cannot start from stall under load. An engine's torque increases rather than decreases with increasing angular velocity.

Figure 2-40b (p. 76) shows a family of **load lines** superposed on the *torque-speed* curve of a PM motor. These load lines represent a time-varying load applied to the driven mechanism. The problem comes from the fact that *as the required load torque increases, the motor must reduce speed to supply it.* Thus, the input speed will vary in

TABLE 2-5

Motor Power Classes

Class	HP
Subfractional	< 1/20
Fractional	1/20 – 1
Integral	> 1

2

(a) Speed–torque characteristic of a PM electric motor

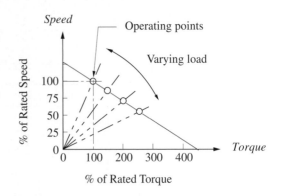

(b) Load lines superposed on speed–torque curve

FIGURE 2-40

DC permanent magnet (PM) electric motor's typical speed-torque characteristic

response to load variations in most motors, regardless of their design.[*] If constant speed is required, this may be unacceptable. Other types of DC motors have either more or less speed sensitivity to load than the PM motor. *A motor is typically selected based on its torque-speed curve.*

SHUNT-WOUND DC MOTORS have a torque speed curve like that shown in Figure 2-41a. Note the flatter slope around the rated torque point (at 100%) compared to Figure 2-40. The shunt-wound motor is less speed-sensitive to load variation in its operating range, but stalls very quickly when the load exceeds its maximum overload capacity of about 250% of rated torque. Shunt-wound motors are typically used on fans and blowers.

SERIES-WOUND DC MOTORS have a torque-speed characteristic like that shown in Figure 2-41b. This type is more speed-sensitive than the shunt or PM configurations. However, its starting torque can be as high as 800% of full-load rated torque. It also does not have any theoretical maximum no-load speed, which makes it tend to run away if the load is removed. Actually, friction and windage losses will limit its maximum speed, which can be as high as 20,000 to 30,000 revolutions per minute (rpm). Overspeed detectors are sometimes fitted to limit its unloaded speed. Series-wound motors are used in sewing machines and portable grinders where their speed variability can be an advantage as it can be controlled, to a degree, with voltage variation. They are also used in heavy-duty applications such as vehicle traction drives where their high starting torque is an advantage. Also their speed sensitivity (large slope) is advantageous in high-load applications as it gives a "soft start" when moving high-inertia loads. The motor's tendency to slow down when the load is applied cushions the shock that would be felt if a large step in torque were suddenly applied to the mechanical elements.

COMPOUND-WOUND DC MOTORS have their field and armature coils connected in a combination of series and parallel. As a result their torque-speed characteristic has aspects of both the shunt-wound and series-wound motors as shown in Figure 2-41c. Their speed sensitivity is greater than a shunt-wound but less than a series-wound motor and it will not run away when unloaded. This feature plus its high starting torque and soft-start capability make it a good choice for cranes and hoists that experience high inertial

---

[*] The synchronous AC motor and the speed-controlled DC motor are exceptions.

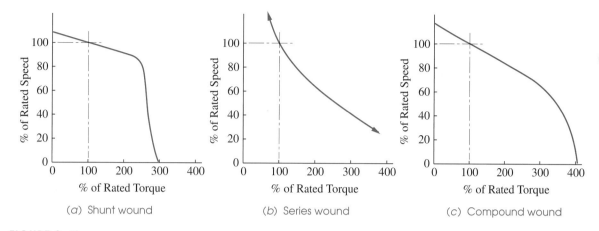

FIGURE 2-41

Torque-speed curves for three types of DC motor

loads and can suddenly lose the load due to cable failure, creating a potential runaway problem if the motor does not have a self-limited no-load speed.

SPEED-CONTROLLED DC MOTORS   If precise speed control is needed, as is often the case in production machinery, another solution is to use a speed-controlled DC motor that operates from a controller that increases and decreases the current to the motor in the face of changing load to try to maintain constant speed. These speed-controlled (typically PM) DC motors will run from an AC source since the controller also converts AC to DC. The cost of this solution is high, however. Another possible solution is to provide a **flywheel** on the input shaft, which will store kinetic energy and help smooth out the speed variations introduced by load variations. Flywheels will be investigated in Chapter 11.

AC MOTORS   are the least expensive way to get continuous rotary motion, and they can be had with a variety of *torque-speed* curves to suit various load applications. They are limited to a few standard speeds that are a function of the AC line frequency (60 Hz in North America, 50 Hz elsewhere). The synchronous motor speed n_s is a function of line frequency f and the number of magnetic poles p present in the rotor.

$$n_s = \frac{120f}{p} \tag{2.17}$$

Synchronous motors "lock on" to the AC line frequency and run exactly at synchronous speed. These motors are used for clocks and timers. Nonsynchronous AC motors have a small amount of slip that makes them lag the line frequency by about 3 to 10%.

Table 2-6 shows the synchronous and nonsynchronous speeds for various AC motor-pole configurations. The most common AC motors have 4 poles, giving nonsynchronous *no-load speeds* of about 1725 rpm, which reflects slippage from the 60-Hz synchronous speed of 1800 rpm.

Figure 2-42 (p. 78) shows typical torque-speed curves for single-phase (1ϕ) and 3-phase (3ϕ) AC motors of various designs. The single-phase shaded pole and permanent

TABLE 2-6
AC Motor Speeds

Poles	Sync rpm	Async rpm
2	3600	3450
4	1800	1725
6	1200	1140
8	900	850
10	720	690
12	600	575

(a) Single phase   (b) Three phase

FIGURE 2-42

Torque-speed curves for single- and three-phase AC motors

split capacitor designs have a starting torque lower than their full-load torque. To boost the start torque, the split-phase and capacitor-start designs employ a separate starting circuit that is cut off by a centrifugal switch as the motor approaches operating speed. The broken curves indicate that the motor has switched from its starting circuit to its running circuit. The NEMA[*] three-phase motor designs B, C, and D in Figure 2-42 differ mainly in their starting torque and in speed sensitivity (slope) near the full-load point.

GEARMOTORS   If different single (as opposed to variable) output speeds than the standard ones of Table 2-6 (p. 77) are needed, a gearbox speed reducer can be attached to the motor's output shaft, or a gearmotor can be purchased that has an integral gearbox. Gearmotors are commercially available in a large variety of output speeds and power ratings. The kinematics of gearbox design are covered in Chapter 9.

SERVOMOTORS   These are fast-response, closed-loop-controlled motors capable of providing a programmed function of acceleration or velocity, providing position control, and of holding a fixed position against a load. **Closed loop** means that *sensors (typically shaft encoders) on the motor or the output device being moved feed back information on its position and velocity.* Circuitry in the motor controller responds to the fed back information by reducing or increasing (or reversing) the current flow (and/or its frequency) to the motor. Precise positioning of the output device is then possible, as is control of the speed and shape of the motor's response to changes in load or input commands. These are relatively expensive devices[†] that are commonly used in applications such as moving the flight control surfaces in aircraft and guided missiles, in numerically controlled machining centers, automated manufacturing machinery, and in controlling robots, for example.

Servomotors are made in both AC and DC configurations, with the AC type currently becoming more popular. These achieve speed control by the controller generating a variable frequency current that the synchronous AC motor locks onto. The controller first rectifies the AC to DC and then "chops" it into the desired frequency, a common method being pulse-width modification. They have high torque capability and a flat torque-speed

[*] National Electrical Manufacturers Association.

[†] Costs of all electronic devices seem to continuously fall as technology advances and motor controllers are no exception.

curve similar to Figure 2-41a (p. 77). Also, they will typically provide as much as three times their continuous rated torque for short periods such as under intermittent overloads. Other advantages of servomotors include their ability to do programmed "soft starts," hold any speed to a close tolerance in the face of variation in the load torque, and make a rapid emergency stop using dynamic braking.

STEPPER MOTORS   These are brushless permanent magnet, variable reluctance, or hybrid-type motors designed to position an output device. Unlike servomotors, they typically run **open loop**, meaning they *receive no feedback as to whether the output device has responded as requested.* Thus, they can get out of phase with the desired program. They will, however, happily sit energized for an indefinite period, holding the output in one position (though they do get hot—100-150°F). Their internal construction consists of a number of magnetic strips arranged around the circumference of both the rotor and stator. When energized, the rotor will move one step, to the next magnet, for each pulse received. Thus, these are **intermittent motion** devices and do not provide continuous rotary motion like other motors. The number of magnetic strips and controller type determine their resolution (typically 200 steps/rev, but a microstepper drive can increase this to 2000 or more steps/rev). They are relatively small compared to AC/DC motors and have low drive torque capacity but have high holding torque. They are moderately expensive and require special controllers.

Air and Hydraulic Motors

These have more limited application than electric motors, simply because they require the availability of a compressed air or hydraulic source. Both of these devices are less energy efficient than the direct electrical to mechanical conversion of electric motors, because of the losses associated with the conversion of the energy first from chemical or electrical to fluid pressure and then to mechanical form. Every energy conversion involves some losses. Air motors find widest application in factories and shops, where high-pressure compressed air is available for other reasons. A common example is the air impact wrench used in automotive repair shops. Although individual air motors and air cylinders are relatively inexpensive, these pneumatic systems are quite expensive when the cost of all the ancillary equipment is included. Hydraulic motors are most often found within machines or systems such as construction equipment (cranes), aircraft, and ships, where high-pressure hydraulic fluid is provided for many purposes. Hydraulic systems are very expensive when the cost of all the ancillary equipment is included.

Air and Hydraulic Cylinders

These are linear actuators (piston in cylinder) that provide a limited stroke, straight-line output from a pressurized fluid flow input of either compressed air or hydraulic fluid (usually oil). They are the method of choice if you need a linear motion as the input. However, they share the same high cost, low efficiency, and complication factors as listed under their air and hydraulic motor equivalents above.

Another problem is that of control. Most motors, left to their own devices, will tend to run at a constant speed. A linear actuator, when subjected to a constant pressure fluid source, typical of most compressors, will respond with more nearly constant acceleration, which means its velocity will increase linearly with time. This can result in severe impact loads on the driven mechanism when the actuator comes to the end of its stroke at maximum velocity. Servovalve control of the fluid flow, to slow the actuator at the end of its stroke, is possible but is quite expensive.

The most common application of fluid power cylinders is in farm and construction equipment such as tractors and bulldozers, where open loop (nonservo) hydraulic cylinders actuate the bucket or blade through linkages. The cylinder and its piston become two of the links (slider and track) in a slider-crank mechanism. See Figure 1-1b (p. 7).

Solenoids

These are electromechanical (AC or DC) linear actuators that share some of the limitations of air cylinders, and they possess a few more of their own. They are *energy inefficient*, are limited to very short strokes (about 2 to 3 cm), develop a force that varies exponentially over the stroke, and deliver high impact loads. They are, however, inexpensive, reliable, and have very rapid response times. They cannot handle much power, and they are typically used as control or switching devices rather than as devices that do large amounts of work on a system.

A common application of solenoids is in camera shutters, where a small solenoid is used to pull the latch and trip the shutter action when you push the button to take the picture. Its nearly instantaneous response is an asset in this application, and very little work is being done in tripping a latch. Another application is in electric door or trunk locking systems in automobiles, where the click of their impact can be clearly heard when you turn the key (or press the button) to lock or unlock the mechanism.

2.20   REFERENCES

1   **Reuleaux, F.** (1963). *The Kinematics of Machinery.* A. B. W. Kennedy, translator. Dover Publications: New York, pp. 29-55.

2   **Gruebler, M.** (1917). *Getriebelehre.* Springer-Verlag: Berlin.

3   **Fang, W. E., and F. Freudenstein.** (1990). "The Stratified Representation of Mechanisms." *Journal of Mechanical Design,* **112**(4), p. 514.

4   **Kim, J. T., and B. M. Kwak.** (1992). "An Algorithm of Topological Ordering for Unique Representation of Graphs." *Journal of Mechanical Design,* **114**(1), p. 103.

5   **Tang, C. S., and T. Liu.** (1993). "The Degree Code—A New Mechanism Identifier." *Journal of Mechanical Design,* **115**(3), p. 627.

6   **Dhararipragada, V. R., et al.** (1994). "A More Direct Method for Structural Synthesis of Simple-Jointed Planar Kinematic Chains." *Proc. of 23rd Biennial Mechanisms Conference,* Minneapolis, MN, p. 507.

7   **Yadav, J. N., et al.** (1995). "Detection of Isomorphism Among Kinematic Chains Using the Distance Concept." *Journal of Mechanical Design,* **117**(4).

8   **Grashof, F.** (1883). *Theoretische Maschinenlehre.* Vol. 2. Voss: Hamburg.

9   **Paul, B.** (1979). "A Reassessment of Grashof's Criterion." *Journal of Mechanical Design,* **101**(3), pp. 515-518.

10   **Barker, C.** (1985). "A Complete Classification of Planar Fourbar Linkages." *Mechanism and Machine Theory,* **20**(6), pp. 535-554.

11   **Ting, K. L.** (1993). "Fully Rotatable Geared Fivebar Linkages." *Proc. of 3rd Applied Mechanisms and Robotics Conference,* Cincinnati, pp. 67-1.

12   **Ting, K. L., and Y. W. Liu**. (1991). "Rotatability Laws for N-Bar Kinematic Chains and Their Proof." *Journal of Mechanical Design*, **113**(1), pp. 32-39.

13   **Shyu, J. H., and K. L. Ting**. (1994). "Invariant Link Rotatability of N-Bar Kinematic Chains." *Journal of Mechanical Design*, **116**(1), p. 343.

14   **Miller, W. S.**, ed. *Machine Design Electrical and Electronics Reference Issue*. Penton Publishing: Cleveland, Ohio. (See also www.machinedesign.com.)

15   **Norton, R. L.** (2006). *Machine Design: An Integrated Approach*, 3ed. Prentice-Hall: Upper Saddle River, NJ.

16   **Howell, L. H.** (2001). *Compliant Mechanisms*. John Wiley & Sons: New York.

17   **Karunamoorthy, S.**, (1998). "Rule Based Number Synthesis for Kinematic Linkage Mechanism With Full Revolute Joints," ASME paper DETC98-MECH-5818.

18   **Hunt, K. H.**, (1978). *Kinematic Geometry of Mechanisms*. Oxford University Press, pp. 18, 39, 84.

2.21   PROBLEMS[†]

*2-1   Find three (or other number as assigned) of the following common devices. Sketch careful kinematic diagrams and find their total degrees of freedom.

   a.   An automobile hood hinge mechanism
   b.   An automobile hatchback lift mechanism
   c.   An electric can opener
   d.   A folding ironing board
   e.   A folding card table
   f.   A folding beach chair
   g.   A baby swing
   h.   A folding baby walker
   i.   A fancy corkscrew as shown in Figure P2-9 (p. 88)
   j.   A windshield wiper mechanism
   k.   A dump truck dump mechanism
   l.   A trash truck dumpster mechanism
   m.   A pickup truck tailgate mechanism
   n.   An automobile jack
   o.   A collapsible auto radio antenna

2-2   How many *DOF* do you have in your wrist and hand combined? Describe them.

*2-3   How many *DOF* do the following joints have?

   a.   Your knee
   b.   Your ankle
   c.   Your shoulder
   d.   Your hip
   e.   Your knuckle

*2-4   How many *DOF* do the following have in their normal environment?

   a.   A submerged submarine   b. An earth-orbiting satellite
   c.   A surface ship   d. A motorcycle (road bike)
   e.   A two-button mouse   f. A computer joystick

*2-5   Are the joints in Problem 2-3 force closed or form closed?

*2-6   Describe the motion of the following items as pure rotation, pure translation, or complex planar motion.

---

* Answers in Appendix F

[†] All problem figures are provided as PDF files, and some are provided as animated AVI and Working Model files, all on the DVD. Filenames are the same as the figure number. Run *Animations.html* to see the animations.

2

a.   A windmill
b.   A bicycle (in the vertical plane, not turning)
c.   A conventional "double-hung" window
d.   The keys on a computer keyboard
e.   The hand of a clock
f.   A hockey puck on the ice
g.   A "casement" window

*2-7   Calculate the mobility of the linkages assigned from Figure P2-1 part 1 and part 2.

*2-8   Identify the items in Figure P2-1 as mechanisms, structures, or preloaded structures.

2-9   Use linkage transformation on the linkage of Figure P2-1a to make it have 1 *DOF*.

2-10   Use linkage transformation on the linkage of Figure P2-1d to make it have 2 *DOF*.

2-11   Use number synthesis to find all the possible link combinations for 2 *DOF*, up to 9 links, to hexagonal order, using only revolute joints.

2-12   Find all valid isomers of the eightbar 1-*DOF* link combinations in Table 2-2 (p. 46) having:

a.   Four binary and four ternary links
b.   Five binaries, two ternaries, and one quaternary link
c.   Six binaries and two quaternary links
d.   Six binaries, one ternary, and one pentagonal link

2-13   Use linkage transformation to create a 1-*DOF* mechanism with two sliding full joints from Stephenson's sixbar linkage in Figure 2-16a (p. 56).

(a)

(b)

(c)

(d)

FIGURE P2-1 Part 1

Linkages for Problems 2-7 to 2-10

2

(e)

bellcrank

ground

tooling

cam

connecting rod

ground   roller
follower

follower arm

piston rod

air in/out

piston (attached to piston rod)

air cylinder

ground

rocker

valve
spring

pushrod

valve

(f)

tappet

cam

rocker pivots   rocker arms

ground   rollers

cams

INT   EXH

(g)   valves

rods are
fixed to
ground

ω_{cam}

barrel cam

cam groove

follower
moves

follower

roller
in
groove

(h)

FIGURE P2-1   Part 2

Linkages for Problems 2-7 to 2-8

2-14   Use linkage transformation to create a 1-*DOF* mechanism with one sliding full joint and a half joint from Stephenson's sixbar linkage in Figure 2-16b (p. 56).

*2-15   Calculate the Grashof condition of the fourbar mechanisms defined below. Build cardboard models of the linkages and describe the motions of each inversion. Link lengths are in inches (or double given numbers for centimeters).

a.   2   4.5   7   9
b.   2   3.5   7   9
c.   2   4.0   6   8

2-16   What type(s) of electric motor would you specify

a.   To drive a load with large inertia.
b.   To minimize variation of speed with load variation.
c.   To maintain accurate constant speed regardless of load variations.

2-17   Describe the difference between a cam-follower (half) joint and a pin joint.

2-18   Examine an automobile hood hinge mechanism of the type described in Section 2.15 (p. 65). Sketch it carefully. Calculate its mobility and Grashof condition. Make a cardboard model. Analyze it with a free-body diagram. Describe how it keeps the hood up.

2-19   Find an adjustable arm desk lamp of the type shown in Figure P2-2. Measure it and sketch it to scale. Calculate its mobility and Grashof condition. Make a cardboard model. Analyze it with a free-body diagram. Describe how it keeps itself stable. Are there any positions in which it loses stability? Why?

2-20   The torque-speed curve for a 1/8 hp permanent magnet (PM) DC motor is shown in Figure P2-3. The rated speed for this fractional horsepower motor is 2500 rpm at a rated voltage of 130 V. Determine:

a,   The rated torque in oz-in (ounce-inches, the U.S. industry standard for fractional hp motors)
b.   The no-load speed
c.   Plot the power-torque curve and determine the maximum power that the motor can deliver.

FIGURE P2-2

Problem 2-19

*2-21   Find the mobility of the mechanisms in Figure P2-4 (p. 86).

2-22   Find the Grashof condition and Barker classifications of the mechanisms in Figure P2-4a, b, and d (p. 86).

2-23   Find the rotatability of each loop of the mechanisms in Figure P2-4e, f, and g (p. 86).

*2-24   Find the mobility of the mechanisms in Figure P2-5 (p. 87).

2-25   Find the mobility of the ice tongs in Figure P2-6 (p. 87).

a.   When operating them to grab the ice block.
b.   When clamped to the ice block but before it is picked up (ice grounded).
c.   When the person is carrying the ice block with the tongs.

*2-26   Find the mobility of the automotive throttle mechanism in Figure P2-7 (p. 87).

*2-27   Sketch a kinematic diagram of the scissors jack shown in Figure P2-8 (p. 88) and determine its mobility. Describe how it works.

2-28   Find the mobility of the corkscrew in Figure P2-9 (p. 88).

---

* Answers in Appendix F

FIGURE P2-3

Torque-speed characteristic of a 1/8 hp, 2500 rpm, PM DC motor for Problem 2-20

2-29   Figure P2-10 (p. 88) shows Watt's sun and planet drive that he used in his steam en-
        gine. The beam 2 is driven in oscillation by the piston of the engine. The planet gear
        is fixed rigidly to link 3 and its center is guided in the fixed track 1. The output rotation
        is taken from the sun gear 4. Sketch a kinematic diagram of this mechanism and deter-
        mine its mobility. Can it be classified by the Barker scheme? If so, what Barker class
        and subclass is it?

2-30   Figure P2-11 (p. 89) shows a bicycle handbrake lever assembly. Sketch a kinematic
        diagram of this device and draw its equivalent linkage. Determine its mobility. *Hint*:
        Consider the flexible cable to be a link.

2-31   Figure P2-12 (p. 89) shows a bicycle brake caliper assembly. Sketch a kinematic dia-
        gram of this device and draw its equivalent linkage. Determine its mobility under two
        conditions.

        a.   Brake pads not contacting the wheel rim.
        b.   Brake pads contacting the wheel rim.

        Hint: Consider the flexible cables to be replaced by forces in this case.

2-32   Find the mobility, the Grashof condition, and the Barker classification of the mecha-
        nism in Figure P2-13 (p. 90).

2-33   The approximate torque-speed curve and its equation for a 1/4 hp shunt-wound DC mo-
        tor are shown in Figure P2-14 (p. 90). The rated speed for this fractional horsepower
        motor is 10 000 rpm at a rated voltage of 130 V. Determine:

        a.   The rated torque in oz-in (ounce-inches, the U.S. industry standard for fractional hp
              motors)
        b.   The no-load speed
        c.   The operating speed range
        d.   Plot the power-torque curve in the operating range and determine the maximum
              power that the motor can deliver in that range.

        _____
        * Answers in Appendix F

2

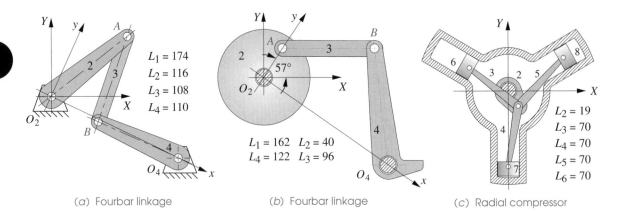

$L_1 = 174$
$L_2 = 116$
$L_3 = 108$
$L_4 = 110$

$L_1 = 162$ $L_2 = 40$
$L_4 = 122$ $L_3 = 96$

$L_2 = 19$
$L_3 = 70$
$L_4 = 70$
$L_5 = 70$
$L_6 = 70$

(a) Fourbar linkage

(b) Fourbar linkage

(c) Radial compressor

$L_1 = 150$ $L_2 = 30$
$L_3 = 150$ $L_4 = 30$

ω_2

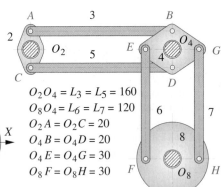

$O_2O_4 = L_3 = L_5 = 160$
$O_8O_4 = L_6 = L_7 = 120$
$O_2A = O_2C = 20$
$O_4B = O_4D = 20$
$O_4E = O_4G = 30$
$O_8F = O_8H = 30$

$L_2 = 63$
$L_3 = 130$
offset = 52

(d) Walking-beam conveyor

(e) Bellcrank mechanism

(f) Offset slider crank

$L_1 = 87$
$L_2 = 49$
$L_3 = 100$
$L_4 = 153$
$L_5 = 100$
$L_6 = 153$

121°

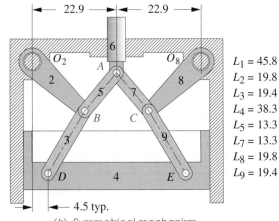

22.9   22.9

$L_1 = 45.8$
$L_2 = 19.8$
$L_3 = 19.4$
$L_4 = 38.3$
$L_5 = 13.3$
$L_7 = 13.3$
$L_8 = 19.8$
$L_9 = 19.4$

4.5 typ.

(g) Drum brake mechanism

(h) Symmetrical mechanism

all dimensions in mm

FIGURE P2-4

Problems 2-21 to 2-23 *Adapted from P. H. Hill and W. P. Rule. (1960). Mechanisms: Analysis and Design, with permission*

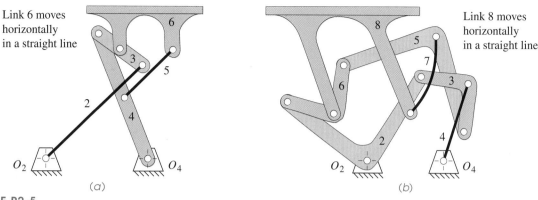

Link 6 moves horizontally in a straight line

Link 8 moves horizontally in a straight line

(a)   (b)

FIGURE P2-5

Problem 2-24 (a) Chebyschev and (b) Sylvester-Kempe straight-line mechanism *Adapted from Kempe, How to Draw a Straight Line, Macmillan: London, 1877*

2-34 Figure P2-15 (p. 91) shows a power hacksaw, used to cut metal. Link 5 pivots at O_5 and its weight forces the sawblade against the workpiece while the linkage moves the blade (link 4) back and forth within link 5 to cut the part. Sketch its kinematic diagram, determine its mobility and its type (i.e., is it a fourbar, a Watt sixbar, a Stephenson sixbar, an eightbar, or what?). Use reverse linkage transformation to determine its pure revolute-jointed equivalent linkage.

*2-35 Figure P2-16 (p. 91) shows a manual press used to compact powdered materials. Sketch its kinematic diagram, determine its mobility and its type (i.e., is it a fourbar, a Watt sixbar, a Stephenson sixbar, an eightbar, or what?). Use reverse linkage transformation to determine its pure revolute-jointed equivalent linkage.

2-36 Sketch the equivalent linkage for the cam and follower mechanism in Figure P2-17 (p. 91) in the position shown. Show that it has the same *DOF* as the original mechanism.

F F

ice

W

FIGURE P2-6

Problem 2-25

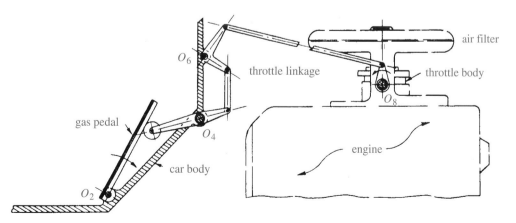

gas pedal

car body

throttle linkage

air filter

throttle body

engine

O_6

O_4

O_2

O_8

FIGURE P2-7

Problem 2-26. *Adapted from P. H. Hill and W. P Rule. (1960) Mechanisms: Analysis and Design, with permission*

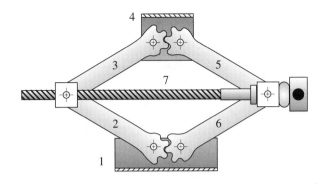

FIGURE P2-8

Problem 2-27

2-37 Describe the motion of the following rides, commonly found at an amusement park, as pure rotation, pure translation, or complex planar motion.

a. A Ferris wheel
b. A "bumper" car
c. A drag racer ride
d. A roller coaster whose foundation is laid out in a straight line
e. A boat ride through a maze
f. A pendulum ride
g. A train ride

2-38 For the mechanism in Figure P2-1a (p. 82), number the links, starting with 1. (Don't forget the "ground" link.) Letter the joints alphabetically, starting with point *A*.

FIGURE P2-9

Problem 2-28

$L_1 = 2.15$
$L_2 = 1.25$
$L_3 = 1.80$
$L_4 = 0.54$

Section A—A

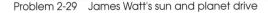

FIGURE P2-10

Problem 2-29   James Watt's sun and planet drive

FIGURE P2-11

Problem 2-30   Bicycle hand brake lever assembly

     a.   Using your link numbers, describe each link as binary, ternary, etc.
     b.   Using your joint letters, determine each joint's order.
     c.   Using your joint letters, determine whether each is a half or full joint.

2-39   Repeat Problem 2-38 for Figure P2-1b (p. 82).

2-40   Repeat Problem 2-38 for Figure P2-1c (p. 82).

2-41   Repeat Problem 2-38 for Figure P2-1d (p. 82).

2-42   Find the mobility, the Grashof condition, and the Barker classification of the oil field
       pump shown in Figure P2-18 (p. 92).

FIGURE P2-12

Problem 2-31   Bicycle brake caliper assembly

$L_1 = 0.92$
$L_2 = 0.27$
$L_3 = 0.50$
$L_4 = 0.60$

FIGURE P2-13

Problem 2-32   Crimping tool

2-43   Find the mobility, the Grashof condition, and the Barker classification of the aircraft overhead bin shown in Figure P2-19 (p. 92). Make a model and investigate its motions.

2-44   Figure P2-20 (p. 93) shows a "Rube Goldberg" mechanism that turns a light switch on when a room door is opened and off when the door is closed. The pivot at O_2 goes through the wall. There are two spring-loaded piston-in-cylinder devices in the assembly. An arrangement of ropes and pulleys inside the room (not shown) transfers the door swing into a rotation of link 2. Door opening rotates link 2 CW, pushing the switch up as shown in the figure, and door closing rotates link 2 CCW, pulling the switch down. Consider the spring-loaded cylinder at the switch to be effectively a single variable-length binary link. Find the mobility of the linkage.

2-45   All the eightbar linkages in Figure 2-11 part 2 (p. 50) have eight possible inversions. Some of these will give motions similar to others. Those that have distinct motions

$$N(T) = -0.1 \frac{N_R}{T_R} T + 1.1 N_R \qquad \text{if } T \le 62.5 \text{ oz-in}$$

$$N(T) = -1.7 \frac{N_R}{T_R} T + 5.1 N_R \qquad \text{if } T > 62.5 \text{ oz-in}$$

FIGURE P2-14

Problem 2-33   Torque-speed characteristic of a 1/4 hp, 10 000 rpm DC motor

Problem 2-34   Power hacksaw *Adapted from P. H. Hill and W. P Rule. (1960). Mechanisms: Analysis and Design*

are called *distinct inversions*. How many distinct inversions does the linkage in row 4, column 1 have?

2-46   Repeat Problem 2-45 for the linkage in row 4, column 2.

2-47   Repeat Problem 2-45 for the linkage in row 4, column 3.

2-48   Find the mobility of the mechanism shown in Figure 3-33 (p. 152).

FIGURE P2-16

Problem 2-35   Powder compacting press *Adapted from P. H. Hill and W. P. Rule. (1960). Mechanisms: Analysis and Design, with permission*

FIGURE P2-17

Problem 2-36

FIGURE P2-18

Problem 2-42   An oil field pump - dimensions in inches

2-49   Find the mobility of the mechanism shown in Figure 3-34 (p. 153).

2-50   Find the mobility of the mechanism shown in Figure 3-35 (p. 153).

2-51   Find the mobility of the mechanism shown in Figure 3-36 (p. 154).

2-52   Find the mobility of the mechanism shown in Figure 3-37b (p. 154).

2-53   Repeat Problem 2-38 for Figure P2-1e (p. 83).

2-54   Repeat Problem 2-38 for Figure P2-1f (p. 83).

2-55   Repeat Problem 2-38 for Figure P2-1g (p. 83).

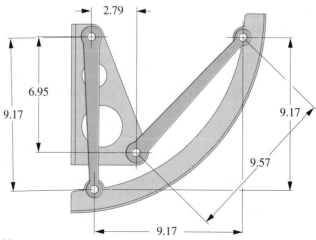

FIGURE P2-19

Problem 2-43   An aircraft overhead bin mechanism - dimensions in inches

FIGURE P2-20

A "Rube Goldberg" light switch actuating mechanism *(Courtesy of Robert Taylor, WPI)*

2-56   For the example linkage shown in Figure 2-4 (p. 36) find the number of links and their respective link orders, the number of joints and their respective orders, and the mobility of the linkage.

2-57   For the linkage shown in Figure 2-5b (p. 37) find the number of joints, their respective orders, and mobility for:

    a.   The condition of a finite load W in the direction shown and a zero load F
    b.   The condition of a finite load W and a finite load F both in the directions shown after link 6 is off the stop.

2-58   Figure P2-21a (p. 94) shows a "Nuremberg scissors" mechanism. Find its mobility.

2-59   Figure P2-21b (p. 94) shows a mechanism. Find its mobility and classify its isomer type.

2-60   Figure P2-21c (p. 94) shows a circular saw mounted on the coupler of a fourbar linkage. The centerline of the sawblade is at a coupler point that moves in an approximate straight line. Draw its kinematic diagram and determine its mobility.

*2-61   Figure P2-21d (p. 94) shows a log transporter. Draw a kinematic diagram of the mechanism, specify the number of links and joints, and then determine its mobility:

    a.   For the transporter wheels locked and no log in the claw.
    b.   For the transporter wheels locked with it lifting a log.
    c.   For the transporter moving a log to a destination in a straight line.

* Answers in Appendix F

2

(a) Nuremberg linkage   (b) Mechanism   (c) Circular saw

(d) Log transporter   (e) Tractor-mounted plow mechanism

FIGURE P2-21

Problems 2-58 to 2-62 (*Some illustrations from E. A. Dijksman, Motion Geometry of Mechanisms, Cambridge Press, London, 1976*)

FIGURE P2-22

Problem 2-63
from Ref. 18. p. 15

*2-62   Figure P2-21e shows a plow mechanism attached to a tractor. Draw its kinematic diagram and find its mobility including the earth as a "link."

    a.   When the tractor is stopped and the turnbuckle is fixed. (*Hint*: Consider the tractor and wheel to be one with the earth.)

    b.   When the tractor is stopped and the turnbuckle is being adjusted. (Same hint.)

    c.   When the tractor is moving and the turnbuckle is fixed. (*Hint*: Add the moving tractor's *DOF* to those found in part a.)

2-63   Figure P2-22 shows a Hart inversor sixbar linkage. (a) Is it a Watt or Stephenson linkage? (b) Determine its inversion, i.e., is it a type I, II, or III?

2-64   Figure P2-23 shows the top view of the partially open doors on one side of an entertainment center cabinet. The wooden doors are hinged to each other and one door is hinged to the cabinet. There is also a ternary, metal link attached to the cabinet and door through pin joints. A spring-loaded piston-in-cylinder device attaches to the ternary link and the cabinet through pin joints. Draw a kinematic diagram of the door system and find the mobility of this mechanism.

2-65   Figure P2-24a shows the seat and seat-back of a reclining chair with the linkage that connects them to the chair frame. Draw its kinematic diagram and determine its mobility with respect to the frame of the chair.

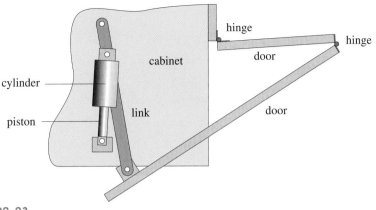

FIGURE P2-23

Problem 2-64

2-66 Figure P2-24b shows the mechanism used to extend the foot support on a reclining chair. Draw its kinematic diagram and determine its mobility with respect to the frame of the chair.

2-67 Figure P2-24b shows the mechanism used to extend the foot support on a reclining chair. Number the links, starting with 1. (*Hint*: Don't forget the "ground" link.) Letter the joints alphabetically, starting with A.

    a.   Using the link numbers, describe each link as binary, ternary, etc.
    b.   Using the joint letters, determine each joint's order.
    c.   Using the joint letters, determine whether each is a half or full joint.

2-68 Figure P2-25 shows a sixbar linkage.

    a.   Is it a Watt or Stephenson linkage?
    b.   Determine its inversion, i.e., is it a type I, II, or III?

FIGURE P2-25

Problem 2-68

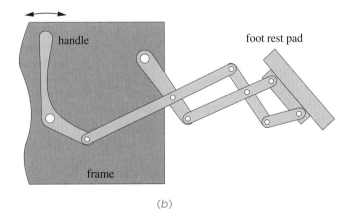

(a)   (b)

FIGURE P2-24

Problems 2-65 to 2-67

GRAPHICAL LINKAGE SYNTHESIS

*Genius is 1% inspiration
and 99% perspiration*
THOMAS A. EDISON

3.0   INTRODUCTION

Most engineering design practice involves a combination of synthesis and analysis. Most engineering courses deal primarily with analysis techniques for various situations. However, one cannot analyze anything until it has been synthesized into existence. Many machine design problems require the creation of a device with particular motion characteristics. Perhaps you need to move a tool from position A to position B in a particular time interval. Perhaps you need to trace out a particular path in space to insert a part into an assembly. The possibilities are endless, but a common denominator is often the need for a linkage to generate the desired motions. So, we will now explore some simple synthesis techniques to enable you to create potential linkage design solutions for some typical kinematic applications.

3.1   SYNTHESIS

QUALITATIVE SYNTHESIS   means *the creation of potential solutions in the absence of a well-defined algorithm that configures or predicts the solution.* Since most real design problems will have many more unknown variables than you will have equations to describe the system's behavior, you cannot simply solve the equations to get a solution.

Nevertheless you must work in this fuzzy context to create a potential solution and to also judge its **quality**. You can then analyze the proposed solution to determine its viability, and iterate between synthesis and analysis, as outlined in the **design process**, until you are satisfied with the result. Several tools and techniques exist to assist you in this process. The traditional tool is the drafting board, on which you lay out, to scale, multiple orthographic views of the design, and investigate its motions by drawing arcs, showing multiple positions, and using transparent, movable overlays. *Computer-aided drafting* (CAD) systems can speed this process to some degree, but you will probably find that the quickest way to get a sense of the quality of your linkage design is to model it, to scale, in cardboard, foam board, or drafting Mylar® and see the motions directly.

Other tools are available in the form of computer programs such as LINKAGES, DYNACAM, ENGINE, and MATRIX (all included with this text), some of which do synthesis, but these are mainly analysis tools. They can analyze a trial mechanism solution so rapidly that their dynamic graphical output gives almost instantaneous visual feedback on the quality of the design. Commercially available programs such as *Solidworks, Pro-Engineer,* and *Working Model* also allow rapid analysis of a proposed mechanical design. The process then becomes one of **qualitative design by successive analysis,** which is really *an iteration between synthesis and analysis.* Very many trial solutions can be examined in a short time using these *computer-aided engineering* (CAE) tools. We will develop the mathematical solutions used in these programs in subsequent chapters in order to provide the proper foundation for understanding their operation. But if you want to try these programs to reinforce some of the concepts in these early chapters, you may do so. Appendix A describes these programs, and the DVD contains a manual for their use. Reference will be made to program features that are germane to topics in each chapter, as they are introduced. Data files for input to these computer programs are also provided on disk for example problems and figures in these chapters. The data filenames are noted near the figure or example. The student is encouraged to open these sample files in the programs in order to observe more dynamic examples than the printed page can provide. These examples can be run by merely accepting the defaults provided for all inputs.

TYPE SYNTHESIS   refers to *the definition of the proper type of mechanism best suited to the problem* and is a form of qualitative synthesis.[*] This is perhaps the most difficult task for the student as it requires some experience and knowledge of the various types of mechanisms that exist and which also may be feasible from a performance and manufacturing standpoint. As an example, assume that the task is to design a device to track the straight-line motion of a part on a conveyor belt and spray it with a chemical coating as it passes by. This has to be done at high, constant speed, with good accuracy and repeatability, and it must be reliable. Moreover, the solution must be inexpensive. Unless you have had the opportunity to see a wide variety of mechanical equipment, you might not be aware that this task could conceivably be accomplished by any of the following devices:

- *A straight-line linkage*
- *A cam and follower*
- *An air cylinder*
- *A hydraulic cylinder*
- *A robot*
- *A solenoid*

[*] A good discussion of type synthesis and an extensive bibliography on the topic can be found in Olson, D. G., et al. (1985). "A Systematic Procedure for Type Synthesis of Mechanisms with Literature Review." *Mechanism and Machine Theory*, **20**(4), pp. 285-295.

3

Each of these solutions, while possible, may not be optimal or even practical. Greater detail needs to be known about the problem to make that judgment, and that detail will come from the research phase of the design process. The straight-line linkage may prove to be too large and to have undesirable accelerations; the cam and follower will be expensive, though accurate and repeatable. The air cylinder itself is inexpensive but is noisy and unreliable. The hydraulic cylinder is more expensive, as is the robot. The solenoid, while cheap, has high impact loads and high impact velocity. So, you can see that the choice of device type can have a significant effect on the quality of the design. A poor choice at the type synthesis stage can create insoluble problems later on. The design might have to be scrapped after completion, at great expense. **Design is essentially an exercise in trade-offs**. Each proposed type of solution in this example has good and bad points. Seldom will there be a clear-cut, obvious solution to a real engineering design problem. It will be your job as a design engineer to balance these conflicting features and find a solution that gives the best trade-off of functionality against cost, reliability, and all other factors of interest. Remember, *an engineer can do, with one dollar, what any fool can do for ten dollars*. Cost is always an important constraint in engineering design.

QUANTITATIVE SYNTHESIS, OR ANALYTICAL SYNTHESIS   means the generation of one or more solutions of a particular type that you know to be suitable to the problem, and more importantly, one for which there is a synthesis algorithm defined. As the name suggests, this type of solution can be quantified, as some set of equations exists that will give a numerical answer. Whether that answer is a good or suitable one is still a matter for the judgment of the designer and requires analysis and iteration to optimize the design. Often the available equations are fewer than the number of potential variables, in which case you must assume some reasonable values for enough unknowns to reduce the remaining set to the number of available equations. Thus some qualitative judgment enters into the synthesis in this case as well. Except for very simple cases, a CAE tool is needed to do quantitative synthesis. Examples of such tools are the programs SYMECH by J. Cook et al.[*] that solves the three-position multibar linkage synthesis problem and LINCAGES,[†] by A. Erdman et al. [1], that solves the four-position fourbar linkage synthesis problem. The computer programs provided with this text also allow you to do three-position **analytical synthesis** and general linkage **design by successive analysis**. The fast computation of these programs allows one to analyze the performance of many trial mechanism designs in a short time and promotes rapid iteration to a better solution.

DIMENSIONAL SYNTHESIS   of a linkage *is the determination of the proportions (lengths) of the links necessary to accomplish the desired motions* and can be a form of quantitative synthesis if an algorithm is defined for the particular problem, but can also be a form of qualitative synthesis if there are more variables than equations. The latter situation is more common for linkages. (Dimensional synthesis of cams is quantitative.) Dimensional synthesis assumes that, through *type synthesis*, you have already determined that a linkage (or a cam) is the most appropriate solution to the problem. This chapter discusses **graphical dimensional synthesis** of linkages in detail. Chapter 5 presents methods of **analytical linkage synthesis**, and Chapter 8 presents **cam synthesis**.

3.2   FUNCTION, PATH, AND MOTION GENERATION

FUNCTION GENERATION   is defined as *the correlation of an input motion with an output motion in a mechanism*. A function generator is conceptually a "black box" that delivers

---

[*] SyMech Inc., 1207 Downey Place, Celebration, FL 34747 415-221-5111 http://www.symech.com

---

[†] Available from Prof. A. Erdman, U. Minn., 111 Church St. SE, Minneapolis, MN 55455 612-625-8580

some predictable output in response to a known input. Historically, before the advent of electronic computers, mechanical function generators found wide application in artillery rangefinders and shipboard gun aiming systems, and many other tasks. They are, in fact, **mechanical analog computers**. The development of inexpensive digital electronic microcomputers for control systems coupled with the availability of compact servo and stepper motors has reduced the demand for these mechanical function generator linkage devices. Many such applications can now be served more economically and efficiently with electromechanical devices.[*] Moreover, the computer-controlled electromechanical function generator is programmable, allowing rapid modification of the function generated as demands change. For this reason, while presenting some simple examples in this chapter and a general, analytical design method in Chapter 5, we will not emphasize mechanical linkage function generators in this text. Note, however, that the cam-follower system, discussed extensively in Chapter 8, is in fact a form of mechanical function generator, and it is typically capable of higher force and power levels per dollar than electromechanical systems.

PATH GENERATION   is defined as *the control of a **point** in the plane such that it follows some prescribed path.* This is typically accomplished with at least four bars, wherein a point on the coupler traces the desired path. Specific examples are presented in the section on coupler curves below. Note that no attempt is made in path generation to control the orientation of the link that contains the point of interest. However, it is common for the timing of the arrival of the point at particular locations along the path to be defined. This case is called *path generation with prescribed timing* and is analogous to function generation in that a particular output function is specified. Analytical path and function generation will be dealt with in Chapter 5.

MOTION GENERATION   is defined as *the control of a **line** in the plane such that it assumes some prescribed set of sequential positions.* Here orientation of the link containing the line is important. This is a more general problem than path generation, and in fact, path generation is a subset of motion generation. An example of a motion generation problem is the control of the "bucket" on a bulldozer. The bucket must assume a set of positions to dig, pick up, and dump the excavated earth. Conceptually, the motion of a line, painted on the side of the bucket, must be made to assume the desired positions. A linkage is the usual solution.

PLANAR MECHANISMS VERSUS SPATIAL MECHANISMS   The above discussion of controlled movement has assumed that the motions desired are planar (2-D). We live in a three-dimensional world, however, and our mechanisms must function in that world. **Spatial mechanisms** *are* 3-D *devices*. Their design and analysis are much more complex than those of **planar mechanisms**, which *are* 2-D *devices*. The study of spatial mechanisms is beyond the scope of this introductory text. Some references for further study are in the bibliography to this chapter. However, the study of planar mechanisms is not as practically limiting as it might first appear since many devices in three dimensions are constructed of multiple sets of 2-D devices coupled together. An example is any folding chair. It will have some sort of linkage in the left side plane that allows folding. There will be an identical linkage on the right side of the chair. These two *XY* planar linkages will be connected by some structure along the *Z* direction, which ties the two planar linkages into a 3-D assembly. Many real mechanisms are arranged in this way, as **duplicate planar linkages**, displaced in the *Z* direction in parallel planes and rigidly connected. When you open the hood of a car, take note of the hood hinge mechanism. It will be duplicated

_____

[*] It is worth noting that the day is long past when a mechanical engineer can be content to remain ignorant of electronics and electromechanics. Virtually all modern machines are controlled by electronic devices. Mechanical engineers must understand their operation.

on each side of the car. The hood and the car body tie the two planar linkages together into a 3-D assembly. Look and you will see many other such examples of assemblies of planar linkages into 3-D configurations. So, the 2-D techniques of synthesis and analysis presented here will prove to be of practical value in designing in 3-D as well.

3.3   LIMITING CONDITIONS

The manual, graphical, dimensional synthesis techniques presented in this chapter and the computerizable, analytical synthesis techniques presented in Chapter 5 are reasonably rapid means to obtain a trial solution to a motion control problem. Once a potential solution is found, it must be evaluated for its quality. There are many criteria that may be applied. In later chapters, we will explore the analysis of these mechanisms in detail. However, one does not want to expend a great deal of time analyzing, in great detail, a design that can be shown to be inadequate by some simple and quick evaluations.

TOGGLE POSITIONS   One important test can be applied within the synthesis procedures described below. You need to check that the linkage can in fact reach all of the specified design positions without encountering a limit position. Linkage synthesis procedures often only provide that the particular positions specified will be obtained. They say nothing about the linkage's behavior between those positions. Figure 3-1a shows a non-Grashof fourbar linkage at its limits of motion called **toggle positions**. *The toggle positions are determined by the **colinearity** of two of the moving links.* C_1D_1 and C_2D_2 (solid lines) are the toggle positions reached when driven from link 2. C_3D_3 and C_4D_4 (dashed lines) are the toggle positions reached when driven from link 4. A fourbar triple-rocker mechanism will have four, and a Grashof double-rocker two, of these toggle positions in which the linkage assumes a triangular configuration. When in a triangular (toggle) position, it will not allow further input motion in one direction from one of its rocker links (either of link 2 from positions C_1D_1 and C_2D_2 or link 4 from positions C_3D_3 and C_4D_4). A different link will then have to be driven to get it out of toggle.

STATIONARY POSITIONS   A Grashof fourbar crank-rocker linkage will also assume two stationary positions as shown in Figure 3-1b, when the shortest link (crank O_2C) is

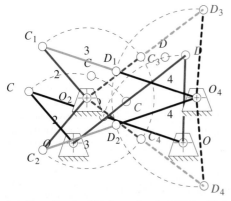

(a) Non-Grashof triple-rocker toggle positions

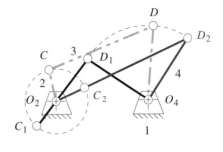

(b) Grashof crank-rocker stationary configurations

FIGURE 3-1

Linkage limit positions

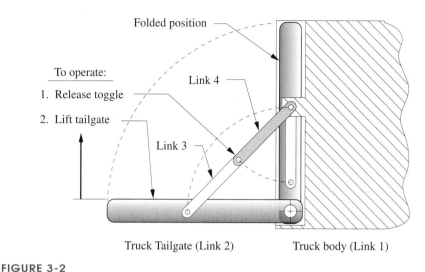

Folded position

To operate:
1. Release toggle
2. Lift tailgate

Link 4

Link 3

Truck Tailgate (Link 2)   Truck body (Link 1)

FIGURE 3-2

Deltoid toggle linkage used to control truck tailgate motion

colinear with the coupler CD (link 3), either *extended colinear* ($O_2C_2D_2$) or *overlapping colinear* ($O_2C_1D_1$). It cannot be *back driven* from the rocker O_4D (link 4) through these colinear positions (which then act as toggles), but when the crank O_2C (link 2) is driven, it will carry through both stationary positions because it is Grashof. Note that the stationary positions define the limits of motion of the driven rocker (link 4), at which its angular velocity will go through zero. Use program LINKAGES to read the data files F03-01A.4br and F03-1b.4br and animate these examples.

After synthesizing a **double- or triple-rocker** solution to a multiposition (motion generation) problem, you **must** check for the presence of toggle positions *between* your design positions. *An easy way to do this is with a model of the linkage.* A CAE tool such as LINKAGES or *Working Model* will also check for this problem. It is important to realize that a toggle condition is only undesirable if it is preventing your linkage from getting from one desired position to the other. In other circumstances the toggle is very useful. It can provide a self-locking feature when a linkage is moved slightly beyond the toggle position and against a fixed stop. Any attempt to reverse the motion of the linkage then causes it merely to jam harder against the stop. It must be manually pulled "over center," out of toggle, before the linkage will move. You have encountered many examples of this application, as in card table or ironing board leg linkages and also pickup truck or station wagon tailgate linkages. An example of such a toggle linkage is shown in Figure 3-2. It happens to be a special-case Grashof linkage in the deltoid configuration (see also Figure 2-17d, p. 57), which provides a locking toggle position when open, and folds on top of itself when closed, to save space. We will analyze the toggle condition in greater detail in a later chapter.

TRANSMISSION ANGLE   Another useful test that can be very quickly applied to a linkage design to judge its quality is the measurement of its transmission angle. This can be done analytically, graphically on the drawing board, or a model for a rough approximation. (Extend the links beyond the pivot to measure the angle.) The **transmission angle**

* The transmission angle as defined by Alt.[2] has limited application. It only predicts the quality of force or torque transmission if the input and output links are pivoted to ground. If the output force is taken from a floating link (coupler), then the transmission angle is of no value. A different index of merit called the joint force index (JFI) is presented in Chapter 11 which discusses force analysis in linkages. (See Section 11.12 on p. 611.) The JFI is useful for situations in which the output link is floating as well as for giving the same kind of information when the output is taken from a link rotating against the ground. However, the JFI requires that a complete force analysis of the linkage be done, whereas the transmission angle is determined from linkage geometry alone.

3

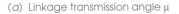

(a) Linkage transmission angle μ   (b) Static forces at a linkage joint

FIGURE 3-3

Transmission angle in the fourbar linkage

* Alt, [2] who defined the
transmission angle, recommended keeping
$\mu_{min} > 40°$. But it can be
argued that at higher speeds,
the momentum of the mov-
ing elements and/or the addition of a flywheel will carry a
mechanism through locations
of poor transmission angle.
The most common example
is the back-driven slider-
crank (as used in internal
combustion engines) which
has $\mu = 0$ twice per revolution. Also, the transmission
angle is only critical in a
fourbar linkage when the
rocker is the output link
on which the working load
impinges. If the working
load is taken by the coupler
rather than by the rocker,
then minimum transmission
angles less than 40° may be
viable. A more definitive
way to qualify a mecha-
nism's dynamic function is
to compute the variation in
its required driving torque.
Driving torque and flywheels
are addressed in Chapter
11. A joint force index (JFI)
can also be calculated. (See
footnote on previous page.)

μ is shown in Figure 3-3a and is defined as *the angle between the output link and the coupler.*[*] It is usually taken as the *absolute value of the acute angle of the pair of angles at the intersection of the two links* and *varies continuously from some minimum to some maximum value as the linkage goes through its range of motion.* It is a measure of the quality of force and velocity transmission at the joint. Note in Figure 3-2 that the linkage cannot be moved from the open position shown by any force applied to the tailgate, link 2, since the transmission angle between links 3 and 4 is zero at that position. But a force applied to link 4 as the input link will move it. The transmission angle is now between links 3 and 2 and is 45 degrees.

Figure 3-3b shows a torque T_2 applied to link 2. Even before any motion occurs, this causes a static, colinear force F_3 to b applied by link 3 to link 4 at point D. Its radial and tangential components F_{34}^r and F_{34}^t are resolved parallel and perpendicular to link 4, respectively. Ideally, we would like all of the force F_{34} to go into producing output torque T_4 on link 4. However, only the tangential component creates torque on link 4. The radial component F_{34}^r provides only tension or compression in link 4. This radial component only increases pivot friction and does not contribute to the output torque. Therefore, the optimum value for the **transmission angle** is 90°. When μ is less than 45° the radial component will be larger than the tangential component. Most machine designers try to keep the **minimum transmission angle above about 40°** to promote smooth running and good force transmission. However, if in your particular design there will be little or no external force or torque applied to link 4, you may be able to get away with even lower values of μ.[*] The transmission angle provides one means to judge the quality of a newly synthesized linkage. If it is unsatisfactory, you can iterate through the synthesis procedure to improve the design. We will investigate the transmission angle in greater detail in later chapters.

3.4   DIMENSIONAL SYNTHESIS

Dimensional synthesis of a linkage is *the determination of the proportions (lengths) of the links necessary to accomplish the desired motions.* This section assumes that, through *type synthesis*, you have determined that a linkage is the most appropriate solution to the problem. Many techniques exist to accomplish this task of **dimensional synthesis of a fourbar linkage**. The simplest and quickest methods are graphical. These work well for up to three design positions. Beyond that number, a numerical, analytical synthesis approach as described in Chapter 5, using a computer, is usually necessary.

Note that the principles used in these graphical synthesis techniques are simply those of **euclidean geometry**. The rules for bisection of lines and angles, properties of parallel and perpendicular lines, and definitions of arcs, etc., are all that is needed to generate these linkages. **Compass**, **protractor**, and **rule** are the only tools needed for graphical linkage synthesis. Refer to any introductory (high school) text on geometry if your geometric theorems are rusty.

Two-Position Synthesis

Two-position synthesis subdivides into two categories: **rocker output** (pure rotation) and **coupler output** (complex motion). Rocker output is most suitable for situations in which a Grashof crank-rocker is desired and is, in fact, a trivial case of *function generation* in which the output function is defined as two discrete angular positions of the rocker. Coupler output is more general and is a simple case of *motion generation* in which two positions of a line in the plane are defined as the output. This solution will frequently lead to a triple-rocker. However, the fourbar triple-rocker can be motor driven by the addition of a **dyad** (twobar chain), which makes the final result a **Watt sixbar** containing a **Grashof fourbar subchain**. We will now explore the synthesis of each of these types of solution for the two-position problem.*

✎⬜EXAMPLE 3-1

Rocker Output - Two Positions with Angular Displacement. (Function Generation)

Problem: Design a fourbar Grashof crank-rocker to give 45° of rocker rotation with equal time forward and back, from a constant speed motor input.

Solution: (See Figure 3-4,† p. 104.)

1   Draw the output link O_4B in both extreme positions, B_1 and B_2 in any convenient location, such that the desired angle of motion θ_4 is subtended.

2   Draw the chord B_1B_2 and extend it in either direction.

3   Select a convenient point O_2 on line B_1B_2 extended.

4   Bisect line segment B_1B_2, and draw a circle of that radius about O_2.

5   Label the two intersections of the circle and B_1B_2 extended, A_1 and A_2.

6   Measure the length of the coupler as A_1 to B_1 or A_2 to B_2.

7   Measure ground length 1, crank length 2, and rocker length 4.

8   Find the Grashof condition. If non-Grashof, redo steps 3 to 8 with O_2 farther from O_4.

---

* A video on "Position Synthesis" showing how to synthesize linkages for two and three positions is included on the book's DVD

---

† This figure is provided as animated AVI and Working Model files on the DVD. Its filename is the same as the figure number.

3

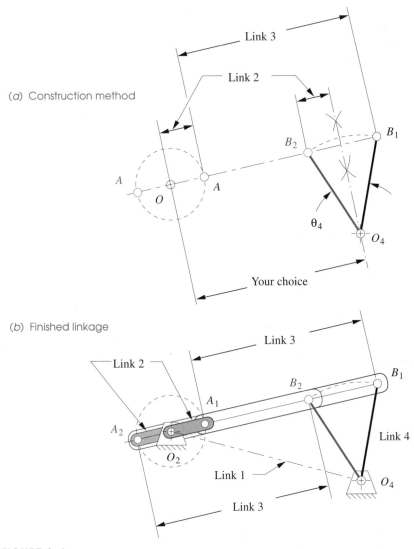

(a) Construction method

(b) Finished linkage

FIGURE 3-4

Two-position function synthesis with rocker output (non-quick-return)

9   Make a model of the linkage and check its function and transmission angles.

10   You can input the file F03-04.4br to program LINKAGES to see this example come alive.

Note several things about this synthesis process. We started with the output end of the system, as it was the only aspect defined in the problem statement. We had to make many quite arbitrary decisions and assumptions to proceed because there were many more variables than we could have provided "equations" for. We are frequently forced to make "free choices" of "a convenient angle or length." These free choices are actually definitions of design parameters. A poor choice will lead to a poor design. Thus these are **qualitative synthesis** approaches and require an iterative process, even for this simple an example. The first solution you reach will probably not be satisfactory, and several attempts (iterations) should be expected to be necessary. As you gain more experience in designing kinematic solutions you will be able to make better choices for these design parameters with fewer iterations. **The value of making a simple model of your design cannot be overstressed!** You will get the *most insight* into your design's quality for the *least effort* by making, articulating, and studying the model. These general observations will hold for most of the linkage synthesis examples presented.

EXAMPLE 3-2

Rocker Output - Two Positions with Complex Displacement. (Motion Generation)

Problem: Design a fourbar linkage to move link CD from position C_1D_1 to C_2D_2.

Solution: (See Figure 3-5,[*] p. 106.)

1   Draw the link CD in its two desired positions, C_1D_1 and C_2D_2, in the plane as shown.

2   Draw construction lines from point C_1 to C_2 and from point D_1 to D_2.

3   Bisect line C_1C_2 and line D_1D_2 and extend their perpendicular bisectors to intersect at O_4. Their intersection is the **rotopole**.

4   Select a convenient radius and draw an arc about the rotopole to intersect both lines O_4C_1 and O_4C_2. Label the intersections B_1 and B_2.

5   Do steps 2 to 8 of Example 3-1 (p. 103) to complete the linkage.

6   Make a model of the linkage and articulate it to check its function and its transmission angles.

Note that Example 3-2 reduces to the method of Example 3-1 (p. 103) once the **rotopole** is found. Thus a link represented by a line in complex motion can be reduced to the simpler problem of pure rotation and moved to any two positions in the plane as the rocker on a fourbar linkage. The next example moves the same link through the same two positions as the coupler of a fourbar linkage.

EXAMPLE 3-3

Coupler Output - Two Positions with Complex Displacement. (Motion Generation)

Problem: Design a fourbar linkage to move the link CD shown from position C_1D_1 to C_2D_2 (with moving pivots at C and D).

Solution: (See Figure 3-6, p. 107.)

1   Draw the link CD in its two desired positions, C_1D_1 and C_2D_2, in the plane as shown.

2   Draw construction lines from point C_1 to C_2 and from point D_1 to D_2.

---
[*] This figure is provided as animated AVI and Working Model files on the DVD. Its filename is the same as the figure number.

3

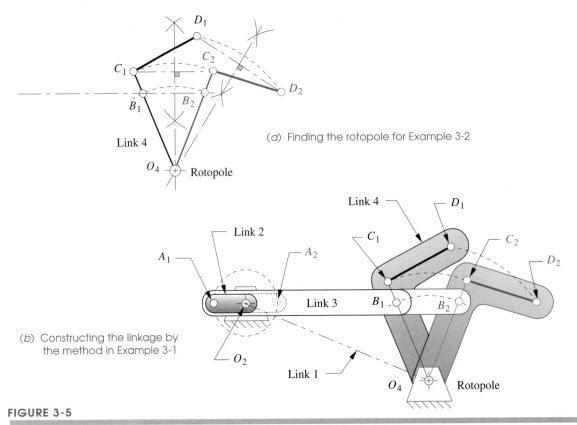

(a) Finding the rotopole for Example 3-2

(b) Constructing the linkage by the method in Example 3-1

FIGURE 3-5

Two-position motion synthesis with rocker output (non-quick-return)

3   Bisect line C_1C_2 and line D_1D_2 and extend the perpendicular bisectors in convenient direc-
    tions. The rotopole will **not** be used in this solution.

4   Select any convenient point on each bisector as the fixed pivots O_2 and O_4, respectively.

5   Connect O_2 with C_1 and call it link 2. Connect O_4 with D_1 and call it link 4.

6   Line C_1D_1 is link 3. Line O_2O_4 is link 1.

7   Check the Grashof condition, and repeat steps 4 to 7 if unsatisfied. Note that any Grashof
    condition is potentially acceptable in this case.

8   Construct a model and check its function to be sure it can get from the initial to final position
    without encountering any limit (toggle) positions.

9   Check the transmission angles.

Input file F03-06.4br to program LINKAGES to see Example 3-3 (p. 105). Note that this
example had nearly the same problem statement as Example 3-2 (p. 105), but the solution
is quite different. Thus a link can also be moved to any two positions in the plane as the

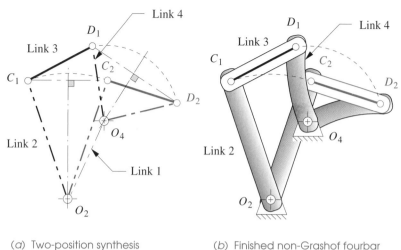

(a) Two-position synthesis   (b) Finished non-Grashof fourbar

FIGURE 3-6

Two-position motion synthesis with coupler output

coupler of a fourbar linkage, rather than as the rocker. However, to limit its motions to those two coupler positions as extrema, two additional links are necessary. These additional links can be designed by the method shown in Example 3-4 and Figure 3-7 (p. 108).

✐ EXAMPLE 3-4

Adding a Dyad (Twobar Chain) to Control Motion in Example 3-3 (p. 105).

Problem: Design a **dyad** to control and limit the extremes of motion of the linkage in Example 3-3 to its two design positions.

Solution: (See Figure 3-7a.)

1   Select a convenient point on link 2 of the linkage designed in Example 3-3. Note that it need not be on the line O_2C_1. Label this point B_1.

2   Draw an arc about center O_2 through B_1 to intersect the corresponding line O_2B_2 in the second position of link 2. Label this point B_2. The chord B_1B_2 provides us with the same problem as in Example 3-1 (p. 103).

3   Do steps 2 to 9 of Example 3-1 to complete the linkage, except add links 5 and 6 and center O_6 rather than links 2 and 3 and center O_2. Link 6 will be the driver crank. The fourbar subchain of links O_6, A_1, B_1, O_2 must be a Grashof crank-rocker.

Note that we have used the approach of Example 3-1 (p. 103) to add a **dyad** to serve as a *driver stage* for our existing fourbar. This results in a **sixbar Watt mechanism** whose first stage is Grashof as shown in Figure 3-7b.[*] Thus we can drive this with a motor on link 6. Note also that we can locate the motor center O_6 anywhere in the plane

(a) Adding a driver dyad to the fourbar chain

(b) The completed Watt sixbar linkage with motor at O_6

(c) An alternate location of the driver dyad with motor at O_6

FIGURE 3-7

Driving a non-Grashof linkage with a dyad (non-quick-return) by judicious choice of point B_1 on link 2. If we had put B_1 below center O_2, the motor would be to the right of links 2, 3, and 4 as shown in Figure 3-7c. There is *an infinity of driver dyads* possible that will drive any double-rocker assemblage of links. Input the files F03-07b.6br and F03-07c.6br to program LINKAGES to see Example 3-4 (p. 107) in motion for these two solutions.

Three-Position Synthesis with Specified Moving Pivots

3

Three-position synthesis allows the definition of three positions of a line in the plane and will create a fourbar linkage configuration to move it to each of those positions. This is a **motion generation** problem. The synthesis technique is a logical extension of the method used in Example 3-3 (p. 105) for two-position synthesis with coupler output. The resulting linkage may be of any Grashof condition and will usually require the addition of a dyad to control and limit its motion to the positions of interest. Compass, protractor, and rule are the only tools needed in this graphical method.

✎ EXAMPLE 3-5

Coupler Output - 3 Positions with Complex Displacement. (Motion Generation)

Problem: Design a fourbar linkage to move the link CD shown from position C_1D_1 to C_2D_2 and then to position C_3D_3. Moving pivots are at C and D. Find the fixed pivot locations.

Solution: (See Figure 3-8, p. 110.)

1   Draw link CD in its three design positions C_1D_1, C_2D_2, C_3D_3 in the plane as shown.

2   Draw construction lines from point C_1 to C_2 and from point C_2 to C_3.

3   Bisect line C_1C_2 and line C_2C_3 and extend their perpendicular bisectors until they intersect. Label their intersection O_2.

4   Repeat steps 2 and 3 for lines D_1D_2 and D_2D_3. Label the intersection O_4.

5   Connect O_2 with C_1 and call it link 2. Connect O_4 with D_1 and call it link 4.

6   Line C_1D_1 is link 3. Line O_2O_4 is link 1.

7   Check the Grashof condition. Any Grashof condition is potentially acceptable in this case.

8   Construct a model and check its function to be sure it can get from initial to final position without encountering any limit (toggle) positions.

9   Construct a driver dyad according to the method in Example 3-4 (p. 107) using an extension of link 3 to attach the dyad.

Note that while a solution is usually obtainable for this case, it is possible that you may not be able to move the linkage continuously from one position to the next without disassembling the links and reassembling them to get them past a limiting position. That

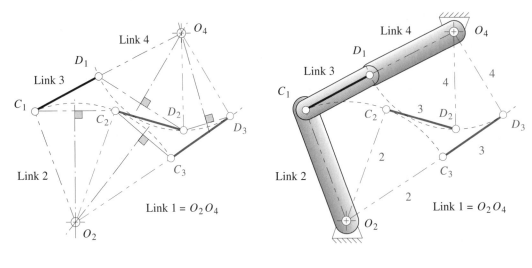

(a) Construction method   (b) Finished non-Grashof fourbar

FIGURE 3-8

Three-position motion synthesis

will obviously be unsatisfactory. In the particular solution presented in Figure 3-8, note that links 3 and 4 are in toggle at position one, and links 2 and 3 are in toggle at position three. In this case we will have to drive link 3 with a driver dyad, since any attempt to drive either link 2 or link 4 will fail at the toggle positions. No amount of torque applied to link 2 at position C_1 will move link 4 away from point D_1, and driving link 4 will not move link 2 away from position C_3. Input the file F03-08.4br to program LINKAGES to see Example 3-5 (p. 109).

Three-Position Synthesis with Alternate Moving Pivots

Another potential problem is the possibility of an undesirable location of the fixed pivots O_2 and O_4 with respect to your packaging constraints. For example, if the fixed pivot for a windshield wiper linkage design ends up in the middle of the windshield, you may want to redesign it. Example 3-6 shows a way to obtain an alternate configuration for the same three-position motion as in Example 3-5 (p. 109). And, the method shown in Example 3-8 (ahead on p. 115) allows you to specify the location of the fixed pivots in advance and then find the locations of the moving pivots on link 3 that are compatible with those fixed pivots.

✏️ EXAMPLE 3-6

Coupler Output - Three Positions with Complex Displacement - Alternate Attachment Points for Moving Pivots. (Motion Generation)

Problem: Design a fourbar linkage to move the link CD shown from position C_1D_1 to C_2D_2 and then to position C_3D_3. Use different moving pivots than CD. Find the fixed pivot locations.

Solution: (See Figure 3-9,[*] p. 112.)

1 Draw the link CD in its three desired positions C_1D_1, C_2D_2, C_3D_3, in the plane as done in Example 3-5 (p. 109).

2 Define new attachment points E_1 and F_1 that have a fixed relationship between C_1D_1 and E_1F_1 within the link. Now use E_1F_1 to define the three positions of the link.

3 Draw construction lines from point E_1 to E_2 and from point E_2 to E_3.

4 Bisect line E_1E_2 and line E_2E_3 and extend the perpendicular bisectors until they intersect. Label the intersection O_2.

5 Repeat steps 2 and 3 for lines F_1F_2 and F_2F_3. Label the intersection O_4.

6 Connect O_2 with E_1 and call it link 2. Connect O_4 with F_1 and call it link 4.

7 Line E_1F_1 is link 3. Line O_2O_4 is link 1.

8 Check the Grashof condition. Note that any Grashof condition is potentially acceptable in this case.

9 Construct a model and check its function to be sure it can get from initial to final position without encountering any limit (toggle) positions. If not, change locations of points E and F and repeat steps 3 to 9.

10 Construct a driver dyad acting on link 2 according to the method in Example 3-4 (p. 107).

Note that the shift of the attachment points on link 3 from CD to EF has resulted in a shift of the locations of fixed pivots O_2 and O_4 as well. Thus they may now be in more favorable locations than they were in Example 3-5 (p. 109). It is important to understand that any two points on link 3, such as E and F, can serve to fully define that link as a rigid body, and that there is an infinity of such sets of points to choose from. While points C and D have some particular location in the plane that is defined by the linkage's function, points E and F can be anywhere on link 3, thus creating an infinity of solutions to this problem.

The solution in Figure 3-9 (p. 112) is different from that of Figure 3-8 in several respects. It avoids the toggle positions and thus can be driven by a dyad acting on one of the rockers, as shown in Figure 3-9c, and the transmission angles are better. However, the toggle positions of Figure 3-8 might actually be of value if a self-locking feature were desired. *Recognize that both of these solutions are to the same problem,* and the solution in Figure 3-8 is just a special case of that in Figure 3-9. Both solutions may be useful. Line CD moves through the same three positions with both designs. There is an infinity of other solutions to this problem waiting to be found as well. Input the file F03-09c.6br to program LINKAGES to see Example 3-6.

Three-Position Synthesis with Specified Fixed Pivots

Even though one can probably find an acceptable solution to the three-position problem by the methods described in the two preceding examples, it can be seen that the designer will have little direct control over the location of the fixed pivots since they are one of

[*] This figure is provided as animated AVI and Working Model files on the DVD. Its filename is the same as the figure number.

3

(a) Alternate attachment points

(b) Three-position synthesis

Link 1 = $O_2 O_4$

(c) Completed Watt sixbar linkage with motor at O_6

FIGURE 3-9

Three-position synthesis with alternate moving pivots

the results of the synthesis process. The fixed pivots need to be located where the ground plane of the package exists and is accessible. It would be preferable if we could define the fixed pivot locations, as well as the three positions of the moving link, and then synthesize the appropriate attachment points, E and F, to the moving link to satisfy these more realistic constraints. The principle of **inversion** can be applied to this problem. Examples 3-5 (p. 109) and 3-6 (p. 110) showed how to find the required fixed pivots for three chosen positions of the moving pivots. Inverting this problem allows specification of the fixed pivot locations and determination of the required moving pivots for those locations. The first step is to find the three positions of the ground plane that correspond to the three desired coupler positions. This is done by **inverting the linkage**[*] as shown in Figure 3-10 (p. 114) and Example 3-7.

✐ EXAMPLE 3-7

Three-Position Synthesis with Specified Fixed Pivots – Inverting the Three-Position Motion Synthesis Problem

Problem: Invert a fourbar linkage which moves the link CD shown from position C_1D_1 to C_2D_2 and then to position C_3D_3. Use specified fixed pivots O_2 and O_4.

Solution: First find the inverted positions of the ground link corresponding to the three coupler positions specified. (See Figure 3-10, p. 114.)

1 Draw the link CD in its three desired positions C_1D_1, C_2D_2, C_3D_3, in the plane, as was done in Example 3-5 (p. 109) and as shown in Figure 3-10a.

2 Draw the ground link O_2O_4 in its desired position in the plane with respect to the first coupler position C_1D_1 as shown in Figure 3-10a.

3 Draw construction arcs from point C_2 to O_2 and from point D_2 to O_2 whose radii define the sides of triangle $C_2O_2D_2$. This defines the relationship of the fixed pivot O_2 to the coupler line CD in the second coupler position as shown in Figure 3-10b.

4 Draw construction arcs from point C_2 to O_4 and from point D_2 to O_4 to define the triangle $C_2O_4D_2$. This defines the relationship of the fixed pivot O_4 to the coupler line CD in the second coupler position as shown in Figure 3-10b.

5 Now transfer this relationship back to the first coupler position C_1D_1 so that the ground plane position $O_2'O_4'$ bears the same relationship to C_1D_1 as O_2O_4 bore to the second coupler position C_2D_2. In effect, you are sliding C_2 along the dotted line C_2-C_1 and D_2 along the dotted line D_2-D_1. By doing this, we have pretended that the ground plane moved from O_2O_4 to $O_2'O_4'$ instead of the coupler moving from C_1D_1 to C_2D_2. We have *inverted* the problem.

6 Repeat the process for the third coupler position as shown in Figure 3-10d and transfer the third relative ground link position to the first, or reference, position as shown in Figure 3-10e.

7 The three inverted positions of the ground plane that correspond to the three desired coupler positions are labeled O_2O_4, $O_2'O_4'$, and $O_2''O_4''$ and have also been renamed E_1F_1, E_2F_2, and E_3F_3 as shown in Figure 3-10f. These correspond to the three coupler positions shown in Figure 3-10a. Note that the original three lines C_1D_1, C_2D_2, and C_3D_3 are not now needed for the linkage synthesis.

---

[*] This method and example were supplied by Mr. Homer D. Eckhardt, Consulting Engineer, Lincoln, MA.

3

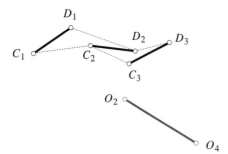

(a) Original coupler three-position
problem with specified pivots

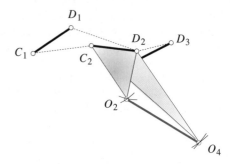

(b) Position of the ground plane relative
to the second coupler position

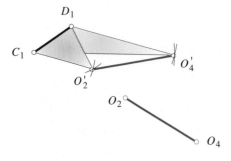

(c) Transferring second ground plane position
to reference location at first position

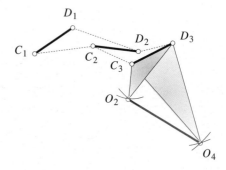

(d) Position of the ground plane relative
to the third coupler position

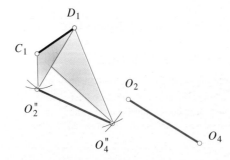

(e) Transferring third ground plane position
to reference location at first position

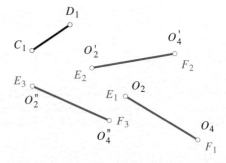

(f) The three inverted positions of the ground plane
corresponding to the original coupler position

FIGURE 3-10

Inverting the three-position motion synthesis problem

We can use these three new lines E_1F_1, E_2F_2, and E_3F_3 to find the attachment points GH (moving pivots) on link 3 that will allow the desired fixed pivots O_2 and O_4 to be used for the three specified output positions. In effect we will now consider the ground link O_2O_4 to be a coupler moving through the inverse of the original three positions, find the "ground pivots" GH needed for that inverted motion, and put them on the real coupler instead. The inversion process done in Example 3-7 (p. 113) and Figure 3-10 has swapped the roles of coupler and ground plane. The remaining task is identical to that done in Example 3-5 (p. 109) and Figure 3-8 (p. 110). The result of the synthesis then must be reinverted to obtain the solution.

EXAMPLE 3-8

Finding the Moving Pivots for Three Positions and Specified Fixed Pivots

Problem: Design a fourbar linkage to move the link CD shown from position C_1D_1 to C_2D_2 and then to position C_3D_3. Use specified fixed pivots O_2 and O_4. Find the required moving pivot locations on the coupler by inversion.

Solution: Using the inverted ground link positions E_1F_1, E_2F_2, and E_3F_3 found in Example 3-7 (p. 113), find the fixed pivots for that inverted motion, then reinvert the resulting linkage to create the moving pivots for the three positions of coupler CD that use the selected fixed pivots O_2 and O_4 as shown in Figure 3-10a (see also Figure 3-11,[*] p. 116).

1   Start with the inverted three positions in the plane as shown in Figures 3-10f and 3-11a. Lines E_1F_1, E_2F_2, and E_3F_3 define the three positions of the inverted link to be moved.

2   Draw construction lines from point E_1 to E_2 and from point E_2 to E_3.

3   Bisect line E_1E_2 and line E_2E_3 and extend the perpendicular bisectors until they intersect. Label the intersection G.

4   Repeat steps 2 and 3 for lines F_1F_2 and F_2F_3. Label the intersection H.

5   Connect G with E_1 and call it link 2. Connect H with F_1 and call it link 4. See Figure 3-11b.

6   In this inverted linkage, line E_1F_1 is the coupler, link 3. Line GH is the "ground" link 1.

7   We must now reinvert the linkage to return to the original arrangement. Line E_1F_1 is really the ground link O_2O_4, and GH is really the coupler. Figure 3-11c shows the reinversion of the linkage in which points G and H are now the moving pivots on the coupler and E_1F_1 has resumed its real identity as ground link O_2O_4. (See Figure 3-10e)

8   Figure 3-11d reintroduces the original line C_1D_1 in its correct relationship to line O_2O_4 at the initial position as shown in the original problem statement in Figure 3-10a. This forms the required coupler plane and defines a minimal shape of link 3.

9   The angular motions required to reach the second and third positions of line CD shown in Figure 3-11e are the same as those defined in Figure 3-11b for the linkage inversion. The angle F_1HF_2 in Figure 3-11b is the same as angle $H_1O_4H_2$ in Figure 3-11e and F_2HF_3 is the same as angle $H_2O_4H_3$. The angular excursions of link 2 retain the same relationship between Figure

---

[*] This figure is provided as animated AVI and Working Model files on the DVD. Its filename is the same as the figure number.

3

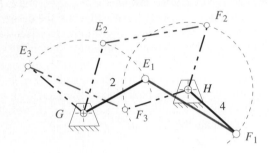

(a) Construction to find "fixed" pivots G and H

(b) The correct inversion of desired linkage

(c) Reinvert to obtain the result

(d) Re-place line CD on link

(e) The three positions (link 4 driving CCW)

FIGURE 3-11

Constructing the linkage for three positions with specified fixed pivots by inversion

3-11b and e as well. The angular motions of links 2 and 4 are the same for both inversions as the link excursions are relative to one another.

10 Check the Grashof condition. Note that any Grashof condition is potentially acceptable in this case provided that the linkage has mobility among all three positions. This solution is a non-Grashof linkage.

11 Construct a model and check its function to be sure it can get from initial to final position without encountering any limit (toggle) positions. In this case links 3 and 4 reach a toggle position between points H_1 and H_2. This means that this linkage cannot be driven from link 2 as it will hang up at that toggle position. It must be driven from link 4.

By inverting the original problem, we have reduced it to a more tractable form that allows a direct solution by the general method of three-position synthesis from Examples 3-5 (p. 109) and 3-6 (p. 110).

Position Synthesis for More Than Three Positions

It should be obvious that the more constraints we impose on these synthesis problems, the more complicated the task becomes to find a solution. When we define more than three positions of the output link, the difficulty increases substantially.

FOUR-POSITION SYNTHESIS does not lend itself as well to manual graphical solutions, though Hall[3] does present one approach. Probably the best approach is that used by Sandor and Erdman[4] and others, which is a quantitative synthesis method and requires a computer to execute it. Briefly, a set of simultaneous vector equations is written to represent the desired four positions of the entire linkage. These are then solved after some free choices of variable values are made by the designer. The computer program LINCAGES[1] by Erdman et al. and the program KINSYN[5] by Kaufman, both provide a convenient and user-friendly computer graphics-based means to make the necessary design choices to solve the four-position problem. See Chapter 5 for further discussion.

3.5 QUICK-RETURN MECHANISMS

Many machine design applications have a need for a difference in average velocity between their "forward" and "return" strokes. Typically some external work is being done by the linkage on the forward stroke, and the return stroke needs to be accomplished as rapidly as possible so that a maximum of time will be available for the working stroke. Many arrangements of links will provide this feature. The only problem is to synthesize the right one!

Fourbar Quick-Return

The linkage synthesized in Example 3-1 (p. 103) is perhaps the simplest example of a fourbar linkage design problem (see Figure 3-4, p. 104, and program LINKAGES disk file F03-04.4br). It is a crank-rocker that gives two positions of the rocker with equal time for the forward stroke and the return stroke. This is called a *non-quick-return* linkage, and it is a special case of the more general **quick-return** case. The reason for its non-quick-return state is the positioning of the crank center O_2 on the chord B_1B_2 extended. This results in equal angles of 180 degrees being swept out by the crank as it drives the rocker from one extreme (toggle position) to the other. If the crank is rotating at constant angular velocity, as it will tend to when motor driven, then each 180 degree sweep, forward and back, will take the same time interval. Try this with your model from Example 3-1 by rotating the crank at uniform velocity and observing the rocker motion and velocity.

If the crank center O_2 is located off the chord B_1B_2 extended, as shown in Figure 3-1b (p. 100) and Figure 3-12, then unequal angles will be swept by the crank between the toggle positions (defined as colinearity of crank and coupler). Unequal angles will give unequal time, when the crank rotates at constant velocity. These angles are labeled α and β in Figure 3-12. Their ratio α/β is called the **time ratio** (T_R) and defines the degree of quick return of the linkage. Note that the term **quick return** is arbitrarily used to describe this kind of linkage. If the crank is rotated in the opposite direction, it will be a **quick-forward** mechanism. Given a completed linkage, it is a trivial task to estimate the time ratio by measuring or calculating the angles α and β. It is a more difficult task to design the linkage for a chosen time ratio. Hall [6] provides a graphical method to synthesize a quick-return Grashof fourbar. T do so we need to compute the values of α an β that will give the specified time ratio. We can write two equations involving α and β and solve them simultaneously.

$$T_R = \frac{\alpha}{\beta} \qquad \alpha + \beta = 360 \qquad \therefore \quad T_R = \frac{\alpha}{360 - \alpha} \qquad (3.1)$$

We also must define a construction angle,

$$\delta = |180 - \alpha| = |180 - \beta| \qquad (3.2)$$

which will be used to synthesize the linkage.

EXAMPLE 3-9

Fourbar Crank-Rocker Quick-Return Linkage for Specified Time Ratio

Problem: Redesign Example 3-1 (p. 103) to provide a time ratio of 1:1.25 with 45° output rocker motion.

Solution: (See Figure 3-12.[*])

1 Draw the output link O_4B in both extreme positions, in any convenient location, such that the desired angle of motion, θ_4, is subtended.

2 Calculate α, β, and δ using equations 3.1 and 3.2. In this example, α = 160°, β = 200°, δ = 20°.

3 Draw a construction line through point B_1 at any convenient angle.

4 Draw a construction line through point B_2 at angle δ from the first line.

5 Label the intersection of the two construction lines O_2.

6 The line O_2O_4 now defines the ground link.

7 Find lengths of crank and coupler by measuring O_2B_1 and O_2B_2 and simultaneously solving:

Coupler + crank = O_2B_1

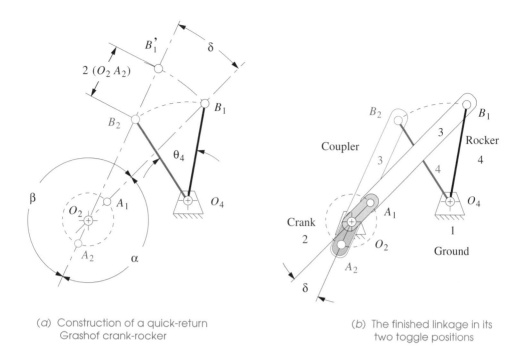

(a) Construction of a quick-return
Grashof crank-rocker

(b) The finished linkage in its
two toggle positions

FIGURE 3-12

Quick-return Grashof fourbar crank-rocker linkage

$$\text{Coupler} - \text{crank} = O_2B_2$$

or you can construct the crank length by swinging an arc centered at O_2 from B_1 to cut line O_2B_2 extended. Label that intersection $B_1{}'$. The line $B_2B_1{}'$ is twice the crank length. Bisect this line segment to measure crank length O_2A_1.

8 Calculate the Grashof condition. If non-Grashof, repeat steps 3 to 8 with O_2 farther from O_4.

9 Make a model of the linkage and articulate it to check its function.

10 Check the transmission angles.

This method works well for time ratios down to about 1:1.5. Beyond that value the transmission angles become poor, and a more complex linkage is needed. Input the file F03-12.4br to program LINKAGES to see Example 3-9.

Sixbar Quick-Return

Larger time ratios, up to about 1:2, can be obtained by designing a sixbar linkage. The strategy here is to first design a fourbar drag link mechanism that has the desired time ratio between its driver crank and its driven or "dragged" crank, and then add a dyad (twobar) output stage, driven by the dragged crank. This dyad can be arranged to have either a rocker or a translating slider as the output link. First the drag link fourbar will be synthesized; then the dyad will be added.

* This figure is provided as animated AVI and Working Model files on the DVD. Its filename is the same as the figure number.

3

✐EXAMPLE 3-10

Sixbar Drag Link Quick-Return Linkage for Specified Time Ratio.

Problem: Provide a time ratio of 1:1.4 with 90° rocker motion.

Solution: (See Figure 3-13.)

1   Calculate α and β using equations 3.1. For this example, $\alpha = 150°$ and $\beta = 210°$.

2   Draw a line of centers XX at any convenient location.

3   Choose a crank pivot location O_2 on line XX. Draw an axis YY perpendicular to XX through O_2.

4   Draw a circle of convenient radius O_2A about center O_2.

5   Lay out angle α with vertex at O_2, symmetrical about quadrant one.

6   Label points A_1 and A_2 at the intersections of the lines subtending angle α and the circle of radius O_2A .

7   Set the compass to a convenient radius AC long enough to cut XX in two places on either side of O_2 when swung from both A_1 and A_2. Label the intersections C_1 and C_2.

8   The line O_2A_1 is the driver crank, link 2, and line A_1C_1 is the coupler, link 3.

9   The distance C_1C_2 is twice the driven (dragged) crank length. Bisect it to find fixed pivot O_4.

10 The line O_2O_4 now defines the ground link. Line O_4C_1 is the driven crank, link 4.

11 Calculate the Grashof condition. If non-Grashof, repeat steps 7 to 11 using a smaller radius in step 7.

12 Invert the method of Example 3-1 (p. 103) to create the output dyad using XX as the chord and O_4C_1 as the driving crank. The points B_1 and B_2 will lie on line XX and be spaced apart a distance $2O_4C_1$. The pivot O_6 will lie on the perpendicular bisector of B_1B_2, at a distance from line XX which subtends the specified output rocker angle.

13 Check the transmission angles.

This linkage provides a quick return when a constant-speed motor is attached to link 2. Link 2 will go through angle α while link 4 (which is dragging the output dyad along) goes through the first 180 degrees, from position C_1 to C_2. Then, while link 2 completes its cycle through β degrees, the output stage will complete another 180 degrees from C_2 to C_1. Since angle β is greater than α, the forward stroke takes longer. Note that the chordal stroke of the output dyad is twice the crank length O_4C_1. This is independent of the angular displacement of the output link which can be tailored by moving the pivot O_6 closer to or farther from the line XX.

The transmission angle at the joint between link 5 and link 6 will be optimized if the fixed pivot O_6 is placed on the perpendicular bisector of the chord B_1B_2 as shown in Figure 3-13a.[*] If a translating output is desired, the slider (link 6) will be located on line XX and

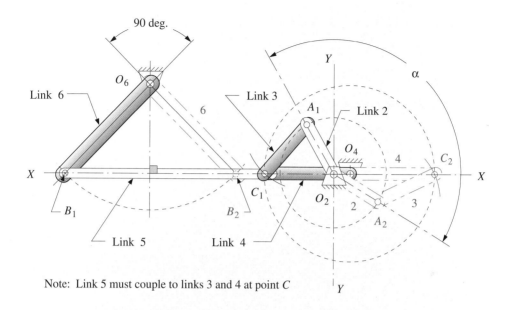

Note: Link 5 must couple to links 3 and 4 at point C

(a) Rocker output sixbar drag link quick-return mechanism

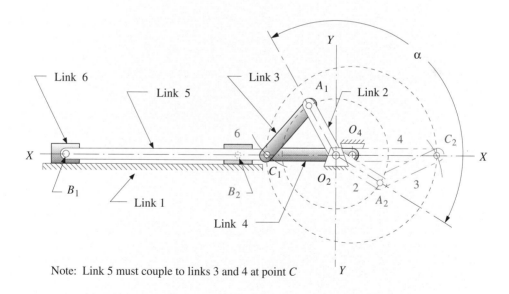

Note: Link 5 must couple to links 3 and 4 at point C

(b) Slider output sixbar drag link quick-return mechanism

FIGURE 3-13

Synthesizing a sixbar drag link quick-return mechanism

will oscillate between B_1 and B_2 as shown in Figure 3-13b. The arbitrarily chosen size of this or any other linkage can be scaled up or down, simply by multiplying all link lengths by the same scale factor. Thus a design made to arbitrary size can be fit to any package. Input the file F03-13a.6br to program LINKAGES to see Example 3-10 (p. 120) in action.

CRANK-SHAPER QUICK RETURN   A commonly used mechanism capable of large time ratios is shown in Figure 3-14.[*] It is often used in metal shaper machines to provide a slow cutting stroke and a quick-return stroke when the tool is doing no work. It is the inversion #2 of the crank-slider mechanism as was shown in Figure 2-13b (p. 52). It is very easy to synthesize this linkage by simply moving the rocker pivot O_4 along the vertical centerline O_2O_4 while keeping the two extreme positions of link 4 tangent to the circle of the crank, until the desired time ratio (α / β) is achieved. Note that the angular displacement of link 4 is then defined as well. Link 2 is the input and link 6 is the output.

Depending on the relative lengths of the links this mechanism is known as a **Whitworth** or **crank-shaper** mechanism. If the ground link is the shortest, then it will behave as a double-crank linkage, or *Whitworth mechanism*, with both pivoted links making full revolutions as shown in Figure 2-13b (p. 52). If the driving crank is the shortest link, then it will behave as a crank-rocker linkage, or *crank-shaper mechanism*, as shown in Figure 3-14. They are the same inversion as the slider block is in complex motion in each case.

3.6   COUPLER CURVES

* This figure is provided as animated AVI and Working Model files on the DVD. Its filename is the same as the figure number.

A **coupler** is the most interesting link in any linkage. It is in complex motion, and thus points on the coupler can have path motions of high degree.[†] In general, the more links, the higher the degree of curve generated, where **degree** here means *the highest power of any term in its equation*. A curve (function) can have *up to* as many intersections (roots)

† In 1876, Kempe [7a] proved his theory that a linkage with only revolute (pin) and prismatic (slider) joints can be found that will trace any algebraic curve of any order or complexity. But the linkage for a particular curve may be excessively complex, may be unable to traverse the curve without encountering limit (toggle) positions, and may even need to be disassembled and reassembled to reach all points on the curve. See the discussion of circuit and branch defects in Section 4.12 (p. 207). Nevertheless this theory points to the potential for interesting motions from the coupler curve.

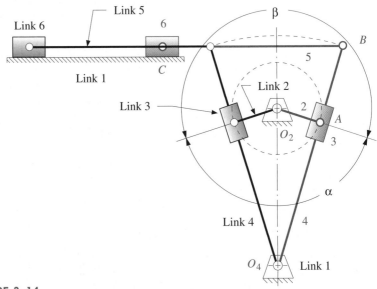

FIGURE 3-14

Crank-shaper quick-return mechanism

with any straight line as the degree of the function. The *fourbar crank-slider* has, in general, fourth-degree coupler curves; the *pin-jointed fourbar*, up to sixth degree.[*] The geared fivebar, the sixbar, and more complicated assemblies all have still higher-degree curves. Wunderlich [7b] derived an expression for the highest degree *m* possible for a coupler curve of a mechanism of *n* links connected with only revolute joints.

$$m = 2 \cdot 3^{(n/2-1)} \tag{3.3}$$

This gives, respectively, degrees of 6, 18, and 54 for the fourbar, sixbar, and eightbar linkage coupler curves. Specific points on their couplers may have degenerate curves of lower degree as, for example, the pin joints between any crank or rocker and the coupler that describes second-degree curves (circles). The parallelogram fourbar linkage has degenerate coupler curves, all of which are circles.

All linkages that possess one or more "floating" coupler links will generate coupler curves. It is interesting to note that these will be closed curves even for non-Grashof linkages. The coupler (or any link) can be extended infinitely in the plane. Figure 3-15[†] (p. 124) shows a fourbar linkage with its coupler extended to include a large number of points, each of which describes a different coupler curve. Note that these points may be anywhere on the coupler, including along line *AB*. There is, of course, an infinity of points on the coupler, each of which generates a different curve.

Coupler curves can be used to generate quite useful path motions for machine design problems. They are capable of *approximating straight lines* and *large circle arcs* with remote centers. Recognize that the coupler curve is a solution to the path generation problem described in Section 3.2 (p. 98). It is not by itself a solution to the motion generation problem, since the attitude or orientation of a line on the coupler is not predicted by the information contained in the path. Nevertheless it is a very useful device, and it can be converted to a parallel motion generator by adding two links as described in the next section. As we shall see, approximate straight-line motions, dwell motions, and more complicated symphonies of timed motions are available from even the simple fourbar linkage and its infinite variety of often surprising coupler curve motions.

CUSPS AND CRUNODES   come in a variety of shapes which can be crudely categorized as shown in Figure 3-16 (p. 124). There is an infinite range of variation between these generalized shapes. Interesting features of some coupler curves are the **cusp** and **crunode**. A **cusp** is *a sharp point on the curve which has the useful property of instantaneous zero velocity*. Note that *the acceleration at the cusp is not zero*. The simplest example of a curve with a cusp is the cycloid curve which is generated by a point on the rim of a wheel rotating on a flat surface. When the point touches the surface, it has the same (zero) velocity as all points on the stationary surface, provided there is pure rolling and no slip between the elements. Anything attached to a cusp point will come smoothly to a stop along one path and then accelerate smoothly away from that point on a different path. The cusp's feature of zero velocity has value in such applications as transporting, stamping and feeding processes. A **crunode** *is a double point that occurs where the coupler curve crosses itself creating multiple loops*. The two slopes (tangents) at a crunode give the point two different velocities, neither of which is zero in contrast to the cusp. In general, a fourbar coupler curve can have up to three real double points,[‡] which may be a combination of cusps and crunodes as can be seen in Figure 3-16.

---

[*] The algebraic equation of the coupler curve is sometimes referred to as a "tricircular sextic" referring respectively to its circularity of 3 (it can contain 3 loops) and its degree of 6. See Chapter 5 for its equation.

[†] This figure is provided as animated AVI and Working Model files on the DVD. Its filename is the same as the figure number.

[‡] Actually, the fourbar coupler curve has 9 double points of which 6 are usually imaginary. However, Fichter and Hunt [8b] point out that some unique configurations of the fourbar linkage (i.e., rhombus parallelograms and those close to this configuration) can have up to 6 real double points which they denote as comprising 3 "proper" and 3 "improper" real double points. For non-special-case Grashof fourbar linkages with minimum transmission angles suitable for engineering applications, only the 3 "proper" double points will appear.

(a) Pseudo ellipse

3

(b) Kidney bean

(c) Banana

(d) Crescent

(e) Single straight

(f) Double straight

FIGURE 3-16 Part 1

A "Cursory Catalog" of coupler curve shapes

---

* This figure is provided as animated AVI and Working Model files on the DVD. Its filename is the same as the figure number.

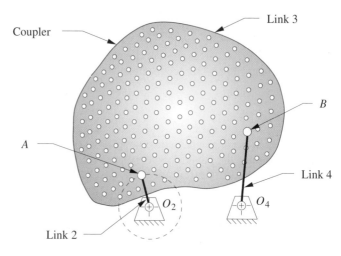

FIGURE 3-15

The fourbar coupler extended to include a large number of coupler points

The *Hrones and Nelson* (H&N) atlas of fourbar coupler curves [8a] is a useful reference which can provide the designer with a starting point for further design and analysis. It contains about 7000 coupler curves and defines the linkage geometry for each of its Grashof crank-rocker linkages. Figure 3-17a* (p. 126) reproduces a page from this book and the entire atlas is reproduced as PDF files on the book DVD. The H&N atlas is logically arranged, with all linkages defined by their link ratios, based on a unit length crank. The coupler is shown as a matrix of fifty coupler points for each linkage geometry, arranged ten to a page. Thus each linkage geometry occupies five pages. Each page contains a schematic "key" in the upper right corner which defines the link ratios.

Figure 3-17b* (p. 126) shows a "fleshed out" linkage drawn on top of the H&N atlas page to illustrate its relationship to the atlas information. The double circles in Figure 3-17a (p. 126) define the fixed pivots. The crank is always of unit length. The ratios of the other link lengths to the crank are given on each page. The actual link lengths can be scaled up or down to suit your package constraints and this will affect the size but not the shape of the coupler curve. Any one of the ten coupler points shown can be used by incorporating it into a triangular coupler link. The location of the chosen coupler point can be scaled from the atlas and is defined within the coupler by the position vector **R** whose constant angle ϕ is measured with respect to the line of centers of the coupler. The H&N coupler curves are shown as dashed lines. Each dash station represents **five degrees** of crank rotation. So, for an assumed constant crank velocity, the dash spacing is proportional to path velocity. The changes in velocity and the quick-return nature of the coupler path motion can be clearly seen from the dash spacing.

One can peruse this linkage atlas resource on the book's DVD and find an approximate solution to virtually any path generation problem. Then one can take the tentative solution from the atlas to a CAE resource such as the LINKAGES program and further refine the design, based on the complete analysis of positions, velocities, and accelerations provided by the program. The only data needed for the LINKAGES program are the four link lengths and the location of the chosen coupler point with respect to the line of centers of the coupler link as shown in Figure 3-17 (p. 126). These parameters can be changed within program LINKAGES to alter and refine the design. Input the file F03-17b.4br to program LINKAGES to animate the linkage shown in that figure. Also see the video "Coupler Curves" on the DVD for more information.

An example of an application of a fourbar linkage to a practical problem is shown in Figure 3-18[*] (p. 127) which is a movie camera (or projector) film advance mechanism. Point O_2 is the crank pivot which is motor driven at constant speed. Point O_4 is the rocker pivot, and points A and B are the moving pivots. Points A, B, and C define the coupler where C is the coupler point of interest. A movie is really a series of still pictures, each "frame" of which is projected for a small fraction of a second on the screen. Between each picture, the film must be moved very quickly from one frame to the next while the shutter is closed to blank the screen. The whole cycle takes only 1/24 of a second. The human eye's response time is too slow to notice the flicker associated with this discontinuous stream of still pictures, so it appears to us to be a continuum of changing images.

The linkage shown in Figure 3-18[*] (p. 127) is cleverly designed to provide the required motion. A hook is cut into the coupler of this fourbar Grashof crank-rocker at point C which generates the coupler curve shown. The hook will enter one of the sprocket holes in the film as it passes point F_1. Notice that the direction of motion of the hook at that point is nearly perpendicular to the film, so it enters the sprocket hole cleanly. It then turns abruptly downward and follows a crudely approximate straight line as it rapidly pulls the film downward to the next frame. The film is separately guided in a straight track called the "gate." The shutter (driven by another linkage from the same driveshaft at O_2) is closed during this interval of film motion, blanking the screen. At point F_2 there is a cusp on the coupler curve which causes the hook to decelerate smoothly to zero velocity in the vertical direction, and then as smoothly accelerate up and out of the sprocket hole. The abrupt transition of direction at the cusp allows the hook to back out of the hole without jarring the film, which would make the image jump on the screen as the shutter opens. The rest of the coupler curve motion is essentially "wasting time" as it proceeds up the back side, to be ready to enter the film again to repeat the process. Input the file F03-18.4br to program LINKAGES to animate the linkage shown in that figure.

Some advantages of using this type of device for this application are that it is very simple and inexpensive (only four links, one of which is the frame of the camera), is extremely reliable, has low friction if good bearings are used at the pivots, and can be reliably timed with the other events in the overall camera mechanism through common shafting from a single motor. There are a myriad of other examples of fourbar coupler curves used in machines and mechanisms of all kinds.

One other example of a very different application is that of the automobile suspension (Figure 3-19, p. 127). Typically, the up and down motions of the car's wheels are controlled by some combination of planar fourbar linkages, arranged in duplicate to provide three-dimensional control as described in Section 3.2 (p. 98). Only a few manufacturers currently use a true spatial linkage in which the links are not arranged in parallel planes. In all cases the wheel assembly is attached to the coupler of the linkage assembly, and its motion is along a set of coupler curves. The orientation of the wheel is also of concern in this case, so this is not strictly a path generation problem. By designing the linkage to

(g) Teardrop

(h) Scimitar

(i) Umbrella

(j) Triple cusp

(k) Figure eight

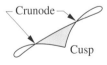

(l) Triple loop

FIGURE 3-16 Part 2

A "Cursory Catalog" of coupler curve shapes

[*] This figure is provided as animated AVI and Working Model files on the DVD. Its filename is the same as the figure number.

3

* The Hrones and Nelson atlas is long out of print but a reproduction is included as PDF files on this book's DVD. A video, "Coupler Curves" is also provided on DVD that describes the curve's properties and shows how to extract the information from the atlas and use it to design a practical mechanism.

Also, a similar volume to the H&N book called the *Atlas of Linkage Design and Analysis Vol. 1: The Four Bar Linkage* is available from Saltire Software, 9725 SW Gemini Drive, Beaverton, OR 97005, (800) 659-1874.

There is also a web site at http://www.cedarville.edu/ dept/eg/kinematics/ccapdf/ fccca.htm created by Prof. Thomas J. Thompson of Cedarville College, which provides an interactive coupler curve atlas that allows the link dimensions to be changed and generates the coupler curves on screen.[21]

Program LINKAGES, included with this text, also allows rapid investigation of coupler curve shapes. For any defined linkage geometry, the program draws the coupler curve. By shift-clicking the mouse pointer on the coupler point and dragging it around, you will see the coupler curve shape instantly update for each new coupler point location. When you release the mouse button, the new linkage geometry is preserved for that curve.

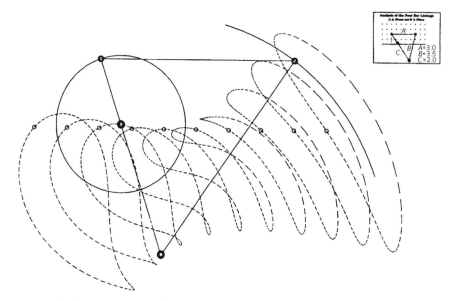

(*a*) A page from the Hrones and Nelson atlas of fourbar coupler curves

Hrones, J. A., and G. L. Nelson (1951). Analysis of the Fourbar Linkage.
MIT Technology Press, Cambridge, MA. Reprinted with permission.

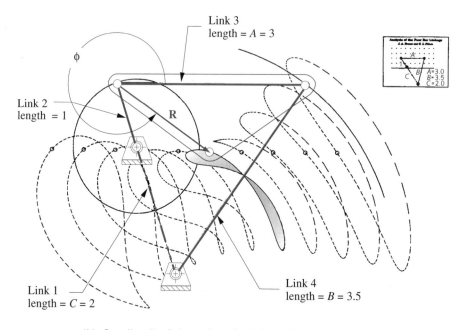

(*b*) Creating the linkage from the information in the atlas

FIGURE 3-17

Selecting a coupler curve and constructing the linkage from the Hrones and Nelson atlas control the paths of multiple points on the wheel (tire contact patch, wheel center, etc.—all of which are points on the same coupler link extended), motion generation is achieved as the coupler has complex motion. Figure 3-19a[*] and b[*] shows parallel planar fourbar linkages suspending the wheels. The coupler curve of the wheel center is nearly a straight line over the small vertical displacement required. This is desirable as the idea is to keep the tire perpendicular to the ground for best traction under all cornering and attitude changes of the car body. This is an application in which a non-Grashof linkage is perfectly acceptable, as full rotation of the wheel in this plane might have some undesirable results and surprise the driver. Limit stops are of course provided to prevent such behavior, so even a Grashof linkage could be used. The springs support the weight of the vehicle and provide a fifth, variable-length "force link" that stabilizes the mechanism as was described in Section 2.15 (p. 65). The function of the fourbar linkage is solely to guide and control the wheel motions. Figure 3-19c shows a true spatial linkage of seven links (including frame and wheel) and nine joints (some of which are ball-and-socket joints) used to control the motion of the rear wheel. These links do not move in parallel planes but rather control the three-dimensional motion of the coupler which carries the wheel assembly.

FIGURE 3-18

Movie camera film-advance mechanism. *From Die Wissenschaftliche und Angenwandte Photographie, Michel, Kurt (ed.). (1955). Vol. 3, Harold Weise, Die Kinematographische Kamera, p. 202, Springer Verlag, OHG, Vienna.* (Input the file F03-18.4br to program FOURBAR to animate this linkage.)

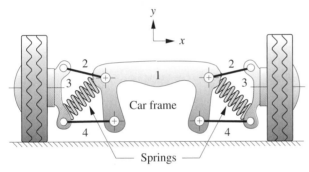

(*a*) Fourbar planar linkages are duplicated in parallel planes, displaced in the *z* direction, behind the links shown

(*b*) Parallel-planar linkage used to control Viper wheel motion (*Courtesy of Chrysler Corporation*)

(*c*) Multilink true spatial linkage used to control rear wheel motion (*Courtesy of Mercedes-Benz of North America Inc.*)

FIGURE 3-19

Linkages used in automotive chassis suspensions

[*] This figure is provided as animated AVI and Working Model files on the DVD. Its filename is the same as the figure number.

SYMMETRICAL LINKAGES COUPLER CURVES   When a fourbar linkage's geometry is such that the coupler and rocker are the same length pin-to-pin, all coupler points that lie on a circle centered on the coupler-rocker joint with radius equal to the coupler length will generate symmetrical coupler curves. Figure 3-20 shows such a linkage, its symmetrical coupler curve, and the locus of all points that will give symmetrical curves. Using the notation of that figure, the criterion for coupler curve symmetry can be stated as:

$$AB = O_4B = BP \tag{3.4}$$

A linkage for which equation 3.4 is true is referred to as a **symmetrical fourbar linkage**. The axis of symmetry of the coupler curve is the line O_4P drawn when the crank O_2A and the ground link O_2O_4 are colinear-extended (i.e., $\theta_2 = 180°$). Symmetrical coupler curves prove to be quite useful as we shall see in the next several sections. Some give good approximations to circular arcs and others give very good approximations to straight lines (over a portion of the coupler curve).

In the general case, nine parameters are needed to define the geometry of a **nonsymmetrical fourbar linkage** with one coupler point.[*] We can reduce this to five as follows. Three parameters can be eliminated by fixing the location and orientation of the ground link. The four link lengths can be reduced to three parameters by normalizing three link lengths to the fourth. The shortest link (the crank, if a Grashof crank-rocker linkage) is usually taken as the reference link, and three link ratios are formed as L_1/L_2, L_3/L_2, L_4/L_2, where L_1 = ground, L_2 = crank, L_3 = coupler, and L_4 = rocker length as shown in Figure 3-20. Two parameters are needed to locate the coupler point: the distance from a convenient reference point on the coupler (either B or A in Figure 3-20) to the coupler point P, and the angle that the line BP (or AP) makes with the line of centers of the coupler AB (either δ or γ). Thus, with a defined ground link, five parameters that will define the geometry of a nonsymmetrical fourbar linkage (using point B as the reference in link 3 and the labels of Figure 3-20) are: L_1/L_2, L_3/L_2, L_4/L_2, BP/L_2, and γ. Note that multiplying these parameters by a scaling factor will change the size of the linkage and its coupler curve but will not change the coupler curve's shape.

A **symmetrical fourbar linkage** with a defined ground link needs only *three parameters* to define its geometry because three of the five nonsymmetrical parameters are now equal per equation 3.4: $L_3/L_2 = L_4/L_2 = BP/L_2$. Three possible parameters to define the geometry of a symmetrical fourbar linkage in combination with equation 3.4 are then: L_1/L_2, L_3/L_2, and γ. Having only three parameters to deal with rather than five greatly simplifies an analysis of the behavior of the coupler curve shape when the linkage geometry is varied. Other relationships for the isosceles-triangle coupler are shown in Figure 3-20. Length AP and angle δ are needed for input of the linkage geometry to program LINKAGES.

Kota [9] did an extensive study of the characteristics of coupler curves of symmetrical fourbar linkages and mapped coupler curve shape as a function of the three linkage parameters defined above. He defined a three-dimensional design space to map the coupler curve shape. Figure 3-21 (p. 130) shows two orthogonal plane sections taken through this design space for particular values of link ratios,[†] and Figure 3-22 (p. 131) shows a schematic of the design space. Though the two cross sections of Figure 3-21 show only a small fraction of the information in the 3-D design space of Figure 3-22, they nevertheless give a sense of the way that variation of the three linkage parameters affects the coupler

[*] The nine independent parameters of a fourbar linkage are: four link lengths, two coordinates of the coupler point with respect to the coupler link, and three parameters that define the location and orientation of the fixed link in the global coordinate system.

[†] Adapted from materials provided by Professor Sridhar Kota, University of Michigan.

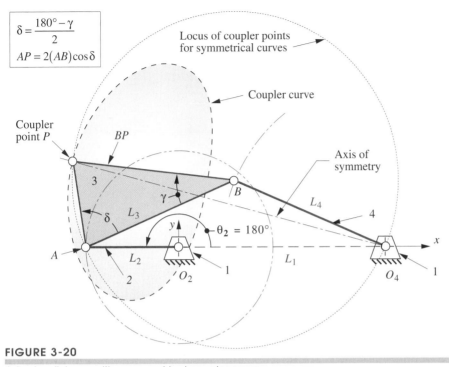

$$\delta = \frac{180° - \gamma}{2}$$

$$AP = 2(AB)\cos\delta$$

Locus of coupler points for symmetrical curves

Coupler curve

Coupler point P

BP

Axis of symmetry

L_4

4

3

γ

B

δ L_3

y

$\theta_2 = 180°$

A

L_2

L_1

x

2

1

O_2

O_4

1

FIGURE 3-20

A fourbar linkage with a symmetrical coupler curve

curve shape. Used in combination with a linkage design tool such as program LINKAGES, these design charts can help guide the designer in choosing suitable values for the linkage parameters to achieve a desired path motion.

GEARED FIVEBAR COUPLER CURVES (Figure 3-23, p. 131) are more complex than the fourbar variety. Because there are three additional, independent design variables in a geared fivebar compared to the fourbar (an additional link ratio, the gear ratio, and the phase angle between the gears), the coupler curves can be of higher degree than those of the fourbar. This means that the curves can be more convoluted, having more cusps and crunodes (loops). In fact, if the gear ratio used is noninteger, the input link will have to make a number of revolutions equal to the factor necessary to make the ratio an integer before the coupler curve pattern will repeat. The Zhang, Norton, Hammond (ZNH) *Atlas of Geared FiveBar Mechanisms* (GFBM) [10] shows typical coupler curves for these linkages limited to symmetrical geometry (e.g., link 2 = link 5 and link 3 = link 4) and gear ratios of ±1 and ±2. A page from the ZNH atlas is reproduced in Figure 3-23 (p. 131). Additional pages are in Appendix E and the entire atlas is on the book DVD. Each page shows the family of coupler curves obtained by variation of the phase angle for a particular set of link ratios and gear ratio. A key in the upper right corner of each page defines the ratios: α = link 3 / link 2, β = link 1 / link 2, λ = gear 5 / gear 2. Symmetry defines links 4 and 5 as noted above. The phase angle ϕ is noted on the axes drawn at each coupler curve and can be seen to have a significant effect on the resulting coupler curve shape.

3

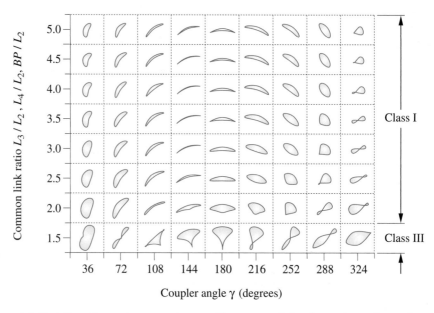

(a) Variation of coupler curve shape with common link ratio and coupler angle for a ground link ratio $L_1 / L_2 = 2.0$

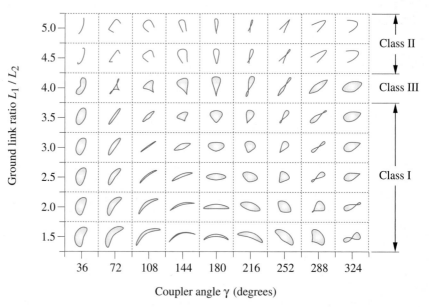

(b) Variation of coupler curve shape with ground link ratio and coupler angle for a common link ratio $L_3 / L_2 = L_4 / L_2 = BP / L_2 = 2.5$

FIGURE 3-21

Coupler curve shapes of symmetrical fourbar linkages *Adapted from reference (9)*

FIGURE 3-22

A three-dimensional map of coupler shapes of symmetrical fourbar linkages (9)

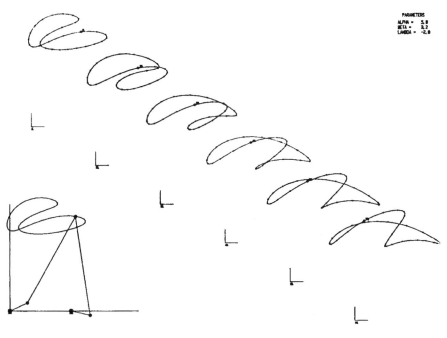

FIGURE 3-23

A page from the Zhang-Norton-Hammond atlas of geared fivebar coupler curves (10)

This reference atlas is intended to be used as a starting point for a geared fivebar linkage design. The link ratios, gear ratio, and phase angle can be input to the program FIVEBAR and then varied to observe the effects on coupler curve shape, velocities, and accelerations. Asymmetry of links can be introduced, and a coupler point location other than the pin joint between links 3 and 4 defined within the LINKAGES program as well. Note that program LINKAGES expects the gear ratio to be in the form gear 2 / gear 5 which is the inverse of the ratio λ in the ZNH atlas.

3.7   COGNATES*

It sometimes happens that a good solution to a linkage synthesis problem will be found that satisfies path generation constraints but which has the fixed pivots in inappropriate locations for attachment to the available ground plane or frame. In such cases, the use of a **cognate** to the linkage may be helpful. The term **cognate** was used by Hartenberg and Denavit [11] to describe *a linkage, of different geometry, which generates the same coupler curve*. Samuel Roberts (1875)[23] and Chebyschev (1878) independently discovered the theorem which now bears their names:

Roberts-Chebyschev Theorem

Three different planar, pin-jointed fourbar linkages will trace identical coupler curves.

Hartenberg and Denavit[11] presented extensions of this theorem to the crank-slider and the sixbar linkages:

Two different planar crank-slider linkages will trace identical coupler curves.†

The coupler-point curve of a planar fourbar linkage is also described by the joint of a dyad of an appropriate sixbar linkage.

Figure 3-24a (p. 134) shows a fourbar linkage for which we want to find the two cognates. The first step is to release the fixed pivots O_A and O_B. While holding the coupler stationary, rotate links 2 and 4 into colinearity with the line of centers (A_1B_1) of link 3 as shown in Figure 3-24b. We can now construct lines parallel to all sides of the links in the original linkage to create the **Cayley diagram**[24] in Figure 3-24c. This schematic arrangement defines the lengths and shapes of links 5 through 10 which belong to the cognates. All three fourbars share the original coupler point P and will thus generate the same path motion on their coupler curves.

In order to find the correct location of the fixed pivot O_C from the Cayley diagram, the ends of links 2 and 4 are returned to the original locations of the fixed pivots O_A and O_B as shown in Figure 3-25a (p. 135). The other links will follow this motion, maintaining the parallelogram relationships between links, and fixed pivot O_C will then be in its proper location on the ground plane. This configuration is called a **Roberts diagram**— three fourbar linkage cognates which share the same coupler curve.

The Roberts diagram can be drawn directly from the original linkage without resort to the Cayley diagram by noting that the parallelograms which form the other cognates are also present in the Roberts diagram and the three couplers are similar triangles. It is also possible to locate fixed pivot O_C directly from the original linkage as shown in Figure 3-25a (p. 135). Construct a similar triangle to that of the coupler, placing its base (AB) between O_A and O_B. Its vertex will be at O_C.

* A video "Cognates" that shows how to find the cognates of a fourbar linkage is provided on the book DVD.

† Dijksman and Smals [25] state that an inverted crank-slider linkage does not possess any cognates.

3

The ten-link Roberts configuration (Cayley's nine plus the ground) can now be articulated up to any toggle positions, and point P will describe the original coupler path which is the same for all three cognates. Point O_C will not move when the Roberts linkage is articulated, proving that it is a ground pivot. The cognates can be separated as shown in Figure 3-25b (p. 135) and any one of the three linkages used to generate the same coupler curve. Corresponding links in the cognates will have the same angular velocity as the original mechanism as defined in Figure 3-25.

Nolle [12] reports on work by Luck [13] (in German) that defines the character of all fourbar cognates and their transmission angles. If the original linkage is a Grashof crank-rocker, then one cognate will be also, and the other will be a Grashof double-rocker. The minimum transmission angle of the crank-rocker cognate will be the same as that of the original crank-rocker. If the original linkage is a Grashof double-crank (drag link), then both cognates will be also and their minimum transmission angles will be the same in pairs that are driven from the same fixed pivot. If the original linkage is a non-Grashof triple-rocker, then both cognates are also triple-rockers.

These findings indicate that cognates of Grashof linkages do not offer improved transmission angles over the original linkage. Their main advantages are the different fixed pivot location and different velocities and accelerations of other points in the linkage. While the coupler path is the same for all cognates, its velocities and accelerations will not generally be the same since each cognate's overall geometry is different.

When the coupler point lies on the line of centers of link 3, the Cayley diagram degenerates to a group of colinear lines. A different approach is needed to determine the geometry of the cognates. Hartenberg and Denavit [11] give the following set of steps to find the cognates in this case. The notation refers to Figure 3-26 (p. 136).

1   Fixed pivot O_C lies on the line of centers $O_A O_B$ extended and divides it in the same ratio as point P divides AB (i.e., $O_C / O_A = PA / AB$).

2   Line $O_A A_2$ is parallel to $A_1 P$ and $A_2 P$ is parallel to $O_A A_1$, locating A_2.

3   Line $O_B A_3$ is parallel to $B_1 P$ and $A_3 P$ is parallel to $O_B B_1$, locating A_3.

4   Joint B_2 divides line $A_2 P$ in the same ratio as point P divides AB. This defines the first cognate $O_A A_2 B_2 O_C$.

5   Joint B_3 divides line $A_3 P$ in the same ratio as point P divides AB. This defines the second cognate $O_B A_3 B_3 O_C$.

The three linkages can then be separated and each will independently generate the same coupler curve. The example chosen for Figure 3-26 (p. 136) is unusual in that the two cognates of the original linkage are identical, mirror-image twins. These are special linkages and will be discussed further in the next section.

Program LINKAGES will automatically calculate the two cognates for any linkage configuration input to it. The velocities and accelerations of each cognate can then be calculated and compared. The program also draws the Cayley diagram for the set of cognates. Input the file F03-24.4br to program LINKAGES to display the Cayley diagram of Figure 3-24 (p. 134). Input the files COGNATE1.4br, COGNATE2.4br, and COGNATE3.4br to animate and view the motion of each cognate shown in Figure 3-25 (p. 135). Their coupler curves (at least those portions that each cognate can reach) will be seen to be identical.

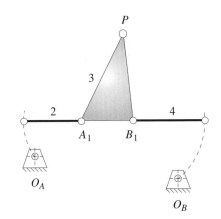

(a) Original fourbar linkage
(cognate #1)

(b) Align links 2 and 4 with coupler

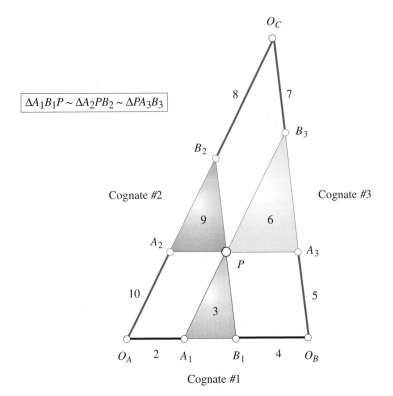

$$\Delta A_1 B_1 P \sim \Delta A_2 P B_2 \sim \Delta P A_3 B_3$$

(c) Construct lines parallel to all sides of the original fourbar linkage to create cognates

FIGURE 3-24

Cayley diagram to find cognates of a fourbar linkage

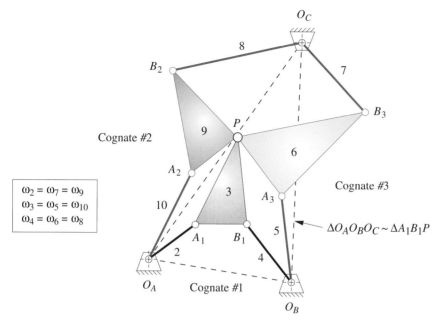

$\Delta O_A O_B O_C \sim \Delta A_1 B_1 P$

$$\omega_2 = \omega_7 = \omega_9$$
$$\omega_3 = \omega_5 = \omega_{10}$$
$$\omega_4 = \omega_6 = \omega_8$$

(a) Return links 2 and 4 to their fixed pivots O_A and O_B.
      Point O_C will assume its proper location

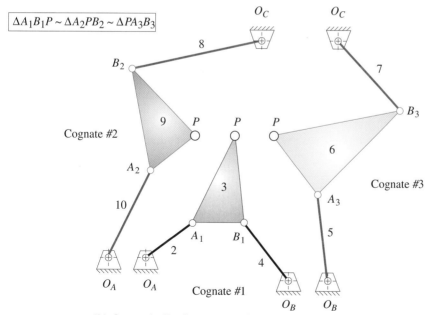

$$\Delta A_1 B_1 P \sim \Delta A_2 P B_2 \sim \Delta P A_3 B_3$$

(b) Separate the three cognates.
     Point P has the same path motion in each cognate

FIGURE 3-25

Roberts diagram of three fourbar cognates

(a) A fourbar linkage and its coupler curve   (b) Cognates of the fourbar linkage

FIGURE 3-26

Finding cognates of a fourbar linkage when its coupler point lies on the line of centers of the coupler

Parallel Motion*

It is quite common to want the output link of a mechanism to follow a particular path without any rotation of the link as it moves along the path. Once an appropriate path motion in the form of a coupler curve and its fourbar linkage have been found, a cognate of that linkage provides a convenient means to replicate the coupler path motion and provide curvilinear translation (i.e., no rotation) of a new output link that follows the coupler path. This is referred to as **parallel motion.** Its design is best described with an example, the result of which will be a Watt II sixbar linkage† that incorporates the original fourbar and parts of one of its cognates. The method shown is as described in Soni.[14]

* A video on constructing "Parallel Motion" linkages is provided on the book DVD.

†Another common method used to obtain parallel motion is to duplicate the same linkage (i.e., the identical cognate), connect them with a parallelogram loop, and remove two redundant links. This results in an eight-link mechanism. See Figure P3-7 on p. 160 for an example of such a mechanism. The method shown here using a different cognate results in a simpler linkage, but either approach will accomplish the desired goal.

✏️ EXAMPLE 3-11

Parallel Motion from a Fourbar Linkage Coupler Curve.

Problem: Design a sixbar linkage for parallel motion over a fourbar linkage coupler path.

Solution: (See Figure 3-27.)

1   Figure 3-27a shows the chosen Grashof crank-rocker fourbar linkage and its coupler curve. The first step is to create the Roberts diagram and find its cognates as shown in Figure 3-27b. The Roberts linkage can be found directly, without resort to the Cayley diagram, as described on p. 133. The fixed center O_C is found by drawing a triangle similar to the coupler triangle A_1B_1P with base O_AO_B.

2   One of a crank-rocker linkage's cognates will also be a crank-rocker (here cognate #3) and

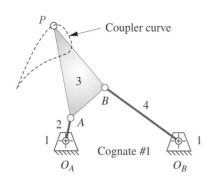

(a) Original fourbar linkage with coupler curve

$\Delta O_A O_B O_C \sim \Delta A_1 B_1 P$

(b) Roberts diagram showing all cognates

$$\omega_2 = \omega_7 = \omega_9$$
$$\omega_3 = \omega_5 = \omega_{10}$$
$$\omega_4 = \omega_6 = \omega_8$$

(c) Cognate #3 shifted with O_C moving to O_A

(d) Redundant link 5 omitted and links 2 & 7 combined leaving a Watt sixbar

FIGURE 3-27

Method to construct a Watt-I sixbar that replicates a coupler path with curvilinear translation (parallel motion) (14)

the other is a Grashof double-rocker (here cognate #2). Discard the double-rocker, keeping the links numbered 2, 3, 4, 5, 6, and 7 in Figure 3-27b (p. 137). Note that links 2 and 7 are the two cranks, and both have the same angular velocity. The strategy is to coalesce these two cranks on a common center (O_A) and then combine them into a single link.

3   Draw the line qq parallel to line O_AO_C and through point O_B as shown in Figure 3-27c.

4   Without allowing links 5, 6, and 7 to rotate, slide them as an assembly along lines O_AO_C and qq until the free end of link 7 is at point O_A. The free end of link 5 will then be at point O_B' and point P on link 6 will be at P'.

5   Add a new link of length O_AO_C between P and P'. This is the *new output link* 8 and all points on it describe the original coupler curve as depicted at points P, P', and P'' in Figure 3-27c.

6   The mechanism in Figure 3-27c has 8 links, 10 revolute joints, and one *DOF*. When driven by **either** crank 2 or 7, all points on link 8 will duplicate the coupler curve of point P.

7   This is an *overclosed linkage* with redundant links. Because links 2 and 7 have the same angular velocity, they can be joined into one link as shown in Figure 3-27d. Then link 5 can be removed and link 6 reduced to a binary link supported and constrained as part of the loop 2, 6, 8, 3. The resulting mechanism is a Watt-I sixbar (see Figure 2-14, p. 54) with the links numbered 1, 2, 3, 4, 6, and 8. Link 8 is in *curvilinear translation* and follows the coupler path of the original point P.[*]

Geared Fivebar Cognates of the Fourbar

Chebyschev also discovered that any fourbar coupler curve can be duplicated with a **geared fivebar mechanism whose gear ratio is plus one,** meaning that the gears turn with the same speed and direction. The geared fivebar's link lengths will be different from those of the fourbar but can be determined directly from the fourbar. Figure 3-28a shows the construction method, as described by Hall [15], to obtain the geared fivebar which will give the same coupler curve as a fourbar. The original fourbar is $O_AA_1B_1O_B$ (links 1, 2, 3, 4). The fivebar is $O_AA_2PB_2O_B$ (links 1, 5, 6, 7, 8). The two linkages share only the coupler point P and fixed pivots O_A and O_B. The fivebar is constructed by simply drawing link 6 parallel to link 2, link 7 parallel to link 4, link 5 parallel to A_1P, and link 8 parallel to B_1P.

A three-gear set is needed to couple links 5 and 8 with a ratio of plus one (gear 5 and gear 8 are the same diameter and have the same direction of rotation, due to the idler gear), as shown in Figure 3-28b. Link 5 is attached to gear 5, as is link 8 to gear 8. This construction technique may be applied to each of the three fourbar cognates, yielding three geared fivebars (which may or may not be Grashof). The three fivebar cognates can actually be seen in the Roberts diagram. Note that in the example shown, a non-Grashof triple-rocker fourbar yields a Grashof fivebar, which can be motor driven. This conversion to a GFBM linkage could be an advantage when the "right" coupler curve has been found on a non-Grashof fourbar linkage, but continuous output through the fourbar's toggle positions is needed. Thus we can see that there are at least seven linkages which will generate the same coupler curve, three fourbars, three GFBMs and one or more sixbars.

Program LINKAGES calculates the equivalent geared fivebar configuration for any fourbar linkage and displays the result. The file F03-28a.4br can be opened in LINKAGES

[*] Another example of a parallel motion sixbar linkage is the Chebyschev straight-line linkage of Figure P2-5a on p. 87. It is a combination of two of the cognates shown in Figure 3-26 (p. 136), assembled by the method described in Example 3-11 (p. 136) and shown in Figure 3-27 (p. 137).

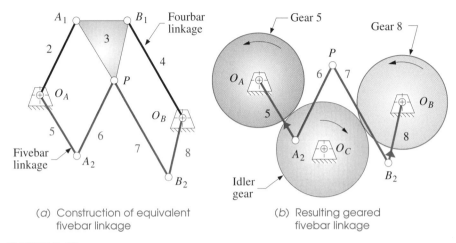

(a) Construction of equivalent   (b) Resulting geared
    fivebar linkage   fivebar linkage

FIGURE 3-28

A geared fivebar linkage cognate of a fourbar linkage

to animate the linkage shown in Figure 3-28a. Then also open the file F03-28b.5br in program LINKAGES to see the motion of the equivalent geared fivebar linkage. Note that the original fourbar linkage is a triple-rocker, so cannot reach all portions of the coupler curve when driven from one rocker. But its geared fivebar equivalent linkage can make a full revolution and traverses the entire coupler path. Program LINKAGES will create the equivalent GFBM of any fourbar linkage.

3.8   STRAIGHT-LINE MECHANISMS

A very common application of coupler curves is the generation of approximate straight lines. Straight-line linkages have been known and used since the time of James Watt in the 18th century. Many kinematicians, such as Watt, Chebyschev, Peaucellier, Kempe, Evans, and Hoeken (as well as others) over a century ago, developed or discovered either approximate or exact straight-line linkages, and their names are associated with those devices to this day. Figure 3-29 (p. 140) shows a collection of the better-known ones, most of which are also provided as animated files on the DVD.

The first recorded application of a coupler curve to a motion problem is that of **Watt's straight-line linkage,** patented in 1784, and shown in Figure 3-29a. Watt devised several straight-line linkages to guide the long-stroke piston of his steam engine at a time when metal-cutting machinery that could create a long, straight guideway did not yet exist.[*] Figure 3-29b shows the Watt linkage used to guide the steam engine piston.[†] This triple-rocker linkage is still used in automobile suspension systems to guide the rear axle up and down in a straight line as well as in many other applications.

Richard Roberts (1789-1864) (not to be confused with Samuel Roberts of the cognates) discovered the **Roberts straight-line linkage** shown in Figure 3-29c. This is a triple-rocker. Other values for AP and BP are possible, but the ones shown give the most

[*] In Watt's time, straight-line motion was dubbed "parallel motion" though we use that term somewhat differently now. James Watt is reported to have told his son, "*Though I am not over anxious after fame, yet I am more proud of the parallel motion than of any other mechanical invention I have made.*" Quoted in Muirhead, J. P. (1854). *The Origin and Progress of the Mechanical Inventions of James Watt*, Vol. 3, London, p. 89.

[†] Note also in Figure 3-29b (and in Figure P2-10 on p. 88) that the driven dyad (links 7 and 8 in Figure 3-29b or 3 and 4 in Figure P2-10, p. 88) are a complicated arrangement of sun and planet gears with the planet axle in a circular track. These have the same effect as the simpler crank and connecting rod. Watt was forced to invent the sun and planet drive to get around James Pickard's 1780 patent on the crankshaft and connecting rod.

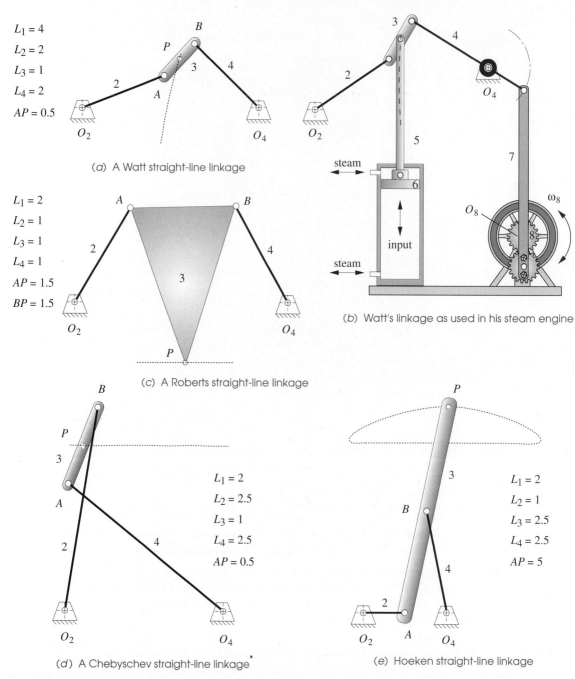

$L_1 = 4$
$L_2 = 2$
$L_3 = 1$
$L_4 = 2$
$AP = 0.5$

(a) A Watt straight-line linkage

$L_1 = 2$
$L_2 = 1$
$L_3 = 1$
$L_4 = 1$
$AP = 1.5$
$BP = 1.5$

(c) A Roberts straight-line linkage

(b) Watt's linkage as used in his steam engine

steam
steam
input
O_8
ω_8

$L_1 = 2$
$L_2 = 2.5$
$L_3 = 1$
$L_4 = 2.5$
$AP = 0.5$

(d) A Chebyschev straight-line linkage[*]

$L_1 = 2$
$L_2 = 1$
$L_3 = 2.5$
$L_4 = 2.5$
$AP = 5$

(e) Hoeken straight-line linkage

FIGURE 3-29 Part 1

Some common and classic approximate straight-line linkages

* The link ratios of the Chebyschev straight-line linkage shown have been reported differently by various authors. The ratios used here are those first reported (in English) by Kempe (1877). But Kennedy (1893) describes the same linkage, reportedly "as Chebyschev demonstrated it at the Vienna Exhibition of 1893" as having the link ratios 1, 3.25, 2.5, 3.25. We will assume the earliest reference by Kempe to be correct as listed in the figure. Both may be correct as Chebyschev recorded several straight-line linkage designs.[20]

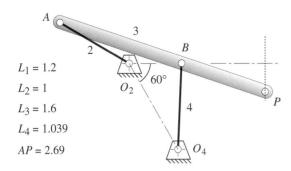

$L_1 = 1.2$
$L_2 = 1$
$L_3 = 1.6$
$L_4 = 1.039$
$AP = 2.69$

(f) Evans approx. straight-line linkage #1

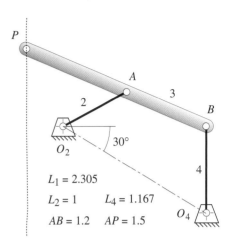

$L_1 = 2.305$
$L_2 = 1$ $L_4 = 1.167$
$AB = 1.2$ $AP = 1.5$

(g) Evans approx. straight-line linkage #2

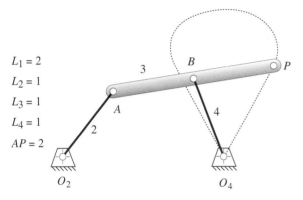

$L_1 = 2$
$L_2 = 1$
$L_3 = 1$
$L_4 = 1$
$AP = 2$

(h) Evans approx. straight-line linkage #3

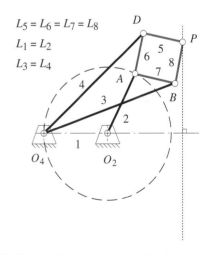

$L_5 = L_6 = L_7 = L_8$
$L_1 = L_2$
$L_3 = L_4$

(j) Peaucellier exact straight-line linkage

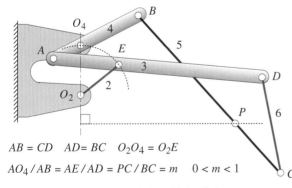

$AB = CD$ $AD = BC$ $O_2O_4 = O_2E$
$AO_4 / AB = AE / AD = PC / BC = m$ $0 < m < 1$

(i) Hart inversor exact straight-line linkage

FIGURE 3-29 Part 2

Approximate and exact straight-line linkages

accurate straight line with a deviation from straight of only 0.04% (0.0004 dec%) of the length of link 2 over the range of $49° < \theta_2 < 69°$.

Chebyschev (1821-1894) also devised many **straight-line linkages**. His well-known Grashof double-rocker is shown in Figure 3-29d.

* Hain [17] (1967) cites the Hoeken reference [16] (1926) for this linkage. Nolle [18] (1974) shows the Hoeken mechanism but refers to it as a Chebyschev crank-rocker without noting its cognate relationship to the Chebyschev double-rocker, which he also shows. It is certainly conceivable that Chebyschev, as one of the creators of the theorem of cognate linkages, would have discovered the "Hoeken" cognate of his own double-rocker. However, this author has been unable to find any mention of its genesis in the English literature other than the ones cited here.

† Peaucellier was a French army captain and military engineer who first proposed his "compas compose" or *compound compass* in 1864 but received no immediate recognition therefor. (He later received the "Prix Montyon" from the Institute of France.) The British-American mathematician James Sylvester reported on it to the *Atheneum Club* in London in 1874. He observed that "*The perfect parallel motion of Peaucellier looks so simple and moves so easily that people who see it at work almost universally express astonishment that it waited so long to be discovered.*" A model of the Peaucellier linkage was passed around the table. The famous physicist Sir William Thomson (later Lord Kelvin) refused to relinquish it, declaring, "*No. I have not had nearly enough of it—it is the most beautiful* (footnotes cont'd. opp. page)

The **Hoeken linkage** [16] in Figure 3-29e (p. 140) is a Grashof crank-rocker, which is a significant practical advantage. In addition, the Hoeken linkage has the feature of very *nearly constant velocity along the center portion of its straight-line motion.* It is interesting to note that the **Hoeken** and **Chebyschev** linkages are cognates of one another.* The cognates shown in Figure 3-26 (p. 136) are the Chebyschev and Hoeken linkages.

Figure 3-29f shows one of **Evans'** many straight-line linkages. It is a triple rocker with a range of input link motion of about 27 to 333° between toggle positions. The portion of coupler curve shown is between 150 and 210° and has a very accurate straight line with a deviation of only 0.25% (0.0025 dec%) of the crank length.

Figure 3-29g shows a second **Evans straight-line linkag**e, also a triple rocker with a range of input link motion of about –81 to +81° between toggle positions. The portion of coupler curve shown is between –40 and 40° and has a long but less accurate straight line with a deviation of 1.5% (0.015 dec%) of the crank length.

Figure 3-29h shows a third **Evans straight-line linkag**e. It is a triple rocker with a range of input link motion of about –75 to +75° between toggle positions. The portion of coupler curve shown is all that is reachable between those limits and has two straight portions. The remainder of the coupler curve is a mirror image making a figure eight.

Some of these straight-line linkages are provided as built-in examples in program LINKAGES. AVI and Working Model files of many of them are also on the DVD. Artobolevsky [20] shows seven Watt, seven Chebyschev, five Roberts, and sixteen Evans straight-line linkages in his Vol. I that include the ones shown here. A quick look in the Hrones and Nelson atlas of coupler curves (on the DVD) will reveal a large number of coupler curves with **approximate straight-line** segments. They are quite common.

To generate an **exact straight line** with only pin joints requires more than four links. At least six links and seven pin joints are needed to generate an exact straight line with a pure revolute-jointed linkage, i.e., a Watt or Stephenson sixbar. Figure 3-29i shows the **Hart inversor exact straight-line sixbar mechanism**. A symmetrical geared fivebar mechanism (Figure 2-21, p. 63), with a gear ratio of –1 and a phase angle of π radians, will also generate an exact straight line at the joint between links 3 and 4. But this linkage is merely a transformed Watt sixbar obtained by replacing one binary link with a higher joint in the form of a gear pair. This geared fivebar's straight-line motion can be seen by opening the file STRAIGHT.5BR in program LINKAGES, and animating the linkage.

Peaucellier† (1864) discovered an **exact straight-line** mechanism of eight bars and six pins, shown in Figure 3-29j.‡ Links 5, 6, 7, 8 form a rhombus of convenient size. Links 3 and 4 can be any convenient but equal lengths. When O_2O_4 exactly equals O_2A, point C generates an *arc of infinite radius*, i.e., **an exact straight line**. By moving the pivot O_2 left or right from the position shown, changing only the length of link 1, this mechanism *will generate true circle arcs with radii much larger than the link lengths*. Other exact straight-line linkages exist as well. See Artobolevsky.[20]

Designing Optimum Straight-Line Fourbar Linkages

Given the fact that an exact straight line can be generated with six or more links using only revolute joints, why use a fourbar approximate straight-line linkage at all? One reason is the desire for simplicity in machine design. The pin-jointed fourbar is the simplest possible 1-*DOF* mechanism. Another reason is that a very good approximation to a true straight line can be obtained with just four links, and this is often "good enough" for the needs of the machine being designed. Manufacturing tolerances will, after all, cause any mechanism's performance to be less than ideal. As the number of links and joints increases, the probability that an exact straight-line mechanism will deliver its theoretical performance in practice is obviously reduced.

There is a real need for straight-line motions in machinery of all kinds, especially in automated production machinery. Many consumer products such as cameras, film, toiletries, razors, and bottles are manufactured, decorated, or assembled on sophisticated and complicated machines that contain a myriad of linkages and cam-follower systems. Traditionally, most of this kind of production equipment has been of the intermittent-motion variety. This means that the product is carried through the machine on a linear or rotary conveyor that stops for any operation to be done on the product, and then indexes the product to the next workstation where it again stops for another operation to be performed. The forces, torque, and power required to accelerate and decelerate the large mass of the conveyor (which is independent of, and typically larger than, the mass of the product) severely limit the speeds at which these machines can be run.

Economic considerations continually demand higher production rates, requiring higher speeds or additional, expensive machines. This economic pressure has caused many manufacturers to redesign their assembly equipment for continuous conveyor motion. When the product is in continuous motion in a straight line and at constant velocity, every workhead that operates on the product must be articulated to chase the product and match both its straight-line path and its constant velocity while performing the task. These factors have increased the need for straight-line mechanisms, including ones capable of near-constant velocity over the straight-line path.

A (near) perfect straight-line motion is easily obtained with a fourbar crank-slider mechanism. Ball-bushings (Figure 2-33, p. 71) and ball-slides (Figure 2-38, p. 73) are available commercially at moderate cost and make this a reasonable, low-friction solution to the straight-line path guidance problem. But, the cost and lubrication problems of a properly guided crank-slider mechanism are still greater than those of a pin-jointed fourbar linkage. Moreover, a crank-slider block has a velocity profile that is nearly sinusoidal (with some harmonic content) and is far from having constant velocity over any of its motion. (See Section 3.10 on p. 151 for a modified crank-slider mechanism that has nearly constant slider velocity for part of its stroke.)

The Hoeken-type linkage offers an optimum combination of straightness and near constant velocity and is a crank-rocker, so it can be motor driven. Its geometry, dimensions, and coupler path are shown in Figure 3-30 (p. 144). This is a symmetrical fourbar linkage. Since the angle γ of line BP is specified and $L_3 = L_4 = BP$, only two link ratios are needed to define its geometry, say L_1 / L_2 and L_3 / L_2. If the crank L_2 is driven at constant angular velocity ω_2, the linear velocity V_x along the straight-line portion Δx of the coupler path will be very close to constant over a significant portion of crank rotation $\Delta\beta$.

A study was done to determine the errors in straightness and constant velocity of the Hoeken-type linkage over various fractions $\Delta\beta$ of the crank cycle as a function of the link ratios. [19] The structural error in position (i.e., straightness) ε_S and the structural error in velocity ε_V are defined using notation from Figure 3-30 as:

(continued from opp. page)
thing I have ever seen in my life."   Source: Strandh, S. (1979). *A History of the Machine.* A&W Publishers: New York, p. 67.
A JAVA applet that animates a Peaucellier cell can be found at: http://math2.math.nthu.edu.tw/jcchuan/java-sketchpad/peau.htm.

3

‡ This figure is provided as animated AVI and Working Model files on the DVD. Its filename is the same as the figure number.

3

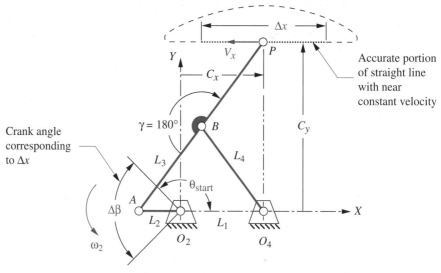

FIGURE 3-30

Hoekens linkage geometry. Linkage shown with P at center of straight-line portion of path

$$\varepsilon_S = \frac{MAX_{i=1}^{n}\left(C_{y_i}\right) - MIN_{i=1}^{n}\left(C_{y_i}\right)}{\Delta x}$$

$$(3.5)^*$$

$$\varepsilon_V = \frac{MAX_{i=1}^{n}\left(V_{x_i}\right) - MIN_{i=1}^{n}\left(V_{x_i}\right)}{\overline{V}_v}$$

* See reference [19] for the
derivation of equations 3.5.

TABLE 3-1   Link Ratios for Smallest Attainable Errors in Straightness and Velocity for Various Crank-Angle Ranges of a Hoeken-Type Fourbar Approximate Straight-Line Linkage (19)

Range of Motion			Optimized for Straightness							Optimized for Constant Velocity						
$\Delta\beta$ (deg)	θ (deg)	% of cycle	Maximum ΔC_y %	ΔV %	V_x ($L_2\,\omega_2$)	L_1 / L_2	L_3 / L_2	$\Delta x / L_2$		Maximum ΔV_x %	ΔC_y %	V_x ($L_2\,\omega_2$)	L_1 / L_2	L_3 / L_2	$\Delta x / L_2$
20	170	5.6%	0.00001%	0.38%	1.725	2.975	3.963	0.601		0.006%	0.137%	1.374	2.075	2.613	0.480
40	160	11.1%	0.00004%	1.53%	1.717	2.950	3.925	1.193		0.038%	0.274%	1.361	2.050	2.575	0.950
60	150	16.7%	0.00027%	3.48%	1.702	2.900	3.850	1.763		0.106%	0.387%	1.347	2.025	2.538	1.411
80	140	22.2%	0.001%	6.27%	1.679	2.825	3.738	2.299		0.340%	0.503%	1.319	1.975	2.463	1.845
100	130	27.8%	0.004%	9.90%	1.646	2.725	3.588	2.790		0.910%	0.640%	1.275	1.900	2.350	2.237
120	120	33.3%	0.010%	14.68%	1.611	2.625	3.438	3.238		1.885%	0.752%	1.229	1.825	2.238	2.600
140	110	38.9%	0.023%	20.48%	1.565	2.500	3.250	3.623		3.327%	0.888%	1.178	1.750	2.125	2.932
160	100	44.4%	0.047%	27.15%	1.504	2.350	3.025	3.933		5.878%	1.067%	1.124	1.675	2.013	3.232
180	90	50.0%	0.096%	35.31%	1.436	2.200	2.800	4.181		9.299%	1.446%	1.045	1.575	1.863	3.456

The structural errors were computed separately for each of nine crank-angle ranges $\Delta\beta$ from 20° to 180°. Table 3-1 shows the link ratios that give the smallest possible structural error in either position or velocity over values of $\Delta\beta$ from 20° to 180°. Note that one cannot attain optimum straightness and minimum velocity error in the same linkage. However, reasonable compromises between the two criteria can be achieved, especially for smaller ranges of crank angle. The errors in both straightness and velocity increase as longer portions of the curve are used (larger $\Delta\beta$). The use of Table 3-1 to design a straight-line linkage will be shown with an example.

✐EXAMPLE 3-12

Designing a Hoeken-Type Straight-Line Linkage

Problem: A 100-mm-long straight-line motion is needed over 1/3 of the total cycle (120° of crank rotation). Determine the dimensions of a Hoeken-type linkage that will

(a) Provide minimum deviation from a straight line. Determine its maximum deviation from constant velocity.
(b) Provide minimum deviation from constant velocity. Determine its maximum deviation from a straight line.

Solution: (See Figure 3-30 and Table 3-1.)

1 Part (a) requires the most accurate straight line. Enter Table 3-1 at the 6th row which is for a crank-angle duration $\Delta\beta$ of the required 120°. The 4th column shows the minimum possible deviation from straight to be 0.01% of the length of the straight-line portion used. For a 100-mm length the absolute deviation will then be 0.01 mm (0.0004 in). The 5th column shows that its velcity error will be 14.68% of the average velocity over the 100-mm length. The absolute value of this velocity error of course depends on the speed of the crank.

2 The linkage dimensions for part (a) are found from the ratios in columns 7, 8, and 9. The crank length required to obtain the 100-mm length of straight line Δx is:

from Table 3-1:
$$\frac{\Delta x}{L_2} = 3.238$$

(a)

$$L_2 = \frac{\Delta x}{3.238} = \frac{100 \text{ mm}}{3.23} = 30.88 \text{ mm}$$

The other link lengths are then:

from Table 3-1:
$$\frac{L_1}{L_2} = 2.625$$

(b)

$$L_1 = 2.625 L_2 = 2.625(30.88 \text{ mm}) = 81.07 \text{ mm}$$

from Table 3-1:
$$\frac{L_3}{L_2} = 3.438$$

(c)

$$L_3 = 3.438 L_2 = 3.438(30.88 \text{ mm}) = 106.18 \text{ mm}$$

The complete linkage is then $L_1 = 81.07, L_2 = 30.88, L_3 = L_4 = BP = 106.18$ mm. The nominal velocity V_x of the coupler point at the center of the straight line ($\theta_2 = 180°$) can be found from the factor in the 6th column which must be multiplied by the crank length L_2 and the crank angular velocity ω_2 in radians per second (rad/sec).

3   Part (b) requires the most accurate velocity. Again enter Table 3-1 (p. 144) at the 6th row which is for a crank angle duration $\Delta\beta$ of the required 120°. The 10th column shows the minimum possible deviation from constant velocity to be 1.885% of the average velocity V_x over the length of the straight-line portion used. The 11th column shows the deviation from straight to be 0.752% of the length of the straight-line portion used. For a 100-mm length the absolute deviation in straightness for this optimum constant velocity linkage will then be 0.75 mm (0.030 in).

4   Link lengths for this mechanism are found in the same way as was done in step 2 except that the link ratios 1.825, 2.238, and 2.600 from columns 13, 14, and 15 are used. The result is $L_1 = 70.19, L_2 = 38.46, L_3 = L_4 = BP = 86.08$ mm. The nominal velocity V_x of the coupler point at the center of the straight line ($\theta_2 = 180°$) can be found from the factor in the 12th column which must be multiplied by the crank length L_2 and the crank angular velocity ω_2 in rad/sec.

5   The first solution (step 2) gives an extremely accurate straight line over a significant part of the cycle. but its 15% deviation in velocity would probably be unacceptable if that factor were considered important. The second solution (step 3) gives less than 2% deviation from constant velocity, which may be viable for a design application. Its 3/4% deviation from straightness, while much greater than the first design, may be acceptable in some situations.

3.9   DWELL MECHANISMS[*]

A common requirement in machine design problems is the need for a dwell in the output motion. A **dwell** is defined as *zero output motion for some nonzero input motion.* In other words, the motor keeps going, but the output link stops moving. Many production machines perform a series of operations which involve feeding a part or tool into a workspace, and then holding it there (in a dwell) while some task is performed. Then the part must be removed from the workspace, and perhaps held in a second dwell while the rest of the machine "catches up" by indexing or performing some other tasks. Cams and followers (Chapter 8) are often used for these tasks because it is trivially easy to create a dwell with a cam. But there is always a trade-off in engineering design, and cams have their problems of high cost and wear as described in Section 2.17 (p. 67).

It is also possible to obtain dwells with "pure" linkages of only links and pin joints, which have the advantage over cams of low cost and high reliability. Dwell linkages are more difficult to design than are cams with dwells. Linkages will usually yield only an approximate dwell but will be much cheaper to make and maintain than cams. Thus they may be worth the effort.

Single-Dwell Linkages

There are two usual approaches to designing single-dwell linkages. Both result in **sixbar mechanisms**, and both require first finding a fourbar with a suitable coupler curve. A **dyad** is then added to provide an output link with the desired dwell characteristic. The first approach to be discussed requires the design or definition of a fourbar with a coupler curve that contains an approximate circle arc portion, which "arc" occupies the desired portion of the input link (crank) cycle designated as the dwell. An atlas of coupler curves is invaluable for this part of the task. Symmetrical coupler curves are also well suited to this task, and the information in Figure 3-21 (p. 130) can be used to find them.

EXAMPLE 3-13

Single-Dwell Mechanism with Only Revolute Joints

Problem: Design a sixbar linkage for 90° rocker motion over 300 crank degrees with dwell for the remaining 60°.

Solution: See Figure 3-31 (p. 148).

1   Search the H&N atlas for a fourbar linkage with a coupler curve having an approximate (pseudo) circle arc portion which occupies 60° of crank motion (12 dashes). The chosen four-bar is shown in Figure 3-31a.

2   Lay out this linkage to scale including the coupler curve and find the approximate center of the chosen coupler curve pseudo-arc using graphical geometric techniques. To do so, draw the chord of the arc and construct its perpendicular bisector as shown in Figure 3-31b. The center will lie on this bisector. Find it by striking arcs with your compass point on the bisector, while adjusting the radius to get the best fit to the coupler curve. Label the arc center D.

3   Your compass should now be set to the approximate radius of the coupler arc. This will be the length of link 5 which is to be attached at the coupler point P.

4   Trace the coupler curve with the compass point, while keeping the compass pencil lead on the perpendicular bisector, and find the extreme location along the bisector that the compass lead will reach. Label this point E.

5   The line segment DE represents the maximum displacement that a link of length PD, attached at P, will reach along the bisector.

6   Construct a perpendicular bisector of the line segment DE, and extend it in a convenient direction.

7   Locate fixed pivot O_6 on the bisector of DE such that lines O_6D and O_6E subtend the desired output angle, in this example, 90°.

8   Draw link 6 from D (or E) through O_6 and extend to any convenient length. This is the output link which will dwell for the specified portion of the crank cycle.

9   Check the transmission angles.

10 Make a model of the linkage and articulate it to check its function.

This linkage dwells because, during the time that the coupler point P is traversing the pseudo-arc portion of the coupler curve, the other end of link 5, attached to P and the same length as the arc radius, is essentially stationary at its other end, which is the arc center.

3

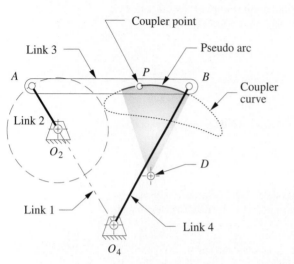

(a) Chosen fourbar crank-rocker with
    pseudo-arc section for 60° of link 2 rotation

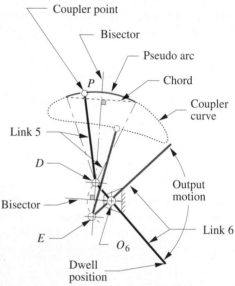

(b) Construction of the output-dwell dyad

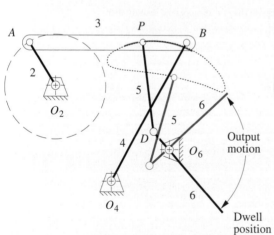

(c) Completed sixbar single-dwell linkage
    with rocker output option

(d) Completed sixbar single-dwell linkage
    with slider output option

FIGURE 3-31

Design of a sixbar single-dwell mechanism with rocker output or slider output, using a pseudo-arc coupler curve

However the dwell at point D will have some "jitter" or oscillation, due to the fact that D is only an approximate center of the pseudo-arc on the sixth-degree coupler curve. When point P leaves the arc portion, it will smoothly drive link 5 from point D to point E, which will in turn rotate the output link 6 through its arc as shown in Figure 3-31c.[*] Note that we can have any angular displacement of link 6 we desire with the same links 2 to 5, as they alone completely define the dwell aspect. Moving pivot O_6 left and right along the bisector of line DE will change the angular displacement of link 6 but not its timing. In fact, a slider block could be substituted for link 6 as shown in Figure 3-31d,[*] and linear translation along line DE with the same timing and dwell at D will result. Input the file F03-31c.6br to program LINKAGES and animate to see the linkage of Example 3-13 (p. 147) in motion. The dwell in the motion of link 6 can be clearly seen in the animation, including the jitter due to its approximate nature.

Double-Dwell Linkages

It is also possible, using a fourbar coupler curve, to create a double-dwell output motion. One approach is the same as that used in the single-dwell of Example 3-11 (p. 136). Now a coupler curve is needed which has *two* approximate circle arcs of the same radius but with different centers, both convex or both concave. A link 5 of length equal to the radius of the two arcs will be added such that it and link 6 will remain nearly stationary at the center of each of the arcs, while the coupler point traverses the circular parts of its path. Motion of the output link 6 will occur only when the coupler point is between those arc portions. Higher-order linkages, such as the geared fivebar, can be used to create multiple-dwell outputs by a similar technique since they possess coupler curves with multiple, approximate circle arcs. See the built-in example double-dwell linkage in program LINKAGES for a demonstration of this approach.

A second approach uses a coupler curve with two approximate straight-line segments of appropriate duration. If a pivoted slider block (link 5) is attached to the coupler at this point, and link 6 is allowed to slide in link 5, it only remains to choose a pivot O_6 at the intersection of the straight-line segments extended. The result is shown in Figure 3-32 (p. 150). While block 5 is traversing the "straight-line" segments of the curve, it will not impart any angular motion to link 6. The approximate nature of the fourbar straight line causes some jitter in these dwells also.

✏️ EXAMPLE 3-14

Double-Dwell Mechanism.

Problem: Design a sixbar linkage for 80° rocker output motion over 20 crank degrees with dwell for 160°, return motion over 140° and second dwell for 40°.

Solution: See Figure 3-32 (p. 150).

1 Search the H&N atlas on the DVD for a linkage with a coupler curve having two approximate straight-line portions. One should occupy 160° of crank motion (32 dashes), and the second 40° of crank motion (8 dashes). This is a wedge-shaped curve as shown in Figure 3-32a.

2 Lay out this linkage to scale including the coupler curve and find the intersection of two tangent lines colinear with the straight segments. Label this point O_6.

[*] This figure is provided as animated AVI and Working Model files on the DVD. Its filename is the same as the figure number.

3

3   Design link 6 to lie along these straight tangents, pivoted at O_6. Provide a slot in link 6 to accommodate slider block 5 as shown in Figure 3-32b.

4   Connect slider block 5 to the coupler point P on link 3 with a pin joint. The finished sixbar is shown in Figure 3-32c.

5   Check the transmission angles.

(a) Fourbar coupler curve with two "straight" segments

(b) Slider dyad for double-dwell

(c) Complete sixbar double-dwell linkage

FIGURE 3-32

Double-dwell sixbar linkage

It should be apparent that these linkage dwell mechanisms have some disadvantages. Besides being difficult to synthesize, they give only approximate dwells which have some jitter on them. Also, they tend to be large for the output motions obtained, so do not package well. The acceleration of the output link can also be very high as in Figure 3-32, when block 5 is near pivot O_6. (Note the large angular displacement of link 6 resulting from a small motion of link 5.)   Nevertheless they may be of value in situations where a completely stationary dwell is not required, and the low cost and high reliability of a linkage are important factors.   Program LINKAGES has both single-dwell and double-dwell example linkages built in.

3.10   OTHER USEFUL LINKAGES[*]

There are many practical machine design problems that can be solved with clever linkage design. One of the best references for these mechanisms is by Hain.[22] Another useful catalog of linkages is the four volumes of Artobolevsky.[20] We will present a few examples from these that we find useful. Some are fourbar linkages, others are Watt's or Stephenson's sixbars, or eightbar linkages. Artobolevsky provides link ratios, but Hain does not. Hain describes their graphical construction, so the dimensions of his linkages shown here are approximate, obtained by scaling his drawings.

Constant Velocity Piston Motion

The fourbar crank-slider linkage is probably the most frequently used linkage in machinery. Every internal combustion (IC) engine and reciprocating compressor has as many of them as it has cylinders. Manufacturing machinery uses them to obtain straight-line motions. In most cases this simple linkage is completely adequate for the application, converting continuous rotary input to oscillating straight-line output. One limitation is lack of control over the slider's velocity profile when the crank is driven with constant angular velocity. Altering the link ratios (crank vs. coupler) has a second-order effect on the shape of the slider's velocity and acceleration curves[†] but it will always be fundamentally a sinusoidal motion. In some cases, a constant or near constant velocity is needed on either the forward or backward stroke of the slider. An example is a piston pump for metering fluids whose flow rate needs to be constant during the delivery stroke. A direct solution is to use a cam to drive the piston with a constant velocity motion rather than use a crank-slider linkage. However, Hain[22] provides a pure linkage solution to this problem that adds a drag-link fourbar stage to the crank-slider with the drag-link geometry chosen to modulate the sinusoidal slider motion to be approximately constant velocity.

Figure 3-33 (p. 152) shows the result, which is effectively a Watt sixbar. Constant angular velocity is input to link 2 of the drag-link stage. It causes its "output" link 4 to have a nonconstant angular velocity that repeats each cycle. This varying angular velocity becomes the "input" to the crank-slider stage 4-5-6 whose input link is now link 4. Thus, the drag link's velocity oscillation effectively "corrects" or modulates the slider velocity to be close to constant on the forward stroke as plotted in the figure. The deviation from constant velocity is < 1% over $240° < \theta_2 < 270°$ and <= 4% over $190° < \theta_2 < 316°$. Its velocity on the return stroke must therefore vary to a greater degree than in the unmodulated linkage. This is an example of the effect of cascading linkages. Each stage's output function becomes the input to the next, and the end result is their mathematical combination, analogous to adding terms to a series. See the *TKSolver* file Dragslider.tkw.

[*] A number of quite interesting linkages can be found at the web site: http://www.mfdabbs.pwp .blueyonder.co.uk/ Maths_Pages/SketchPad_ Files/Mechanical_Linkages/ Mechanical_Linkages.html where you will find drawings of linkages and their coupler curves plus links to downloadable JAVA applets that animate the linkage motions. A collection of such applets is at: http:// www.dynamicgeometry. com/General_Resources/ Links.html

[†] This topic is addressed in depth in Chapter 13.

$$L_1 = 1.000$$
$$L_2 = 2.170$$
$$L_3 = 2.067$$
$$L_4 = 2.310$$
$$L_5 = 5.400$$

FIGURE 3-33

Approximate constant velocity, drag link driven slider-crank sixbar mechanism[22]

In addition to metering fluids, this linkage has application in situations where a part must be picked up from a nest on the stationary ground plane and transferred to a conveyor that is moving at constant velocity. The slider has points of zero velocity at each extreme of motion, exact straight line motion in both directions, and a long region of approximately constant velocity. Note however, that the Hoeken straight-line linkage of Section 3.8 (p. 139) gives a nearly exact straight line with close to constant velocity using only four links and four pin joints rather than the six links and slider track needed here. Hoeken's linkage is also useful for the pick-and-place-at-constant-velocity application.

Large Angular Excursion Rocker Motion[*]

It is often desired to obtain a rocking motion through a large angle with continuous rotary input. A Grashof fourbar crank-rocker linkage is limited to about 120° of rocker oscillation if the transmission angles are to be kept above 30°. A rocker oscillation of 180° would obviously take the transmission angle to zero and also create a Barker Class III linkage with change points, an unacceptable solution. To get a larger oscillation than about 120° with good transmission angles requires six links. Hain[22] designed such a linkage (shown in Figure 3-34) as a Stephenson III sixbar that gives 180° of rocker motion with continuous rotation of the input crank. It is a non-quick-return linkage in which 180° of input crank rotation corresponds to the full oscillation of the output rocker.

An even larger rocker output of about 212° is obtained from the Watt II sixbar linkage shown in Figure 3-35. This mechanism is used to oscillate the agitator in some washing machines. The motor drives toothed crank 2 through a pinion P. Crank 2 oscillates rocker 4 through 102° via coupler 3. Rocker 4 serves as the input to rocker 6 through coupler 5. Rocker 6 is attached to the agitator in the washtub. The minimum transmission angles are 36° in stage 1 (links 2-3-4) and 23° in stage 2 (links 4-5-6).

Hain[22] also created a remarkable eightbar linkage that gives ±360° of oscillatory rocker motion from continuous unidirectional rotation of the input crank! This linkage,

[*] The linkages shown in Figures 3-34 and 3-35 can be animated in program LINKAGES by opening the files F03-34.6br and F03-35.6br, respectively.

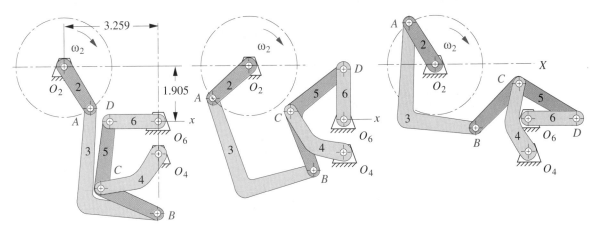

(a) Extreme CCW position of link 6   (b) Mid position of link 6   (c) Extreme CW position of link 6

$O_4O_6 = 1.00$	$L_3 = AB = 4.248$	$L_6 = 1.542$	$DB = 3.274$
$L_2 = 1.556$	$L_4 = 2.125$	$CD = 2.158$	$\angle CDB \quad 36°$

FIGURE 3-34

Stephenson's III sixbar with 180° oscillation of link 6 when crank 2 revolves fully (*Source: Hain*[22], *pp. 448-450*)

shown in Figure 3-36 (p. 154), has a minimum transmission angle of 30°. Slight changes to this linkage's geometry will give more or less than ±360° of output crank oscillation.

Remote Center Circular Motion

When a rotary motion is needed but the center of that rotation is not available to mount the pivot of a crank, a linkage can be used to describe an approximate or exact circular motion "in the air" remote from the fixed and moving pivots of the linkage. Artobolevsky[20]

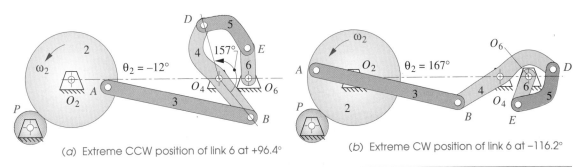

(a) Extreme CCW position of link 6 at +96.4°   (b) Extreme CW position of link 6 at −116.2°

$L_2 = 1.000$	$L_3 = 3.800$	$L_5 = 1.286$	$L_6 = 0.771$	$O_4B = 1.286$	$O_4D = 1.429$	$O_2O_4 = 3.857$	$O_2O_6 = 4.643$

Link 6 rotates through 212.8° and back with every revolution of link 2

FIGURE 3-35

Washing machine agitator mechanism - constant speed motor drives link 2 and agitator is oscillated by link 6 at O_6

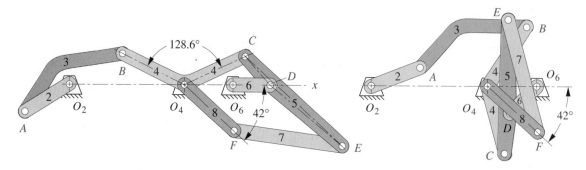

(a) First extreme position, $\theta_2 = 209°$ (b) Second extreme position, $\theta_2 = 19°$

$O_2O_4 = 1.00$	$L_2 = 0.450$	$L_3 = 0.990$	$L_6 = 0.325$	$L_7 = 0.938$	$L_8 = 0.572$	$CD = 0.325$	$CE = 1.145$
	$DE = 0.823$	$O_4O_6 = 0.419$		$O_4B = O_4C = 0.590$		$\angle CDE = 173°$	

FIGURE 3-36

Eightbar linkage with ±360° oscillatory rotation of link 8 when crank 2 revolves fully (*Source: Hain*[22], *pp. 368-370*)

shows ten such mechanisms, two of which are reproduced in Figure 3-37. Figure 3-37a shows a Chebyschev fourbar approximate circle-tracing linkage. When the crank rotates CCW, point *P* traces a circle of the same diameter CW. Figure 3-37b shows a Delone exact circle-tracing sixbar linkage that contains a pantograph cell (*B-C-D-O_4*) that causes point *P* to mimic the motion of point *A*, giving an exact 1:1 replication of the circular motion of *A* about O_4, but rotating in the opposite direction. If a link were added between O_P and *P*, it would rotate at the same speed but in the opposite direction to link 2. Thus this linkage could be substituted for a pair of external gears (gearset) with a 1:1 ratio (see Chapter 9 for information on gears).

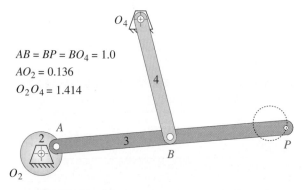

$AB = BP = BO_4 = 1.0$
$AO_2 = 0.136$
$O_2O_4 = 1.414$

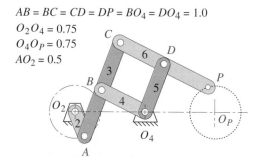

$AB = BC = CD = DP = BO_4 = DO_4 = 1.0$
$O_2O_4 = 0.75$
$O_4O_P = 0.75$
$AO_2 = 0.5$

(a) Chebyschev approximate circle-tracing fourbar   (b) Delone exact circle-tracing sixbar

FIGURE 3-37

Circle generating mechanisms (*Source: Artobolevsky*[20], *Vol. 1, pp. 450-451*)

3.11 REFERENCES

1 **Erdman, A. G., and J. E. Gustafson**. (1977). "LINCAGES: Linkage INteractive Computer Analysis and Graphically Enhanced Synthesis." ASME Paper: 77-DTC-5.

2 **Alt, H.** (1932). "Der Übertragungswinkel und seine Bedeutung für das Konstruieren periodischer Getriebe (The Transmission Angle and Its Importance for the Design of Periodic Mechanisms)." *Werkstatttechnik*, **26**, pp. 61-64.

3 **Hall, A. S.** (1961). *Kinematics and Linkage Design.* Waveland Press: Prospect Heights, IL, p. 146.

4 **Sandor, G. N., and A. G. Erdman**. (1984). *Advanced Mechanism Design: Analysis and Synthesis.* Vol. 2. Prentice-Hall: Upper Saddle River, NJ, pp. 177-187.

5 **Kaufman, R. E.** (1978). "Mechanism Design by Computer." *Machine Design*, October 26, 1978, pp. 94-100.

6 **Hall, A. S.** (1961). *Kinematics and Linkage Design.* Waveland Press: Prospect Heights, IL, pp. 33-34.

7a **Kempe, A. B.** (1876). "On a General Method of Describing Plane Curves of the Nth Degree by Linkwork." *Proceedings London Mathematical Society*, **7**, pp. 213-216.

7b **Wunderlich, W.** (1963). "Höhere Koppelkurven." *Österreichisches Ingenieur Archiv*, **XVII**(3), pp. 162-165.

8a **Hrones, J. A., and G. L. Nelson**. (1951). *Analysis of the Fourbar Linkage.* MIT Technology Press: Cambridge, MA.

8b **Fichter, E. F., and K. H. Hunt**. (1979). "The Variety, Cognate Relationships, Class, and Degeneration of the Coupler Curves of the Planar 4R Linkage." *Proc. of 5th World Congress on Theory of Machines and Mechanisms*, Montreal, pp. 1028-1031.

9 **Kota, S.** (1992). "Automatic Selection of Mechanism Designs from a Three-Dimensional Design Map." *Journal of Mechanical Design*, **114**(3), pp. 359-367.

10 **Zhang, C., R. L. Norton, and T. Hammond**. (1984). "Optimization of Parameters for Specified Path Generation Using an Atlas of Coupler Curves of Geared Five-Bar Linkages." *Mechanism and Machine Theory*, **19**(6), pp. 459-466.

11 **Hartenberg, R. S., and J. Denavit**. (1959). "Cognate Linkages." *Machine Design*, April 16, 1959, pp. 149-152.

12 **Nolle, H.** (1974). "Linkage Coupler Curve Synthesis: A Historical Review - II. Developments after 1875." *Mechanism and Machine Theory*, **9**, 1974, pp. 325-348.

13 **Luck, K.** (1959). "Zur Erzeugung von Koppelkurven viergliedriger Getriebe." *Maschinenbautechnik (Getriebetechnik)*, **8**(2), pp. 97-104.

14 **Soni, A. H.** (1974). *Mechanism Synthesis and Analysis.* Scripta, McGraw-Hill: New York, pp. 381-382.

15 **Hall, A. S.** (1961). *Kinematics and Linkage Design.* Waveland Press: Prospect Heights, IL, p. 51.

16 **Hoeken, K.** (1926). "Steigerung der Wirtschaftlichkeit durch zweckmäßige." *Anwendung der Getriebelehre Werkstatttechnik.*

17 **Hain, K.** (1967). *Applied Kinematics.* D. P. Adams, translator. McGraw-Hill: New York, pp. 308-309.

18 **Nolle, H.** (1974). "Linkage Coupler Curve Synthesis: A Historical Review - I. Developments up to 1875." *Mechanism and Machine Theory*, **9**, pp. 147-168.

19 **Norton, R. L.** (1999). "In Search of the 'Perfect' Straight Line and Constant Velocity Too." *Proc. 6th Applied Mechanisms and Robotics Conference*, Cincinnati, OH.

20 **Artobolevsky, I. I.** (1975). *Mechanisms in Modern Engineering Design.* N. Weinstein, translator. Vols. I to IV. MIR Publishers: Moscow. pp. 431-447.

21   **Thompson, T. J.** (2000). "A Web-Based Atlas of Coupler Curves and Centrodes of Fourbar Mechanisms." *Proc. of SCSI Conf. on Simulation and Multimedia in Eng. Education*, San Diego, pp. 209-213.

22   **Hain, K.** (1967). *Applied Kinematics.* D. P. Adams, translator. McGraw-Hill: New York.

23   **Roberts, S.** (1875). "Three-bar Motion in Plane Space." *Proc. Lond. Math. Soc.*, **7**, pp. 14-23.

24   **Cayley, A.** (1876). "On Three-bar Motion." *Proc. Lond. Math. Soc.*, **7**, pp. 136-166.

25   **Dijksman, E. A. and T. J. M. Smals.** (2000). "How to Exchange Centric Inverted Slider Cranks with λ-formed Fourbar Linkages." *Mechanism and Machine Theory*, **35**, pp. 305-327.

3.12   BIBLIOGRAPHY

*For additional information on **type synthesis**, the following are recommended:*

Artobolevsky, I. I. (1975). *Mechanisms in Modern Engineering Design.* N. Weinstein, translator. Vols. I to IV. MIR Publishers: Moscow.

Chironis, N. P., ed. (1965). *Mechanisms, Linkages, and Mechanical Controls.* McGraw-Hill: New York.

Chironis, N. P., ed. (1966). *Machine Devices and Instrumentation.* McGraw-Hill: New York.

Chironis, N. P., and N. Sclater, eds. (1996). *Mechanisms and Mechanical Devices Sourcebook.* McGraw-Hill: New York.

Jensen, P. W. (1991). *Classical and Modern Mechanisms for Engineers and Inventors.* Marcel Dekker: New York.

Jones, F., H. Horton, and J. Newell. (1967). *Ingenious Mechanisms for Engineers.* Vols. I to IV. Industrial Press: New York.

Olson, D. G., et al. (1985). "A Systematic Procedure for Type Synthesis of Mechanisms with Literature Review." *Mechanism and Machine Theory*, **20**(4), pp. 285-295.

Tuttle, S. B. (1967). *Mechanisms for Engineering Design.* John Wiley & Sons: New York.

*For additional information on **dimensional linkage synthesis**, the following are recommended:*

Djiksman, E. A. (1976). *Motion Geometry of Mechanisms.* Cambridge University Press: London.

Hain, K. (1967). *Applied Kinematics.* D. P. Adams, translator. McGraw-Hill: New York, p. 399.

Hall, A. S. (1961). *Kinematics and Linkage Design.* Waveland Press: Prospect Heights, IL.

Hartenberg, R. S., and J. Denavit. (1964). *Kinematic Synthesis of Linkages.* McGraw-Hill: New York.

McCarthy, J. M. (2000). *Geometric Design of Linkages.* Springer-Verlag: New York.

Molian, S. (1982). *Mechanism Design: An Introductory Text.* Cambridge University Press: Cambridge.

Sandor, G. N., and A. G. Erdman. (1984). *Advanced Mechanism Design: Analysis and Synthesis.* Vol. 2. Prentice-Hall: Upper Saddle River, NJ.

Tao, D. C. (1964). *Applied Linkage Synthesis.* Addison-Wesley: Reading, MA.

*For information on **spatial linkages**, the following are recommended:*

Haug, E. J. (1989). *Computer Aided Kinematics and Dynamics of Mechanical Systems.* Allyn and Bacon: Boston.

Nikravesh, P. E. (1988). *Computer Aided Analysis of Mechanical Systems.* Prentice-Hall: Upper Saddle River, NJ.

Suh, C. H., and C. W. Radcliffe. (1978). *Kinematics and Mechanism Design.* John Wiley & Sons: New York.

3.13   PROBLEMS[†]

*3-1   Define the following examples as path, motion, or function generation cases.

    a.   A telescope aiming (star tracking) mechanism

    b.   A backhoe bucket control mechanism

    c.   A thermostat adjusting mechanism

    d.   A computer printer head moving mechanism

    e.   An XY plotter pen control mechanism

3-2   Design a fourbar Grashof crank-rocker for 90° of output rocker motion with no quick return. (See Example 3-1, p. 103.) Build a model and determine the toggle positions and the minimum transmission angle from the model.

*3-3   Design a fourbar mechanism to give the two positions shown in Figure P3-1 of output rocker motion with no quick return. (See Example 3-2, p. 105.) Build a model and determine the toggle positions and the minimum transmission angle from the model.

3-4   Design a fourbar mechanism to give the two positions shown in Figure P3-1 of coupler motion. (See Example 3-3, p. 105.) Build a model and determine the toggle positions and the minimum transmission angle from the model. Add a driver dyad. (See Example 3-4, p. 107.)

*3-5   Design a fourbar mechanism to give the three positions of coupler motion with no quick return shown in Figure P3-2. (See also Example 3-5, p. 109.) Ignore the points O_2 and O_4 shown. Build a model and determine the toggle positions and the minimum transmission angle from the model. Add a driver dyad. (See Example 3-4, p. 107.)

*3-6   Design a fourbar mechanism to give the three positions shown in Figure P3-2 using the fixed pivots O_2 and O_4 shown. Build a model and determine the toggle positions and the minimum transmission angle from the model. Add a driver dyad.

3-7   Repeat Problem 3-2 with a quick-return time ratio of 1:1.4. (See Example 3-9, p. 118.)

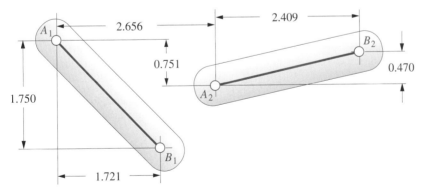

FIGURE P3-1

Problems 3-3 to 3-4

* Answers in Appendix F.

[†] All problem figures are provided as PDF files, and some are also provided as animated AVI and Working Model files; all are on the DVD. PDF filenames are the same as the figure number. Run the file *Animations. html* to access and run the animations.

FIGURE P3-2

Problems 3-5 to 3-6

[*]3-8 Design a sixbar drag link quick-return linkage for a time ratio of 1:2 and output rocker motion of 60°.

3-9   Design a crank-shaper quick-return mechanism for a time ratio of 1:3 (see Figure 3-14, p. 122).

[*]3-10   Find the two cognates of the linkage in Figure 3-17 (p. 126). Draw the Cayley and Roberts diagrams. Check your results with program LINKAGES.

3-11   Find the three equivalent geared fivebar linkages for the three fourbar cognates in Figure 3-25a (p. 135). Check your results by comparing the coupler curves with program LINKAGES.

3-12   Design a sixbar single-dwell linkage for a dwell of 90° of crank motion, with an output rocker motion of 45°.

3-13   Design a sixbar double-dwell linkage for a dwell of 90° of crank motion, with an output rocker motion of 60°, followed by a second dwell of about 60° of crank motion.

3-14   Figure P3-3 shows a treadle-operated grinding wheel driven by a fourbar linkage. Make a model of the linkage to any convenient scale. Find its minimum transmission angles from the model. Comment on its operation. Will it work? If so, explain how it does.

* Answers in Appendix F.

3-15   Figure P3-4 shows a non-Grashof fourbar linkage that is driven from link O_2A. All
        dimensions are in centimeters (cm).

        a.   Find the transmission angle at the position shown.
        b.   Find the toggle positions in terms of angle AO_2O_4.
        c.   Find the maximum and minimum transmission angles over its range of motion by
              graphical techniques.
        d.   Draw the coupler curve of point P over its range of motion.

3-16   Draw the Roberts diagram for the linkage in Figure P3-4 and find its two cognates. Are
        they Grashof or non-Grashof?

3-17   Design a Watt I sixbar to give parallel motion that follows the coupler path of point P
        of the linkage in Figure P3-4.

3-18   Add a driver dyad to the solution of Problem 3-17 to drive it over its possible range of
        motion with no quick return. (The result will be an eightbar linkage.)

3-19   Design a pin-jointed linkage that will guide the forks of the fork lift truck in Figure
        P3-5 up and down in an approximate straight line over the range of motion shown.
        Arrange the fixed pivots so they are close to some part of the existing frame or body of
        the truck.

3-20   Figure P3-6 (p. 160) shows a "V-link" off-loading mechanism for a paper roll con-
        veyor. Design a pin-jointed linkage to replace the air cylinder driver that will rotate the
        rocker arm and V-link through the 90° motion shown. Keep the fixed pivots as close
        to the existing frame as possible. Your fourbar linkage should be Grashof and be in
        toggle at each extreme position of the rocker arm.

3-21   Figure P3-7 (p. 160) shows a walking-beam transport mechanism that uses a fourbar
        coupler curve, replicated with a parallelogram linkage for parallel motion. Note the
        duplicate crank and coupler shown ghosted in the right half of the mechanism—they
        are redundant and have been removed from the duplicate fourbar linkage. Using the

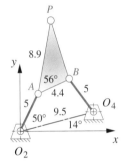

FIGURE P3-3

Problem 3-14. Treadle-
operated grinding
wheel

FIGURE P3-4

Problems 3-15 to 3-18

fork lift truck

FIGURE P3-5

Problem 3-19

FIGURE P3-6

Problem 3-20

FIGURE P3-8

Problem 3-23
Loom laybar drive

same fourbar driving stage (links L_1, L_2, L_3, L_4 with coupler point P), design a Watt I sixbar linkage that will drive link 8 in the same parallel motion using two fewer links.

*3-22 Find the maximum and minimum transmission angles of the fourbar driving stage (links L_1, L_2, L_3, L_4) in Figure P3-7 to graphical accuracy.

*3-23 Figure P3-8 shows a fourbar linkage used in a power loom to drive a comblike reed against the thread, "beating it up" into the cloth. Determine its Grashof condition and its minimum and maximum transmission angles to graphical accuracy.

3-24 Draw the Roberts diagram and find the cognates of the linkage in Figure P3-9.

3-25 Find the equivalent geared fivebar mechanism cognate of the linkage in Figure P3-9.

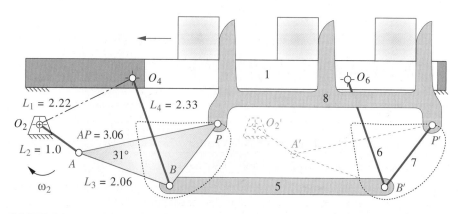

FIGURE P3-7

Problems 3-21 to 3-22 Straight-line walking-beam eightbar transport mechanism

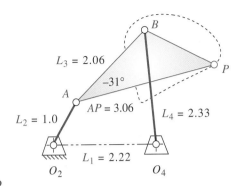

FIGURE P3-9

Problems 3-24 to 3-27

3-26   Use the linkage in Figure P3-9 to design an eightbar double-dwell mechanism that has a rocker output through 45°.

3-27   Use the linkage in Figure P3-9 to design an eightbar double-dwell mechanism that has a slider output stroke of 5 crank units.

3-28   Use two of the cognates in Figure 3-26b (p. 136) to design a Watt I sixbar parallel motion mechanism that carries a link through the same coupler curve at all points. Comment on its similarities to the original Roberts diagram.

3-29   Find the cognates of the Watt straight-line mechanism in Figure 3-29a (p. 140).

3-30   Find the cognates of the Roberts straight-line mechanism in Figure 3-29c (p. 140).

*3-31   Design a Hoeken straight-line linkage to give minimum error in velocity over 22% of the cycle for a 15-cm-long straight-line motion. Specify all linkage parameters.

3-32   Design a Hoeken straight-line linkage to give minimum error in straightness over 39% of the cycle for a 20-cm-long straight-line motion. Specify all linkage parameters.

3-33   Design a linkage that will give a symmetrical "kidney bean" shaped coupler curve as shown in Figure 3-16 (pp. 124 and 125). Use the data in Figure 3-21 (p. 130) to determine the required link ratios and generate the coupler curve with program LINKAGES.

3-34   Repeat Problem 3-33 for a "double straight" coupler curve.

3-35   Repeat problem 3-33 for a "scimitar" coupler curve with two distinct cusps. Show that there are (or are not) true cusps on the curve by using program LINKAGES. (*Hint*: Think about the definition of a cusp and how you can use the program's data to show it.)

*3-36   Find the Grashof condition, inversion, any limit positions, and the extreme values of the transmission angle (to graphical accuracy) of the linkage in Figure P3-10.

3-37   Draw the Roberts diagram and find the cognates of the linkage in Figure P3-10.

3-38   Find the three geared fivebar cognates of the linkage in Figure P3-10.

*3-39   Find the Grashof condition, any limit positions, and the extreme values of the transmission angle (to graphical accuracy) of the linkage in Figure P3-11.

3-40   Draw the Roberts diagram and find the cognates of the linkage in Figure P3-11.

* Answers in Appendix F.

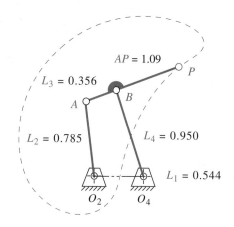

FIGURE P3-10

Problems 3-36 to 3-38

3-41   Find the three geared fivebar cognates of the linkage in Figure P3-11.

*3-42   Find the Grashof condition, any limit positions, and the extreme values of the transmission angle (to graphical accuracy) of the linkage in Figure P3-12.

3-43   Draw the Roberts diagram and find the cognates of the linkage in Figure P3-12.

3-44   Find the three geared fivebar cognates of the linkage in Figure P3-12.

3-45   Prove that the relationships between the angular velocities of various links in the Roberts diagram as shown in Figure 3-25 (p. 135) are true.

3-46   Design a fourbar linkage to move the object in Figure P3-13 from position 1 to 2 using points A and B for attachment. Add a driver dyad to limit its motion to the range of positions designed making it a sixbar. All fixed pivots should be on the base.

3-47   Design a fourbar linkage to move the object in Figure P3-13 from position 2 to 3 using points A and B for attachment. Add a driver dyad to limit its motion to the range of positions designed making it a sixbar. All fixed pivots should be on the base.

FIGURE P3-11

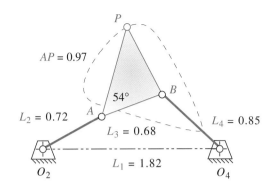

FIGURE P3-12

Problems 3-42 to 3-44

3-48 Design a fourbar linkage to move the object in Figure P3-13 through the three positions shown using points *A* and *B* for attachment. Add a driver dyad to limit its motion to the range of positions designed, making it a sixbar. All fixed pivots should be on the base.

3-49 Design a fourbar linkage to move the object in Figure P3-14 from position 1 to 2 using points *A* and *B* for attachment. Add a driver dyad to limit its motion to the range of positions designed making it a sixbar. All fixed pivots should be on the base.

3-50 Design a fourbar linkage to move the object in Figure P3-14 from position 2 to 3 using points *A* and *B* for attachment. Add a driver dyad to limit its motion to the range of positions designed making it a sixbar. All fixed pivots should be on the base.

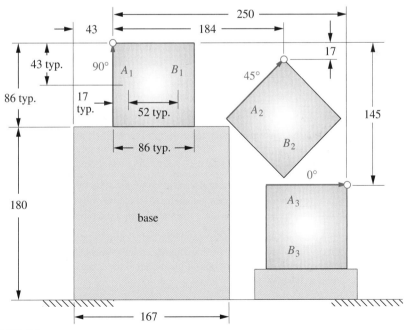

FIGURE P3-13

Problems 3-46 to 3-48

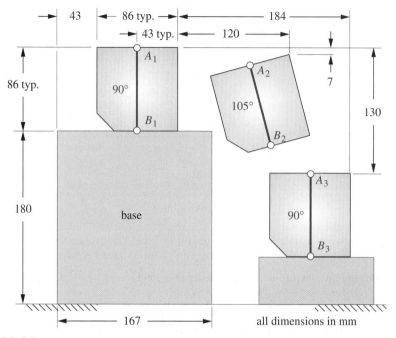

FIGURE P3-14

Problems 3-49 to 3-51

3-51   Design a fourbar linkage to move the object in Figure P3-14 through the three positions shown using points A and B for attachment. Add a driver dyad to limit its motion to the range of positions designed, making it a sixbar. All fixed pivots should be on the base.

3-52   Design a fourbar linkage to move the object in Figure P3-15 from position 1 to 2 using points A and B for attachment. Add a driver dyad to limit its motion to the range of positions designed, making it a sixbar. All fixed pivots should be on the base.

3-53   Design a fourbar linkage to move the object in Figure P3-15 from position 2 to 3 using points A and B for attachment. Add a driver dyad to limit its motion to the range of positions designed, making it a sixbar. All fixed pivots should be on the base.

3-54   Design a fourbar linkage to move the object in Figure P3-15 through the three positions shown using points A and B for attachment. Add a driver dyad to limit its motion to the range of positions designed, making it a sixbar. All fixed pivots should be on the base.

3-55   Design a fourbar mechanism to move the link shown in Figure P3-16 from position 1 to position 2. Ignore the third position and the fixed pivots O_2 and O_4 shown. Build a model and add a driver dyad to limit its motion to the range of positions designed, making it a sixbar.

3-56   Design a fourbar mechanism to move the link shown in Figure P3-16 from position 2 to position 3. Ignore the first position and the fixed pivots O_2 and O_4 shown. Build a model and add a driver dyad to limit its motion to the range of positions designed, making it a sixbar.

FIGURE P3-15

Problems 3-52 to 3-54

3-57 Design a fourbar mechanism to give the three positions shown in Figure P3-16. Ignore the fixed pivots O_2 and O_4 shown. Build a model and add a driver dyad to limit its motion to the range of positions designed, making it a sixbar.

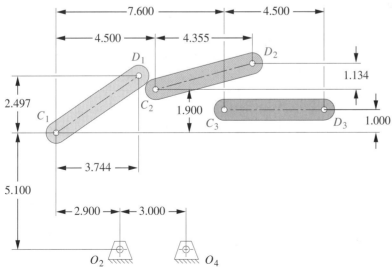

FIGURE P3-16

Problems 3-55 to 3-58

FIGURE P3-17

Problems 3-59 to 3-62

3-58   Design a fourbar mechanism to give the three positions shown in Figure P3-16 using the fixed pivots O_2 and O_4 shown. (See Example 3-7, p. 113.) Build a model and add a driver dyad to limit its motion to the range of positions designed, making it a sixbar.

3-59   Design a fourbar mechanism to move the link shown in Figure P3-17 from position 1 to position 2. Ignore the third position and the fixed pivots O_2 and O_4 shown. Build a model and add a driver dyad to limit its motion to the range of positions designed, making it a sixbar.

3-60   Design a fourbar mechanism to move the link shown in Figure P3-17 from position 2 to position 3. Ignore the first position and the fixed pivots O_2 and O_4 shown. Build a model and add a driver dyad to limit its motion to the range of positions designed, making it a sixbar.

3-61   Design a fourbar mechanism to give the three positions shown in Figure P3-17. Ignore the fixed pivots O_2 and O_4 shown. Build a model and add a driver dyad to limit its motion to the range of positions designed, making it a sixbar.

3-62   Design a fourbar mechanism to give the three positions shown in Figure P3-17 using the fixed pivots O_2 and O_4 shown. (See Example 3-7, p. 113.) Build a model and add a driver dyad to limit its motion to the range of positions designed, making it a sixbar.

3-63   Design a fourbar mechanism to move the link shown in Figure P3-18 from position 1 to position 2. Ignore the third position and the fixed pivots O_2 and O_4 shown. Build a model and add a driver dyad to limit its motion to the range of positions designed, making it a sixbar.

3-64   Design a fourbar mechanism to move the link shown in Figure P3-18 from position 2 to position 3. Ignore the first position and the fixed pivots O_2 and O_4 shown. Build

FIGURE P3-18

Problems 3-63 to 3-66

a model and add a driver dyad to limit its motion to the range of positions designed, making it a sixbar.

3-65 Design a fourbar mechanism to give the three positions shown in Figure P3-18. Ignore the fixed pivots O_2 and O_4 shown. Build a model and add a driver dyad to limit its motion to the range of positions designed, making it a sixbar.

3-66 Design a fourbar mechanism to give the three positions shown in Figure P3-18 using the fixed pivots O_2 and O_4 shown. (See Example 3-7, p. 113.) Build a model and add a driver dyad to limit its motion to the range of positions designed, making it a sixbar.

3-67 Design a fourbar Grashof crank-rocker for 120° of output rocker motion with a quick-return time ratio of 1:1.2. (See Example 3-9, p. 118.)

3-68 Design a fourbar Grashof crank-rocker for 100° of output rocker motion with a quick-return time ratio of 1:1.5. (See Example 3-9, p. 118.)

3-69 Design a fourbar Grashof crank-rocker for 80° of output rocker motion with a quick-return time ratio of 1:1.33. (See Example 3-9, p. 118.)

3-70 Design a sixbar drag link quick-return linkage for a time ratio of 1:4 and output rocker motion of 50°.

3-71   Design a crank shaper quick-return mechanism for a time ratio of 1:2.5 (See Figure 3-14, p. 122).

3-72   Design a sixbar, single-dwell linkage for a dwell of 70° of crank motion, with an output rocker motion of 30° using a symmetrical fourbar linkage having the following parameters: ground link ratio = 2.0, common link ratio = 2.0, and coupler angle $\gamma = 40°$. (See Example 3-13, p. 147.)

3-73   Design a sixbar, single-dwell linkage for a dwell of 100° of crank motion, with an output rocker motion of 50° using a symmetrical fourbar linkage having the following parameters: ground link ratio = 2.0, common link ratio = 2.5, and coupler angle $\gamma = 60°$. (See Example 3-13, p. 147.)

3-74   Design a sixbar, single-dwell linkage for a dwell of 80° of crank motion, with an output rocker motion of 45° using a symmetrical fourbar linkage having the following parameters: ground link ratio = 2.0, common link ratio = 1.75, and coupler angle $\gamma = 70°$. (See Example 3-13, p. 147.)

3-75   Using the method of Example 3-11 (p. 136), show that the sixbar Chebyschev straight-line linkage of Figure P2-5 (p. 87) is a combination of the fourbar Chebyschev straight-line linkage of Figure 3-29d and its Hoeken cognate of Figure 3-29e (p. 140). See also Figure 3-26 (p. 136) for additional information useful to this solution. Graphically construct the Chebyschev sixbar parallel motion linkage of Figure P2-5a (p. 87) from its two fourbar linkage constituents and build a physical or computer model of the result.

3-76   Design a driver dyad to drive link 2 of the Evans straight-line linkage in Figure 3-29f (p. 141) from 150° to 210°. Make a model of the resulting sixbar linkage and trace the coupler curve.

3-77   Design a driver dyad to drive link 2 of the Evans straight-line linkage in Figure 3-29g (p. 141) from –40° to 40°. Make a model of the resulting sixbar linkage and trace the coupler curve.

3-78   Figure 6 on page *ix* of the Hrones and Nelson atlas of fourbar coupler curves (on the book's DVD) shows a 50-point coupler that was used to generate the curves in the atlas. Using the definition of the vector **R** given in Figure 3-17b (p. 126) of the text, determine the 10 possible pairs of values of ϕ and R for the first row of points above the horizontal axis if the grid point spacing is one-half the length of the unit crank.

*3-79   The first set of 10 coupler curves on page 1 of the Hrones and Nelson atlas of fourbar coupler curves (on the book's DVD) has $A = B = C = 1.5$. Model this linkage with program LINKAGES using the coupler point farthest to the left in the row shown on page 1 and plot the resulting coupler curve. Note that you will have to orient link 1 at the proper angle in LINKAGES to get the curve as shown in the atlas.

3-80   Repeat problem 3-79 for the set of coupler curves on page 17 of the Hrones and Nelson atlas (see page 32 of the PDF file on the book's DVD) which has $A = 1.5$, $B = C = 3.0$, using the coupler point farthest to the right in the row shown.

3-81   Repeat problem 3-79 for the set of coupler curves on page 21 of the Hrones and Nelson atlas (see page 36 of the PDF file on the book's DVD) which has $A = 1.5$, $B = C = 3.5$, using the second coupler point from the right end in the row shown.

FIGURE P3-19

Problems 3-84 to 3-87

3-82 Repeat problem 3-79 for the set of coupler curves on page 34 of the Hrones and Nelson atlas (see page 49 of the PDF file on the book's DVD) which has $A = 2.0$, $B = 1.5$, $C = 2.0$, using the coupler point farthest to the right in the row shown.

3-83 Repeat problem 3-79 for the set of coupler curves on page 115 of the Hrones and Nelson atlas (see page 130 of the PDF file on the book's DVD) which has $A = 2.5$, $B = 1.5$, $C = 2.5$, using the fifth coupler point from the left end in the row shown.

3-84 Design a fourbar mechanism to move the link shown in Figure P3-19 from position 1 to position 2. Ignore the third position and the fixed pivots O_2 and O_4 shown. Build a cardboard model that demonstrates the required movement.

3-85 Design a fourbar mechanism to move the link shown in Figure P3-19 from position 2 to position 3. Ignore the first position and the fixed pivots O_2 and O_4 shown. Build a cardboard model that demonstrates the required movement.

3-86 Design a fourbar mechanism to give the three positions shown in Figure P3-19. Ignore the points O_2 and O_4 shown. Build a cardboard model that has stops to limit its motion to the range of positions designed.

3-87 Design a fourbar mechanism to give the three positions shown in Figure P3-19 using the fixed pivots O_2 and O_4 shown. See Example 3-7 (p. 113). Build a cardboard model that has stops to limit its motion to the range of positions designed.

3.14 PROJECTS

These larger-scale project statements deliberately lack detail and structure and are loosely defined. Thus, they are similar to the kind of "identification of need" problem statement commonly encountered in engineering practice. It is left to the student to structure the problem through background research and to create a clear goal statement and set of performance specifications before attempting to design a solution. This design process is spelled out in Chapter 1 and should be followed in all of these examples. These projects can be done as an exercise in mechanism synthesis alone or can be revisited and thoroughly analyzed by the methods presented in later chapters as well. All results should be documented in a professional engineering report.

P3-1   The tennis coach needs a better tennis ball server for practice. This device must fire a sequence of standard tennis balls from one side of a standard tennis court over the net such that they land and bounce within each of the three court areas defined by the court's white lines. The order and frequency of a ball's landing in any one of the three court areas must be random. The device should operate automatically and unattended except for the refill of balls. It should fire 50 balls between reloads. The timing of ball releases should vary. For simplicity, a motor driven pin-jointed linkage design is preferred.

P3-2   A quadriplegic patient has lost all motion except that of her head. She can only move a small "mouth stick" to effect a switch closure. She was an avid reader before her injury and would like again to be able to read standard hardcover books without the need of a person to turn pages for her. Thus, a reliable, simple, and inexpensive automatic page turner is needed. The book may be placed in the device by an assistant. It should accommodate as wide a range of book sizes as possible. Book damage is to be avoided and safety of the user is paramount.

P3-3   Grandma's off her rocker again! Junior's run down to the Bingo parlor to fetch her, but we've got to do something about her rocking chair before she gets back. She's been complaining that her arthritis makes it too painful to push the rocker. So, for her 100th birthday in 2 weeks, we're going to surprise her with a new, automated, motorized rocking chair. The only constraints placed on the problem are that the device must be safe and must provide interesting and pleasant motions, similar to those of her present *Boston* rocker, to all parts of the occupant's body. Since simplicity is the mark of good design, a linkage solution with only full pin joints is preferred.

P3-4   The local amusement park's business is suffering as a result of the proliferation of computer game parlors. They need a new and more exciting ride which will attract new customers. The only constraints are that it must be safe, provide excitement, and not subject the occupants to excessive accelerations or velocities. Also it must be as compact as possible, since space is limited. Continuous rotary input and full pin joints are preferred.

P3-5   The student section of ASME is sponsoring a spring fling on campus. They need a mechanism for their "Dunk the Professor" booth which will carry the unfortunate (untenured) volunteer into and out of the water tub. The contestants will provide the inputs to a multiple-*DOF* mechanism. If they know their kinematics, they can provide a combination of inputs which will dunk the victim.

P3-6   The National House of Flapjacks wants to automate its flapjack production. It needs a mechanism that will automatically flip the flapjacks "on the fly" as they travel through the griddle on a continuously moving conveyor. This mechanism must track the constant velocity of the conveyor, pick up a pancake, flip it over, and place it back onto the conveyor.

P3-7   Many varieties and shapes of computer video monitors now exist. Their long-term use leads to eyestrain and body fatigue. There is a need for an adjustable stand which will hold the video monitor and the separate keyboard at any position the user deems comfortable. The computer's central processor unit (CPU) can be remotely located. This device should be freestanding to allow use with a comfortable chair, couch, or lounge of the user's choice. It should not require the user to assume the conventional "seated at a desk" posture to use the computer. It must be stable in all positions and safely support the equipment's weight.

P3-8   Most small boat trailers must be submerged in the water to launch or retrieve the boat. This greatly reduces the life of the trailer, especially in salt water. A need exists for a trailer that will remain on dry land while it launches or retrieves the boat. No part of the trailer should get wet. User safety is of greatest concern, as is protection of the boat from damage.

P3-9   The "Save the Skeet" foundation has requested a more humane skeet launcher be designed. While they have not yet succeeded in passing legislation to prevent the wholesale slaughter of these little devils, they are concerned about the inhumane aspects of the large accelerations imparted to the skeet as it is launched into the sky for the sportsman to shoot it down. The need is for a skeet launcher that will smoothly accelerate the clay pigeon onto its desired trajectory.

P3-10   The coin-operated "kid bouncer" machines found outside supermarkets typically provide a very unimaginative rocking motion to the occupant. There is a need for a superior "bouncer" which will give more interesting motions while remaining safe for small children.

P3-11   Horseback riding is a very expensive hobby or sport. There is a need for a horseback riding simulator to train prospective riders sans the expensive horse. This device should provide similar motions to the occupant as she would feel in the saddle under various gaits such as a walk, canter, or gallop. A more advanced version might contain jumping motions as well. User safety is most important.

P3-12   The nation is on a fitness craze. Many exercise machines have been devised. There is still room for improvement to these devices. They are typically designed for the young, strong athlete. There is also a need for an ergonomically optimum exercise machine for the older person who needs gentler exercise.

P3-13   A paraplegic patient needs a device to get himself from his wheelchair into the Jacuzzi with no assistance. He has good upper body and arm strength. Safety is paramount.

P3-14   The army has requested a mechanical walking device to test army boots for durability. It should mimic a person's walking motion and provide forces similar to an average soldier's foot.

P3-15   NASA wants a zero-g machine for astronaut training. It must carry one person and provide a negative 1-g acceleration for as long as possible.

P3-16   The Amusement Machine Co. Inc. wants a portable "whip" ride that will give two or four passengers a thrilling but safe ride, and which can be pulled behind a pickup truck from one location to another.

P3-17   The Air Force has requested a pilot training simulator which will give potential pilots exposure to g forces similar to those they will experience in dogfight maneuvers.

P3-18   Cheers needs a better "mechanical bull" simulator for its "yuppie" bar in Boston. It must give a thrilling "bucking bronco" ride but be safe.

P3-19   Despite the improvements in handicap access, many curbs block wheelchairs from
         public places. Design an attachment for a conventional wheelchair which will allow it
         to get up over a curb.

P3-20   A carpenter needs a dumping attachment to fit in her pickup truck so she can dump
         building materials. She can't afford to buy a dump truck.

P3-21   The carpenter in Project P3-20 wants an inexpensive lift gate designed to fit her full-
         sized pickup truck, in order to lift and lower heavy cargo to the truck bed.

P3-22   The carpenter in Project P3-20 is very demanding (and lazy). She also wants a device
         to lift sheet rock and blueboard into place on ceilings or walls to hold it while she nails
         it on.

P3-23   Click and Clack, the tappet brothers, need a better transmission jack for their Good
         News Garage. This device should position a transmission under a car (on a lift) and
         allow it to be maneuvered into place safely and quickly.

P3-24   A paraplegic who was an avid golfer before his injury wants a mechanism to allow him
         to stand up in his wheelchair in order to once again play golf. It must not interfere with
         normal wheelchair use, though it could be removed from the chair when he is not golfing.

P3-25   A wheelchair lift is needed to raise the wheelchair and person 3 ft from the garage
         floor to the level of the first floor of the house. Safety, reliability, and cost are of major
         concern.

P3-26   A paraplegic needs a mechanism that can be installed on a full-size 3-door pickup
         truck that will lift the wheelchair into the area behind the driver's seat. This person
         has excellent upper body strength and, with the aid of specially installed handles on the
         truck, can get into the cab from the chair. The truck can be modified as necessary to
         accommodate this task. For example, attachment points can be added to its structure
         and the back seat of the truck can be removed if necessary.

P3-27   There is demand for a better *baby transport device*. Many such devices are on the mar-
         ket. Some are called carriages, others strollers. Some are convertible to multiple uses.
         Our marketing survey data so far seems to indicate that the customers want portability
         (i.e., foldability), light weight, one-handed operation, and large wheels. Some of these
         features are present in existing devices. We need a better design that more completely
         meets the needs of the customer. The device must be stable, effective, and safe for the
         baby and the operator. Full joints are preferred to half joints and simplicity is the mark
         of good design. A linkage solution with manual input is desired.

P3-28   A boat owner has requested that we design a lift mechanism to automatically move a
         1000-lb, 15-ft boat from a cradle on land to the water. A seawall protects the owner's
         yard, and the boat cradle sits above the seawall. The tidal variation is 4 ft and the
         seawall is 3 ft above the high tide mark. Your mechanism will be attached to land and
         move the boat from its stored position on the cradle to the water and return it to the
         cradle. The device must be safe and easy to use and not overly expensive.

P3-29   The landfills are full! We're about to be up to our ears in trash! The world needs a bet-
         ter trash compactor. It should be simple, inexpensive, quiet, compact, and safe. It can
         either be manually powered or motorized, but manual operation is preferred to keep the
         cost down. The device must be stable, effective, and safe for the operator.

P3-30   A small contractor needs a mini-dumpster attachment for his pickup truck. He has
         made several trash containers which are 4 ft x 4 ft x 3.5 ft high. The empty container

3

weighs 150 lb. He needs a mechanism which he can attach to his fleet of standard, full-size pickup trucks (Chevrolet, Ford, or Dodge). This mechanism should be able to pick up the full trash container from the ground, lift it over the closed tailgate of the truck, dump its contents into the truck bed, and then return it empty to the ground. He would like not to tip his truck over in the process. The mechanism should store permanently on the truck in such a manner as to allow the normal use of the pickup truck at all other times. You may specify any means of attachment of your mechanism to the container and to the truck.

P3-31   As a feast day approaches, the task of inserting the leaves in the dining room table presents itself. Typically, the table leaves are stored in some forgotten location, and when found have to be carried to the table and manually placed in it. Wouldn't it be better if the leaves (leaf) were stored within the table itself and were automatically inserted into place when the table was opened? The only constraints imposed on the problem are that the device must be simple to use, preferably using the action of opening the table halves as the actuating motion. That is, as you pull the table open, the stored leaf should be carried by the mechanism of your design into its proper place in order to extend the dining surface. Thus, a linkage solution with manual input is desired and full joints are preferred to half joints, though either may be used.

P3-32   Small sailboats often are not "self-bailing," meaning that they accumulate rainwater and can sink at the mooring if not manually bailed (emptied of water) after a rainstorm. These boats usually do not have a power source such as a battery aboard. Design a mechanism that can be quickly attached to, detached from, and stored in a 20-foot-long daysailer, and that will use wave action (boat rocking) as the input to a bilge pump to automatically keep the boat dry and afloat when left at a mooring.

P3-33   A machine uses several 200 kg servomotors that bolt underneath the machine's bedplate, which is 0.75 m above the floor. The machine frame has a 400 mm square front opening through which the motor can be inserted. It must be extended 0.5 m horizontally to its installed location. Design a mechanism to transport the motor from the stockroom to the machine, insert it under the machine and raise it 200 mm into position. Your mechanism also should be capable of removing a motor from the machine.

P3-34   A paraplegic client has requested that we design a mechanism to attach to his wheelchair that will store his backpack behind his seatback. This mechanism must also bring the backpack around toward the front of the chair so that he can access its contents. He has some use of his upper body and so can do something to cause your mechanism to move. It should safely lock itself in place when stowed behind the seatback. It should not upset the chair's stabilty or limit its mobility.

P3-35   In an effort to reduce chronic back injury among janitorial staff, our client, Ready Refuse, has requested that we design a mechanism to safely lift an office size rectangular trash or recycling container and dump it into a large rolling trash barrel. The mechanism needs to be motorized to dump the smaller container automatically. To operate the system, the user will roll the large trash barrel up to the rectangular container, which is sitting on the floor, and press a button that will cause the mechanism move through the required motion and dump the contents of the container into the large barrel. The grip design between your mechanism and the top lip of the rectangular trash container is being designed by another team at Widgets Perfected Inc.; assume that it works. Your task is to design the motorized mechanism that dumps the container without spilling the contents outside of the large trash barrel. [*]

[*] Problem courtesy of Prof. James Van de Ven, WPI.

Chapter **4**

POSITION ANALYSIS

Theory is the distilled essence of practice
RANKINE

4.0   INTRODUCTION

Once a tentative mechanism design has been **synthesized**, it must then be **analyzed**. A principal goal of kinematic analysis is to determine the accelerations of all the moving parts in the assembly. **Dynamic forces** are proportional to acceleration, from Newton's second law. We need to know the dynamic forces in order to calculate the **stresses** in the components. The design engineer must ensure that the proposed mechanism or machine will not fail under its operating conditions. Thus the stresses in the materials must be kept well below allowable levels. To calculate the stresses, we need to know the static and dynamic forces on the parts. To calculate the dynamic forces, we need to know the **accelerations**. In order to calculate the accelerations, we must first find the **positions** of all the links or elements in the mechanism for each increment of input motion, and then differentiate the position equations versus time to find **velocities**, and then differentiate again to obtain the expressions for acceleration. For example, in a simple Grashof fourbar linkage, we would probably want to calculate the positions, velocities, and accelerations of the output links (coupler and rocker) for perhaps every two degrees (180 positions) of input crank position for one revolution of the crank.

This can be done by any of several methods. We could use a **graphical approach** to determine the position, velocity, and acceleration of the output links for all 180 positions of interest, or we could **derive the general equations** of motion for any position, differentiate for velocity and acceleration, and then solve these **analytical expressions** for our 180 (or more) crank locations. A computer will make this latter task much more palatable. If we choose to use the graphical approach to analysis, we will have to do an independent graphical solution for each of the positions of interest. None of the information obtained graphically for the first position will be applicable to the second position or to any others. In contrast, once the analytical solution is derived for a particular mechanism, it can be quickly solved (with a computer) for all positions. If you want information for more than 180 positions, it only means you will have to wait longer for the computer to generate those data. The derived equations are the same. So, have another cup of coffee while the computer crunches the numbers! In this chapter, we will present and derive analytical solutions to the position analysis problem for various planar mechanisms. We will also discuss graphical solutions which are useful for checking your analytical results. In Chapters 6 and 7 we will do the same for velocity and acceleration analysis of planar mechanisms.

It is interesting to note that **graphical position analysis** of linkages is a truly trivial exercise, while the algebraic approach to position analysis is much more complicated. If you can draw the linkage to scale, you have then solved the position analysis problem graphically. It only remains to measure the link angles on the scale drawing to protractor accuracy. But the converse is true for velocity and especially for acceleration analysis. Analytical solutions for these are less complicated to derive than is the analytical position solution. However, graphical velocity and acceleration analysis becomes quite complex and difficult. Moreover, the graphical vector diagrams must be redone *de novo* (meaning literally *from new*) for each of the linkage positions of interest. This is a very tedious exercise and was the only practical method available in the days *B.C.* (*Before Computer*), not so long ago. The proliferation of inexpensive microcomputers in recent years has truly revolutionized the practice of engineering. As a graduate engineer, you will never be far from a computer of sufficient power to solve this type of problem and may even have one in your pocket. Thus, in this text we will emphasize analytical solutions which are easily solved with a microcomputer. The computer programs provided with this text use the same analytical techniques as derived in the text.

Geez Joe, - now I wish I took that programming course!

4.1   COORDINATE SYSTEMS

Coordinate systems and reference frames exist for the convenience of the engineer who defines them. In the next chapters we will provide our systems with multiple coordinate systems as needed, to aid in understanding and solving the problem. We will denote one of these as the *global* or *absolute* coordinate system, and the others will be *local* coordinate systems within the global framework. The global system is often taken to be attached to Mother Earth, though it could as well be attached to another ground plane such as the frame of an automobile. If our goal is to analyze the motion of a windshield wiper blade, we may not care to include the gross motion of the automobile in the analysis. In that case a global coordinate system (GCS—denoted as X,Y) attached to the car would be useful, and we could consider it to be an **absolute** coordinate system. Even if we use the earth as an absolute reference frame, we must realize that it is not stationary either, and as such is not very useful as a reference frame for a space probe. Though we will speak of absolute positions, velocities, and accelerations, keep in mind that ultimately, until we discover some stationary point in the universe, all motions are really relative. The term **inertial reference frame** is used to denote *a system which itself has no acceleration*. All angles in this text will be measured according to the *right-hand rule*. That is, **counterclockwise angles**, angular velocities, and angular accelerations *are positive in sign*.

Local coordinate systems are typically attached to a link at some point of interest. This might be a pin joint, a center of gravity, or a line of centers of a link. These local coordinate systems may be either rotating or nonrotating as we desire. If we want to measure the angle of a link as it rotates in the global system, we probably will want to attach a local nonrotating coordinate system (LNCS—denoted as x, y) to some point on the link (say a pin joint). This nonrotating system will move with its origin on the link but remains always parallel to the global system. If we want to measure some parameters within a link, independent of its rotation, then we will want to construct a local rotating coordinate system (LRCS—denoted as x', y') along some line on the link. This system will both move and rotate with the link in the global system. Most often we will need to have both types of local coordinate systems (LNCS and LRCS) on our moving links to do a complete analysis. Obviously we must define the angles and/or positions of these moving, local coordinate systems in the global system at all positions of interest.

4.2   POSITION AND DISPLACEMENT

Position

The **position** of a point in the plane can be defined by the use of a **position vector** as shown in Figure 4-1. The choice of **reference axes** is arbitrary and is selected to suit the observer. Figure 4-1a shows a point in the plane defined in a global coordinate system and Figure 4-1b shows the same point defined in a local coordinate system with its origin coincident with the global system A two-dimensional vector has two attributes, which can be expressed in either *polar* or *cartesian* coordinates. The **polar form** provides the magnitude and the angle of the vector. The **cartesian form** provides the X and Y components of the vector. Each form is directly convertible into the other by[*]

4

[*] Note that a two-argument arctangent function must be used to obtain angles in all four quadrants. The single-argument arctangent function found in most calculators and computer programming languages returns angle values in only the first and fourth quadrants. You can calculate your own two-argument arctangent function very easily by testing the sign of the x component of the arguments and, if x is minus, adding π radians or 180° to the result obtained from the available single-argument arctangent function.

For example (in Fortran):

```
FUNCTION Atan2( x, y )
IF x <> 0 THEN Q = y / x
Temp = ATAN(Q)
IF x < 0 THEN
    Atan2 = Temp + 3.14159
ELSE
    Atan2 = Temp
END IF
RETURN
END
```

The above code assumes that the language used has a built-in single-argument arctangent function called ATAN(x) which returns an angle between $\pm\pi/2$ radians when given a signed argument representing the value of the tangent of that angle.

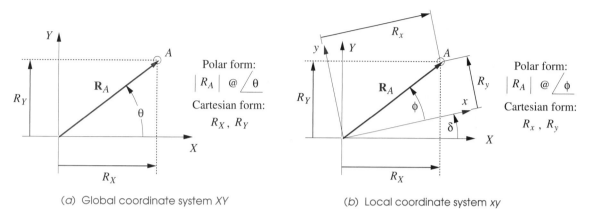

FIGURE 4-1

A position vector in the plane - expressed in both global and local coordinates

the Pythagorean theorem:

$$R_A = \sqrt{R_X^2 + R_Y^2}$$

and trigonometry: (4.0a)

$$\theta = \arctan\left(\frac{R_Y}{R_X}\right)$$

Equations 4.0a are shown in global coordinates but could as well be expressed in local coordinates.

Coordinate Transformation

It is often necessary to transform the coordinates of a point defined in one system to co-ordinates in another. If the system's origins are coincident as shown in Figure 4-1b and the required transformation is a rotation, it can be expressed in terms of the original coordinates and the signed angle δ between the coordinate systems. If the position of point A in Figure 4-1b is expressed in the local xy system as R_x, R_y, and it is desired to transform its coordinates to R_X, R_Y in the global XY system, the equations are:

$$R_X = R_x \cos\delta - R_y \sin\delta$$
$$R_Y = R_x \sin\delta + R_y \cos\delta$$ (4.0b)

Displacement

Displacement of a point is the change in its position and can be defined as *the straight-line distance between the initial and final position of a point which has moved in the reference frame.* Note that displacement is not necessarily the same as the path length which the point may have traveled to get from its initial to final position. Figure 4-2a shows a point in two positions, A and B. The curved line depicts the path along which the point traveled. The position vector $\mathbf{R}_{BA}$ defines the displacement of the point B with respect to

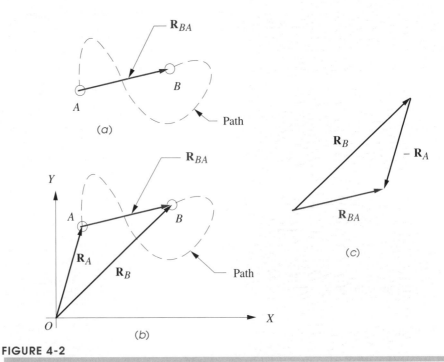

FIGURE 4-2

Position difference and relative position

point A. Figure 4-2b defines this situation more rigorously and with respect to a reference frame XY. The notation $\mathbf{R}$ will be used to denote a position vector. The vectors $\mathbf{R}_A$ and $\mathbf{R}_B$ define, respectively, the absolute positions of points A and B with respect to this *global XY* reference frame. The vector $\mathbf{R}_{BA}$ denotes the difference in position, or the *displacement*, between A and B. This can be expressed as the *position difference equation:*

$$\mathbf{R}_{BA} = \mathbf{R}_B - \mathbf{R}_A \qquad (4.1a)$$

This expression is read: *The position of B with respect to A is equal to the (absolute) position of B minus the (absolute) position of A*, where *absolute* means with respect to the origin of the *global* reference frame. This expression could also be written as:

$$\mathbf{R}_{BA} = \mathbf{R}_{BO} - \mathbf{R}_{AO} \qquad (4.1b)$$

with the second subscript O denoting the origin of the XY reference frame. When a position vector is rooted at the origin of the reference frame, it is customary to omit the second subscript. It is understood, in its absence, to be the origin. Also, a vector referred to the origin, such as $\mathbf{R}_A$, is often called an absolute vector. This means that it is taken with respect to a reference frame which is assumed to be stationary, e.g., *the ground*. It is important to realize, however, that the ground is usually also in motion in some larger frame of reference. Figure 4-2c shows a graphical solution to equations 4.1.

In our example of Figure 4-2, we have tacitly assumed so far that this point, which is first located at A and later at B, is, in fact, the same particle, moving within the reference frame. It could be, for example, one automobile moving along the road from A to B. With that assumption, it is conventional to refer to the vector $\mathbf{R}_{BA}$ as a **position difference**. There is, however, another situation which leads to the same diagram and equation but needs a different name. Assume now that points A and B in Figure 4-2b represent not the same particle but two independent particles moving in the same reference frame, as perhaps two automobiles traveling on the same road. The vector equations 4.1 and the diagram in Figure 4-2b still are valid, but we now refer to $\mathbf{R}_{BA}$ as a **relative position**, or **apparent position**. We will use the *relative position* term here. A more formal way to distinguish between these two cases is as follows:

CASE 1: *One body in two successive positions* => ***position difference***

CASE 2: *Two bodies simultaneously in separate positions* => ***relative position***

This may seem a rather fine point to distinguish, but the distinction will prove useful, and the reasons for it more clear, when we analyze velocities and accelerations, especially when we encounter (Case 2 type) situations in which the two bodies occupy the same position at the same time but have different motions.

4.3 TRANSLATION, ROTATION, AND COMPLEX MOTION

So far we have been dealing with a particle, or point, in plane motion. It is more interesting to consider the motion of a **rigid body**, or link, which involves both the position of a point on the link and the orientation of a line on the link, sometimes called the **POSE** of the link. Figure 4-3a (p. 180) shows a link AB denoted by a position vector $\mathbf{R}_{BA}$. An axis system has been set up at the root of the vector, at point A, for convenience.

Translation

Figure 4-3b shows link AB moved to a new position $A'B'$ by translation through the displacement AA' or BB' which are equal, i.e., $\mathbf{R}_{A'A} = \mathbf{R}_{B'B}$.

A definition of *translation* is:

All points on the body have the same displacement.

As a result the link retains its angular orientation. Note that the translation need not be along a straight path. The curved lines from A to A' and B to B' are the **curvilinear translation** path of the link. There is no rotation of the link if these paths are parallel. If the path happens to be straight, then it will be the special case of **rectilinear translation**, and the path and the displacement will be the same.

Rotation

Figure 4-3c shows the same link AB moved from its original position at the origin by rotation through an angle. Point A remains at the origin, but B moves through the position difference vector $\mathbf{R}_{B'B} = \mathbf{R}_{B'A} - \mathbf{R}_{BA}$.

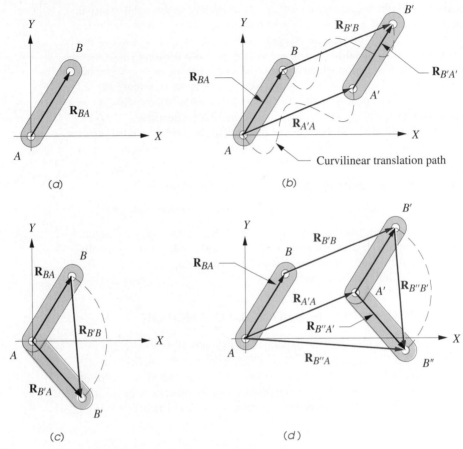

FIGURE 4-3

Translation, rotation, and complex motion

A definition of *rotation* is:

Different points in the body undergo different displacements and thus there is a displacement difference between any two points chosen.

The link now changes its angular orientation in the reference frame, and all points have different displacements.

Complex Motion

The general case of **complex motion** is the sum of the translation and rotation components. Figure 4-3d shows the same link moved through both a translation and a rotation. Note that the order in which these two components are added is immaterial. The resulting complex displacement will be the same whether you first rotate and then translate or vice versa. This is so because the two factors are independent. The total complex displacement of point *B* is defined by the following expression:

Total displacement = translation component + rotation component

$$\mathbf{R}_{B''B} = \mathbf{R}_{B'B} + \mathbf{R}_{B''B'} \qquad (4.1c)$$

The new absolute position of point B referred to the origin at A is:

$$\mathbf{R}_{B''A} = \mathbf{R}_{A'A} + \mathbf{R}_{B''A'} \qquad (4.1d)$$

Note that the above two formulas are merely applications of the position difference equation 4.1a (p. 178). See also Section 2.2 (p. 31) for definitions and discussion of *rotation, translation,* and *complex motion.* These motion states can be expressed as the following theorems.

Theorems

Euler's theorem:

The general displacement of a rigid body with one point fixed is a rotation about some axis.

This applies to pure rotation as defined above and in Section 2.2 (p. 31). Chasles (1793-1880) provided a corollary to Euler's theorem now known as:

Chasles' theorem:[6] *

Any displacement of a rigid body is equivalent to the sum of a translation of any one point on that body and a rotation of the body about an axis through that point.

This describes complex motion as defined above and in Section 2.2. Note that equation 4.1c is an expression of Chasles' theorem.

4.4   GRAPHICAL POSITION ANALYSIS OF LINKAGES

For any one-*DOF* linkage, such as a fourbar, only one parameter is needed to completely define the positions of all the links. The parameter usually chosen is the angle of the input link. This is shown as θ_2 in Figure 4-4 (p. 182). We want to find θ_3 and θ_4. The link lengths are known. Note that we will consistently number the ground link as 1 and the driver link as 2 in these examples.

The graphical analysis of this problem is trivial and can be done using only high-school geometry. If we draw the linkage carefully to scale with rule, compass, and protractor in a particular position (given θ_2), then it is only necessary to measure the angles of links 3 and 4 with the protractor. Note that all link angles are measured from a positive X axis. In Figure 4-4, a *local xy* axis system, parallel to the *global XY* system, has been created at point A to measure θ_3. The accuracy of this graphical solution will be limited by our care and drafting ability and by the crudity of the protractor used. Nevertheless, a very rapid approximate solution can be found for any one position.

Figure 4-5 (p. 182) shows the construction of the graphical position solution. The four link lengths a, b, c, d and the angle θ_2 of the input link are given. First, the ground link (1) and the input link (2) are drawn to a convenient scale such that they intersect at the origin O_2 of the global XY coordinate system with link 2 placed at the input angle θ_2. Link 1 is drawn along the X axis for convenience. The compass is set to the scaled length of link 3, and an arc of that radius is swung about the end of link 2 (point A). Then the

4

FIGURE 4-4

Measurement of angles in the fourbar linkage

compass is set to the scaled length of link 4, and a second arc is swung about the end of link 1 (point O_4). These two arcs will have two intersections at B and B' that define the two solutions to the position problem for a fourbar linkage which can be assembled in two configurations, called circuits, labeled open and crossed in Figure 4-5. Circuits in linkages will be discussed in a later section.

The angles of links 3 and 4 can be measured with a protractor. One circuit has angles θ_3 and θ_4, the other $\theta_{3'}$ and $\theta_{4'}$. A graphical solution is only valid for the particular value of input angle used. For each additional position analysis we must completely redraw the linkage. This can become burdensome if we need a complete analysis at every 1- or 2-degree increment of θ_2. In that case we will be better off to derive an analytical solution for θ_3 and θ_4 that can be solved by computer.

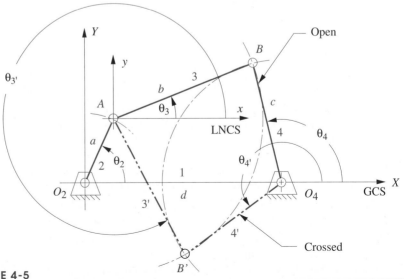

FIGURE 4-5

Graphical position solution to the open and crossed configurations of the fourbar linkage

4.5   ALGEBRAIC POSITION ANALYSIS OF LINKAGES

The same procedure that was used in Figure 4-5 to solve geometrically for the intersections B and B' and angles of links 3 and 4 can be encoded into an algebraic algorithm. The coordinates of point A are found from

$$A_x = a\cos\theta_2$$

$$A_y = a\sin\theta_2$$

(4.2a)

The coordinates of point B are found using the equations of circles about A and O_4.

$$b^2 = \left(B_x - A_x\right)^2 + \left(B_y - A_y\right)^2$$

(4.2b)

$$c^2 = \left(B_x - d\right)^2 + B_y^2$$

(4.2c)

which provide a pair of simultaneous equations in B_x and B_y.

Subtracting equation 4.2c from 4.2b gives an expression for B_x.

$$B_x = \frac{a^2 - b^2 + c^2 - d^2}{2\left(A_x - d\right)} - \frac{2A_y B_y}{2\left(A_x - d\right)} = S - \frac{2A_y B_y}{2\left(A_x - d\right)}$$

(4.2d)

Substituting equation 4.2d into 4.2c gives a quadratic equation in B_y which has two solutions corresponding to those in Figure 4-5.

$$B_y^2 + \left(S - \frac{A_y B_y}{A_x - d} - d\right)^2 - c^2 = 0$$

(4.2e)

This can be solved with the familiar expression for the roots of a quadratic equation,

$$B_y = \frac{-Q \pm \sqrt{Q^2 - 4PR}}{2P}$$

(4.2f)

where:

$$P = \frac{A_y^2}{\left(A_x - d\right)^2} + 1 \qquad\qquad Q = \frac{2A_y\left(d - S\right)}{A_x - d}$$

$$R = (d - S)^2 - c^2 \qquad\qquad S = \frac{a^2 - b^2 + c^2 - d^2}{2\left(A_x - d\right)}$$

Note that the solutions to this equation set can be real or imaginary. If the latter, it indicates that the links cannot connect at the given input angle or at all. Once the two values of B_y are found (if real), they can be substituted into equation 4.2d to find their corresponding x components. The link angles for this position can then be found from

$$\theta_3 = \tan^{-1}\left(\frac{B_y - A_y}{B_x - A_x}\right)$$

$$\theta_4 = \tan^{-1}\left(\frac{B_y}{B_x - d}\right)$$

(4.2g)

A two-argument arctangent function must be used to solve equations 4.2g since the angles can be in any quadrant. Equations 4.2 can be encoded in any computer language or equation solver, and the value of θ_2 varied over the linkage's usable range to find all corresponding values of the other two link angles.

Vector Loop Representation of Linkages

An alternate approach to linkage position analysis creates a vector loop (or loops) around the linkage as first proposed by Raven.[9] This approach offers some advantages in the synthesis of linkages which will be addressed in Chapter 5. The links are represented as **position vectors**. Figure 4-6 shows the same fourbar linkage as in Figure 4-4 (p. 182), but the links are now drawn as position vectors that form a vector loop. This loop closes on itself, making the sum of the vectors around the loop zero. The lengths of the vectors are the link lengths, which are known. The current linkage position is defined by the input angle θ_2 as it is a one-*DOF* mechanism. We want to solve for the unknown angles θ_3 and θ_4. To do so we need a convenient notation to represent the vectors.

Complex Numbers as Vectors

There are many ways to represent vectors. They may be defined in **polar coordinates**, by their *magnitude* and *angle*, or in **cartesian coordinates** as x and y components. These forms are of course easily convertible from one to the other using equations 4.0a. The position vectors in Figure 4-6 can be represented as any of these expressions:

Polar form	Cartesian form	
$R@\angle\theta$	$r\cos\theta\,\hat{\mathbf{i}}+r\sin\theta\,\hat{\mathbf{j}}$	(4.3a)
$re^{j\theta}$	$r\cos\theta+jr\sin\theta$	(4.3b)

Equation 4.3a uses **unit vectors** to represent the x and y vector component directions in the cartesian form. Figure 4-7 shows the unit vector notation for a position vector. Equation 4.3b uses **complex number notation** wherein the X direction component is called the *real portion* and the Y direction component is called the *imaginary portion*. This unfortunate term *imaginary* comes about because of the use of the notation j to represent the square root of minus one, which of course cannot be evaluated numerically.

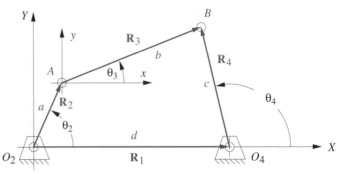

FIGURE 4-6

Position vector loop for a fourbar linkage

However, this *imaginary* number is used in a **complex number** as an **operator**, *not as a value*. Figure 4-8a (p. 186) shows the **complex plane** in which the *real* axis represents the *X*-directed component of the vector in the plane, and the *imaginary* axis represents the *Y*-directed component of the same vector. So, any term in a complex number which has no *j* operator is an *x* component, and a *j* indicates a *y* component.

Note in Figure 4-8b (p. 186) that each multiplication of the vector $\mathbf{R}_A$ by the operator *j* results in a *counterclockwise rotation* of the vector through 90 degrees. The vector $\mathbf{R}_B = j\mathbf{R}_A$ is directed along the *positive imaginary* or *j* axis. The vector $\mathbf{R}_C = j^2 \mathbf{R}_A$ is directed along the *negative real* axis because $j^2 = -1$ and thus $\mathbf{R}_C = -\mathbf{R}_A$. In similar fashion, $\mathbf{R}_D = j^3 \mathbf{R}_A = -j\mathbf{R}_A$ and this component is directed along the *negative j axis*.

One advantage of using this complex number notation to represent planar vectors comes from the **Euler identity**:

$$e^{\pm j\theta} = \cos\theta \pm j\sin\theta \tag{4.4a}$$

Any two-dimensional vector can be represented by the compact polar notation on the left side of equation 4.4a. There is no easier function to differentiate or integrate, since it is its own derivative:

$$\frac{de^{j\theta}}{d\theta} = je^{j\theta} \tag{4.4b}$$

We will use this **complex number notation** for vectors to develop and derive the equations for position, velocity, and acceleration of linkages.

The Vector Loop Equation for a Fourbar Linkage

The directions of the position vectors in Figure 4-6 are chosen so as to define their angles where we desire them to be measured. By definition, *the angle of a vector is always measured at its root, **not at its head***. We would like angle θ_4 to be measured at the fixed pivot O_4, so vector $\mathbf{R}_4$ is arranged to have its root at that point. We would like to measure angle θ_3 at the point where links 2 and 3 join, so vector $\mathbf{R}_3$ is rooted there. A similar logic dictates the arrangement of vectors $\mathbf{R}_1$ and $\mathbf{R}_2$. Note that the *X* (*real*) axis is taken for convenience along link 1 and the origin of the global coordinate system is taken at point

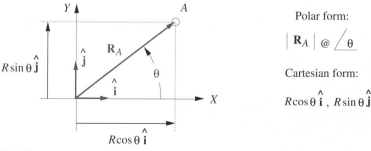

Polar form:

$$\left| \mathbf{R}_A \right| @ \underline{\diagdown\,\theta}$$

Cartesian form:

$$R\cos\theta\,\hat{\mathbf{i}}, \; R\sin\theta\,\hat{\mathbf{j}}$$

FIGURE 4-7

Unit vector notation for position vectors

4

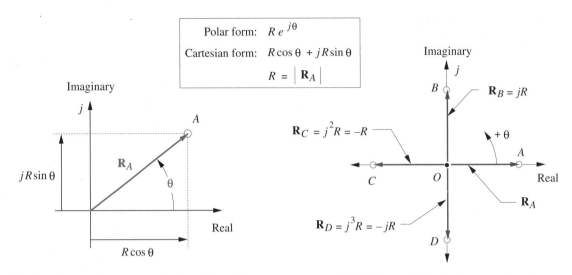

(a) Complex number representation of a position vector   (b) Vector rotations in the complex plane

FIGURE 4-8

Complex number representation of vectors in the plane

O_2, the root of the input link vector $\mathbf{R}_2$. These choices of vector directions and senses, as indicated by their arrowheads, lead to this vector loop equation:

$$\mathbf{R}_2 + \mathbf{R}_3 - \mathbf{R}_4 - \mathbf{R}_1 = 0 \qquad (4.5a)$$

An **alternate notation** for these position vectors is to use the labels of the points at the vector **tips** and **roots** (*in that order*) as subscripts. The second subscript is conventionally omitted if it is the origin of the global coordinate system (point O_2):

$$\mathbf{R}_A + \mathbf{R}_{BA} - \mathbf{R}_{BO_4} - \mathbf{R}_{O_4} = 0 \qquad (4.5b)$$

Next, we substitute the complex number notation for each position vector. To simplify the notation and minimize the use of subscripts, we will denote the scalar lengths of the four links as a, b, c, and d. These are so labeled in Figure 4-6 (p. 184). The equation then becomes:

$$a e^{j\theta_2} + b e^{j\theta_3} - c e^{j\theta_4} - d e^{j\theta_1} = 0 \qquad (4.5c)$$

These are three forms of the same vector equation, and as such can be solved for two unknowns. There are four variables in this equation, namely the four link angles. The link lengths are all constant in this particular linkage. Also, the value of the angle of link 1 is fixed (at zero) since this is the ground link. The *independent variable* is θ_2 which we will control with a motor or other driver device. That leaves the angles of link 3 and 4 to be found. We need algebraic expressions which define θ_3 and θ_4 as functions only of the constant link lengths and the one input angle, θ_2. These expressions will be of the form:

$$\theta_3 = f\{a, b, c, d, \theta_2\}$$

$$\theta_4 = g\{a, b, c, d, \theta_2\}$$

(4.5d)

To solve the polar form, vector equation 4.5c, we must substitute the *Euler equivalents* (equation 4.4a, p. 185) for the $e^{j\theta}$ terms, and then separate the resulting cartesian form vector equation into two scalar equations which can be solved simultaneously for θ_3 and θ_4. Substituting equation 4.4a into equation 4.5c:

$$a(\cos\theta_2 + j\sin\theta_2) + b(\cos\theta_3 + j\sin\theta_3) - c(\cos\theta_4 + j\sin\theta_4) - d(\cos\theta_1 + j\sin\theta_1) = 0 \quad (4.5e)$$

This equation can now be separated into its real and imaginary parts and each set to zero.

real part (*x* component):

$$a\cos\theta_2 + b\cos\theta_3 - c\cos\theta_4 - d\cos\theta_1 = 0$$

but: $\theta_1 = 0$, so: (4.6a)

$$a\cos\theta_2 + b\cos\theta_3 - c\cos\theta_4 - d = 0$$

imaginary part (*y* component):

$$ja\sin\theta_2 + jb\sin\theta_3 - jc\sin\theta_4 - jd\sin\theta_1 = 0$$

but: $\theta_1 = 0$, and the j's divide out, so: (4.6b)

$$a\sin\theta_2 + b\sin\theta_3 - c\sin\theta_4 = 0$$

The scalar equations 4.6a and 4.6b can now be solved simultaneously for θ_3 and θ_4. To solve this set of two simultaneous trigonometric equations is straightforward but tedious. Some substitution of trigonometric identities will simplify the expressions. The first step is to rewrite equations 4.6a and 4.6b so as to isolate one of the two unknowns on the left side. We will isolate θ_3 and solve for θ_4 in this example.

$$b\cos\theta_3 = -a\cos\theta_2 + c\cos\theta_4 + d \quad (4.6c)$$
$$b\sin\theta_3 = -a\sin\theta_2 + c\sin\theta_4 \quad (4.6d)$$

Now square both sides of equations 4.6c and 4.6d and add them:

$$b^2\left(\sin^2\theta_3 + \cos^2\theta_3\right) = (-a\sin\theta_2 + c\sin\theta_4)^2 + (-a\cos\theta_2 + c\cos\theta_4 + d)^2 \quad (4.7a)$$

Note that the quantity in parentheses on the left side is equal to 1, eliminating θ_3 from the equation, leaving only θ_4 which can now be solved for.

$$b^2 = (-a\sin\theta_2 + c\sin\theta_4)^2 + (-a\cos\theta_2 + c\cos\theta_4 + d)^2 \quad (4.7b)$$

Expand this expression and collect terms.

$$b^2 = a^2 + c^2 + d^2 - 2ad\cos\theta_2 + 2cd\cos\theta_4 - 2ac(\sin\theta_2\sin\theta_4 + \cos\theta_2\cos\theta_4) \quad (4.7c)$$

Divide through by $2ac$ and rearrange to get:

$$\frac{d}{a}\cos\theta_4 - \frac{d}{c}\cos\theta_2 + \frac{a^2 - b^2 + c^2 + d^2}{2ac} = \sin\theta_2 \sin\theta_4 + \cos\theta_2 \cos\theta_4 \qquad (4.7d)$$

To further simplify this expression, the constants $K_1, K_2,$ and K_3 are defined in terms of the constant link lengths in equation 4.7d:

$$K_1 = \frac{d}{a} \qquad K_2 = \frac{d}{c} \qquad K_3 = \frac{a^2 - b^2 + c^2 + d^2}{2ac} \qquad (4.8a)$$

and:

$$K_1 \cos\theta_4 - K_2 \cos\theta_2 + K_3 = \cos\theta_2 \cos\theta_4 + \sin\theta_2 \sin\theta_4 \qquad (4.8b)$$

If we substitute the identity $\cos(\theta_2 - \theta_4) = \cos\theta_2 \cos\theta_4 + \sin\theta_2 \sin\theta_4$, we get the form known as Freudenstein's equation.

$$K_1 \cos\theta_4 - K_2 \cos\theta_2 + K_3 = \cos(\theta_2 - \theta_4) \qquad (4.8c)$$

In order to reduce equation 4.8b to a more tractable form for solution, it will be useful to substitute the *half-angle identities* which will convert the $\sin\theta_4$ and $\cos\theta_4$ terms to $\tan\theta_4$ terms:

$$\sin\theta_4 = \frac{2\tan\left(\dfrac{\theta_4}{2}\right)}{1 + \tan^2\left(\dfrac{\theta_4}{2}\right)}; \qquad \cos\theta_4 = \frac{1 - \tan^2\left(\dfrac{\theta_4}{2}\right)}{1 + \tan^2\left(\dfrac{\theta_4}{2}\right)} \qquad (4.9)$$

This results in the following simplified form, where the link lengths and known input value (θ_2) terms have been collected as constants A, B, and C.

$$A \tan^2\left(\frac{\theta_4}{2}\right) + B \tan\left(\frac{\theta_4}{2}\right) + C = 0 \qquad (4.10a)$$

where:

$$A = \cos\theta_2 - K_1 - K_2 \cos\theta_2 + K_3$$
$$B = -2\sin\theta_2$$
$$C = K_1 - (K_2 + 1)\cos\theta_2 + K_3$$

Note that equation 4.10a is quadratic in form, and the solution is:

$$\tan\left(\frac{\theta_4}{2}\right) = \frac{-B \pm \sqrt{B^2 - 4AC}}{2A} \qquad (4.10b)$$

$$\theta_{4_{1,2}} = 2\arctan\left(\frac{-B \pm \sqrt{B^2 - 4AC}}{2A}\right)$$

Equation 4.10b has two solutions, obtained from the ± conditions on the radical. These two solutions, as with any quadratic equation, may be of three types: *real and equal, real and unequal, complex conjugate*. If the discriminant under the radical is negative, then the solution is complex conjugate, which simply means that the link lengths chosen are not capable of connection for the chosen value of the input angle θ_2. This can occur either when the link lengths are completely incapable of connection in any position or, in a non-Grashof linkage, when the input angle is beyond a toggle limit position. There is then no real solution for that value of input angle θ_2. Excepting this situation, the solution will usually be real and unequal, meaning there are two values of θ_4 corresponding to any one value of θ_2. These are referred to as the **crossed** and **open** configurations of the linkage and also as the two **circuits** of the linkage.[*] In the fourbar linkage, the minus solution gives θ_4 for the open configuration and the positive solution gives θ_4 for the crossed configuration.

Figure 4-5 (p. 182) shows both crossed and open solutions for a Grashof crank-rocker linkage. The terms crossed and open are based on the assumption that the input link 2, for which θ_2 is defined, is placed in the first quadrant (i.e., $0 < \theta_2 < \pi/2$). A Grashof linkage is then defined as **crossed** if the two links adjacent to the shortest link cross one another, and as **open** if they do not cross one another in this position. Note that the configuration of the linkage, either crossed or open, is solely dependent upon the way that the links are assembled. You cannot predict, based on link lengths alone, which of the solutions will be the desired one. In other words, you can obtain either solution with the same linkage by simply taking apart the pin which connects links 3 and 4 in Figure 4-5 (p. 182), and moving those links to the only other positions at which the pin will again connect them. In so doing, you will have switched from one position solution, or **circuit**, to the other.

The solution for angle θ_3 is essentially similar to that for θ_4. Returning to equations 4.6, we can rearrange them to isolate θ_4 on the left side.

$$c \cos\theta_4 = a\cos\theta_2 + b\cos\theta_3 - d \qquad (4.6e)$$
$$c \sin\theta_4 = a\sin\theta_2 + b\sin\theta_3 \qquad (4.6f)$$

Squaring and adding these equations will eliminate θ_4. The resulting equation can be solved for θ_3 as was done above for θ_4, yielding this expression:

$$K_1 \cos\theta_3 + K_4 \cos\theta_2 + K_5 = \cos\theta_2 \cos\theta_3 + \sin\theta_2 \sin\theta_3 \qquad (4.11a)$$

The constant K_1 is the same as defined in equation 4.8b, and K_4 and K_5 are:

$$K_4 = \frac{d}{b} \qquad\qquad K_5 = \frac{c^2 - d^2 - a^2 - b^2}{2ab} \qquad (4.11b)$$

This also reduces to a quadratic form:

$$D \tan^2\left(\frac{\theta_3}{2}\right) + E \tan\left(\frac{\theta_3}{2}\right) + F = 0 \qquad (4.12)$$

where
$$D = \cos\theta_2 - K_1 + K_4 \cos\theta_2 + K_5$$
$$E = -2\sin\theta_2$$
$$F = K_1 + (K_4 - 1)\cos\theta_2 + K_5$$

and the solution is:

* See Section 4-13 (p. 208) for a more complete discussion of circuits and branches in linkages.

$$\theta_{3_{1,2}} = 2 \arctan\left(\frac{-E \pm \sqrt{E^2 - 4DF}}{2D}\right) \tag{4.13}$$

As with the angle θ_4, this also has two solutions, corresponding to the crossed and open circuits of the linkage, as shown in Figure 4-5 (p. 182).

✐ EXAMPLE 4-1

Position Analysis of a Fourbar Linkage with the Vector Loop Method.

Problem: Given a fourbar linkage with the link lengths $L_1 = d = 100$ mm, $L_2 = a = 40$ mm, $L_3 = b = 120$ mm, $L_4 = c = 80$ mm. For $\theta_2 = 40°$ find all possible values of θ_3 and θ_4.

Solution: See Figure 4-6 (p. 184) for nomenclature.

1   Using equation 4.8a, calculate the link ratios K_1, K_2 and K_3.

$$K_1 = \frac{d}{a} = \frac{100}{40} = 2.5$$

$$K_2 = \frac{d}{c} = \frac{100}{80} = 1.25 \tag{a}$$

$$K_3 = \frac{a^2 - b^2 + c^2 + d^2}{2ac} = \frac{40^2 - 120^2 + 80^2 + 100^2}{2(40)(80)} = 0.562$$

2   Use these link ratios to find the intermediate parameters A, B, and C from equation 4.10a.

$$A = \cos\theta_2 - K_1 - K_2 \cos\theta_2 + K_3 = \cos(40°) - 2.5 - 1.25\cos(40°) + 0.562 = -2.129$$
$$B = -2\sin\theta_2 = -2\sin(40°) = -1.286 \tag{b}$$
$$C = K_1 - (K_2 + 1)\cos\theta_2 + K_3 = 2.5 - (1.25 + 1)\cos(40°) + 0.562 = 1.339$$

3   Use equation 4.10b to find θ_4 for both the open and crossed configurations.

$$\theta_{4_{open}} = 2\arctan\left(\frac{-B - \sqrt{B^2 - 4AC}}{2A}\right) = 2\arctan\left(\frac{1.286 - \sqrt{-1.286^2 - 4(-2.129)(1.339)}}{2(-2.129)}\right)$$

$$= 57.33°$$
$$\tag{c}$$
$$\theta_{4_{crossed}} = 2\arctan\left(\frac{-B + \sqrt{B^2 - 4AC}}{2A}\right) = 2\arctan\left(\frac{1.286 + \sqrt{-1.286^2 - 4(-2.129)(1.339)}}{2(-2.129)}\right)$$

$$= -98.01°$$

4   Use equation 4.11b to find the ratios K_4 and K_5.

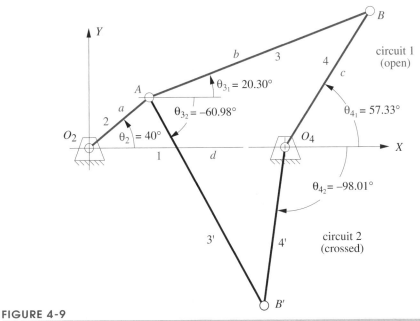

FIGURE 4-9

Solution to Example 4-1

$$K_4 = \frac{d}{b} = \frac{100}{120} = 0.833$$

$$K_5 = \frac{c^2 - d^2 - a^2 - b^2}{2ab} = \frac{80^2 - 100^2 - 40^2 - 120^2}{2(40)(120)} = -2.042 \qquad (d)$$

5 Use equation 4.12 to find the intermediate parameters D, E, and F.

$$D = \cos\theta_2 - K_1 + K_4 \cos\theta_2 + K_5 = \cos(40°) - 2.5 + 0.833(40°) - 2.042 = -3.137$$
$$E = -2\sin\theta_2 = -2\sin(40°) = -1.286 \qquad (e)$$
$$F = K_1 + (K_4 - 1)\cos\theta_2 + K_5 = 2.5 + (0.833 - 1)\cos(40°) - 2.042 = 0.331$$

6 Use equation 4.13 to find θ_3 for both the open and crossed configurations.

$$\theta_{3_{open}} = 2\arctan\left(\frac{-E - \sqrt{E^2 - 4DF}}{2D}\right) = 2\arctan\left(\frac{1.286 - \sqrt{-1.286^2 - 4(-3.137)(0.331)}}{2(-3.137)}\right)$$

$$= 20.30° \qquad (f)$$

$$\theta_{3_{crossed}} = 2\arctan\left(\frac{-E + \sqrt{E^2 - 4DF}}{2D}\right) = 2\arctan\left(\frac{1.286 + \sqrt{-1.286^2 - 4(-3.137)(0.331)}}{2(-3.137)}\right)$$

$$= -60.98°$$

7 The solution is shown in Figure 4-9.

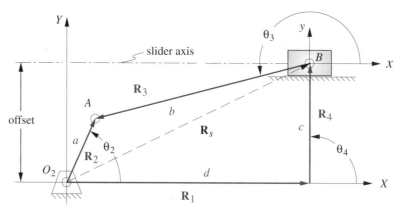

FIGURE 4-10

Position vector loop for a fourbar crank-slider or slider-crank linkage

4.6   THE FOURBAR CRANK-SLIDER POSITION SOLUTION

The same vector loop approach as used for the pure pin-jointed fourbar can be applied to a linkage containing sliders. Figure 4-10 shows an offset fourbar crank-slider linkage, inversion #1. The term **offset** means that *the slider axis extended does not pass through the crank pivot.* This is the general case. (The nonoffset crank-slider linkages shown in Figure 2-13 (p. 52) are the special cases.) This linkage could be represented by only three position vectors, $\mathbf{R}_2$, $\mathbf{R}_3$, and $\mathbf{R}_s$, but one of them ($\mathbf{R}_s$) will be a vector of varying magnitude and angle. It will be easier to use four vectors, $\mathbf{R}_1$, $\mathbf{R}_2$, $\mathbf{R}_3$, and $\mathbf{R}_4$ with $\mathbf{R}_1$ arranged parallel to the axis of sliding and $\mathbf{R}_4$ perpendicular. In effect the pair of vectors $\mathbf{R}_1$ and $\mathbf{R}_4$ are orthogonal components of the position vector $\mathbf{R}_s$ from the origin to the slider.

It simplifies the analysis to arrange one coordinate axis parallel to the axis of sliding. The variable-length, constant-direction vector $\mathbf{R}_1$ then represents the slider position with magnitude d. The vector $\mathbf{R}_4$ is orthogonal to $\mathbf{R}_1$ and defines the constant magnitude **offset** of the linkage. Note that for the special-case, nonoffset version, the vector $\mathbf{R}_4$ will be zero and $\mathbf{R}_1 = \mathbf{R}_s$. The vectors $\mathbf{R}_2$ and $\mathbf{R}_3$ complete the vector loop. The coupler's position vector $\mathbf{R}_3$ is placed with its root at the slider which then defines its angle θ_3 at point B. This particular arrangement of position vectors leads to a vector loop equation similar to the pin-jointed fourbar example:

$$\mathbf{R}_2 - \mathbf{R}_3 - \mathbf{R}_4 - \mathbf{R}_1 = 0 \tag{4.14a}$$

Compare equation 4.14a to equation 4.5a (p. 186) and note that the only difference is the sign of $\mathbf{R}_3$. This is due solely to the somewhat arbitrary choice of the sense of the position vector $\mathbf{R}_3$ in each case. The angle θ_3 must always be measured at the root of vector $\mathbf{R}_3$, and in this example it will be convenient to have that angle θ_3 at the joint labeled B. Once these arbitrary choices are made it is crucial that the resulting algebraic signs be carefully observed in the equations, or the results will be completely erroneous. Letting the vector magnitudes (link lengths) be represented by a, b, c, d as shown, we can substitute the complex number equivalents for the position vectors.

$$a e^{j\theta_2} - b e^{j\theta_3} - c e^{j\theta_4} - d e^{j\theta_1} = 0 \tag{4.14b}$$

Substitute the Euler equivalents:

$$a(\cos\theta_2 + j\sin\theta_2) - b(\cos\theta_3 + j\sin\theta_3)$$
$$-c(\cos\theta_4 + j\sin\theta_4) - d(\cos\theta_1 + j\sin\theta_1) = 0 \qquad (4.14c)$$

Separate the real and imaginary components:

real part (x component):

$$a\cos\theta_2 - b\cos\theta_3 - c\cos\theta_4 - d\cos\theta_1 = 0$$

but: $\theta_1 = 0$, so: $a\cos\theta_2 - b\cos\theta_3 - c\cos\theta_4 - d = 0 \qquad (4.15a)$

imaginary part (y component):

$$ja\sin\theta_2 - jb\sin\theta_3 - jc\sin\theta_4 - jd\sin\theta_1 = 0$$

but: $\theta_1 = 0$, and the $j's$ divide out, so: $\qquad\qquad (4.15b)$

$$a\sin\theta_2 - b\sin\theta_3 - c\sin\theta_4 = 0$$

We want to solve equations 4.15 simultaneously for the two unknowns, link length d and link angle θ_3. The independent variable is crank angle θ_2. Link lengths a and b, the offset c, and angle θ_4 are known. But note that since we set up the coordinate system to be parallel and perpendicular to the axis of the slider block, the angle θ_1 is zero and θ_4 is 90°. Equation 4.15b can be solved for θ_3 and the result substituted into equation 4.15a to solve for d. The solution is:

$$\theta_{3_1} = \arcsin\left(\frac{a\sin\theta_2 - c}{b}\right) \qquad (4.16a)$$

$$d = a\cos\theta_2 - b\cos\theta_3 \qquad (4.16b)$$

Note that there are again two valid solutions corresponding to the two circuits of the linkage. The arcsine function is multivalued. Its evaluation will give a value between ±90° representing only one circuit of the linkage. The value of d is dependent on the calculated value of θ_3. The value of θ_3 for the second circuit of the linkage can be found from:

$$\theta_{3_2} = \arcsin\left(-\frac{a\sin\theta_2 - c}{b}\right) + \pi \qquad (4.17)$$

✐ EXAMPLE 4-2

Position Analysis of a Fourbar Crank-Slider Linkage with the Vector Loop Method.

Problem: Given a fourbar crank-slider linkage with the link lengths $L_2 = a = 40$ mm, $L_3 = b = 120$ mm, *offset* $= c = -20$ mm. For $\theta_2 = 60°$ find all possible values of θ_3 and slider position d.

Solution: (See Figure 4-10 for nomenclature.)

1   Using equation 4.16a, calculate the link coupler angle θ_3 for the open configuration.

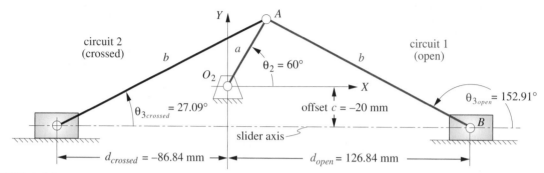

FIGURE 4-11

Solution to Example 4-2

$$\theta_{3_{open}} = \arcsin\left(\frac{a\sin\theta_2 - c}{b}\right) = \arcsin\left(\frac{40\sin(60°) - (-20)}{120}\right) = 152.91° \qquad (a)$$

2   Using equation 4.16b and the result from step 1, calculate slider position d for open linkage.

$$d = a\cos\theta_2 - b\cos\theta_3 = 40\cos(60°) - 120\cos(152.91°) = 126.84 \text{ mm} \qquad (b)$$

3   Using equation 4.17, calculate the link coupler angle θ_3 for the crossed configuration.

$$\theta_{3_{crossed}} = \arcsin\left(-\frac{a\sin\theta_2 - c}{b}\right) + \pi = \arcsin\left(-\frac{40\sin(60°) - (-20)}{120}\right) + \pi = 27.09° \qquad (c)$$

4   Using equation 4.16b and the result from step 3, calculate slider position d for crossed linkage.

$$d = a\cos\theta_2 - b\cos\theta_3 = 40\cos(60°) - 120\cos(27.09°) = -86.84 \text{ mm} \qquad (d)$$

5   Note that θ_3 is measured at the slider end of the coupler as shown in Figure 4-11.

4.7   THE FOURBAR SLIDER-CRANK POSITION SOLUTION

The *fourbar* **slider-crank** *linkage* has the same geometry as the *fourbar* **crank-slider** *linkage* that was analyzed in the previous section. The name change indicates that it will be driven with the slider as input and the crank as output. This is sometimes referred to as a "back-driven" crank-slider. We will use the term **slider-crank** to define it as slider-driven. This is a very commonly used linkage configuration. Every internal-combustion piston engine has as many of these as it has cylinders. The vector loop is as shown in Figure 4-10 (p. 192) and the vector loop equation is identical to equation 4.14a. But now we must solve this equation for θ_2 as a function of slider position d.

Start with equation 4.14a, make the substitutions of equation 4.14b and the simplifications of equations 4.15 to get the same simultaneous equation set:

$$a\cos\theta_2 - b\cos\theta_3 - c\cos\theta_4 - d = 0 \qquad (4.15a)$$
$$a\sin\theta_2 - b\sin\theta_3 - c\sin\theta_4 = 0 \qquad (4.15b)$$

but

$$\theta_4 = 90° \quad \therefore \quad \sin\theta_4 = 1, \quad \cos\theta_4 = 0$$

so

$$a\cos\theta_2 - b\cos\theta_3 - d = 0 \qquad (4.18a)$$
$$a\sin\theta_2 - b\sin\theta_3 - c = 0 \qquad (4.18b)$$

As was done in the fourbar linkage solution, isolate the θ_3 terms on one side, square both equations, and add them to eliminate θ_3.

$$b\cos\theta_3 = a\cos\theta_2 - d$$
$$b\sin\theta_3 = a\sin\theta_2 - c$$

square:
$$b^2 \cos^2\theta_3 = (a\cos\theta_2 - d)^2$$
$$b^2 \sin^2\theta_3 = (a\sin\theta_2 - c)^2$$

add:
$$b^2\left(\sin^2\theta_3 + \cos^2\theta_3\right) = (a\cos\theta_2 - d)^2 + (a\sin\theta_2 - c)^2$$
$$b^2 = (a\cos\theta_2 - d)^2 + (a\sin\theta_2 - c)^2$$
$$b^2 = a^2\cos^2\theta_2 - 2ad\cos\theta_2 + d^2 + a^2\sin^2\theta_2 - 2ac\sin\theta_2 + c^2$$
$$b^2 = a^2\left(\sin^2\theta_2 + \cos^2\theta_2\right) - 2ad\cos\theta_2 - 2ac\sin\theta_2 + c^2 + d^2$$
$$a^2 - b^2 + c^2 + d^2 - 2ac\sin\theta_2 - 2ad\cos\theta_2 = 0 \qquad (4.19)$$

To simplify, create some constant parameters:

let
$$K_1 = a^2 - b^2 + c^2 + d^2, \quad K_2 = -2ac, \quad K_3 = -2ad$$
then
$$K_1 + K_2\sin\theta_2 + K_3\cos\theta_2 = 0 \qquad (4.20)$$

As we did for the fourbar linkage, substitute the tangent half-angle identities (equation 4.9) for $\sin\theta_2$ and $\cos\theta_2$ to get the equation in terms of one trigonometric function.

$$K_1 + K_2\left(\frac{2\tan\frac{\theta_2}{2}}{1 + \tan^2\frac{\theta_2}{2}}\right) + K_3\left(\frac{1 - \tan^2\frac{\theta_2}{2}}{1 + \tan^2\frac{\theta_2}{2}}\right) = 0$$

simplify
$$(K_1 - K_3)\tan^2\frac{\theta_2}{2} + 2K_2\tan\frac{\theta_2}{2} + (K_1 + K_3) = 0$$

let
$$A = K_1 - K_3, \quad B = 2K_2, \quad C = K_1 + K_3$$

then
$$A\tan^2\frac{\theta_2}{2} + B\tan\frac{\theta_2}{2} + C = 0$$

and
$$\theta_{2_{1,2}} = 2\arctan\left(\frac{-B \pm \sqrt{B^2 - 4AC}}{2A}\right) \qquad (4.21)$$

Once θ_2 is known for a given value of d, θ_3 can be found from either equation 4.18a or 4.18b.

Note that there are two solutions to equation 4.21 representing the two branches of the linkage on the circuit to which the given value of slider position d applies.[*] The equation will fail when the backdriven slider-crank is at either top dead center (TDC) or bottom dead center (BDC). These are indeterminate change points between the branches at which the mathematics cannot predict which branch the linkage will go to next. A real slider-crank linkage can only make a full revolution of the crank if there is some stored energy in the crank to carry it through the dead centers twice per revolution. This is why you must spin a piston engine to start it and why they typically have a flywheel attached to the crankshaft to provide the angular momentum needed to pass through TDC and BDC.

4

[*] The crank-slider and slider-crank linkage both have two circuits or configurations in which they can be independently assembled, sometimes called open and crossed. Because effective link 4 is always perpendicular to the slider axis, it is parallel to itself on both circuits. This results in the two circuits being mirror images of one another, mirrored about a line through the crank pivot and perpendicular to the slide axis. Thus, the choice of value of slider position d in the calculation of the slider-crank linkage determines which circuit is being analyzed. But, because of the change points at TDC and BDC, the slider-crank has two branches on each circuit, and the two solutions obtained from equation 4.21 represent the two branches on the one circuit being analyzed. In contrast, the crank-slider has only one branch per circuit because when the crank is driven, it can make a full revolution and there are no change points to separate branches. See Section 4.13 (p. 208) for a more complete discussion of circuits and branches in linkages.

✏️ **EXAMPLE 4-3**

Position Analysis of a Fourbar Slider-Crank Linkage with the Vector Loop Method

Problem: Given a fourbar slider-crank linkage with the link lengths $L_2 = a = 40$ mm, $L_3 = b = 120$ mm, $offset = c = -20$ mm. For $d = 100$ mm, find all possible values of θ_2 and θ_3 on the circuit defined by the given value of d.

Solution: See Figure 4-9 (p. 191) for nomenclature.

1   Find the TDC and BDC positions of the linkage.

$$d_{BDC} = b - a = 120 - 40 = 80 \text{ mm}$$
$$d_{TDC} = b + a = 120 + 40 = 160 \text{ mm}$$

(a)

The requested position of $d = 100$ mm is within the range of motion of the slider-crank linkage and is neither TDC nor BDC, so equations 4.20 and 4.21 can be used.

2   Find the intermediate parameters needed from equations 4.20 and 4.21.

$$K_1 = a^2 - b^2 + c^2 + d^2 = 40^2 - 120^2 + (-20)^2 + 100^2 = -2400$$
$$K_2 = -2ac = -2(40)(-20) = 1600$$
$$K_3 = -2ad = -2(40)(100) = -8000$$
$$A = K_1 - K_3 = -2400 - (-8000) = 5600$$
$$B = 2K_2 = 2(1600) = 3200$$
$$C = K_1 + K_3 = -2400 + (-8000) = -10400$$

(b)

3   Find the two values of θ_2 from equation 4.21.

$$\theta_{2_1} = 2\tan^{-1}\left(\frac{-B + \sqrt{B^2 - 4AC}}{2A}\right) = 2\tan^{-1}\left(\frac{-3200 + \sqrt{3200^2 - 4(5600)(-10400)}}{2(5600)}\right) = 95.798°$$

(c)

$$\theta_{2_2} = 2\tan^{-1}\left(\frac{-B - \sqrt{B^2 - 4AC}}{2A}\right) = 2\tan^{-1}\left(\frac{-3200 - \sqrt{3200^2 - 4(5600)(-10400)}}{2(5600)}\right) = -118.418°$$

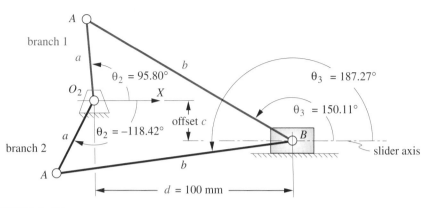

FIGURE 4-12

Solution to Example 4-3

4   Find the two values of θ_3 from either equation 4.16a or 4.17. Calculate θ_3 with both equations for one value of θ_2 and then use equation 4.16b with that result to determine which of the two equations gives the correct value of d to match the circuit of this linkage. Then use that equation with each of the θ_2 values to get the correct values of θ_3 for each branch of this circuit. This example needs equation 4.17 for its circuit.

$$\theta_{3_1} = \sin^{-1}\left(-\frac{a\sin\theta_{2_1} - c}{b}\right) + \pi = \sin^{-1}\left(-\frac{40\sin(95.798°) - (-20)}{120}\right) + \pi = 150.113°$$

$$\hspace{10cm} (d)$$

$$\theta_{3_2} = \cos^{-1}\left(\frac{a\sin\theta_{2_2} - c}{b}\right) + \pi = \cos^{-1}\left(\frac{40\sin(-118.418°) - (-20)}{120}\right) + \pi = 187.267°$$

5   The solution is shown in Figure 4-12.

4.8   AN INVERTED CRANK-SLIDER POSITION SOLUTION

Figure 4-13a* (p. 198) shows inversion #3 of the common fourbar crank-slider linkage in which the sliding joint is between links 3 and 4 at point B. This is shown as an **offset** crank-slider mechanism. The slider block has pure rotation with its center offset from the slide axis. (Figure 2-13c, p. 52, shows the nonoffset version of this linkage in which the vector $\mathbf{R}_4$ is zero.)

The global coordinate system is again taken with its origin at input crank pivot O_2 and the positive X axis along link 1, the ground link. A local axis system has been placed at point B in order to define θ_3. Note that there is a fixed angle γ within link 4 which defines the slot angle with respect to that link.

In Figure 4-13b (p. 198), the links have been represented as position vectors having senses consistent with the coordinate systems that were chosen for convenience in defining the link angles. This particular arrangement of position vectors leads to the same vector loop equation as the previous crank-slider example.

* This figure is provided as animated AVI and Working Model files on the DVD. Its filename is the same as the figure number.

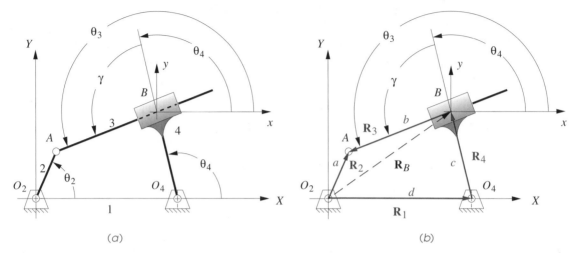

FIGURE 4-13

Inversion #3 of the slider-crank fourbar linkage

Equations 4.14 and 4.15 (pp. 192–193) apply to this inversion as well. Note that the absolute position of point B is defined by vector $\mathbf{R}_B$ which varies in both magnitude and direction as the linkage moves. We choose to represent $\mathbf{R}_B$ as the vector difference $\mathbf{R}_2 - \mathbf{R}_3$ in order to use the actual links as the position vectors in the loop equation.

All slider linkages will have at least one link whose effective length between joints will vary as the linkage moves. In this example the length of link 3 between points A and B, designated as b, will change as it passes through the slider block on link 4. Thus the value of b will be one of the variables to be solved for in this inversion. Another variable will be θ_4, the angle of link 4. Note however, that we also have an unknown in θ_3, the angle of link 3. This is a total of three unknowns. Equations 4.15 can only be solved for two unknowns. Thus we require another equation to solve the system. There is a fixed relationship between angles θ_3 and θ_4, shown as γ in Figure 4-10 (p. 192), which gives the equation:

$$\theta_3 = \theta_4 + \gamma \qquad (4.22)$$

Repeating equations 4.15 and renumbering them for the reader's convenience:

$$a\cos\theta_2 - b\cos\theta_3 - c\cos\theta_4 - d = 0 \qquad (4.23a)$$
$$a\sin\theta_2 - b\sin\theta_3 - c\sin\theta_4 = 0 \qquad (4.23b)$$

These have only two unknowns and can be solved simultaneously for θ_4 and b. Equation 4.23b can be solved for link length b and substituted into equation 4.23a.

$$b = \frac{a\sin\theta_2 - c\sin\theta_4}{\sin\theta_3} \qquad (4.24a)$$

$$a\cos\theta_2 - \frac{a\sin\theta_2 - c\sin\theta_4}{\sin\theta_3}\cos\theta_3 - c\cos\theta_4 - d = 0 \qquad (4.24b)$$

Substitute equation 4.22 and after some algebraic manipulation, equation 4.24 can be reduced to:

$$P\sin\theta_4 + Q\cos\theta_4 + R = 0$$

where   (4.25)

$$P = a\sin\theta_2 \sin\gamma + (a\cos\theta_2 - d)\cos\gamma$$
$$Q = -a\sin\theta_2 \cos\gamma + (a\cos\theta_2 - d)\sin\gamma$$
$$R = -c\sin\gamma$$

Note that the factors P, Q, R are constant for any input value of θ_2. To solve this for θ_4, it is convenient to substitute the tangent half angle identities (equation 4.9, p. 188) for the $\sin\theta_4$ and $\cos\theta_4$ terms. This will result in a quadratic equation in $\tan(\theta_4/2)$ which can be solved for the two values of θ_4.

$$P\frac{2\tan\left(\dfrac{\theta_4}{2}\right)}{1+\tan^2\left(\dfrac{\theta_4}{2}\right)} + Q\frac{1-\tan^2\left(\dfrac{\theta_4}{2}\right)}{1+\tan^2\left(\dfrac{\theta_4}{2}\right)} + R = 0 \qquad (4.26a)$$

This reduces to:

$$(R-Q)\tan^2\left(\frac{\theta_4}{2}\right) + 2P\tan\left(\frac{\theta_4}{2}\right) + (Q+R) = 0$$

let

$$S = R - Q, \qquad T = 2P, \qquad U = Q + R$$

then

$$S\tan^2\left(\frac{\theta_4}{2}\right) + T\tan\left(\frac{\theta_4}{2}\right) + U = 0 \qquad (4.26b)$$

and the solution is:

$$\theta_{4_{1,2}} = 2\arctan\left(\frac{-T \pm \sqrt{T^2 - 4SU}}{2S}\right) \qquad (4.26c)$$

As was the case with the previous examples, this also has a crossed and an open solution represented by the plus and minus signs on the radical. Note that we must also calculate the values of link length b for each θ_4 by using equation 4.24a. The coupler angle θ_3 is found from equation 4.22 (p. 198).

4.9   LINKAGES OF MORE THAN FOUR BARS

With some exceptions,* the same approach as shown here for the fourbar linkage can be used for any number of links in a closed-loop configuration. More complicated linkages may have multiple loops which will lead to more equations to be solved simultaneously and may require an iterative solution. Alternatively, Wampler [10] presents a new, general, noniterative method for the analysis of planar mechanisms containing any number of rigid links connected by rotational and/or translational joints.

* Waldron and Sreenivasan[1] report that the common solution methods for position analysis are not general, i.e., are not extendable to n-link mechanisms. Conventional position analysis methods, such as those used here, rely on the presence of a fourbar loop in the mechanism that can be solved first, followed by a decomposition of the remaining links into a series of dyads. Not all mechanisms contain fourbar loops. (One eightbar, 1-DOF linkage contains no fourbar loops—see the 16th isomer at lower right in Figure 2-11d on p. 50). Even if there is a fourbar loop, its pivots may not be grounded, requiring that the linkage be inverted to start the solution. Also, if the driving joint is not in the fourbar loop, then interpolation is needed to solve for link positions.

The Geared Fivebar Linkage

Another example, which can be reduced to two equations in two unknowns, is the **geared fivebar linkage**, which was introduced in Section 2.14 (p. 62) and is shown in Figure 4-14a and program LINKAGES disk file F04-11.5br. The vector loop for this linkage is shown in Figure 4-14b. It obviously has one more position vector than the fourbar. Its vector loop equation is:

$$\mathbf{R}_2 + \mathbf{R}_3 - \mathbf{R}_4 - \mathbf{R}_5 - \mathbf{R}_1 = 0 \tag{4.27a}$$

Note that the vector senses are again chosen to suit the analyst's desires to have the vector angles defined at a convenient end of the respective link. Equation 4.27b substitutes the complex polar notation for the position vectors in equation 4-23a, using a, b, c, d, f to represent the scalar lengths of the links as shown in Figure 4-14.

$$a e^{j\theta_2} + b e^{j\theta_3} - c e^{j\theta_4} - d e^{j\theta_5} - f e^{j\theta_1} = 0 \tag{4.27b}$$

Note also that this vector loop equation has three unknown variables in it, namely the angles of links 3, 4, and 5. (The angle of link 2 is the input, or independent, variable, and link 1 is fixed with constant angle.) Since a two-dimensional vector equation can only be solved for two unknowns, we will need another equation to solve this system. Because this is a geared fivebar linkage, there exists a relationship between the two geared links, here links 2 and 5. Two factors determine how link 5 behaves with respect to link 2, namely, the **gear ratio** λ and the **phase angle** ϕ. The relationship is:

$$\theta_5 = \lambda \theta_2 + \phi \tag{4.27c}$$

This allows us to express θ_5 in terms of θ_2 in equation 4.27b and reduce the unknowns to two by substituting equation 4.27c into equation 4.27b.

$$a e^{j\theta_2} + b e^{j\theta_3} - c e^{j\theta_4} - d e^{j(\lambda\theta_2 + \phi)} - f e^{j\theta_1} = 0 \tag{4.28a}$$

Note that the gear ratio λ is the ratio of the diameters of the gears connecting the two links ($\lambda = dia_2 / dia_5$), and the phase angle ϕ is the *initial angle* of link 5 with respect to link 2. When link 2 is at zero degrees, link 5 is at the **phase angle** ϕ. Equation 4.27c defines the relationship between θ_2 and θ_5. Both λ and ϕ are design parameters selected by the design engineer along with the link lengths. With these parameters defined, the only unknowns left in equation 4.28 are θ_3 and θ_4.

The behavior of the geared fivebar linkage can be modified by changing the link lengths, the gear ratio, or the phase angle. The phase angle can be changed simply by lifting the gears out of engagement, rotating one gear with respect to the other, and re-engaging them. Since links 2 and 5 are rigidly attached to gears 2 and 5, respectively, their relative angular rotations will be changed also. It is this fact that results in different positions of links 3 and 4 with any change in phase angle. The coupler curve's shapes will also change with variation in any of these parameters as can be seen in Figure 3-23 (p. 131) and in Appendix E.

The procedure for solution of this vector loop equation is the same as that used for the fourbar linkage:

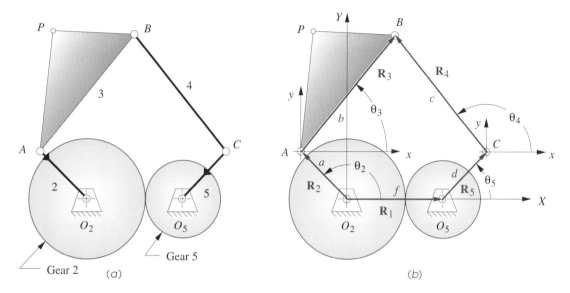

FIGURE 4-14

The geared fivebar linkage and its vector loop

1 Substitute the Euler equivalent (equation 4.4a, p. 185) into each term in the vector loop equation 4.28a.

$$a(\cos\theta_2 + j\sin\theta_2) + b(\cos\theta_3 + j\sin\theta_3) - c(\cos\theta_4 + j\sin\theta_4)$$
$$-d\left[\cos(\lambda\theta_2 + \phi) + j\sin(\lambda\theta_2 + \phi)\right] - f(\cos\theta_1 + j\sin\theta_1) = 0 \qquad (4.28b)$$

2 Separate the real and imaginary parts of the cartesian form of the vector loop equation.

$$a\cos\theta_2 + b\cos\theta_3 - c\cos\theta_4 - d\cos(\lambda\theta_2 + \phi) - f\cos\theta_1 = 0 \qquad (4.28c)$$
$$a\sin\theta_2 + b\sin\theta_3 - c\sin\theta_4 - d\sin(\lambda\theta_2 + \phi) - f\sin\theta_1 = 0 \qquad (4.28d)$$

3 Rearrange to isolate one unknown (either θ_3 or θ_4) in each scalar equation. Note that θ_1 is zero.

$$b\cos\theta_3 = -a\cos\theta_2 + c\cos\theta_4 + d\cos(\lambda\theta_2 + \phi) + f \qquad (4.28e)$$
$$b\sin\theta_3 = -a\sin\theta_2 + c\sin\theta_4 + d\sin(\lambda\theta_2 + \phi) \qquad (4.28f)$$

4 Square both equations and add them to eliminate one unknown, say θ_3.

$$b^2 = 2c\left[d\cos(\lambda\theta_2 + \phi) - a\cos\theta_2 + f\right]\cos\theta_4$$
$$+ 2c\left[d\sin(\lambda\theta_2 + \phi) - a\sin\theta_2\right]\sin\theta_4$$
$$+ a^2 + c^2 + d^2 + f^2 - 2af\cos\theta_2$$
$$- 2d(a\cos\theta_2 - f)\cos(\lambda\theta_2 + \phi)$$
$$- 2ad\sin\theta_2 \sin(\lambda\theta_2 + \phi) \qquad (4.28g)$$

5   Substitute the tangent half-angle identities (equation 4.9, p. 188) for the sine and cosine terms and manipulate the resulting equation in the same way as was done for the fourbar linkage in order to solve for θ_4.

$$A = 2c\left[d\cos\left(\lambda\theta_2 + \phi\right) - a\cos\theta_2 + f\right]$$

$$B = 2c\left[d\sin\left(\lambda\theta_2 + \phi\right) - a\sin\theta_2\right]$$

$$C = a^2 - b^2 + c^2 + d^2 + f^2 - 2af\cos\theta_2$$
$$\qquad - 2d\left(a\cos\theta_2 - f\right)\cos\left(\lambda\theta_2 + \phi\right) - 2ad\sin\theta_2\sin\left(\lambda\theta_2 + \phi\right)$$

$$D = C - A, \qquad\qquad E = 2B, \qquad\qquad F = A + C$$

$$\theta_{4_{1,2}} = 2\arctan\left(\frac{-E \pm \sqrt{E^2 - 4DF}}{2D}\right) \qquad\qquad (4.28\text{h})$$

6   Repeat steps 3 to 5 for the other unknown angle θ_3.

$$G = 2b\left[a\cos\theta_2 - d\cos\left(\lambda\theta_2 + \phi\right) - f\right]$$

$$H = 2b\left[a\sin\theta_2 - d\sin\left(\lambda\theta_2 + \phi\right)\right]$$

$$K = a^2 + b^2 - c^2 + d^2 + f^2 - 2af\cos\theta_2$$
$$\qquad - 2d\left(a\cos\theta_2 - f\right)\cos\left(\lambda\theta_2 + \phi\right)$$
$$\qquad - 2ad\sin\theta_2\sin\left(\lambda\theta_2 + \phi\right)$$

$$L = K - G; \quad M = 2H; \quad N = G + K$$

$$\theta_{3_{1,2}} = 2\arctan\left(\frac{-M \pm \sqrt{M^2 - 4LN}}{2L}\right) \qquad\qquad (4.28\text{i})$$

Note that these derivation steps are essentially identical to those for the pin-jointed fourbar linkage once θ_2 is substituted for θ_5 using equation 4.27c (p. 200).

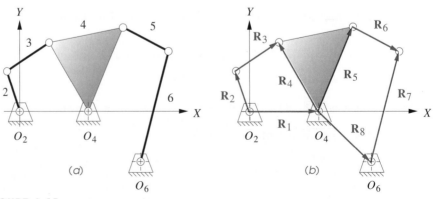

FIGURE 4-15

Watt's sixbar linkage and vector loop

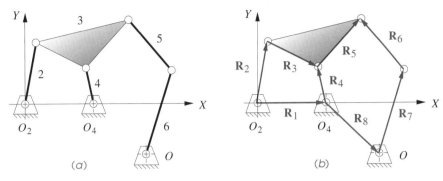

FIGURE 4-16

Stephenson's sixbar linkage and vector loops

Sixbar Linkages

WATT'S SIXBAR   is essentially two fourbar linkages in series, as shown in Figure 4-15a, and can be analyzed as such. Two vector loops are drawn as shown in Figure 4-15b. These vector loop equations can be solved in succession with the results of the first loop applied as input to the second loop. Note that there is a constant angular relationship between vectors $\mathbf{R}_4$ and $\mathbf{R}_5$ within link 4. The solution for the fourbar linkage (equations 4.10 and 4.13, pp. 188 and 189, respectively) is simply applied twice in this case. Depending on the inversion of the Watts linkage being analyzed, there may be two four-link loops or one four-link and one five-link loop. (See Figure 2-14, p. 54.) In either case, if the four-link loop is analyzed first, there will not be more than two unknown link angles to be found at one time.

STEPHENSON'S SIXBAR   is a more complicated mechanism to analyze. Two vector loops can be drawn, but depending on the inversion being analyzed, either one or both loops will have five links[*] and three unknown angles as shown in Figure 4-13a and b (p. 198). However, the two loops will have at least one nonground link in common and so a solution can be found. In the other cases an iterative solution such as a Newton-Raphson method (see Section 4.14, p. 210) must be used to find the roots of the equations. Program LINKAGES is limited to the inversions which allow a closed-form solution, one of which is shown in Figure 4-16, and it does not do the iterative solution.

4.10   POSITION OF ANY POINT ON A LINKAGE

Once the angles of all the links are found, it is simple and straightforward to define and calculate the position of any point on any link for any input position of the linkage. Figure 4-17 shows a fourbar linkage whose coupler, link 3, is enlarged to contain a coupler point P. The crank and rocker have also been enlarged to show points S and U which might represent the centers of gravity of those links. We want to develop algebraic expressions for the positions of these (or any) points on the links.

---

[*] See footnote on p. 199.

FIGURE 4-17

Positions of points on the links

To find the position of point S, draw a position vector from the fixed pivot O_2 to point S. This vector $\mathbf{R}_{SO_2}$ makes an angle δ_2 with the vector $\mathbf{R}_{AO_2}$. This angle δ_2 is completely defined by the geometry of link 2 and is constant. The position vector for point S is then:

$$\mathbf{R}_{SO_2} = \mathbf{R}_S = se^{j(\theta_2 + \delta_2)} = s\left[\cos(\theta_2 + \delta_2) + j\sin(\theta_2 + \delta_2)\right] \qquad (4.29)$$

The position of point U on link 4 is found in the same way, using the angle δ_4 which is a constant angular offset within the link. The expression is:

$$\mathbf{R}_{UO_4} = ue^{j(\theta_4 + \delta_4)} = u\left[\cos(\theta_4 + \delta_4) + j\sin(\theta_4 + \delta_4)\right] \qquad (4.30)$$

The position of point P on link 3 can be found from the addition of two position vectors $\mathbf{R}_A$ and $\mathbf{R}_{PA}$. Vector $\mathbf{R}_A$ is already defined from our analysis of the link angles in equation 4.5 (p. 186). Vector $\mathbf{R}_{PA}$ is the relative position of point P with respect to point A. Vector $\mathbf{R}_{PA}$ is defined in the same way as $\mathbf{R}_S$ or $\mathbf{R}_U$, using the internal link offset angle δ_3 and the position angle of link 3, θ_3.

$$\mathbf{R}_{PA} = pe^{j(\theta_3 + \delta_3)} = p\left[\cos(\theta_3 + \delta_3) + j\sin(\theta_3 + \delta_3)\right] \qquad (4.31a)$$

$$\mathbf{R}_P = \mathbf{R}_A + \mathbf{R}_{PA} \qquad (4.31b)$$

Compare equation 4.31b with equations 4.1 (p. 178). Equation 4.31b is the position difference equation.

4.11 TRANSMISSION ANGLES

The transmission angle was defined in Section 3.3 (p. 100) for a fourbar linkage. That definition is repeated here for your convenience.

The **transmission angle** μ is shown in Figure 3-3a (p. 102) and is defined as *the angle between the output link and the coupler*. It is usually taken as the absolute value of the acute angle of the pair of angles at the intersection of the two links and varies continuously from some minimum to some maximum value as the linkage goes through its range of motion. It is a measure of the quality of force transmission at the joint.[*]

We will expand that definition here to represent the angle between any two links in a linkage, as a linkage can have many transmission angles. The angle between any output link and the coupler which drives it is a transmission angle. Now that we have developed the analytic expressions for the angles of all the links in a mechanism, it is easy to define the transmission angle algebraically. It is merely the difference between the angles of the two joined links through which we wish to pass some force or velocity. For our fourbar linkage example it will be the difference between θ_3 and θ_4. By convention we take the absolute value of the difference and force it to be an acute angle.

$$\theta_{trans} = |\theta_3 - \theta_4|$$

$$\text{if} \quad \theta_{trans} > \frac{\pi}{2} \qquad \text{then} \quad \mu = \pi - \theta_{trans} \quad \text{else} \quad \mu = \theta_{trans} \qquad (4.32)$$

This computation can be done for any joint in a linkage by using the appropriate link angles.

Extreme Values of the Transmission Angle

For a Grashof crank-rocker fourbar linkage the minimum value of the transmission angle will occur when the crank is colinear with the ground link as shown in Figure 4-18. The values of the transmission angle in these positions are easily calculated from the law of cosines since the linkage is then in a triangular configuration. The sides of the two triangles are link 3, link 4, and either the sum or difference of links 1 and 2. Depending on the linkage geometry, the minimum value of the transmission angle μ_{min} will occur either when links 1 and 2 are *colinear and overlapping* as shown in Figure 4-18a or when links 1 and 2 are *colinear and nonoverlapping* as shown in Figure 4-18b. Using notation consistent with Section 4.5 (p. 183) and Figure 4-7 (p. 185) we will label the links:

[*] The transmission angle has limited application. It only predicts the quality of force or torque transmission if the input and output links are pivoted to ground. If the output force is taken from a floating link (coupler), then the transmission angle is of no value. A different index of merit called the joint force index (JFI) is presented in Chapter 11 which discusses force analysis in linkages. (See Section 11.12 p. 611.) The JFI is useful for situations in which the output link is floating as well as giving the same kind of information when the output is taken from a link rotating against the ground. However, the JFI requires a complete force analysis of the linkage be done whereas the transmission angle is determined from linkage geometry alone.

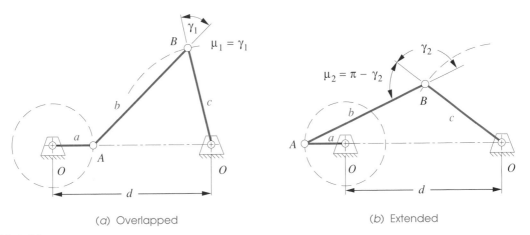

(a) Overlapped   (b) Extended

FIGURE 4-18

The minimum transmission angle in the Grashof crank-rocker fourbar linkage occurs in one of two positions

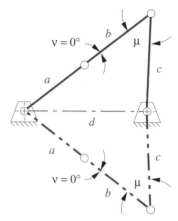

(a) Toggle positions for links b and c (b) Toggle positions for links a and b

FIGURE 4-19

Non-Grashof triple-rocker linkages in toggle

a = link 2, b = link 3, c = link 4, d = link 1

For the overlapping case (Figure 4-15a, p. 202) the cosine law gives

$$\mu_1 = \gamma_1 = \arccos\left[\frac{b^2 + c^2 - (d-a)^2}{2bc}\right] \qquad (4.33a)$$

and for the extended case, the cosine law gives

$$\mu_2 = \pi - \gamma_2 = \pi - \arccos\left[\frac{b^2 + c^2 - (d+a)^2}{2bc}\right] \qquad (4.33b)$$

The minimum transmission angle μ_{min} in a Grashof crank-rocker linkage is then the smaller of μ_1 and μ_2.

For a **Grashof double-rocker** linkage the transmission angle can vary from 0 to 90 degrees because the coupler can make a full revolution with respect to the other links. For a **non-Grashof triple-rocker** linkage the transmission angle will be zero degrees in the toggle positions which occur when the output rocker c and the coupler b are colinear as shown in Figure 4-19a. In the other toggle positions when input rocker a and coupler b are colinear (Figure 4-19b), the transmission angle can be calculated from the cosine law as:

when $\nu = 0,$

$$\mu = \arccos\left[\frac{(a+b)^2 + c^2 - d^2}{2c(a+b)}\right] \qquad (4.34)$$

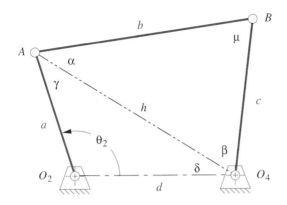

FIGURE 4-20

Finding the crank angle corresponding to the toggle positions

This is not the smallest value that the transmission angle μ can have in a triple-rocker, as that will obviously be zero. Of course, when analyzing any linkage, the transmission angles can easily be computed and plotted for all positions using equation 4.32. Program LINKAGES does this. The student should investigate the variation in transmission angle for the example linkages in those programs. Disk file F04-15.4br can be opened in program LINKAGES to observe that linkage in motion.

4.12 TOGGLE POSITIONS

The input link angles which correspond to the toggle positions (stationary configurations) of the **non-Grashof triple-rocker** can be calculated by the following method, using trigonometry. Figure 4-20 shows a non-Grashof fourbar linkage in a general position. A construction line h has been drawn between points A and O_4. This divides the quadrilateral loop into two triangles, O_2AO_4 and ABO_4. Equation 4.35 uses the cosine law to express the transmission angle μ in terms of link lengths and the input link angle θ_2.

$$h^2 = a^2 + d^2 - 2ad\cos\theta_2$$

also:

$$h^2 = b^2 + c^2 - 2bc\cos\mu$$

so:

$$a^2 + d^2 - 2ad\cos\theta_2 = b^2 + c^2 - 2bc\cos\mu$$

and:

$$\cos\mu = \frac{b^2 + c^2 - a^2 - d^2}{2bc} + \frac{ad}{bc}\cos\theta_2 \qquad (4.35)$$

To find the maximum and minimum values of input angle θ_2, we can differentiate equation 4.35, form the derivative of θ_2 with respect to μ, and set it equal to zero.

$$\frac{d\theta_2}{d\mu} = \frac{bc}{ad}\frac{\sin\mu}{\sin\theta_2} = 0 \qquad (4.36)$$

The link lengths a, b, c, d are never zero, so this expression can only be zero when $\sin\mu$ is zero. This will be true when angle μ in Figure 4-20 is either zero or 180°. This is consistent with the definition of toggle given in Section 3.3 (p. 100). If μ is zero or

180° then cos μ will be ±1. Substituting these two values for cos μ into equation 4.35 will give a solution for the value of θ_2 between zero and 180° which corresponds to the toggle position of a triple-rocker linkage when driven from one rocker.

$$\cos\mu = \frac{b^2 + c^2 - a^2 - d^2}{2bc} + \frac{ad}{bc}\cos\theta_2 = \pm 1$$

or:

$$\cos\theta_2 = \frac{a^2 + d^2 - b^2 - c^2}{2ad} \pm \frac{bc}{ad} \qquad (4.37)$$

and:

$$\theta_{2_{toggle}} = \arccos\left(\frac{a^2 + d^2 - b^2 - c^2}{2ad} \pm \frac{bc}{ad}\right) \qquad 0 \le \theta_{2_{toggle}} \le \pi$$

One of these ± cases will produce an argument for the arccosine function which lies between ±1. The toggle angle which is in the first or second quadrant can be found from this value. The other toggle angle will then be the negative of the one found, due to the mirror symmetry of the two toggle positions about the ground link as shown in Figure 4-19 (p. 206). Program LINKAGES computes the values of these toggle angles for any non-Grashof linkage.

4.13   CIRCUITS AND BRANCHES IN LINKAGES

In Section 4.5 (p. 183) it was noted that the fourbar linkage position problem has two solutions which correspond to the two circuits of the linkage. This section will explore the topics of circuits and branches in linkages in greater detail.

Chase and Mirth[2] define a **circuit** in a linkage as "*all possible orientations of the links that can be realized without disconnecting any of the joints*" and a **branch** as "*a continuous series of positions of the mechanism on a circuit between two stationary configurations . . . The stationary configurations divide a circuit into a series of branches.*" A linkage may have one or more circuits each of which may contain one or more branches. The number of circuits corresponds to the number of solutions possible from the position equations for the linkage.

Circuit defects are fatal to linkage operation, but branch defects are not. A mechanism that must change circuits to move from one desired position to the other (referred to as a **circuit defect**) is not useful as it cannot do so without disassembly and reassembly. A mechanism that changes branch when moving from one circuit to another (referred to as a **branch defect**) may or may not be usable depending on the designer's intent.

The tailgate linkage shown in Figure 3-2 (p. 101) is an example of a linkage with a deliberate branch defect in its range of motion (actually at the limit of its range of motion). The toggle position (stationary configuration) that it reaches with the tailgate fully open serves to hold it open. But the user can move it out of this stationary configuration by rotating one of the links out of toggle. Folding chairs and tables often use a similar scheme as do fold-down seats in automobiles.

Another example of a common linkage with a branch defect is the slider-crank linkage (crankshaft, connecting rod, and slider driving) used in every piston engine and shown

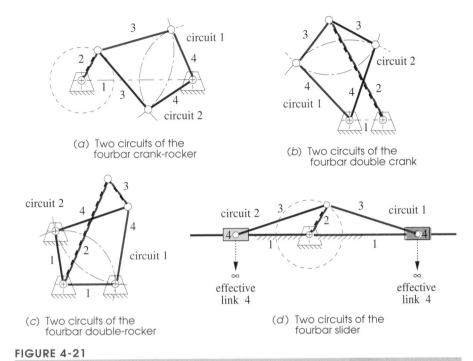

(a) Two circuits of the
    fourbar crank-rocker

(b) Two circuits of the
    fourbar double crank

(c) Two circuits of the
    fourbar double-rocker

(d) Two circuits of the
    fourbar slider

FIGURE 4-21

Circuits of the fourbar linkage

TABLE 4-1
Circuits & Branches
In the Fourbar Linkage

Fourbar Linkage Type	Number of Circuits	Branches per Circuit
Non-Grashof triple-rocker	1	2
Grashof* crank-rocker	2	1
Grashof* double-crank	2	1
Grashof* double-rocker	2	2
Grashof* rocker-crank	2	2

* Valid only for non-special-case Grashof linkages.

TABLE 4-2
Circuits & Branches
In the Fourbar Slider

Fourbar Slider Type	Number of Circuits	Branches per Circuit
Crank-slider	2	1
Slider-crank	2	2

in Figure 13-3 (p. 663). This linkage has two toggle positions (top and bottom dead center) giving it two branches within one revolution of its crank. It works nevertheless because it is carried through these stationary configurations by the angular momentum of the rotating crank and its attached flywheel. One penalty is that the engine must be spun to start it in order to build sufficient momentum to carry it through these toggle positions.

The Watt sixbar linkage can have four circuits, and the Stephenson sixbar can have either four or six circuits depending on which link is driving. Eightbar linkages can have as many as 16 or 18 circuits, not all of which may be real, however.[2]

The number of circuits and branches in the fourbar linkage depends on its Grashof condition and the inversion used. A non-Grashof, triple-rocker fourbar linkage has only one circuit but has two branches. All Grashof fourbar linkages have two circuits, but the number of branches per circuit differs with the inversion. The crank-rocker and double-crank have only one branch within each circuit. The double-rocker and rocker-crank have two branches within each circuit. Table 4-1 summarizes these relationships.[2] Table 4-2 shows the circuits and branches for the two configurations of the fourbar slider linkage. Figure 4-21 shows the circuits for the Grashof fourbar linkage and the fourbar slider.

Any solution for the position of a linkage must take into account the number of possible circuits that it contains. A closed-form solution, if available, will contain all the circuits. An iterative solution such as is described in the next section will only yield the position data for one circuit, and it may not be the one you expect.

4.14   NEWTON-RAPHSON SOLUTION METHOD

The solution methods for position analysis shown so far in this chapter are all of "closed form," meaning that they provide the solution with a direct, noniterative approach.[*]   In some situations, particularly with multiloop mechanisms, a closed-form solution may not be attainable. Then an alternative approach is needed, and the Newton-Raphson method (sometimes just called Newton's method) provides one that can solve sets of simultaneous nonlinear equations. Any iterative solution method requires that one or more guess values be provided to start the computation. It then uses the guess values to obtain a new solution that may be closer to the correct one. This process is repeated until it converges to a solution close enough to the correct one for practical purposes. However, there is no guarantee that an iterative method will converge at all. It may diverge, taking successive solutions further from the correct one, especially if the initial guess is not sufficiently close to the real solution.

Though we will need to use the multidimensional (Newton-Raphson) version of Newton's method for these linkage problems, it is easier to understand how the algorithm works by first discussing the one-dimensional Newton method for finding the roots of a single nonlinear function in one independent variable. Then we will discuss the multidimensional Newton-Raphson method.

One-Dimensional Root-Finding (Newton's Method)

A nonlinear function may have multiple roots, where a root is defined as the intersection of the function with any straight line. Typically the zero axis of the independent variable is the straight line for which we desire the roots. Take, for example, a cubic polynomial which will have three roots, with either one or all three being real.

$$y = f(x) = -x^3 - 2x^2 + 50x + 60 \qquad (4.38)$$

There is a closed-form solution for the roots of a cubic function[†] which allows us to calculate in advance that the roots of this particular cubic are all real and are $x = -7.562$, -1.177, and 6.740.

Figure 4-22 shows this function plotted over a range of x. In Figure 4-22a, an initial guess value of $x_1 = 1.8$ is chosen. Newton's algorithm evaluates the function for this guess value, finding y_1. The value of y_1 is compared to a user-selected tolerance (say 0.001) to see if it is close enough to zero to call x_1 the root. If not, then the slope (m) of the function at x_1, y_1 is calculated either by using an analytic expression for the derivative of the function or by doing a numerical differentiation (less desirable). The equation of the tangent line is then evaluated to find its intercept at x_2 which is used as a new guess value. The above process is repeated, finding y_2; testing it against the user selected tolerance; and, if it is too large, calculating another tangent line whose x intercept is used as a new guess value. This process is repeated until the value of the function y_i at the latest x_i is close enough to zero to satisfy the user.

The Newton algorithm described above can be expressed algebraically (in pseudocode) as shown in equation 4.39. The function for which the roots are sought is $f(x)$, and its derivative is $f'(x)$. The slope m of the tangent line is equal to $f'(x)$ at the current point x_i, y_i.

[*] Kramer [3] states that "In theory, any nonlinear algebraic system of equations can be manipulated into the form of a single polynomial in one unknown. The roots of this polynomial can then be used to determine all unknowns in the system. However, if the derived polynomial is greater than degree four, factoring and/or some form of iteration are necessary to obtain the roots. In general, systems that have more than a fourth degree polynomial associated with the eliminant of all but one variable must be solved by iteration. However, if factoring of the polynomial into terms of degree four or less is possible, all roots may be found without iteration. Therefore the only truly symbolic solutions are those that can be factored into terms of fourth degree or less. This is the formal definition of a closed form solution."

[†] Viete's method from "De Emendatione" by Francois Viete (1615) as described in reference [4].

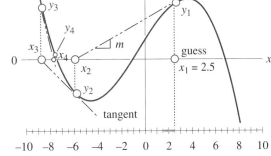

(a) A guess of x = 1.8 converges to the root at x = -1.177   (b) A guess of x = 2.5 converges to the root at x = -7.562

FIGURE 4-22

Newton-Raphson method of solution for roots of nonlinear functions

step 1 $y_i = f(x_i)$

step 2   IF $y_i \leq tolerance$ THEN STOP

step 3 $m = f'(x_i)$

step 4 $x_{i+1} = x_i - \dfrac{y_i}{m}$

step 5 $y_{i+1} = f(x_{i+1})$

step 6   IF $y_{i+1} \leq tolerance$ THEN STOP

          ELSE $x_i = x_{i+1}$: $y_i = y_{i+1}$:   GOTO step 1   (4.39)

If the initial guess value is close to a root, this algorithm will converge rapidly to the solution. However, it is quite sensitive to the initial guess value. Figure 4-22b shows the result of a slight change in the initial guess from $x_1 = 1.8$ to $x_1 = 2.5$. With this slightly different guess it converges to another root. Note also that if we choose an initial guess of $x_1 = 3.579$ which corresponds to a local maximum of this function, the tangent line will be horizontal and will not intersect the x axis at all. The method fails in this situation. Can you suggest a value of x_1 that would cause it to converge to the root at $x = 6.74$?

So this method has its drawbacks. It may fail to converge. It may behave chaotically.* It is sensitive to the guess value. It also is incapable of distinguishing between multiple circuits in a linkage. The circuit solution it finds is dependent on the initial guess. It requires that the function be differentiable, and the derivative as well as the function must be evaluated at every step. Nevertheless, it is the method of choice for functions whose derivatives can be efficiently evaluated and which are continuous in the region of the root. Furthermore, it is about the only choice for systems of nonlinear equations.

* Kramer[3] points out that "the Newton Raphson algorithm can exhibit chaotic behavior when there are multiple solutions to kinematic constraint equations. . . . Newton Raphson has no mechanism for distinguishing between the two solutions" (circuits). He does an experiment with just two links, exactly analogous to finding the angles of the coupler and rocker in the fourbar linkage position problem, and finds that the initial guess values need to be quite close to the desired solution (one of the two possible circuits) to avoid divergence or chaotic oscillation between the two solutions.

Multidimensional Root-Finding (Newton-Raphson Method)

The one-dimensional Newton method is easily extended to multiple, simultaneous, non-linear equation sets and is then called the Newton-Raphson method. First, let's generalize the expression developed for the one-dimensional case in step 4 of equation 4.39. Refer also to Figure 4-22 (p. 211).

$$x_{i+1} = x_i - \frac{y_i}{m} \qquad \text{or} \qquad m\left(x_{i+1} - x_i\right) = -y_i$$

but: $\qquad y_i = f(x_i) \qquad m = f'(x_i) \qquad x_{i+1} - x_i = \Delta x$

substituting: $\qquad f'(x_i) \cdot \Delta x = -f(x_i)$ $\qquad\qquad$ (4.40)

Here a Δx term is introduced which will approach zero as the solution converges. The Δx term rather than y_i will be tested against a selected tolerance in this case. Note that this form of the equation avoids the division operation which is acceptable in a scalar equation but impossible with a matrix equation.

A multidimensional problem will have a set of equations of the form

$$\begin{bmatrix} f_1(x_1, x_2, x_3, \ldots, x_n) \\ f_2(x_1, x_2, x_3, \ldots, x_n) \\ \vdots \qquad\qquad \vdots \\ f_n(x_1, x_2, x_3, \ldots, x_n) \end{bmatrix} = \mathbf{B} \qquad\qquad (4.41)$$

where the set of equations constitutes a vector, here called $\mathbf{B}$.

Partial derivatives are required to obtain the slope terms

$$\begin{bmatrix} \dfrac{\partial f_1}{\partial x_1} & \dfrac{\partial f_1}{\partial x_2} & \cdots & \dfrac{\partial f_1}{\partial x_n} \\ \vdots & \vdots & & \vdots \\ \dfrac{\partial f_n}{\partial x_1} & \dfrac{\partial f_n}{\partial x_2} & \cdots & \dfrac{\partial f_n}{\partial x_n} \end{bmatrix} = \mathbf{A} \qquad\qquad (4.42)$$

which form the *Jacobian matrix* of the system, here called $\mathbf{A}$.

The error terms are also a vector, here called $\mathbf{X}$.

$$\begin{bmatrix} \Delta x_1 \\ \Delta x_2 \\ \vdots \\ \Delta x_n \end{bmatrix} = \mathbf{X} \qquad\qquad (4.43)$$

Equation 4.40 then becomes a matrix equation for the multidimensional case.

$$\mathbf{AX} = -\mathbf{B} \qquad\qquad (4.44)$$

Equation 4.44 can be solved for $\mathbf{X}$ either by matrix inversion or by Gaussian elimination. The values of the elements of $\mathbf{A}$ and $\mathbf{B}$ are calculable for any assumed (guess) values of the variables. A criterion for convergence can be taken as the sum of the error vector $\mathbf{X}$ at each iteration where the sum approaches zero at a root.

Let's set up this Newton-Raphson solution for the fourbar linkage.

Newton-Raphson Solution for the Fourbar Linkage

The vector loop equation of the fourbar linkage, separated into its real and imaginary parts (equations 4.6a and 4.6b, p. 187) provides the set of functions that define the two unknown link angles θ_3 and θ_4. The link lengths, a, b, c, d, and the input angle θ_2 are given.

4

$$f_1 = a\cos\theta_2 + b\cos\theta_3 - c\cos\theta_4 - d = 0$$
$$f_2 = a\sin\theta_2 + b\sin\theta_3 - c\sin\theta_4 = 0 \tag{4.45a}$$

$$\mathbf{B} = \begin{bmatrix} a\cos\theta_2 + b\cos\theta_3 - c\cos\theta_4 - d \\ a\sin\theta_2 + b\sin\theta_3 - c\sin\theta_4 \end{bmatrix} \tag{4.45b}$$

The error vector is:

$$\mathbf{X} = \begin{bmatrix} \Delta\theta_3 \\ \Delta\theta_4 \end{bmatrix} \tag{4.46}$$

$$\mathbf{A} = \begin{bmatrix} \dfrac{\partial f_1}{\partial\theta_3} & \dfrac{\partial f_1}{\partial\theta_4} \\ \dfrac{\partial f_2}{\partial\theta_3} & \dfrac{\partial f_2}{\partial\theta_4} \end{bmatrix} = \begin{bmatrix} -b\sin\theta_3 & c\sin\theta_4 \\ b\cos\theta_3 & -c\cos\theta_4 \end{bmatrix} \tag{4.47}$$

This matrix is known as the **Jacobian** of the system, and, in addition to its usefulness in this solution method, it also tells something about the solvability of the system. The system of equations for position, velocity, and acceleration (in all of which the Jacobian appears) can only be solved if the value of the determinant of the Jacobian is nonzero.

Substituting equations 4.45b, 4.46, and 4.47 into equation 4.44 gives:

$$\begin{bmatrix} -b\sin\theta_3 & c\sin\theta_4 \\ b\cos\theta_3 & -c\cos\theta_4 \end{bmatrix} \begin{bmatrix} \Delta\theta_3 \\ \Delta\theta_4 \end{bmatrix} = -\begin{bmatrix} a\cos\theta_2 + b\cos\theta_3 - c\cos\theta_4 - d \\ a\sin\theta_2 + b\sin\theta_3 - c\sin\theta_4 \end{bmatrix} \tag{4.48}$$

To solve this matrix equation, guess values will have to be provided for θ_3 and θ_4 and the two equations then solved simultaneously for $\Delta\theta_3$ and $\Delta\theta_4$. For a larger system of equations, a matrix reduction algorithm will need to be used. For this simple system in two unknowns, the two equations can be solved by combination and reduction. The test described above which compares the sum of the values of $\Delta\theta_3$ and $\Delta\theta_4$ to a selected tolerance must be applied after each iteration to determine if a root has been found.

Equation Solvers

Some commercially available equation solver software packages include the ability to do a Newton-Raphson iterative solution on sets of nonlinear simultaneous equations. *TKSolver*[*] and *Mathcad*[†] are examples. *TKSolver* automatically invokes its Newton-

[*] Universal Technical Systems, 1220 Rock St. Rockford, IL 61101, USA. (800) 435-7887

[†] PTC Inc., 140 Kendrick St., Needham, MA 02494 (781) 370-5000

4

Raphson solver when it cannot directly solve the presented equation set, provided that enough guess values have been supplied for the unknowns. These equation solver tools are quite convenient in that the user need only supply the equations for the system in "raw" form such as equation 4.45a. It is not necessary to arrange them into the Newton-Raphson algorithm as shown in the previous section. Lacking such a commercial equation solver, you will have to write your own computer code to program the solution as described above. Reference [5] is a useful aid in this regard. The DVD included with this text contains example *TKSolver* files for the solution of this fourbar position problem as well as others.

4.15 REFERENCES

1 **Waldron, K. J., and S. V. Sreenivasan**. (1996). "A Study of the Solvability of the Position Problem for Multi-Circuit Mechanisms by Way of Example of the Double Butterfly Linkage." *Journal of Mechanical Design*, **118**(3), p. 390.

2 **Chase, T. R., and J. A. Mirth.** (1993). "Circuits and Branches of Single-Degree-of-Freedom Planar Linkages." *Journal of Mechanical Design*, **115**, p. 223.

3 **Kramer, G.** (1992). *Solving Geometric Constraint Systems: A Case Study in Kinematics.* MIT Press: Cambridge, MA, pp. 155-158.

4 **Press, W. H., et al.** (1986). *Numerical Recipes: The Art of Scientific Computing.* Cambridge University Press: Cambridge, pp. 145-146.

5 Ibid., pp. 254-273.

6 **Chasles, M.** (1830). "Note Sur les Proprietes Generales du Systeme de Deux Corps Semblables entr'eux (Note on the general properties of a system of two similar bodies in combination)." *Bulletin de Sciences Mathematiques, Astronomiques Physiques et Chimiques,* Baron de Ferussac, Paris, pp. 321-326.

7 **Ceccarelli, M.** (2000). "Screw Axis Defined by Giulio Mozzi in 1763 and Early Studies on Helicoidal Motion." *Mechanism and Machine Theory*, **35**, pp. 761-770.

8 **Mozzi, G.** (1763). *Discorso matematico sopra il rotamento momentaneo dei corpi (Mathematical Treatise on the temporally revolving of bodies).*

9 **Raven, F. H.** (1958). "Velocity and Acceleration Analysis of Plane and Space Mechanisms by Means of Independent-Position Equations." *Trans ASME*, **25**, pp. 1-6.

10 **Wampler, C. W.** (1999). "Solving the Kinematics of Planar Mechanisms." *Journal of Mechanical Design,* **121**(3), pp. 387-391.

4.16 PROBLEMS‡

4-1 A position vector is defined as having a length equal to your height in inches (or centimeters). The tangent of its angle is defined as your weight in pounds (or kilograms) divided by your age in years. Calculate the data for this vector and:

  a.   Draw the position vector to scale on cartesian axes.
  b.   Write an expression for the position vector using unit vector notation.
  c.   Write an expression for the position vector using complex number notation, in both polar and cartesian forms.

4-2 A particle is traveling along an arc of 6.5-in radius. The arc center is at the origin of a coordinate system. When the particle is at position *A*, its position vector makes a

‡ All problem figures are provided as PDF files, and some are also provided as animated AVI and Working Model files; all are on the DVD. PDF filenames are the same as the figure number. Run the file *Animations. html* to access and run the animations.

TABLE P4-1   Data for Problems 4-6, 4-7 and 4-13 to 4-15[‡]

Row	Link 1	Link 2	Link 3	Link 4	θ_2
a	6	2	7	9	30
b	7	9	3	8	85
c	3	10	6	8	45
d	8	5	7	6	25
e	8	5	8	6	75
f	5	8	8	9	15
g	6	8	8	9	25
h	20	10	10	10	50
i	4	5	2	5	80
j	20	10	5	10	33
k	4	6	10	7	88
l	9	7	10	7	60
m	9	7	11	8	50
n	9	7	11	6	120

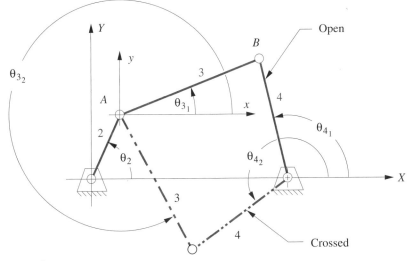

FIGURE P4-1[‡]

Problems 4-6 to 4-7. General configuration and terminology for the fourbar linkage

45° angle with the X axis. At position B, its vector makes a 75° angle with the X axis.
Draw this system to some convenient scale and:

a. Write an expression for the particle's position vector in position A using complex number notation, in both polar and cartesian forms.

b. Write an expression for the particle's position vector in position B using complex number notation, in both polar and cartesian forms.

c. Write a vector equation for the position difference between points B and A. Substitute the complex number notation for the vectors in this equation and solve for the position difference numerically.

d. Check the result of part c with a graphical method.

‡ These problem figures are provided as PDF files, and some are also provided as animated AVI and Working Model files; all are on the DVD. PDF filenames are the same as the figure number. Run the file *Animations.html* to access and run the animations.

TABLE P4-2   Data for Problems 4-9 to 4-10‡

Row	Link 2	Link 3	Offset	θ_2
a	1.4	4	1	45
b	2	6	-3	60
c	3	8	2	-30
d	3.5	10	1	120
e	5	20	-5	225
f	3	13	0	100
g	7	25	10	330

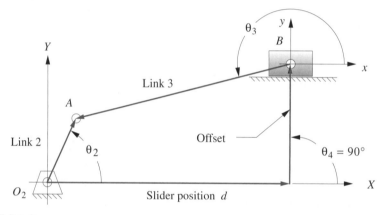

FIGURE P4-2

Problems 4-9, 4-10, 4-60, 4-61 Open configuration and terminology for a fourbar slider linkage

† These problems are suited to solution using *Mathcad, Matlab,* or *TKSolver* equation solver programs. In most cases, your solution can be checked with program LINKAGES.

‡ These problem figures are provided as PDF files, and some are also provided as animated AVI and Working Model files; all are on the DVD. PDF filenames are the same as the figure number. Run the file *Animations.html* to access and run the animations.

4-3 Repeat problem 4-2 considering points A and B to represent separate particles, and find their relative position.

4-4 Repeat Problem 4-2 with the particle's path defined as being along the line $y = -2x + 10$.

4-5 Repeat Problem 4-3 with the path of the particle defined as being along the curve $y = -2x^2 - 2x + 10$.

*4-6 The link lengths and the value of θ_2 for some fourbar linkages are defined in Table P4-1 (p. 215). The linkage configuration and terminology are shown in Figure P4-1 (p. 215). For the rows assigned, draw the linkage to scale and graphically find all possible solutions (both open and crossed) for angles θ_3 and θ_4. Determine the Grashof condition.

*†4-7 Repeat Problem 4-6 except solve by the vector loop method.

4-8 Expand equation 4.7b and prove that it reduces to equation 4.7c (p. 187).

*4-9 The link lengths and the value of θ_2 and offset for some fourbar crank-slider linkages are defined in Table P4-2. The linkage configuration and terminology are shown in Figure P4-2. For the rows assigned, draw the linkage to scale and graphically find all possible solutions (both open and crossed) for angle θ_3 and slider position d.

TABLE P4-3   Data for Problems 4-11 to 4-12

Row	Link 1	Link 2	Link 4	γ	θ_2
a	6	2	4	90	30
b	7	9	3	75	85
c	3	10	6	45	45
d	8	5	3	60	25
e	8	4	2	30	75
f	5	8	8	90	150

4

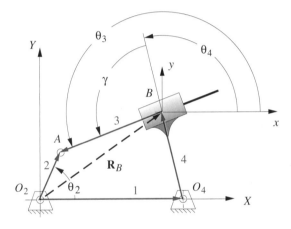

FIGURE P4-3

Problems 4-11 to 4-12 Terminology for inversion #3 of the fourbar crank-slider linkage

*†4-10   Repeat Problem 4-9 except solve by the vector loop method.

*4-11   The link lengths and the value of θ_2 and γ for some inverted fourbar crank-slider linkages are defined in Table P4-3. The linkage configuration and terminology are shown in Figure P4-3. For the rows assigned, draw the linkage to scale and graphically find both open and crossed solutions for angles θ_3 and θ_4 and vector $\mathbf{R}_B$.

*†4-12   Repeat Problem 4-11 except solve by the vector loop method.

*†4-13   Find the transmission angles of the linkages in the assigned rows in Table P4-1 (p. 215).

*†4-14   Find the minimum and maximum values of the transmission angle for all the Grashof crank-rocker linkages in Table P4-1 (p. 215).

*†4-15   Find the input angles corresponding to the toggle positions of the non-Grashof linkages in Table P4-1 (p. 215). (For this problem, ignore the values of θ_2 given in the table.)

*4-16   The link lengths, gear ratio (λ), phase angle (φ), and the value of θ_2 for some geared fivebar linkages are defined in Table P4-4 (p. 218). The linkage configuration and terminology are shown in Figure P4-4 (p. 218). For the rows assigned, draw the linkage to scale and graphically find all possible solutions for angles θ_3 and θ_4.

*†4-17   Repeat Problem 4-16 except solve by the vector loop method.

* Answers in Appendix F.

† These problems are suited to solution using *Mathcad*, *Matlab*, or *TKSolver* equation solver programs. In most cases, your solution can be checked with program LINKAGES.

TABLE P4-4   Data for Problems 4-16 to 4-17

Row	Link 1	Link 2	Link 3	Link 4	Link 5	λ	ϕ	θ_2
a	6	1	7	9	4	2	30	60
b	6	5	7	8	4	-2.5	60	30
c	3	5	7	8	4	-0.5	0	45
d	4	5	7	8	4	-1	120	75
e	5	9	11	8	8	3.2	-50	-39
f	10	2	7	5	3	1.5	30	120
g	15	7	9	11	4	2.5	-90	75
h	12	8	7	9	4	-2.5	60	55
i	9	7	8	9	4	-4	120	100

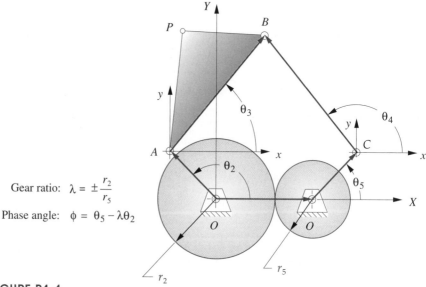

Gear ratio: $\quad \lambda = \pm \dfrac{r_2}{r_5}$

Phase angle: $\quad \phi = \theta_5 - \lambda\theta_2$

FIGURE P4-4

Problems 4-16 to 4-17 Open configuration and terminology for the geared fivebar linkage

4-18   Figure P4-5 shows the mechanisms for the following problems each of which refers to the part of the figure having the same letter. Reference all calculated angles to the global XY axes.

    a.   The angle between the X and x axes is 25°. Find the angular displacement of link 4 when link 2 rotates clockwise from the position shown (+37°) to horizontal (0°). How does the transmission angle vary and what is its minimum between those two positions? Find the toggle positions of this linkage in terms of the angle of link 2.

    b.   Find and plot the angular position of links 3 and 4 and the transmission angle as a function of the angle of link 2 as it rotates through one revolution.

    c.   Find and plot the position of any one piston as a function of the angle of crank 2 as it rotates through one revolution. Once one piston's motion is defined, find the

† These problems are suited to solution using *Mathcad, Matlab,* or *TKSolver* equation solver programs. In most cases, your solution can be checked with program LINKAGES.

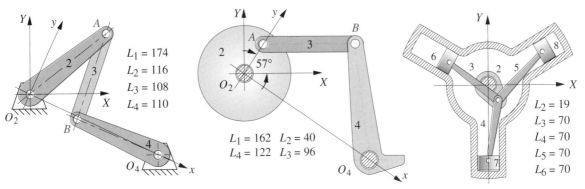

$$L_1 = 174$$
$$L_2 = 116$$
$$L_3 = 108$$
$$L_4 = 110$$

(a) Fourbar linkage

$57°$

$$L_1 = 162 \quad L_2 = 40$$
$$L_4 = 122 \quad L_3 = 96$$

(b) Fourbar linkage

$$L_2 = 19$$
$$L_3 = 70$$
$$L_4 = 70$$
$$L_5 = 70$$
$$L_6 = 70$$

(c) Radial compressor

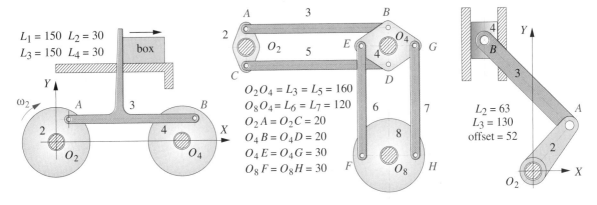

$$L_1 = 150 \quad L_2 = 30$$
$$L_3 = 150 \quad L_4 = 30$$

box

ω_2

(d) Walking-beam conveyor

$$O_2O_4 = L_3 = L_5 = 160$$
$$O_8O_4 = L_6 = L_7 = 120$$
$$O_2A = O_2C = 20$$
$$O_4B = O_4D = 20$$
$$O_4E = O_4G = 30$$
$$O_8F = O_8H = 30$$

(e) Bellcrank mechanism

$$L_2 = 63$$
$$L_3 = 130$$
$$offset = 52$$

(f) Offset slider-crank

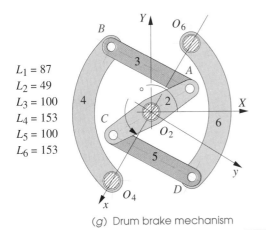

$$L_1 = 87$$
$$L_2 = 49$$
$$L_3 = 100$$
$$L_4 = 153$$
$$L_5 = 100$$
$$L_6 = 153$$

(g) Drum brake mechanism

22.9 22.9

$$L_1 = 45.8$$
$$L_2 = 19.8$$
$$L_3 = 19.4$$
$$L_4 = 38.3$$
$$L_5 = 13.3$$
$$L_7 = 13.3$$
$$L_8 = 19.8$$
$$L_9 = 19.4$$

4.5 typ.

(h) Symmetrical mechanism

all dimensions in mm

FIGURE P4-5

Mechanisms for Problem 4-18 *Adapted from P. H. Hill and W. P. Rule. (1960). Mechanisms: Analysis and Design, with permission*

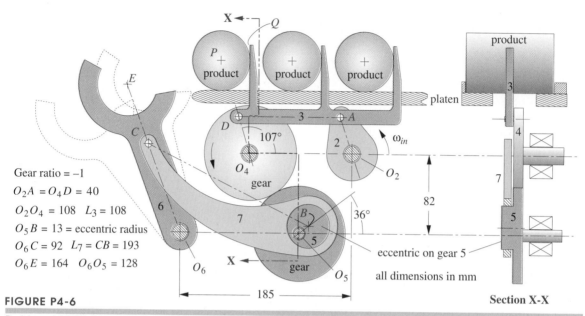

Gear ratio = –1
$O_2A = O_4D = 40$
$O_2O_4 = 108$ $L_3 = 108$
$O_5B = 13$ = eccentric radius
$O_6C = 92$ $L_7 = CB = 193$
$O_6E = 164$ $O_6O_5 = 128$

FIGURE P4-6   **Section X-X**

Problem 4-19 Walking-beam indexer with pick-and-place mechanism *Adapted from P. H. Hill and W. P. Rule. (1960).*
Mechanisms: Analysis and Design, with permission

motions of the other two pistons and their phase relationship to the first piston.

d.   Find the total angular displacement of link 3 and the total stroke of the box as link 2 makes a complete revolution.

e.   Determine the ratio of angular displacement between links 8 and 2 as a function of angular displacement of input crank 2. Plot the transmission angle at point *B* for one revolution of crank 2. Comment on the behavior of this linkage. Can it make a full revolution as shown?

f.   Find and plot the displacement of piston 4 and the angular displacement of link 3 as a function of the angular displacement of crank 2.

g.   Find and plot the angular displacement of link 6 versus the angle of input link 2 as it is rotated from the position shown (+30°) to a vertical position (+90°). Find the toggle positions of this linkage in terms of the angle of link 2.

h.   Find link 4's maximum displacement vertically downward from the position shown. What will the angle of input link 2 be at that position?

†4-19   For one revolution of driving link 2 of the walking-beam indexing and pick-and-place mechanism in Figure P4-6, find the horizontal stroke of link 3 for the portion of their motion where its tips are above the top of the platen. Express the stroke as a percentage of the crank length O_2A. What portion of a revolution of link 2 does this stroke correspond to? Also find the total angular displacement of link 6 over one revolution of link 2. The vertical distance from O_2 to the top of the platen is 64 mm. The vertical distance from line *AD* to the top left corner *Q* of the leftmost pusher finger is 73 mm. The horizontal distance from point *A* to *Q* is 95 mm.

†4-20   Figure P4-7 shows a power hacksaw, used to cut metal. Link 5 pivots at O_5 and its weight forces the sawblade against the workpiece while the linkage moves the blade (link 4) back and forth on link 5 to cut the part. It is an offset crank-slider mechanism. The dimensions are shown in the figure. For one revolution of driving link 2 of the

* Answers in Appendix F.

† These problems are suited to solution using *Mathcad, Matlab,* or *TKSolver* equation solver programs. In most cases, your solution can be checked with program LINKAGES.

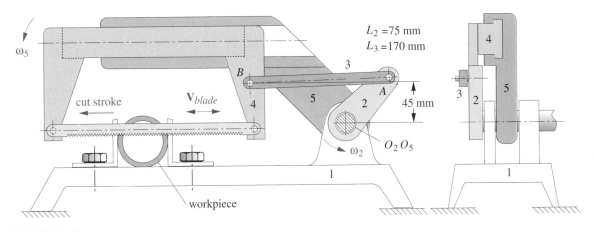

FIGURE P4-7

Problem 4-20   Power hacksaw *Adapted from P. H. Hill and W. P. Rule. (1960). Mechanisms: Analysis and Design, with permission*

hacksaw mechanism on the cutting stroke, find and plot the horizontal stroke of the sawblade as a function of the angle of link 2.

*†4-21   For the linkage in Figure P4-8, find its limit (toggle) positions in terms of the angle of link O_2A referenced to the line of centers O_2O_4 when driven from link O_2A. Then calculate and plot the *xy* coordinates of coupler point P between those limits, referenced to the line of centers O_2O_4.

†4-22   For the walking-beam mechanism of Figure P4-9, calculate and plot the *x* and *y* components of the position of the coupler point P for one complete revolution of the crank O_2A. *Hint*: Calculate them first with respect to the ground link O_2O_4 and then transform them into the global *XY* coordinate system (i.e., horizontal and vertical in the figure). Scale the figure for any additional information needed.

FIGURE P4-8

Problem 4-21

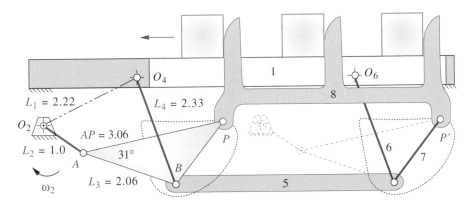

FIGURE P4-9

Problem 4-22   Straight-line walking-beam eightbar transport mechanism

* Answers in Appendix F.

† These problems are suited to solution using *Mathcad, Matlab,* or *TKSolver* equation solver programs. In most cases, your solution can be checked with program LINKAGES.

FIGURE P4-10

Problem 4-23

ω_{in} coupler
         8.375"   reed

crank   rocker
2"   7.187"

ground
9.625"
@ –43°

FIGURE P4-11

Problem 4-24

*†4-23   For the linkage in Figure P4-10, calculate and plot the angular displacement of links 3 and 4 and the path coordinates of point P with respect to the angle of the input crank O_2A for one revolution.

†4-24   For the linkage in Figure P4-11, calculate and plot the angular displacement of links 3 and 4 with respect to the angle of the input crank O_2A for one revolution.

*†4-25   For the linkage in Figure P4-12, find its limit (toggle) positions in terms of the angle of link O_2A referenced to the line of centers O_2O_4 when driven from link O_2A. Then calculate and plot the angular displacement of links 3 and 4 and the path coordinates of point P with respect to the angle of the input crank O_2A over its possible range of motion referenced to the line of centers O_2O_4.

*†4-26   For the linkage in Figure P4-13, find its limit (toggle) positions in terms of the angle of link O_2A referenced to the line of centers O_2O_4 when driven from link O_2A. Then calculate and plot the angular displacement of links 3 and 4 and the path coordinates of point P between those limits, with respect to the angle of the input crank O_2A over its possible range of motion referenced to the line of centers O_2O_4.

* Answers in Appendix F.

---

† These problems are suited to solution using *Mathcad, Matlab,* or *TKSolver* equation solver programs. In most cases, your solution can be checked with program LINKAGES.

FIGURE P4-12

Problem 4-25

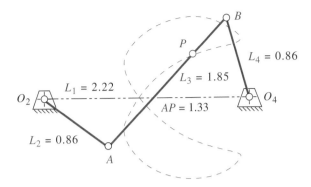

FIGURE P4-13

Problems 4-26 to 4-27

†4-27 For the linkage in Figure P4-13, find its limit (toggle) positions in terms of the angle of link O_4B referenced to the line of centers O_4O_2 when driven from link O_4B. Then calculate and plot the angular displacement of links 2 and 3 and the path coordinates of point P between those limits, with respect to the angle of the input crank O_4B over its possible range of motion referenced to the line of centers O_4O_2.

†4-28 For the rocker-crank linkage in Figure P4-14, find the maximum angular displacement possible for the treadle link (to which force F is applied). Determine the toggle positions. How does this work? Explain why the grinding wheel is able to fully rotate despite the presence of toggle positions when driven from the treadle. How would you get it started if it were in a toggle position?

**†4-29 For the linkage in Figure P4-15, find its limit (toggle) positions in terms of the angle of link O_2A referenced to the line of centers O_2O_4 when driven from link O_2A. Then calculate and plot the angular displacement of links 3 and 4 and the path coordinates of point P between those limits, with respect to the angle of the input crank O_2A over its possible range of motion referenced to the line of centers O_2O_4.

**†4-30 For the linkage in Figure P4-15, find its limit (toggle) positions in terms of the angle of link O_4B referenced to the line of centers O_4O_2 when driven from link O_4B. Then calculate and plot the angular displacement of links 2 and 3 and the path coordinates of

FIGURE P4-14

Problem 4-28

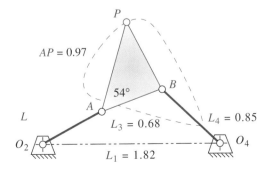

FIGURE P4-15

Problems 4-29 to 4-30

* Answers in Appendix F.

† These problems are suited to solution using *Mathcad*, *Matlab*, or *TKSolver* equation solver programs. In most cases, your solution can be checked with program LINKAGES.

point *P* between those limits, with respect to the angle of the input crank O_4B over its possible range of motion referenced to the line of centers O_4O_2.

*†4-31   Write a computer program (or use an equation solver such as *Mathcad, Matlab,* or *TKSolver*) to find the roots of $y = 9x^2 + 50x - 40$. Hint: Plot the function to determine good guess values.

†4-32   Write a computer program (or use an equation solver such as *Mathcad, Matlab,* or *TKSolver*) to find the roots of $y = -x^3 - 4x^2 + 80x - 40$. Hint: Plot the function to determine good guess values.

†4-33   Figure 4-18 (p. 227) plots the cubic function from equation 4.38. Write a computer program (or use an equation solver such as *Mathcad, Matlab,* or *TKSolver* to solve the matrix equation) to investigate the behavior of the Newton-Raphson algorithm as the initial guess value is varied from $x = 1.8$ to 2.5 in steps of 0.1. Determine the guess value at which the convergence switches roots. Explain this root-switching phenomenon based on your observations from this exercise.

†4-34   Write a computer program or use an equation solver such as *Mathcad, Matlab,* or *TKSolver* to calculate and plot the angular position of link 4 and the position of slider 6 in Figure 3-33 (p. 152) as a function of the angle of input link 2.

†4-35   Write a computer program or use an equation solver such as *Mathcad, Matlab,* or *TKSolver* to calculate and plot the transmission angles at points *B* and *C* of the linkage in Figure 3-33 (p. 152) as a function of the angle of input link 2.

†4-36   Write a computer program or use an equation solver such as *Mathcad, Matlab,* or *TKSolver* to calculate and plot the path of the coupler point of the straight-line linkage shown in Figure 3-29f (p. 141). (Use LINKAGES to check your result.)

†4-37   Write a computer program or use an equation solver such as *Mathcad, Matlab,* or *TKSolver* to calculate and plot the angular position of link 6 in Figure 3-34 (p. 153) as a function of the angle of input link 2.

†4-38   Write a computer program or use an equation solver such as *Mathcad, Matlab,* or *TK-Solver* to calculate and plot the transmission angles at points *B, C,* and *D* of the linkage in Figure 3-34 (p. 153) as a function of the angle of input link 2.

†4-39   Write a computer program or use an equation solver such as *Mathcad, Matlab,* or *TKSolver* to calculate and plot the path of the coupler point of the straight-line linkage shown in Figure 3-29g (p. 141). (Use LINKAGES to check your result.)

†4-40   Write a computer program or use an equation solver such as *Mathcad, Matlab,* or *TKSolver* to calculate and plot the angular position of link 6 in Figure 3-35 (p. 153) as a function of the angle of input link 2.

†4-41   Write a computer program or use an equation solver such as *Mathcad, Matlab,* or *TK-Solver* to calculate and plot the transmission angles at points *B, C,* and *D* of the linkage in Figure 3-35 (p. 153) as a function of the angle of input link 2.

4-42   Write a computer program or use an equation solver such as *Mathcad, Matlab,* or *TKSolver* to calculate and plot the path of the coupler point of the straight-line linkage shown in Figure 3-29h (p. 141). (Use LINKAGES to check your result.)

†4-43   Write a computer program or use an equation solver such as *Mathcad, Matlab,* or *TKSolver* to calculate and plot the angular position of link 8 in Figure 3-36 (p. 154) as a function of the angle of input link 2.

† These problems are suited to solution using *Mathcad, Matlab,* or *TKSolver* equation solver programs. In most cases, the solution can be checked with program LINKAGES.

†4-44 Write a computer program or use an equation solver such as *Mathcad, Matlab*, or *TKSolver* to calculate and plot the transmission angles at points B, C, D, E, and F of the linkage in Figure 3-36 (p. 154) as a function of the angle of input link 2.

†4-45 Model the linkage shown in Figure 3-37a (p. 154) in LINKAGES. Export the coupler curve coordinates to EXCEL and calculate the error function versus a true circle.

†4-46 Write a computer program or use an equation solver such as *Mathcad, Matlab*, or *TKSolver* to calculate and plot the path of point P in Figure 3-37a (p. 154) as a function of the angle of input link 2. Also plot the variation (error) in the path of point P versus that of point A, i.e., how close to a perfect circle is point P's path.

†4-47 Write a computer program or use an equation solver such as *Mathcad, Matlab*, or *TK-Solver* to calculate and plot the transmission angles at point B of the linkage in Figure 3-37a (p. 154) as a function of the angle of input link 2.

†4-48 Figure 3-29f (p. 141) shows Evan's approximate straight-line linkage #1. Determine the range of motion of link 2 for which point P varies no more than 0.0025 from the straight line $x = 1.690$ in a coordinate system with origin at O_2 and its x axis rotated 60° from O_2O_4.

†4-49 Write a computer program or use an equation solver such as *Mathcad, Matlab*, or *TKSolver* to calculate and plot the path of point P in Figure 3-37b (p. 154) as a function of the angle of input link 2.

†4-50 Write a computer program or use an equation solver such as *Mathcad, Matlab*, or *TK-Solver* to calculate and plot the transmission angles at points B, C, and D of the linkage in Figure 3-37b (p. 154) as a function of the angle of input link 2.

†4-51 Figure 3-29g (p. 141) shows Evan's approximate straight-line linkage #2. Determine the range of motion of link 2 for which point P varies no more than 0.005 from the straight line $x = -0.500$ in a coordinate system with origin at O_2 and its x axis rotated 30° from O_2O_4.

4-52 For the linkage in Figure P4-16 (p. 226), what are the angles that link 2 makes with the positive X axis when links 2 and 3 are in toggle positions?

4-53 The coordinates of the point P_1 on link 4 in Figure P4-16 (p. 226) are (114.68, 33.19) with respect to the xy coordinate system when link 2 is in the position shown. When link 2 is in another position the coordinates of P_2 with respect to the xy system are (100.41, 43.78). Calculate the coordinates of P_1 and P_2 in the XY system for the two positions of link 2. What is the salient feature of the coordinates of P_1 and P_2 in the XY system?

†4-54 Write a computer program or use an equation solver such as *Mathcad, Matlab*, or *TKSolver* to calculate and plot the angular position of link 4 with respect to the XY coordinate frame and the transmission angle at point B of the linkage in Figure P4-16 (p. 226) as a function of the angle of link 2 with respect to the XY frame.

4-55 For the linkage in Figure P4-17 (p. 226), calculate the maximum CW rotation of link 2 from the position shown, which is at –26° with respect to the local xy coordinate system. What angles do link 3 and link 4 rotate through for that excursion of link 2?

†4-56 Write a computer program or use an equation solver such as *Mathcad, Matlab*, or *TK-Solver* to calculate and plot the position of the coupler point P of the linkage in Figure P4-17 (p. 226) with respect to the XY coordinate system as a function of the angle of link 2 with respect to the XY system. The position of the coupler point P on link 3 with respect to point A is: $p = 15.00$, $\delta_3 = 0°$.

† Note that these can be long problems to solve and may be more appropriate for a project assignment than an overnight problem. In most cases, the solution can be checked with program LINKAGES.

FIGURE P4-16

Problems 4-52 to 4-54   An oil field pump - dimensions in inches

4-57   For the linkage in Figure P4-17, calculate the coordinates of the point P in the XY coordinate system if its coordinates in the xy system are (12.816, 10.234).

†4-58   The elliptical trammel in Figure P4-18 must be driven by rotating link 3 in a full circle. Derive analytical expressions for the positions of points A, B, and a point C on link 3 midway between A and B as a function of θ_3 and the length AB of link 3. Use a vector loop equation. (*Hint*: Place the global origin off the mechanism, preferably below and to the left and use a total of 5 vectors.) Code your solution in an equation solver

† Note that these can be long problems to solve and may be more appropriate for a project assignment than an overnight problem. In most cases, the solution can be checked with program LINKAGES.

FIGURE P4-17

Problems 4-55 to 4-57   An aircraft overhead bin mechanism - dimensions in inches such as *Mathcad*, *Matlab*, or *TKSolver* to calculate and plot the path of point C for one revolution of link 3.

†4-59 Figure P4-19 shows a mechanism commonly used as a cabinet door hinge. Write a computer program or use an equation solver such as *Mathcad*, *Matlab*, or *TKSolver* to calculate and plot the angular position of link 6 in Figure P4-19 as a function of the angle of input link 2. $O_2O_4 = AB = BC = DE = 1$. $O_2A = O_4B = BE = CD = 1.75$. $O_4C = AE = 2.60$. *Hint: Because the linkage geometry is simple and symmetrical, the analysis can be done with simple trigonometry.*

4-60 The link lengths, offset, and value of d for some fourbar slider-crank linkages are defined in Table P4-5. The linkage configuration and terminology are shown in Figure P4-2 (p. 216). For the rows assigned, draw the linkage to scale and graphically find all possible solutions (both open and crossed) for angles θ_2 and θ_3.

4-61 Repeat Problem 4-60 except solve by the vector loop method.

FIGURE P4-18

Elliptical trammel - Problem 4-58

TABLE P4-5 Data for Problems 4-60 to 4-61‡

Row	Link 2	Link 3	Offset	d
a	1.4	4	1	2.5
b	2	6	−3	5
c	3	8	2	8
d	3.5	10	1	−8
e	5	20	−5	15
f	3	13	0	−12
g	7	25	10	25

‡ Drawings of these linkages are in the *PDF Problem Workbook* folder on the DVD.

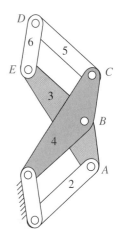

FIGURE P4-19

Problem 4-59

---

† Note that these can be long problems to solve and may be more appropriate for a project assignment than an overnight problem. In most cases, the solution can be checked with program LINKAGES.

Chapter 5

ANALYTICAL LINKAGE SYNTHESIS

Imagination is more important than knowledge
ALBERT EINSTEIN

5.0   INTRODUCTION

With the fundamentals of position analysis established, we can now use these techniques to **synthesize linkages** for specified output positions **analytically**. The synthesis techniques presented in Chapter 3 were strictly graphical and somewhat intuitive. The **analytical synthesis** procedure is algebraic rather than graphical and is less intuitive. However, its algebraic nature makes it quite suitable for computerization. These analytical synthesis methods were originated by Sandor[1] and further developed by his students Erdman,[2] Kaufman[3], and Loerch et al.[4,5]

5.1   TYPES OF KINEMATIC SYNTHESIS

Erdman and Sandor[6] define three types of kinematic synthesis, **function**, **path**, and **motion generation**, which were discussed in Section 3.2 (p. 98). Brief definitions are repeated here for your convenience.

FUNCTION GENERATION   is defined as *the correlation of an **input function** with an **output function** in a mechanism.* Typically, a double-rocker or crank-rocker is the result, with pure rotation input and pure rotation output. A slider-crank linkage can be a function generator as well, driven from either end, i.e., rotation in and translation out or vice versa.

PATH GENERATION   is defined as *the control of a **point** in the plane such that it follows some prescribed path.* This is typically accomplished with a fourbar crank-rocker or double-rocker, wherein a point on the coupler traces the desired output path. No attempt is made in path generation to control the orientation of the link which contains the point of

228

interest. The coupler curve is made to pass through a set of desired output points. However, it is common for the timing of the arrival of the coupler point at particular locations
along the path to be defined. This case is called *path generation with prescribed timing*
and is analogous to function generation in that a particular output function is specified.

MOTION GENERATION is defined as *the control of a **line** in the plane such that it
assumes some sequential set of prescribed positions*. Here orientation of the link containing the line is important. This is typically accomplished with a fourbar crank-rocker
or double-rocker, wherein a point on the coupler traces the desired output path and the
linkage also controls the angle of the coupler link containing the output line of interest.

5.2   TWO-POSITION SYNTHESIS FOR ROCKER OUTPUT

Example 3-1 (p. 103) showed a simple graphical technique for synthesis of a non-quick-
return, Grashof fourbar linkage to drive a rocker through an angle. This technique was
employed in later examples (e.g., 3-2, 3-4, 3-6) to construct a driver dyad to move a synthesized fourbar linkage through its desired range of motion, thus creating a Watt sixbar
linkage. The rocker excursion cannot exceed 180° theoretically but should be limited to
about 120° practically, which will give minimum transmission angles of 30°. The same
dyad synthesis procedure can be done analytically and will prove to be of value in combination with the other synthesis techniques presented in this chapter.

Figure 5-0 (p. 230) shows the same problem as Figure 3-4 (p. 104) with generic an-
notation suitable for analytical determination of link lengths for the driver dyad. Link 4
(which might represent the input link to the next stage of the resulting Watt sixbar) is here
the output link to be driven by a dyad consisting of links 2 and 3, whose lengths, along
with that of the ground link 1 and its pivot location O_2, are to be determined. The pivot
location O_4 (defined in any convenient coordinate system XY), the initial angle θ_4, and
the excursion angle β are given. The procedure is as follows:[*]

First choose a suitable location on link 4 to attach link 3, here labeled B_1 and B_2 in
its extreme locations. This defines R_4, the length of link 4. These points can be defined
in the chosen coordinate system as:

$$B_{1_x} = O_{4_x} + R_4 \cos(\theta_4) \qquad\qquad B_{1_y} = O_{4_y} + R_4 \sin(\theta_4)$$
$$\text{(5.0a)}$$
$$B_{2_x} = O_{4_x} + R_4 \cos(\theta_4 + \beta) \qquad\qquad B_{2_y} = O_{4_y} + R_4 \sin(\theta_4 + \beta)$$

The vector $\mathbf{M}$ is the position difference between vectors $\mathbf{R}_{B2}$ and $\mathbf{R}_{B1}$

$$\mathbf{M} = \mathbf{R}_{B_2} - \mathbf{R}_{B_1} \qquad\qquad \text{(5.0b)}$$

The parametric equation for line $\mathbf{L}$ can be written as:

$$\mathbf{L}(u) = \mathbf{R}_{B_1} + u\mathbf{M} \qquad\qquad -\infty \le u \le \infty \qquad\qquad \text{(5.0c)}$$

We want the resulting linkage to be a Class 1 Grashof crank rocker. We can achieve
this by placing the crank pivot O_2 suitably far from B_1 along line $\mathbf{L}$. Let $M = |\mathbf{M}|$. It will
be a Class 2 (non-Grashof) linkage when $B_1O_2 < M$, become Class 3 (Grashof with change
points) when $B_1O_2 = M$, be Class 1 when $B_1O_2 > M$, and will again approach Class 3 when
$B_1O_2 \gg M$. A reasonable range for B_1O_2 seems to be two to three times M.

* This procedure was
provided by Prof. Pierre Larochelle of Florida Institute
of Technology.

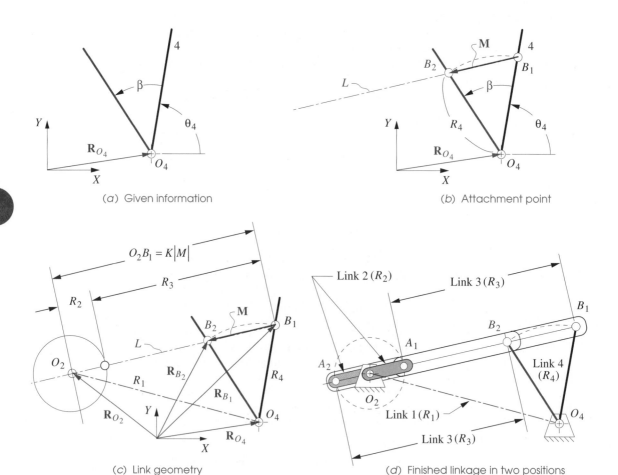

(a) Given information

(b) Attachment point

(c) Link geometry

(d) Finished linkage in two positions

FIGURE 5-0

Analytical two-position synthesis with rocker output (non-quick-return)

$$\text{Let: } \mathbf{R}_{O_2} = \mathbf{R}_{B_1} \pm K\mathbf{M} \qquad\qquad 2 < K < 3 \qquad\qquad (5.0d)$$

As shown in Example 3-1 (p. 103), the length of the crank must be half the length of vector $\mathbf{M}$:

$$R_2 = 0.5\,|\mathbf{M}| = R_4 \sin(\beta/2) \qquad\qquad (5.0e)$$

where β is in radians. Link 3 can be found by subtracting R_2 from the magnitude of $\mathbf{R}_{B1}-\mathbf{R}_{O2}$ and link 1 is found by subtracting $\mathbf{R}_{O2}$ from $\mathbf{R}_{O4}$.

$$R_3 = \left|\mathbf{R}_{B_1} - \mathbf{R}_{O_2}\right| - R_2; \qquad\qquad R_1 = \left|\mathbf{R}_{O_4} - \mathbf{R}_{O_2}\right| \qquad\qquad (5.0f)$$

This algorithm will result in a Grashof crank rocker mechanism that drives the rocker through the specified angle with no quick return.

5.3   PRECISION POINTS

The *points, or positions, prescribed for successive locations of the output (coupler or rocker) link in the plane* are generally referred to as **precision points** or **precision positions**. The number of precision points which can be synthesized is limited by the number of equations available for solution. The fourbar linkage can be synthesized by closed-form methods for up to five precision points for motion or path generation with prescribed timing (coupler output) and up to seven points for function generation (rocker output). Synthesis for two or three precision points is relatively straightforward, and each of these cases can be reduced to a system of linear simultaneous equations easily solved on a calculator. The four or more position synthesis problems involve the solution of nonlinear, simultaneous equation systems, and so are more complicated to solve, requiring a computer.

Note that these analytical synthesis procedures provide a solution which will be able to "be at" the specified precision points, but no guarantee is provided regarding the linkage's behavior between those precision points. It is possible that the resulting linkage will be incapable of moving from one precision point to another due to the presence of a toggle position or other constraint. This situation is actually no different than that of the graphical synthesis cases in Chapter 3, wherein there was also the possibility of a toggle position between design points. In fact, these analytical synthesis methods are just an alternate way to solve the same multiposition synthesis problems. One should still build a simple cardboard model of the synthesized linkage to observe its behavior and check for the presence of problems, even if the synthesis was performed by an esoteric analytical method.

5.4   TWO-POSITION MOTION GENERATION BY ANALYTICAL SYNTHESIS

Figure 5-1 (p. 232) shows a fourbar linkage in one position with a coupler point located at a first precision position P_1. It also indicates a second precision position (point P_2) to be achieved by the rotation of the input rocker, link 2, through an as yet unspecified angle β_2. Note also that the angle of the coupler link 3 at each of the precision positions is defined by the angles of the position vectors $\mathbf{Z}_1$ and $\mathbf{Z}_2$. The angle ϕ corresponds to the angle θ_3 of link 3 in its first position. This angle is unknown at the start of the synthesis and will be found. The angle α_2 represents the angular change of link 3 from position one to position two. This angle is defined in the problem statement.

It is important to realize that the linkage as shown in the figure is schematic. Its dimensions are unknown at the outset and are to be found by this synthesis technique. Thus, for example, the length of the position vector $\mathbf{Z}_1$ as shown is not indicative of the final length of that edge of link 3, nor are the lengths (W, Z, U, V) or angles ($\theta, \phi, \sigma, \psi$) of any of the links as shown predictive of the final result.

The problem statement is:

Design a fourbar linkage which will move a line on its coupler link such that a point P on that line will be first at P_1 and later at P_2 and will also rotate the line through an angle α_2 between those two precision positions. Find the lengths and angles of the four links and the coupler link dimensions A_1P_1 and B_1P_1 as shown in Figure 5-1.

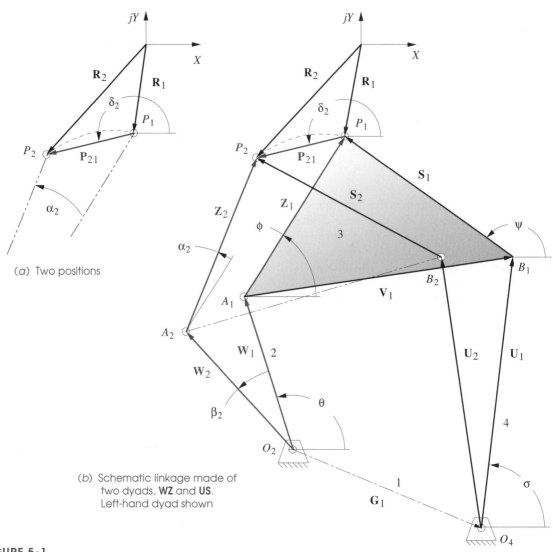

FIGURE 5-1

Two-position analytical synthesis

The two-position analytical motion synthesis procedure is as follows:

Define the two desired precision positions in the plane with respect to an arbitrarily chosen global coordinate system *XY* using position vectors $\mathbf{R}_1$ and $\mathbf{R}_2$ as shown in Figure 5-1a. The change in angle α_2 of vector $\mathbf{Z}$ is the rotation required of the coupler link. Note that the position difference vector $\mathbf{P}_{21}$ defines the displacement of the output motion of point *P* and is defined as:

$$\mathbf{P}_{21} = \mathbf{R}_2 - \mathbf{R}_1 \tag{5.1}$$

The dyad $\mathbf{W}_1\mathbf{Z}_1$ defines the left half of the linkage. The dyad $\mathbf{U}_1\mathbf{S}_1$ defines the right half of the linkage. Note that $\mathbf{Z}_1$ and $\mathbf{S}_1$ are both embedded in the rigid coupler (link 3), and both of these vectors will undergo the same rotation through angle α_2 from position 1 to position 2. The pin-to-pin length and angle of link 3 (vector $\mathbf{V}_1$) is defined in terms of vectors $\mathbf{Z}_1$ and $\mathbf{S}_1$.

$$\mathbf{V}_1 = \mathbf{Z}_1 - \mathbf{S}_1 \tag{5.2a}$$

The ground link 1 is also definable in terms of the two dyads.

$$\mathbf{G}_1 = \mathbf{W}_1 + \mathbf{V}_1 - \mathbf{U}_1 \tag{5.2b}$$

Thus if we can define the two dyads $\mathbf{W}_1$, $\mathbf{Z}_1$, and $\mathbf{U}_1$, $\mathbf{S}_1$, we will have defined a linkage that meets the problem specifications.

We will first solve for the left side of the linkage (vectors $\mathbf{W}_1$ and $\mathbf{Z}_1$) and later use the same procedure to solve for the right side (vectors $\mathbf{U}_1$ and $\mathbf{S}_1$). To solve for $\mathbf{W}_1$ and $\mathbf{Z}_1$ we need only write a vector loop equation around the loop which includes both positions P_1 and P_2 for the left-side dyad. We will go clockwise around the loop, starting with $\mathbf{W}_2$.

$$\mathbf{W}_2 + \mathbf{Z}_2 - \mathbf{P}_{21} - \mathbf{Z}_1 - \mathbf{W}_1 = 0 \tag{5.3}$$

Now substitute the complex number equivalents for the vectors.

$$we^{j(\theta+\beta_2)} + ze^{j(\phi+\alpha_2)} - p_{21}e^{j\delta_2} - ze^{j\phi} - we^{j\theta} = 0 \tag{5.4}$$

The sums of angles in the exponents can be rewritten as products of terms.

$$we^{j\theta}e^{j\beta_2} + ze^{j\phi}e^{j\alpha_2} - p_{21}e^{j\delta_2} - ze^{j\phi} - we^{j\theta} = 0 \tag{5.5a}$$

Simplifying and rearranging:

$$we^{j\theta}\left(e^{j\beta_2} - 1\right) + ze^{j\phi}\left(e^{j\alpha_2} - 1\right) = p_{21}e^{j\delta_2} \tag{5.5b}$$

Note that the lengths of vectors $\mathbf{W}_1$ and $\mathbf{W}_2$ are the same magnitude w because they represent the same rigid link in two different positions. The same can be said about vectors $\mathbf{Z}_1$ and $\mathbf{Z}_2$ whose common magnitude is z.

Equations 5.5 are vector equations, each of which contains two scalar equations and so can be solved for two unknowns. The two scalar equations can be revealed by substituting Euler's identity (equation 4.4a, p. 185) and separating the real and imaginary terms as was done in Section 4.5 (p. 187).

real part:

$$\left[w\cos\theta\right](\cos\beta_2 - 1) - \left[w\sin\theta\right]\sin\beta_2$$
$$+\left[z\cos\phi\right](\cos\alpha_2 - 1) - \left[z\sin\phi\right]\sin\alpha_2 = p_{21}\cos\delta_2 \tag{5.6a}$$

imaginary part (with complex operator j divided out):

$$\left[w\sin\theta\right](\cos\beta_2 - 1) + \left[w\cos\theta\right]\sin\beta_2$$
$$+\left[z\sin\phi\right](\cos\alpha_2 - 1) + \left[z\cos\phi\right]\sin\alpha_2 = p_{21}\sin\delta_2 \tag{5.6b}$$

There are eight variables in these two equations: w, θ, β_2, z, ϕ, α_2, p_{21}, and δ_2. We can only solve for two. Three of the eight are defined in the problem statement, namely α_2, p_{21}, and δ_2. Of the remaining five, w, θ, β_2, z, ϕ, we are forced to choose three as "free choices" (assumed values) in order to solve for the other two.

One strategy is to assume values for the three angles, θ, β_2, ϕ, on the premise that we may want to specify the orientation θ, ϕ of the two link vectors $\mathbf{W}_1$ and $\mathbf{Z}_1$ to suit packaging constraints, and also specify the angular excursion β_2 of link 2 to suit some driving constraint. This choice also has the advantage of leading to a set of equations that are linear in the unknowns and are thus easy to solve. For this solution, the equations can be simplified by setting the assumed and specified terms to be equal to some constants.

In equation 5.6a (p. 233), let:

$$A = \cos\theta(\cos\beta_2 - 1) - \sin\theta\sin\beta_2$$
$$B = \cos\phi(\cos\alpha_2 - 1) - \sin\phi\sin\alpha_2 \qquad (5.7a)$$
$$C = p_{21}\cos\delta_2$$

and in equation 5.6b (p. 233) let:

$$D = \sin\theta(\cos\beta_2 - 1) + \cos\theta\sin\beta_2$$
$$E = \sin\phi(\cos\alpha_2 - 1) + \cos\phi\sin\alpha_2 \qquad (5.7b)$$
$$F = p_{21}\sin\delta_2$$

then:

$$Aw + Bz = C$$
$$Dw + Ez = F \qquad (5.7c)$$

and solving simultaneously,

$$w = \frac{CE - BF}{AE - BD}; \qquad\qquad z = \frac{AF - CD}{AE - BD} \qquad (5.7d)$$

A second strategy is to assume a length z and angle ϕ for vector $\mathbf{Z}_1$ and the angular excursion β_2 of link 2 and then solve for the vector $\mathbf{W}_1$. This is a commonly used approach. Note that the terms in square brackets in each of equations 5.6 are respectively the x and y components of the vectors $\mathbf{W}_1$ and $\mathbf{Z}_1$.

$$W_{1_x} = w\cos\theta; \qquad\qquad Z_{1_x} = z\cos\phi$$
$$W_{1_y} = w\sin\theta; \qquad\qquad Z_{1_y} = z\sin\phi \qquad (5.8a)$$

Substituting in equation 5.6,

$$W_{1_x}(\cos\beta_2 - 1) - W_{1_y}\sin\beta_2$$
$$+ Z_{1_x}(\cos\alpha_2 - 1) - Z_{1_y}\sin\alpha_2 = p_{21}\cos\delta_2$$

$$W_{1_y}(\cos\beta_2 - 1) + W_{1_x}\sin\beta_2$$
$$+ Z_{1_y}(\cos\alpha_2 - 1) + Z_{1_x}\sin\alpha_2 = p_{21}\sin\delta_2 \qquad (5.8b)$$

Z_{1x} and Z_{1y} are known from equation 5.8a with z and ϕ assumed as free choices. To further simplify the expression, combine other known terms as:

$$A = \cos\beta_2 - 1; \qquad B = \sin\beta_2; \qquad C = \cos\alpha_2 - 1$$
$$D = \sin\alpha_2; \qquad E = p_{21}\cos\delta_2; \qquad F = p_{21}\sin\delta_2$$

(5.8c)

substituting,

$$AW_{1_x} - BW_{1_y} + CZ_{1_x} - DZ_{1_y} = E$$
$$AW_{1_y} + BW_{1_x} + CZ_{1_y} + DZ_{1_x} = F$$

(5.8d)

and the solution is:

$$W_{1_x} = \frac{A\left(-CZ_{1_x} + DZ_{1_y} + E\right) + B\left(-CZ_{1_y} - DZ_{1_x} + F\right)}{-2A}$$

$$W_{1_y} = \frac{A\left(-CZ_{1_y} - DZ_{1_x} + F\right) + B\left(CZ_{1_x} - DZ_{1_y} - E\right)}{-2A}$$

(5.8e)

Either of these strategies results in the definition of a left dyad $\mathbf{W}_1\mathbf{Z}_1$ and its pivot locations which will provide the motion generation specified.

We must repeat the process for the right-hand dyad, $\mathbf{U}_1\mathbf{S}_1$. Figure 5-2 (p. 236) highlights the two positions $\mathbf{U}_1\mathbf{S}_1$ and $\mathbf{U}_2\mathbf{S}_2$ of the right dyad. Vector $\mathbf{U}_1$ is initially at angle σ and moves through angle γ_2 from position 1 to 2. Vector $\mathbf{S}_1$ is initially at angle ψ. Note that the rotation of vector $\mathbf{S}$ from $\mathbf{S}_1$ to $\mathbf{S}_2$ is through the same angle α_2 as vector $\mathbf{Z}$, since they are in the same link. A vector loop equation similar to equation 5.3 (p. 233) can be written for this dyad.

$$\mathbf{U}_2 + \mathbf{S}_2 - \mathbf{P}_{21} - \mathbf{S}_1 - \mathbf{U}_1 = 0$$

(5.9a)

Rewrite in complex variable form and collect terms.

$$ue^{j\sigma}\left(e^{j\gamma_2} - 1\right) + se^{j\psi}\left(e^{j\alpha_2} - 1\right) = p_{21}e^{j\delta_2}$$

(5.9b)

When this is expanded and the proper angles substituted, the x and y component equations become:

real part:

$$u\cos\sigma\left(\cos\gamma_2 - 1\right) - u\sin\sigma\sin\gamma_2$$
$$+ s\cos\psi\left(\cos\alpha_2 - 1\right) - s\sin\psi\sin\alpha_2 = p_{21}\cos\delta_2$$

(5.10a)

imaginary part (with complex operator j divided out):

$$u\sin\sigma\left(\cos\gamma_2 - 1\right) + u\cos\sigma\sin\gamma_2$$
$$+ s\sin\psi\left(\cos\alpha_2 - 1\right) + s\cos\psi\sin\alpha_2 = p_{21}\sin\delta_2$$

(5.10b)

Compare equations 5.10 to equations 5.6 (p. 233).

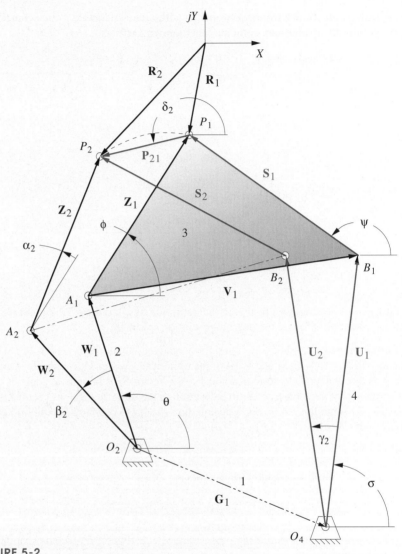

FIGURE 5-2

Right-side dyad shown in two positions

The same first strategy can be applied to equations 5.10 (p. 235) as was used for equations 5.6 (p. 233) to solve for the magnitudes of vectors $\mathbf{U}$ and $\mathbf{S}$, assuming values for angles σ, ψ, and γ_2. The quantities p_{21}, δ_2, and α_2 are defined from the problem statement as before.

In equation 5.10a let:

$$A = \cos\sigma\left(\cos\gamma_2 - 1\right) - \sin\sigma\sin\gamma_2$$
$$B = \cos\psi\left(\cos\alpha_2 - 1\right) - \sin\psi\sin\alpha_2 \qquad (5.11a)$$
$$C = p_{21}\cos\delta_2$$

and in equation 5.10b (p. 235) let:

$$D = \sin\sigma\left(\cos\gamma_2 - 1\right) + \cos\sigma\sin\gamma_2$$
$$E = \sin\psi\left(\cos\alpha_2 - 1\right) + \cos\psi\sin\alpha_2 \qquad (5.11b)$$
$$F = p_{21}\sin\delta_2$$

then:

$$Au + Bs = C$$
$$Du + Es = F \qquad (5.11c)$$

and solving simultaneously,

$$u = \frac{CE - BF}{AE - BD}; \qquad\qquad s = \frac{AF - CD}{AE - BD} \qquad (5.11d)$$

If the second strategy is used, assuming angle γ_2 and the magnitude and direction of vector $\mathbf{S}_1$ (which will define link 3), the result will be:

$$U_{1_x} = u\cos\sigma; \qquad\qquad S_{1_x} = s\cos\psi$$
$$U_{1_y} = u\sin\sigma; \qquad\qquad S_{1_y} = s\sin\psi \qquad (5.12a)$$

Substitute in equation 5.10:

$$U_{1_x}\left(\cos\gamma_2 - 1\right) - U_{1_y}\sin\gamma_2$$
$$+ S_{1_x}\left(\cos\alpha_2 - 1\right) - S_{1_y}\sin\alpha_2 = p_{21}\cos\delta_2 \qquad (5.12b)$$
$$U_{1_y}\left(\cos\gamma_2 - 1\right) + U_{1_x}\sin\gamma_2$$
$$+ S_{1_y}\left(\cos\alpha_2 - 1\right) + S_{1_x}\sin\alpha_2 = p_{21}\sin\delta_2$$

Let: $\qquad A = \cos\gamma_2 - 1; \qquad B = \sin\gamma_2; \qquad C = \cos\alpha_2 - 1$
$$D = \sin\alpha_2; \qquad E = p_{21}\cos\delta_2; \qquad F = p_{21}\sin\delta_2 \qquad (5.12c)$$

Substitute in equation 5.12b,

$$AU_{1_x} - BU_{1_y} + CS_{1_x} - DS_{1_y} = E$$
$$AU_{1_y} + BU_{1_x} + CS_{1_y} + DS_{1_x} = F \qquad (5.12d)$$

and the solution is:

$$U_{1_x} = \frac{A\left(-CS_{1_x} + DS_{1_y} + E\right) + B\left(-CS_{1_y} - DS_{1_x} + F\right)}{-2A}$$
$$\qquad\qquad\qquad\qquad\qquad\qquad\qquad\qquad (5.12e)$$
$$U_{1_y} = \frac{A\left(-CS_{1_y} - DS_{1_x} + F\right) + B\left(CS_{1_x} - DS_{1_y} - E\right)}{-2A}$$

Note that there are infinities of possible solutions to this problem because we may choose any set of values for the three free choices of variables in this two-position case. Technically there is an infinity of solutions for each free choice. Three choices then give infinity cubed solutions! But since infinity is defined as a number larger than the largest number you can think of, infinity cubed is not any more impressively large than just plain infinity. While not strictly correct mathematically, we will, for simplicity, refer to all of these cases as having "an infinity of solutions," regardless of the power to which infinity may be raised as a result of the derivation. There are plenty of solutions to pick from, at any rate. *Unfortunately, not all will work.* Some will have circuit, branch, or order (CBO) defects such as toggle positions between the precision points. Others will have poor transmission angles or poor pivot locations or overlarge links. Design judgment is still most important in selecting the assumed values for your free choices. Despite their name you must pay for those "free choices" later. Make a model!

5.5   COMPARISON OF ANALYTICAL AND GRAPHICAL TWO-POSITION SYNTHESIS

Note that in the **graphical solution** to this two-position synthesis problem (in Example 3-3, p. 105 and Figure 3-6, p. 107), we also had to make *three free choices* to solve the problem. The identical two-position synthesis problem from Figure 3-6 is reproduced in Figure 5-3. The approach taken in Example 3-3 used the two points A and B as the attachments for the moving pivots. Figure 5-3a shows the graphical construction used to find the fixed pivots O_2 and O_4. For the analytical solution we will use those points A and B as the joints of the two dyads **WZ** and **US**. These dyads meet at point P, which is the precision point. The relative position vector $\mathbf{P}_{21}$ defines the displacement of the precision point.

Note that in the graphical solution, we implicitly defined the left dyad vector **Z** by locating attachment points A and B on link 3 as shown in Figure 5-3a. This defined the two variables, z and ϕ. We also implicitly chose the value of w by selecting an arbitrary location for pivot O_2 on the perpendicular bisector. When that third choice was made, the remaining two unknowns, angles β_2 and θ, were solved for graphically at the same time, because the geometric construction was in fact a graphical "computation" for the solution of the simultaneous equations 5.8a (p. 234).

The graphical and analytical methods represent two alternate solutions to the same problem. All of these problems can be solved both analytically and graphically. One method can provide a good check for the other. We will now solve this problem analytically and correlate the results with the graphical solution from Chapter 3.

✎ EXAMPLE 5-1

Two-Position Analytical Motion Synthesis.

Problem: Design a fourbar linkage to move the link APB shown from position $A_1P_1B_1$ to $A_2P_2.B_2$.

Solution: (See Figure 5-3.)

1   Draw the link APB in its two desired positions, $A_1P_1B_1$ and $A_2P_2B_2$, to scale in the plane as shown.

2   Measure or calculate the values of the magnitude and angle of vector $\mathbf{P}_{21}$, namely, p_{21} and δ_2. In this example they are:

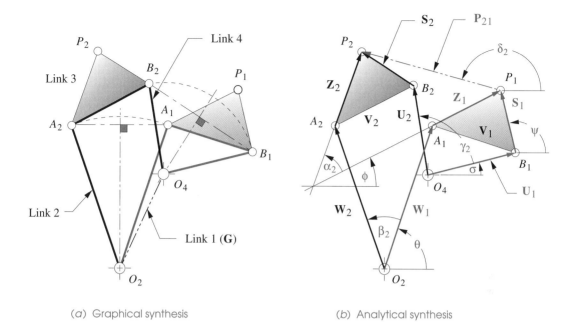

(a) Graphical synthesis   (b) Analytical synthesis

FIGURE 5-3

Two-position motion synthesis with coupler output

$$p_{21} = 2.416; \qquad \delta_2 = 165.2°$$

3   Measure or calculate the value of the change in angle, α_2, of vector $\mathbf{Z}$ from position 1 to position 2. In this example it is:

$$\alpha_2 = 43.3°$$

4   The three values in steps 2 and 3 are the only ones defined in the problem statement. We must assume three additional "free choices" to solve the problem. Method two (see equations 5.8, p. 234) chooses the length z and angle ϕ of vector $\mathbf{Z}$ and β_2, the change in angle of vector $\mathbf{W}$. In order to obtain the same solution as the graphical method produced in Figure 5-3a (from the infinities of solutions available), we will choose those values consistent with the graphical solution.

$$z = 1.298; \qquad \phi = 26.5°; \qquad \beta_2 = 38.4°$$

5   Substitute these six values in equations 5.8 and obtain:

$$w = 2.467 \qquad \theta = 71.6°$$

6   Compare these to the graphical solution;

$$w = 2.48 \qquad \theta = 71°$$

which is a reasonable match given graphical accuracy. This vector $\mathbf{W}_1$ is link 2 of the fourbar.

7   Repeat the procedure for the link-4 side of the linkage. The free choices will now be:

$$s = 1.035; \qquad \psi = 104.1°; \qquad \gamma_2 = 85.6°$$

8   Substitute these three values along with the original three values from steps 2 and 3 in equations 5.12 (p. 237) and obtain:

$$u = 1.486 \qquad \sigma = 15.4°$$

9   Compare these to the graphical solution:

$$u = 1.53 \qquad \sigma = 14°$$

These are a reasonable match for graphical accuracy. Vector $\mathbf{U}_1$ is link 4 of the fourbar.

10   Line A_1B_1 is link 3 and can be found from equation 5.2a (p. 233). Line O_2O_4 is link 1 and can be found from equation 5.2b (p. 233).

11   Check the Grashof condition, and repeat steps 4 to 7 if unsatisfied. Note that any Grashof condition is potentially acceptable in this case.

12   Construct a model in CAD or cardboard and check its function to be sure it can get from initial to final position without encountering any limit (toggle) positions.

13   Check transmission angles.

Open the file E05-01.4br in program LINKAGES to see Example 5-1.

5.6   SIMULTANEOUS EQUATION SOLUTION

These methods of analytical synthesis lead to sets of linear simultaneous equations. The two-position synthesis problem results in two simultaneous equations which can be solved by direct substitution. The three-position synthesis problem will lead to a system of four simultaneous linear equations and will require a more complicated method of solution. A convenient approach to the solution of sets of linear simultaneous equations is to put them in a standard matrix form and use a numerical matrix solver to obtain the answers. Matrix solvers are built into most engineering and scientific pocket calculators. Some spreadsheet packages and equation solvers will also do a matrix solution.

As an example, consider the following set of simultaneous equations:

$$
\begin{aligned}
-2x_1 \; -x_2 \; +x_3 &= -1 \\
x_1 \; +x_2 \; +x_3 &= 6 \\
3x_1 \; +x_2 \; -x_3 &= 2
\end{aligned}
\qquad (5.13a)
$$

A system this small can be solved longhand by the elimination method, but we will put it in matrix form to show the general approach which will work regardless of the number of equations. The equations 5.13a can be written as the product of two matrices set equal to a third matrix.

$$\begin{bmatrix} -2 & -1 & 1 \\ 1 & 1 & 1 \\ 3 & 1 & -1 \end{bmatrix} \times \begin{bmatrix} x_1 \\ x_2 \\ x_3 \end{bmatrix} = \begin{bmatrix} -1 \\ 6 \\ 2 \end{bmatrix} \qquad (5.13b)$$

We will refer to these matrices as **A**, **B**, and **C**,

$$\begin{bmatrix} \mathbf{A} \end{bmatrix} \times \begin{bmatrix} \mathbf{B} \end{bmatrix} = \begin{bmatrix} \mathbf{C} \end{bmatrix} \qquad (5.13c)$$

where **A** is the matrix of coefficients of the unknowns, **B** is a column vector of the unknown terms and **C** is a column vector of the constant terms. When matrix **A** is multiplied by **B**, the result will be the same as the left sides of equation 5.13a. See any text on linear algebra such as reference [7] for a discussion of the procedure for matrix multiplication.

If equation 5.13c were a scalar equation,

$$ab = c \qquad (5.14a)$$

rather than a vector (matrix) equation, it would be very easy to solve it for the unknown b when a and c are known. We would simply divide c by a to find b.

$$b = \frac{c}{a} \qquad (5.14b)$$

Unfortunately, division is not defined for matrices, so another approach must be used. Note that we could also express the division in equation 5.14b as:

$$b = a^{-1}c \qquad (5.14c)$$

If the equations to be solved are linearly independent, then we can find the inverse of matrix **A** and multiply it by matrix **C** to find **B**. The inverse of a matrix is defined as that matrix which when multiplied by the original matrix yields the identity matrix. The **identity matrix** is a square matrix with ones on the main diagonal and zeros everywhere else. The inverse of a matrix is denoted by adding a superscript of negative one to the symbol for the original matrix.

$$\begin{bmatrix} \mathbf{A} \end{bmatrix}^{-1} \times \begin{bmatrix} \mathbf{A} \end{bmatrix} = \begin{bmatrix} \mathbf{I} \end{bmatrix} = \begin{bmatrix} 1 & 0 & 0 \\ 0 & 1 & 0 \\ 0 & 0 & 1 \end{bmatrix} \qquad (5.15)$$

Not all matrices will possess an inverse. The determinant of the matrix must be nonzero for an inverse to exist. The class of problems dealt with here will yield matrices which have inverses provided that all data are correctly calculated for input to the matrix and represent a real physical system. The calculation of the terms of the inverse for a matrix is a complicated numerical process which requires a computer or preprogrammed pocket calculator to invert any matrix of significant size. A Gauss-Jordan-elimination numerical method is usually used to find an inverse. For our simple example in equation 5.13 the inverse of matrix **A** is found to be:

$$\begin{bmatrix} -2 & -1 & 1 \\ 1 & 1 & 1 \\ 3 & 1 & -1 \end{bmatrix}^{-1} = \begin{bmatrix} 1.0 & 0.0 & 1.0 \\ -2.0 & 0.5 & -1.5 \\ 1.0 & 0.5 & 0.5 \end{bmatrix} \qquad (5.16)$$

If the inverse of matrix $\mathbf{A}$ can be found, we can solve equations 5.13 (p. 240) for the unknowns $\mathbf{B}$ by multiplying both sides of the equation by the inverse of $\mathbf{A}$. Note that unlike scalar multiplication, matrix multiplication is not commutative; i.e., $\mathbf{A} \times \mathbf{B}$ is not equal to $\mathbf{B} \times \mathbf{A}$. We will premultiply each side of the equation by the inverse.

but:

$$[\mathbf{A}]^{-1} \times [\mathbf{A}] \times [\mathbf{B}] = [\mathbf{A}]^{-1} \times [\mathbf{C}]$$

$$[\mathbf{A}]^{-1} \times [\mathbf{A}] = [\mathbf{I}] \qquad\qquad (5.17)$$

so:

$$[\mathbf{B}] = [\mathbf{A}]^{-1} \times [\mathbf{C}]$$

The product of $\mathbf{A}$ and its inverse on the left side of the equation is equal to the identity matrix $\mathbf{I}$. Multiplying by the identity matrix is equivalent, in scalar terms, to multiplying by one, so it has no effect on the result. Thus the unknowns can be found by premultiplying the inverse of the coefficient matrix $\mathbf{A}$ times the matrix of constant terms $\mathbf{C}$.

This method of solution works no matter how many equations are present as long as the inverse of $\mathbf{A}$ can be found and enough computer memory and/or time is available to do the computation. Note that it is not actually necessary to find the inverse of matrix $\mathbf{A}$ to solve the set of equations. The Gauss-Jordan algorithm which finds the inverse can also be used to directly solve for the unknowns $\mathbf{B}$ by assembling the $\mathbf{A}$ and $\mathbf{C}$ matrices into an **augmented matrix** of n rows and $n + 1$ columns. The added column is the $\mathbf{C}$ vector. This approach requires fewer calculations, so it is faster and more accurate. The augmented matrix for this example is:

$$\begin{bmatrix} -2 & -1 & 1 & \vdots & -1 \\ 1 & 1 & 1 & \vdots & 6 \\ 3 & 1 & -1 & \vdots & 2 \end{bmatrix} \qquad\qquad (5.18a)$$

The Gauss-Jordan algorithm manipulates this augmented matrix until it is in the form shown below, in which the left, square portion has been reduced to the identity matrix and the rightmost column contains the values of the column vector of unknowns. In this case the results are $x_1 = 1$, $x_2 = 2$, and $x_3 = 3$ which are the correct solution to the original equations 5.13 (p. 240).

$$\begin{bmatrix} 1 & 0 & 0 & \vdots & 1 \\ 0 & 1 & 0 & \vdots & 2 \\ 0 & 0 & 1 & \vdots & 3 \end{bmatrix} \qquad\qquad (5.18b)$$

The program MATRIX, supplied with this text, solves these problems with this Gauss-Jordan elimination method and operates on the augmented matrix without actually finding the inverse of $\mathbf{A}$ in explicit form. See Appendix A for instructions on running program MATRIX. For a review of matrix algebra see reference [7].

5.7   THREE-POSITION MOTION GENERATION BY ANALYTICAL SYNTHESIS

The same approach of defining two dyads, one at each end of the fourbar linkage, as used for two-position motion synthesis can be extended to three, four, and five positions in the plane. The three-position motion synthesis problem will now be addressed. Figure 5-4 (p. 244) shows a fourbar linkage in one general position with a coupler point located at its first precision position P_1. Second and third precision positions (points P_2 and P_3) are also shown. These are to be achieved by the rotation of the input rocker, link 2, through as yet unspecified angles β_2 and β_3. Note also that the angles of the coupler link 3 at each of the precision positions are defined by the angles of the position vectors Z_1, Z_2, and Z_3. The linkage shown in the figure is schematic. Its dimensions are unknown at the outset and are to be found by this synthesis technique. Thus, for example, the length of the position vector Z_1 as shown is not indicative of the final length of that edge of link 3 nor are the lengths or angles of any of the links shown predictive of the final result.

The problem statement is:

Design a fourbar linkage which will move a line on its coupler link such that a point P on that line will be first at P_1, later at P_2, and still later at P_3, and also will rotate the line through an angle α_2 between the first two precision positions and through an angle α_3 between the first and third precision positions. Find the lengths and angles of the four links and the coupler link dimensions A_1P_1 and B_1P_1 as shown in Figure 5-4.

The three-position analytical motion synthesis procedure is as follows:

For convenience, we will place the global coordinate system XY at the first precision point P_1. We define the other two desired precision positions in the plane with respect to this global system as shown in Figure 5-4. The position difference vectors P_{21}, drawn from P_1 to P_2, and P_{31}, drawn from P_1 to P_3, have angles δ_2 and δ_3, respectively. The position difference vectors P_{21} and P_{31} define the displacements of the output motion of point P from point 1 to 2 and from 1 to 3, respectively.

The dyad $W_1\ Z_1$ defines the left half of the linkage. The dyad $U_1\ S_1$ defines the right half of the linkage. Vectors Z_1 and S_1 are both embedded in the rigid coupler (link 3), and both will undergo the same rotations, through angle α_2 from position 1 to position 2 and through angle α_3 from position 1 to position 3. The pin-to-pin length and angle of link 3 (vector V_1) are defined in terms of vectors Z_1 and S_1 as in equation 5.2a (p. 233). The ground link is defined by equation 5.2b as before.

As we did in the two-position case, we will first solve for the left side of the linkage (vectors W_1 and Z_1) and later use the same procedure to solve for the right side (vectors U_1 and S_1). To solve for W_1 and Z_1 we need to now write **two vector loop equations**, one around the loop which includes positions P_1 and P_2 and the second one around the loop which includes positions P_1 and P_3 (see Figure 5-4). We will go clockwise around the first loop for motion from position 1 to 2, starting with W_2, and then write the second loop equation for motion from position 1 to 3 starting with W_3.

$$W_2 + Z_2 - P_{21} - Z_1 - W_1 = 0$$
$$W_3 + Z_3 - P_{31} - Z_1 - W_1 = 0$$

$$(5.19)$$

Substituting the complex number equivalents for the vectors.

$$we^{j(\theta+\beta_2)} + ze^{j(\phi+\alpha_2)} - p_{21}e^{j\delta_2} - ze^{j\phi} - we^{j\theta} = 0$$
$$we^{j(\theta+\beta_3)} + ze^{j(\phi+\alpha_3)} - p_{31}e^{j\delta_3} - ze^{j\phi} - we^{j\theta} = 0$$

$$(5.20)$$

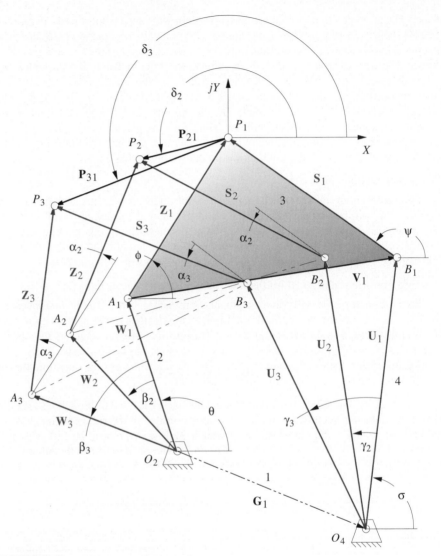

FIGURE 5-4

Three-position analytical synthesis

Rewriting the sums of angles in the exponents as products of terms.

$$we^{j\theta}e^{j\beta_2} + ze^{j\phi}e^{j\alpha_2} - p_{21}e^{j\delta_2} - ze^{j\phi} - we^{j\theta} = 0$$

$$we^{j\theta}e^{j\beta_3} + ze^{j\phi}e^{j\alpha_3} - p_{31}e^{j\delta_3} - ze^{j\phi} - we^{j\theta} = 0$$

(5.21a)

Simplifying and rearranging:

$$we^{j\theta}\left(e^{j\beta_2} - 1\right) + ze^{j\phi}\left(e^{j\alpha_2} - 1\right) = p_{21}e^{j\delta_2}$$

$$we^{j\theta}\left(e^{j\beta_3} - 1\right) + ze^{j\phi}\left(e^{j\alpha_3} - 1\right) = p_{31}e^{j\delta_3}$$

(5.21b)

The magnitude w of vectors $\mathbf{W}_1$, $\mathbf{W}_2$, and $\mathbf{W}_3$ is the same in all three positions because it represents the same line in a rigid link. The same can be said about vectors $\mathbf{Z}_1$, $\mathbf{Z}_2$, and $\mathbf{Z}_3$ whose common magnitude is z.

Equations 5.21 are a set of two vector equations, each of which contains two scalar equations. This set of four equations can be solved for four unknowns. The scalar equations can be revealed by substituting Euler's identity (equation 4.4a, p. 185) and separating the real and imaginary terms as was done in the two-position example above.

real part:

$$w\cos\theta(\cos\beta_2 - 1) - w\sin\theta\sin\beta_2$$
$$+ z\cos\phi(\cos\alpha_2 - 1) - z\sin\phi\sin\alpha_2 = p_{21}\cos\delta_2 \qquad (5.22a)$$

$$w\cos\theta(\cos\beta_3 - 1) - w\sin\theta\sin\beta_3$$
$$+ z\cos\phi(\cos\alpha_3 - 1) - z\sin\phi\sin\alpha_3 = p_{31}\cos\delta_3 \qquad (5.22b)$$

imaginary part (with complex operator j divided out):

$$w\sin\theta(\cos\beta_2 - 1) + w\cos\theta\sin\beta_2$$
$$+ z\sin\phi(\cos\alpha_2 - 1) + z\cos\phi\sin\alpha_2 = p_{21}\sin\delta_2 \qquad (5.22c)$$

$$w\sin\theta(\cos\beta_3 - 1) + w\cos\theta\sin\beta_3$$
$$+ z\sin\phi(\cos\alpha_3 - 1) + z\cos\phi\sin\alpha_3 = p_{31}\sin\delta_3 \qquad (5.22d)$$

There are **twelve variables** in these four equations 5.22: w, θ, β_2, β_3, z, ϕ, α_2, α_3, p_{21}, p_{31}, δ_2, and δ_3. **We can solve for only four.** Six of them are defined in the problem statement, namely α_2, α_3, p_{21}, p_{31}, δ_2, and δ_3. Of the remaining six, w, θ, β_2, β_3, z, ϕ, **we must choose two as free choices** (assumed values) in order to solve for the other four. One strategy is to assume values for the two angles, β_2 and β_3, on the premise that we may want to specify the angular excursions of link 2 to suit some driving constraint. (This choice also has the benefit of leading to a set of linear equations for simultaneous solution.)

This leaves the magnitudes and angles of vectors $\mathbf{W}$ and $\mathbf{Z}$ to be found (w, θ, z, ϕ). To simplify the solution, we can substitute the following relationships to obtain the x and y components of the two unknown vectors $\mathbf{W}$ and $\mathbf{Z}$, rather than their polar coordinates.

$$W_{1_x} = w\cos\theta; \qquad\qquad Z_{1_x} = z\cos\phi$$
$$\qquad\qquad\qquad\qquad\qquad\qquad\qquad\qquad\qquad\qquad (5.23)$$
$$W_{1_y} = w\sin\theta; \qquad\qquad Z_{1_y} = z\sin\phi$$

Substituting equations 5.23 into 5.22 we obtain:

$$W_{1_x}(\cos\beta_2 - 1) - W_{1_y}\sin\beta_2$$
$$+ Z_{1_x}(\cos\alpha_2 - 1) - Z_{1_y}\sin\alpha_2 = p_{21}\cos\delta_2 \qquad (5.24a)$$

$$W_{1_x}(\cos\beta_3 - 1) - W_{1_y}\sin\beta_3$$
$$+ Z_{1_x}(\cos\alpha_3 - 1) - Z_{1_y}\sin\alpha_3 = p_{31}\cos\delta_3 \qquad (5.24b)$$

$$W_{1_y}\left(\cos\beta_2 - 1\right) + W_{1_x}\sin\beta_2$$
$$+ Z_{1_y}\left(\cos\alpha_2 - 1\right) + Z_{1_x}\sin\alpha_2 = p_{21}\sin\delta_2 \tag{5.24c}$$

$$W_{1_y}\left(\cos\beta_3 - 1\right) + W_{1_x}\sin\beta_3$$
$$+ Z_{1_y}\left(\cos\alpha_3 - 1\right) + Z_{1_x}\sin\alpha_3 = p_{31}\sin\delta_3 \tag{5.24d}$$

These are four equations in the four unknowns W_{1x}, W_{1y}, Z_{1x}, and Z_{1y}. By setting the coefficients which contain the assumed and specified terms equal to some constants, we can simplify the notation and obtain the following solutions.

$$
\begin{array}{lll}
A = \cos\beta_2 - 1; & B = \sin\beta_2; & C = \cos\alpha_2 - 1 \\
D = \sin\alpha_2; & E = p_{21}\cos\delta_2; & F = \cos\beta_3 - 1 \\
G = \sin\beta_3; & H = \cos\alpha_3 - 1; & K = \sin\alpha_3 \\
L = p_{31}\cos\delta_3; & M = p_{21}\sin\delta_2; & N = p_{31}\sin\delta_3
\end{array} \tag{5.25}
$$

Substituting equations 5.25 in 5.24 to simplify:

$$AW_{1_x} - BW_{1_y} + CZ_{1_x} - DZ_{1_y} = E \tag{5.26a}$$
$$FW_{1_x} - GW_{1_y} + HZ_{1_x} - KZ_{1_y} = L \tag{5.26b}$$
$$BW_{1_x} + AW_{1_y} + DZ_{1_x} + CZ_{1_y} = M \tag{5.26c}$$
$$GW_{1_x} + FW_{1_y} + KZ_{1_x} + HZ_{1_y} = N \tag{5.26d}$$

This system can be put into standard matrix form:

$$
\begin{bmatrix}
A & -B & C & -D \\
F & -G & H & -K \\
B & A & D & C \\
G & F & K & H
\end{bmatrix}
\times
\begin{bmatrix}
W_{1_x} \\
W_{1_y} \\
Z_{1_x} \\
Z_{1_y}
\end{bmatrix}
=
\begin{bmatrix}
E \\
L \\
M \\
N
\end{bmatrix} \tag{5.27}
$$

This is the general form of equation 5.13c (p. 241). The vector of unknowns **B** can be solved for by premultiplying the inverse of the coefficient matrix **A** by the constant vector **C** or by forming the augmented matrix as in equation 5.18 (p. 242). For any numerical problem, the inverse of a 4 × 4 matrix can be found with many pocket calculators. The computer program MATRIX, supplied with this text, will also solve the augmented matrix equation.

Equations 5.25 and 5.26 solve the three-position synthesis problem for the left-hand side of the linkage using any pair of assumed values for β_2 and β_3. We must repeat the above process for the right-hand side of the linkage to find vectors **U** and **S**. Figure 5-4 (p. 244) also shows the three positions of the **US** dyad, and the angles σ, γ_2, γ_3, ψ, α_2, and α_3, which define those vector rotations for all three positions. The solution derivation for the right-side dyad, **US**, is identical to that just done for the left dyad **WZ**. The angles and vector labels are the only difference. The vector loop equations are:

$$\mathbf{U}_2 + \mathbf{S}_2 - \mathbf{P}_{21} - \mathbf{S}_1 - \mathbf{U}_1 = 0$$
$$\mathbf{U}_3 + \mathbf{S}_3 - \mathbf{P}_{31} - \mathbf{S}_1 - \mathbf{U}_1 = 0 \tag{5.28}$$

Substituting, simplifying, and rearranging,

$$ue^{j\sigma}\left(e^{j\gamma_2}-1\right)+se^{j\psi}\left(e^{j\alpha_2}-1\right)=p_{21}e^{j\delta_2}$$

$$ue^{j\sigma}\left(e^{j\gamma_3}-1\right)+se^{j\psi}\left(e^{j\alpha_3}-1\right)=p_{31}e^{j\delta_3}$$

(5.29)

The solution requires that two free choices be made. We will assume values for the angles γ_2 and γ_3. Note that α_2 and α_3 are the same as for dyad **WZ**. We will, in effect, solve for angles σ and ψ by finding the x and y components of the vectors **U** and **S**. The solution is:

$A=\cos\gamma_2-1;$	$B=\sin\gamma_2;$	$C=\cos\alpha_2-1$
$D=\sin\alpha_2;$	$E=p_{21}\cos\delta_2;$	$F=\cos\gamma_3-1$
$G=\sin\gamma_3;$	$H=\cos\alpha_3-1;$	$K=\sin\alpha_3$
$L=p_{31}\cos\delta_3;$	$M=p_{21}\sin\delta_2;$	$N=p_{31}\sin\delta_3$

(5.30)

$$AU_{1_x}-BU_{1_y}+CS_{1_x}-DS_{1_y}=E \tag{5.31a}$$

$$FU_{1_x}-GU_{1_y}+HS_{1_x}-KS_{1_y}=L \tag{5.31b}$$

$$BU_{1_x}+AU_{1_y}+DS_{1_x}+CS_{1_y}=M \tag{5.31c}$$

$$GU_{1_x}+FU_{1_y}+KS_{1_x}+HS_{1_y}=N \tag{5.31d}$$

Equations 5.31 can be solved using the approach of equations 5.27 and 5.18 (p. 242), by changing W to U and Z to S and using the definitions of the constants given in equation 5.30 in equation 5.27.

It should be apparent that there are infinities of solutions to this three-position synthesis problem as well. An inappropriate selection of the two free choices could lead to a solution which has circuit, branch, or order problems in moving among all specified positions. Thus we must check the function of the solution synthesized by this or any other method. A simple model is the quickest check.

5.8   COMPARISON OF ANALYTICAL AND GRAPHICAL THREE-POSITION SYNTHESIS

Figure 5-5 (p. 249) shows the same three-position synthesis problem as was done graphically in Example 3-6 (p. 110). Compare this figure to Figure 3-9 (p. 112). The labeling has been changed to be consistent with the notation in this chapter. The points P_1, P_2, and P_3 correspond to the three points labeled D in the earlier figure. Points A_1, A_2, and A_3 correspond to points E; points B_1, B_2, and B_3 correspond to points F. The old line AP becomes the present **Z** vector. Point P is the coupler point which will go through the specified precision points, P_1, P_2, and P_3. Points A and B are the attachment points for the rockers (links 2 and 4, respectively) on the coupler (link 3). We wish to solve for the coordinates of vectors **W**, **Z**, **U**, and **S**, which define not only the lengths of those links but also the locations of the fixed pivots O_2 and O_4 in the plane and the lengths of links 3 and 1. Link 1 is defined as vector **G** in Figure 5-4 (p. 244) and can be found from equation 5.2b (p. 233). Link 3 is vector **V** found from equation 5.2a (p. 233).

Four free choices must be made to constrain the problem to a particular solution out of the infinities of solutions available. In this case the values of link angles β_2, β_3, γ_2, and γ_3 have been chosen to be the same values as those which were found in the graphical solution to Example 3-6 (p. 110) in order to obtain the same solution as a check and comparison. Recall that in doing the graphical three-position synthesis solution to this same problem we in fact also had to make four free choices. These were the x,y coordinates of the moving pivot locations E and F in Figure 3-9 (p. 112) which correspond in concept to our four free choices of link angles here.

Example 3-5 (p. 109) also shows a graphical solution to this same problem resulting from the free choice of the x,y coordinates of points C and D on the coupler for the moving pivots (see Figure 3-8, p. 110 and Example 3-5, p. 109). We found some problems with toggle positions in that solution and redid it using points E and F as moving pivots in Example 3-6 (p. 110) and Figure 3-9 (p. 112). In effect the graphical three-position synthesis solution presented in Chapter 3 is directly analogous to the analytical solution presented here. For this analytical approach we choose to select the link angles β_2, β_3, γ_2, and γ_3 rather than the moving pivot locations E and F in order to force the resulting equations to be linear in the unknowns. The graphical solution done in the earlier examples is actually a solution of simultaneous nonlinear equations.

✎ EXAMPLE 5-2

Three-Position Analytical Motion Synthesis.

Problem: Design a fourbar linkage to move the link APB shown from position $A_1P_1B_1$ to $A_2P_2B_2$ and then to position $A_3P_3B_3$.

Solution: (See Figure 5-5.)

1   Draw the link APB in its three desired positions, $A_1P_1B_1$, $A_2P_2B_2$, and $A_3P_3B_3$ to scale in the plane as shown in the figure.

2   The three positions are then defined with respect to a global origin positioned at the first precision point P_1. The given data are the magnitudes and angles of the position difference vectors between precision points:

$$p_{21} = 2.798 \qquad \delta_2 = -31.19° \qquad p_{31} = 3.919 \qquad \delta_3 = -16.34°$$

3   The angle changes of the coupler between precision points are:

$$\alpha_2 = -45° \qquad\qquad \alpha_3 = 9.3°$$

4   The free choices assumed for the link angles are:

$$\beta_2 = 342.3° \qquad \beta_3 = 324.8° \qquad \gamma_2 = 30.9° \qquad \gamma_3 = 80.6°$$

These defined variables and the free choices are also listed on the figure.

5   Once the free choices of link angles are made, the terms for the matrices of equation 5.27 (p. 246) can be defined by solving equation 5.25 (p. 246) for the first dyad of the linkage and equation 5.30 (p. 247) for the second dyad of the linkage. For this example they evaluate to:

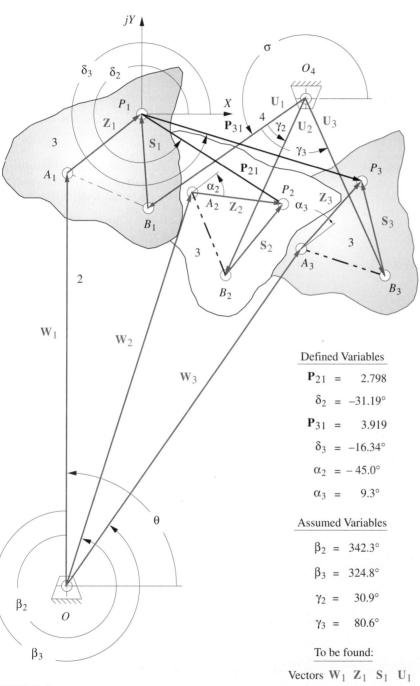

Defined Variables

$\mathbf{P}_{21}$ = 2.798

δ_2 = −31.19°

$\mathbf{P}_{31}$ = 3.919

δ_3 = −16.34°

α_2 = − 45.0°

α_3 = 9.3°

Assumed Variables

β_2 = 342.3°

β_3 = 324.8°

γ_2 = 30.9°

γ_3 = 80.6°

To be found:

Vectors $\mathbf{W}_1$ $\mathbf{Z}_1$ $\mathbf{S}_1$ $\mathbf{U}_1$

FIGURE 5-5

Data needed for three-position analytical synthesis

First dyad (**WZ**):

$A = -0.0473$	$B = -0.3040$	$C = -0.2929$	$D = -0.7071$
$E = 2.3936$	$F = -0.1829$	$G = -0.5764$	$H = -0.0131$
$K = 0.1616$	$L = 3.7607$	$M = -1.4490$	$N = -1.1026$

Second dyad (**US**):

$A = -0.1419$	$B = 0.5135$	$C = -0.2929$	$D = -0.7071$
$E = 2.3936$	$F = -0.8367$	$G = 0.9866$	$H = -0.0131$
$K = 0.1616$	$L = 3.7607$	$M = -1.4490$	$N = -1.1026$

6   Program MATRIX is used to solve this matrix equation once with the values from equation 5.25 (p. 246) inserted to get the coordinates of vectors **W** and **Z**, and a second time with values from equation 5.31 (p. 247) in the matrix to get the coordinates of vectors **U** and **S**. The calculated coordinates of the link vectors from equations 5.25 to 5.31 are:

$W_x = 0.055$	$W_y = 6.832$	$Z_x = 1.179$	$Z_y = 0.940$

Link $2 = w = 6.832$

$U_x = -2.628$	$U_y = -1.825$	$S_x = -0.109$	$S_y = 1.487$

Link $4 = u = 3.2$

7   Equation 5.2a (p. 233) is used to find link 3.

$$V_x = Z_x - S_x = 1.179 - (-0.109) = 1.288$$

$$V_y = Z_y - S_y = 0.940 - 1.487 = -0.547$$

Link $3 = v = 1.399$

8   The ground link is found from equation 5.2b (p. 233)

$$G_x = W_x + V_x - U_x = 0.055 + 1.288 - (-2.628) = 3.971$$

$$G_y = W_y + V_y - U_y = 6.832 - 0.547 - (-1.825) = 8.110$$

Link $1 = g = 9.03$

9   The appropriate vector components are added together to get the locations of fixed pivots O_2 and O_4 with respect to the global origin at precision point P_1. See Figures 5-4 (p. 244) and 5-5 (p. 249).

$$O_{2x} = -Z_x - W_x = -1.179 - 0.055 = -1.234$$

$$O_{2y} = -Z_y - W_y = -0.940 - 6.832 = -7.772$$

$$O_{4x} = -S_x - U_x = -(-0.109) - (-2.628) = 2.737$$

$$O_{4y} = -S_y - U_y = -1.487 - (-1.825) = 0.338$$

Table 5-1 shows the linkage parameters as synthesized by this method. These agree with the solution found in Example 3-6 (p. 110) within its graphical accuracy. Open the files E05-02a.mtr and E05-02b.mtr into program MATRIX to compute these results.

This problem can also be solved with program LINKAGES using the same method as derived in Section 5.7 (p. 242). Though the derivation was done in terms of the polar coordinates of the position difference vectors $\mathbf{P}_{21}$ and $\mathbf{P}_{31}$, it was considered more convenient to supply the cartesian coordinates of these vectors to program LINKAGES. (It is generally more accurate to measure x,y coordinates from a sketch of the desired positions than to measure angles with a protractor.) Thus the program requests the rectangular coordinates of $\mathbf{P}_{21}$ and $\mathbf{P}_{31}$. For this example they are:

$$p_{21x} = 2.394 \qquad p_{21y} = -1.449 \qquad p_{31x} = 3.761 \qquad p_{31y} = -1.103$$

The angles α_2 and α_3 must be measured from the diagram and supplied, in degrees. These six items constitute the set of "givens." **Note that these data are all *relative information* relating the second and third positions to the first.** No information about their absolute locations is needed. The global reference system can be taken to be anywhere in the plane. We took it to be at the first precision point position P_1 for convenience. The free choices β_2 and β_3 for the first dyad and γ_2, γ_3 for the second dyad must also be input to program LINKAGES as they also were to program MATRIX.

Program LINKAGES then solves the matrix equation 5.27 (p. 246) once with the values from equation 5.25 (p. 246) inserted to get the coordinates of vectors $\mathbf{W}$ and $\mathbf{Z}$, and a second time with values from equation 5.31 (p. 247) in the matrix to get the coordinates of vectors $\mathbf{U}$ and $\mathbf{S}$. Equations 5.2 (p. 233) are then solved to find links 1 and 3, and the appropriate vector components are added together to get the locations of fixed pivots O_2 and O_4. The link lengths are returned to the main part of program LINKAGES so that other linkage parameters can be calculated and the linkage animated.

Note that there are two ways to assemble any fourbar linkage, open and crossed (see Figure 4-5, p. 182), and this analytical synthesis technique gives no information on which

TABLE 5-1   Results of Analytical Synthesis for Example 5-2

Link Number	Analytical Solution Length Calculated (in)	Graphical Solution Length from Fig. 3-9 (in)
1	9.03	8.9
2	6.83	6.7
3	1.40	1.5
4	3.20	3.2
Coupler Pt. =	1.51 @ 61.31 degrees	1.5 @ 61 degrees
Open/Crossed =	CROSSED	CROSSED
Start Alpha2 =	0 rad/sec^2	
Start Omega2 =	1 rad/sec	
Start Theta2 =	29 degrees	
Final Theta2 =	11 degrees	
Delta Theta2 =	-9 degrees	

mode of assembly is necessary to get the desired solution. Thus you may have to try both modes of assembly in program LINKAGES to find the correct one after determining the proper link lengths with this method.

The finished linkage is the same as the one in Figure 3-9c (p. 112) that shows a driver dyad added to move links 2, 3, and 4 through the three precision points. You may open the file E05-02.4br in program LINKAGES to see the motions of the analytically synthesized fourbar solution. The linkage will move through the three positions defined in the problem statement. The file F03-09c.6br may also be opened in program LINKAGES to see the full motion of the finished sixbar linkage.

5.9   SYNTHESIS FOR A SPECIFIED FIXED PIVOT LOCATION

In Example 3-8 (p. 115) we used graphical synthesis techniques and inversion to create a fourbar linkage for three-position motion generation with specified fixed pivot locations. This is a commonly encountered problem as the available locations for fixed pivots in most machines are quite limited. Loerch et al.[4], show how we can use these analytical synthesis techniques to find a linkage with specified fixed pivots and three output positions for motion generation. In effect we will now take as our four free choices the x and y coordinates of the two fixed pivots instead of the angles of the links. This approach will lead to a set of nonlinear equations containing transcendental functions of the unknown angles.

Figure 5-6 shows the $\mathbf{WZ}$ dyad in three positions. Because we want to relate the fixed pivots of vectors $\mathbf{W}$ and $\mathbf{U}$ to our precision points, we will place the origin of our global axis system at precision point P_1. A position vector $\mathbf{R}_1$ can then be drawn from the root of vector $\mathbf{W}_1$ to the global origin at P_1, $\mathbf{R}_2$ to P_2, and $\mathbf{R}_3$ to P_3. The vector $-\mathbf{R}_1$ defines the location of the fixed pivot in the plane with respect to the global origin at P_1.

We will subsequently have to repeat this procedure for three positions of vector $\mathbf{U}$ at the right end of the linkage as we did with the three-position solution in Section 5.8 (p. 247). The procedure is presented here in detail only for the left end of the linkage (vectors $\mathbf{W}$, $\mathbf{Z}$). It is left to the reader to substitute $\mathbf{U}$ for $\mathbf{W}$ and $\mathbf{S}$ for $\mathbf{Z}$ in equations 5.32 to generate the solution for the right side.

We can write the vector loop equation for each precision position:

$$\mathbf{W}_1 + \mathbf{Z}_1 = \mathbf{R}_1$$
$$\mathbf{W}_2 + \mathbf{Z}_2 = \mathbf{R}_2 \qquad (5.32a)$$
$$\mathbf{W}_3 + \mathbf{Z}_3 = \mathbf{R}_3$$

Substitute the complex number equivalents for the vectors $\mathbf{W}_i$ and $\mathbf{Z}_i$:

$$we^{j\theta} + ze^{j\phi} = \mathbf{R}_1$$
$$we^{j(\theta+\beta_2)} + ze^{j(\phi+\alpha_2)} = \mathbf{R}_2 \qquad (5.32b)$$
$$we^{j(\theta+\beta_3)} + ze^{j(\phi+\alpha_3)} = \mathbf{R}_3$$

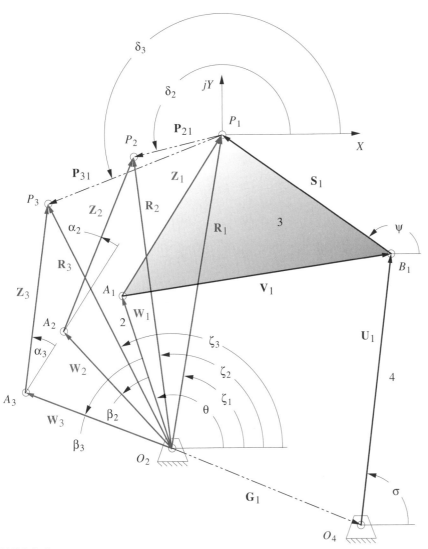

FIGURE 5-6

Three-position synthesis of a linkage with specified fixed pivot locations

Expand:

$$we^{j\theta} + ze^{j\phi} = \mathbf{R}_1$$
$$we^{j\theta}e^{j\beta_2} + ze^{j\phi}e^{j\alpha_2} = \mathbf{R}_2 \qquad (5.32c)$$
$$we^{j\theta}e^{j\beta_3} + ze^{j\phi}e^{j\alpha_3} = \mathbf{R}_3$$

Note that:

$$\mathbf{W} = we^{j\theta}; \qquad \mathbf{Z} = ze^{j\phi} \qquad (5.32d)$$

and:
$$\mathbf{W} + \mathbf{Z} = \mathbf{R}_1$$
$$\mathbf{W}e^{j\beta_2} + \mathbf{Z}e^{j\alpha_2} = \mathbf{R}_2 \qquad (5.32e)$$
$$\mathbf{W}e^{j\beta_3} + \mathbf{Z}e^{j\alpha_3} = \mathbf{R}_3$$

Previously, we chose β_2 and β_3 and solved for the vectors $\mathbf{W}$ and $\mathbf{Z}$. Now we wish to, in effect, specify the x, y components of the fixed pivot O_2 ($-R_{1x}$, $-R_{1y}$) as our two free choices. This leaves β_2 and β_3 to be solved for. These angles are contained in transcendental expressions in the equations. Note that, if we assumed values for β_2 and β_3 as before, there could only be a solution for $\mathbf{W}$ and $\mathbf{Z}$ if the determinant of the augmented matrix of coefficients of equations 5.32e were equal to zero.

$$\begin{bmatrix} 1 & 1 & \mathbf{R}_1 \\ e^{j\beta_2} & e^{j\alpha_2} & \mathbf{R}_2 \\ e^{j\beta_3} & e^{j\alpha_3} & \mathbf{R}_3 \end{bmatrix} = 0 \qquad (5.33a)$$

Expand this determinant about the first column which contains the present unknowns β_2 and β_3:

$$\left(\mathbf{R}_3 e^{j\alpha_2} - \mathbf{R}_2 e^{j\alpha_3}\right) + e^{j\beta_2}\left(\mathbf{R}_1 e^{j\alpha_3} - \mathbf{R}_3\right) + e^{j\beta_3}\left(\mathbf{R}_2 - \mathbf{R}_1 e^{j\alpha_2}\right) = 0 \qquad (5.33b)$$

To simplify, let:

$$A = \mathbf{R}_3 e^{j\alpha_2} - \mathbf{R}_2 e^{j\alpha_3}$$
$$B = \mathbf{R}_1 e^{j\alpha_3} - \mathbf{R}_3 \qquad (5.33c)$$
$$C = \mathbf{R}_2 - \mathbf{R}_1 e^{j\alpha_2}$$

then:

$$A + Be^{j\beta_2} + Ce^{j\beta_3} = 0 \qquad (5.33d)$$

Equation 5.33d expresses the summation of vectors around a closed loop. Angles β_2 and β_3 are contained within transcendental expressions making their solution cumbersome. The procedure is similar to that used for the analysis of the fourbar linkage in Section 4.5 (p. 183). Substitute the complex number equivalents for all vectors in equation 5.33d. Expand using the Euler identity (equation 4.4a, p. 185). Separate real and imaginary terms to get two simultaneous equations in the two unknowns β_2 and β_3. Square these expressions and add them to eliminate one unknown. Simplify the resulting mess and substitute the tangent half angle identities to get rid of the mixture of sines and cosines. It will ultimately reduce to a quadratic equation in the tangent of half the angle sought, here β_3. β_2 can then be found by back substituting β_3 in the original equations. The results are:[*]

* Note that a two-argument arctangent function must be used to obtain the proper quadrants for angles β_2 and β_3. Also, the minus signs in numerator and denominator of the equation for β_2 look like they could be canceled, but should not be. They are needed to determine the correct quadrant of β_2 in the two-argument arctangent function.

$$\beta_3 = 2\arctan\left(\frac{K_2 \pm \sqrt{K_1^2 + K_2^2 - K_3^2}}{K_1 + K_3}\right)$$

$$\beta_2 = \arctan\left[\frac{-(A_3 \sin\beta_3 + A_2 \cos\beta_3 + A_4)}{-(A_5 \sin\beta_3 + A_3 \cos\beta_3 + A_6)}\right] \qquad (5.34a)$$

where:
$$K_1 = A_2 A_4 + A_3 A_6$$
$$K_2 = A_3 A_4 + A_5 A_6 \qquad\qquad (5.34b)$$
$$K_3 = \frac{A_1^2 - A_2^2 - A_3^2 - A_4^2 - A_6^2}{2}$$

and:
$$A_1 = -C_3{}^2 - C_4{}^2; \qquad\qquad A_2 = C_3 C_6 - C_4 C_5$$
$$A_3 = -C_4 C_6 - C_3 C_5; \qquad\qquad A_4 = C_2 C_3 + C_1 C_4 \qquad (5.34c)$$
$$A_5 = C_4 C_5 - C_3 C_6; \qquad\qquad A_6 = C_1 C_3 - C_2 C_4$$

$$C_1 = R_3 \cos(\alpha_2 + \zeta_3) - R_2 \cos(\alpha_3 + \zeta_2)$$
$$C_2 = R_3 \sin(\alpha_2 + \zeta_3) - R_2 \sin(\alpha_3 + \zeta_2)$$
$$C_3 = R_1 \cos(\alpha_3 + \zeta_1) - R_3 \cos\zeta_3$$
$$C_4 = -R_1 \sin(\alpha_3 + \zeta_1) + R_3 \sin\zeta_3 \qquad (5.34d)$$
$$C_5 = R_1 \cos(\alpha_2 + \zeta_1) - R_2 \cos\zeta_2$$
$$C_6 = -R_1 \sin(\alpha_2 + \zeta_1) + R_2 \sin\zeta_2$$

The ten variables in these equations are: α_2, α_3, β_2, β_3, ζ_1, ζ_2, ζ_3, R_1, R_2, and R_3. The constants C_1 to C_6 are defined in terms of the eight known variables, R_1, R_2, R_3, ζ_1, ζ_2, and ζ_3 (which are the magnitudes and angles of position vectors $\mathbf{R}_1$, $\mathbf{R}_2$, and $\mathbf{R}_3$) and the angles α_2 and α_3 that define the change in angle of the coupler. See Figure 5-6 (p. 253) for depictions of these variables.

Note in equation 5.34a (p. 254) that there are two solutions for each angle (just as there were to the position analysis of the fourbar linkage in Section 4.5 and Figure 4-5, pp. 182 - 183). One solution in this case will be a trivial one wherein $\beta_2 = \alpha_2$ and $\beta_3 = \alpha_3$. The nontrivial solution is the one desired.

This procedure is then repeated, solving equations 5.34 for the right-hand end of the linkage using the desired location of fixed pivot O_4 to calculate the necessary angles γ_2 and γ_3 for link 4.

We have now reduced the problem to that of three-position synthesis without specified pivots as described in Section 5.7 (p. 242) and Example 5-2 (p. 248). In effect we have found the particular values of β_2, β_3, γ_2, and γ_3 which correspond to the solution that uses the desired fixed pivots. The remaining task is to solve for the values of W_x, W_y, Z_x, Z_y using equations 5.25 through 5.31 (pp. 246 - 247).

EXAMPLE 5-3

Three-Position Analytical Synthesis with Specified Fixed Pivots.

Problem: Design a fourbar linkage to move the line AP shown from position $A_1 P_1$ to $A_2 P_2$ and then to position $A_3 P_3$ using fixed pivots O_2 and O_4 in the locations specified.

Solution: (See Figure 5-7.)

1   Draw the link AP in its three desired positions, $A_1 P_1$, $A_2 P_2$, and $A_3 P_3$ to scale in the plane as

shown in Figure 5-7. The three positions are defined with respect to a global origin positioned at the first precision point P_1. The given data are specified in parts 2 to 4 below.

2 The position difference vectors between precision points are:

$$P_{21x} = -0.244 \qquad P_{21y} = 0.013 \qquad P_{31x} = -0.542 \qquad P_{31y} = 0.029$$

3 The angle changes of the coupler between precision points are:

$$\alpha_2 = -11.34° \qquad \alpha_3 = -22.19°$$

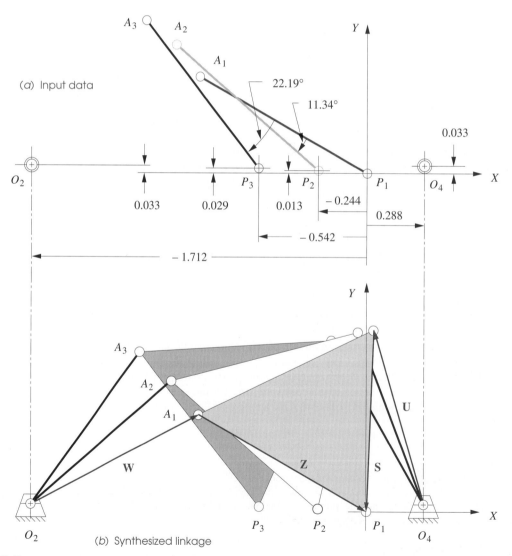

(a) Input data

(b) Synthesized linkage

FIGURE 5-7

Three-position synthesis example for specified fixed pivots

4   The assumed free choices are the desired fixed pivot locations.

$$O_{2x} = -1.712 \qquad O_{2y} = 0.033 \qquad O_{4x} = 0.288 \qquad O_{4y} = 0.033$$

5   Solve equations 5.34 (p. 254) twice, once using the O_2 pivot location coordinates and again using the O_4 pivot location coordinates.

For pivot O_2:

$$C_1 = -0.205 \qquad C_2 = 0.3390 \qquad C_3 = 0.4028$$
$$C_4 = 0.6731 \qquad C_5 = 0.2041 \qquad C_6 = 0.3490$$
$$A_1 = -0.6152 \qquad A_2 = 0.0032 \qquad A_3 = -0.3171$$
$$A_4 = -0.0017 \qquad A_5 = -0.0032 \qquad A_6 = -0.3108$$
$$K_1 = 0.0986 \qquad K_2 = 0.0015 \qquad K_3 = 0.0907$$

The values found for the link angles to match this choice of fixed pivot location O_2 are:

$$\beta_2 = 11.96° \qquad\qquad \beta_3 = 23.96°$$

For pivot O_4:

$$C_1 = -0.3144 \qquad C_2 = -0.0231 \qquad C_3 = 0.5508$$
$$C_4 = -0.0822 \qquad C_5 = 0.2431 \qquad C_6 = -0.0443$$
$$A_1 = -0.3102 \qquad A_2 = -0.0044 \qquad A_3 = -0.1376$$
$$A_4 = 0.0131 \qquad A_5 = 0.0044 \qquad A_6 = -0.1751$$
$$K_1 = 0.0240 \qquad K_2 = -0.0026 \qquad K_3 = 0.0232$$

The values found for the link angles to match this choice of fixed pivot location O_4 are:

$$\gamma_2 = 2.78° \qquad\qquad \gamma_3 = 9.96°$$

6   At this stage, the problem has been reduced to the same one as in the previous section; i.e., find the linkage given the free choices of the above angles β_2, β_3, γ_2, γ_3, using equations 5.25 through 5.31 (pp. 246 - 247). The data needed for the remaining calculations are those given in steps 2, 3, and 5 of this example, namely:

for dyad 1:

$$P_{21x} \qquad P_{21y} \qquad P_{31x} \qquad P_{31y} \qquad a_2 \qquad a_3 \qquad b_2 \qquad b_3$$

for dyad 2:

$$P_{21x} \qquad P_{21y} \qquad P_{31x} \qquad P_{31y} \qquad a_2 \qquad a_3 \qquad g_2 \qquad g_3$$

See Example 5-2 (p. 248) and Section 5.7 (p. 242) for the procedure. A matrix solving calculator, *Mathcad, TKSolver, Matlab*, program MATRIX, or program LINKAGES will solve this and compute the coordinates of the link vectors:

$$W_x = 0.866 \qquad W_y = 0.500 \qquad Z_x = 0.846 \qquad Z_y = -0.533$$

$$U_x = -0.253 \qquad U_y = 0.973 \qquad S_x = -0.035 \qquad S_y = -1.006$$

7   The link lengths are computed as was done in Example 5-2 (p. 248) and are shown in Table 5-2.

This example can be opened in program LINKAGES from the file E05-03.4br and animated.

5.10   CENTER-POINT AND CIRCLE-POINT CIRCLES

It would be quite convenient if we could find the loci of all possible solutions to the three-position synthesis problem, as we would then have an overview of the potential locations of the ends of the vectors $\mathbf{W}$, $\mathbf{Z}$, $\mathbf{U}$, and $\mathbf{S}$. Loerch et al.[5] show that by holding one of the free choices (say β_2) at an arbitrary value, and then solving equations 5.25 and 5.26 (p. 246) while iterating the other free choice (β_3) through all possible values from 0 to 2π, a circle will be generated. This circle is the locus of all possible locations of the root of vector $\mathbf{W}$ (for the particular value of β_2 used). The root of the vector $\mathbf{W}$ is the location of the fixed pivot or *center* O_2. Thus, this circle is called a **center-point circle**. The vector $\mathbf{N}$ in Figure 5-8 defines points on the *center-point* circle with respect to the global coordinate system which is placed at precision point P_1 for convenience.

If the same thing is done for vector $\mathbf{Z}$, holding α_2 constant at some arbitrary value and iterating α_3 from 0 to 2π, another circle will be generated. This circle is the locus of all possible locations of the root of vector $\mathbf{Z}$ for the chosen value of α_2. Because the root of vector $\mathbf{Z}$ is joined to the tip of vector $\mathbf{W}$ and $\mathbf{W}$'s tip describes a circle about pivot O_2 in the finished linkage, this locus is called the **circle-point circle**. Vector $(-\mathbf{Z})$ defines points on the *circle-point* circle with respect to the global coordinate system.

The x,y components of vectors $\mathbf{W}$ and $\mathbf{Z}$ are defined by equations 5.25 and 5.26 (p. 246). Negating the x,y components of $\mathbf{Z}$ will give the coordinates of points on the circle-point circle for any assumed value of α_2 as angle α_3 is iterated from 0 to 2π. The x,y components of $\mathbf{N} = -\mathbf{Z} - \mathbf{W}$ define points on the O_2 center-point circle for any assumed value of β_2 as β_3 is iterated through 0 to 2π. Vector $\mathbf{W}$ is calculated using angles β_2 and β_3, and vector $\mathbf{Z}$ using angles α_2 and α_3, both from equations 5.25 and 5.26.

For the right-hand dyad, there will also be separate center-point circles and circle-point circles. The x,y components of $\mathbf{M} = -\mathbf{S} - \mathbf{U}$ define points on the O_4 center-point circle for any assumed value of γ_2 as γ_3 is iterated through 0 to 2π. (See Figure 5-8 and also Figure 5-4, p. 244.) Negating the x,y components of $\mathbf{S}$ will give the coordinates of points on the circle-point circle for any assumed value of α_2 as α_3 is iterated through 0 to 2π. Vector $\mathbf{U}$ is calculated using angles γ_2 and γ_3, and vector $\mathbf{S}$ using angles α_2 and α_3, both from equations 5.30 and 5.31 (p. 247).

Note that there is still an infinity of solutions because we are choosing the value of one angle arbitrarily. Thus there will be an **infinite number of sets of center-point and circle-point circles**. A computer program can be of help in choosing a linkage design which has pivots in convenient locations. Program LINKAGES, provided with this text, will calculate the solutions to the analytical synthesis equations derived in this section, for user-selected values of all the free choices needed for three-position synthesis, both with and without specification of fixed pivot locations. Information about the computer program LINKAGES is in Appendix A.

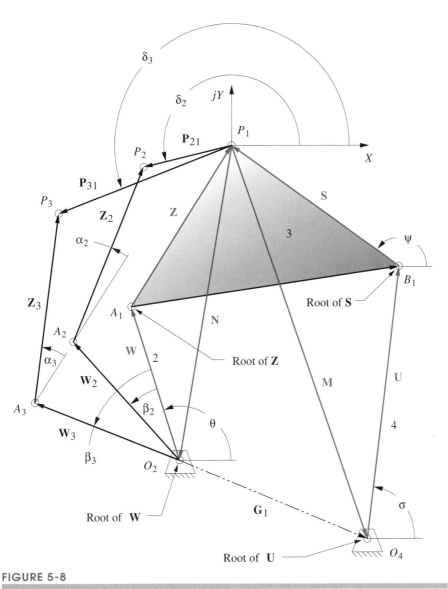

FIGURE 5-8

Definition of vectors to define center-point and circle-point circles

Figure 5-9 (p. 260) shows the circle-point and center-point circles for the Chebyschev straight-line linkage for choices of $\beta_2 = 26°$, $\alpha_2 = 97.41°$, $\alpha_3 = 158.18°$ for the left dyad and $\gamma_2 = 36°$, $\alpha_2 = 97.41°$, $\alpha_3 = 158.18°$ for the right dyad. In this example the two larger circles are the center-point circles which define the loci of possible fixed pivot locations O_2 and O_4. The smaller two circles define the loci of possible moving pivot locations I_{23} and I_{34}. Note that the coordinate system has its origin at the reference precision point, in this case P_1, from which we measured all parameters used in the analysis. These circles define the pivot loci of all possible linkages which will reach the three precision points P_1,

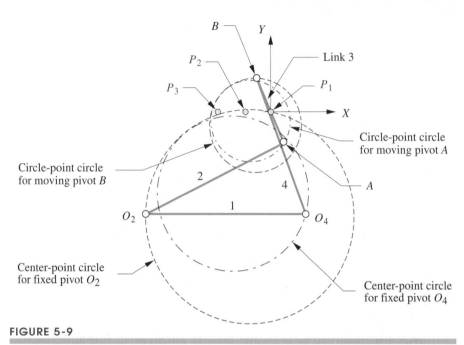

FIGURE 5-9

Circle-point and center-point circles and a linkage that reaches the precision points

P_2, and P_3 that were specified for particular choices of angles β_2, γ_2, and α_2. An example linkage is drawn on the diagram to illustrate one possible solution.

5.11 FOUR- AND FIVE-POSITION ANALYTICAL SYNTHESIS

The same techniques derived above for two- and three-position synthesis can be extended to four and five positions by writing more vector loop equations, one for each precision point. To facilitate this we will now put the vector loop equations in a more general form, applicable to any number of precision positions. Figure 5-4 (p. 244) will still serve to illustrate the notation for the general solution. The angles α_2, α_3, β_2, β_3, γ_2, and γ_3 will now be designated as α_k, β_k, and γ_k, $k = 2$ to n, where k represents the precision position and $n = 2, 3, 4,$ or 5 represents the total number of positions to be solved for. The vector loop general equation set then becomes:

$$\mathbf{W}_k + \mathbf{Z}_k - \mathbf{P}_{k1} - \mathbf{Z}_1 - \mathbf{W}_1 = 0, \qquad k = 2 \text{ to } n \qquad (5.35a)$$

which, after substituting the complex number forms and simplifying, becomes:

$$we^{j\theta}\left(e^{j\beta_k} - 1\right) + ze^{j\phi}\left(e^{j\alpha_k} - 1\right) = p_{k1}e^{j\delta_k}, \qquad k = 2 \text{ to } n \qquad (5.35b)$$

This can be put in a more compact form by substituting vector notation for those terms to which it applies, let:

$$\mathbf{W} = we^{j\theta}; \qquad\qquad \mathbf{Z} = ze^{j\phi}; \qquad\qquad \mathbf{P}_{k1} = p_{k1}e^{j\delta_k} \qquad (5.35c)$$

then:

$$\mathbf{W}\left(e^{j\beta_k}-1\right)+\mathbf{Z}\left(e^{j\alpha_k}-1\right)=\mathbf{P}_{k1}e^{j\delta_k}, \qquad k=2 \text{ to } n \qquad (5.35d)$$

Equation 5.35d is called the **standard form equation** by Erdman and Sandor.[6] By substituting the values of α_k, β_k, and δ_k, in equation 5.35d for all the precision positions desired, the requisite set of simultaneous equations can be written for the left dyad of the linkage. The standard form equation applies to the right-hand dyad **US** as well, with appropriate changes to variable names as required.

$$\mathbf{U}\left(e^{j\beta_k}-1\right)+\mathbf{S}\left(e^{j\alpha_k}-1\right)=\mathbf{P}_{k1}e^{j\delta_k}, \qquad k=2 \text{ to } n \qquad (5.35e)$$

The number of resulting equations, variables, and free choices for each value of n is shown in Table 5-3 (after Erdman and Sandor). They provide solutions for the four- and five-position problems in reference [6]. The circle-point and center-point circles of the three-position problem become cubic curves, called **Burmester curves**, in the four-position problem. Erdman et al.'s commercially available computer program LINCAGES[8] solves the **four-position problem** in an interactive way, allowing users to select center and circle pivot locations on their Burmester curve loci, which are drawn on the graphics screen of the computer.

5.12 ANALYTICAL SYNTHESIS OF A PATH GENERATOR WITH PRESCRIBED TIMING

The approach derived above for motion generation synthesis is also applicable to the case of **path generation with prescribed timing**. In path generation, the precision points are to be reached, but the angle of a line on the coupler is not of concern. Instead, the timing at which the coupler reaches the precision point is specified in terms of input rocker angle β_2. In the three-position motion generation problem we specified the angles α_2 and α_3 of vector $\mathbf{Z}$ in order to control the angle of the coupler. Here we instead want to specify angles β_2 and β_3 of the input rocker, to define the timing. Before, the free choices were β_2 and β_3. Now they will be α_2 and α_3. In either case, all four angles are either specified or assumed as free choices and the solution is identical. Figure 5-4 (p. 244) and equations 5.25, 5.26, 5.30, and 5.31 (pp. 246 - 247) apply to this case as well. This case can be extended to as many as five precision points as shown in Table 5-3.

TABLE 5-3   Number of Variables and Free Choices for Analytical Precision Point Motion and Timed Path Synthesis.[7]

No. of Positions (n)	No. of Scalar Variables	No. of Scalar Equations	No. of Prescribed Variables	No. of Free Choices	No. of Available Solutions
2	8	2	3	3	∞^3
3	12	4	6	2	∞^2
4	16	6	9	1	∞^1
5	20	8	12	0	Finite

5.13 ANALYTICAL SYNTHESIS OF A FOURBAR FUNCTION GENERATOR

A similar process to that used for the synthesis of path generation with prescribed timing can be applied to the problem of function generation. In this case we do not care about motion of the coupler at all. In a fourbar function generator, the coupler exists only to **couple** the input link to the output link. Figure 5-10 shows a fourbar linkage in three positions. Note that the coupler, link 3, is merely a line from point A to point P. Point P can be thought of as a coupler point which happens to coincide with the pin joint between links 3 and 4. As such it will have simple arc motion, pivoting about O_4, rather than, for example, the higher-order path motion of the coupler point P_1 in Figure 5-4 (p. 244).

Our **function generator** uses *link 2 as the input link and takes the output from link 4*. The "**function**" generated is the **relationship between the angles of link 2 and link 4** for the specified three-position positions, P_1, P_2, and P_3. These are located in the plane with respect to an arbitrary global coordinate system by position vectors $\mathbf{R}_1$, $\mathbf{R}_2$, and $\mathbf{R}_3$. The function is:

$$\gamma_k = f\left(\beta_k\right), \qquad k = 1, 2, \ , n; \qquad n \le 7 \qquad (5.36)$$

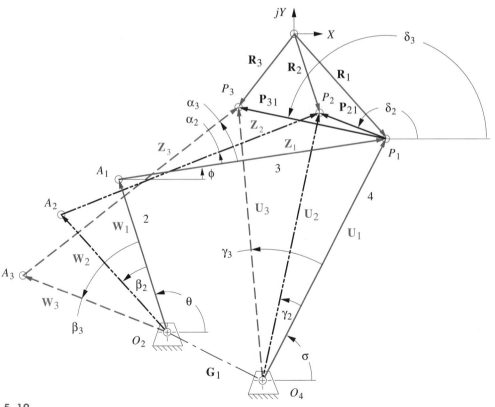

FIGURE 5-10

Analytical synthesis of a fourbar function generator

This is *not* a continuous function. The relationship holds only for the discrete points (k) specified.

To synthesize the lengths of the links needed to satisfy equation 5.36, we will write vector loop equations around the linkage in pairs of positions, as was done for the previous examples. However, we now wish to include both link 2 and link 4 in the loop, since link 4 is the output. See Figure 5-10.

$$\mathbf{W}_2 + \mathbf{Z}_2 - \mathbf{U}_2 + \mathbf{U}_1 - \mathbf{Z}_1 - \mathbf{W}_1 = 0$$
$$\mathbf{W}_3 + \mathbf{Z}_3 - \mathbf{U}_3 + \mathbf{U}_1 - \mathbf{Z}_1 - \mathbf{W}_1 = 0$$

$$(5.37a)$$

rearranging:

$$\mathbf{W}_2 + \mathbf{Z}_2 - \mathbf{Z}_1 - \mathbf{W}_1 = \mathbf{U}_2 - \mathbf{U}_1$$
$$\mathbf{W}_3 + \mathbf{Z}_3 - \mathbf{Z}_1 - \mathbf{W}_1 = \mathbf{U}_3 - \mathbf{U}_1$$

$$(5.37b)$$

but,

$$\mathbf{P}_{21} = \mathbf{U}_2 - \mathbf{U}_1$$
$$\mathbf{P}_{31} = \mathbf{U}_3 - \mathbf{U}_1$$

$$(5.37c)$$

substituting:

$$\mathbf{W}_2 + \mathbf{Z}_2 - \mathbf{Z}_1 - \mathbf{W}_1 = \mathbf{P}_{21}$$
$$\mathbf{W}_3 + \mathbf{Z}_3 - \mathbf{Z}_1 - \mathbf{W}_1 = \mathbf{P}_{31}$$

$$(5.37d)$$

$$we^{j(\theta+\beta_2)} + ze^{j(\phi+\alpha_2)} - ze^{j\phi} - we^{j\theta} = p_{21}e^{j\delta_2}$$
$$we^{j(\theta+\beta_3)} + ze^{j(\phi+\alpha_3)} - ze^{j\phi} - we^{j\theta} = p_{31}e^{j\delta_3}$$

$$(5.37e)$$

Note that equations 5.37d and 5.37e are identical to equations 5.19 and 5.20 (p. 243) derived for the three-position motion generation case and can also be put into Erdman's *standard form*[6] of equation 5.35 (p. 260) for the *n*-position case. The twelve variables in equation 5.37e are the same as those in equation 5.20 (p. 243): w, θ, β_2, β_3, z, ϕ, α_2, α_3, p_{21}, p_{31}, δ_2, and δ_3.

For the three-position function generation case the solution procedure then can be the same as that described by equations 5.20 through 5.27 (pp. 243 - 246) for the motion synthesis problem. In other words, the solution equations are the same for **all three types** of kinematic synthesis, *function generation, motion generation*, and *path generation with prescribed timing*. This is why Erdman and Sandor called equation 5.35 the **standard form equation**. To develop the data for the function generation solution, expand equation 5.37b (p. 263):

$$we^{j(\theta+\beta_2)} + ze^{j(\phi+\alpha_2)} - ze^{j\phi} - we^{j\theta} = ue^{j(\sigma+\gamma_2)} - ue^{j\sigma}$$
$$we^{j(\theta+\beta_3)} + ze^{j(\phi+\alpha_3)} - ze^{j\phi} - we^{j\theta} = ue^{j(\sigma+\gamma_3)} - ue^{j\sigma}$$

$$(5.37f)$$

There are also **twelve variables** in equation 5.37f: w, θ, z, ϕ, α_2, α_3, β_2, β_3, u, σ, γ_2, and γ_3. We can solve for any four. Four angles, β_2, β_3, γ_2, and γ_3 are specified from the function to be generated in equation 5.36 (p. 262). This leaves **four free choices**. In the function generation problem it is often convenient to define the length of the output

TABLE 5-4   Number of Variables and Free Choices for Function Generation
               Synthesis.[7]

No. of Positions (n)	No. of Scalar Variables	No. of Scalar Equations	No. of Prescribed Variables	No. of Free Choices	No. of Available Solutions
2	8	2	1	5	∞^5
3	12	4	4	4	∞^4
4	16	6	7	3	∞^3
5	20	8	10	2	∞^2
6	24	10	13	1	∞^1
7	28	12	16	0	Finite

rocker, u, and its initial angle σ to suit the package constraints. Thus, selecting the components u and σ of vector $\mathbf{U}_1$ can provide two convenient free choices of the four required.

With u, σ, γ_2, and γ_3 known, $\mathbf{U}_2$ and $\mathbf{U}_3$ can be found. Vectors $\mathbf{P}_{21}$ and $\mathbf{P}_{31}$ can then be found from equation 5.37c (p. 263). Six of the unknowns in equation 5.37e are then defined, namely, β_2, β_3, p_{21}, p_{31}, δ_2, and δ_3. Of the remaining six (w, θ, z, ϕ, α_2, α_3), we must assume values for two more as free choices in order to solve for the remaining four. We will assume values (free choices) for the two angles, α_2 and α_3 (as was done for path generation with prescribed timing) and solve equations 5.37e for the components of $\mathbf{W}$ and $\mathbf{Z}$ (w, θ, z, ϕ). We have now reduced the problem to that of Section 5.7 (p. 242) and Example 5-2 (p. 248). See equations 5.20 through 5.27 (pp. 243 - 246) for the solution.

Having chosen vector $\mathbf{U}_1$ (u, σ) as a free choice in this case, we only have to solve for one dyad, $\mathbf{WZ}$. Though we arbitrarily choose the length of vector $\mathbf{U}_1$, the resulting function generator linkage can be scaled up or down to suit packaging constraints without affecting the input/output relation defined in equation 5.36 (p. 262), because it is a function of angles only. This fact is not true for the motion or path generation cases, as scaling them will change the absolute coordinates of the path or motion output precision points which were specified in the problem statement.

Table 5-4 shows the relationships between number of positions, variables, free choices, and solutions for the function generation case. Note that up to seven angular output positions can be solved for with this method.

5.14   OTHER LINKAGE SYNTHESIS METHODS

Many other techniques for the synthesis of linkages to provide a prescribed motion have been created or discovered in recent years. Most of these approaches are somewhat involved and many are mathematically complicated. Only a few allow a closed-form solution; most require an iterative numerical solution. Most address the path synthesis problem with or without concern for prescribed timing. As Erdman and Sandor point out, the path, motion, and function generation problems are closely related.[6]

Space does not permit a complete exposition of even one of these approaches in this text. We choose instead to present brief synopses of a number of synthesis methods along with complete references to their full descriptions in the engineering and scientific literature. The reader interested in a detailed account of any method listed may consult the referenced papers which can be obtained through any university library or large public library. Also, some of the authors of these methods may make copies of their computer code available to interested parties.

Table 5-5 (p. 266) summarizes some of the existing fourbar linkage synthesis methods and for each one lists the method type, the maximum number of positions synthesized, the approach, special features, and a bibliographic reference (see the end of this chapter for the complete reference). The list in Table 5-5 is not exhaustive; other methods than these also exist.

The listed methods are divided into three types labeled **precision**, **equation**, and **optimized** (first column of Table 5-5). By **precision** (from precision point) is meant a method, such as the ones described in previous sections of this chapter, that attempts to find a solution that will pass exactly through the desired (precision) points but may deviate from the desired path between these points. Precision point methods are limited to matching a number of points equal to the number of independently adjustable parameters that define the mechanism. For a fourbar linkage, this is nine.[*] (Higher-order linkages with more links and joints will have a larger number of possible precision points.)

For up to 5 precision points in the fourbar linkage, the equations can be solved in closed form without iteration. (The four-point solution is used as a tool to solve for 5 positions in closed form, but for 6 points or more the nonlinear equations are difficult to handle.) For 6 to 9 precision points an iterative method is needed to solve the equation set. There can be problems of nonconvergence, or convergence to singular or imaginary solutions, when iterating nonlinear equations. Regardless of the number of points solved for, the solution found may be unusable due to circuit, branch, or order (CBO) defects. A circuit defect means that the linkage must be disassembled and reassembled to reach some positions, and a branch defect means that a toggle position is encountered between successive positions (see Section 4.13, p. 208). An order defect means that the points are all reachable on the same branch but are encountered in the wrong order.

The type labeled **equation** in Table 5-5 refers to methods that solve the tricircular, trinodal sextic coupler curve to find a linkage that will generate an entire coupler curve that closely approximates a set of desired points on the curve.

The type labeled **optimized** in Table 5-5 refers to an iterative optimization procedure that attempts to minimize an **objective function** that can be defined in many ways, such as the least-squares deviation between the calculated and desired coupler point positions, for example. The calculated points are found by solving a set of equations that define the behavior of the linkage geometry, using assumed initial values for the linkage parameters. A set of inequality constraints that limit the range of variation of parameters such as link length ratios, Grashof condition, or transmission angle may also be included in the calculation. New values of linkage parameters are generated with each iteration step according to the particular optimization scheme used. The closest achievable fit between the calculated solution points and the desired points is sought, defined as minimization of the chosen objective function. None of the desired points will be exactly matched by these methods, but for most engineering tasks this is an acceptable result.

---

[*] The nine independent parameters of a fourbar linkage are: four link lengths, two coordinates of the coupler point with respect to the coupler link, and three parameters that define the location and orientation of the fixed link in the global coordinate system.

TABLE 5-5   Some Methods for the Analytic Synthesis of Linkages

Type	Max Pos.	Approach	Special Features	Bibliography	References
Precision	4	Loop equations—closed form	Linear equations extendable to five positions	Freudenstein (1959) Sandor (1959) Erdman (1981)	1, 2, 4, 5, 6, 8, 10
Precision	5	Loop equations—Newton-Raphson	Uses displacement matrix	Suh (1967)	11
Precision	5	Loop equations—continuation	Specified fixed pivots, specified moving pivots	Morgan (1990) Subbian (1991)	14, 15, 16, 17
Precision	7	Closed form 5 pt.—iterative to 7 pt.	Extendable to Watt I sixbar	Tylaska (1994)	19, 20
Precision	9	Loop equations—Newton-Raphson	Exhaustive solution	Morgan (1987) Wampler (1992)	12, 13, 18
Equation	10	Coupler curve eqn.	Iterative solution	Blechschmidt (1986)	21
Equation	15	Coupler curve eqn.	Builds on Blechschmidt	Ananthasuresh (1993)	22
Optimized	N	Loop equations—least squares	Specified fixed pivots, control force and torque	Fox (1966)	24
Optimized	N	Loop equations—various criteria	Path or function generation	Youssef (1975)	25
Optimized	N	Least squares on linear equations	Prescribed timing, rapid convergence	Nolle (1971)	9
Optimized	N	Selective precision synthesis (SPS)	Relaxes precision requirements	Kramer (1975)	26, 27
Optimized	N	SPS + fuzzy logic	Extends Kramer's SPS	Krishnamurthi (1993)	28
Optimized	N	Quasi-precision pos.	Builds on Kramer	Mirth (1994)	29
Optimized	3 or 4	Loop equations and dynamic criteria	Kinematics and dynamic forces and torques	Conte (1975) Kakatsios (1987)	30, 31, 32
Optimized	N	Loop equations—least squares	Avoids branch problems, rapid convergence	Angeles (1988)	33
Optimized	N	Energy method	FEA approach	Aviles (1994)	34
Optimized	N	Genetic algorithm	Whole curve synthesis	Fang (1994)	35
Optimized	N	Fourier descriptors	Whole curve synthesis	Ullah (1996)	36, 37
Optimized	N	Neural network	Whole curve synthesis	Vasilio (1998)	38
Optimized	2, 3, or 4	Loop equations—various criteria	Automatic generation CBO defect-free	Bawab (1997)	39
Optimized	N	Approximate—continuation	All solutions—no initial guess required	Liu (1999)	40

Optimization methods allow larger numbers of points to be specified than do the precision methods, limited only by available computer time and numerical roundoff error. Table 5-5 shows a variety of optimization schemes ranging from the mundane (least squares) to the esoteric (fuzzy logic, genetic algorithms). All require a computer-programmed solution. Most can be run on current desktop computers in reasonably short times. Each different optimization approach has advantages and disadvantages in respect to convergence, accuracy, reliability, complexity, speed, and computational burden. Convergence often depends on a good choice of initial assumptions (guess values) for the linkage parameters. Some methods, if they converge at all, do so to a local minimum (only one of many possible solutions), and it may not be the best one for the task.

Precision Point Methods

Table 5-5 shows several precision point synthesis methods. Some of these are based on original work by Freudenstein and Sandor.[10] Sandor [1] and Erdman [2], [6] developed this approach into the "standard form" which is described in detail in this chapter. This method yields closed-form solutions for 2, 3, and 4 precision positions and is extendable to 5 positions. It suffers from the possible circuit, branch, and order (CBO) defects common to all precision point methods.

The method of Suh and Radcliffe [11] is quite similar to that of Freudenstein and others [1], [2], [6], [10] but leads to a set of simultaneous nonlinear equations which are solved for up to 5 positions using the Newton-Raphson numerical method (see Section 4.14, p. 210). This approach adds to the usual CBO problems the possibilities of nonconvergence, or convergence to singular or imaginary solutions.

Recent developments in the mathematical theory of polynomials have created new methods of solution called **continuation methods** (also called **homotopy methods**) which do not suffer from the same convergence problems as other methods and can also determine all the solutions of the equations starting from any set of assumed values. [12], [13] Continuation methods are a general solution to this class of problem and are reliable and fast enough to allow multiple designs to be investigated in a reasonable time (typically measured in CPU **hours** on a powerful computer).

Several researchers have developed solutions for the 5- to 9-precision point problem using this technique. Morgan and Wampler [14] solved the fourbar linkage 5-point problem with specified fixed pivots completely and found a maximum of 36 real solutions. Subbian and Flugrad [15] used specified moving pivots for the 5-point problem, extended the 5-point method to sixbar linkages,[16] and also synthesized eightbar and geared fivebar mechanisms for 6 and 7 precision points using continuation methods. [17]

Only the continuation method has yet been able to completely solve the fourbar linkage 9-precision-point problem and yield all its possible solutions. Wampler, Morgan, and Sommese [18] used a combination of analytical equation reduction and numerical continuation methods to exhaustively compute all possible nondegenerate, generic solutions to the 9-point problem.* They proved that there is a maximum of 4326 distinct, nondegenerate linkages (occurring in 1442 sets of cognate triples) that will potentially solve a generic 9-precision-point fourbar problem. Their method does not eliminate physically impossible (complex link) linkages or those with CBO defects. These still have to be removed by examination of the various solutions. They also solved four examples and found the maximum number of linkages with real link lengths that generated these particular 9-point paths to be, respectively, 21, 45, 64, and 120 cognate triples. Computation times ranged from 69 to 321 CPU minutes on an IBM 3090 for these four examples.

Tylaska and Kazerounian [19], [20] took a different approach and devised a method that synthesizes a fourbar linkage for up to 7 precision points and also synthesized a Watt I sixbar linkage for up to six body guidance (motion specification) positions with control

---

* The authors report that this calculation took 332 CPU hours on an IBM 3081 computer.

over locations of some ground and moving pivots. Their method yields the entire set of solutions for any set of design data and is an improvement over iterative methods that are sensitive to initial guesses. It is less computationally intensive than continuation methods.

Coupler Curve Equation Methods

Blechschmidt and Uicker [21] and Ananthasuresh and Kota [22] used the algebraic coupler curve equation rather than a vector loop approach to calculate the coupler point path. The equation of the coupler curve is a tricircular, trinodal sextic of 15 terms. Beyer[41] gives one form of the coupler curve equation as:*

$$a^2\left[(x-k)^2+y^2\right]\left(x^2+y^2+b^2-r^2\right)^2 -2ab\left[\left(x^2+y^2-kx\right)\cos\gamma+ky\sin\gamma\right]$$
$$\left(x^2+y^2+b^2-r^2\right)\left[(x-k)^2+y^2+a^2-R^2\right]+b^2\left(x^2+y^2\right)$$
$$\left[(x-k)^2+y^2+a^2-R^2\right]^2 -4a^2b^2\left[\left(x^2+y^2-kx\right)\sin\gamma-ky\cos\gamma\right]^2=0$$
(5.38)

Nolle [23] states that:

> The coupler curve equation itself is very complex and as far as is known in the study of mechanics (or for that matter elsewhere) no other mathematical result has been found having algebraic characteristics matching those of the coupler curve.

Its solution is quite involved and requires iteration. Blechschmidt and Uicker's approach [21] chose coordinates for 10 points on the desired curve. Ananthasuresh and Kota used 15 points with some trial and error required in their selection. The advantage of these coupler curve equation approaches is that they define the entire curve which can be plotted and examined for suitability and defects prior to calculating the link dimensions, which requires significant additional computing time.

Optimization Methods

The methods listed as **optimized** in Table 5-5 (p. 266) are a diverse group and some have little in common except the goal of finding a linkage that will generate a desired path. All allow a theoretically unlimited number of design points to be specified, but making N too large will increase the computation time and may not improve the result. One inherent limitation of optimization methods is that they may converge to a local minimum near the starting conditions. The result may not be as good as other minima located elsewhere in the N-space of the variables. Finding the global optimum is possible but more difficult and time consuming.

Perhaps the earliest application (1966) of optimization techniques to this fourbar linkage path synthesis problem is that of Fox and Willmert [24] in which they minimized the area between the desired and calculated curves subject to a number of equality and inequality constraints. They controlled link lengths to be positive and less than some maximum, controlled for Grashof condition, limited forces and torques, and restricted the locations of the fixed pivots. They used Powell's method to find the minimum of the objective function.

Youssef et al.[25], used sum of squares, sum of absolute values, or area error criteria to minimize the objective function. They accommodated path and function generation

---

* Beyer's linkage geometry notation is different than that used in this book. Beyer's labeling for the equation, as shown by Hall,[42] is:

for single-loop (fourbar) or multiloop linkages with both pin and slider joints. They allowed constraints to be imposed on the allowable ranges of link lengths and angles, any of which also may be held constant during the iteration. An example of an optimization done with this method for 19 evenly spaced points around a desired fourbar coupler path is shown in Figure 5-11.[25] Another example of this method is the 10-bar crank-slider linkage in Figure 5-12 [25] (p. 270) which also shows the desired and actual coupler curve generated by point P for 24 points corresponding to equal increments of input crank angle.

Nolle and Hunt [9] derived analytical expressions that lead to a set of ten linear simultaneous nonhomogeneous equations whose solution gives values for all the independent variables. They used a least squares approach to the optimization and also allowed specified timing of the input crank to each position on the coupler. Because their equations are linear, convergence is rapid requiring only about one second per iteration.

Kramer and Sandor [26], [27] described a variant on the precision point technique which they call **selective precision synthesis** (SPS). It relaxes the requirement that the curve pass exactly through the precision points by defining "accuracy neighborhoods" around each point. The size of these tolerance zones can be different for each point, and more than nine points can be used. They point out that exact correspondence to a set of points is often not necessary in engineering applications and even if achieved theoretically would be compromised by manufacturing tolerances.

The SPS approach is suitable to any linkage constructible from dyads or triads and so can accommodate sixbar and geared fivebar linkages as well as fourbars. Fourbar function, motion, or path generation (with prescribed timing) can all be synthesized, using the standard form approach which considers all three forms equivalent in terms of equation formulation. Spatial mechanisms can also be accommodated. The solutions are stable and less sensitive to small changes in the data than precision point methods. Krishnamurthi et al.[28], extended the SPS approach by using fuzzy set theory which gives a mechanism path as close to the specified points as is possible for a given start point; but it is sensitive to start point selection and may find local optima rather than global.

Mirth [29] provided a variation on Kramer's SPS technique called quasi-precision position synthesis which uses three precision positions and N quasi positions which are defined as tolerance zones. This approach retains the computational advantages of the Burmester (precision point) approach while also allowing the specification of a larger number of points to improve and refine the design.

Conte et al.[30], and Kakatsios and Tricamo [31], [32] described methods to satisfy a small number of precision points and simultaneously optimize the linkage's dynamic characteristics. The link lengths are controlled to reasonable size, the Grashof condition is constrained, and the input torque, dynamic bearing and reaction forces, and shaking moments are simultaneously minimized.

Many of the optimization methods listed above use some form of inequality constraints to limit the allowable values of design parameters such as link lengths and transmission angles. These constraints often cause problems that lead to nonconvergence, or to CBO defects. Angeles et al.[33] described an unconstrained nonlinear least-squares-method that avoids these problems. Continuation methods are employed, and good convergence is claimed with no branch defects.

desired curve ··········
actual curve ─────

(a) Coupler curve

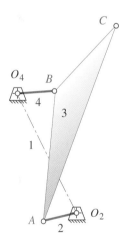

(b) Synthesized linkage

FIGURE 5-11

Linkage synthesized to generate a desired coupler curve by an optimization method

Reproduced from "Optimal Kinematic Synthesis of Planar Linkage Mechanisms" [25] with the kind permission of Professional Engineering Publishing, Bury St. Edmunds, UK.

5

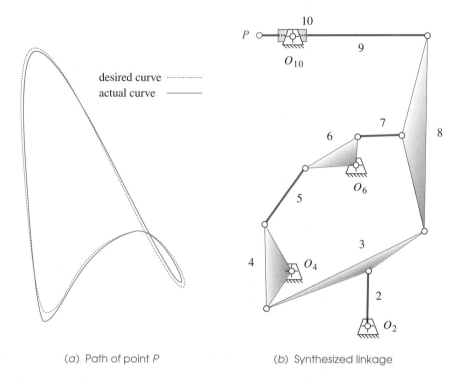

(a) Path of point P   (b) Synthesized linkage

FIGURE 5-12

Example of synthesis of a 10 link mechanism to generate a coupler path Reproduced from Youssef et al. (1975), "Optimal Kinematic Synthesis of Planar Linkage Mechanisms" [25] with the kind permission of Professional Engineering Publishing, Bury St. Edmunds, UK.

Aviles et al.[34] proposed a novel approach to the linkage synthesis problem that uses the elastic energy that would be stored in the links if they were allowed to deform elastically such that the coupler point reaches the desired location. The objective function is defined as the minimum energy condition in the set of deformed links which of course will occur when their rigid body positions most closely approach the desired path. This is essentially a finite element method approach that considers each link to be a bar element. Newton's method is used for the iteration and, in this case, converges to a minimum even when the initial guess is far from a solution.

Fang [35] described an unusual approach to linkage synthesis using genetic algorithms. Genetic algorithms emulate the way that living organisms adapt to nature. Initially, a population of random "organisms" is generated that represents the system to be optimized. This takes the form of a bit string, analogous to a cell's chromosomes, which is called the first generation. Two operations are performed on a given population, called crossover and mutation. Crossover combines part of the "genetic code" of a "father" organism with part of the code of a "mother" organism. Mutation changes values of the genetic code at random points in the bit string. An objective function is created that expresses the "fitness" of the organism for the desired task. Each successive generation is produced by selecting the organisms that best fit the task. The population "evolves" through generations until a termination criterion is met based on the objective function.

Some advantages of this approach are that it searches from population to population rather than point to point, and this makes it less likely to be trapped at local optima. The population also preserves a number of valid solutions rather than converging to only one. The disadvantage is long computation times due to the large number of objective function evaluations required. Nevertheless it is more efficient than random walk or exhaustive search algorithms. All other optimization approaches listed here deal only with dimensional synthesis, but genetic algorithms can also deal with type synthesis.

Ullah and Kota [36], [37] separated the linkage synthesis problem into two steps. The first step seeks an acceptable match for the shape of the desired curve without regard to the size, orientation, or location of the curve in space. Once a curve of suitable shape and its associated linkage are found, the result can be translated, rotated, and scaled as desired. This approach simplifies the optimization task compared to the algorithms that seek a structural optimization that includes size, orientation, and location of the coupler curve all at once in the objective function. Fourier descriptors are used to characterize the shape of the curve as is done in many pattern matching applications such as for automated robotic assembly tasks. A stochastic global optimization algorithm is used which avoids unwanted convergence to suboptimal local minima.

Vasiliu and Yannou [38] also focus solely on the shape of the desired path, approximating it with five terms of a Fourier series. They use an artificial *neural network* approach to synthesize a linkage to generate the approximate curve shape. A neural network is a graph of *input neurons* that represent the shape of the path and *output neurons* that represent the dimensional parameters of the linkage. The network is "taught" to properly relate the output to the input with various algorithms. Learning time was 30 hours for 14 000 iterations for their example, so this method is computer intensive. The matching of their resulting linkage curve shape to the desired curve is less accurate than that of the method shown in Figures 5-11 (p. 269) and 5-12.

Bawab et al. [39] described an approach that will automatically (within the software program) synthesize a fourbar linkage for two, three, or four positions using Burmester theory and eliminate all solutions having CBO defects. Limits on link length ratios and transmission angle are specified and the objective function is based on these criteria with weighting factors applied. Regions in the plane within which the fixed or moving pivots must be located may also be specified.

Liu and Yang [40] proposed a method for finding all solutions to the approximate synthesis problem for function generation, rigid body guidance, and path generation with timing, using a combination of continuation methods and optimization. Their approach does not require an initial guess, and all possible solutions can be obtained with relatively short computational times.

5.15 REFERENCES

1   **Sandor, G. N.** (1959). "A General Complex Number Method for Plane Kinematic Synthesis with Applications." Ph.D. Thesis, Columbia University, University Microfilms, Ann Arbor, MI.

5

2   **Erdman, A. G.** (1981). "Three and Four Precision Point Kinematic Synthesis of Planar Linkages." *Mechanism and Machine Theory*, **16**, pp. 227-245.

3   **Kaufman, R. E.** (1978). "Mechanism Design by Computer." *Machine Design*, October 26, 1978, pp. 94-100.

4   **Loerch, R. J., et al.** (1975). "Synthesis of Fourbar Linkages with Specified Ground Pivots." *Proc. of 4th Applied Mechanisms Conference*, Chicago, IL, pp. 10.1-10.6.

5   **Loerch, R. J., et al.** (1979). "On the Existence of Circle-Point and Center-Point Circles for Three Position Dyad Synthesis." *Journal of Mechanical Design*, **101**(3), pp. 554-562.

6   **Erdman, A. G., and G. N. Sandor**. (1997). *Mechanism Design: Analysis and Synthesis.* Vol. 1, 3d ed., and *Advanced Mechanism Design, Analysis and Synthesis*, Vol. 2 (1984). Prentice-Hall: Upper Saddle River, NJ.

7   **Jennings, A.** (1977). *Matrix Computation for Engineers and Scientists.* John Wiley & Sons: New York.

8   **Erdman, A. G.,** and **J. E. Gustafson**. (1977). "LINCAGES: Linkage INteractive Computer Analysis and Graphically Enhanced Synthesis." ASME Paper: 77-DTC-5.

9   **Nolle, H., and K. H. Hunt**. (1971). "Optimum Synthesis of Planar Linkages to Generate Coupler Curves." *Journal of Mechanisms*, **6**, pp. 267-287.

10 **Freudenstein, F., and G. N. Sandor.** (1959). "Synthesis of Path Generating Mechanisms by Means of a Programmed Digital Computer." *ASME Journal for Engineering in Industry*, **81**, p. 2.

11 **Suh, C. H., and C. W. Radcliffe**. (1966). "Synthesis of Planar Linkages With Use of the Displacement Matrix." ASME Paper: 66-MECH-19, 9 pp.

12 **Morgan, A. P., and A. J. Sommese**. (1987). "Computing All Solutions to Polynomial Systems Using Homotopy Continuation." *Applied Mathematics and Computation*, **24**, pp. 115-138.

13 **Morgan, A. P.** (1987). *Solving Polynomial Systems Using Continuation for Scientific and Engineering Problems.* Prentice-Hall: Upper Saddle River, NJ.

14 **Morgan, A. P., and C. W. Wampler**. (1990). "Solving a Planar Fourbar Design Problem Using Continuation." *Journal of Mechanical Design*, **112**(4), p. 544.

15 **Subbian, T., and J. D. R. Flugrad**. (1991). "Fourbar Path Generation Synthesis by a Continuation Method." *Journal of Mechanical Design*, **113**(1), p. 63.

16 **Subbian, T., and J. D. R. Flugrad**. (1993). "Five Position Triad Synthesis with Applications to Four and Sixbar Mechanisms." *Journal of Mechanical Design*, **115**(2), p. 262.

17 **Subbian, T., and J. D. R. Flugrad**. (1994). "Six and Seven Position Triad Synthesis Using Continuation Methods." *Journal of Mechanical Design*, **116**(2), p. 660.

18 **Wampler, C. W., et al.** (1992). "Complete Solution of the Nine-Point Path Synthesis Problem for Fourbar Linkages." *Journal of Mechanical Design*, **114**(1), p. 153.

19 **Tylaska, T., and K. Kazerounian**. (1994). "Synthesis of Defect-Free Sixbar Linkages for Body Guidance Through Up to Six Finitely Separated Positions." *Proc. of 23rd Biennial Mechanisms Conference*, Minneapolis, MN, p. 369.

20 **Tylaska, T., and K. Kazerounian**. (1993). "Design of a Six Position Body Guidance Watt I Sixbar Linkage and Related Concepts." *Proc. of 3rd Applied Mechanisms and Robotics Conference*, Cincinnati, p. 93-1.

21 **Blechschmidt, J. L., and J. J. Uicker**. (1986). "Linkage Synthesis Using Algebraic Curves." *J. Mechanisms, Transmissions, and Automation in Design*, **108** (December 1986), pp. 543-548.

22 **Ananthasuresh, G. K., and S. Kota**. (1993). "A Renewed Approach to the Synthesis of Fourbar Linkages for Path Generation via the Coupler Curve Equation." *Proc. of 3rd Applied Mechanisms and Robotics Conference*, Cincinnati, p. 83-1.

23   **Nolle, H.** (1975). "Linkage Coupler Curve Synthesis: A Historical Review - III. Spatial Synthesis and Optimization." *Mechanism and Machine Theory*, **10**, 1975, pp. 41-55.

24   **Fox, R. L., and K. D. Willmert.** (1967). "Optimum Design of Curve-Generating Linkages with Inequality Constraints." *Journal of Engineering for Industry (*Feb. 1967), pp. 144-152.

25   **Youssef, A. H., et al.** (1975). "Optimal Kinematic Synthesis of Planar Linkage Mechanisms." *I. Mech. E.*, pp. 393-398.

26   **Kramer, S. N., and G. N. Sandor.** (1975). "Selective Precision Synthesis—A General Method of Optimization for Planar Mechanisms." *Trans. ASME J. Eng. for Industry*, **97B**(2), pp. 689-701.

27   **Kramer, S. N.** (1987). "Selective Precision Synthesis—General Design Method for Planar and Spatial Mechanisms." *Proc. of 7th World Congress on Theory of Machines and Mechanisms*, Seville, Spain.

28   **Krishnamurthi, S., et al.** (1993). "Fuzzy Synthesis of Mechanisms." *Proc. of 3rd Applied Mechanisms and Robotics Conference*, Cincinnati, p. 94-1.

29   **Mirth, J. A.** (1994). "Quasi-Precision Position Synthesis of Fourbar Linkages." *Proc. of 23rd Biennial Mechanisms Conference*, Minneapolis, MN, p. 215.

30   **Conte, F. L., et al.** (1975). "Optimum Mechanism Design Combining Kinematic and Dynamic-Force Considerations." *Journal of Engineering for Industry (*May 1975), pp. 662-670.

31   **Kakatsios, A. J., and S. J. Tricamo.** (1987). "Precision Point Synthesis of Mechanisms with Optimal Dynamic Characteristics." *Proc. of 7th World Congress on the Theory of Machines and Mechanisms*, Seville, Spain, pp. 1041-1046.

32   **Kakatsios, A. J., and S. J. Tricamo.** (1986). "Design of Planar Rigid Body Guidance Mechanisms with Simultaneously Optimized Kinematic and Dynamic Characteristics." ASME Paper: 86-DET-142.

33   **Angeles, J., et al.** (1988). "An Unconstrained Nonlinear Least-Square Method of Optimization of RRRR Planar Path Generators." *Mechanism and Machine Theory*, **23**(5), pp. 343-353.

34   **Aviles, R., et al.** (1994). "An Energy-Based General Method for the Optimum Synthesis of Mechanisms." *Journal of Mechanical Design*, **116**(1), p. 127.

35   **Fang, W. E.** (1994). "Simultaneous Type and Dimensional Synthesis of Mechanisms by Genetic Algorithms." *Proc. of 23rd Biennial Mechanisms Conference*, Minneapolis, MN, p. 36.

36   **Ullah, I., and S. Kota.** (1994). "A More Effective Formulation of the Path Generation Mechanism Synthesis Problem." *Proc. of 23rd Biennial Mechanisms Conference*, Minneapolis, MN, p. 239.

37   **Ullah, I., and S. Kota.** (1996). "Globally-Optimal Synthesis of Mechanisms for Path Generation Using Simulated Annealing and Powell's Method." *Proc. of ASME Design Engineering Conference*, Irvine, CA, pp. 1-8.

38   **Vasiliu, A., and B. Yannou.** (1998). "Dimensional Synthesis of Planar Path Generator Linkages Using Neural Networks." *Mechanism and Machine Theory*, **32**(65).

39   **Bawab, S., et al.** (1997). "Automatic Synthesis of Crank Driven Fourbar Mechanisms for Two, Three, or Four Position Motion Generation." *Journal of Mechanical Design*, **119**(June), pp. 225-231.

40   **Liu, A. X., and T. L. Yang.** (1999). "Finding All Solutions to Unconstrained Nonlinear Optimization for Approximate Synthesis of Planar Linkages Using Continuation Method." *Journal of Mechanical Design*, **121**(3), pp. 368-374.

41   **Beyer, R.** (1963). *The Kinematic Synthesis of Mechanisms.* McGraw-Hill: New York. p. 254.

42   **Hall, A. S.** (1961). *Kinematics and Linkage Design.* Prentice-Hall: Englewood Cliffs, NJ, p. 49.

5

5.16   PROBLEMS‡

Note that all three-position synthesis problems below may be done using a matrix solving calculator, equation solver such as Mathcad, Matlab, or TKSolver, program MATRIX, or program LINKAGES. Two-position synthesis problems can be done with a four-function calculator.

5-1   Redo Problem 3-3 (p. 157) using the analytical methods of this chapter.

5-2   Redo Problem 3-4 (p. 157) using the analytical methods of this chapter.

5-3   Redo Problem 3-5 (p. 157) using the analytical methods of this chapter.

5-4   Redo Problem 3-6 (p. 157) using the analytical methods of this chapter.

5-5   See Project P3-8 (p. 171). Define three positions of the boat and analytically synthesize a linkage to move through them.

5-6   See Project P3-30 (p. 172). Define three positions of the dumpster and analytically synthesize a linkage to move through them. The fixed pivots must be located on the existing truck.

5-7   See Project P3-7 (p. 171). Define three positions of the computer monitor and analytically synthesize a linkage to move through them. The fixed pivots must be located on the floor or wall.

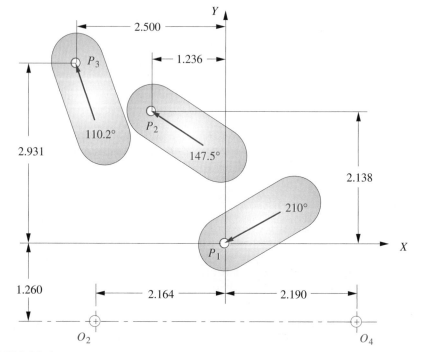

FIGURE P5-1

Data for Problems 5-8 to 5-11

*†5-8 Design a linkage to carry the body in Figure P5-1 through the two positions P_1 and P_2 at the angles shown in the figure. Use analytical synthesis without regard for the fixed pivots shown. Hint: Try the free choice values $z = 1.075$, $\phi = 204.4°$, $\beta_2 = -27°$; $s = 1.24$, $\psi = 74°$, $\gamma_2 = -40°$.

†5-9 Design a linkage to carry the body in Figure P5-1 through the two positions P_2 and P_3 at the angles shown in the figure. Use analytical synthesis without regard for the fixed pivots shown. Hint: First try a rough graphical solution to create realistic values for free choices.

†5-10 Design a linkage to carry the body in Figure P5-1 through the three positions P_1, P_2, and P_3 at the angles shown in the figure. Use analytical synthesis without regard for the fixed pivots shown. Hint: Try the free choice values $\beta_2 = 30°$, $\beta_3 = 60°$; $\gamma_2 = -10°$, $\gamma_3 = 25°$.

*†5-11 Design a linkage to carry the body in Figure P5-1 through the three positions P_1, P_2, and P_3 at the angles shown in the figure. Use analytical synthesis and design it for the fixed pivots shown.

†5-12 Design a linkage to carry the body in Figure P5-2 through the two positions P_1 and P_2 at the angles shown in the figure. Use analytical synthesis without regard for the fixed pivots shown. Hint: Try the free choice values $z = 2$, $\phi = 150°$, $\beta_2 = 30°$; $s = 3$, $\psi = -50°$, $\gamma_2 = 40°$.

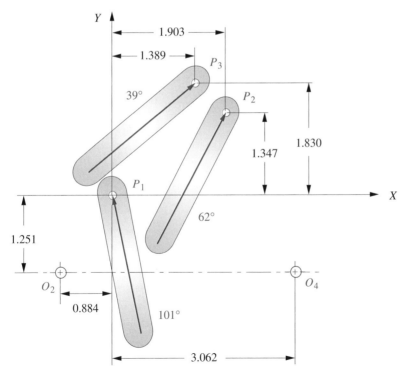

FIGURE P5-2

Data for Problems 5-12 to 5-15

* Answers in Appendix F.

† These problems are suited to solution using *Mathcad*, or *TKSolver* equation solver programs. In most cases, your solution can be checked with program LINKAGES.

†5-13   Design a linkage to carry the body in Figure P5-2 (p. 275) through the two positions P_2 and P_3 at the angles shown in the figure. Use analytical synthesis without regard for the fixed pivots shown. Hint: First try a rough graphical solution to create realistic values for free choices.

†5-14   Design a linkage to carry the body in Figure P5-2 (p. 275) through the three positions P_1, P_2, and P_3 at the angles shown in the figure. Use analytical synthesis without regard for the fixed pivots shown.

*†5-15   Design a linkage to carry the body in Figure P5-2 (p. 275) through the three positions P_1, P_2, and P_3 at the angles shown in the figure. Use analytical synthesis and design it for the fixed pivots shown.

†5-16   Design a linkage to carry the body in Figure P5-3 through the two positions P_1 and P_2 at the angles shown in the figure. Use analytical synthesis without regard for the fixed pivots shown.

†5-17   Design a linkage to carry the body in Figure P5-3 through the two positions P_2 and P_3 at the angles shown in the figure. Use analytical synthesis without regard for the fixed pivots shown.

†5-18   Design a linkage to carry the body in Figure P5-3 through the three positions P_1, P_2, and P_3 at the angles shown in the figure. Use analytical synthesis without regard for the fixed pivots shown.

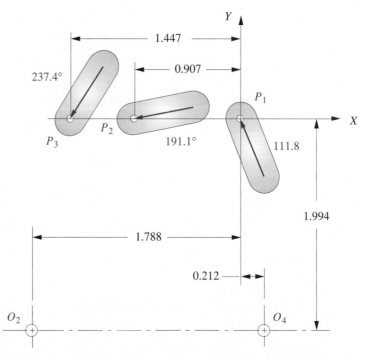

* Answers in Appendix F.

† These problems are suited to solution using *Mathcad*, or *TKSolver* equation solver programs. In most cases, your solution can be checked with program LINKAGES.

FIGURE P5-3

Data for Problems 5-16 to 5-20

†5-19 Design a linkage to carry the body in Figure P5-3 through the three positions P_1, P_2, and P_3 at the angles shown in the figure. Use analytical synthesis and design it for the fixed pivots shown.

†5-20 Write a program to generate and plot the circle-point and center-point circles for Problem 5-19 using an equation solver or any programming language.

†5-21 Design a fourbar linkage to carry the box in Figure P5-4 from position 1 to 2 without regard for the fixed pivot locations shown. Use points A and B for your attachment points. Determine the range of the transmission angle. The fixed pivots should be on the base.

†5-22 Design a fourbar linkage to carry the box in Figure P5-4 from position 1 to 3 without regard for the fixed pivots shown. Use points A and B for your attachment points. Determine the range of the transmission angle. The fixed pivots should be on the base.

†5-23 Design a fourbar linkage to carry the box in Figure P5-4 from position 2 to 3 without regard for the fixed pivots shown. Use points A and B for your attachment points. Determine the range of the transmission angle. The fixed pivots should be on the base.

†5-24 Design a fourbar linkage to carry the box in Figure P5-4 through the three positions shown in their numbered order without regard for the fixed pivots shown. Determine the range of the transmission angle. Use any points on the object as your attachment points. The fixed pivots should be on the base.

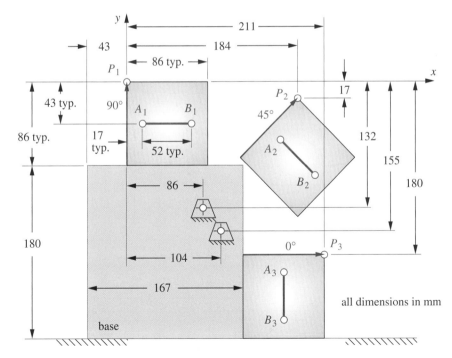

FIGURE P5-4

Data for Problems 5-21 to 5-26

* Answers in Appendix F.

† These problems are suited to solution using *Mathcad,* or *TKSolver* equation solver programs. In most cases, your solution can be checked with program LINKAGES.

†5-25   Design a fourbar linkage to carry the box in Figure P5-4 (p. 277) through the three positions shown in their numbered order without regard for the fixed pivots shown. Use points *A* and *B* for your attachment points. Determine the range of the transmission angle.

*†5-26   Design a fourbar linkage to carry the box in Figure P5-4 (p. 277) through the three positions shown in their numbered order using the fixed pivots shown. Determine the range of the transmission angle.

†5-27   Design a fourbar linkage to carry the object in Figure P5-5 through the three positions shown in their numbered order without regard for the fixed pivots shown. Use any points on the object as attachment points. The fixed pivots should be on the base. Determine the range of the transmission angle.

†5-28   Design a fourbar linkage to carry the object in Figure P5-5 through the three positions shown in their numbered order without regard for the fixed pivots shown. Use points *A* and *B* for your attachment points. Determine the range of the transmission angle.

†5-29   Design a fourbar linkage to carry the object in Figure P5-5 through the three positions shown in their numbered order using the fixed pivots shown. Determine the range of the transmission angle.

†5-30   To the linkage solution from Problem 5-29, add a driver dyad with a crank to control the motion of your fourbar so that it cannot move beyond positions one and three.

* Answers in Appendix F.

† These problems are suited to solution using *Mathcad,* or *TKSolver* equation solver programs. In most cases, your solution can be checked with program LINKAGES.

FIGURE P5-5

Data for Problems 5-27 to 5-30

FIGURE P5-6

Data for Problems 5-31 to 5-33

[†]5-31   Design a fourbar linkage to carry the object in Figure P5-6 through the three positions shown in their numbered order without regard for the fixed pivots shown. Use points A and B for your attachment points. Determine the range of the transmission angle.

[†]5-32   Design a fourbar linkage to carry the object in Figure P5-6 through the three positions shown in their numbered order without regard for the fixed pivots shown. Use any points on the object as attachment points. The fixed pivots should be on the base. Determine the range of the transmission angle.

^{*†}5-33   Design a fourbar linkage to carry the object in Figure P5-6 through the three positions shown in their numbered order using the fixed pivots shown. Determine the range of the transmission angle.

[†]5-34   Design a fourbar linkage to carry the bolt in Figure P5-7 (p. 280) from positions 1 to 2 to 3 without regard to the fixed pivots shown. The bolt is fed into the gripper in the z direction (into the paper). The gripper grabs the bolt, and your linkage moves it to position 3 to be inserted into the hole. A second degree of freedom within the gripper assembly (not shown) pushes the bolt into the hole. Extend the gripper assembly as necessary to include the moving pivots. The fixed pivots should be on the base. Hint: Try guess values of $\beta_2 = 70°$, $\beta_3 = 140°$, $\gamma_2 = -5°$, $\gamma_3 = -49°$.

---

* Answers in Appendix F.

[†] These problems are suited to solution using *Mathcad*, or *TKSolver* equation solver programs. In most cases, your solution can be checked with program LINKAGES.

FIGURE P5-7

Data for Problems 5-34 to 5-36

*†5-35   Design a fourbar linkage to carry the bolt in Figure P5-7 from positions 1 to 2 to 3
using the fixed pivot locations shown. Extend the gripper assembly as necessary to
include the moving pivots. See Problem 5-34 for more information.

FIGURE P5-8

Problem 5-37

FIGURE P5-9

Data for problems 5-38 to 5-40 and 5-47

5-36   To the linkage solution from Problem 5-35, add a driver dyad with a crank to control the motion of your fourbar so that it cannot move beyond positions one and three.

5-37   Figure P5-8 shows an off-loading mechanism for paper rolls. The V-link is rotated through 90° by an air-driven fourbar slider-crank linkage. Design a pin-jointed fourbar linkage to replace the existing off-loading station and perform essentially the same function. Choose three positions of the roll including its two end positions and synthesize a substitute mechanism. Use a link similar to the existing V-link as one of your links. Add a driver dyad to limit its motion to the range desired.

†5-38   Design a fourbar linkage to carry the object in Figure P5-9 through the three positions shown in their numbered order without regard for the fixed pivots shown. Use points C and D for your attachment points. Determine the range of the transmission angle.

†5-39   Design a fourbar linkage to carry the object in Figure P5-9 through the three positions shown in their numbered order without regard for the fixed pivots shown. Use any points on the object as attachment points. Determine the range of the transmission angle.

†5-40   Design a fourbar linkage to carry the object in Figure P5-9 through the three positions shown in their numbered order using the fixed pivots shown. Determine the range of the transmission angle.

†5-41   Repeat Problem 5-38 using the data shown in Figure P5-10 (p. 282) instead.

†5-42   Repeat Problem 5-39 using the data shown in Figure P5-10 instead.

†5-43   Repeat Problem 5-40 using the data shown in Figure P5-10 instead.

†5-44   Repeat Problem 5-38 using the data shown in Figure P5-11 (p. 282) instead.

†5-45   Repeat Problem 5-39 using the data shown in Figure P5-11 instead.

* Answers in Appendix F.

† These problems are suited to solution using *Mathcad,* or *TKSolver* equation solver programs. In most cases, your solution can be checked with program LINKAGES.

FIGURE P5-10

Data for problems 5-41 to 5-43 and 5-48

FIGURE P5-11

Data for problems 5-44 to 5-46 and 5-49

†5-46   Repeat Problem 5-40 (p. 281) using the data shown in Figure P5-11 instead.

†5-47   Write a program to generate and plot the circle-point and center-point circles for Problem 5-40 using an equation solver or any programming language.

†5-48   Repeat Problem 5-47 using the data from Problem 5-43 (p. 281) instead.

†5-49   Repeat Problem 5-47 using the data from Problem 5-46 instead.

†5-50   In Example 5-2 (p. 248) the precision points and rotation angles are specified while the input and output rotation angles β and γ are free choices. Using the choices given for β_2 and γ_2, determine the radii and center coordinates of the center-point circles for O_2 and O_4. Plot those circles (or portions of them) and show that the choices of β_3 and γ_3 give a solution that falls on the center-point circles. You can get some verification of your circle calculations by using program LINKAGES.

5-51   Design a driver dyad to move link 2 of Example 5-1 (p. 238) from position 1 to position 2 and return.

5-52   Design a driver dyad to move link 2 of Example 5-2 (p. 248) from position 1 to position 3 and return.

5-53   Design a driver dyad to move link 2 of Example 5-3 (p. 255) from position 1 to position 3 and return.

5-54   Design a fourbar linkage to carry the object in Figure P5-12 (p. 284) from position 1 to 2 using points C and D for your attachment points. The fixed pivots should be within the indicated area.

5-55   Design a fourbar linkage to carry the object in Figure P5-12 (p. 284) from position 1 to 3 using points C and D for your attachment points. The fixed pivots should be within the indicated area.

5-56   Design a fourbar linkage to carry the object in Figure P5-12 (p. 284) from position 2 to 3 using points C and D for your attachment points. The fixed pivots should be within the indicated area.

5-57   Design a fourbar linkage to carry the object in Figure P5-12 (p. 284) through the three positions shown in their numbered order using points C and D for your attachment points. The fixed pivots should be within the indicated area.

† These problems are suited to solution using *Mathcad*, or *TKSolver* equation solver programs. In most cases, your solution can be checked with program LINKAGES.

5

Problems 5-54 to 5-57

Chapter **6**

VELOCITY ANALYSIS

The faster I go, the behinder I get
ANON. PENN. DUTCH

6.0 INTRODUCTION

Once a position analysis is done, the next step is to determine the velocities of all links and points of interest in the mechanism. We need to know the velocities in our mechanism or machine, both to calculate the stored kinetic energy from $mV^2/2$, and also as a step on the way to the determination of the link's accelerations that are needed for the dynamic force calculations. Many methods and approaches exist to find velocities in mechanisms. We will examine only a few of these methods here. We will first develop manual graphical methods, which are often useful as a check on the more complete and accurate analytical solution. We will also investigate the properties of the instant center of velocity which can shed much light on a mechanism's velocity behavior with very little effort. Finally, we will derive the analytical solution for the fourbar and inverted crank-slider as examples of the general vector loop equation solution to velocity analysis problems. From these calculations we will be able to establish some indices of merit to judge our designs while they are still on the drawing board (or in the computer).

6.1 DEFINITION OF VELOCITY

Velocity is defined as *the rate of change of position with respect to time*. Position (**R**) is a vector quantity and so is velocity. Velocity can be **angular** or **linear**. **Angular velocity** will be denoted as ω and **linear velocity** as **V**.

$$\omega = \frac{d\theta}{dt}; \qquad \mathbf{V} = \frac{d\mathbf{R}}{dt} \qquad (6.1)$$

Figure 6-1 (p. 286) shows a link *PA* in pure rotation, pivoted at point *A* in the *xy* plane. Its position is defined by the position vector $\mathbf{R}_{PA}$. We are interested in the velocity of

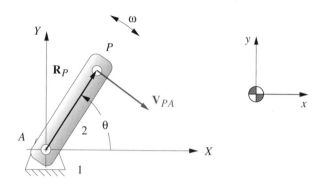

FIGURE 6-1

A link in pure rotation

point P when the link is subjected to an angular velocity ω. If we represent the position vector $\mathbf{R}_{PA}$ as a complex number in polar form,

$$\mathbf{R}_{PA} = pe^{j\theta} \tag{6.2}$$

where p is the scalar length of the vector. We can easily differentiate it to obtain:

$$\mathbf{V}_{PA} = \frac{d\mathbf{R}_{PA}}{dt} = p\,je^{j\theta}\frac{d\theta}{dt} = p\omega\,je^{j\theta} \tag{6.3}$$

Compare the right side of equation 6.3 to the right side of equation 6.2. Note that as a result of the differentiation, the velocity expression has been multiplied by the (constant) complex operator j. This causes a rotation of this velocity vector through 90 degrees with respect to the original position vector. (See also Figure 4-8b, p. 186.) This 90-degree rotation is positive, or counterclockwise. However, the velocity expression is also multiplied by ω, which may be either positive or negative. As a result, the velocity vector will be **rotated 90 degrees** from the angle θ of the position vector **in a direction dictated by the sign of ω.** This is just mathematical verification of what you already knew, namely that *velocity is always in a direction perpendicular to the radius of rotation and is tangent to the path of motion* as shown in Figure 6-1.

Substituting the Euler identity (equation 4.4a, p. 185) into equation 6.3 gives us the real and imaginary (or x and y) components of the velocity vector.

$$\mathbf{V}_{PA} = p\omega\,j(\cos\theta + j\sin\theta) = p\omega(-\sin\theta + j\cos\theta) \tag{6.4}$$

Note that the sine and cosine terms have swapped positions between the real and imaginary terms, due to multiplying by the j coefficient. This is evidence of the 90-degree rotation of the velocity vector versus the position vector. The former x component has become the y component, and the former y component has become a minus x component. Study Figure 4-8b (p. 186) to review why this is so.

The velocity $\mathbf{V}_{PA}$ in Figure 6-1 can be referred to as an **absolute velocity** since it is referenced to A, which is the origin of the global coordinate axes in that system. As such, we could have referred to it as $\mathbf{V}_P$, with the absence of the second subscript implying

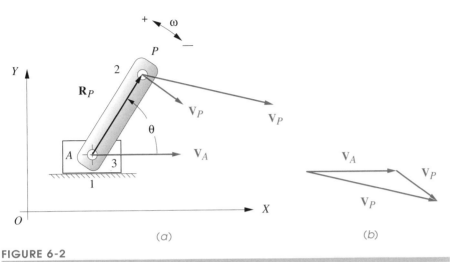

FIGURE 6-2

Velocity difference

reference to the global coordinate system. Figure 6-2a shows a different and slightly more complicated system in which the pivot A is no longer stationary. It has a known linear velocity $\mathbf{V}_A$ as part of the translating carriage, link 3. If ω is unchanged, the velocity of point P versus A will be the same as before, but $\mathbf{V}_{PA}$ can no longer be considered an absolute velocity. It is now a **velocity difference** and **must** carry the second subscript as $\mathbf{V}_{PA}$. The absolute velocity $\mathbf{V}_P$ must now be found from the **velocity difference equation** whose graphical solution is shown in Figure 6-2b:

$$\mathbf{V}_{PA} = \mathbf{V}_P - \mathbf{V}_A \tag{6.5a}$$

rearranging:

$$\mathbf{V}_P = \mathbf{V}_A + \mathbf{V}_{PA} \tag{6.5b}$$

Note the similarity of equations 6.5 to the **position difference equation** 4.1 (p. 177).

Figure 6-3 (p. 288) shows two independent bodies P and A, which could be two automobiles, moving in the same plane. If their independent velocities $\mathbf{V}_P$ and $\mathbf{V}_A$ are known, their **relative velocity** $\mathbf{V}_{PA}$ can be found from equations 6.5 arranged algebraically as:

$$\mathbf{V}_{PA} = \mathbf{V}_P - \mathbf{V}_A \tag{6.6}$$

The graphical solution to this equation is shown in Figure 6-3b. Note that it is similar to Figure 6-2b except for a different vector being the resultant.

As we did for position analysis, we give these two cases different names despite the fact that the same equation applies. Repeating the definition from Section 4.2 (p. 179), modified to refer to velocity:

CASE 1:   *Two points in the same body* => **velocity difference**

CASE 2:   *Two points in different bodies* => **relative velocity**

We will find use for this semantic distinction when we analyze both linkage velocities and the velocity of slip later in this chapter.

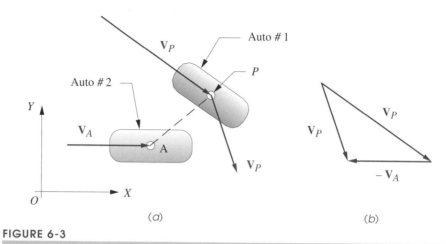

FIGURE 6-3

Relative velocity

6.2   GRAPHICAL VELOCITY ANALYSIS

Before programmable calculators and computers became universally available to engineers, graphical methods were the only practical way to solve these velocity analysis problems. With some practice and with proper tools such as a drafting machine or CAD package, one can fairly rapidly solve for the velocities of particular points in a mechanism for any one input position by drawing vector diagrams. However, it is a tedious process if velocities for many positions of the mechanism are to be found, because each new position requires a completely new set of vector diagrams be drawn. Very little of the work done to solve for the velocities at position 1 carries over to position 2, etc. Nevertheless, this method still has more than historical value as it can provide a quick check on the results from a computer program solution. Such a check needs only be done for a few positions to prove the validity of the program. Also, graphical solutions provide the beginning student some visual feedback on the solution that can help develop an understanding of the underlying principles. It is principally for this last reason that graphical solutions are included in this text even in this "age of the computer."

To solve any velocity analysis problem graphically, we need only two equations, 6.5 (p. 287) and 6.7 (which is merely the scalar form of equation 6.3, p. 286):

$$|\mathbf{V}| = v = r\omega \tag{6.7}$$

Note that the scalar equation 6.7 defines only the **magnitude** (v) of the velocity of any point on a body that is in pure rotation. In a graphical CASE 1 analysis, the **direction** of the vector due to the rotation component must be understood from equation 6.3 to be perpendicular to the radius of rotation. Thus, if the center of rotation is known, the direction of the velocity component due to that rotation is known and its sense will be consistent with the angular velocity ω of the body.

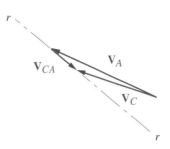

(b) Velocity diagram for points A and B

(c) Velocity diagram for points A and C

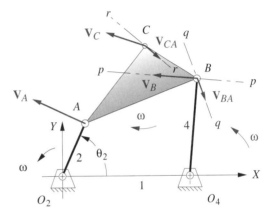

(a) Linkage showing velocity of point A

(d) Linkage showing velocities of points A, B, and C

FIGURE 6-4

Graphical solution for velocities in a pin-jointed linkage

Figure 6-4 shows a fourbar linkage in a particular position. We wish to solve for the angular velocities of links 3 and 4 (ω_3, ω_4) and the linear velocities of points A, B, and C ($\mathbf{V}_A$, $\mathbf{V}_B$, $\mathbf{V}_C$). Point C represents any general point of interest. Perhaps C is a coupler point. The solution method is valid for any point on any link. To solve this problem we need to know the *lengths of all the links*, the *angular positions of all the links*, and the *instantaneous input velocity of any one driving link or driving point*. Assuming we have designed this linkage, we will know or can measure the link lengths. We must also first do a **complete position analysis** to find the link angles θ_3 and θ_4 given the input link's position θ_2. This can be done by any of the methods in Chapter 4. In general we must solve these problems in stages, first for link positions, then for velocities, and finally for accelerations. For the following example, we will assume that a complete position analysis has been done and that the input is to link 2 with known θ_2 and ω_2 for this one "freeze frame" position of the moving linkage.

✍EXAMPLE 6-1

Graphical Velocity Analysis for One Position of Linkage.

Problem: Given θ_2, θ_3, θ_4, ω_2, find ω_3, ω_4, $\mathbf{V}_A$, $\mathbf{V}_B$, $\mathbf{V}_C$ by graphical methods.

Solution: (See Figure 6-4, p. 289.)

1   Start at the end of the linkage about which you have the most information. Calculate the magnitude of the velocity of point A using scalar equation 6.7 (p. 288).

$$v_a = (AO_2)\omega_2 \tag{a}$$

2   Draw the velocity vector $\mathbf{V}_A$ with its length equal to its magnitude v_A at some convenient scale with its root at point A and its direction perpendicular to the radius AO_2. Its sense is the same as that of ω_2 as shown in Figure 6-4a.

3   Move next to a point about which you have some information. Note that the direction of the velocity of point B is predictable since it is pivoting in pure rotation about point O_4. Draw the construction line pp through point B perpendicular to BO_4, to represent the direction of $\mathbf{V}_B$ as shown in Figure 6-4a.

4   Write the velocity difference vector equation 6.5 (p. 287) for point B versus point A.

$$\mathbf{V}_B = \mathbf{V}_A + \mathbf{V}_{BA} \tag{b}$$

We will use point A as the reference point to find $\mathbf{V}_B$ because A is in the same link as B and we have already solved for $\mathbf{V}_A$. Any two-dimensional vector equation can be solved for two unknowns. Each term has two parameters, namely magnitude and direction. There are then potentially six unknowns in this equation, two per term. We must know four of them to solve it. We know both magnitude and direction of $\mathbf{V}_A$ and the direction of $\mathbf{V}_B$. We need to know one more parameter.

5   The term $\mathbf{V}_{BA}$ represents the velocity of B with respect to A. If we assume that the link BA is rigid, then there can be no component of $\mathbf{V}_{BA}$ that is directed along the line BA, because point B cannot move toward or away from point A without shrinking or stretching the rigid link! Therefore, the direction of $\mathbf{V}_{BA}$ must be perpendicular to the line BA. Draw construction line qq through point B and perpendicular to BA to represent the direction of $\mathbf{V}_{BA}$, as shown in Figure 6-4a.

6   Now the vector equation can be solved graphically by drawing a vector diagram as shown in Figure 6-4b. Either drafting tools or a CAD package is necessary for this step. First draw velocity vector $\mathbf{V}_A$ carefully to some scale, maintaining its direction. (It is drawn twice size in the figure.) The equation in step 4 says to add $\mathbf{V}_{BA}$ to $\mathbf{V}_A$, so draw a line parallel to line qq across the tip of $\mathbf{V}_A$. The resultant, or left side of the equation, must close the vector diagram, from the tail of the first vector drawn ($\mathbf{V}_A$) to the tip of the last, so draw a line parallel to pp across the tail of $\mathbf{V}_A$. The intersection of these lines parallel to pp and qq defines the lengths of $\mathbf{V}_B$ and $\mathbf{V}_{BA}$. The senses of the vectors are determined from reference to the equation. $\mathbf{V}_A$ was added to $\mathbf{V}_{BA}$, so they must be arranged tip to tail. $\mathbf{V}_B$ is the resultant, so it must be from the tail of the first to the tip of the last. The resultant vectors are shown in Figure 6-4b and d.

7   The angular velocities of links 3 and 4 can be calculated from equation 6.7 (p. 288):

$$\omega_4 = \frac{v_B}{BO_4} \qquad \text{and} \qquad \omega_3 = \frac{v_{BA}}{BA} \qquad\qquad (c)$$

Note that the velocity difference term $\mathbf{V}_{BA}$ represents the rotational component of velocity of link 3 due to ω_3. This must be true if point B cannot move toward or away from point A. The only velocity difference they can have, one to the other, is due to rotation of the line connecting them. You may think of point B on the line BA rotating about point A as a center, or point A on the line AB rotating about B as a center. The rotational velocity ω of any body is a "free vector" that has no particular point of application to the body. It exists everywhere on the body.

8   Finally we can solve for $\mathbf{V}_C$, again using equation 6.5 (p. 287). We select any point in link 3 for which we know the absolute velocity to use as the reference, such as point A.

$$\mathbf{V}_C = \mathbf{V}_A + \mathbf{V}_{CA} \qquad\qquad (d)$$

In this case, we can calculate the magnitude of $\mathbf{V}_{CA}$ from equation 6.7 as we have already found ω_3,

$$v_{ca} = c\omega_3 \qquad\qquad (e)$$

Since both $\mathbf{V}_A$ and $\mathbf{V}_{CA}$ are known, the vector diagram can be directly drawn as shown in Figure 6-4c (p. 289). $\mathbf{V}_C$ is the resultant that closes the vector diagram. Figure 6-4d shows the calculated velocity vectors on the linkage diagram. Note that the velocity difference vector $\mathbf{V}_{CA}$ is perpendicular to line CA (along line rr) for the same reasons as discussed in step 7 above.

6

The above example contains some interesting and significant principles that deserve further emphasis. Equation 6.5a is repeated here for discussion.

$$\mathbf{V}_P = \mathbf{V}_A + \mathbf{V}_{PA} \qquad\qquad (6.5a)$$

This equation represents the *absolute* velocity V_P of some general point P referenced to the origin of the global coordinate system. The right side defines it as the sum of the absolute velocity V_A of some other reference point A in the same system and the velocity difference (or relative velocity) V_{PA} of point P versus point A. This equation could also be written:

Velocity = translation component + rotation component

These are the same two components of motion defined by Chasles' theorem, and introduced for displacement in Section 4.3 (p. 181). Chasles' theorem holds for velocity as well. These two components of motion, translation and rotation, are independent of one another. If either is zero in a particular example, the complex motion will reduce to one of the special cases of pure translation or pure rotation. When both are present, the total velocity is merely their vector sum.

Let us review what was done in Example 6-1 in order to extract the general strategy for solution of this class of problem. We started at the input side of the mechanism, as that is where the driving angular velocity is defined. We first looked for a point (A) for which the motion was pure rotation so that one of the terms in equation 6.5 (p. 287) would be zero. (We could as easily have looked for a point in pure translation to bootstrap the solution.) We then solved for the absolute velocity of that point ($\mathbf{V}_A$) using equations 6.5 and 6.7 (pp. 287-288). *(Steps 1 and 2)*

We then used the point (*A*) just solved for as a reference point to define the translation component in equation 6.5 written for a new point (*B*). Note that we needed to choose a second point (*B*) that was in the same rigid body as the reference point (*A*) which we had already solved and about which we could predict some aspect of the new point's (*B*'s) velocity. In this example, we knew the direction of the velocity $\mathbf{V}_B$. In general this condition will be satisfied by any point on a link that is jointed to ground (as is link 4). In this example, we could not have solved for point *C* until we solved for *B*, because point *C* is on a floating link for which point we do not yet know the velocity direction. *(Steps 3 and 4)*

To solve the equation for the second point (*B*), we also needed to recognize that the rotation component of velocity is directed perpendicular to the line connecting the two points in the link (*B* and *A* in the example). You **will always know the direction of the rotation component** in equation 6.5 **if it represents a velocity difference** (CASE 1) **situation**. *If the rotation component relates two points in the* **same rigid body,** *then that velocity difference component is always perpendicular to the line connecting those two points* (see Figure 6-2, p. 287). This will be true regardless of the two points selected. But, *this is not true in a* CASE 2 *situation* (see Figure 6-3, p. 288). *(Steps 5 and 6)*

Once we found the absolute velocity ($\mathbf{V}_B$) of a second point on the same link (CASE 1) we could solve for the angular velocity of that link. (Note that points *A* and *B* are both on link 3 and the velocity of point O_4 is zero.) Once the angular velocities of all the links were known, we could solve for the linear velocity of any point (such as *C*) in any link using equation 6.5. To do so, we had to understand the concept of angular velocity as a **free vector**, meaning that it exists everywhere on the link at any given instant. It has no particular center. *It has an infinity of potential centers.* The link simply *has an angular velocity*, just as does a frisbee thrown and spun across the lawn.

All points on a *frisbee*, if spinning while flying, obey equation 6.5. Left to its own devices, the frisbee will spin about its center of gravity (*CG*), which is close to the center of its circular shape. But if you are an expert frisbee player (and have rather pointed fingers), you can imagine catching that flying frisbee between your two index fingers in some off-center location (not at the *CG*), such that the frisbee continues to spin about your fingertips. In this somewhat far-fetched example of championship frisbee play, you will have taken the translation component of the frisbee's motion to zero, but its independent rotation component will still be present. Moreover, it will now be spinning about a different center (your fingers) than it was in flight (its *CG*). Thus this **free vector** of angular velocity (ω) is happy to attach itself to any point on the body. The body still has the same ω, regardless of the assumed center of rotation. It is this property that allows us to solve equation 6.5 for literally **any point** on a rigid body in complex motion **referenced to any other point** on that body. *(Steps 7 and 8)*

6.3   INSTANT CENTERS OF VELOCITY[*]

[*] See the video *Instant Centers and Centrodes* on the book's DVD.

The definition of an **instant center** of velocity is *a point, common to two bodies in plane motion, which point has the same instantaneous velocity in each body.* Instant centers are sometimes also called *centros* or *poles*. Since it takes two bodies or links to create an instant center (*IC*), we can easily predict the quantity of instant centers to expect from any collection of links. The combination formula for *n* things taken *r* at a time is:

$$C = \frac{n(n-1)(n-2)\cdots(n-r+1)}{r!} \qquad (6.8a)$$

For our case *r* = 2 and it reduces to:

$$C = \frac{n(n-1)}{2} \qquad (6.8b)$$

From equation 6.8b we can see that a fourbar linkage has 6 instant centers, a sixbar has 15, and an eightbar has 28.

Figure 6-5 (p. 294) shows a fourbar linkage in an arbitrary position. It also shows a **linear graph**[†] that is useful for keeping track of which *ICs* have been found. This particular graph can be created by drawing a circle on which we mark off as many points as there are links in our assembly. We will then draw a line between the dots representing the link pairs each time we find an instant center. The resulting linear graph is the set of lines connecting the dots. It does not include the circle that was used only to place the dots. This graph is actually a geometric solution to equation 6.8b, since connecting all the points in pairs gives all the possible combinations of points taken two at a time.

Some *ICs* can be found by inspection, using only the definition of the instant center. Note in Figure 6-5a that the four pin joints each satisfy the definition. They clearly must have the same velocity in both links at all times. These have been labeled $I_{1,2}$, $I_{2,3}$, $I_{3,4}$, and $I_{1,4}$. The order of the subscripts is immaterial. Instant center $I_{2,1}$ is the same as $I_{1,2}$. These pin-joint *ICs* are sometimes called "permanent" instant centers as they remain in the same location for all positions of the linkage. In general, instant centers will move to new locations as the linkage changes position, thus the adjective *instant*. In this fourbar example there are two more *ICs* to be found. It will help to use the Aronhold-Kennedy theorem,[‡] also called *Kennedy's rule*,[3] to locate them.

Kennedy's rule:

*Any three bodies in plane motion will have exactly three instant centers, and **they will lie on the same straight line***.

The first part of this rule is just a restatement of equation 6.8b for *n* = 3. It is the second clause in this rule that is most useful. Note that this rule does **not** require that the three bodies be connected in any way. We can use this rule, in conjunction with the linear graph, to find the remaining *ICs* that are not obvious from inspection. Figure 6.5b shows the construction necessary to find instant center $I_{1,3}$. Figure 6-5c shows the construction necessary to find instant center $I_{2,4}$. The following example describes the procedure in detail.

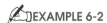EXAMPLE 6-2

Finding All Instant Centers for a Fourbar Linkage.

Problem: Given a fourbar linkage in one position, find all *ICs* by graphical methods.

6

[†] Note that this *graph* is not a plot of points on an *x, y* coordinate system. Rather it is a *linear graph* from the fascinating branch of mathematics called *graph theory*, which is itself a branch of topology. Linear graphs are often used to depict interrelationships between various phenomena. They have many applications in kinematics especially as a way to classify linkages and to find isomers.

[‡] Discovered independently by Aronhold in Germany, in 1872, and by Kennedy in England, in 1886. Kennedy[3] states in his preface, "The theorem of the three virtual (instant) centers … was first given, I believe, by Aronhold, although its previous publication was unknown to me until some years after I had given it in my lectures." It tends to be attributed to Kennedy in the English-speaking world and to Aronhold in the German-speaking world.

Solution: (See Figure 6-5 and the video *Instant Centers and Centrodes* on DVD.)

1   Draw a circle with all links numbered around the circumference as shown in Figure 6-5a.

2   Locate as many *ICs* as possible by inspection. All pin joints will be permanent *ICs*. Connect the link numbers on the circle to create a linear graph and record those *ICs* found, as shown in Figure 6-5a.

3   Identify a link combination on the linear graph for which the *IC* has not been found, and draw a dotted line connecting those two link numbers. Identify two triangles on the graph that each contain the dotted line and whose other two sides are solid lines representing *ICs* already

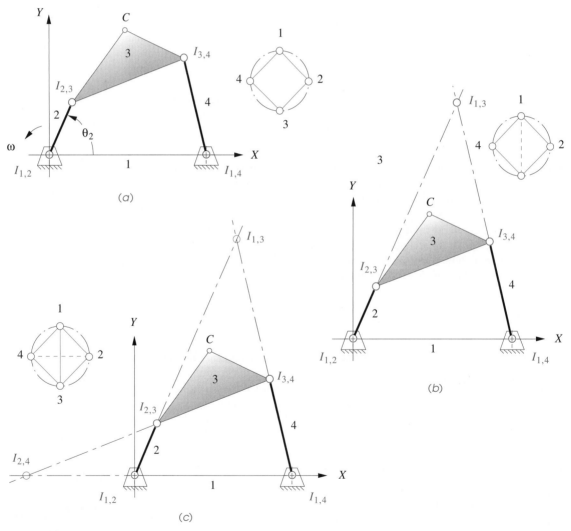

FIGURE 6-5

Locating instant centers in the pin-jointed linkage

found. On the graph in Figure 6-5b, link numbers 1 and 3 have been connected with a dotted line. This line forms one triangle with sides 13, 34, 14 and another with sides 13, 23, 12. These triangles define trios of *ICs* that obey **Kennedy's rule**. Thus *ICs* 13, 34, and 14 **must lie on the same straight line**. Also *ICs* 13, 23 and 12 will **lie on a different straight line**.

4   On the linkage diagram draw a line through the two known *ICs* that form a trio with the unknown *IC*. Repeat for the other trio. In Figure 6-5b, a line has been drawn through $I_{1,2}$ and $I_{2,3}$ and extended. $I_{1,3}$ must lie on this line. Another line has been drawn through $I_{1,4}$ and $I_{3,4}$ and extended to intersect the first line. By Kennedy's rule, instant center $I_{1,3}$ must also lie on this line, so their intersection is $I_{1,3}$.

5   Connect link numbers 2 and 4 with a dotted line on the linear graph as shown in Figure 6-5c. This line forms one triangle with sides 24, 23, 34 and another with sides 24, 12, 14. These sides represent trios of *ICs* that obey Kennedy's rule. Thus *ICs* 24, 23, and 34 must lie on the same straight line. Also *ICs* 24, 12, and 14 lie on a different straight line.

6   On the linkage diagram draw a line through the two known *ICs* that form a trio with the unknown *IC*. Repeat for the other trio. In Figure 6-5c, a line has been drawn through $I_{1,2}$ and $I_{1,4}$ and extended. $I_{2,4}$ must lie on this line. Another line has been drawn through $I_{2,3}$ and $I_{3,4}$ and extended to intersect the first line. By Kennedy's rule, instant center $I_{2,4}$ must also lie on this line, so their intersection is $I_{2,4}$.

7   If there were more links, this procedure would be repeated until all *ICs* were found.

The presence of slider joints makes finding the instant centers a little more subtle as is shown in the next example. Figure 6-6a (p. 296) shows a **fourbar crank-slider linkage**. Note that there are only three pin joints in this linkage. All pin joints are *permanent instant centers*. But the joint between links 1 and 4 is a rectilinear, sliding full joint. A sliding joint is kinematically equivalent to an infinitely long link, "pivoted" at infinity. Figure 6-6b shows a nearly equivalent pin-jointed version of the crank-slider in which link 4 is a very long rocker. Point *B* now swings through a shallow arc that is nearly a straight line. It is clear in Figure 6-6b (p. 296) that, in this linkage, $I_{1,4}$ is at pivot O_4. Now imagine increasing the length of this long, link 4 rocker even more. In the limit, link 4 approaches infinite length, the pivot O_4 approaches infinity along the line that was originally the long rocker, and the arc motion of point *B* approaches a straight line. Thus, *a slider joint will have its instant center at infinity along a line perpendicular to the direction of sliding* as shown in Figure 6-6a.

✏️EXAMPLE 6-3

Finding All Instant Centers for a Crank-Slider Linkage.

Problem: Given a crank-slider linkage in one position, find all *ICs* by graphical methods.

Solution: (See Figure 6-7, p. 297, and the video *Instant Centers and Centrodes* on DVD.)

1   Draw a circle with all links numbered around the circumference as shown in Figure 6-7a.

2   Locate all *ICs* possible by inspection. All pin joints will be permanent *ICs*. The slider joint's

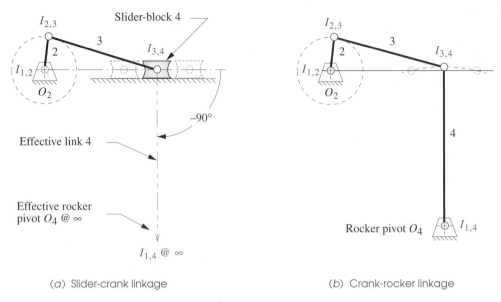

FIGURE 6-6

A rectilinear slider's instant center is at infinity

instant center will be at infinity along a line perpendicular to the axis of sliding. Connect the link numbers on the circle to create a linear graph and record those *ICs* found, as shown in Figure 6-7a.

3 Identify a link combination on the linear graph for which the *IC* has not been found, and draw a dotted line connecting those two link numbers. Identify two triangles on the graph that each contain the dotted line and whose other two sides are solid lines representing *ICs* already found. In the graph on Figure 6-7b, link numbers 1 and 3 have been connected with a dotted line. This line forms one triangle with sides 13, 34, 14 and another with sides 13, 23, 12. These sides represent trios of *ICs* that obey Kennedy's rule. Thus *ICs* 13, 34, and 14 must lie on the same straight line. Also *ICs* 13, 23, and 12 lie on a different straight line.

4 On the linkage diagram draw a line through the two known *ICs* that form a trio with the unknown *IC*. Repeat for the other trio. In Figure 6-7b, a line has been drawn from $I_{1,2}$ through $I_{2,3}$ and extended. $I_{1,3}$ must lie on this line. Another line has been drawn from $I_{1,4}$ (at infinity) through $I_{3,4}$ and extended to intersect the first line. By Kennedy's rule, instant center $I_{1,3}$ must also lie on this line, so their intersection is $I_{1,3}$.

5 Connect link numbers 2 and 4 with a dotted line on the graph as shown in Figure 6-7c. This line forms one triangle with sides 24, 23, 34 and another with sides 24, 12, 14. These sides also represent trios of *ICs* that obey Kennedy's rule. Thus *ICs* 24, 23, and 34 must lie on the same straight line. Also *ICs* 24, 12, and 14 lie on a different straight line.

6 On the linkage diagram draw a line through the two known *ICs* that form a trio with the unknown *IC*. Repeat for the other trio. In Figure 6-7c, a line has been drawn from $I_{1,2}$ to intersect $I_{1,4}$ and extended. Note that the only way to "intersect" $I_{1,4}$ at infinity is to draw a line parallel to the line $I_{3,4} I_{1,4}$ since all parallel lines intersect at infinity. Instant center $I_{2,4}$

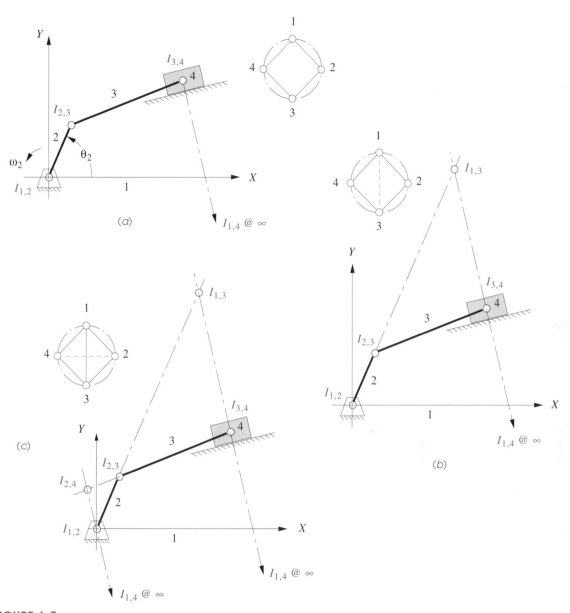

FIGURE 6-7

Locating instant centers in the slider-crank linkage

must lie on this parallel line. Another line has been drawn through $I_{2,3}$ and $I_{3,4}$ and extended to intersect the first line. By Kennedy's rule, instant center $I_{2,4}$ must also lie on this line, so their intersection is $I_{2,4}$.

7 If there were more links, this procedure would be repeated until all *ICs* were found.

The procedure in this slider example is identical to that used in the pin-jointed fourbar, except that it is complicated by the presence of instant centers located at infinity.

In Section 2.9 and Figure 2-10c (p. 47) we showed that a cam-follower mechanism is really a fourbar linkage in disguise. As such it will also possess instant centers. The presence of the half joint in this, or any linkage, makes the location of the instant centers a little more complicated. We have to recognize that the instant center between any two links will be along a line that is perpendicular to the *relative velocity* vector between the links at the half joint, as shown in the following example. Figure 6-8 shows the same cam-follower mechanism as in Figure 2-10c (p. 47). The effective links 2, 3, and 4 are also shown.

✎ EXAMPLE 6-4

Finding All Instant Centers for a Cam-Follower Mechanism.

Problem: Given a cam and follower in one position, find all *IC*s by graphical methods.

Solution: (See Figure 6-8.)

1   Draw a circle with all links numbered around the circumference as shown in Figure 6-8b. In this case there are only three links and thus only three *IC*s to be found as shown by equation 6.8. Note that the links are numbered 1, 2, and 4. The missing link 3 is the variable-length effective coupler.

2   Locate all *IC*s possible by inspection. All pin joints will be permanent *IC*s. The two fixed pivots $I_{1,2}$ and $I_{1,4}$ are the only pin joints here. Connect the link numbers on the circle to create a linear graph and record those *IC*s found, as shown in Figure 6-8b. The only link combination on the linear graph for which the *IC* has not been found is $I_{2,4}$, so draw a dotted line connecting those two link numbers.

3   Kennedy's rule says that all three *IC*s must lie on the same straight line; thus the remaining instant center $I_{2,4}$ must lie on the line $I_{1,2}$ $I_{1,4}$ extended. Unfortunately in this example, we have too few links to find a second line on which $I_{2,4}$ must lie.

4   On the linkage diagram draw a line through the two known *IC*s that form a trio with the unknown *IC*. In Figure 6-8c, a line has been drawn from $I_{1,2}$ through $I_{1,4}$ and extended. This is, of course, link 1. By Kennedy's rule, $I_{2,4}$ must lie on this line.

5   Looking at Figure 6-8c that shows the effective links of the equivalent fourbar linkage for this position, we can extend effective link 3 until it intersects link 1 extended. Just as in the "pure" fourbar linkage, instant center 2,4 lies on the intersection of links 1 and 3 extended (see Example 6-2, p. 293).

6   Figure 6-8d shows that it is not necessary to construct the effective fourbar linkage to find $I_{2,4}$. Note that the **common tangent** to links 2 and 4 at their contact point (the half joint) has been drawn. This line is also called the **axis of slip** because it is the line along that all relative (slip) velocity will occur between the two links. Thus the velocity of link 4 versus 2, $\mathbf{V}_{42}$, is directed along the axis of slip. Instant center $I_{2,4}$ must therefore lie along a line perpendicular to the common tangent, called the **common normal.** Note that this line is the same as the effective link 3 line in Figure 6-8c.

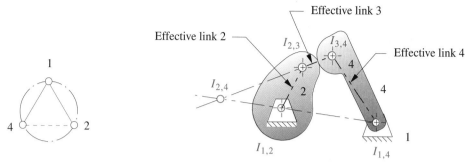

(b) The linkage graph

(c) The instantaneously equivalent "effective linkage"

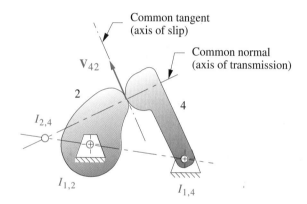

(a) The cam and follower

(d) Finding $I_{2,4}$ without using the effective linkage

FIGURE 6-8

Locating instant centers in the cam-follower mechanism

6.4 VELOCITY ANALYSIS WITH INSTANT CENTERS

Once the *ICs* have been found, they can be used to do a very rapid graphical velocity analysis of the linkage. Note that, depending on the particular position of the linkage being analyzed, some of the *ICs* may be very far removed from the links. For example, if links 2 and 4 are nearly parallel, their extended lines will intersect at a point far away and not be practically available for velocity analysis. Figure 6-9 (p. 300) shows the same linkage as Figure 6-5 (p. 294) with $I_{1,3}$ located and labeled. From the definition of the instant center, both links sharing the instant center will have identical velocity at that point. Instant center $I_{1,3}$ involves the coupler (link 3), which is in complex motion, and the ground link 1, which is stationary. All points on link 1 have zero velocity in the global coordinate system, which is embedded in link 1. Therefore, $I_{1,3}$ must have zero velocity at this instant. If $I_{1,3}$ has zero velocity, then it can be considered to be an instantaneous "fixed pivot" about which link 3 is in pure rotation with respect to link 1. A moment later, $I_{1,3}$ will move to a new location and link 3 will be "pivoting" about a new instant center.

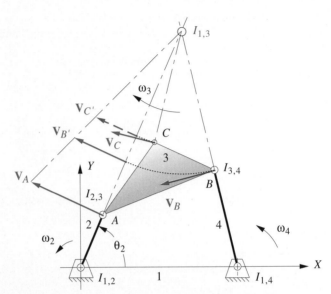

FIGURE 6-9

Velocity analysis using instant centers

The velocity of point A is shown on Figure 6-9. The magnitude of $\mathbf{V}_A$ can be computed from equation 6.7 (p. 288). Its direction and sense can be determined by inspection as was done in Example 6-1 (p. 290). Note that point A is also instant center $I_{2,3}$. It has the same velocity as part of link 2 and as part of link 3. Since link 3 is effectively pivoting about $I_{1,3}$ at this instant, the angular velocity ω_3 can be found by rearranging equation 6.7:

$$\omega_3 = \frac{v_A}{\left(AI_{1,3}\right)} \tag{6.9a}$$

Once ω_3 is known, the magnitude of $\mathbf{V}_B$ can also be found from equation 6.7:

$$v_B = \left(BI_{1,3}\right)\omega_3 \tag{6.9b}$$

Once $\mathbf{V}_B$ is known, ω_4 can also be found from equation 6.7:

$$\omega_4 = \frac{v_B}{\left(BO_4\right)} \tag{6.9c}$$

Finally, the magnitude of $\mathbf{V}_C$ (or the velocity of any other point on the coupler) can be found from equation 6.7:

$$v_C = \left(CI_{1,3}\right)\omega_3 \tag{6.9d}$$

Note that equations 6.7 and 6.9 provide only the **scalar magnitude** of these velocity vectors. We have to determine their **direction** from the information in the scale diagram (Figure 6-9). Since we know the location of $I_{1,3}$, which is an instantaneous "fixed" pivot for link 3, all of that link's absolute velocity vectors for this instant will be **perpendicular** to their radii from $I_{1,3}$ **to the point in question**. $\mathbf{V}_B$ and $\mathbf{V}_C$ can be seen to be perpendicular to their radii from $I_{1,3}$. Note that $\mathbf{V}_B$ is also perpendicular to the radius from O_4 because B is also pivoting about that point as part of link 4.

A rapid graphical solution to equations 6.9 is shown in the figure. Arcs centered at $I_{1,3}$ are swung from points B and C to intersect line $AI_{1,3}$. The magnitudes of velocities $\mathbf{V}_{B'}$ and $\mathbf{V}_{C'}$ are found from the vectors drawn perpendicular to that line at the intersections of the arcs and line $AI_{1,3}$. The lengths of the vectors are defined by the line from the tip of $\mathbf{V}_A$ to the instant center $I_{1,3}$. These vectors can then be slid along their arcs back to points B and C, maintaining their tangency to the arcs.

Thus, we have in only a few steps found all the same velocities that were found in the more tedious method of Example 6-1 (p. 290). The instant center method is a quick graphical method to analyze velocities, but it will only work if the instant centers are in reachable locations for the particular linkage position analyzed. However, the graphical method using the velocity difference equation shown in Example 6-1 will always work, regardless of linkage position.

Angular Velocity Ratio

The **angular velocity ratio** m_V is defined as *the output angular velocity divided by the input angular velocity*. For a fourbar mechanism this is expressed as:

$$m_V = \frac{\omega_4}{\omega_2} \tag{6.10}$$

We can derive this ratio for any linkage by constructing a **pair of effective links** as shown in Figure 6-10a (p. 302). The definition of **effective link pairs** is *two lines, mutually parallel, drawn through the fixed pivots and intersecting the coupler extended*. These are shown as O_2A' and O_4B' in Figure 6-10a. Note that there is an infinity of possible effective link pairs. They must be parallel to one another but may make any angle with link 3. In the figure they are shown perpendicular to link 3 for convenience in the derivation to follow. The angle between links 2 and 3 is shown as ν. The transmission angle between links 3 and 4 is μ. We will now derive an expression for the angular velocity ratio using these effective links, the actual link lengths, and angles ν and μ.

From geometry:

$$O_2A' = (O_2A)\sin\nu \qquad\qquad O_4B' = (O_4B)\sin\mu \tag{6.11a}$$

From equation 6.7 (p. 288)

$$V_{A'} = (O_2A')\omega_2 \tag{6.11b}$$

The component of velocity $V_{A'}$ lies along the link AB. Just as with a two-force member in which a force applied at one end transmits only its component that lies along the link to the other end, this velocity component can be transmitted along the link to point B. This is sometimes called the **principle of transmissibility**. We can then equate these components at either end of the link.

$$V_{A'} = V_{B'} \tag{6.11c}$$

(a)

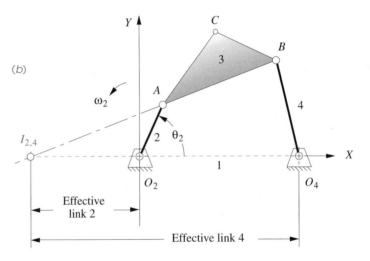

(b)

FIGURE 6-10

Effective links and the angular velocity ratio

Then:

$$O_2A'\omega_2 = O_4B'\omega_4 \qquad\qquad (6.11d)$$

rearranging:

$$\frac{\omega_4}{\omega_2} = \frac{O_2A'}{O_4B'} \qquad\qquad (6.11e)$$

and substituting:

$$\frac{\omega_4}{\omega_2} = \frac{O_2A\sin\nu}{O_4B\sin\mu} = m_V \qquad\qquad (6.11f)$$

Note in equation 6.11f that as angle ν goes through zero, the angular velocity ratio will be zero regardless of the values of ω_2 or the link lengths, and thus ω_4 will be zero. When angle ν is zero, links 2 and 3 will be colinear and thus be in their toggle positions. We learned in Section 3.3 (p. 100) that the limiting positions of link 4 are defined by these toggle conditions. We should expect that the velocity of link 4 will be zero when it has come to the end of its travel. An even more interesting situation obtains if we allow angle μ to go to zero. Equation 6.11f shows that ω_4 **will go to infinity** when $\mu = 0$, regardless of the values of ω_2 or the link lengths. We clearly cannot allow μ to reach zero. In fact, we learned in Section 3.3 that we should keep this transmission angle μ above about 40 degrees to maintain good quality of motion and force transmission.[*]

Figure 6-10b shows the same linkage as in Figure 6-10a, but the effective links have now been drawn so that they are not only parallel but are also colinear, and thus lie on top of one another. Both intersect the extended coupler at the same point, which is instant center $I_{2,4}$. So, A' and B' of Figure 6-10a are now coincident at $I_{2,4}$. This allows us to write an equation for the **angular velocity ratio** in terms of the distances from the fixed pivots to instant center $I_{2,4}$.

$$m_V = \frac{\omega_4}{\omega_2} = \frac{O_2 I_{2,4}}{O_4 I_{2,4}} \tag{6.11g}$$

Thus, the instant center $I_{2,4}$ can be used to determine the **angular velocity ratio**.

Mechanical Advantage

The power P in a mechanical system can be defined as the dot or scalar product of the force vector $\mathbf{F}$ and the velocity vector $\mathbf{V}$ at any point:

$$P = \mathbf{F} \cdot \mathbf{V} = F_x V_x + F_y V_y \tag{6.12a}$$

For a rotating system, power P becomes the product of torque T and angular velocity ω that, in two dimensions, have the same (z) direction:

$$P = T\omega \tag{6.12b}$$

The power flows through a passive system and:

$$P_{in} = P_{out} + losses \tag{6.12c}$$

Mechanical efficiency can be defined as:

$$\varepsilon = \frac{P_{out}}{P_{in}} \tag{6.12d}$$

Linkage systems can be very efficient if they are well made with low friction bearings on all pivots. Losses are often less than 10%. For simplicity in the following analysis we will assume that the losses are zero (i.e., a conservative system). Then, letting T_{in} and ω_{in} represent input torque and angular velocity, and T_{out} and ω_{out} represent output torque and angular velocity,

[*] This limitation on transmission angle is only critical if the output load is applied to a link that is pivoted to ground (i.e., to link 4 in the case of a fourbar linkage). If the load is applied to a floating link (e.g., a coupler), then other measures of the quality of force transmission than the transmission angle are more appropriate, as discussed in Chapter 11, Section 11.12 (p. 611) where the joint force index (JFI) is defined.

$$P_{in} = T_{in}\omega_{in}$$

$$P_{out} = T_{out}\omega_{out} \tag{6.12e}$$

Note that the **torque ratio** ($m_T = T_{out}/T_{in}$) is the inverse of the angular velocity ratio.

Mechanical advantage (m_A) can be defined as:

$$m_A = \frac{F_{out}}{F_{in}} \tag{6.13a}$$

Assuming that the input and output forces are applied at some radii r_{in} and r_{out}, perpendicular to their respective force vectors,

$$F_{out} = \frac{T_{out}}{r_{out}}$$

$$F_{in} = \frac{T_{in}}{r_{in}} \tag{6.13b}$$

substituting equations 6.13b in 6.13a gives an expression in terms of torque.

$$m_A = \left(\frac{T_{out}}{T_{in}}\right)\left(\frac{r_{in}}{r_{out}}\right) \tag{6.13c}$$

Substituting equation 6.12f in 6.13c gives

$$m_A = \left(\frac{\omega_{in}}{\omega_{out}}\right)\left(\frac{r_{in}}{r_{out}}\right) \tag{6.13d}$$

and substituting equation 6.11f (p. 302) gives

$$m_A = \left(\frac{O_4 B \sin\mu}{O_2 A \sin\nu}\right)\left(\frac{r_{in}}{r_{out}}\right) \tag{6.13e}$$

See Figure 6-11 and compare equation 6.13e to equation 6.11f (p. 302) and its discussion under **angular velocity ratio** (p. 301). Equation 6.13e shows that for any choice of r_{in} and r_{out}, the mechanical advantage responds to changes in angles ν and μ in opposite fashion to that of the angular velocity ratio. If the transmission angle μ goes to zero (which we don't want it to do), the mechanical advantage also goes to zero regardless of the amount of input force or torque applied. But, when angle ν goes to zero (which it can and does, twice per cycle in a Grashof linkage), the mechanical advantage becomes infinite! This is the principle of a rock-crusher mechanism as shown in Figure 6-11. A quite moderate force applied to link 2 can generate a huge force on link 4 to crush the rock. Of course, we cannot expect to achieve the theoretical output of infinite force or torque magnitude, as the strengths of the links and joints will limit the maximum forces and torques obtainable. Another common example of a linkage that takes advantage of this theoretically infinite mechanical advantage at the toggle position is a ViseGrip locking pliers (see Figure P6-21, p. 341).

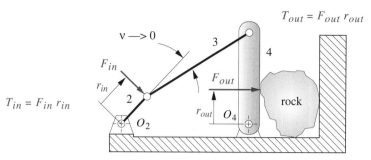

FIGURE 6-11

"Rock-crusher" toggle mechanism

These two ratios, **angular velocity ratio** and **mechanical advantage**, provide useful, dimensionless **indices of merit** by which we can judge the relative quality of various linkage designs that may be proposed as solutions.

Using Instant Centers in Linkage Design

In addition to providing a quick numerical velocity analysis, instant center analysis more importantly gives the designer a remarkable overview of the linkage's global behavior. It is quite difficult to mentally visualize the complex motion of a "floating" coupler link even in a simple fourbar linkage, unless you build a model or run a computer simulation. Because this complex coupler motion in fact reduces to an instantaneous pure rotation about the instant center $I_{1,3}$, finding that center allows the designer to visualize the motion of the coupler as a pure rotation. One can literally *see* the motion and the directions of velocities of any points of interest by relating them to the instant center. It is only necessary to draw the linkage in a few positions of interest, showing the instant center locations for each position.

Figure 6-12 (p. 306) shows a practical example of how this visual, qualitative analysis technique could be applied to the design of an automobile rear suspension system. Most automobile suspension mechanisms are either fourbar linkages or fourbar cranksliders, with the wheel assembly carried on the coupler (as was also shown in Figure 3-19, p. 127). Figure 6-12a shows a rear suspension design from a domestic car of 1970s vintage that was later redesigned because of a disturbing tendency to "bump steer," i.e., turn the rear axle when hitting a bump on one side of the car. The figure is a view looking from the center of the car outward, showing the fourbar linkage that controls the up and down motion of one side of the rear axle and one wheel. Links 2 and 4 are pivoted to the frame of the car which is link 1. The wheel and axle assembly is rigidly attached to the coupler, link 3. Thus the wheel assembly has complex motion in the vertical plane. Ideally, one would like the wheel to move up and down in a straight vertical line when hitting a bump. Figure 6-12b shows the motion of the wheel and the new instant center ($I_{1,3}$) location for the situation when one wheel has hit a bump. The velocity vector for the center of the wheel in each position is drawn perpendicular to its radius from $I_{1,3}$. You can see that the wheel center has a significant horizontal component of motion as it moves up over the bump. This horizontal component causes the wheel center on that side of the car to move forward while it moves upward, thus turning the axle (about a

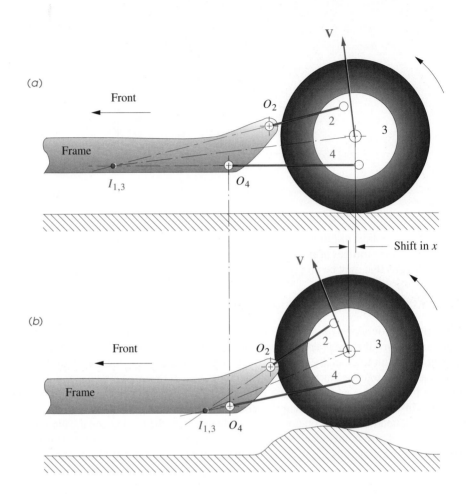

FIGURE 6-12

"Bump steer" due to shift in instant center location

vertical axis) and steering the car with the rear wheels in the same way that you steer a toy wagon. Viewing the path of the instant center over some range of motion gives a clear picture of the behavior of the coupler link. The undesirable behavior of this suspension linkage system could have been predicted from this simple instant center analysis before ever building the mechanism.

Another practical example of the effective use of instant centers in linkage design is shown in Figure 6-13, which is an optical adjusting mechanism used to position a mirror and allow a small amount of rotational adjustment. [1] A more detailed account of this design case study [2] is provided in Chapter 1. The designer, K. Towfigh, recognized that $I_{1,3}$ at point E is an instantaneous "fixed pivot" and will allow very small pure rotations about that point with very small translational error. He then designed a one-piece, plastic fourbar linkage whose "pin joints" are thin webs of plastic that flex to allow slight rotation. This is termed a **compliant linkage,**[*] one that uses elastic deformations of the links

The Mechanism

The Final Product of Keivan Towfigh

FIGURE 6-13

An optical adjustment compliant linkage *Reproduced from reference (2) with permission*

as hinges instead of pin joints. He then placed the mirror on the coupler at $I_{1,3}$. Even the fixed link 1 is the same piece as the "movable links" and has a small set screw to provide the adjustment. A simple and elegant design.

6.5 CENTRODES†

Figure 6-14 (p. 309) illustrates the fact that the successive positions of an instant center (or **centro**) form a path of their own. *This path, or locus, of the instant center is called the* **centrode**. Since there are two links needed to create an instant center, there will be two centrodes associated with any one instant center. These are formed by projecting the path of the instant center first on one link and then on the other. Figure 6-14a shows the locus of instant center $I_{1,3}$ as projected onto link 1. Because link 1 is stationary, or fixed, this is called the **fixed centrode**. By temporarily inverting the mechanism and fixing link 3 as the ground link, as shown in Figure 6-14b, we can move link 1 as the coupler and project the locus of $I_{1,3}$ onto link 3. In the original linkage, link 3 was the moving coupler, so this is called the **moving centrode**. Figure 6-14c shows the original linkage with both fixed and moving centrodes superposed.

The definition of the instant center says that both links have the same velocity at that point, at that instant. Link 1 has zero velocity everywhere, as does the fixed centrode. So, as the linkage moves, the moving centrode must roll against the fixed centrode without slipping. If you cut the fixed and moving centrodes out of metal, as shown in Figure 6-14d, and roll the moving centrode (which is link 3) against the fixed centrode (which is link 1), the complex motion of link 3 will be identical to that of the original linkage. *All of the coupler curves of points on link 3 will have the same path shapes as in the original linkage.* We now have, in effect, a "linkless" fourbar linkage, really one composed of two bodies that have these centrode shapes rolling against one another. Links 2 and 4 have

* See also Section 2.16 (p. 65) for more information on compliant mechanisms.

---

† See the video *Instant Centers and Centrodes* on the book's DVD.

been eliminated. Note that the example shown in Figure 6-14 is a non-Grashof fourbar. The lengths of its centrodes are limited by the double-rocker toggle positions.

All instant centers of a linkage will have centrodes.[*] If the links are directly connected by a joint, such as $I_{2,3}$, $I_{3,4}$, $I_{1,2}$, and $I_{1,4}$, their fixed and moving centrodes will degenerate to a point at that location on each link. The most interesting centrodes are those involving links not directly connected to one another such as $I_{1,3}$ and $I_{2,4}$. If we look at the double-crank linkage in Figure 6-15a (p. 310) in which links 2 and 4 both revolve fully, we see that the centrodes of $I_{1,3}$ form closed curves. The motion of link 3 with respect to link 1 could be duplicated by causing these two centrodes to roll against one another without slipping. Note that there are two loops to the moving centrode. Both must roll on the single-loop fixed centrode to complete the motion of the equivalent double-crank linkage.

We have so far dealt largely with the instant center $I_{1,3}$. Instant center $I_{2,4}$ involves two links that are each in pure rotation and not directly connected to one another. If we use a special-case Grashof linkage with the links crossed (sometimes called an **antiparallelogram** linkage), the centrodes of $I_{2,4}$ become ellipses as shown in Figure 6-15b. To guarantee no slip, it will probably be necessary to put meshing teeth on each centrode. We then will have a pair of elliptical, **noncircular gears**, or *gearset*, which gives the *same output motion as the original double-crank linkage* and will have the *same variations in the angular velocity ratio and mechanical advantage as the linkage* had. Thus we can see that *gearsets are also just fourbar linkages in disguise*. Noncircular gears find much use in machinery, such as printing presses, where rollers must be speeded and slowed with some pattern during each cycle or revolution. More complicated shapes of noncircular gears are analogous to cams and followers in that the equivalent fourbar linkage must have variable-length links. **Circular gears** are just a special case of noncircular gears that give a **constant angular velocity ratio** and are widely used in all machines. Gears and gearsets will be dealt with in greater detail in Chapter 10.

In general, centrodes of crank-rockers and double- or triple-rockers will be open curves with asymptotes. Centrodes of double-crank linkages will be closed curves. Program LINKAGES will calculate and draw the fixed and moving centrodes for any linkage input to it. Open the files F06-14.4br, F06-15a.4br, and F06-15b.4br into program LINKAGES to see the centrodes of these linkages drawn as the linkages rotate.

A "Linkless" Linkage

A common example of a mechanism made of centrodes is shown in Figure 6-16a (p. 310). You have probably rocked in a *Boston* or *Hitchcock* rocking chair and experienced the soothing motions that it delivers to your body. You may have also rocked in a *platform rocker* as shown in Figure 6-16b and noticed that its motion did not feel as soothing.

There are good kinematic reasons for the difference. The platform rocker has a fixed pin joint between the seat and the base (floor). Thus all parts of your body are in pure rotation along concentric arcs. You are in effect riding on the rocker of a linkage.

The Boston rocker has a shaped (curved) base, or "runners," which rolls against the floor. These runners are usually *not* circular arcs. They have a higher-order curve contour. They are, in fact, **moving centrodes**. The floor is the **fixed centrode**. When one is rolled against the other, the chair and its occupant experience coupler curve motion. Every part

---

[*] Since instant centers are called *poles* as well as *centros*, centrodes are sometimes also called *polodes*. We will use the *centro* and *centrode* nomenclature in this text.

(a) The fixed centrode

(b) The moving centrode

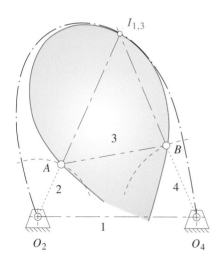

(c) The centrodes in contact

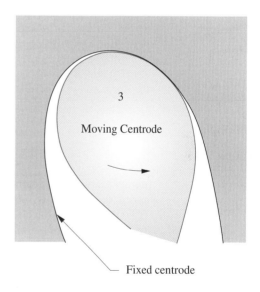

(d) Roll the moving centrode against the fixed centrode to produce the same coupler motion as the original linkage

FIGURE 6-14

Open-loop fixed and moving centrodes (or polodes) of a fourbar linkage

of your body travels along a different sixth-order coupler curve that provides smooth accelerations and velocities and feels better than the cruder second-order (circular) motion of the platform rocker. Our ancestors, who carved these rocking chairs, probably had never heard of fourbar linkages and centrodes, but they knew intuitively how to create comfortable motions.

(a) Closed-loop centrodes of $I_{1,3}$
    for a Grashof double-crank linkage

(b) Ellipsoidal centrodes of $I_{2,4}$
    for a special-case Grashof
    anti-parallelogram linkage

FIGURE 6-15

Closed-loop fixed and moving centrodes

(a) Boston rocker

(b) Platform rocker

FIGURE 6-16

Some rocking chairs
use centrodes of a
fourbar linkage

Cusps

Another example of a centrode that you probably use frequently is the path of the tire on your car or bicycle. As your tire rolls against the road without slipping, the road becomes a fixed centrode and the circumference of the tire is the moving centrode. The tire is, in effect, the coupler of a linkless fourbar linkage. All points on the contact surface of the tire move along cycloidal coupler curves and pass through a cusp of zero velocity when they reach the fixed centrode at the road surface as shown in Figure 6-17a. All other points on the tire and wheel assembly travel along coupler curves that do not have cusps. This last fact is a clue to a means to identify coupler points that will have cusps in their coupler curve. *If a coupler point is chosen to be on the moving centrode at one extreme of its path motion (i.e., at one of the positions of $I_{1,3}$), then it will have a cusp in its coupler curve.* Figure 6-17b shows a coupler curve of such a point, drawn with program LINKAGES. The right end of the coupler path touches the moving centrode and as a result has a cusp at that point. So, if you desire a cusp in your coupler motion, many are available. Simply choose a coupler point on the moving centrode of link 3. Open the file F06-17b.4br in program LINKAGES to animate that linkage with its coupler curve or centrodes. Note in Figure 6-14 (p. 309) that choosing any location of instant center $I_{1,3}$ on the coupler as the coupler point will provide a cusp at that point.

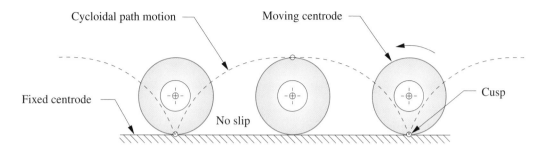

Cycloidal path motion   Moving centrode

Fixed centrode

No slip

Cusp

(a) Cycloidal motion of a circular, moving centrode rolling on a straight, fixed centrode

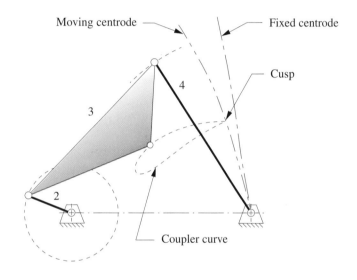

Moving centrode   Fixed centrode

Cusp

4

3

2

Coupler curve

(b) Coupler curve cusps exist only on the moving centrode

FIGURE 6-17

Examples of centrodes

6.6   VELOCITY OF SLIP

When there is a sliding joint between two links and neither one is the ground link, the velocity analysis is more complicated. Figure 6-18 (p. 312) shows an inversion of the fourbar crank-slider mechanism in which the sliding joint is floating, i.e., not grounded. To solve for the velocity at the sliding joint A, we have to recognize that there is more than one point A at that joint. There is a point A as part of link 2 (A_2), a point A as part of link 3 (A_3), and a point A as part of link 4 (A_4). This is a CASE 2 situation in which we have at least two points belonging to different links but occupying the same location at a given instant. Thus, the **relative velocity** equation 6.6 (p. 287) will apply. We can usually solve for the velocity of at least one of these points directly from the known input information using equation 6.7 (p. 288). It and equation 6.6 are all that is needed to solve

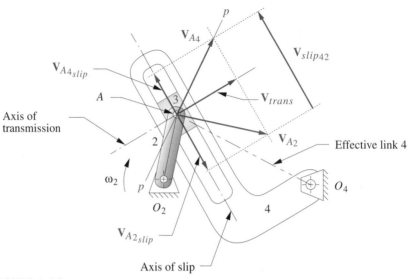

FIGURE 6-18

Velocity of slip and velocity of transmission (note that the applied ω is negative as shown)

for everything else. In this example link 2 is the driver, and θ_2 and ω_2 are given for the "freeze frame" position shown. We wish to solve for ω_4, the angular velocity of link 4, and also for the velocity of slip at the joint labeled A.

In Figure 6-18 the **axis of slip** is shown to be tangent to the slider motion and is the line along which all sliding occurs between links 3 and 4. The **axis of transmission** is defined to be perpendicular to the axis of slip and pass through the slider joint at A. This *axis of transmission is the* **only line** *along which we can transmit motion or force across the slider joint, except for friction.* We will assume friction to be negligible in this example. Any force or velocity vector applied to point A can be resolved into two components along these two axes that provide a *translating and rotating, local coordinate system* for analysis at the joint. The component along the axis of transmission will do useful work at the joint. But, the component along the axis of slip does no work, except *friction work.*

✏ EXAMPLE 6-5

Graphical Velocity Analysis at a Sliding Joint.

Problem: Given θ_2, θ_3, θ_4, ω_2, find ω_3, ω_4, $\mathbf{V}_A$, by graphical methods.

Solution: (See Figure 6-18.)

1   Start at the end of the linkage for which you have the most information. Calculate the magnitude of the velocity of **point A as part of link 2** (A_2) using scalar equation 6.7 (p. 288).

$$v_{A_2} = (AO_2)\omega_2 \tag{a}$$

2   Draw the velocity vector $\mathbf{V}_{A2}$ with its length equal to its magnitude v_{A2} at some convenient scale and with its root at point A and its direction perpendicular to the radius AO_2. Its sense is the same as that of ω_2 as is shown in Figure 6-18.

3   Draw the **axis of slip** and **axis of transmission** through point A.

4   Project $\mathbf{V}_{A2}$ onto the axis of slip and onto the axis of transmission to create the components $\mathbf{V}_{A2slip}$ and $\mathbf{V}_{trans}$ of $\mathbf{V}_{A2}$ on the axes of slip and transmission, respectively. Note that the **transmission component** is shared by all true velocity vectors at this point, as it is the only component that can transmit across the joint.

5   Note that link 3 is pin-jointed to link 2, so $\mathbf{V}_{A3} = \mathbf{V}_{A2}$.

6   Note that the direction of the velocity of point $\mathbf{V}_{A4}$ is predictable since all points on link 4 are pivoting in pure rotation about point O_4. Draw the line pp through point A and perpendicular to the effective link 4, AO_4. Line pp is the direction of velocity $\mathbf{V}_{A4}$.

7   Construct the true magnitude of velocity vector $\mathbf{V}_{A4}$ by extending the projection of the **transmission component** $\mathbf{V}_{trans}$ until it intersects line pp.

8   Project $\mathbf{V}_{A4}$ onto the axis of slip to create the **slip component** $\mathbf{V}_{A4slip}$.

9   Write the relative velocity vector equation 6.6 (p. 287) for the **slip components** of point A_2 versus point A_4.

$$V_{slip42} = V_{A_4 slip} - V_{A_2 slip} \qquad (b)$$

10 The angular velocities of links 3 and 4 are identical because they share the slider joint and must rotate together. They can be calculated from equation 6.7 (p. 288):

$$\omega_4 = \omega_3 = \frac{V_{A_4}}{AO_4} \qquad (c)$$

Instant center analysis also can be used to solve sliding-joint velocity problems.

✏️ EXAMPLE 6-6

Graphical Velocity Analysis of a Cam and Follower.

Problem: Given θ_2, ω_2, find ω_3, by graphical methods.

Solution: (See Figure 6-19, p. 314.)

1   Construct the effective radius of the cam $R_{2\,eff}$ at the instantaneous point of contact with the follower for this position (point A in the figure). Its length is distance O_2A. Calculate the magnitude of the velocity of point A as part of link 2 (A_2) using scalar equation 6.7 (p. 288).

$$v_{A_2} = (AO_2)\omega_2 \qquad (a)$$

2 Draw the velocity vector $\mathbf{V}_{A2}$ with its length equal to its magnitude v_{A2} at some convenient scale and with its root at point A and its direction perpendicular to the radius O_2A. Its sense is the same as that of ω_2 as is shown in Figure 6-19.

3 Construct the axis of slip (common tangent to cam and follower) and its normal, the axis of transmission, as shown in Figure 6-19.

4 Project $\mathbf{V}_{A2}$ onto the axis of transmission to create the component $\mathbf{V}_{trans}$. Note that the **transmission component** is shared by all true velocity vectors at this point, as it is the only component that can transmit across the joint.

5 Project $\mathbf{V}_{A2}$ onto the axis of slip to create the **slip component** $\mathbf{V}_{A2slip}$.

6 Note that the direction of the velocity of point $\mathbf{V}_{A3}$ is predictable since all points on link 3 are pivoting in pure rotation about point O_3. Construct the effective radius of the follower $R_{3\,eff}$ at the instantaneous point of contact with the follower for this position (point A in the figure). Its length is distance O_3A.

7 Construct a line in the direction of $\mathbf{V}_{A3}$ perpendicular to $R_{3\,eff}$. Construct the true magnitude of velocity vector $\mathbf{V}_{A3}$ by extending the projection of the transmission component $\mathbf{V}_{trans}$ until it intersects the line of $\mathbf{V}_{A3}$.

8 Project $\mathbf{V}_{A3}$ onto the axis of slip to create the **slip component** $\mathbf{V}_{A3slip}$.

9 The total slip velocity at A is the vector difference between the two slip components. Write the relative velocity vector equation 6.6 (p. 287) for the slip components of point A_3 versus A_2.

$$V_{slip32} = V_{A3slip} - V_{A2slip} \qquad\qquad (b)$$

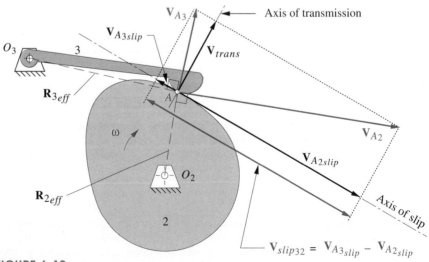

FIGURE 6-19

Graphical velocity analysis of a cam and follower

10 The angular velocity of link 3 can be calculated from equation 6.7:

$$\omega_3 = \frac{V_{A_3}}{AO_3} \qquad\qquad (c)$$

The above examples show how mechanisms with sliding or half joints can be solved graphically for velocities at one position. In the next section, we will develop the general solution using algebraic equations to solve similar problems.

6.7 ANALYTICAL SOLUTIONS FOR VELOCITY ANALYSIS

The Fourbar Pin-Jointed Linkage

The vector-loop position equations for the fourbar pin-jointed linkage were derived in Section 4.5 (pp. 185-190). The linkage was shown in Figure 4-6 (p. 184) and is shown again in Figure 6-20 (p. 316) on which we also show an input angular velocity ω_2 applied to link 2. This ω_2 can be a time-varying input velocity. The vector loop equation is shown in equations 4.5a and 4.5c, repeated here for your convenience.

$$\mathbf{R}_2 + \mathbf{R}_3 - \mathbf{R}_4 - \mathbf{R}_1 = 0 \qquad\qquad (4.5a)$$

As before, we substitute the complex number notation for the vectors, denoting their scalar lengths as a, b, c, d as shown in Figure 6-20a.

$$a e^{j\theta_2} + b e^{j\theta_3} - c e^{j\theta_4} - d e^{j\theta_1} = 0 \qquad\qquad (4.5c)$$

To get an expression for velocity, differentiate equation 4.5c with respect to time.

$$ja e^{j\theta_2} \frac{d\theta_2}{dt} + jb e^{j\theta_3} \frac{d\theta_3}{dt} - jc e^{j\theta_4} \frac{d\theta_4}{dt} = 0 \qquad\qquad (6.14a)$$

But,

$$\frac{d\theta_2}{dt} = \omega_2; \qquad \frac{d\theta_3}{dt} = \omega_3; \qquad \frac{d\theta_4}{dt} = \omega_4 \qquad\qquad (6.14b)$$

and:

$$ja\omega_2 e^{j\theta_2} + jb\omega_3 e^{j\theta_3} - jc\omega_4 e^{j\theta_4} = 0 \qquad\qquad (6.14c)$$

Note that the θ_1 term has dropped out because that angle is a constant, and thus its derivative is zero. Note also that equation 6.14 is, in fact, the **relative velocity** or **velocity difference equation**.

$$\mathbf{V}_A + \mathbf{V}_{BA} - \mathbf{V}_B = 0 \qquad\qquad (6.15a)$$

where:

$$\mathbf{V}_A = ja\omega_2 e^{j\theta_2}$$

$$\mathbf{V}_{BA} = jb\omega_3 e^{j\theta_3} \qquad\qquad (6.15b)$$

$$\mathbf{V}_B = jc\omega_4 e^{j\theta_4}$$

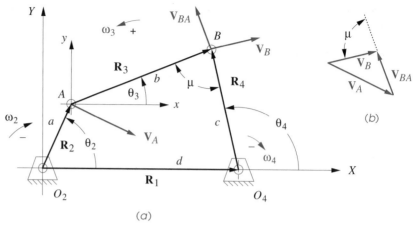

FIGURE 6-20

Position vector loop for a fourbar linkage showing velocity vectors for a negative (cw) ω_2

Compare equations 6.15 to equations 6.3, 6.5, and 6.6 (pp. 286-287). This equation is solved graphically in the vector diagram of Figure 6-20b. Note the transmission angle μ drawn between links 3 and 4 and also between V_B and V_{BA}. This shows an alternate way to define the transmission angle using the velocity vectors at point B.

We now need to solve equation 6.14 for ω_3 and ω_4, knowing the input velocity ω_2, the link lengths, and all link angles. Thus the position analysis derived in Section 4.5 (p. 183) must be done first to determine the link angles before this velocity analysis can be completed. We wish to solve equation 6.14 to get expressions in this form:

$$\omega_3 = f\left(a, b, c, d, \theta_2, \theta_3, \theta_4, \omega_2\right) \qquad \omega_4 = g\left(a, b, c, d, \theta_2, \theta_3, \theta_4, \omega_2\right) \qquad (6.16)$$

The strategy of solution will be the same as was done for the position analysis. First, substitute the Euler identity from equation 4.4a (p. 185) in each term of equation 6.14c:

$$ja\omega_2 \left(\cos\theta_2 + j\sin\theta_2\right) + jb\omega_3 \left(\cos\theta_3 + j\sin\theta_3\right)$$
$$- jc\omega_4 \left(\cos\theta_4 + j\sin\theta_4\right) = 0 \qquad (6.17a)$$

Multiply through by the operator j:

$$a\omega_2 \left(j\cos\theta_2 + j^2\sin\theta_2\right) + b\omega_3 \left(j\cos\theta_3 + j^2\sin\theta_3\right)$$
$$- c\omega_4 \left(j\cos\theta_4 + j^2\sin\theta_4\right) = 0 \qquad (6.17b)$$

The cosine terms have become the imaginary, or y-directed terms, and because $j^2 = -1$, the sine terms have become real or x-directed.

$$a\omega_2 \left(-\sin\theta_2 + j\cos\theta_2\right) + b\omega_3 \left(-\sin\theta_3 + j\cos\theta_3\right)$$
$$- c\omega_4 \left(-\sin\theta_4 + j\cos\theta_4\right) = 0 \qquad (6.17c)$$

We can now separate this vector equation into its two components by collecting all real and all imaginary terms separately:

real part (x component):

$$-a\omega_2 \sin\theta_2 - b\omega_3 \sin\theta_3 + c\omega_4 \sin\theta_4 = 0 \qquad (6.17d)$$

imaginary part (y component):

$$a\omega_2 \cos\theta_2 + b\omega_3 \cos\theta_3 - c\omega_4 \cos\theta_4 = 0 \qquad (6.17e)$$

Note that the j's have canceled in equation 6.17e. We can solve these two equations, 6.17d and 6.17e, simultaneously by direct substitution to get:

$$\omega_3 = \frac{a\omega_2}{b} \frac{\sin(\theta_4 - \theta_2)}{\sin(\theta_3 - \theta_4)} \qquad (6.18a)$$

$$\omega_4 = \frac{a\omega_2}{c} \frac{\sin(\theta_2 - \theta_3)}{\sin(\theta_4 - \theta_3)} \qquad (6.18b)$$

Once we have solved for ω_3 and ω_4, we can then solve for the linear velocities by substituting the Euler identity into equations 6.15,

$$\mathbf{V}_A = ja\omega_2 (\cos\theta_2 + j\sin\theta_2) = a\omega_2 (-\sin\theta_2 + j\cos\theta_2) \qquad (6.19a)$$

$$\mathbf{V}_{BA} = jb\omega_3 (\cos\theta_3 + j\sin\theta_3) = b\omega_3 (-\sin\theta_3 + j\cos\theta_3) \qquad (6.19b)$$

$$\mathbf{V}_B = jc\omega_4 (\cos\theta_4 + j\sin\theta_4) = c\omega_4 (-\sin\theta_4 + j\cos\theta_4) \qquad (6.19c)$$

where the real and imaginary terms are the x and y components, respectively. Equations 6.18 and 6.19 provide a complete solution for the angular velocities of the links and the linear velocities of the joints in the pin-jointed fourbar linkage. Note that there are also two solutions to this velocity problem, corresponding to the open and crossed circuits of the linkage. They are found by the substitution of the open or crossed circuit values of θ_3 and θ_4 obtained from equations 4.10 (p. 188) and 4.12-4.13 (pp. 189-190) into equations 6.18 and 6.19. Figure 6-20a shows the open circuit.

✐ EXAMPLE 6-7

Velocity Analysis of a Fourbar Linkage with the Vector Loop Method.

Problem: Given a fourbar linkage with the link lengths $L_1 = d = 100$ mm, $L_2 = a = 40$ mm, $L_3 = b = 120$ mm, $L_4 = c = 80$ mm. For $\theta_2 = 40°$ and $\omega_2 = 25$ rad/sec find the values of ω_3 and ω_4, V_A, V_{BA}, and V_B for the open circuit of the linkage. Use the angles found for the same linkage and position in Example 4-1 (p. 190).

Solution: (See Figure 6-20 for nomenclature.)

1 Example 4-1 found the link angles for the open circuit of this linkage to be $\theta_3 = 20.298°$ and $\theta_4 = 57.325°$.

2   Use these angles and equations 6.18 (p. 317) to find ω_3 and ω_4 for the open circuit.

$$\omega_3 = \frac{a\omega_2}{b}\frac{\sin(\theta_4-\theta_2)}{\sin(\theta_3-\theta_4)} = \frac{40(25)}{120}\frac{\sin(57.325°-40°)}{\sin(20.298°-57.325°)} = -4.121 \text{ rad/sec}$$

$$\qquad\qquad\qquad (a)$$

$$\omega_4 = \frac{a\omega_2}{c}\frac{\sin(\theta_2-\theta_3)}{\sin(\theta_4-\theta_3)} = \frac{40(25)}{80}\frac{\sin(40°-20.298°)}{\sin(57.325°-20.298°)} = 6.998 \text{ rad/sec}$$

3   Use the angular velocities and equations 6.19 (p. 317) to find the linear velocities of points A and B.

$$\mathbf{V}_A = a\omega_2\left(-\sin\theta_2 + j\cos\theta_2\right)$$
$$\quad = 40(25)\left(-\sin 40° + j\cos 40°\right) = -642.79 + j766.04$$
$$\mathbf{V}_{A_x} = -642.79; \quad \mathbf{V}_{A_y} = 766.04; \quad \mathbf{V}_{A_{mag}} = 1000 \text{ mm/sec}; \quad \mathbf{V}_{A_{ang}} = 130° \qquad (b)$$

$$\mathbf{V}_{BA} = b\omega_3\left(-\sin\theta_3 + j\cos\theta_3\right)$$
$$\quad = 120(-4.121)\left(-\sin 20.298° + j20.298°\right) = 171.55 - j463.80$$
$$\mathbf{V}_{BA_x} = 171.55; \quad \mathbf{V}_{BA_y} = -463.80; \quad \mathbf{V}_{BA_{mag}} = 494.51 \text{ mm/sec}; \quad \mathbf{V}_{BA_{ang}} = -69.70° \quad (c)$$

$$\mathbf{V}_B = c\omega_4\left(-\sin\theta_4 + j\cos\theta_4\right)$$
$$\quad = 80(6.998)\left(-\sin 57.325 + j\cos 57.325\right) = -471.242 + j302.243$$
$$\mathbf{V}_{B_x} = -471.242; \quad \mathbf{V}_{B_y} = 302.243; \quad \mathbf{V}_{B_{mag}} = 599.84 \text{ mm/sec}; \quad \mathbf{V}_{B_{ang}} = 147.33° \quad (d)$$

4   As an exercise, repeat the above process to find the velocities for the crossed circuit of the linkage.

The Fourbar Crank-Slider

The position equations for the fourbar offset crank-slider linkage (inversion #1) were derived in Section 4.6 (p. 192). The linkage was shown in Figure 4-10 (p. 192) and is shown again in Figure 6-21a on which we also show an input angular velocity ω_2 applied to link 2. This ω_2 can be a time-varying input velocity. The vector loop equation 4.14 is repeated here for your convenience.

$$\mathbf{R}_2 - \mathbf{R}_3 - \mathbf{R}_4 - \mathbf{R}_1 = 0 \qquad\qquad (4.14a)$$

$$ae^{j\theta_2} - be^{j\theta_3} - ce^{j\theta_4} - de^{j\theta_1} = 0 \qquad\qquad (4.14b)$$

Differentiate equation 4.14b with respect to time noting that a, b, c, θ_1, and θ_4 are constant but the length of link d varies with time in this inversion.

$$ja\omega_2 e^{j\theta_2} - jb\omega_3 e^{j\theta_3} - \dot{d} = 0 \qquad\qquad (6.20a)$$

The term $\dot{d}$ is the linear velocity of the slider block. Equation 6.20a is the velocity difference equation 6.5 (p. 287) and can be written in that form.

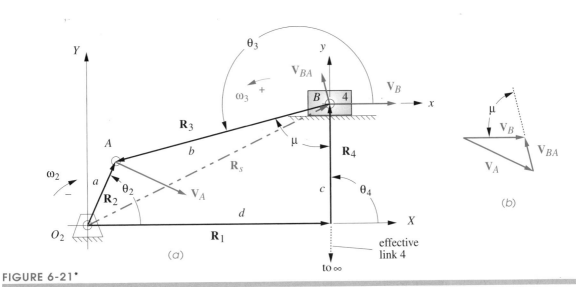

FIGURE 6-21*

Position vector loop for a fourbar slider-crank linkage showing velocity vectors for a negative (CW) ω_2

$$V_A - V_{AB} - V_B = 0$$

or:
$$V_A = V_B + V_{AB}$$

but:
$$V_{AB} = -V_{BA}$$

then:
$$V_B = V_A + V_{BA} \qquad (6.20b)$$

Equation 6.20 is identical in form to equations 6.5 and 6.15a (pp. 287 and 315). Note that because we arranged the position vector R_3 in Figure 4-10 (p. 192) and Figure 6-21 with its root at point B, directed from B to A, its derivative represents the velocity difference of point A with respect to point B, the opposite of that in the previous fourbar example. Compare this also to equation 6.15b noting that its vector R_3 is directed from A to B. Figure 6-21b shows the vector diagram of the graphical solution to equation 6.20b.

Substitute the Euler equivalent, equation 4.4a (p. 185), in equation 6.20a,

$$ja\omega_2\left(\cos\theta_2 + j\sin\theta_2\right) - jb\omega_3\left(\cos\theta_3 + j\sin\theta_3\right) - \dot{d} = 0 \qquad (6.21a)$$

simplify,

$$a\omega_2\left(-\sin\theta_2 + j\cos\theta_2\right) - b\omega_3\left(-\sin\theta_3 + j\cos\theta_3\right) - \dot{d} = 0 \qquad (6.21b)$$

and separate into real and imaginary components.

real part (x component):

$$-a\omega_2\sin\theta_2 + b\omega_3\sin\theta_3 - \dot{d} = 0 \qquad (6.21c)$$

imaginary part (y component):

$$a\omega_2\cos\theta_2 - b\omega_3\cos\theta_3 = 0 \qquad (6.21d)$$

* Note the transmission angle μ in Figure 6-21a drawn between link 3 and effective link 4 as previously defined. It is also shown drawn between vectors V_B and V_{BA} in Figure 6-21b, indicating an alternate way to define the transmission angle as *the acute angle between the absolute velocity and velocity difference vectors* at a point such as B. This approach does not require construction of the slider's effective link 4 to determine the transmission angle.

These are two simultaneous equations in the two unknowns, $\dot{d}$ and ω_3. Equation 6.21d (p. 319) can be solved for ω_3 and substituted into 6.21c to find $\dot{d}$.

$$\omega_3 = \frac{a \cos\theta_2}{b \cos\theta_3}\omega_2 \qquad (6.22a)$$

$$\dot{d} = -a\omega_2 \sin\theta_2 + b\omega_3 \sin\theta_3 \qquad (6.22b)$$

The absolute velocity of point A and the velocity difference of point A versus point B are found from equation 6.20:

$$\mathbf{V}_A = a\omega_2\left(-\sin\theta_2 + j\cos\theta_2\right) \qquad (6.23a)$$

$$\mathbf{V}_{AB} = b\omega_3\left(-\sin\theta_3 + j\cos\theta_3\right) \qquad (6.23b)$$

$$\mathbf{V}_{BA} = -\mathbf{V}_{AB} \qquad (6.23c)$$

6

✎ EXAMPLE 6-8

Velocity Analysis of a Fourbar Crank-Slider Linkage with the Vector Loop Method.

Problem: Given a fourbar crank-slider linkage with the link lengths $L_2 = a = 40$ mm, $L_3 = b = 120$ mm, *offset* $= c = -20$ mm. For $\theta_2 = 60°$ and $\omega_2 = -30$ rad/sec, find ω_3 and linear velocities of points A and B for the open circuit. Use the angles and positions found for the same linkage and its link 2 position in Example 4-2 (p. 193).

Solution: (See Figure 6-21, p. 319, for nomenclature.)

1 Example 4-2 found angle $\theta_3 = 152.91°$ and slider position $d = 126.84$ mm for the open circuit.

2 Using equation 6.22a and the data from step 1, calculate the coupler angular velocity ω_3.

$$\omega_3 = \frac{a \cos\theta_2}{b \cos\theta_3}\omega_2 = \frac{40}{120}\frac{\cos 60°}{\cos 152.91°}(-30) = 5.616 \text{ rad/sec} \qquad (a)$$

3 Using equation 6.22b and the data from steps 1 and 2, calculate the slider velocity $\dot{d}$.

$$\dot{d} = -a\omega_2 \sin\theta_2 + b\omega_3 \sin\theta_3 = -40(-30)\sin 60° + 120(5.616)\sin 152.91° = 1346 \text{ mm/sec} \quad (b)$$

4 Using equation 6.23 and the result from step 2, calculate the linear velocities V_A and V_{BA}.

$$\mathbf{V}_A = a\omega_2\left(-\sin\theta_2 + j\cos\theta_2\right) = 40(-30)\left(-\sin 60° + j\cos 60°\right) = 1039.23 - j600$$
$$\mathbf{V}_{A_x} = 1039.23; \quad \mathbf{V}_{A_y} = -600; \quad \mathbf{V}_{A_{mag}} = 1200 \text{ mm/sec}; \quad \mathbf{V}_{A_{ang}} = -30° \qquad (c)$$

$$\mathbf{V}_{AB} = b\omega_3\left(-\sin\theta_3 + j\cos\theta_3\right)$$
$$\mathbf{V}_{AB} = 120(5.616)\left(-\sin 152.91° + j\cos 152.91°\right) = -306.86 - j600$$
$$\mathbf{V}_{BA} = -\mathbf{V}_{AB} = 306.86 + j600$$
$$\mathbf{V}_{BA_x} = 306.86; \quad \mathbf{V}_{BA_y} = 600; \quad \mathbf{V}_{BA_{mag}} = 673.92 \text{ mm/sec}; \quad \mathbf{V}_{BA_{ang}} = 62.91° \qquad (d)$$

The Fourbar Slider-Crank

The *fourbar slider-crank linkage* has the same geometry as the *fourbar crank-slider linkage* that was analyzed in the previous section. The name change indicates that it will be driven with the slider as input and the crank as output. This is sometimes referred to as a "back-driven" crank-slider. We will use the term **slider-crank** to define it as slider-driven. This is a very commonly used linkage configuration. Every internal-combustion, piston engine has as many of these as it has cylinders. The vector loop is as shown in Figure 6-21 and the vector loop equation is identical to that of the crank-slider (equation 4.14a, p. 192). The derivation for θ_2 as a function of slider position d was done in Section 4-7 (p. 194). Now we want to solve for ω_2 as a function of slider velocity $\dot{d}$ and the known link lengths and angles.

We can start with equations 6.21c and d, which also apply to this linkage:

$$-a\omega_2 \sin\theta_2 + b\omega_3 \sin\theta_3 - \dot{d} = 0 \qquad (6.21c)$$

$$a\omega_2 \cos\theta_2 - b\omega_3 \cos\theta_3 = 0 \qquad (6.21d)$$

Solve equation 6.21d for ω_3 in terms of ω_2.

$$\omega_3 = \frac{a\omega_2 \cos\theta_2}{b\cos\theta_3} \qquad (6.24a)$$

Substitute equation 6.24a for ω_3 in equation 6.21c and solve for ω_2.

$$\omega_2 = \frac{\dot{d}\cos\theta_3}{a\left(\cos\theta_2 \sin\theta_3 - \sin\theta_2 \cos\theta_3\right)} \qquad (6.24b)$$

The circuit of the linkage depends on the value of d chosen and the angular velocities will be for the circuit represented by the values of θ_2 and θ_3 used from equation 4.21 (p. 195).*

* The crank-slider and slider-crank linkage both have two circuits or configurations in which they can be independently assembled, sometimes called open and crossed. Because effective link 4 is always perpendicular to the slider axis, it is parallel to itself on both circuits. This results in the two circuits being mirror images of one another, mirrored about a line through the crank pivot and perpendicular to the slide axis. Thus, the choice of value of slider position d in the calculation of the slider-crank linkage determines which circuit is being analyzed. But, because of the change points at TDC and BDC, the slider crank has two branches on each circuit and the two solutions obtained from equation 4.21 represent the two branches on the one circuit being analyzed. In contrast, the crank-slider has only one branch per circuit because when the crank is driven, it can make a full revolution and there are no change points to separate branches. See Section 4.13 (p. 208) for a more complete discussion of circuits and branches in linkages.

✒️ EXAMPLE 6-9

Velocity Analysis of a Fourbar Slider-Crank Linkage with the Vector Loop Method.

Problem: Given a fourbar slider-crank linkage with the link lengths $L_2 = a = 40$ mm, $L_3 = b = 120$ mm, *offset* = $c = -20$ mm. For $d = 100$ mm and $\dot{d} = 1200$ mm/sec, find ω_2 and ω_3 for both branches of one circuit of the linkage. Use the angles found for the same linkage in Example 4-3 (p. 196).

Solution: (See Figure 6-21, p. 319 for nomenclature.)

1 Example 4-3 found angles $\theta_{2_1} = 95.798°$, $\theta_{3_1} = 150.113°$ for branch 1 and $\theta_{2_2} = -118.418°$, $\theta_{3_2} = 187.267°$ for branch 2 of this linkage.

2 Using equation 6.24b and the data from step 1, calculate the crank angular velocity ω_{2_1}.

$$\omega_{2_1} = \frac{\dot{d}\cos\theta_{3_1}}{a\left(\cos\theta_{2_1} \sin\theta_{3_1} - \sin\theta_{2_1} \cos\theta_{3_1}\right)}$$

$$= \frac{1200\cos 150.113°}{40\left(\cos 95.798° \sin 150.113° - \sin 95.798° \cos 150.113°\right)} = -32.023 \text{ rad/sec} \qquad (a)$$

6

3   Using equation 6.24a (p. 321) and data from steps 1 and 2, calculate coupler angular velocity ω_{3_1}.

$$\omega_{3_1} = \frac{a\omega_{2_1}\cos\theta_{2_1}}{b\cos\theta_{3_1}} = \frac{40(-32.023)\cos 95.798°}{120\cos 150.113°} = -1.244 \text{ rad/sec} \tag{b}$$

4   Example 4-3 (p. 196) found $\theta_{2_2} = -118.418°$ and $\theta_{3_2} = 187.267°$ for branch 2 of this linkage.

5   Using equation 6.24b and the data from step 2, calculate the crank angular velocity ω_{2_2}.

$$\omega_{2_2} = \frac{\dot{d}\cos\theta_{3_2}}{a\left(\cos\theta_{2_2}\sin\theta_{3_2} - \sin\theta_{2_2}\cos\theta_{3_2}\right)}$$

$$= \frac{1200\cos(187.267°)}{40\left[\cos(-118.418°)\sin(187.267°) - \sin(-118.418°)\cos(187.267°)\right]} = 36.639 \text{ rad/sec} \tag{c}$$

6   Using equation 6.24a and the data from steps 3 and 4, calculate coupler angular velocity ω_{3_2}.

$$\omega_{3_2} = \frac{a\omega_{2_2}\cos\theta_{2_2}}{b\cos\theta_{3_2}} = \frac{40(36.639)\cos(-118.418°)}{120\cos(187.267°)} = 5.859 \text{ rad/sec} \tag{d}$$

The Fourbar Inverted Crank-Slider

The position equations for the fourbar inverted crank-slider linkage were derived in Section 4.8 (p. 197). The linkage was shown in Figure 4-13 (p. 198) and is shown again in Figure 6-22 on which we also show an input angular velocity ω_2 applied to link 2. This ω_2 can vary with time. The vector loop equations 4.14 repeated on p. 318 are valid for this linkage as well.

All slider linkages will have at least one link whose effective length between joints varies as the linkage moves. In this inversion the length of link 3 between points A and B, designated as b, will change as it passes through the slider block on link 4. To get an expression for velocity, differentiate equation 4.14b with respect to time noting that a, c, d, and θ_1 are constant and b varies with time.

$$ja\omega_2 e^{j\theta_2} - jb\omega_3 e^{j\theta_3} - \dot{b}e^{j\theta_3} - jc\omega_4 e^{j\theta_4} = 0 \tag{6.25a}$$

The value of db/dt will be one of the variables to be solved for in this case and is the $\dot{b}$ term in the equation. Another variable will be ω_4, the angular velocity of link 4. Note, however, that we also have an unknown in ω_3, the angular velocity of link 3. This is a total of three unknowns. Equation 6.25a can only be solved for two unknowns. Thus we require another equation to solve the system. There is a fixed relationship between angles θ_3 and θ_4, shown as γ in Figure 6-22 and defined in equation 4.22, repeated here:

$$\theta_3 = \theta_4 + \gamma \tag{4.22}$$

Differentiate it with respect to time to obtain:

$$\omega_3 = \omega_4 \tag{6.25b}$$

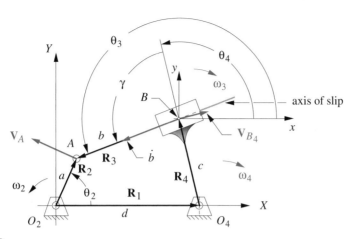

FIGURE 6-22

Velocity analysis of inversion #3 of the slider-crank fourbar linkage

We wish to solve equation 6.25a to get expressions in this form:

$$\omega_3 = \omega_4 = f\left(a, b, c, d, \theta_2, \theta_3, \theta_4, \omega_2\right)$$

$$\frac{db}{dt} = \dot{b} = g\left(a, b, c, d, \theta_2, \theta_3, \theta_4, \omega_2\right)$$

(6.26)

Substitution of the Euler identity (equation 4.4a, p. 185) into equation 6.25a yields:

$$ja\,\omega_2\left(\cos\theta_2 + j\sin\theta_2\right) - jb\,\omega_3\left(\cos\theta_3 + j\sin\theta_3\right)$$
$$-\dot{b}\left(\cos\theta_3 + j\sin\theta_3\right) - jc\,\omega_4\left(\cos\theta_4 + j\sin\theta_4\right) = 0$$

(6.27a)

Multiply by the operator j and substitute ω_4 for ω_3 from equation 6.25b:

$$a\,\omega_2\left(-\sin\theta_2 + j\cos\theta_2\right) - b\,\omega_4\left(-\sin\theta_3 + j\cos\theta_3\right)$$
$$-\dot{b}\left(\cos\theta_3 + j\sin\theta_3\right) - c\,\omega_4\left(-\sin\theta_4 + j\cos\theta_4\right) = 0$$

(6.27b)

We can now separate this vector equation into its two components by collecting all real and all imaginary terms separately:

real part (x component):

$$-a\,\omega_2\sin\theta_2 + b\,\omega_4\sin\theta_3 - \dot{b}\cos\theta_3 + c\,\omega_4\sin\theta_4 = 0$$

(6.28a)

imaginary part (y component):

$$a\,\omega_2\cos\theta_2 - b\,\omega_4\cos\theta_3 - \dot{b}\sin\theta_3 - c\,\omega_4\cos\theta_4 = 0$$

(6.28b)

Collect terms and rearrange equations 6.28 to isolate one unknown on the left side.

$$\dot{b}\cos\theta_3 = -a\,\omega_2\sin\theta_2 + \omega_4\left(b\sin\theta_3 + c\sin\theta_4\right)$$

(6.29a)

$$\dot{b}\sin\theta_3 = a\,\omega_2\cos\theta_2 - \omega_4\left(b\cos\theta_3 + c\cos\theta_4\right)$$

(6.29b)

Either equation can be solved for $\dot{b}$ and the result substituted in the other. Solving equation 6.29a (p. 323):

$$\dot{b} = \frac{-a\omega_2\sin\theta_2 + \omega_4\left(b\sin\theta_3 + c\sin\theta_4\right)}{\cos\theta_3} \tag{6.30a}$$

Substitute in equation 6.29b (p. 323) and simplify:

$$\omega_4 = \frac{a\omega_2\cos\left(\theta_2 - \theta_3\right)}{b + c\cos\left(\theta_4 - \theta_3\right)} \tag{6.30b}$$

Equation 6.30a provides the **velocity of slip** at point B. Equation 6.30b gives the **angular velocity** of link 4. Note that we can substitute $-\gamma = \theta_4 - \theta_3$ from equation 4.18 (for an open linkage) into equation 6.30b to further simplify it. Note that $\cos(-\gamma) = \cos(\gamma)$.

$$\omega_4 = \frac{a\omega_2\cos\left(\theta_2 - \theta_3\right)}{b + c\cos\gamma} \tag{6.30c}$$

The **velocity of slip** from equation 6.30a is always directed along the **axis of slip** as shown in Figure 6-22 (p. 323). There is also a component orthogonal to the axis of slip called the **velocity of transmission**. This lies along the **axis of transmission** which is the only line along which any useful work can be transmitted across the sliding joint. All energy associated with motion along the slip axis is converted to heat and lost.

The absolute linear velocity of point A is found from equation 6.23a (p. 320). We can find the absolute velocity of point B on link 4 since ω_4 is now known. From equation 6.15b (p. 315):

$$\mathbf{V}_{B_4} = jc\omega_4 e^{j\theta_4} = c\omega_4\left(-\sin\theta_4 + j\cos\theta_4\right) \tag{6.31a}$$

The velocity of transmission is the component of V_{b4} normal to the axis of slip. The absolute velocity of point B on link 3 is found from equation 6.5 (p. 287) as

$$\mathbf{V}_{B_3} = \mathbf{V}_{B_4} + \mathbf{V}_{B_{34}} = \mathbf{V}_{B_4} + \mathbf{V}_{slip_{34}} \tag{6.31b}$$

6.8   VELOCITY ANALYSIS OF THE GEARED FIVEBAR LINKAGE

The position loop equation for the geared fivebar mechanism was derived in Section 4.8 (p. 197) and is repeated here. See Figure P6-4 (p. 331) for notation.

$$ae^{j\theta_2} + be^{j\theta_3} - ce^{j\theta_4} - de^{j\theta_5} - fe^{j\theta_1} = 0 \tag{4.27b}$$

Differentiate this with respect to time to get an expression for velocity.

$$a\omega_2 je^{j\theta_2} + b\omega_3 je^{j\theta_3} - c\omega_4 je^{j\theta_4} - d\omega_5 je^{j\theta_5} = 0 \tag{6.32a}$$

Substitute the Euler equivalents:

$$a\omega_2 j\left(\cos\theta_2 + j\sin\theta_2\right) + b\omega_3 j\left(\cos\theta_3 + j\sin\theta_3\right)$$
$$-c\omega_4 j\left(\cos\theta_4 + j\sin\theta_4\right) - d\omega_5 j\left(\cos\theta_5 + j\sin\theta_5\right) = 0 \tag{6.32b}$$

Note that the angle θ_5 is defined in terms of θ_2, the gear ratio λ, and the phase angle ϕ.

$$\theta_5 = \lambda \theta_2 + \phi \qquad (4.27c)$$

Differentiate with respect to time:

$$\omega_5 = \lambda \omega_2 \qquad (6.32c)$$

Since a complete position analysis must be done before a velocity analysis, we will assume that the values of θ_5 and ω_5 have been found and will leave these equations in terms of θ_5 and ω_5.

Separating the real and imaginary terms in equation 6.32b:

real: $\qquad -a\omega_2 \sin\theta_2 - b\omega_3 \sin\theta_3 + c\omega_4 \sin\theta_4 + d\omega_5 \sin\theta_5 = 0 \qquad (6.32d)$

imaginary: $\qquad a\omega_2 \cos\theta_2 + b\omega_3 \cos\theta_3 - c\omega_4 \cos\theta_4 - d\omega_5 \cos\theta_5 = 0 \qquad (6.32e)$

The only two unknowns are ω_3 and ω_4. Either equation 6.32d or 6.32e can be solved for one unknown and the result substituted in the other. The solution for ω_3 is:

$$\omega_3 = -\frac{2\sin\theta_4 \left[a\omega_2 \sin(\theta_2 - \theta_4) + d\omega_5 \sin(\theta_4 - \theta_5) \right]}{b\left[\cos(\theta_3 - 2\theta_4) - \cos\theta_3 \right]} \qquad (6.33a)$$

The angular velocity ω_4 can be found from equation 6.32d using ω_3.

$$\omega_4 = \frac{a\omega_2 \sin\theta_2 + b\omega_3 \sin\theta_3 - d\omega_5 \sin\theta_5}{c\sin\theta_4} \qquad (6.33b)$$

With all link angles and angular velocities known, the linear velocities of the pin joints can be found from:

$$\mathbf{V}_A = a\omega_2 \left(-\sin\theta_2 + j\cos\theta_2 \right) \qquad (6.33c)$$

$$\mathbf{V}_{BA} = b\omega_3 \left(-\sin\theta_3 + j\cos\theta_3 \right) \qquad (6.33d)$$

$$\mathbf{V}_C = d\omega_5 \left(-\sin\theta_5 + j\cos\theta_5 \right) \qquad (6.33e)$$

$$\mathbf{V}_B = \mathbf{V}_A + \mathbf{V}_{BA} \qquad (6.33f)$$

6.9 VELOCITY OF ANY POINT ON A LINKAGE

Once the angular velocities of all the links are found it is easy to define and calculate the velocity of *any point on any link* for any input position of the linkage. Figure 6-23 shows the fourbar linkage with its coupler, link 3, enlarged to contain a coupler point P. The crank and rocker have also been enlarged to show points S and U which might represent the centers of gravity of those links. We want to develop algebraic expressions for the velocities of these (or any) points on the links.

To find the velocity of point S, draw the position vector from the fixed pivot O_2 to point S. This vector, $\mathbf{R}_{SO_2}$ makes an angle δ_2 with the vector $\mathbf{R}_{AO_2}$. The angle δ_2 is completely defined by the geometry of link 2 and is constant. The position vector for point S is then:

$$\mathbf{R}_{SO_2} = \mathbf{R}_S = se^{j(\theta_2 + \delta_2)} = s\left[\cos(\theta_2 + \delta_2) + j\sin(\theta_2 + \delta_2) \right] \qquad (4.29)$$

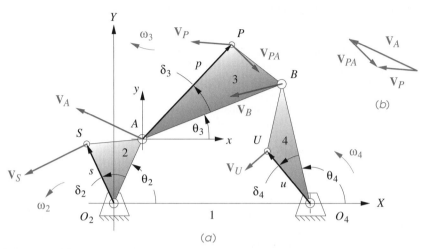

FIGURE 6-23

Finding the velocities of points on the links

Differentiate this position vector to find the velocity of that point.

$$\mathbf{V}_S = jse^{j(\theta_2+\delta_2)}\omega_2 = s\omega_2\left[-\sin(\theta_2+\delta_2)+j\cos(\theta_2+\delta_2)\right] \tag{6.34}$$

The position of point U on link 4 is found in the same way, using the angle δ_4 which is a constant angular offset within the link. The expression is:

$$\mathbf{R}_{UO_4} = ue^{j(\theta_4+\delta_4)} = u\left[\cos(\theta_4+\delta_4)+j\sin(\theta_4+\delta_4)\right] \tag{4.30}$$

Differentiate this position vector to find the velocity of that point.

$$\mathbf{V}_U = jue^{j(\theta_4+\delta_4)}\omega_4 = u\omega_4\left[-\sin(\theta_4+\delta_4)+j\cos(\theta_4+\delta_4)\right] \tag{6.35}$$

The velocity of point P on link 3 can be found from the addition of two velocity vectors, such as $\mathbf{V}_A$ and $\mathbf{V}_{PA}$. $\mathbf{V}_A$ is already defined from our analysis of the link velocities. $\mathbf{V}_{PA}$ is the velocity difference of point P with respect to point A. Point A is chosen as the reference point because angle θ_3 is defined in a LNCS and angle δ_3 is defined in a LRCS whose origins are both at A. Position vector $\mathbf{R}_{PA}$ is defined in the same way as $\mathbf{R}_S$ or $\mathbf{R}_U$ using the internal link offset angle δ_3 and the angle of link 3, θ_3. This was done in equations 4.31 (repeated here).

$$\mathbf{R}_{PA} = pe^{j(\theta_3+\delta_3)} = p\left[\cos(\theta_3+\delta_3)+j\sin(\theta_3+\delta_3)\right] \tag{4.31a}$$

$$\mathbf{R}_P = \mathbf{R}_A + \mathbf{R}_{PA} \tag{4.31b}$$

Differentiate equations 4.31 to find the velocity of point P.

$$\mathbf{V}_{PA} = jpe^{j(\theta_3+\delta_3)}\omega_3 = p\omega_3\left[-\sin(\theta_3+\delta_3)+j\cos(\theta_3+\delta_3)\right] \tag{6.36a}$$

$$\mathbf{V}_P = \mathbf{V}_A + \mathbf{V}_{PA} \tag{6.36b}$$

Please compare equations 6.36 with equations 6.5 (p. 287) and 6.15 (p. 315). It is, again, the velocity difference equation.

Note that if, for example, you wished to derive an equation for the velocity of a coupler point P on the crank-slider linkage as set up in Figure 6-21 (p. 319), or the inverted crank-slider of Figure 6-22 (p. 323), both of which have the vector for link 3 defined with its root at point B rather than at point A, you might want to use point B as the reference point rather than point A, making equation 6.36b become:

$$V_P = V_{B_3} + V_{PB_3} \qquad (6.36c)$$

Angle θ_3 would then be defined in a LNCS at point B, and δ_3 in a LRCS at point B.

6.10 REFERENCES

1   **Towfigh, K.** (1969). "The Fourbar Linkage as an Adjustment Mechanism." *Proc. of Applied Mechanism Conference*, Tulsa, OK, pp. 27-1 to 27-4.

2   **Wood, G. A.** (1977). "Educating for Creativity in Engineering." *Proc. of ASEE 85th Annual Conference*, University of North Dakota, pp. 1-13.

3   **Kennedy, A. B. W.** (1893). *Mechanics of Machinery*. Macmillan, London, pp. vii, 73.

6.11 PROBLEMS‡

6-1   Use the relative velocity equation and solve graphically or analytically.

    a.   A ship is steaming due north at 20 knots (nautical miles per hour). A submarine is laying in wait 1/2 mile due west of the ship. The sub fires a torpedo on a course of 85 degrees. The torpedo travels at a constant speed of 30 knots. Will it strike the ship? If not, by how many nautical miles will it miss?

    b.   A plane is flying due south at 500 mph at 35,000 ft altitude, straight and level. A second plane is initially 40 miles due east of the first plane, also at 35,000 feet altitude, flying straight and level and traveling at 550 mph. Determine the compass angle at which the second plane would be on a collision course with the first. How long will it take for the second plane to catch the first?

6-2   A point is at a 6.5 in radius on a body in pure rotation with $\omega = 100$ rad/sec. The rotation center is at the origin of a coordinate system. When the point is at position A, its position vector makes a $45°$ angle with the X axis. At position B, its position vector makes a $75°$ angle with the X axis. Draw this system to some convenient scale and:

    a.   Write an expression for the particle's velocity vector in position A using complex number notation, in both polar and cartesian forms.

    b.   Write an expression for the particle's velocity vector in position B using complex number notation, in both polar and cartesian forms.

    c.   Write a vector equation for the velocity difference between points B and A. Substitute the complex number notation for the vectors in this equation and solve for the position difference numerically.

    d.   Check the result of part c with a graphical method.

6-3   Repeat problem 6-2 considering points A and B to be on separate bodies rotating about the origin with ω's of -50 (A) and $+75$ rad/sec (B). Find their relative velocity.

*6-4   A general fourbar linkage configuration and its notation are shown in Figure P6-1. The link lengths, coupler point location, and the values of θ_2 and ω_2 for the same fourbar

‡ All problem figures are provided as PDF files, and some are also provided as animated Working Model files; all are on the DVD. PDF filenames are the same as the figure number. Run the file *Animations.html* to access and run the animations.

TABLE P6-1   Data for Problems 6-4 to 6-5[‡]

Row	Link 1	Link 2	Link 3	Link 4	θ_2	ω_2	R_{pa}	δ_3
a	6	2	7	9	30	10	6	30
b	7	9	3	8	85	-12	9	25
c	3	10	6	8	45	-15	10	80
d	8	5	7	6	25	24	5	45
e	8	5	8	6	75	-50	9	300
f	5	8	8	9	15	-45	10	120
g	6	8	8	9	25	100	4	300
h	20	10	10	10	50	-65	6	20
i	4	5	2	5	80	25	9	80
j	20	10	5	10	33	25	1	0
k	4	6	10	7	88	-80	10	330
l	9	7	10	7	60	-90	5	180
m	9	7	11	8	50	75	10	90
n	9	7	11	6	120	15	15	60

[‡] Drawings of these linkages are in the *PDF Problem Workbook* folder on the DVD.

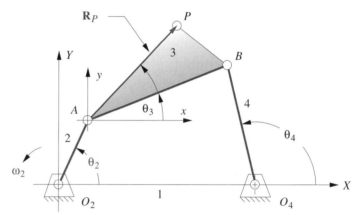

FIGURE P6-1

Configuration and terminology for the pin-jointed fourbar linkage of Problems 6-4 to 6-5

linkages as used for position analysis in Chapter 4 are redefined in Table P6-1, which is the same as Table P4-1 (p. 215). *For the row(s) assigned*, draw the linkage to scale and find the velocities of the pin joints A and B and of instant centers $I_{1,3}$ and $I_{2,4}$ using a graphical method. Then calculate ω_3 and ω_4 and find the velocity of point P.

*[†]6-5   Repeat Problem 6-4 using an analytical method. Draw the linkage to scale and label it before setting up the equations.

*6-6   The general linkage configuration and terminology for an offset fourbar crank-slider linkage are shown in Figure P6-2. The link lengths and the values of θ_2 and ω_2 are defined in Table P6-2. *For the row(s) assigned*, draw the linkage to scale and find the velocities of the pin joints A and B and the velocity of slip at the sliding joint using a graphical method.

*[†]6-7   Repeat Problem 6-6 using an analytical method. Draw the linkage to scale and label it before setting up the equations.

TABLE P6-2   Data for Problems 6-6 to 6-7[‡]

Row	Link 2	Link 3	Offset	θ_2	ω_2
a	1.4	4	1	45	10
b	2	6	-3	60	-12
c	3	8	2	-30	-15
d	3.5	10	1	120	24
e	5	20	-5	225	-50
f	3	13	0	100	-45
g	7	25	10	330	100

[‡] Drawings of these linkages are in the *PDF Problem Workbook* folder on the DVD.

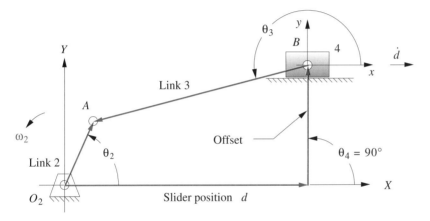

FIGURE P6-2

Configuration and terminology for Problems 6-6, 6-7, 6-110, 6-111

*6-8   The general linkage configuration and terminology for an inverted fourbar crank-slider linkage are shown in Figure P6-3. The link lengths and the values of θ_2, ω_2, and γ are defined in Table P6-3. *For the row(s) assigned*, draw the linkage to scale and find the velocities of points A and B and velocity of slip at the sliding joint using a graphical method.

*[†]6-9   Repeat Problem 6-8 using an analytical method. Draw the linkage to scale and label it before setting up the equations.

*6-10   The general linkage configuration and terminology for a geared fivebar linkage are shown in Figure P6-4. The link lengths, gear ratio (λ), phase angle (ϕ), and the values of θ_2 and ω_2 are defined in Table P6-4. *For the row(s) assigned*, draw the linkage to scale and find ω_3 and ω_4 using a graphical method.

*[†]6-11   Repeat Problem 6-10 using an analytical method. Draw the linkage to scale and label it before setting up the equations.

6-12   Find all the instant centers of the linkages shown in Figure P6-5 (p. 332).

6-13   Find all the instant centers of the linkages shown in Figure P6-6 (p. 332).

6-14   Find all the instant centers of the linkages shown in Figure P6-7 (p. 333).

6

* Answers in Appendix F.

[†] These problems are suited to solution using *Mathcad*, *Matlab*, or *TKSolver* equation solver programs.

TABLE P6-3   Data for Problems 6-8 to 6-9

Row	Link 1	Link 2	Link 4	γ	θ_2	ω_2
a	6	2	4	90	30	10
b	7	9	3	75	85	-15
c	3	10	6	45	45	24
d	8	5	3	60	25	-50
e	8	4	2	30	75	-45
f	5	8	8	90	150	100

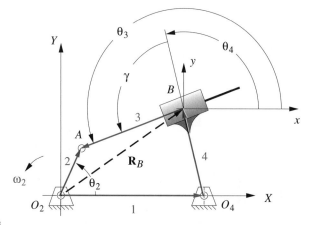

FIGURE P6-3

Configuration and terminology for Problems 6-8 to 6-9

6-15   Find all the instant centers of the linkages shown in Figure P6-8 (p. 335).

*6-16   The linkage in Figure P6-5a (p. 332) has $O_2A = 0.8$, $AB = 1.93$, $AC = 1.33$, and offset = 0.38 in. The crank angle in the position shown is 34.3° and angle $BAC = 38.6°$. Find ω_3, V_A, V_B, and V_C for the position shown for $\omega_2 = 15$ rad/sec in the direction shown.

   a.   Using the velocity difference graphical method.
   b.   Using the instant center graphical method.
   †c.   Using an analytical method.

6-17   The linkage in Figure P6-5c (p. 332) has $I_{12}A = 0.75$, $AB = 1.5$, and $AC = 1.2$ in. The effective crank angle in the position shown is 77° and angle $BAC = 30°$. Find ω_3, ω_4, V_A, V_B, and V_C for the position shown for $\omega_2 = 15$ rad/sec in direction shown.

   a.   Using the velocity difference graphical method.
   b.   Using the instant center graphical method.
   †c.   Using an analytical method. (Hint: Create an effective linkage for the position shown and analyze as a pin-jointed fourbar.)

6-18   The linkage in Figure P6-5f (p. 332) has $AB = 1.8$ and $AC = 1.44$ in. The angle of AB in the position shown is 128° and angle $BAC = 49°$. The slider at B is at an angle of 59°. Find ω_3, V_A, V_B, and V_C for the position shown for $V_A = 10$ in/sec in the direction shown.

   a.   Using the velocity difference graphical method.
   b.   Using the instant center graphical method.
   †c.   Using an analytical method.

* Answers in Appendix F.

† These problems are suited to solution using *Mathcad*, *Matlab*, or *TKSolver* equation solver programs.

TABLE P6-4   Data for Problems 6-10 to 6-11

Row	Link 1	Link 2	Link 3	Link 4	Link 5	λ	ϕ	ω_2	θ_2
a	6	1	7	9	4	2	30	10	60
b	6	5	7	8	4	-2.5	60	-12	30
c	3	5	7	8	4	-0.5	0	-15	45
d	4	5	7	8	4	-1	120	24	75
e	5	9	11	8	8	3.2	-50	-50	-39
f	10	2	7	5	3	1.5	30	-45	120
g	15	7	9	11	4	2.5	-90	100	75
h	12	8	7	9	4	-2.5	60	-65	55
i	9	7	8	9	4	-4	120	25	100

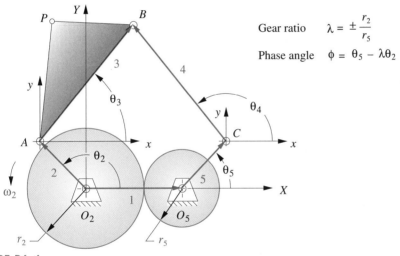

Gear ratio $\quad \lambda = \pm \dfrac{r_2}{r_5}$

Phase angle $\quad \phi = \theta_5 - \lambda\theta_2$

FIGURE P6-4

Configuration and terminology for problems 6-10 and 6-11

6-19   The cam-follower in Figure P6-5d (p. 332) has $O_2A = 0.853$ in. Find $\mathbf{V}_4$, $\mathbf{V}_{trans}$, and $\mathbf{V}_{slip}$ for the position shown with $\omega_2 = 20$ rad/sec in the direction shown.

6-20   The cam-follower in Figure P6-5e (p. 332) has $O_2A = 0.980$ in and $O_3A = 1.344$ in. Find ω_3, $\mathbf{V}_{trans}$, and $\mathbf{V}_{slip}$ for the position shown for $\omega_2 = 10$ rad/sec in the direction shown.

6-21   The linkage in Figure P6-6b (p. 332) has $L_1 = 61.9$, $L_2 = 15$, $L_3 = 45.8$, $L_4 = 18.1$, $L_5 = 23.1$ mm. θ_2 is 68.3° in the xy coordinate system, which is at $-23.3°$ in the XY coordinate system. The X component of O_2C is 59.2 mm. For the position shown, find the velocity ratio $V_{I_{5,6}} / V_{I_{2,3}}$ and the mechanical advantage from link 2 to link 6.

    a.   Using the velocity difference graphical method.
    b.   Using the instant center graphical method.

6-22   Repeat Problem 6-21 for the mechanism in Figure P6-6d, which has the dimensions: $L_2 = 15$, $L_3 = 40.9$, $L_5 = 44.7$ mm. θ_2 is 24.2° in the XY coordinate system.

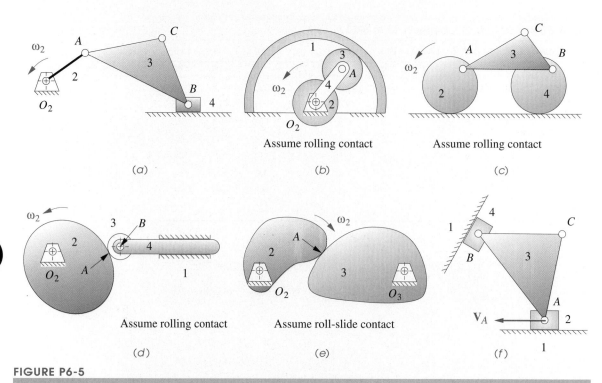

(a)

(b)
Assume rolling contact

(c)
Assume rolling contact

(d)
Assume rolling contact

(e)
Assume roll-slide contact

(f)

FIGURE P6-5

Velocity analysis and instant center problems. Problems 6-12 and 6-16 to 6-20

(a)

(b)

(c)

(d)

FIGURE P6-6

Problems 6-13, 6-21, and 6-22

FIGURE P6-7

Problems 6-14 and 6-23. *From R. T. Hinkle, Problems in Kinematics, Prentice-Hall. Englewood Cliffs, NJ, 1954*

†6-23 Generate and draw the fixed and moving centrodes of links 1 and 3 for the linkage in Figure P6-7a.

6-24 The linkage in Figure P6-8a (p. 335) has link 1 at –25° and O_2A at 37° in the global *XY* coordinate system. Find ω_4, $\mathbf{V}_A$, and $\mathbf{V}_B$ in the global coordinate system for the position shown if $\omega_2 = 15$ rad/sec CW. Use the velocity difference graphical method. (Print the figure from its PDF file on the DVD and draw on it.)

6-25 The linkage in Figure P6-8a (p. 335) has link 1 at –25° and O_2A at 37° in the global *XY* coordinate system. Find ω_4, $\mathbf{V}_A$, and $\mathbf{V}_B$ in the global coordinate system for the position shown if $\omega_2 = 15$ rad/sec CW. Use the instant center graphical method. (Print the figure from its PDF file on the DVD and draw on it.)

†6-26 The linkage in Figure P6-8a has $\theta_2 = 62°$ in the local *x'y'* coordinate system. The angle between the *X* and *x* axes is 25°. Find ω_4, $\mathbf{V}_A$, and $\mathbf{V}_B$ in the local coordinate system for the position shown if $\omega_2 = 15$ rad/sec CW. Use an analytical method.

† These problems are suited to solution using *Mathcad, Matlab,* or *TKSolver* equation solver programs.

†6-27 For the linkage in Figure P6-8a, write a computer program or use an equation solver to find and plot ω_4, $\mathbf{V}_A$, and $\mathbf{V}_B$ in the local coordinate system for the maximum range of motion that this linkage allows if $\omega_2 = 15$ rad/sec CW.

6-28 The linkage in Figure P6-8b has link 1 at $-36°$ and link 2 at $57°$ in the global XY coordinate system. Find ω_4, $\mathbf{V}_A$, and $\mathbf{V}_B$ in the global coordinate system for the position shown if $\omega_2 = 20$ rad/sec CCW. Use the velocity difference graphical method. (Print the figure from its PDF file on the DVD and draw on it.)

6-29 The linkage in Figure P6-8b has link 1 at $-36°$ and link 2 at $57°$ in the global XY coordinate system. Find ω_4, $\mathbf{V}_A$, and $\mathbf{V}_B$ in the global coordinate system for the position shown if $\omega_2 = 20$ rad/sec CCW. Use the instant center graphical method. (Print the figure from its PDF file on the DVD and draw on it.)

†6-30 The linkage in Figure P6-8b has link 1 at $-36°$ and link 2 at $57°$ in the global XY coordinate system. Find ω_4, $\mathbf{V}_A$, and $\mathbf{V}_B$ in the global coordinate system for the position shown if $\omega_2 = 20$ rad/sec CCW. Use an analytical method.

†6-31 The linkage in Figure P6-8b has link 1 at $-36°$ in the global XY coordinate system. Write a computer program or use an equation solver to find and plot ω_4, $\mathbf{V}_A$, and $\mathbf{V}_B$ in the local coordinate system for the maximum range of motion that this linkage allows if $\omega_2 = 20$ rad/sec CCW.

6-32 The offset crank-slider linkage in Figure P6-8f has link 2 at $51°$ in the global XY coordinate system. Find $\mathbf{V}_A$ and $\mathbf{V}_B$ in the global coordinate system for the position shown if $\omega_2 = 25$ rad/sec CW. Use the velocity difference graphical method. (Print the figure from its PDF file on the DVD and draw on it.)

6-33 The offset crank-slider linkage in Figure P6-8f has link 2 at $51°$ in the global XY coordinate system. Find $\mathbf{V}_A$ and $\mathbf{V}_B$ in the global coordinate system for the position shown if $\omega_2 = 25$ rad/sec CW. Use the instant center graphical method. (Print the figure from its PDF file on the DVD and draw on it.)

†6-34 The offset crank-slider linkage in Figure P6-8f has link 2 at $51°$ in the global XY coordinate system. Find $\mathbf{V}_A$ and $\mathbf{V}_B$ in the global coordinate system for the position shown if $\omega_2 = 25$ rad/sec CW. Use an analytical method.

†6-35 For the offset crank-slider linkage in Figure P6-8f , write a computer program or use an equation solver to find and plot $\mathbf{V}_A$ and $\mathbf{V}_B$ in the global coordinate system for the maximum range of motion that this linkage allows if $\omega_2 = 25$ rad/sec CW.

6-36 The linkage in Figure P6-8d has link 2 at $58°$ in the global XY coordinate system. Find $\mathbf{V}_A$, $\mathbf{V}_B$, and $\mathbf{V}_{box}$ in the global coordinate system for the position shown if $\omega_2 = 30$ rad/sec CW. Use the velocity difference graphical method. (Make a copy of the figure from its PDF file on the DVD and draw on it.)

†6-37 The linkage in Figure P6-8d has link 2 at $58°$ in the global XY coordinate system. Find $\mathbf{V}_A$, $\mathbf{V}_B$, and $\mathbf{V}_{box}$ in the global coordinate system for the position shown if $\omega_2 = 30$ rad/sec CW. Use an analytical method.

†6-38 For the linkage in Figure P6-8d, write a computer program or use an equation solver to find and plot $\mathbf{V}_A$, $\mathbf{V}_B$, and $\mathbf{V}_{box}$ in the global coordinate system for the maximum range of motion that this linkage allows if $\omega_2 = 30$ rad/sec CW.

6-39 The linkage in Figure P6-8g has the local xy axis at $-119°$ and O_2A at $29°$ in the global XY coordinate system. Find ω_4, $\mathbf{V}_A$, and $\mathbf{V}_B$ in the global coordinate system for the position shown if $\omega_2 = 15$ rad/sec CW. Use the velocity difference graphical method. (Make a copy of the figure from its PDF file on the DVD and draw on it.)

(a) Fourbar linkage

$L_1 = 174$
$L_2 = 116$
$L_3 = 108$
$L_4 = 110$

(b) Fourbar linkage

$57°$

$L_1 = 162$ $L_2 = 40$
$L_4 = 122$ $L_3 = 96$

(c) Radial compressor

$L_2 = 19$
$L_3 = 70$
$L_4 = 70$
$L_5 = 70$
$L_6 = 70$

6

$L_1 = 150$ $L_2 = 30$
$L_3 = 150$ $L_4 = 30$

box

(d) Walking-beam conveyor

$O_2O_4 = L_3 = L_5 = 160$
$O_8O_4 = L_6 = L_7 = 120$
$O_2A = O_2C = 20$
$O_4B = O_4D = 20$
$O_4E = O_4G = 30$
$O_8F = O_8H = 30$

(e) Bellcrank mechanism

$L_2 = 63$
$L_3 = 130$
offset $= 52$

(f) Offset slider-crank

$L_1 = 87$
$L_2 = 49$
$L_3 = 100$
$L_4 = 153$
$L_5 = 100$
$L_6 = 153$

$121°$

(g) Drum brake mechanism

22.9 22.9

4.5 typ.

$L_1 = 45.8$
$L_2 = 19.8$
$L_3 = 19.4$
$L_4 = 38.3$
$L_5 = 13.3$
$L_7 = 13.3$
$L_8 = 19.8$
$L_9 = 19.4$

(h) Symmetrical mechanism

all dimensions in mm

FIGURE P6-8

Problems 6-15 and 6-24 to 6-45 *Adapted from P. H. Hill and W. P. Rule. (1960). Mechanisms: Analysis and Design, with permission*

6-40   The linkage in Figure P6-8g (p. 335) has the local xy axis at $-119°$ and O_2A at $29°$ in the global XY coordinate system. Find ω_4, $\mathbf{V}_A$, and $\mathbf{V}_B$ in the global coordinate system for the position shown if $\omega_2 = 15$ rad/sec CW. Use the instant center graphical method. (Make a copy of the figure from its PDF file on the DVD and draw on it.)

†6-41   The linkage in Figure P6-8g (p. 335) has the local xy axis at $-119°$ and O_2A at $29°$ in the global XY coordinate system. Find ω_4, $\mathbf{V}_A$, and $\mathbf{V}_B$ in the global coordinate system for the position shown if $\omega_2 = 15$ rad/sec CW. Use an analytical method.

†6-42   The linkage in Figure P6-8g (p. 335) has the local xy axis at $-119°$ in the global XY coordinate system. Write a computer program or use an equation solver to find and plot ω_4, $\mathbf{V}_A$, and $\mathbf{V}_B$ in the local coordinate system for the maximum range of motion that this linkage allows if $\omega_2 = 15$ rad/sec CW.

6-43   The 3-cylinder radial compressor in Figure P6-8c (p. 335) has its cylinders equispaced at $120°$. Find the piston velocities $\mathbf{V}_6$, $\mathbf{V}_7$, $\mathbf{V}_8$ with the crank at $-53°$ using a graphical method if $\omega_2 = 15$ rad/sec CW. (Make a copy of the figure from its PDF file on the DVD and draw on it.)

†6-44   The 3-cylinder radial compressor in Figure P6-8c (p. 335) has its cylinders equispaced at $120°$. Find the piston velocities $\mathbf{V}_6$, $\mathbf{V}_7$, $\mathbf{V}_8$ with the crank at $-53°$ using an analytical method if $\omega_2 = 15$ rad/sec CW.

†6-45   The 3-cylinder radial compressor in Figure P6-8c (p. 335) has its cylinders equispaced at $120°$. Write a program or use an equation solver to find and plot the piston velocities $\mathbf{V}_6$, $\mathbf{V}_7$, $\mathbf{V}_8$ for one revolution of the crank if $\omega_2 = 15$ rad/sec CW.

6-46   Figure P6-9 shows a linkage in one position. Find the instantaneous velocities of points A, B, and P if link O_2A is rotating CW at 40 rad/sec.

**6-47   Figure P6-10 shows a linkage and its coupler curve. Write a computer program or use an equation solver to calculate and plot the magnitude and direction of the velocity of the coupler point P at $2°$ increments of crank angle for $\omega_2 = 100$ rpm. Check your result with program LINKAGES.

**6-48   Figure P6-11 shows a linkage that operates at 500 crank rpm. Write a computer program or use an equation solver to calculate and plot the magnitude and direction of the velocity of point B at $2°$ increments of crank angle. Check the result with program LINKAGES.

**6-49   Figure P6-12 shows a linkage and its coupler curve. Write a computer program or use an equation solver to calculate and plot the magnitude and direction of the velocity of the coupler point P at $2°$ increments of crank angle for $\omega_2 = 20$ rpm over the maximum range of motion possible. Check your result with program LINKAGES.

†6-50   Figure P6-13 (p. 338) shows a linkage and its coupler curve. Write a computer program or use an equation solver to calculate and plot the magnitude and direction of the velocity of the coupler point P at $2°$ increments of crank angle for $\omega_2 = 80$ rpm over the maximum range of motion possible. Check your result with program LINKAGES.

**6-51   Figure P6-14 (p. 338) shows a linkage and its coupler curve. Write a computer program or use an equation solver to calculate and plot the magnitude and direction of the velocity of the coupler point P at $2°$ increments of crank angle for $\omega_2 = 80$ rpm over the maximum range of motion possible. Check your result with program LINKAGES.

†6-52   Figure P6-15 (p. 338) shows a power hacksaw, used to cut metal. Link 5 pivots at O_5 and its weight forces the sawblade against the workpiece while the linkage moves the blade (link 4) back and forth on link 5 to cut the part. It is an offset crank-slider

FIGURE P6-9

Problem 6-46

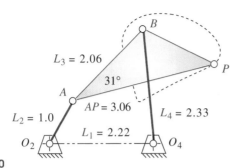

FIGURE P6-10

Problem 6-47 A fourbar linkage with a double straight-line coupler curve

mechanism with the dimensions shown in the figure. Draw an equivalent linkage diagram; then calculate and plot the velocity of the sawblade with respect to the piece being cut over one revolution of the crank at 50 rpm.

†6-53 Figure P6-16 (p. 339) shows a walking-beam indexing and pick-and-place mechanism that can be analyzed as two fourbar linkages driven by a common crank. The link lengths are given in the figure. The phase angle between the two crankpins on links 4 and 5 is given. The product cylinders being pushed have 60-mm diameters. The point of contact between the left vertical finger and the leftmost cylinder in the position shown is 58 mm at 80° versus the left end of the parallelogram's coupler (point D). Calculate and plot the absolute velocities of points E and P and the relative velocity between points E and P for one revolution of gear 2.

†6-54 Figure P6-17 (p. 339) shows a paper roll off-loading mechanism driven by an air cylinder. In the position shown, $AO_2 = 1.1$ m at 178° and O_4A is 0.3 m at 226°. $O_2O_4 = 0.93$ m at 163°. The V-links are rigidly attached to O_4A. The air cylinder is retracted at a constant velocity of 0.2 m/sec. Draw a kinematic diagram of the mechanism, write the necessary equations, and calculate and plot the angular velocity of the paper roll and the linear velocity of its center as it rotates through 90° CCW from the position shown.

FIGURE P6-11

Problem 6-48 Loom laybar drive

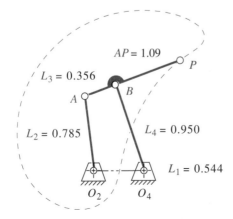

FIGURE P6-12

Problem 6-49

---

† These problems are suited to solution using *Mathcad*, *Matlab*, or *TKSolver* equation solver programs.

FIGURE P6-13

Problem 6-50

FIGURE P6-14

Problem 6-51

FIGURE P6-15

Problem 6-52　Power hacksaw　*Adapted from P. H. Hill and W. P. Rule. (1960). Mechanisms: Analysis and Design, with permission*

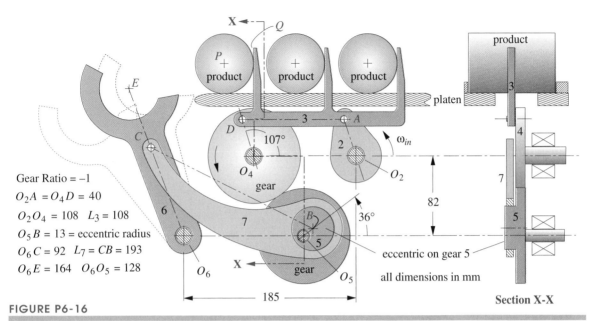

FIGURE P6-16

Problem 6-53   Walking-beam indexer with pick-and-place mechanism   *Adapted from P. H. Hill and W. P. Rule. (1960). Mechanisms: Analysis and Design, with permission*

†6-55   Figure P6-18 (p. 340) shows a powder compaction mechanism.

    a.   Calculate its mechanical advantage for the position shown.

    b.   Calculate and plot its mechanical advantage as a function of the angle of link AC as it rotates from 15 to 60°.

†6-56   Figure P6-19 (p. 340) shows a walking-beam mechanism. Calculate and plot the velocity V_{out} for one revolution of the input crank 2 rotating at 100 rpm.

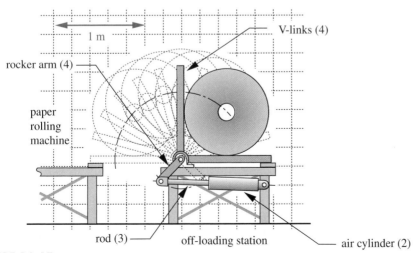

FIGURE P6-17

Problem 6-54

† These problems are suited to solution using *Mathcad*, *Matlab*, or *TKSolver* equation solver programs.

$AB = 105$ @ $44°$

$AC = 301$ @ $44°$

$BD = 172$

All lengths in mm

FIGURE P6-18

Problem 6-55 *From P. H. Hill and W. P. Rule. (1960). Mechanisms: Analysis and Design, with permission*

[†]6-57   Figure P6-20 shows a crimping tool.
   a.   Calculate its mechanical advantage for the position shown.
   b.   Calculate and plot its mechanical advantage as a function of the angle of link AB as it rotates from 60 to 45°.

[†]6-58   Figure P6-21 shows a locking pliers. Calculate its mechanical advantage for the position shown. Scale the diagram for any needed dimensions.

[†]6-59   Figure P6-22 (p. 342) shows a fourbar toggle clamp used to hold a workpiece in place by clamping it at D. $O_2A = 70$, $O_2C = 138$, $AB = 35$, $O_4B = 34$, $O_4D = 82$, and $O_2O_4 = 48$ mm. At the position shown, link 2 is at 104°. Toggle occurs when link 2 reaches 90°.
   a.   Calculate its mechanical advantage for the position shown.
   b.   Calculate and plot its mechanical advantage as a function of the angle of link AB as link 2 rotates from 120 to 90°.

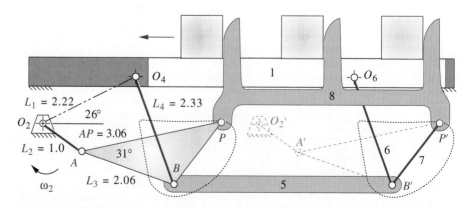

[†] These problems are suited to solution using *Mathcad, Matlab,* or *TKSolver* equation solver programs.

FIGURE P6-19

Problem 6-56 Straight-line walking-beam eightbar transport mechanism

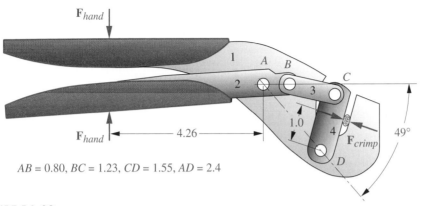

F_{hand}

1 A B

2 3 C

1.0 4

F_{hand} |———— 4.26 ————| F_{crimp} 49°

D

$AB = 0.80, BC = 1.23, CD = 1.55, AD = 2.4$

FIGURE P6-20

Problem 6-57 A crimping tool

†6-60   Figure P6-23 (p. 342) shows a surface grinder. The workpiece is oscillated under the spinning 90-mm-diameter grinding wheel by the crank-slider linkage which has a 22-mm crank, a 157-mm connecting rod, and a 40-mm offset. The crank turns at 120 rpm, and the grinding wheel at 3450 rpm. Calculate and plot the velocity of the grinding wheel contact point relative to the workpiece over one revolution of the crank.

6-61   Figure P6-24 (p. 343) shows an inverted crank-slider mechanism. Link 2 is 2.5 in long. The distance O_4A is 4.1 in and O_2O_4 is 3.9 in. Find ω_2, ω_3, ω_4, V_{A4}, V_{trans}, and V_{slip} for the position shown with $V_{A2} = 20$ in/sec in the direction shown.

*†6-62   Figure P6-25 (p. 343) shows a drag link mechanism with dimensions. Write the necessary equations, and solve them to calculate the angular velocity of link 4 for an input of $\omega_2 = 1$ rad/sec. Comment on uses for this mechanism.

†6-63   Figure P6-25 (p. 343) shows a drag link mechanism with dimensions. Write the necessary equations, and solve them to calculate and plot the centrodes of instant center $I_{2,4}$.

* Answers in Appendix F.

† These problems are suited to solution using *Mathcad, Matlab,* or *TKSolver* equation solver programs.

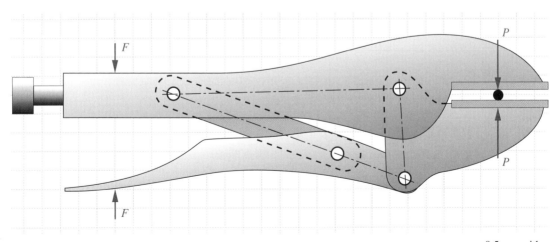

P

F

P

F

0.5-cm grid

FIGURE P6-21

Problem 6-58

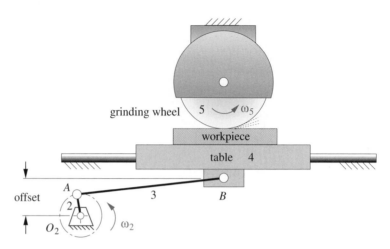

FIGURE P6-22

Problem 6-59

FIGURE P6-23

Problem 6-60 A surface grinder

6-64   Figure P6-26 (p. 344) shows a mechanism with dimensions. Use a graphical method to calculate the velocities of points A, B, and C and the velocity of slip for the position shown. $\omega_2 = 20$ rad/sec.

*6-65   Figure P6-27 (p. 344) shows a cam and follower. Distance $O_2A = 1.89$ in and $O_3B = 1.645$ in. Find the velocities of points A and B, the velocity of transmission, velocity of slip, and ω_3 if $\omega_2 = 50$ rad/sec. Use a graphical method.

6-66   Figure P6-28 (p. 345) shows a quick-return mechanism with dimensions. Use a graphical method to calculate the velocities of points A, B, and C and the velocity of slip for the position shown. $\omega_2 = 10$ rad/sec.

†6-67   Figure P6-29 (p. 345) shows a drum pedal mechanism. $O_2A = 100$ mm at 162° and rotates to 171° at A'. $O_2O_4 = 56$ mm, $AB = 28$ mm, $AP = 124$ mm, and $O_4B = 64$ mm. The distance from O_4 to F_{in} is 48 mm. Find and plot the mechanical advantage and the velocity ratio of the linkage over its range of motion. If the input velocity V_{in} is a constant magnitude of 3 m/sec, and F_{in} is constant at 50 N, find the output velocity and output force over the range of motion and the power in.

6-68   Figure 3-33 (p. 152) shows a sixbar slider-crank linkage. Find all its instant centers in the position shown.

†6-69   Calculate and plot the centrodes of instant center I_{24} of the linkage in Figure 3-33 (p. 152) so that a pair of noncircular gears can be made to replace the driver dyad 23.

6-70   Find the velocity of the slider in Figure 3-33 (p. 152) for the position shown if $\theta_2 = 110°$ with respect to the global X axis assuming $\omega_2 = 1$ rad/sec CW.

       a.   Using a graphical method.
       b.   Using the method of instant centers.
       c.   Using an analytical method.†

†6-71   Write a computer program or use an equation solver such as *Mathcad, Matlab*, or *TKSolver* to calculate and plot the angular velocity of link 4 and the linear velocity of slider 6 in the sixbar crank-slider linkage of Figure 3-33 (p. 152) as a function of the

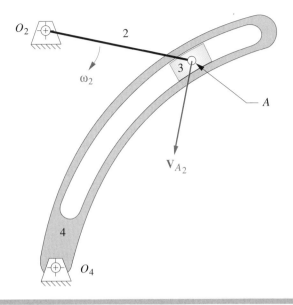

FIGURE P6-24

Problem 6-61 *From P. H. Hill and W. P. Rule. (1960). Mechanisms: Analysis and Design, with permission*

angle of input link 2 for a constant $\omega_2 = 1$ rad/sec CW. Plot V_c both as a function of θ_2 and separately as a function of slider position as shown in the figure. Find the percent deviation from constant velocity over $240° < \theta_2 < 270°$ and over $190° < \theta_2 < 315°$.

6-72 Figure 3-34 (p. 153) shows Stephenson's sixbar mechanism. Find all its instant centers in the position shown:

a. In part (*a*) of the figure.
b. In part (*b*) of the figure.
c. In part (*c*) of the figure.

6-73 Find the angular velocity of link 6 of the linkage in Figure 3-34 part (*b*) (p. 153) for the position shown ($\theta_6 = 90°$ with respect to the *x* axis) assuming $\omega_2 = 10$ rad/sec CW.

a. Using a graphical method.

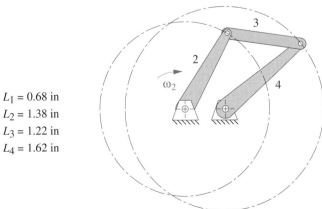

$L_1 = 0.68$ in
$L_2 = 1.38$ in
$L_3 = 1.22$ in
$L_4 = 1.62$ in

FIGURE P6-25

Problems 6-62 and 6-63 *From P. H. Hill and W. P. Rule. (1960). Mechanisms: Analysis and Design*

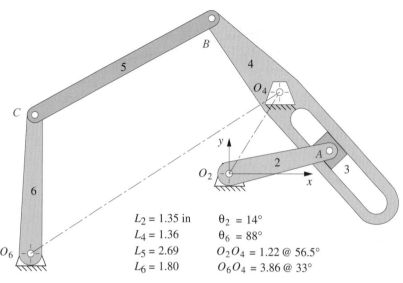

$L_2 = 1.35$ in $\theta_2 = 14°$
$L_4 = 1.36$ $\theta_6 = 88°$
$L_5 = 2.69$ $O_2O_4 = 1.22 @ 56.5°$
$L_6 = 1.80$ $O_6O_4 = 3.86 @ 33°$

FIGURE P6-26

Problems 6-64, 6-106, 6-107 *From P. H. Hill and W. P. Rule. (1960). Mechanisms: Analysis and Design*

    b.   Using the method of instant centers.
    c.   Using an analytical method.[†]

†6-74   Write a computer program or use an equation solver such as *Mathcad, Matlab*, or *TK-Solver* to calculate and plot the angular velocity of link 6 in the sixbar linkage of Figure 3-34 (p. 153) as a function of θ_2 for a constant $\omega_2 = 1$ rad/sec CW.

6-75   Figure 3-35 (p. 153) shows Stephenson's sixbar mechanism. Find all its instant centers in the position shown:

    a.   In part (*a*) of the figure.
    b.   In part (*b*) of the figure.

6-76   Find the angular velocity of link 6 of the linkage in Figure 3-35 (p. 153) with $\theta_2 = 90°$ assuming $\omega_2 = 10$ rad/sec CCW.

    a.   Using a graphical method (use a compass and straightedge to draw the the linkage with link 2 at 90°).

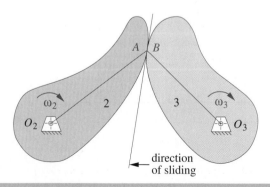

† These problems are suited to solution using *Mathcad, Matlab*, or *TKSolver* equation solver programs.

FIGURE P6-27

Problem 6-65 *From P. H. Hill and W. P. Rule. (1960). Mechanisms: Analysis and Design*

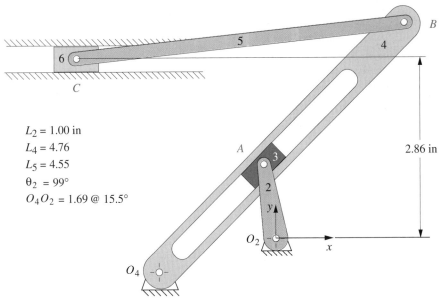

$L_2 = 1.00$ in
$L_4 = 4.76$
$L_5 = 4.55$
$\theta_2 = 99°$
$O_4O_2 = 1.69$ @ $15.5°$

FIGURE P6-28

Problems 6-66, 6-108. 6-109 From P. H. Hill and W. P. Rule. (1960). Mechanisms: Analysis and Design

FIGURE P6-29

Problem 6-67

b. Using the method of instant centers (use a compass and straightedge to draw the the linkage with link 2 at 90°).

c. Using an analytical method.[†]

†6-77 Write a computer program or use an equation solver such as *Mathcad, Matlab*, or *TK-Solver* to calculate and plot the angular velocity of link 6 in the sixbar linkage of Figure 3-35 (p. 153) as a function of θ_2 for a constant $\omega_2 = 1$ rad/sec CCW.

6-78 Figure 3-36 (p. 154) shows an eightbar mechanism. Find all its instant centers in the position shown in part (*a*) of the figure.

†6-79 Write a computer program or use an equation solver such as *Mathcad, Matlab*, or *TK-Solver* to calculate and plot the angular velocity of link 8 in the linkage of Figure 3-36 (p. 154) as a function of θ_2 for a constant $\omega_2 = 1$ rad/sec CCW.

†6-80 Write a computer program or use an equation solver such as *Mathcad, Matlab*, or *TKSolver* to calculate and plot magnitude and direction of the velocity of point *P* in Figure 3-37a (p. 154) as a function of θ_2. Also calculate and plot the velocity of point *P* versus point *A*.

†6-81 Write a computer program or use an equation solver such as *Mathcad, Matlab*, or *TK-Solver* to calculate the percent error of the deviation from a perfect circle for the path of point *P* in Figure 3-37a (p. 154).

†6-82 Repeat problem 6-80 for the linkage in Figure 3-37b (p. 154).

6-83 Find all instant centers of the linkage in Figure P6-30 (p. 346) in the position shown.

6-84 Find the angular velocities of links 3 and 4 and the linear velocities of points *A*, *B* and P_1 in the *XY* coordinate system for the linkage in Figure P6-30 in the position shown. Assume that $\theta_2 = 45°$ in the *XY* coordinate system and $\omega_2 = 10$ rad/sec. The coordinates of the point P_1 on link 4 are (114.68, 33.19) with respect to the *xy* coordinate system.

---

† These problems are suited to solution using *Mathcad, Matlab*, or *TKSolver* equation solver programs.

   a.   Using a graphical method.
   b.   Using the method of instant centers.
   c.   Using an analytical method.[*]

†6-85   Using the data from Problem 6-84, write a computer program or use an equation solver such as *Mathcad, Matlab*, or *TKSolver* to calculate and plot magnitude and direction of the absolute velocity of point P_1 in Figure P6-30 as a function of θ_2.

6-86   Find all instant centers of the linkage in Figure P6-31 in the position shown.

6-87   Find the angular velocities of links 3 and 4, and the linear velocity of point P in the *XY* coordinate system for the linkage in Figure P6-31 in the position shown. Assume that $\theta_2 = -94.121°$ in the *XY* coordinate system and $\omega_2 = 1$ rad/sec. The position of the coupler point P on link 3 with respect to point A is: $p = 15.00$, $\delta_3 = 0°$.

   a.   Using a graphical method.
   b.   Using the method of instant centers.
   c.   Using an analytical method.†

6-88   Figure P6-32 shows a fourbar double slider known as an elliptical trammel. Find all its instant centers in the position shown.

6-89   The elliptical trammel in Figure P6-32 must be driven by rotating link 3 in a full circle. Points on line *AB* describe ellipses. Find and draw (manually or with a computer) the fixed and moving centrodes of instant center I_{13}. (Hint: These are called the Cardan circles.)

6-90   Derive analytical expressions for the velocities of points A and B in Figure P6-32 as a function of θ_3, ω_3, and the length AB of link 3. Use a vector loop equation.

6-91   The linkage in Figure P6-33a (p. 348) has link 2 at 120° in the global *XY* coordinate system. Find ω_6 and $\mathbf{V}_D$ in the global coordinate system for the position shown if $\omega_2 = 10$ rad/sec CCW. Use the velocity difference graphical method. (Print the figure from its PDF file on the DVD and draw on it.)

FIGURE P6-30

Problems 6-83 to 6-85 An oil field pump - dimensions in inches

FIGURE P6-31

Problems 6-86 and 6-87 An aircraft overhead bin mechanism - dimensions in inches

6-92   The linkage in Figure P6-33a (p. 348) has link 2 at 120° in the global XY coordinate system. Find ω_6 and V_D in the global coordinate system for the position shown if ω_2 = 10 rad/sec CCW. Use the instant center graphical method. (Print the figure from its PDF file on the DVD and draw on it.)

6-93   The linkage in Figure P6-33a has link 2 at 120° in the global XY coordinate system. Find ω_6 and V_D in the global coordinate system for the position shown if ω_2 = 10 rad/sec CCW. Use an analytical method.

6-94   The linkage in Figure P6-33b has link 3 perpendicular to the X axis and links 2 and 4 are parallel to each other. Find ω_3, V_A, V_B, and V_P if ω_2 = 15 rad/sec CW. Use the velocity difference graphical method. (Print the figure's PDF file from the DVD and draw on it.)

6-95   The linkage in Figure P6-33b has link 3 perpendicular to the X axis and links 2 and 4 are parallel to each other. Find ω_3, V_A, V_B, and V_P if ω_2 = 15 rad/sec CW. Use the instant center graphical method. (Print the figure from its PDF file on the DVD and draw on it.)

6-96   The linkage in Figure P6-33b has link 3 perpendicular to the X axis and links 2 and 4 are parallel to each other. Find ω_3, V_A, V_B, and V_P if ω_2 = 15 rad/sec CW. Use an analytical method.

6-97   The crosshead linkage shown in Figure P6-33c has 2 *DOF* with inputs at crossheads 2 and 5. Find instant centers $I_{1,3}$ and $I_{1,4}$.

6-98   The crosshead linkage shown in Figure P6-33c has 2 *DOF* with inputs at crossheads 2 and 5. Find V_B, V_{P3}, and V_{P4} if the crossheads are each moving toward the origin of the XY coordinate system with a speed of 20 in/sec. Use a graphical method of your choice. (Print the figure from its PDF file on the DVD and draw on it.)

6-99   The linkage in Figure P6-33d has the path of slider 6 perpendicular to the global X axis and link 2 aligned with the global X axis. Find V_A in the position shown if the velocity of the slider is 20 in/sec downward. Use the velocity difference graphical method. (Print the figure from its PDF file on the DVD and draw on it.)

FIGURE P6-32

Elliptical trammel - Problems 6-88 to 6-90

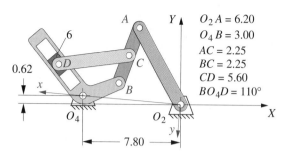

$O_2A = 6.20$
$O_4B = 3.00$
$AC = 2.25$
$BC = 2.25$
$CD = 5.60$
$BO_4D = 110°$

0.62

7.80

(a) Sixbar linkage

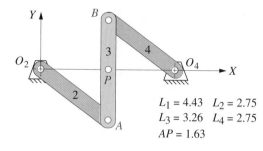

$L_1 = 4.43$　$L_2 = 2.75$
$L_3 = 3.26$　$L_4 = 2.75$
$AP = 1.63$

(b) Fourbar linkage

$AB = 34.32$
$BC = 50.4$
$AP_3 = 31.5$
$BP_3 = 22.2$
$BP_4 = 41.52$
$CP_4 = 27$

59.5

57

(c) Dual crosshead mechanism

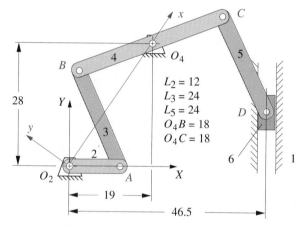

$L_2 = 12$
$L_3 = 24$
$L_5 = 24$
$O_4B = 18$
$O_4C = 18$

28

19

46.5

(d) Sixbar linkage

$L_2 = 5$　$L_3 = 5$　$L_5 = 15$　$BC = 8$
$O_2O_4 = 2.5$　$O_4B = 6$　$O_4C = 6$

(e) Drag link slider-crank

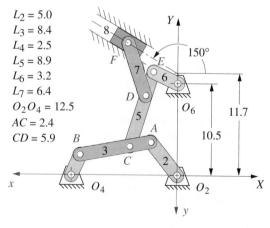

$L_2 = 5.0$
$L_3 = 8.4$
$L_4 = 2.5$
$L_5 = 8.9$
$L_6 = 3.2$
$L_7 = 6.4$
$O_2O_4 = 12.5$
$AC = 2.4$
$CD = 5.9$

150°

11.7

10.5

(f) Eightbar mechanism

FIGURE P6-33

Problems 6-91 to 6-103

6-100   The linkage in Figure P6-33d has the path of slider 6 perpendicular to the global X axis and link 2 aligned with the global X axis. Find $\mathbf{V}_A$ in the position shown if the velocity of the slider is 20 in/sec downward. Use the instant center graphical method. (Print the figure from its PDF file on the DVD and draw on it.)

6-101   For the linkage of Figure P6-33e, write a computer program or use an equation solver to find and plot $\mathbf{V}_D$ in the global coordinate system for one revolution of link 2 if $\omega_2 = 10$ rad/sec CW.

6-102   For the linkage of Figure P6-33f, locate and identify all instant centers.

6-103   The linkage of Figure P6-33f has link 2 at 130° in the global XY coordinate system. Find $\mathbf{V}_D$ in the global coordinate system for the position shown if $\omega_2 = 15$ rad/sec CW. Use any graphical method. (Print the figure from its PDF file on the DVD and draw on it.)

6-104   For the linkage of Figure P6-34, locate and identify all instant centers. $O_2O_4 = AB = BC = DE = 1$. $O_2A = O_4B = BE = CD = 1.75$. $O_4C = AE = 2.60$.

6-105   For the linkage of Figure P6-34, show that $I_{1,6}$ is stationary for all positions of the linkage. $O_2O_4 = AB = BC = DE = 1$. $O_2A = O_4B = BE = CD = 1.75$. $O_4C = AE = 2.60$.

6-106   Figure P6-26 (p. 344) shows a mechanism with dimensions. Use a graphical method to determine the velocities of points A and B, and the velocity of slip for the position shown if $\omega_2 = 24$ rad/sec CW. Ignore links 5 and 6.

6-107   Repeat Problem 6-106 using an analytical method.

6-108   Figure P6-28 (p. 345) shows a quick-return mechanism with dimensions. Use a graphical method to determine the velocities of points A and B and the velocity of slip for the position shown if $\omega_2 = 16$ rad/sec CCW. Ignore links 5 and 6.

6-109   Repeat Problem 6-108 using an analytical method.

6-110   The general linkage configuration and terminology for an offset fourbar slider-crank linkage are shown in Figure P6-2 (p. 329). The link lengths and the values of d and $\dot{d}$ are defined in Table P6-5. For the row(s) assigned, find the velocity of the pin joint A and the angular velocity of the crank using a graphical method.

6-111   The general linkage configuration and terminology for an offset fourbar slider-crank linkage are shown in Figure P6-2 (p. 329). The link lengths and the values of d and $\dot{d}$ are defined in Table P6-5. For the rows assigned, find the velocity of pin joint A and the angular velocity of the crank using the analytic method. Draw the linkage to scale and label it before setting up the equations.

FIGURE P6-34

Problems 6-104 & 6-105

6

TABLE P6-5   Data for Problems 6-110 to 6-111[‡]

Row	Link 2	Link 3	Offset	d	$\dot{d}$
a	1.4	4	1	2.5	10
b	2	6	-3	5	-12
c	3	8	2	8	-15
d	3.5	10	1	-8	24
e	5	20	-5	15	-50
f	3	13	0	-12	-45
g	7	25	10	25	100

[‡] Drawings of these linkages are in the *PDF Problem Workbook* folder on the DVD.

Chapter 7

ACCELERATION ANALYSIS

Take it to warp five, Mr. Sulu
CAPTAIN KIRK

7.0 INTRODUCTION

Once a velocity analysis is done, the next step is to determine the accelerations of all links and points of interest in the mechanism or machine. We need to know the accelerations to calculate the dynamic forces from $\mathbf{F} = m\mathbf{a}$. The dynamic forces will contribute to the stresses in the links and other components. Many methods and approaches exist to find accelerations in mechanisms. We will examine only a few of these methods here. We will first develop a manual graphical method, which is often useful as a check on the more complete and accurate analytical solution. Then we will derive the analytical solution for accelerations in the fourbar and inverted crank-slider linkages as examples of the general vector loop equation solution to acceleration analysis problems.

7.1 DEFINITION OF ACCELERATION

Acceleration is defined as *the rate of change of velocity with respect to time*. Velocity $(\mathbf{V}, \omega)$ is a vector quantity and so is acceleration. Accelerations can be **angular** or **linear**. **Angular acceleration** will be denoted as α and **linear acceleration** as $\mathbf{A}$.

$$\alpha = \frac{d\omega}{dt}; \qquad \mathbf{A} = \frac{d\mathbf{V}}{dt} \qquad (7.1)$$

Figure 7-1 shows a link PA in pure rotation, pivoted at point A in the xy plane. We are interested in the acceleration of point P when the link is subjected to an angular velocity ω and an angular acceleration α, which need not have the same sense. The link's position is defined by the position vector $\mathbf{R}$, and the velocity of point P is $\mathbf{V}_{PA}$. These vectors were defined in equations 6.2 and 6.3 (p. 286) which are repeated here for convenience. (See also Figure 6-1, p. 286.)

$$\mathbf{R}_{PA} = pe^{j\theta} \qquad (6.2)$$

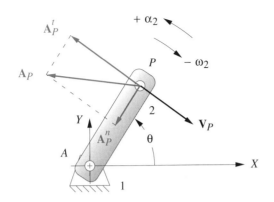

FIGURE 7-1

Acceleration of a link in pure rotation with a positive (CCW) α_2 and a negative (CW) ω_2

$$V_{PA} = \frac{d\mathbf{R}_{PA}}{dt} = p\,je^{j\theta}\,\frac{d\theta}{dt} = p\omega\,je^{j\theta} \qquad (6.3)$$

where p is the scalar length of the vector $\mathbf{R}_{PA}$. We can easily differentiate equation 6.3 to obtain an expression for the acceleration of point P:

$$\mathbf{A}_{PA} = \frac{d\mathbf{V}_{PA}}{dt} = \frac{d\left(p\omega\,je^{j\theta}\right)}{dt}$$

$$\mathbf{A}_{PA} = j\,p\left(e^{j\theta}\,\frac{d\omega}{dt} + \omega\,je^{j\theta}\,\frac{d\theta}{dt}\right) \qquad (7.2)$$

$$\mathbf{A}_{PA} = p\alpha\,je^{j\theta} - p\omega^2\,e^{j\theta}$$

$$\mathbf{A}_{PA} = \mathbf{A}_{PA}^{t} + \mathbf{A}_{PA}^{n}$$

Note that there are two functions of time in equation 6.3, θ and ω. Thus there are two terms in the expression for acceleration, the tangential component of acceleration involving α, and the normal (or centripetal) component $\mathbf{A}_A^n$ involving ω^2. As a result of the differentiation, the tangential component is multiplied by the (constant) complex operator j. This causes a rotation of this acceleration vector through $90°$ with respect to the original position vector. (See also Figure 4-8b, p. 186.) This $90°$ rotation is nominally positive, or counterclockwise (CCW). However, the tangential component is also multiplied by α, which may be either positive or negative. As a result, the tangential component of acceleration will be **rotated 90°** from the angle θ of the *position vector* **in a direction dictated by the sign of** α. This is just mathematical verification of what you already knew, namely that *tangential acceleration is always in a direction perpendicular to the radius of rotation and is thus tangent to the path of motion* as shown in Figure 7-1. The normal, or centripetal, acceleration component is multiplied by j^2, or -1. This directs *the centripetal component at 180° to the angle θ of the original position vector,* i.e., toward the center (centripetal means *toward the center*). The total acceleration $\mathbf{A}_{PA}$ of point P is the vector sum of the tangential $\mathbf{A}_A^t$ and normal $\mathbf{A}_A^n$ components as shown in Figure 7-1 and equation 7.2.

Substituting the Euler identity (equation 4.4a, p. 185) into equations 7.2 gives us the real and imaginary (or x and y) components of the acceleration vector.

$$A_{PA} = p\alpha(-\sin\theta + j\cos\theta) - p\omega^2(\cos\theta + j\sin\theta) \tag{7.3}$$

The acceleration A_{PA} in Figure 7-1 can be referred to as an **absolute acceleration** since it is referenced to A, which is the origin of the global coordinate axes in that system. As such, we could have referred to it as A_P, with the absence of the second subscript implying reference to the global coordinate system.

Figure 7-2a shows a different and slightly more complicated system in which the pivot A is no longer stationary. It has a known linear acceleration A_A as part of the translating carriage, link 3. If α is unchanged, the acceleration of point P versus A will be the same as before, but A_{PA} can no longer be considered an absolute acceleration. It is now an **acceleration difference** and **must** carry the second subscript as A_{PA}. The absolute acceleration A_P must now be found from the **acceleration difference** equation whose graphical solution is shown in Figure 7-2b:

$$A_P = A_A + A_{PA}$$

$$\left(A_P^t + A_P^n\right) = \left(A_A^t + A_A^n\right) + \left(A_{PA}^t + A_{PA}^n\right) \tag{7.4}$$

Note the similarity of equations 7.4 to the **velocity difference equation** (equation 6.5, p. 287). Note also that the solution for A_P in equation 7.4 can be found by adding either the resultant vector A_{PA} or its normal and tangential components A_{PA}^n and A_{PA}^t to the vector A_A in Figure 7-2b. The vector A_A has a zero normal component in this example because link 3 is in pure translation.

Figure 7-3 shows two independent bodies P and A, which could be two automobiles, moving in the same plane. Auto #1 is turning and accelerating into the path of auto #2, that is decelerating to avoid a crash. If their independent accelerations A_P and A_A are known, their **relative acceleration** A_{PA} can be found from equation 7.4 arranged algebraically as:

$$A_{PA} = A_P - A_A \tag{7.5}$$

The graphical solution to this equation is shown in Figure 7-3b.

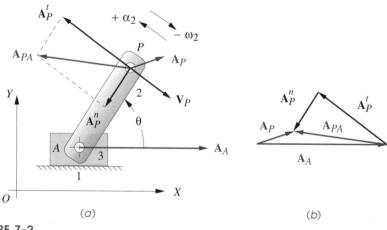

(a)   (b)

FIGURE 7-2

Acceleration difference in a system with a positive (CCW) α_2 and a negative (CW) ω_2

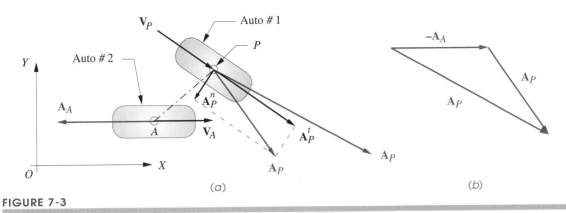

FIGURE 7-3

Relative acceleration

As we did for velocity analysis, we give these two cases different names despite the fact that the same equation applies. Repeating the definition from Section 6.1 (p. 285), modified to refer to acceleration:

CASE 1:   Two points in the same body => *acceleration difference*

CASE 2:   Two points in different bodies => *relative acceleration*

7.2   GRAPHICAL ACCELERATION ANALYSIS

The comments made in regard to graphical velocity analysis in Section 6.2 (p. 288) apply as well to graphical acceleration analysis. Historically, graphical methods were the only practical way to solve these acceleration analysis problems. With some practice, and with proper tools such as a drafting machine, drafting instruments, or a CAD package, one can fairly rapidly solve for the accelerations of particular points in a mechanism for any one input position by drawing vector diagrams. However, if accelerations for many positions of the mechanism are to be found, each new position requires a completely new set of vector diagrams be drawn. Very little of the work done to solve for the accelerations at position 1 carries over to position 2, etc. This is an even more tedious process than that for graphical velocity analysis because there are more components to draw. Nevertheless, this method still has more than historical value as it can provide a quick check on the results from a computer program solution. Such a check only needs to be done for a few positions to prove the validity of the program.

To solve any acceleration analysis problem graphically, we need only three equations, equation 7.4 and equations 7.6 (which are merely the scalar magnitudes of the terms in equation 7.2, p. 351):

$$\left| \mathbf{A}^t \right| = A^t = r\alpha$$

$$ (7.6) $$

$$\left| \mathbf{A}^n \right| = A^n = r\omega^2$$

Note that the scalar equations 7.6 define only the **magnitudes** (A^t, A^n) of the components of acceleration of any point in rotation. In a CASE 1 graphical analysis, the **directions** of the vectors due to the centripetal and tangential components of the acceleration difference must be understood from equation 7.2 to be perpendicular to and along the radius of rotation, respectively. Thus, if the center of rotation is known or assumed, the directions of the acceleration difference components due to that rotation are known and their senses will be consistent with the angular velocity ω and angular acceleration α of the body.

Figure 7-4 shows a fourbar linkage in one particular position. We wish to solve for the angular accelerations of links 3 and 4 (α_3, α_4) and the linear accelerations of points A, B, and C ($\mathbf{A}_A$, $\mathbf{A}_B$, $\mathbf{A}_C$). Point C represents any general point of interest such as a coupler point. The solution method is valid for any point on any link. To solve this problem we need to know the *lengths of all the links*, the *angular positions of all the links*, the *angular velocities of all the links*, and the *instantaneous input acceleration of any one driving link or driving point*. Assuming that we have designed this linkage, we will know or can measure the link lengths. We must also first do a **complete position and velocity analysis** to find the link angles θ_3 and θ_4 and angular velocities ω_3 and ω_4 given the input link's position θ_2, input angular velocity ω_2, and input acceleration α_2. This can be done by any of the methods in Chapters 4 and 6. In general we must solve these problems in stages, first for link positions, then for velocities, and finally for accelerations. For the following example, we will assume that a complete position and velocity analysis has been done and that the input is to link 2 with known θ_2, ω_2, and α_2 for this one "freeze-frame" position of the moving linkage.

✏️ EXAMPLE 7-1

Graphical Acceleration Analysis for One Position of a Fourbar Linkage.

Problem: Given θ_2, θ_3, θ_4, ω_2, ω_3, ω_4, α_2, find α_3, α_4, $\mathbf{A}_A$, $\mathbf{A}_B$, $\mathbf{A}_P$ by graphical methods.

Solution: (See Figure 7-4.)

1 Start at the end of the linkage about which you have the most information. Calculate the magnitudes of the centripetal and tangential components of acceleration of point A using scalar equations 7.6.

$$A_A^n = (AO_2)\omega_2^2; \qquad\qquad\qquad A_A^t = (AO_2)\alpha_2 \qquad\qquad (a)$$

2 On the linkage diagram, Figure 7-4a, draw the acceleration component vectors $\mathbf{A}_A^n$ and $\mathbf{A}_A^t$ with their lengths equal to their magnitudes at some convenient scale. Place their roots at point A with their directions respectively along and perpendicular to the radius AO_2. The sense of $\mathbf{A}_A^t$ is defined by that of α_2 (according to the right-hand rule), and the sense of $\mathbf{A}_A^n$ is the opposite of that of the position vector $\mathbf{R}_A$ as shown in Figure 7-4a.

3 Move next to a point about which you have some information, such as B on link 4. Note that the directions of the tangential and normal components of acceleration of point B are predictable since this link is in pure rotation about point O_4. Draw the construction line pp through point B perpendicular to BO_4, to represent the direction of $\mathbf{A}_B^t$ as shown in Figure 7-4a.

4 Write the acceleration difference vector equation 7.4 (p. 352) for point B versus point A.

$$\mathbf{A}_B = \mathbf{A}_A + \mathbf{A}_{BA} \qquad\qquad (b)$$

Substitute the normal and tangential components for each term:

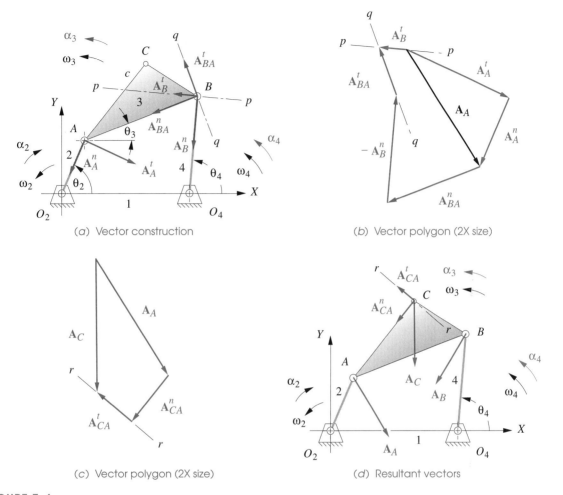

FIGURE 7-4

Graphical solution for acceleration in a pin-jointed linkage with a negative (CW) α_2 and a positive (CCW) ω_2

$$\left(\mathbf{A}_B^t + \mathbf{A}_B^n\right) = \left(\mathbf{A}_A^t + \mathbf{A}_A^n\right) + \left(\mathbf{A}_{BA}^t + \mathbf{A}_{BA}^n\right) \tag{c}$$

We will use point A as the reference point to find $\mathbf{A}_B$ because A is in the same link as B and we have already solved for $\mathbf{A}_A^t$ and $\mathbf{A}_A^n$. Any two-dimensional vector equation can be solved for two unknowns. Each term has two parameters, namely magnitude and direction. There are then potentially twelve unknowns in this equation, two per term. We must know ten of them to solve it. We know both the magnitudes and directions of $\mathbf{A}_A^t$ and $\mathbf{A}_A^n$ and the directions of $\mathbf{A}_B^t$ and $\mathbf{A}_B^n$ that are along line pp and line BO_4, respectively. We can also calculate the magnitude of $\mathbf{A}_B^n$ from equation 7.6 (p. 353) since we know ω_4. This provides seven known values. We need to know three more parameters to solve the equation.

5   The term $\mathbf{A}_{BA}$ represents the acceleration difference of B with respect to A. This has two components. The normal component $\mathbf{A}_{BA}^n$ is directed along the line BA because we are us-ing point A as the reference center of rotation for the free vector ω_3, and its magnitude can be

calculated from equation 7.6. The direction of $\mathbf{A}_{BA}^{t}$ must then be perpendicular to the line BA. Draw construction line qq through point B and perpendicular to BA to represent the direction of $\mathbf{A}_{BA}^{t}$ as shown in Figure 7-4a (p. 355). The calculated magnitude and direction of component $\mathbf{A}_{BA}^{n}$ and the known direction of $\mathbf{A}_{BA}^{t}$ provide the needed additional three parameters.

6   Now the vector equation can be solved graphically by drawing a vector diagram as shown in Figure 7-4b. Either drafting tools or a CAD package is necessary for this step. The strategy is to first draw all vectors for which we know both magnitude and direction, being careful to arrange their senses according to equation 7.4 (p. 352).

First draw acceleration vectors ($\mathbf{A}_{A}^{t}$) and ($\mathbf{A}_{A}^{n}$) tip to tail, carefully to some scale, maintaining their directions. (They are drawn twice size in the figure.) Note that the sum of these two components is the vector $\mathbf{A}_{A}$. The equation in step 4 says to add $\mathbf{A}_{BA}$ to $\mathbf{A}_{A}$. We know $\mathbf{A}_{BA}^{n}$, so we can draw that component at the end of $\mathbf{A}_{A}$. We also know $\mathbf{A}_{B}^{n}$, but this component is on the left side of equation 7.4, so we must subtract it. Draw the negative (opposite sense) of $\mathbf{A}_{B}^{n}$ at the end of $\mathbf{A}_{BA}^{n}$.

This exhausts our supply of components for which we know both magnitude and direction. Our two remaining knowns are the directions of $\mathbf{A}_{B}^{t}$ and $\mathbf{A}_{BA}^{t}$ that lie along the lines pp and qq, respectively. Draw a line parallel to line qq across the tip of the vector representing *minus* $\mathbf{A}_{B}^{n}$. The resultant, or left side of the equation, must close the vector diagram, from the tail of the first vector drawn ($\mathbf{A}_{A}$) to the tip of the last, so draw a line parallel to pp across the tail of $\mathbf{A}_{A}$. The intersection of these lines parallel to pp and qq defines the lengths of $\mathbf{A}_{B}^{t}$ and $\mathbf{A}_{BA}^{t}$. The senses of these vectors are determined from reference to equation 7.4. Vector $\mathbf{A}_{A}$ was added to $\mathbf{A}_{BA}$, so their components must be arranged tip to tail. Vector $\mathbf{A}_{B}$ is the resultant, so its component $\mathbf{A}_{B}^{t}$ must be from the tail of the first to the tip of the last. The resultant vectors are shown in Figure 7-4b and d (p. 355).

7   The angular accelerations of links 3 and 4 can be calculated from equation 7.6 (p. 353):

$$\alpha_4 = \frac{A_B^t}{BO_4} \qquad\qquad \alpha_3 = \frac{A_{BA}^t}{BA} \qquad\qquad (d)$$

Note that the acceleration difference term $\mathbf{A}_{BA}^{t}$ represents the rotational component of acceleration of link 3 due to α_3. The rotational acceleration α of any body is a "**free vector**" which has no particular point of application to the body. It exists everywhere on the body.

8   Finally we can solve for $\mathbf{A}_C$ using equation 7.4 (p. 352) again. We select any point in link 3 for which we know the absolute velocity to use as the reference, such as point A.

$$\mathbf{A}_C = \mathbf{A}_A + \mathbf{A}_{CA} \qquad\qquad (e)$$

In this case, we can calculate the magnitude of $\mathbf{A}_{CA}^{t}$ from equation 7.6 (p. 353) as we have already found α_3,

$$A_{CA}^t = c\alpha_3 \qquad\qquad (f)$$

The magnitude of the component $\mathbf{A}_{CA}^{n}$ can be found from equation 7.6 using ω_3.

$$A_{CA}^n = c\omega_3^2 \qquad\qquad (g)$$

Since both $\mathbf{A}_A$ and $\mathbf{A}_{CA}$ are known, the vector diagram can be directly drawn as shown in Figure 7-4c (p. 355). Vector $\mathbf{A}_C$ is the resultant that closes the vector diagram. Figure 7-4d shows the calculated acceleration vectors on the linkage diagram.

The above example contains some interesting and significant principles that deserve further emphasis. Equations 7.4 are repeated here for discussion.

$$\mathbf{A}_P = \mathbf{A}_A + \mathbf{A}_{PA}$$

$$\left(\mathbf{A}_P^t + \mathbf{A}_P^n\right) = \left(\mathbf{A}_A^t + \mathbf{A}_A^n\right) + \left(\mathbf{A}_{PA}^t + \mathbf{A}_{PA}^n\right) \tag{7.4}$$

These equations represent the *absolute* acceleration of some general point P referenced to the origin of the global coordinate system. The right side defines it as the sum of the absolute acceleration of some other reference point A in the same system and the acceleration difference (or relative acceleration) of point P versus point A. These terms are then further broken down into their normal (centripetal) and tangential components that have definitions as shown in equation 7.2 (p. 351).

Let us review what was done in Example 7-1 in order to extract the general strategy for solution of this class of problem. We started at the input side of the mechanism, as that is where the driving angular acceleration α_2 was defined. We first looked for a point (A) for which the motion was pure rotation. We then solved for the absolute acceleration of that point ($\mathbf{A}_A$) using equations 7.4 and 7.6 (pp. 352, 353) by breaking $\mathbf{A}_A$ into its normal and tangential components. *(Steps 1 and 2)*

We then used the point (A) just solved for as a reference point to define the translation component in equation 7.4 written for a new point (B). Note that we needed to choose a second point (B) in the same rigid body as the reference point (A) that we had already solved, and about which we could predict some aspect of the new point's (B's) acceleration components. In this example, we knew the direction of the component $\mathbf{A}_B^t$, though we did not yet know its magnitude. We could also calculate both magnitude and direction of the centripetal component, $\mathbf{A}_B^n$, since we knew ω_4 and the link length. In general this situation will obtain for any point on a link that is jointed to ground (as is link 4). In this example, we could not have solved for point C until we solved for B, because point C is on a floating link for which we do not yet know the angular acceleration or absolute acceleration direction. *(Steps 3 and 4)*

To solve the equation for the second point (B), we also needed to recognize that the tangential component of the acceleration difference $\mathbf{A}_{BA}^t$ is always directed perpendicular to the line connecting the two related points in the link (B and A in the example). In addition, you will always know the magnitude and direction of the centripetal acceleration components in equation 7.4 (p. 352) *if it represents an acceleration difference* (CASE 1) **situation**. *If the two points are in the **same rigid** body, then that acceleration difference centripetal component has a magnitude of $r\omega^2$ and is always directed along the line connecting the two points, pointing toward the reference point as the center* (see Figure 7-2, p. 352). These observations will be true regardless of the two points selected. But, *note this is not true in a* CASE 2 *situation* as shown in Figure 7-3a (p. 353) where the normal component of acceleration of auto #2 is **not** directed along the line connecting points A and P. *(Steps 5 and 6)*

Once we found the absolute acceleration of point B ($\mathbf{A}_B$) we could solve for α_4, the angular acceleration of link 4 using the tangential component of $\mathbf{A}_B$ in equation (d). Because points A and B are both on link 3, we could also determine the angular acceleration of link 3 using the tangential component of the acceleration difference $\mathbf{A}_{BA}$ between points B and A, in equation (d). Once the angular accelerations of all the links were known, we could then solve for the linear acceleration of any point (such as C) in any link using equation 7.4 (p. 352). To do so, we had to understand the concept of angular acceleration as a **free vector**, which means that it exists everywhere on the link at any given instant. It has no particular center. *It has an infinity of potential centers.* The link simply *has an angular acceleration.* It is this property that allows us to solve equation 7.4 for literally **any point** on a rigid body in complex motion **referenced to any other point** on that body. *(Steps 7 and 8)*

7.3   ANALYTICAL SOLUTIONS FOR ACCELERATION ANALYSIS

The Fourbar Pin-Jointed Linkage

The position equations for the fourbar pin-jointed linkage were derived in Section 4.5 (pp. 185-189). The linkage was shown in Figure 4-6 (p. 184) and is shown again in Figure 7-5a on which we also show an input angular acceleration α_2 applied to link 2. This input angular acceleration α_2 may vary with time. The vector loop equation was shown in equations 4.5a and c, repeated here for your convenience.

$$\mathbf{R}_2 + \mathbf{R}_3 - \mathbf{R}_4 - \mathbf{R}_1 = 0 \tag{4.5a}$$

As before, we substitute the complex number notation for the vectors, denoting their scalar lengths as a, b, c, d as shown in Figure 7-5.

$$ae^{j\theta_2} + be^{j\theta_3} - ce^{j\theta_4} - de^{j\theta_1} = 0 \tag{4.5c}$$

In Section 6.7 (p. 315), we differentiated equation 4.5c versus time to get an expression for velocity which is repeated here.

$$ja\omega_2 e^{j\theta_2} + jb\omega_3 e^{j\theta_3} - jc\omega_4 e^{j\theta_4} = 0 \tag{6.14c}$$

We will now differentiate equation 6.14c versus time to obtain an expression for accelerations in the linkage. Each term in equation 6.14c contains two functions of time, θ and ω. Differentiating with the chain rule in this example will result in two terms in the acceleration expression for each term in the velocity equation.

$$\left(j^2 a\omega_2^2 e^{j\theta_2} + ja\alpha_2 e^{j\theta_2}\right) + \left(j^2 b\omega_3^2 e^{j\theta_3} + jb\alpha_3 e^{j\theta_3}\right) - \left(j^2 c\omega_4^2 e^{j\theta_4} + jc\alpha_4 e^{j\theta_4}\right) = 0 \tag{7.7a}$$

Simplifying and grouping terms:

$$\left(a\alpha_2 je^{j\theta_2} - a\omega_2^2 e^{j\theta_2}\right) + \left(b\alpha_3 je^{j\theta_3} - b\omega_3^2 e^{j\theta_3}\right) - \left(c\alpha_4 je^{j\theta_4} - c\omega_4^2 e^{j\theta_4}\right) = 0 \tag{7.7b}$$

Compare the terms grouped in parentheses with equations 7.2 (p. 351). Equation 7.7 contains the tangential and normal components of the accelerations of points A and B and of the acceleration difference of B to A. Note that these are the same relationships that we used to solve this problem graphically in Section 7.2 (p. 353). Equation 7.7 is, in fact, the **acceleration difference equation** 7.4 (p. 352) which, with the labels used here, is:

$$\mathbf{A}_A + \mathbf{A}_{BA} - \mathbf{A}_B = 0 \tag{7.8a}$$

where:

$$\mathbf{A}_A = \left(\mathbf{A}_A^t + \mathbf{A}_A^n\right) = \left(a\alpha_2 je^{j\theta_2} - a\omega_2^2 e^{j\theta_2}\right)$$

$$\mathbf{A}_{BA} = \left(\mathbf{A}_{BA}^t + \mathbf{A}_{BA}^n\right) = \left(b\alpha_3 je^{j\theta_3} - b\omega_3^2 e^{j\theta_3}\right) \tag{7.8b}$$

$$\mathbf{A}_B = \left(\mathbf{A}_B^t + \mathbf{A}_B^n\right) = \left(c\alpha_4 je^{j\theta_4} - c\omega_4^2 e^{j\theta_4}\right)$$

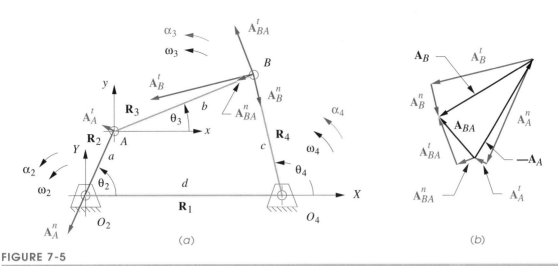

FIGURE 7-5

Position vector loop for a fourbar linkage showing acceleration vectors

The vector diagram in Figure 7-5b shows these components and is a graphical solution to equation 7.8a. The vector components are also shown acting at their respective points on Figure 7-5a.

We now need to solve equation 7.7 for α_3 and α_4, knowing the input angular acceleration α_2, the link lengths, all link angles, and angular velocities. Thus, the position analysis derived in Section 4.5 (p. 183) and the velocity analysis from Section 6.7 (p. 315) must be done first to determine the link angles and angular velocities before this acceleration analysis can be completed. We wish to solve equations 7.8 to get expressions in this form:

$$\alpha_3 = f\left(a, b, c, d, \theta_2, \theta_3, \theta_4, \omega_2, \omega_3, \omega_4, \alpha_2\right) \qquad (7.9a)$$

$$\alpha_4 = g\left(a, b, c, d, \theta_2, \theta_3, \theta_4, \omega_2, \omega_3, \omega_4, \alpha_2\right) \qquad (7.9b)$$

The strategy of solution will be the same as was done for the position and velocity analysis. First, substitute the Euler identity from equation 4.4a (p. 185) in each term of equation 7.7:

$$\left[a\alpha_2\, j\left(\cos\theta_2 + j\sin\theta_2\right) - a\omega_2^2\left(\cos\theta_2 + j\sin\theta_2\right)\right]$$
$$+\left[b\alpha_3\, j\left(\cos\theta_3 + j\sin\theta_3\right) - b\omega_3^2\left(\cos\theta_3 + j\sin\theta_3\right)\right] \qquad (7.10a)$$
$$-\left[c\alpha_4\, j\left(\cos\theta_4 + j\sin\theta_4\right) - c\omega_4^2\left(\cos\theta_4 + j\sin\theta_4\right)\right] = 0$$

Multiply by the operator j and rearrange:

$$\left[a\alpha_2\left(-\sin\theta_2 + j\cos\theta_2\right) - a\omega_2^2\left(\cos\theta_2 + j\sin\theta_2\right)\right]$$
$$+\left[b\alpha_3\left(-\sin\theta_3 + j\cos\theta_3\right) - b\omega_3^2\left(\cos\theta_3 + j\sin\theta_3\right)\right] \qquad (7.10b)$$
$$-\left[c\alpha_4\left(-\sin\theta_4 + j\cos\theta_4\right) - c\omega_4^2\left(\cos\theta_4 + j\sin\theta_4\right)\right] = 0$$

We can now separate this vector equation into its two components by collecting all real and all imaginary terms separately:

real part (x component):

$$-a\alpha_2 \sin\theta_2 - a\omega_2^2 \cos\theta_2 - b\alpha_3 \sin\theta_3 - b\omega_3^2 \cos\theta_3 + c\alpha_4 \sin\theta_4 + c\omega_4^2 \cos\theta_4 = 0 \qquad (7.11a)$$

imaginary part (y component):

$$a\alpha_2 \cos\theta_2 - a\omega_2^2 \sin\theta_2 + b\alpha_3 \cos\theta_3 - b\omega_3^2 \sin\theta_3 - c\alpha_4 \cos\theta_4 + c\omega_4^2 \sin\theta_4 = 0 \qquad (7.11b)$$

Note that the j's have canceled in equation 7.11b. We can solve equations 7.11a and 7.11b simultaneously to get:

$$\alpha_3 = \frac{CD - AF}{AE - BD} \qquad (7.12a)$$

$$\alpha_4 = \frac{CE - BF}{AE - BD} \qquad (7.12b)$$

where:

$$
\begin{aligned}
A &= c\sin\theta_4 \\
B &= b\sin\theta_3 \\
C &= a\alpha_2 \sin\theta_2 + a\omega_2^2 \cos\theta_2 + b\omega_3^2 \cos\theta_3 - c\omega_4^2 \cos\theta_4 \\
D &= c\cos\theta_4 \\
E &= b\cos\theta_3 \\
F &= a\alpha_2 \cos\theta_2 - a\omega_2^2 \sin\theta_2 - b\omega_3^2 \sin\theta_3 + c\omega_4^2 \sin\theta_4
\end{aligned}
\qquad (7.12c)
$$

Once we have solved for α_3 and α_4, we can then solve for the linear accelerations by substituting the Euler identity into equations 7.8b (p. 358),

$$\mathbf{A}_A = a\alpha_2 \left(-\sin\theta_2 + j\cos\theta_2\right) - a\omega_2^2 \left(\cos\theta_2 + j\sin\theta_2\right)$$
$$\mathbf{A}_{A_x} = -a\alpha_2 \sin\theta_2 - a\omega_2^2 \cos\theta_2 \qquad \mathbf{A}_{A_y} = a\alpha_2 \cos\theta_2 - a\omega_2^2 \sin\theta_2 \qquad (7.13a)$$

$$\mathbf{A}_{BA} = b\alpha_3 \left(-\sin\theta_3 + j\cos\theta_3\right) - b\omega_3^2 \left(\cos\theta_3 + j\sin\theta_3\right)$$
$$\mathbf{A}_{BA_x} = -b\alpha_3 \sin\theta_3 - b\omega_3^2 \cos\theta_3 \qquad \mathbf{A}_{BA_y} = b\alpha_3 \cos\theta_3 - b\omega_3^2 \sin\theta_3 \qquad (7.13b)$$

$$\mathbf{A}_B = c\alpha_4 \left(-\sin\theta_4 + j\cos\theta_4\right) - c\omega_4^2 \left(\cos\theta_4 + j\sin\theta_4\right)$$
$$\mathbf{A}_{B_x} = -c\alpha_4 \sin\theta_4 - c\omega_4^2 \cos\theta_4 \qquad \mathbf{A}_{B_y} = c\alpha_4 \cos\theta_4 - c\omega_4^2 \sin\theta_4 \qquad (7.13c)$$

where the real and imaginary terms are the x and y components, respectively. Equations 7.12 and 7.13 provide a complete solution for the angular accelerations of the links and the linear accelerations of the joints in the pin-jointed fourbar linkage.

EXAMPLE 7-2

Acceleration Analysis of a Fourbar Linkage with the Vector Loop Method.

Problem: Given a fourbar linkage with the link lengths $L_1 = d = 100$ mm, $L_2 = a = 40$ mm, $L_3 = b = 120$ mm, $L_4 = c = 80$ mm. For $\theta_2 = 40°$, $\omega_2 = 25$ rad/sec, and $\alpha_2 = 15$ rad/sec^2 find the values of α_3 and α_4, A_A, A_{BA}, and A_B for the open circuit of the linkage. Use the angles and angular velocities found for the same linkage and position in Example 6-7 (p. 317).

Solution: (See Figure 7-5, p. 359, for nomenclature.)

1   Example 4-1 (p. 190) found the link angles for the open circuit of this linkage in this position to be $\theta_3 = 20.298°$ and $\theta_4 = 57.325°$. Example 6-7 (p. 317) found the angular velocities at this position to be $\omega_3 = -4.121$ and $\omega_4 = 6.998$ rad/sec.

2   Use these angles, angular velocities, and equations 7.12 to find α_3 and α_4 for the open circuit. First find the parameters in equation 7.12c.

$$A = c\sin\theta_4 = 80\sin 57.325° = 67.340$$

$$B = b\sin\theta_3 = 120\sin 20.298° = 41.628$$

$$C = a\alpha_2 \sin\theta_2 + a\omega_2^2 \cos\theta_2 + b\omega_3^2 \cos\theta_3 - c\omega_4^2 \cos\theta_4$$

$$= 40(15)\sin 40° + 40(25)^2 \cos 40° + 120(-4.121)^2 \cos 20.298° - 80(6.998)^2 \cos 57.325°$$

$$= 19\,332.98$$

$$D = c\cos\theta_4 = 80\cos 57.325° = 43.190 \qquad\qquad (a)$$

$$E = b\cos\theta_3 = 120\cos 20.298° = 112.548$$

$$F = a\alpha_2 \cos\theta_2 - a\omega_2^2 \sin\theta_2 - b\omega_3^2 \sin\theta_3 + c\omega_4^2 \sin\theta_4$$

$$= 40(15)\cos 40° - 40(25)^2 \sin 40° - 120(-4.121)^2 \sin 20.298° + 80(6.998)^2 \sin 57.325°$$

$$= -13\,019.25$$

3   Then find α_3 and α_4 with equations 7.12a and b.

$$\alpha_3 = \frac{CD - AF}{AE - BD} = \frac{19\,332.98(43.190) - 67.340(-13\,019.25)}{67.340(112.548) - 41.628(43.190)} = 269.089 \text{ rad/sec}^2 \qquad (b)$$

$$\alpha_4 = \frac{CE - BF}{AE - BD} = \frac{19\,332.98(112.548) - 41.628(-13\,019.25)}{67.340(112.548) - 41.628(43.190)} = 470.134 \text{ rad/sec}^2 \qquad (c)$$

4   Use equations 7.13 to find the linear accelerations of points A and B.

$$\mathbf{A}_{A_x} = -a\alpha_2 \sin\theta_2 - a\omega_2^2 \cos\theta_2 = -40(15)\sin 40° - 40(25)^2 \cos 40° = -19.537 \text{ m/sec}^2$$
$$\mathbf{A}_{A_y} = a\alpha_2 \cos\theta_2 - a\omega_2^2 \sin\theta_2 = 40(15)\cos 40° - 40(25)^2 \sin 40° = -15.617 \text{ m/sec}^2 \qquad (d)$$

$$\mathbf{A}_{BA_x} = -b\alpha_3 \sin\theta_3 - b\omega_3^2 \cos\theta_3$$
$$= -120(269.089)\sin 20.298° - 120(-4.121)^2 \cos 20.298° = -14,237 \text{ m/sec}^2 \qquad (e)$$

$$\mathbf{A}_{BA_y} = b\alpha_3 \cos\theta_3 - b\omega_3^2 \sin\theta_3$$
$$= 120(269.089)\cos 20.298° - 120(-4.121)^2 \sin 20.298° = 32.617 \text{ m/sec}^2$$

$$\mathbf{A}_{B_x} = -c\alpha_4 \sin\theta_4 - c\omega_4^2 \cos\theta_4$$
$$= -80(470.134)\sin 57.325° - 80(6.998)^2 \cos 57.325° = -33.774 \text{ m/sec}^2 \qquad (f)$$

$$\mathbf{A}_{B_y} = c\alpha_4 \cos\theta_4 - c\omega_4^2 \sin\theta_4$$
$$= 80(470.134)\cos 57.325° - 80(6.998)^2 \sin 57.325° = 17.007 \text{ m/sec}^2$$

The Fourbar Crank-Slider

The first inversion of the offset crank-slider has its slider block sliding against the ground plane as shown in Figure 7-6a. Its accelerations can be solved for in similar manner as was done for the pin-jointed fourbar.

The position equations for the fourbar offset crank-slider linkage (inversion #1) were derived in Section 4.6 (p. 191). The linkage was shown in Figures 4-9 (p. 191) and 6-21 (p. 319) and is shown again in Figure 7-6a on which we also show an input angular acceleration α_2 applied to link 2. This α_2 can be a time-varying input acceleration. The vector loop equations 4.14 are repeated here for your convenience.

$$\mathbf{R}_2 - \mathbf{R}_3 - \mathbf{R}_4 - \mathbf{R}_1 = 0 \qquad (4.14a)$$

$$a e^{j\theta_2} - b e^{j\theta_3} - c e^{j\theta_4} - d e^{j\theta_1} = 0 \qquad (4.14b)$$

In Section 6.7 (p. 315) we differentiated equation 4.14b with respect to time noting that a, b, c, θ_1, and θ_4 are constant but the length of link d varies with time in this inversion.

$$ja\omega_2 e^{j\theta_2} - jb\omega_3 e^{j\theta_3} - \dot{d} = 0 \qquad (6.20a)$$

The term $\dot{d}$ is the linear velocity of the slider block. Equation 6.20a is the velocity difference equation.

We now will differentiate equation 6.20a with respect to time to get an expression for acceleration in this inversion of the crank-slider mechanism.

$$\left(ja\alpha_2 e^{j\theta_2} + j^2 a\omega_2^2 e^{j\theta_2} \right) - \left(jb\alpha_3 e^{j\theta_3} + j^2 b\omega_3^2 e^{j\theta_3} \right) - \ddot{d} = 0 \qquad (7.14a)$$

Simplifying:

$$\left(a\alpha_2 je^{j\theta_2} - a\omega_2^2 e^{j\theta_2} \right) - \left(b\alpha_3 je^{j\theta_3} - b\omega_3^2 e^{j\theta_3} \right) - \ddot{d} = 0 \qquad (7.14b)$$

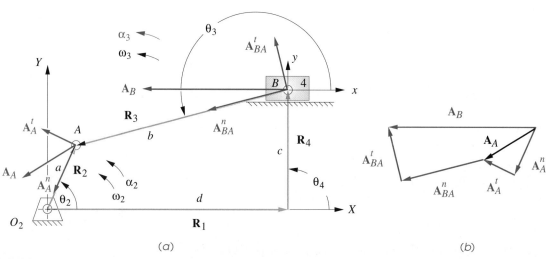

(a)   (b)

FIGURE 7-6

Position vector loop for a fourbar slider-crank linkage showing acceleration vectors

Note that equation 7.14 is again the acceleration difference equation:

$$\mathbf{A}_A - \mathbf{A}_{AB} - \mathbf{A}_B = 0$$
$$\mathbf{A}_{BA} = -\mathbf{A}_{AB} \qquad\qquad (7.15a)$$
$$\mathbf{A}_B = \mathbf{A}_A + \mathbf{A}_{BA}$$

$$\mathbf{A}_A = \left(\mathbf{A}_A^t + \mathbf{A}_A^n\right) = \left(a\alpha_2\, je^{j\theta_2} - a\omega_2^2\, e^{j\theta_2}\right)$$
$$\mathbf{A}_{BA} = \left(\mathbf{A}_{BA}^t + \mathbf{A}_{BA}^n\right) = \left(b\alpha_3\, je^{j\theta_3} - b\omega_3^2\, e^{j\theta_3}\right) \qquad\qquad (7.15b)$$
$$\mathbf{A}_B = \mathbf{A}_B^t = \ddot{d}$$

In this mechanism, link 4 is in pure translation and so has zero ω_4 and zero α_4. The acceleration of link 4 has only a "tangential" component of acceleration along its path.

The two unknowns in the vector equation 7.14 are the angular acceleration of link 3, α_3, and the linear acceleration of link 4, $\ddot{d}$. To solve for them, substitute the Euler identity,

$$a\alpha_2\left(-\sin\theta_2 + j\cos\theta_2\right) - a\omega_2^2\left(\cos\theta_2 + j\sin\theta_2\right)$$
$$- b\alpha_3\left(-\sin\theta_3 + j\cos\theta_3\right) + b\omega_3^2\left(\cos\theta_3 + j\sin\theta_3\right) - \ddot{d} = 0 \qquad\qquad (7.16a)$$

and separate the real (x) and imaginary (y) components:

real part (x component):

$$-a\alpha_2\sin\theta_2 - a\omega_2^2\cos\theta_2 + b\alpha_3\sin\theta_3 + b\omega_3^2\cos\theta_3 - \ddot{d} = 0 \qquad\qquad (7.16b)$$

imaginary part (y component):

$$a\alpha_2\cos\theta_2 - a\omega_2^2\sin\theta_2 - b\alpha_3\cos\theta_3 + b\omega_3^2\sin\theta_3 = 0 \qquad\qquad (7.16c)$$

Equation 7.16c can be solved directly for α_3 and the result substituted in equation 7.16b to find $\ddot{d}$.

$$\alpha_3 = \frac{a\alpha_2\cos\theta_2 - a\omega_2^2\sin\theta_2 + b\omega_3^2\sin\theta_3}{b\cos\theta_3} \qquad\qquad (7.16d)$$

$$\ddot{d} = -a\alpha_2\sin\theta_2 - a\omega_2^2\cos\theta_2 + b\alpha_3\sin\theta_3 + b\omega_3^2\cos\theta_3 \qquad\qquad (7.16e)$$

The other linear accelerations can be found from equation 7.15b and are shown in the vector diagram of Figure 7-6b.

EXAMPLE 7-3

Acceleration Analysis of a Fourbar Crank-Slider Linkage with a Vector Loop Method.

Problem: Given a fourbar crank-slider linkage with the link lengths $L_2 = a = 40$ mm, $L_3 = b = 120$ mm, *offset* $= c = -20$ mm. For $\theta_2 = 60°$, $\omega_2 = -30$ rad/sec, and $\alpha_2 = 20$ rad/sec^2, find α_3 and linear acceleration of the slider for the open circuit. Use the angles, positions, and angular velocities found for the same linkage in Examples 4-2 (p. 193) and 6-8 (p. 320).

Solution:　　(See Figure 7-6, p. 362, for nomenclature.)

1　Example 4-2 (p. 193) found angle $\theta_3 = 152.91°$ and slider position $d = 126.84$ mm for the open circuit. Example 6-8 (p. 320) found the the coupler angular velocity ω_3 to be 5.616 rad/sec.

2　Using equation 7.16d (p. 363) and the data from step 1, calculate the coupler angular acceleration α_3.

$$\alpha_3 = \frac{a\alpha_2 \cos\theta_2 - a\omega_2^2 \sin\theta_2 + b\omega_3^2 \sin\theta_3}{b\cos\theta_3}$$

$$= \frac{40(20)\cos 60° - 40(-30)^2 \sin 60° + 120(5.616)^2 \sin 152.91°}{120\cos 152.91°} = 271.94 \text{ rad/sec}^2 \quad (a)$$

3　Using equation 7.16e and the data from steps 1 and 3, calculate the slider acceleration $\ddot{d}$.

$$\ddot{d} = -a\alpha_2 \sin\theta_2 - a\omega_2^2 \cos\theta_2 + b\alpha_3 \sin\theta_3 + b\omega_3^2 \cos\theta_3$$

$$= -40(20)\sin 60° - 40(-30)^2 \cos 60° + 120(271.94)\sin 152.91° + 120(5.616)^2 \cos 152.91°$$

$$= -7.203 \text{ m/sec}^2 \quad (b)$$

---

* The crank-slider and slider-crank linkage both have two circuits or configurations in which they can be independently assembled, sometimes called open and crossed. Because effective link 4 is always perpendicular to the slider axis, it is parallel to itself on both circuits. This results in the two circuits being mirror images of one another, mirrored about a line through the crank pivot and perpendicular to the slide axis. Thus, the choice of value of slider position d in the calculation of the slider-crank linkage determines which circuit is being analyzed. But, because of the change points at TDC and BDC, the slider-crank has two branches on each circuit and the two solutions obtained from equation 4.21 represent the two branches on the one circuit being analyzed. In contrast, the crank-slider has only one branch per circuit because when the crank is driven, it can make a full revolution and there are no change points to separate branches. See Section 4.13 (p. 207) for a more complete discussion of circuits and branches in linkages.

The Fourbar Slider-Crank

The *fourbar* **slider-crank** *linkage* has the same geometry as the *fourbar* **crank-slider** *linkage* that was analyzed in the previous section. The name change indicates that it will be driven with the slider as input and the crank as output. This is sometimes referred to as a "back-driven" crank-slider. We will use the term **slider-crank** to define it as slider-driven. This is a very commonly used linkage configuration. Every internal-combustion, piston engine has as many of these as it has cylinders. The vector loop is as shown in Figure 7-6 (p. 362) and the vector loop equation is identical to that of the crank-slider (equation 4.14a). The derivation for θ_2 and ω_2 as a function of slider position d and slider velocity $\dot{d}$ were done, respectively, in Sections 4-7 (p. 194) and 6-7 (p. 321). Now we want to solve for α_2 and α_3 as a function of slider acceleration $\ddot{d}$ and the known lengths, angles, and angular velocities of the links.

We can start with equations 7.16b and c, which also apply to this linkage:

$$-a\alpha_2 \sin\theta_2 - a\omega_2^2 \cos\theta_2 + b\alpha_3 \sin\theta_3 + b\omega_3^2 \cos\theta_3 - \ddot{d} = 0 \qquad (7.16b)$$

$$a\alpha_2 \cos\theta_2 - a\omega_2^2 \sin\theta_2 - b\alpha_3 \cos\theta_3 + b\omega_3^2 \sin\theta_3 = 0 \qquad (7.16c)$$

Solve equation 7.16c for α_3 in terms of α_2.

$$\alpha_3 = \frac{a\alpha_2 \cos\theta_2 - a\omega_2^2 \sin\theta_2 + b\omega_3^2 \sin\theta_3}{b\cos\theta_3} \qquad (7.17a)$$

Substitute equation 7.17a for α_3 in equation 7.16b and solve for α_2.

$$\alpha_2 = \frac{a\omega_2^2\left(\cos\theta_2 \cos\theta_3 + \sin\theta_2 \sin\theta_3\right) - b\omega_3^2 + \ddot{d}\cos\theta_3}{a\left(\cos\theta_2 \sin\theta_3 - \sin\theta_2 \cos\theta_3\right)} \qquad (7.17b)$$

The circuit of the linkage depends on the value of d chosen and the angular accelerations will be for the branch represented by the values of θ_2 and θ_3 used from equation 4.21.*

✎□EXAMPLE 7-4

Acceleration Analysis of a Fourbar Slider-Crank Linkage with a Vector Loop Method.

Problem: Given a fourbar slider-crank linkage with the link lengths $L_2 = a = 40$ mm, $L_3 = b = 120$ mm, *offset* $= c = -20$ mm. For $d = 100$ mm and $\ddot{d} = 900$ mm/sec^2, find α_2 and α_3 for both branches of one circuit of the linkage. Use the angles and angular velocities found for the same linkage in Example 4-3 (p. 196) and Example 6-9 (p. 319), respectively.

Solution: (See Figure 7-6 for nomenclature.)

1   Example 4-3 (p. 196) found angles $\theta_{2_1} = 95.80°$, $\theta_{3_1} = 150.11°$ for branch 1 of this linkage. Example 6-9 (p. 321) found the the angular velocities to be $\omega_{2_1} = -32.023$ and $\omega_{3_1} = -1.244$ rad/sec for branch 1.

2   Using equation 7.17b and the data from step 1, calculate the crank angular acceleration α_{2_1}.

$$\alpha_{2_1} = \frac{a\omega_{2_1}^2\left(\cos\theta_{2_1}\,\cos\theta_{3_1} + \sin\theta_{2_1}\,\sin\theta_{3_1}\right) - b\omega_{3_1}^2 + \ddot{d}\cos\theta_{3_1}}{a\left(\cos\theta_{2_1}\,\sin\theta_{3_1} - \sin\theta_{2_1}\,\cos\theta_{3_1}\right)}$$

$$\alpha_{2_1} = \frac{40(-32.023)^2\left(\cos 95.80°\cos 150.11° + \sin 95.80°\sin 150.11°\right) - 120(-1.244)^2 + 900\cos 150.11°}{40\left(\cos 95.80°\sin 150.11° - \sin 95.80°\cos 150.11°\right)}$$

$$\alpha_{2_1} = 706.753 \text{ rad/sec}^2 \qquad\qquad (a)$$

3   Using equation 7.17a and the data from steps 1 and 2, calculate the coupler angular acceleration α_{3_1}.

$$\alpha_{3_1} = \frac{a\alpha_{2_1}\cos\theta_{2_1} - a\omega_{2_1}^2\sin\theta_2 + b\omega_{3_1}^2\sin\theta_{3_1}}{b\cos\theta_{3_1}}$$

$$\alpha_{3_1} = \frac{40(-706.753)\cos 95.80° - 40(-32.023)^2\sin 95.80° + 120(-1.244)^2\sin 150.11°}{120\cos 150.11°}$$

$$\alpha_{3_1} = 418.804 \text{ rad/sec}^2 \qquad\qquad (b)$$

4   Example 4-3 found angles $\theta_{2_2} = -118.42°$, $\theta_{3_2} = 187.27°$ for branch 2 of this linkage. Example 6-9 (p. 321) found the the angular velocities to be $\omega_{2_2} = 36.64$ and $\omega_{3_2} = 5.86$ rad/sec for branch 2. Using equation 7.17b and the data from step 3, calculate the crank angular acceleration α_{2_2} for branch 2.

$$\alpha_{2_2} = \frac{a\omega_{2_2}^2\left(\cos\theta_{2_2}\,\cos\theta_{3_2} + \sin\theta_{2_2}\,\sin\theta_{3_2}\right) - b\omega_{3_2}^2 + \ddot{d}\cos\theta_{3_2}}{a\left(\cos\theta_{2_2}\,\sin\theta_{3_2} - \sin\theta_{2_2}\,\cos\theta_{3_2}\right)}$$

$$\alpha_{2_2} = \frac{40(36.64)^2\left(\cos -118.42°\cos 187.27° + \sin -118.42°\sin 187.27°\right) - 120(5.86)^2 + 900\cos 187.27°}{40\left[\cos(-118.42°)\sin 187.27° - \sin(-118.42°)\cos 187.27°\right]}$$

$$\alpha_{2_2} = -809.801 \text{ rad/sec}^2 \qquad\qquad (c)$$

7

5   Using equation 7.17a (p. 364) and the data from steps 3 and 4, calculate the coupler angular acceleration α_{3_2}.

$$\alpha_{3_2} = \frac{a\alpha_{2_2}\cos\theta_{2_2} - a\omega_{2_2}^2\sin\theta_2 + b\omega_{3_2}^2\sin\theta_{3_2}}{b\cos\theta_{3_2}}$$

$$\alpha_{3_2} = \frac{40(-809.801)\cos-118.42° - 40(36.64)^2\sin-118.42° + 120(5.859)^2\sin187.27°}{120\cos187.27°}$$

$$\alpha_{3_2} = -521.852 \text{ rad/sec}^2 \tag{d}$$

Coriolis Acceleration

The examples used for acceleration analysis above have involved only pin-jointed linkages or the inversion of the crank-slider in which the slider block has no rotation. When a sliding joint is present on a rotating link, an additional component of acceleration will be present, called the **Coriolis component**, after its discoverer. Figure 7-7a shows a simple, two-link system consisting of a link with a radial slot, and a slider block free to slip within that slot.

The instantaneous location of the block is defined by a position vector ($\mathbf{R}_P$) referenced to the global origin at the link center. *This vector is both rotating and changing length as the system moves.* As shown this is a two-degree-of-freedom system. The **two inputs to the system** are the angular acceleration (α) of the link and the relative linear slip velocity ($\mathbf{V}_{Pslip}$) of the block versus the disk. The angular velocity ω is a result of the time history of the angular acceleration. The situation shown, with a counterclockwise α and a clockwise ω, implies that earlier in time the link had been accelerated up to a clockwise angular velocity and is now being slowed down. The transmission component of velocity ($\mathbf{V}_{Ptrans}$) is a result of the ω of the link acting at the radius $\mathbf{R}_P$ whose magnitude is p.

We show the situation in Figure 7-7 at one instant of time. However, the equations to be derived will be valid for all time. We want to determine the acceleration at the center of the block (P) under this combined motion of rotation and sliding. To do so we first write the expression for the position vector $\mathbf{R}_P$ that locates point P.

$$\mathbf{R}_P = pe^{j\theta_2} \tag{7.18a}$$

Note that there are two functions of time in equation 7.17, p and θ. When we differentiate versus time we get two terms in the velocity expression:

$$\mathbf{V}_P = p\omega_2 je^{j\theta_2} + \dot{p}e^{j\theta_2} \tag{7.18b}$$

These are the transmission component and the slip component of velocity.

$$\mathbf{V}_P = \mathbf{V}_{P_{trans}} + \mathbf{V}_{P_{slip}} \tag{7.18c}$$

The $p\omega$ term is the transmission component and is directed at 90 degrees to the axis of slip that, in this example, is coincident with the position vector $\mathbf{R}_P$. The $\dot{p}$ term is the **slip component** and is directed along the **axis of slip** in the same direction as the position vector in this example. Their vector sum is $\mathbf{V}_P$ as shown in Figure 7-7a.

To get an expression for acceleration, we must differentiate equation 7.18 versus time. Note that the transmission component has **three** functions of time in it, p, ω, and θ.

The chain rule will yield three terms for this one term. The slip component of velocity contains two functions of time, p and θ, yielding two terms in the derivative for a total of five terms, two of which turn out to be the same.

$$\mathbf{A}_P = \left(p\alpha_2 je^{j\theta_2} + p\omega_2^2 j^2 e^{j\theta_2} + \dot{p}\omega_2 je^{j\theta_2} \right) + \left(\dot{p}\omega_2 je^{j\theta_2} + \ddot{p}e^{j\theta_2} \right) \qquad (7.19a)$$

Simplifying and collecting terms:

$$\mathbf{A}_P = p\alpha_2 je^{j\theta_2} - p\omega_2^2 e^{j\theta_2} + 2\dot{p}\omega_2 je^{j\theta_2} + \ddot{p}e^{j\theta_2} \qquad (7.19b)$$

These terms represent the following components:

$$\mathbf{A}_P = \mathbf{A}_{P_{\text{tangential}}} + \mathbf{A}_{P_{\text{normal}}} + \mathbf{A}_{P_{\text{coriolis}}} + \mathbf{A}_{P_{\text{slip}}} \qquad (7.19c)$$

Note that the Coriolis term has appeared in the acceleration expression as a result of the differentiation simply because the length of the vector p is a function of time. The Coriolis component magnitude is twice the product of the velocity of slip (equation 7.18) and the angular velocity of the link containing the slider slot. Its direction is rotated 90 degrees from that of the original position vector $\mathbf{R}_P$ either clockwise or counterclockwise depending on the sense of ω.[*] (Note that we chose to align the position vector $\mathbf{R}_P$ with the axis of slip in Figure 7-7 which can always be done regardless of the location of the center of rotation—also see Figure 7-6 (p. 362) where $\mathbf{R}_1$ is aligned with the axis of slip.) All four components from equations 7.19 are shown acting on point P in Figure 7-7b. The total acceleration $\mathbf{A}_P$ is the vector sum of the four terms as shown in Figure 7-7c. Note that the normal acceleration term in equation 7.19b is negative in sign, so it becomes a subtraction when substituted in equation 7.19c.

This Coriolis component of acceleration will always be present when there is a velocity of slip associated with any member that also has an angular velocity. In the absence of either of those two factors the Coriolis component will be zero. You have probably experienced Coriolis acceleration if you have ever ridden on a carousel or merry-go-round. If you attempted to walk radially from the outside to the inside (or vice versa)

[*] This approach works in the 2-D case. Coriolis acceleration is the cross product of 2ω and the velocity of slip. The cross product operation will define its magnitude, direction, and sense in the 3-D case.

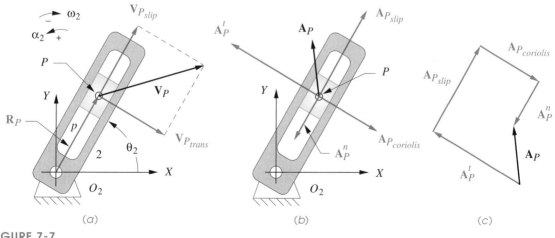

(a)   (b)   (c)

FIGURE 7-7

The Coriolis component of acceleration shown in a system with a positive (CCW) α_2 and a negative (CW) ω_2

while the carousel was turning, you were thrown sideways by the inertial force due to the Coriolis acceleration. You were the *slider block* in Figure 7-7 (p. 367), and your *slip velocity* combined with the rotation of the carousel created the Coriolis component. As you walked from a large radius to a smaller one, your tangential velocity had to change to match that of the new location of your foot on the spinning carousel. Any change in velocity requires an acceleration to accomplish. It was the *"ghost of Coriolis"* that pushed you sideways on that carousel.

Another example of the Coriolis component is its effect on weather systems. Large objects that exist in the earth's lower atmosphere, such as hurricanes, span enough area to be subject to significantly different velocities at their northern and southern extremities. The atmosphere turns with the earth. The earth's surface tangential velocity due to its angular velocity varies from zero at the poles to a maximum of about 1000 mph at the equator. The winds of a storm system are attracted toward the low pressure at its center. These winds have a slip velocity with respect to the surface, which in combination with the earth's ω creates a Coriolis component of acceleration on the moving air masses. This Coriolis acceleration causes the inrushing air to rotate about the center, or "eye" of the storm system. This rotation will be counterclockwise in the northern hemisphere and clockwise in the southern hemisphere. The movement of the entire storm system from south to north also creates a Coriolis component that will tend to deviate the storm's track eastward, though this effect is often overridden by the forces due to other large air masses such as high-pressure systems that can deflect a storm. These complicated factors make it difficult to predict a large storm's true track.

Note that in the analytical solution presented here, the Coriolis component will be accounted for automatically as long as the differentiations are correctly done. However, when doing a graphical acceleration analysis one must be on the alert to recognize the presence of this component, calculate it, and include it in the vector diagrams when its two constituents $\mathbf{V}_{slip}$ and ω are both nonzero.

The Fourbar Inverted Crank-Slider

The position equations for the fourbar inverted crank-slider linkage were derived in Section 4.7 (p. 194). The linkage was shown in Figures 4-10 (p. 192) and 6-22 (p. 323) and is shown again in Figure 7-8a (p. 369) on which we also show an input angular acceleration α_2 applied to link 2. This α_2 can vary with time. The vector loop equations 4.14 (p. 362) are valid for this linkage as well.

All slider linkages will have at least one link whose effective length between joints varies as the linkage moves. In this inversion the length of link 3 between points A and B, designated as b, will change as it passes through the slider block on link 4. In Section 6.7 (p. 315) we got an expression for velocity, by differentiating equation 4.14b with respect to time noting that a, c, d, and θ_1 are constant and b, θ_3, and θ_4 vary with time.

$$ja\omega_2 e^{j\theta_2} - jb\omega_3 e^{j\theta_3} - \dot{b}e^{j\theta_3} - jc\omega_4 e^{j\theta_4} = 0 \qquad (6.25a)$$

Differentiating this with respect to time will give an expression for accelerations in this inversion of the crank-slider mechanism.

$$\left(ja\alpha_2 e^{j\theta_2} + j^2 a\omega_2^2 e^{j\theta_2}\right) - \left(jb\alpha_3 e^{j\theta_3} + j^2 b\omega_3^2 e^{j\theta_3} + j\dot{b}\omega_3 e^{j\theta_3}\right)$$
$$- \left(\ddot{b}e^{j\theta_3} + j\dot{b}\omega_3 e^{j\theta_3}\right) - \left(jc\alpha_4 e^{j\theta_4} + j^2 c\omega_4^2 e^{j\theta_4}\right) = 0 \qquad (7.20a)$$

Simplifying and collecting terms:

$$\left(a\alpha_2\, je^{j\theta_2} - a\omega_2^2 e^{j\theta_2}\right) - \left(b\alpha_3\, je^{j\theta_3} - b\omega_3^2 e^{j\theta_3} + 2\dot{b}\omega_3\, je^{j\theta_3} + \ddot{b}e^{j\theta_3}\right)$$
$$- \left(c\alpha_4\, je^{j\theta_4} - c\omega_4^2 e^{j\theta_4}\right) = 0 \qquad (7.20b)$$

Equation 7.20 is in fact the acceleration difference equation (equation 7.4, p. 352) and can be written in that notation as shown in equations 7.21.

$$\mathbf{A}_A - \mathbf{A}_{AB} - \mathbf{A}_B = 0$$

but:
$$\mathbf{A}_{BA} = -\mathbf{A}_{AB} \qquad (7.21a)$$

and:
$$\mathbf{A}_B = \mathbf{A}_A + \mathbf{A}_{BA}$$

$$\mathbf{A}_A = \mathbf{A}_{A_{\text{tangential}}} + \mathbf{A}_{A_{\text{normal}}}$$

$$\mathbf{A}_{AB} = \mathbf{A}_{AB_{\text{tangential}}} + \mathbf{A}_{AB_{\text{normal}}} + \mathbf{A}_{AB_{\text{coriolis}}} + \mathbf{A}_{AB_{\text{slip}}} \qquad (7.21b)$$

$$\mathbf{A}_B = \mathbf{A}_{B_{\text{tangential}}} + \mathbf{A}_{B_{\text{normal}}}$$

$$\mathbf{A}_{A_{\text{tangential}}} = a\alpha_2\, je^{j\theta_2} \qquad\qquad \mathbf{A}_{A_{\text{normal}}} = -a\omega_2^2 e^{j\theta_2}$$

$$\mathbf{A}_{B_{\text{tangential}}} = c\alpha_4\, je^{j\theta_4} \qquad\qquad \mathbf{A}_{B_{\text{normal}}} = -c\omega_4^2 e^{j\theta_4}$$

$$\mathbf{A}_{AB_{\text{tangential}}} = b\alpha_3\, je^{j\theta_3} \qquad\qquad \mathbf{A}_{AB_{\text{normal}}} = -b\omega_3^2 e^{j\theta_3} \qquad (7.21c)$$

$$\mathbf{A}_{AB_{\text{coriolis}}} = 2\dot{b}\omega_3\, je^{j\theta_3} \qquad\qquad \mathbf{A}_{AB_{\text{slip}}} = \ddot{b}e^{j\theta_3}$$

Because this sliding link also has an angular velocity, there will be a nonzero Coriolis component of acceleration at point B which is the $2\dot{b}$ term in equation 7.20. Since a complete velocity analysis was done before doing this acceleration analysis, the Coriolis component can be readily calculated at this point, knowing both ω and $\mathbf{V}_{slip}$ from the velocity analysis.

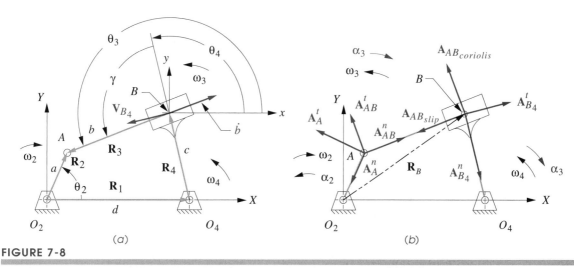

FIGURE 7-8

Acceleration analysis of inversion #3 of the fourbar slider-crank driven with a positive (CCW) α_2 and a negative (CW) ω_2

The $\ddot{b}$ term in equations 7.20b and 7.21c (p. 369) is the *slip component of accelera-tion*. This is one of the variables to be solved for in this acceleration analysis. Another variable to be solved for is α_4, the angular acceleration of link 4. Note, however, that we also have an unknown in α_3, the angular acceleration of link 3. This is a total of three unknowns. Equation 7.20 can only be solved for two unknowns. Thus we require another equation to solve the system. There is a fixed relationship between angles θ_3 and θ_4, shown as γ in Figure 7-8 (p. 369) and defined in equation 4.22, repeated here:

$$\theta_3 = \theta_4 + \gamma \tag{4.22}$$

Differentiate it twice with respect to time to obtain:

$$\omega_3 = \omega_4; \qquad\qquad \alpha_3 = \alpha_4 \tag{7.22}$$

We wish to solve equation 7.20 to get expressions in this form:

$$\alpha_3 = \alpha_4 = f\left(a, b, \dot{b}, c, d, \theta_2, \theta_3, \theta_4, \omega_2, \omega_3, \omega_4, \alpha_2\right) \tag{7.23a}$$

$$\frac{d^2 b}{dt^2} = \ddot{b} = g\left(a, b, \dot{b}, c, d, \theta_2, \theta_3, \theta_4, \omega_2, \omega_3, \omega_4, \alpha_2\right) \tag{7.23b}$$

Substitution of the Euler identity (equation 4.4a, p. 185) into equation 7.20 yields:

$$
\begin{aligned}
a\alpha_2 \, j\left(\cos\theta_2 + j\sin\theta_2\right) &- a\omega_2^2\left(\cos\theta_2 + j\sin\theta_2\right) \\
- b\alpha_3 \, j\left(\cos\theta_3 + j\sin\theta_3\right) &+ b\omega_3^2\left(\cos\theta_3 + j\sin\theta_3\right) \\
- 2\dot{b}\omega_3 \, j\left(\cos\theta_3 + j\sin\theta_3\right) &- \ddot{b}\left(\cos\theta_3 + j\sin\theta_3\right) \\
- c\alpha_4 \, j\left(\cos\theta_4 + j\sin\theta_4\right) &+ c\omega_4^2\left(\cos\theta_4 + j\sin\theta_4\right) = 0
\end{aligned}
\tag{7.24a}
$$

Multiply by the operator j and substitute α_4 for α_3 from equation 7.22:

$$
\begin{aligned}
a\alpha_2 \left(-\sin\theta_2 + j\cos\theta_2\right) &- a\omega_2^2\left(\cos\theta_2 + j\sin\theta_2\right) \\
- b\alpha_4 \left(-\sin\theta_3 + j\cos\theta_3\right) &+ b\omega_3^2\left(\cos\theta_3 + j\sin\theta_3\right) \\
- 2\dot{b}\omega_3 \left(-\sin\theta_3 + j\cos\theta_3\right) &- \ddot{b}\left(\cos\theta_3 + j\sin\theta_3\right) \\
- c\alpha_4 \left(-\sin\theta_4 + j\cos\theta_4\right) &+ c\omega_4^2\left(\cos\theta_4 + j\sin\theta_4\right) = 0
\end{aligned}
\tag{7.24b}
$$

We can now separate this vector equation 7.24b into its two components by collecting all real and all imaginary terms separately:

real part (x component):

$$
\begin{aligned}
-a\alpha_2 \sin\theta_2 - a\omega_2^2 \cos\theta_2 &+ b\alpha_4 \sin\theta_3 + b\omega_3^2 \cos\theta_3 \\
&+ 2\dot{b}\omega_3 \sin\theta_3 - \ddot{b}\cos\theta_3 + c\alpha_4 \sin\theta_4 + c\omega_4^2 \cos\theta_4 = 0
\end{aligned}
\tag{7.25a}
$$

imaginary part (y component):

$$
\begin{aligned}
a\alpha_2 \cos\theta_2 - a\omega_2^2 \sin\theta_2 &- b\alpha_4 \cos\theta_3 + b\omega_3^2 \sin\theta_3 \\
&- 2\dot{b}\omega_3 \cos\theta_3 - \ddot{b}\sin\theta_3 - c\alpha_4 \cos\theta_4 + c\omega_4^2 \sin\theta_4 = 0
\end{aligned}
\tag{7.25b}
$$

Note that the j's have canceled in equation 7.25b. We can solve equations 7.25 simultaneously for the two unknowns, α_4 and $\ddot{b}$. The solution is:

$$\alpha_4 = \frac{a\left[\alpha_2 \cos(\theta_3 - \theta_2) + \omega_2^2 \sin(\theta_3 - \theta_2)\right] + c\omega_4^2 \sin(\theta_4 - \theta_3) - 2\dot{b}\omega_3}{b + c\cos(\theta_3 - \theta_4)} \tag{7.26a}$$

$$\ddot{b} = -\frac{\left\{\begin{array}{c} a\omega_2^2\left[b\cos(\theta_3 - \theta_2) + c\cos(\theta_4 - \theta_2)\right] + a\alpha_2\left[b\sin(\theta_2 - \theta_3) - c\sin(\theta_4 - \theta_2)\right] \\ + 2\dot{b}c\omega_4 \sin(\theta_4 - \theta_3) - \omega_4^2\left[b^2 + c^2 + 2bc\cos(\theta_4 - \theta_3)\right] \end{array}\right\}}{b + c\cos(\theta_3 - \theta_4)} \tag{7.26b}$$

Equation 7.26a provides the **angular acceleration** of link 4. Equation 7.26b provides the **acceleration of slip** at point B. Once these variables are solved for, the linear accelerations at points A and B in the linkage of Figure 7-8 (p. 369) can be found by substituting the Euler identity into equations 7.21 (p. 369).

$$\mathbf{A}_A = a\alpha_2\left(-\sin\theta_2 + j\cos\theta_2\right) - a\omega_2^2\left(\cos\theta_2 + j\sin\theta_2\right) \tag{7.27a}$$

$$\mathbf{A}_{BA} = b\alpha_3\left(\sin\theta_3 - j\cos\theta_3\right) + b\omega_3^2\left(\cos\theta_3 + j\sin\theta_3\right)$$
$$+ 2\dot{b}\omega_3\left(\sin\theta_3 - j\cos\theta_3\right) - \ddot{b}\left(\cos\theta_3 + j\sin\theta_3\right) \tag{7.27b}$$

$$\mathbf{A}_B = -c\alpha_4\left(\sin\theta_4 - j\cos\theta_4\right) - c\omega_4^2\left(\cos\theta_4 + j\sin\theta_4\right) \tag{7.27c}$$

These components of these vectors are shown in Figure 7-8b.

7.4 ACCELERATION ANALYSIS OF THE GEARED FIVEBAR LINKAGE

The velocity equation for the geared fivebar mechanism was derived in Section 6.8 (p. 324) and is repeated here. See Figure P7-4 (p. 383) for notation.

$$a\omega_2 je^{j\theta_2} + b\omega_3 je^{j\theta_3} - c\omega_4 je^{j\theta_4} - d\omega_5 je^{j\theta_5} = 0 \tag{6.32a}$$

Differentiate this with respect to time to get an expression for acceleration.

$$\left(a\alpha_2 je^{j\theta_2} - a\omega_2^2 e^{j\theta_2}\right) + \left(b\alpha_3 je^{j\theta_3} - b\omega_3^2 e^{j\theta_3}\right)$$
$$-\left(c\alpha_4 je^{j\theta_4} - c\omega_4^2 e^{j\theta_4}\right) - \left(d\alpha_5 je^{j\theta_5} - d\omega_5^2 e^{j\theta_5}\right) = 0 \tag{7.28a}$$

Substitute the Euler equivalents:

$$a\alpha_2\left(-\sin\theta_2 + j\cos\theta_2\right) - a\omega_2^2\left(\cos\theta_2 + j\sin\theta_2\right)$$
$$+ b\alpha_3\left(-\sin\theta_3 + j\cos\theta_3\right) - b\omega_3^2\left(\cos\theta_3 + j\sin\theta_3\right)$$
$$- c\alpha_4\left(-\sin\theta_4 + j\cos\theta_4\right) + c\omega_4^2\left(\cos\theta_4 + j\sin\theta_4\right)$$
$$- d\alpha_5\left(-\sin\theta_5 + j\cos\theta_5\right) + d\omega_5^2\left(\cos\theta_5 + j\sin\theta_5\right) = 0 \tag{7.28b}$$

Note that the angle θ_5 is defined in terms of θ_2, the gear ratio λ, and the phase angle ϕ. This relationship and its derivatives are:

$$\theta_5 = \lambda\theta_2 + \phi; \qquad \omega_5 = \lambda\omega_2; \qquad \alpha_5 = \lambda\alpha_2 \tag{7.28c}$$

Since a complete position and velocity analysis must be done before an acceleration analysis, we will assume that the values of θ_5 and ω_5 have been found and will leave these equations in terms of θ_5, ω_5, and α_5.

Separating the real and imaginary terms in equation 7.28b:

real:

$$-a\alpha_2 \sin\theta_2 - a\omega_2^2 \cos\theta_2 - b\alpha_3 \sin\theta_3 - b\omega_3^2 \cos\theta_3$$
$$+ c\alpha_4 \sin\theta_4 + c\omega_4^2 \cos\theta_4 + d\alpha_5 \sin\theta_5 + d\omega_5^2 \cos\theta_5 = 0 \qquad (7.28d)$$

imaginary:

$$a\alpha_2 \cos\theta_2 - a\omega_2^2 \sin\theta_2 + b\alpha_3 \cos\theta_3 - b\omega_3^2 \sin\theta_3$$
$$- c\alpha_4 \cos\theta_4 + c\omega_4^2 \sin\theta_4 - d\alpha_5 \cos\theta_5 + d\omega_5^2 \sin\theta_5 = 0 \qquad (7.28e)$$

The only two unknowns are α_3 and α_4. Either equation 7.28d or 7.28e can be solved for one unknown and the result substituted in the other. The solution for α_3 is:

$$\alpha_3 = \frac{\begin{bmatrix} -a\alpha_2 \sin(\theta_2 - \theta_4) - a\omega_2^2 \cos(\theta_2 - \theta_4) \\ -b\omega_3^2 \cos(\theta_3 - \theta_4) + d\omega_5^2 \cos(\theta_5 - \theta_4) \\ +d\alpha_5 \sin(\theta_5 - \theta_4) + c\omega_4^2 \end{bmatrix}}{b\sin(\theta_3 - \theta_4)} \qquad (7.29a)$$

and angle α_4 is:

$$\alpha_4 = \frac{\begin{bmatrix} a\alpha_2 \sin(\theta_2 - \theta_3) + a\omega_2^2 \cos(\theta_2 - \theta_3) \\ -c\omega_4^2 \cos(\theta_3 - \theta_4) - d\omega_5^2 \cos(\theta_3 - \theta_5) \\ +d\alpha_5 \sin(\theta_3 - \theta_5) + b\omega_3^2 \end{bmatrix}}{c\sin(\theta_4 - \theta_3)} \qquad (7.29b)$$

With all link angles, angular velocities, and angular accelerations known, the linear accelerations for the pin joints can be found from:

$$\mathbf{A}_A = a\alpha_2 \left(-\sin\theta_2 + j\cos\theta_2\right) - a\omega_2^2 \left(\cos\theta_2 + j\sin\theta_2\right) \qquad (7.29c)$$

$$\mathbf{A}_{BA} = b\alpha_3 \left(-\sin\theta_3 + j\cos\theta_3\right) - b\omega_3^2 \left(\cos\theta_3 + j\sin\theta_3\right) \qquad (7.29d)$$

$$\mathbf{A}_C = c\alpha_5 \left(-\sin\theta_5 + j\cos\theta_5\right) - c\omega_5^2 \left(\cos\theta_5 + j\sin\theta_5\right) \qquad (7.29e)$$

$$\mathbf{A}_B = \mathbf{A}_A + \mathbf{A}_{BA} \qquad (7.29f)$$

7.5   ACCELERATION OF ANY POINT ON A LINKAGE

Once the angular accelerations of all the links are found it is easy to define and calculate the acceleration of *any point on any link* for any input position of the linkage. Figure 7-9 shows the fourbar linkage with its coupler, link 3, enlarged to contain a coupler point P. The crank and rocker have also been enlarged to show points S and U which might represent the centers of gravity of those links. We want to develop algebraic expressions for the accelerations of these (or any) points on the links.

To find the acceleration of point S, draw the position vector from the fixed pivot O_2 to point S. This vector $\mathbf{R}_{SO_2}$ makes an angle δ_2 with the vector $\mathbf{R}_{AO_2}$. This angle δ_2 is completely defined by the geometry of link 2 and is constant. The position vector for point S is then:

$$\mathbf{R}_{SO_2} = \mathbf{R}_S = se^{j(\theta_2 + \delta_2)} = s\left[\cos(\theta_2 + \delta_2) + j\sin(\theta_2 + \delta_2)\right] \tag{4.29}$$

We differentiated this position vector in Section 6.9 (p. 325) to find the velocity of that point. The equation is repeated here for your convenience.

$$\mathbf{V}_S = jse^{j(\theta_2 + \delta_2)}\omega_2 = s\omega_2\left[-\sin(\theta_2 + \delta_2) + j\cos(\theta_2 + \delta_2)\right] \tag{6.34}$$

We can differentiate again versus time to find the acceleration of point S.

$$\begin{aligned}
\mathbf{A}_S &= s\alpha_2\, je^{j(\theta_2 + \delta_2)} - s\omega_2^2\, e^{j(\theta_2 + \delta_2)} \\
&= s\alpha_2\left[-\sin(\theta_2 + \delta_2) + j\cos(\theta_2 + \delta_2)\right] \\
&\quad - s\omega_2^2\left[\cos(\theta_2 + \delta_2) + j\sin(\theta_2 + \delta_2)\right]
\end{aligned} \tag{7.30}$$

The position of point U on link 4 is found in the same way, using the angle δ_4 which is a constant angular offset within the link. The expression is:

$$\mathbf{R}_{UO_4} = ue^{j(\theta_4 + \delta_4)} = u\left[\cos(\theta_4 + \delta_4) + j\sin(\theta_4 + \delta_4)\right] \tag{4.30}$$

We differentiated this position vector in Section 6.9 (p. 325) to find the velocity of that point. The equation is repeated here for your convenience.

$$\mathbf{V}_U = jue^{j(\theta_4 + \delta_4)}\omega_4 = u\omega_4\left[-\sin(\theta_4 + \delta_4) + j\cos(\theta_4 + \delta_4)\right] \tag{6.35}$$

We can differentiate again versus time to find the acceleration of point U.

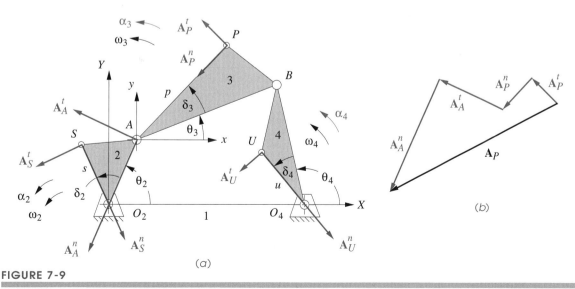

FIGURE 7-9

Finding the acceleration of any point on any link

$$A_U = u\alpha_4\, je^{j(\theta_4 + \delta_4)} - u\omega_4^2\, e^{j(\theta_4 + \delta_4)}$$
$$= u\alpha_4\left[-\sin(\theta_4 + \delta_4) + j\cos(\theta_4 + \delta_4)\right] \quad\quad (7.31)$$
$$- u\omega_4^2\left[\cos(\theta_4 + \delta_4) + j\sin(\theta_4 + \delta_4)\right]$$

The acceleration of point P on link 3 can be found from the addition of two acceleration vectors, such as A_A and A_{PA}. Vector A_A is already defined from our analysis of the link accelerations. A_{PA} is the acceleration difference of point P with respect to point A. Point A is chosen as the reference point because angle θ_3 is defined at a local coordinate system whose origin is at A. Position vector R_{PA} is defined in the same way as R_U or R_S, using the internal link offset angle δ_3 and the angle of link 3, θ_3. We previously analyzed this position vector and differentiated it in Section 6.9 (p. 325) to find the velocity difference of that point with respect to point A. Those equations are repeated here for your convenience.

$$R_{PA} = pe^{j(\theta_3 + \delta_3)} = p\left[\cos(\theta_3 + \delta_3) + j\sin(\theta_3 + \delta_3)\right] \quad\quad (4.31a)$$
$$R_P = R_A + R_{PA} \quad\quad (4.31b)$$

$$V_{PA} = jpe^{j(\theta_3 + \delta_3)}\omega_3 = p\omega_3\left[-\sin(\theta_3 + \delta_3) + j\cos(\theta_3 + \delta_3)\right] \quad\quad (6.36a)$$
$$V_P = V_A + V_{PA} \quad\quad (6.36b)$$

We can differentiate equation 6.36 again versus time to find A_{PA}, the acceleration of point P versus A. This vector can then be added to the vector A_A already found to define the absolute acceleration A_P of point P.

$$A_P = A_A + A_{PA} \quad\quad (7.32a)$$

where:

$$A_{PA} = p\alpha_3\, je^{j(\theta_3 + \delta_3)} - p\omega_3^2\, e^{j(\theta_3 + \delta_3)}$$
$$= p\alpha_3\left[-\sin(\theta_3 + \delta_3) + j\cos(\theta_3 + \delta_3)\right] \quad\quad (7.32b)$$
$$- p\omega_3^2\left[\cos(\theta_3 + \delta_3) + j\sin(\theta_3 + \delta_3)\right]$$

Compare equation 7.32 with equation 7.4 (p. 352). It is again the acceleration difference equation. Note that this equation applies to **any point** on **any link** at any position for which the positions and velocities are defined. It is a general solution for any rigid body.[*]

7.6   HUMAN TOLERANCE OF ACCELERATION

[*] The video *Fourbar Linkage Virtual Laboratory* on the book's DVD shows the measured acceleration of the coupler point on an actual linkage mechanism and also discusses the reasons for differences between the measured values and those calculated with equation 7.32. The measured data are also provided on the DVD.

It is interesting to note that the human body does not sense velocity, except with the eyes, but is very sensitive to acceleration. Riding in an automobile, in the daylight, one can see the scenery passing by and have a sense of motion. But, traveling at night in a commercial airliner at a 500 mph constant velocity, we have no sensation of motion as long as the flight is smooth. What we will sense in this situation is any change in velocity due to atmospheric turbulence, takeoffs, or landings. The semicircular canals in the inner ear are sensitive accelerometers that report to us on any accelerations that we experience. You have no doubt also experienced the sensation of acceleration when riding in an elevator and starting, stopping, or turning in an automobile. Accelerations produce dynamic forces on physical systems, as expressed in Newton's second law, $F = ma$. Force is proportional to acceleration, for a constant mass. The dynamic forces produced within the human body in response to acceleration can be harmful if excessive. The human body is, after all, not rigid. It is a loosely coupled bag of water and tissue, most of which is quite internally mobile. Accelerations in the headward or footward directions will tend to either starve or flood the brain with blood as this liquid responds to Newton's law and effectively moves within the body in a direction opposite to the imposed acceleration as it lags the motion of the skeleton. Lack of blood supply to the brain causes blackout; excess blood supply causes redout. Either results in death if sustained for a long enough period.

A great deal of research has been done, largely by the military and NASA, to determine the limits of human tolerance to sustained accelerations in various directions. Figure 7-10 shows data developed from such tests. [1] The units of linear acceleration were defined in Table 1-4 (p. 19) as in/sec^2, ft/sec^2, or m/sec^2. Another common unit for acceleration is the g, defined as the acceleration due to gravity, which on earth at sea level is approximately 386 in/sec^2, 32.2 ft/sec^2, or 9.8 m/sec^2. The g is a very convenient unit to use for accelerations involving the human as we live in a 1 g environment. Our weight, felt on our feet or buttocks, is defined by our mass times the acceleration due to gravity or mg. Thus an imposed acceleration of 1 g above the baseline of our gravity, or 2 g's, will be felt as a doubling of our weight. At 6 g's we would feel six times as heavy as normal and would have great difficulty even moving our arms against that acceleration. Figure 7-10 shows that the body's tolerance of acceleration is a function of its direction versus the body, its magnitude, and its duration. Note also that the data used for this chart were developed from tests on young, healthy military personnel in prime physical condition. The general population, children and elderly in particular, should not be expected to be able to withstand such high levels of acceleration. Since much machinery is designed for human use, these acceleration tolerance data should be of great interest and value to the machine designer. Several references dealing with these human factors data are provided in the bibliography to Chapter 1 (p. 27).

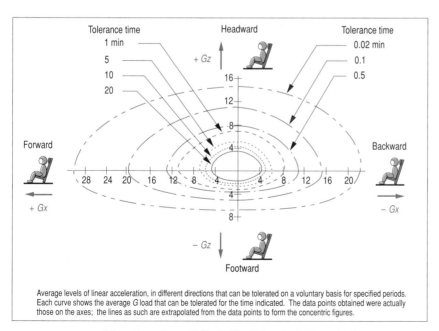

Average levels of linear acceleration, in different directions that can be tolerated on a voluntary basis for specified periods. Each curve shows the average G load that can be tolerated for the time indicated. The data points obtained were actually those on the axes; the lines as such are extrapolated from the data points to form the concentric figures.

(Adapted from reference [1], Fig. 17-17, p. 505, reprinted with permission)

FIGURE 7-10

Human tolerance of acceleration

TABLE 7-1	Acceleration Levels Commonly Encountered in Human Activities
Gentle acceleration in an automobile	+0.1 g
Commercial jet aircraft on takeoff	+0.3 g
Hard acceleration in an automobile	+0.5 g
Panic stop in an automobile	–0.7 g
Fast cornering in a sports car (e.g., BMW, Porsche, Ferrari)	+0.9 g
Formula 1 race car	+2.0 g, –4.0 g
Roller coasters (various)	±3.5 to ±6.5 g[*]
NASA Space Shuttle on takeoff	+4.0 g
Top fuel dragster with drogue chute (>300 mph in 1/4 mile)	±4.5 g
Military jet fighter (e.g., F-15, F-16, F-22—note: pilot wears a G-suit)	±9.0 g

Another useful benchmark when designing machinery for human occupation is to attempt to relate the magnitudes of accelerations that you commonly experience to the calculated values for your potential design. Table 7-1 lists some approximate levels of acceleration, in g's, that humans can experience in everyday life. Your own experience of these will help you develop a "feel" for the values of acceleration that you encounter in designing machinery intended for human occupation.

Acceleration levels in machinery that does not carry humans is limited only by considerations of the stresses in its parts. These stresses are often generated in large part by the dynamic forces due to accelerations. The range of acceleration values in such machinery is so wide that it is not possible to comprehensively define any design guidelines for acceptable or unacceptable levels of acceleration. If the moving mass is small, then very large numerical values of acceleration are reasonable. If the mass is large, the dynamic stresses that the materials can sustain may limit the allowable accelerations to low values. Unfortunately, the designer usually does not know how much acceleration is too much in a design until completing it to the point of calculating stresses in the parts. This usually requires a fairly complete and detailed design. If the stresses turn out to be too high and are due to dynamic forces, then the only recourse is to iterate back through the design process and reduce the accelerations and/or masses in the design. This is one reason that the design process is a circular and not a linear one.

As one point of reference, the acceleration of the piston in a small, four-cylinder economy car engine (about 1.5-L displacement) at idle speed is about 40 g's. At highway speeds the piston acceleration can be as high as 700 g's. At the engine's top speed of 6000 rpm the peak piston acceleration is 2000 g's! As long as you're not riding on the piston, this is acceptable. These engines last a long time in spite of the high accelerations their components experience. One key factor is the choice of proper part geometry and use of low-mass, high-strength, high-stiffness materials for the moving parts to minimize dynamic forces at high acceleration and enable the parts to tolerate the applied stresses.

7.7   JERK

No, not you! The **time derivative of acceleration** is called *jerk, pulse,* or *shock.* The name is apt, as it conjures the proper image of this phenomenon. **Jerk** is *the time rate of*

---

[*] Some U.S. state laws currently limit roller coaster accelerations to a maximum of 5.0 to 5.4 g.

change of acceleration. Force is proportional to acceleration. Rapidly changing acceleration means a rapidly changing force. Rapidly changing forces tend to "jerk" the object
about! You have probably experienced this phenomenon when riding in an automobile.
If the driver is inclined to "jackrabbit" starts and accelerates violently away from the
traffic light, you will suffer from large jerk because your acceleration will go from zero
to a large value quite suddenly. But, when Jeeves, the chauffeur, is driving the Rolls, he
always attempts to minimize jerk by accelerating gently and smoothly, so that Madame
is entirely unaware of the change.

Controlling and minimizing jerk in machine design is often of interest, especially if
low vibration is desired. Large magnitudes of jerk will tend to excite the natural frequencies of vibration of the machine or structure to which it is attached and cause increased
vibration and noise levels. Jerk control is of greater interest in the design of cams than of
linkages, and we will investigate it in greater detail in Chapter 8 on cam design.

The procedure for calculating the jerk in a linkage is a straightforward extension of
the methods shown for acceleration analysis. Let angular jerk be represented by:

$$\varphi = \frac{d\alpha}{dt} \qquad (7.33a)$$

and linear jerk by:

$$\mathbf{J} = \frac{d\mathbf{A}}{dt} \qquad (7.33b)$$

To solve for jerk in a fourbar linkage, for example, the vector loop equation for acceleration (equation 7.7, p. 358) is differentiated versus time. Refer to Figure 7-5 (p.
359) for notation.

$$-a\omega_2^3 je^{j\theta_2} - 2a\omega_2\alpha_2 e^{j\theta_2} + a\alpha_2\omega_2 j^2 e^{j\theta_2} + a\varphi_2 je^{j\theta_2}$$
$$- b\omega_3^3 je^{j\theta_3} - 2b\omega_3\alpha_3 e^{j\theta_3} + b\alpha_3\omega_3 j^2 e^{j\theta_3} + b\varphi_3 je^{j\theta_3}$$
$$+ c\omega_4^3 je^{j\theta_4} + 2c\omega_4\alpha_4 e^{j\theta_4} - c\alpha_4\omega_4 j^2 e^{j\theta_4} - c\varphi_4 je^{j\theta_4} = 0 \qquad (7.34a)$$

Collect terms and simplify:

$$-a\omega_2^3 je^{j\theta_2} - 3a\omega_2\alpha_2 e^{j\theta_2} + a\varphi_2 je^{j\theta_2}$$
$$- b\omega_3^3 je^{j\theta_3} - 3b\omega_3\alpha_3 e^{j\theta_3} + b\varphi_3 je^{j\theta_3}$$
$$+ c\omega_4^3 je^{j\theta_4} + 3c\omega_4\alpha_4 e^{j\theta_4} - c\varphi_4 je^{j\theta_4} = 0 \qquad (7.34b)$$

Substitute the Euler identity and separate into x and y components:

real part (x component):

$$a\omega_2^3 \sin\theta_2 - 3a\omega_2\alpha_2 \cos\theta_2 - a\varphi_2 \sin\theta_2$$
$$+ b\omega_3^3 \sin\theta_3 - 3b\omega_3\alpha_3 \cos\theta_3 - b\varphi_3 \sin\theta_3$$
$$- c\omega_4^3 \sin\theta_4 + 3c\omega_4\alpha_4 \cos\theta_4 + c\varphi_4 \sin\theta_4 = 0 \qquad (7.35a)$$

imaginary part (y component):

$$-a\omega_2^3 \cos\theta_2 - 3a\omega_2\alpha_2 \sin\theta_2 + a\varphi_2 \cos\theta_2$$
$$- b\omega_3^3 \cos\theta_3 - 3b\omega_3\alpha_3 \sin\theta_3 + b\varphi_3 \cos\theta_3$$
$$+ c\omega_4^3 \cos\theta_4 + 3c\omega_4\alpha_4 \sin\theta_4 - c\varphi_4 \cos\theta_4 = 0 \qquad (7.35b)$$

These can be solved simultaneously for φ_3 and φ_4, which are the only unknowns. The driving angular jerk, φ_2, if nonzero, must be known in order to solve the system. All the other factors in equations 7.35 (p. 377) are defined or have been calculated from the position, velocity, and acceleration analyses. To simplify these expressions we will set the known terms to temporary constants.

In equation 7.35a, let:

$$A = a\omega_2^3 \sin\theta_2 \qquad D = b\omega_3^3 \sin\theta_3 \qquad G = 3c\omega_4\alpha_4 \cos\theta_4$$
$$B = 3a\omega_2\alpha_2 \cos\theta_2 \qquad E = 3b\omega_3\alpha_3 \cos\theta_3 \qquad H = c\sin\theta_4 \qquad (7.36a)$$
$$C = a\varphi_2 \sin\theta_2 \qquad F = c\omega_4^3 \sin\theta_4 \qquad K = b\sin\theta_3$$

Equation 7.35a then reduces to:

$$\varphi_3 = \frac{A - B - C + D - E - F + G + H\varphi_4}{K} \qquad (7.36b)$$

Note that equation 7.36b defines angle φ_3 in terms of angle φ_4. We will now simplify equation 7.35b and substitute equation 7.36b into it.

In equation 7.35b, let:

$$L = a\omega_2^3 \cos\theta_2 \qquad P = b\omega_3^3 \cos\theta_3 \qquad S = c\omega_4^3 \cos\theta_4$$
$$M = 3a\omega_2\alpha_2 \sin\theta_2 \qquad Q = 3b\omega_3\alpha_3 \sin\theta_3 \qquad T = 3c\omega_4\alpha_4 \sin\theta_4 \qquad (7.37a)$$
$$N = a\varphi_2 \cos\theta_2 \qquad R = b\cos\theta_3 \qquad U = c\cos\theta_4$$

Equation 7.35b then reduces to:

$$R\varphi_3 - U\varphi_4 - L - M + N - P - Q + S + T = 0 \qquad (7.37b)$$

Substituting equation 7.36b in equation 7.35b:

$$R\left(\frac{A - B - C + D - E - F + G + H\varphi_4}{K}\right) - U\varphi_4 - L - M + N - P - Q + S + T = 0 \qquad (7.38)$$

The solution is:

$$\varphi_4 = \frac{KN - KL - KM - KP - KQ + AR - BR - CR + DR - ER - FR + GR + KS + KT}{KU - HR} \qquad (7.39)$$

The result from equation 7.39 can be substituted into equation 7.36b to find φ_3. Once the angular jerk values are found, the linear jerk at the pin joints can be found from:

$$J_A = -a\omega_2^3 je^{j\theta_2} - 3a\omega_2\alpha_2 e^{j\theta_2} + a\varphi_2 je^{j\theta_2}$$
$$J_{BA} = -b\omega_3^3 je^{j\theta_3} - 3b\omega_3\alpha_3 e^{j\theta_3} + b\varphi_3 je^{j\theta_3} \qquad (7.40)$$
$$J_B = -c\omega_4^3 je^{j\theta_4} - 3c\omega_4\alpha_4 e^{j\theta_4} + c\varphi_4 je^{j\theta_4}$$

The same approach as used in Section 7.5 (p. 372) to find the acceleration of any point on any link can be used to find the linear jerk at any point.

$$J_P = J_A + J_{PA} \qquad (7.41)$$

The jerk difference equation 7.41 can be applied to any point on any link if we let P represent any arbitrary point on any link and A represent any reference point on the same link for which we know the value of the jerk vector. Note that if you substitute equations 7.40 into 7.41, you will get equation 7.34 (p. 377).

7.8 LINKAGES OF *N* BARS

The same analysis techniques presented here for position, velocity, acceleration, and jerk, using the fourbar and fivebar linkage as the examples, can be extended to more complex assemblies of links. Multiple vector loop equations can be written around a linkage of arbitrary complexity. The resulting vector equations can be differentiated and solved simultaneously for the variables of interest. In some cases, the solution will require simultaneous solution of a set of nonlinear equations. A root-finding algorithm such as the Newton-Raphson method will be needed to solve these more complicated cases. A computer is necessary. An equation solver software package such as *TKSolver* or *Mathcad* that will do an iterative root-finding solution will be a useful aid to the solution of any of these analysis problems, including the examples shown here.

7.9 REFERENCE

1 **Sanders, M. S., and E. J. McCormick**. (1987). *Human Factors in Engineering and Design*, 6th ed., McGraw-Hill Co., New York. p. 505.

7.10 PROBLEMS§

7-1 A point at a 6.5-in radius is on a body that is in pure rotation with $\omega = 100$ rad/sec and a constant $\alpha = -500$ rad/sec² at point A. The rotation center is at the origin of a coordinate system. When the point is at position A, its position vector makes a 45° angle with the X axis. It takes 0.01 sec to reach point B. Draw this system to some convenient scale, calculate the θ and ω of position B, and:

   a. Write an expression for the particle's acceleration vector in position A using complex number notation, in both polar and cartesian forms.

   b. Write an expression for the particle's acceleration vector in position B using complex number notation, in both polar and cartesian forms.

   c. Write a vector equation for the acceleration difference between points B and A. Substitute the complex number notation for the vectors in this equation and solve for the acceleration difference numerically.

   d. Check the result of part c with a graphical method.

7-2 In problem 7-1 let A and B represent points on separate, rotating bodies both having the given ω and α at $t = 0$, $\theta_A = 45°$, and $\theta_B = 120°$. Find their relative acceleration.

‡*7-3 The link lengths, coupler point location, and the values of θ_2, ω_2, and α_2 for the same fourbar linkages as used for position and velocity analysis in Chapters 4 and 6 are redefined in Table P7-1 (p. 380), which is the same as Table P6-1 (p. 328). The general linkage configuration and terminology are shown in Figure P7-1 (p. 380). *For the row(s) assigned,* draw the linkage to scale and graphically find the accelerations of points A and B. Then calculate α_3 and α_4 and the acceleration of point P.

*†7-4 Repeat problem 7-3, solving by the analytical vector loop method of Section 7.3 (p. 358).

*7-5 The link lengths and offset and the values of θ_2, ω_2, and α_2 for some noninverted, offset fourbar crank-slider linkages are defined in Table P7-2 (p. 381). The general linkage configuration and terminology are shown in Figure P7-2 (p. 381). *For the row(s) assigned*, draw the linkage to scale and graphically find the accelerations of the pin joints A and B and the acceleration of slip at the sliding joint.

*†7-6 Repeat problem 7-5 using an analytical method.

7

§ See footnote on next page.

TABLE P7-1   Data for Problems 7-3, 7-4 and 7-11‡

Row	Link 1	Link 2	Link 3	Link 4	θ_2	ω_2	α_2	R_{pa}	δ_3
a	6	2	7	9	30	10	0	6	30
b	7	9	3	8	85	− 12	5	9	25
c	3	10	6	8	45	− 15	−10	10	80
d	8	5	7	6	25	24	− 4	5	45
e	8	5	8	6	75	− 50	10	9	300
f	5	8	8	9	15	− 45	50	10	120
g	6	8	8	9	25	100	18	4	300
h	20	10	10	10	50	− 65	25	6	20
i	4	5	2	5	80	25	− 25	9	80
j	20	10	5	10	33	25	− 40	1	0
k	4	6	10	7	88	− 80	30	10	330
l	9	7	10	7	60	− 90	20	5	180
m	9	7	11	8	50	75	− 5	10	90
n	9	7	11	6	120	15	− 65	15	60

‡ Drawings of these linkages are in the *PDF Problem Workbook* folder on the DVD.

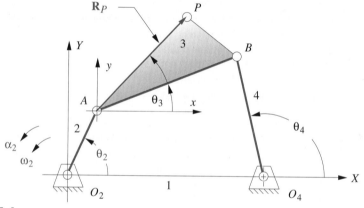

FIGURE P7-1

Configuration and terminology for Problems 7-3, 7-4 and 7-11

*†7-7   The link lengths and the values of θ_2, ω_2, and γ for some inverted fourbar crank-slider linkages are defined in Table P7-3 (p. 382). The general linkage configuration and terminology are shown in Figure P7-3 (p. 382). *For the row(s) assigned*, find accelerations of the pin joints *A* and the acceleration of slip at the sliding joint. Solve by the analytical vector loop method of Section 7.3 (p. 358) for the open configuration of the linkage.

*†7-8   Repeat problem 7-7 for the crossed configuration of the linkage.

*7-9   The link lengths, gear ratio (λ), phase angle (ϕ), and the values of θ_2, ω_2, and α_2 for some geared fivebar linkages are defined in Table P7-4 (p. 373). The general linkage configuration and terminology are shown in Figure P7-4 (p. 373). *For the row(s) assigned*, find α_3 and α_4 and the linear acceleration of point *P*.

†7-10   An automobile driver took a curve too fast. The car spun out of control about its center of gravity (*CG*) and slid off the road in a northeasterly direction. The friction of the

TABLE P7-2   Data for Problems 7-5 and 7-6[‡]

Row	Link 2	Link 3	Offset	θ_2	ω_2	α_2
a	1.4	4	1	45	10	0
b	2	6	-3	60	-12	5
c	3	8	2	-30	-15	-10
d	3.5	10	1	120	24	-4
e	5	20	-5	225	-50	10
f	3	13	0	100	-45	50
g	7	25	10	330	100	18

[‡] Drawings of these linkages are in the *PDF Problem Workbook* folder on the DVD.

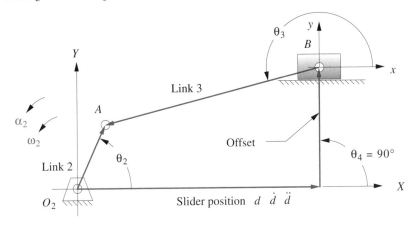

FIGURE P7-2

Configuration and terminology for problems 7-5 to 7-6 and 7-58

skidding tires provided a 0.25 *g* linear deceleration. The car rotated at 100 rpm. When the car hit the tree head-on at 30 mph, it took 0.1 sec to come to rest.

  a. What was the acceleration experienced by the child seated on the middle of the rear seat, 2 ft behind the car's *CG*, just prior to impact?

  b. What force did the 100-lb child exert on her seatbelt harness as a result of the acceleration, just prior to impact?

  c. Assuming a constant deceleration during the 0.1 sec of impact, what was the magnitude of the average deceleration felt by the passengers in that interval?

[†]7-11 For the row(s) assigned in Table P7-1, find the angular jerk of links 3 and 4 and the linear jerk of the pin joint between links 3 and 4 (point *B*). Assume an angular jerk of zero on link 2. The linkage configuration and terminology are shown in Figure P7-11 (p. 387).

[*†]7-12 You are riding on a carousel that is rotating at a constant 12 rpm. It has an inside radius of 4 ft and an outside radius of 12 ft. You begin to run from the inside to the outside along a radius. Your peak velocity with respect to the carousel is 4 mph and occurs at a radius of 8 ft. What are your maximum Coriolis acceleration magnitude and its direction with respect to the carousel?

7-13 The linkage in Figure P7-5a (p. 384) has $O_2A = 0.8$, $AB = 1.93$, $AC = 1.33$, and *offset* = 0.38 in. The crank angle in the position shown is 34.3° and angle $BAC = 38.6°$.

* Answers in Appendix F.

† These problems are suited to solution using *Mathcad*, *Matlab*, or *TKSolver* equation solver programs.

TABLE P7-3   Data for Problems 7-7 to 7-8

Row	Link 1	Link 2	Link 4	γ	θ_2	ω_2	α_2
a	6	2	4	90	30	10	− 25
b	7	9	3	75	85	−15	− 40
c	3	10	6	45	45	24	30
d	8	5	3	60	25	− 50	20
e	8	4	2	30	75	− 45	− 5
f	5	8	8	90	150	100	− 65

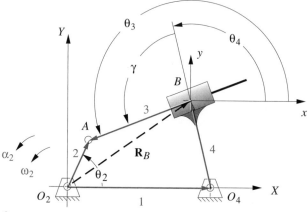

FIGURE P7-3

Configuration and terminology for Problems 7-8 to 7-9 and 7-59

Find α_3, $\mathbf{A}_A$, $\mathbf{A}_B$, and $\mathbf{A}_C$ for the position shown for $\omega_2 = 15$ rad/sec and $\alpha_2 = 10$ rad/sec^2 in directions shown.

    a. Using the acceleration difference graphical method.
   †b. Using an analytical method.

7-14 The linkage in Figure P7-5b has $I_{12}A = 0.75$, $AB = 1.5$, and $AC = 1.2$ in. The effective crank angle in the position shown is 77° and angle $BAC = 30°$. Find α_3, $\mathbf{A}_A$, $\mathbf{A}_B$, and $\mathbf{A}_C$ for the position shown for $\omega_2 = 15$ rad/sec and $\alpha_2 = 10$ rad/sec^2 in the directions shown.

    a. Using the acceleration difference graphical method.
   †b. Using an analytical method. (Hint: Create an effective linkage for the position shown and analyze it as a pin-jointed fourbar.)

7-15 The linkage in Figure P7-5c (p. 384) has $AB = 1.8$ and $AC = 1.44$ in. The angle of AB in the position shown is 128° and angle $BAC = 49°$. The slider at B is at an angle of 59°. Find α_3, $\mathbf{A}_B$, and $\mathbf{A}_C$ for the position shown for $\mathbf{V}_A = 10$ in/sec and $\mathbf{A}_A = 15$ in/sec^2 in the directions shown.

    a. Using the acceleration difference graphical method.
   †b. Using an analytical method.

†7-16 The linkage in Figure P7-6a (p. 384) has $O_2A = 5.6$, $AB = 9.5$, $O_4C = 9.5$, $L_1 = 38.8$ mm. θ_2 is 135° in the xy coordinate system. Write the vector loop equations; differentiate them, and do a complete position, velocity, and acceleration analysis of the linkage. Assume $\omega_2 = 10$ rad/sec and $\alpha_2 = 20$ rad/sec^2.

† These problems are suited to solution using *Mathcad, Matlab,* or *TKSolver* equation solver programs.

TABLE P7-4   Data for Problem 7-9

Row	Link 1	Link 2	Link 3	Link 4	Link 5	λ	ϕ	θ_2	ω_2	α_2	R_{pa}	δ_3
a	6	1	7	9	4	2.0	30	60	10	0	6	30
b	6	5	7	8	4	−2.5	60	30	−12	5	9	25
c	3	5	7	8	4	−0.5	0	45	−15	−10	10	80
d	4	5	7	8	4	−1.0	120	75	24	−4	5	45
e	5	9	11	8	8	3.2	−50	−39	−50	10	9	300
f	10	2	7	5	3	1.5	30	120	−45	50	10	120
g	15	7	9	11	4	2.5	−90	75	100	18	4	300
h	12	8	7	9	4	−2.5	60	55	−65	25	6	20
i	9	7	8	9	4	−4.0	120	100	25	−25	9	80

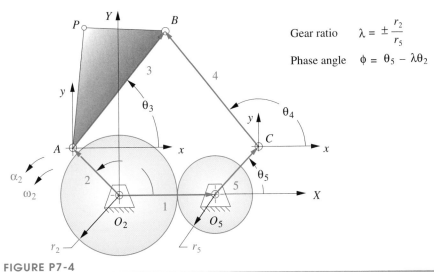

Gear ratio $\quad \lambda = \pm \dfrac{r_2}{r_5}$

Phase angle $\quad \phi = \theta_5 - \lambda\theta_2$

FIGURE P7-4

Configuration and terminology for problems 7-9 and 7-60

†7-17   Repeat Problem 7-16 for the linkage shown in Figure P7-6b (p. 384) which has the dimensions: $L_1 = 61.9$, $L_2 = 15$, $L_3 = 45.8$, $L_4 = 18.1$, $L_5 = 23.1$ mm. θ_2 is 68.3° in the xy coordinate system, which is at −23.3° in the XY coordinate system. The X component of O_2C is 59.2 mm.

†7-18   Repeat Problem 7-16 for the linkage shown in Figure P7-6c (p. 384) which has the dimensions: $O_2A = 11.7$, $O_2C = 20$, $L_3 = 25$, $L_5 = 25.9$ mm. Point B is offset 3.7 mm from the x_1 axis and point D is offset 24.7 mm from the x_2 axis. θ_2 is at 13.3° in the x_2y_2 coordinate system.

†7-19   Repeat Problem 7-16 for the linkage shown in Figure P7-6d (p. 384) which has the dimensions: $L_2 = 15$, $L_3 = 40.9$, $L_5 = 44.7$ mm. θ_2 is 24.2° in the XY coordinate system.

†7-20   Figure P7-7 (p. 384) shows a sixbar linkage with $O_2B = 1$, $BD = 1.5$, $DC = 3.5$, $DO_6 = 3$, and $h = 1.3$ in. Find the angular acceleration of link 6 if ω_2 is a constant 1 rad/sec.

*7-21   The linkage in Figure P7-8a (p. 386) has link 1 at −25° and link 2 at 37° in the global XY coordinate system. Find α_4, $\mathbf{A}_A$, and $\mathbf{A}_B$ in the global coordinate system for the position shown if $\omega_2 = 15$ rad/sec CW and $\alpha_2 = 25$ rad/sec² CCW. Use the acceleration difference graphical method. (Print the figure from its PDF file on the DVD and draw on it.)

* Answers in Appendix F.

† These problems are suited to solution using *Mathcad, Matlab,* or *TKSolver* equation solver programs.

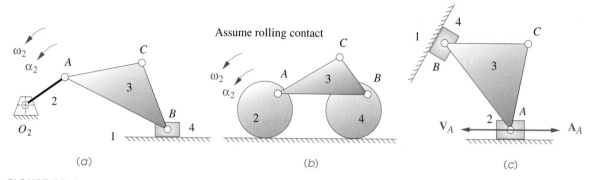

(a)   *(b)*   *(c)*

FIGURE P7-5

Problems 7-13 to 7-15

(a)   *(b)*

(c)   *(d)*

FIGURE P7-6

Problems 7-16 to 7-19

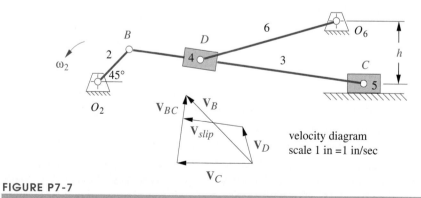

velocity diagram
scale 1 in =1 in/sec

FIGURE P7-7

Problem 7-20. *Courtesy of Prof. J. M. Vance, Iowa State University*

†7-22   The linkage in Figure P7-8a (p. 386) has link 1 at $-25°$ and link 2 at $37°$ in the global XY coordinate system. Find α_4, $\mathbf{A}_A$, and $\mathbf{A}_B$ in the global coordinate system for the position shown if $\omega_2 = 15$ rad/sec CW and $\alpha_2 = 25$ rad/sec^2 CCW. Use an analytical method.

†7-23   At $t = 0$, the non-Grashof linkage in Figure P7-8a (p. 386) has link 1 at $-25°$ and link 2 at $37°$ in the global XY coordinate system and $\omega_2 = 0$. Write a computer program or use an equation solver to find and plot ω_4, α_4, $\mathbf{V}_A$, $\mathbf{A}_A$, $\mathbf{V}_B$, and $\mathbf{A}_B$ in the local coordinate system for the maximum range of motion that this linkage allows if $\alpha_2 = 15$ rad/sec CW constant.

*7-24   The linkage in Figure P7-8b (p. 386) has link 1 at $-36°$ and link 2 at $57°$ in the global XY coordinate system. Find α_4, $\mathbf{A}_A$, and $\mathbf{A}_B$ in the global coordinate system for the position shown if $\omega_2 = 20$ rad/sec CCW, constant. Use the acceleration difference graphical method. (Print the figure from its PDF file on the DVD and draw on it.)

†7-25   The linkage in Figure P7-8b (p. 386) has link 1 at $-36°$ and link 2 at $57°$ in the global XY coordinate system. Find α_4, $\mathbf{A}_A$, and $\mathbf{A}_B$ in the global coordinate system for the position shown if $\omega_2 = 20$ rad/sec CCW, constant. Use an analytical method.

†7-26   For the linkage in Figure P7-8b (p. 386), write a computer program or use an equation solver to find and plot α_4, $\mathbf{A}_A$, and $\mathbf{A}_B$ in the local coordinate system for the maximum range of motion that this linkage allows if $\omega_2 = 20$ rad/sec CCW, constant.

7-27   The offset crank-slider linkage in Figure P7-8f (p. 386) has link 2 at $51°$ in the global XY coordinate system. Find $\mathbf{A}_A$ and $\mathbf{A}_B$ in the global coordinate system for the position shown if $\omega_2 = 25$ rad/sec CW, constant. Use the acceleration difference graphical method. (Print the figure from its PDF file on the DVD and draw on it.)

*†7-28   The offset crank-slider linkage in Figure P7-8f (p. 386) has link 2 at $51°$ in the global XY coordinate system. Find $\mathbf{A}_A$ and $\mathbf{A}_B$ in the global coordinate system for the position shown if $\omega_2 = 25$ rad/sec CW, constant. Use an analytical method.

†7-29   For the offset crank-slider linkage in Figure P7-8f (p. 386), write a computer program or use an equation solver to find and plot $\mathbf{A}_A$ and $\mathbf{A}_B$ in the global coordinate system for the maximum range of motion that this linkage allows if $\omega_2 = 25$ rad/sec CW, constant.

7-30   The linkage in Figure P7-8d (p. 386) has link 2 at $58°$ in the global XY coordinate system. Find $\mathbf{A}_A$, $\mathbf{A}_B$, and $\mathbf{A}_{box}$ (the acceleration of the box) in the global coordinate system for the position shown if $\omega_2 = 30$ rad/sec CW, constant. Use the acceleration difference graphical method. (Print the figure from its PDF file on the DVD and draw on it.)

†7-31   The linkage in Figure P7-8d (p. 386) has link 2 at $58°$ in the global XY coordinate system. Find $\mathbf{A}_A$, $\mathbf{A}_B$, and $\mathbf{A}_{box}$ (the acceleration of the box) in the global coordinate system for the position shown if $\omega_2 = 30$ rad/sec CW, constant. Use an analytical method.

†7-32   For the linkage in Figure P7-8d (p. 386), write a computer program or use an equation solver to find and plot $\mathbf{A}_A$, $\mathbf{A}_B$, and $\mathbf{A}_{box}$ (the acceleration of the box) in the global coordinate system for the maximum range of motion that this linkage allows if $\omega_2 = 30$ rad/sec CW, constant.

7-33   The linkage in Figure P7-8g (p. 386) has the local xy axis at $-119°$ and O_2A at $29°$ in the global XY coordinate system. Find α_4, $\mathbf{A}_A$, and $\mathbf{A}_B$ in the global coordinate system for the position shown if $\omega_2 = 15$ rad/sec CW, constant. Use the acceleration difference graphical method. (Print the figure from its PDF file on the DVD and draw on it.)

†7-34   The linkage in Figure P7-8g (p. 386) has the local xy axis at $-119°$ and O_2A at $29°$ in the global XY coordinate system. Find α_4, $\mathbf{A}_A$, and $\mathbf{A}_B$ in the global coordinate system for the position shown if $\omega_2 = 15$ rad/sec CW and $\alpha_2 = 10$ rad/sec CCW, constant. Use an analytical method.

*   Answers in Appendix F.

†   These problems are suited to solution using *Mathcad, Matlab,* or *TKSolver* equation solver programs.

(a) Fourbar linkage

$L_1 = 174$
$L_2 = 116$
$L_3 = 108$
$L_4 = 110$

(b) Fourbar linkage

$L_1 = 162 \quad L_2 = 40$
$L_4 = 122 \quad L_3 = 96$

(c) Radial compressor

$L_2 = 19$
$L_3 = 70$
$L_4 = 70$
$L_5 = 70$
$L_6 = 70$

(d) Walking-beam conveyor

$L_1 = 150 \quad L_2 = 30$
$L_3 = 150 \quad L_4 = 30$

(e) Bellcrank mechanism

$O_2O_4 = L_3 = L_5 = 160$
$O_8O_4 = L_6 = L_7 = 120$
$O_2A = O_2C = 20$
$O_4B = O_4D = 20$
$O_4E = O_4G = 30$
$O_8F = O_8H = 30$

(f) Offset slider-crank

$L_2 = 63$
$L_3 = 130$
offset = 52

(g) Drum brake mechanism

$L_1 = 87$
$L_2 = 49$
$L_3 = 100$
$L_4 = 153$
$L_5 = 100$
$L_6 = 153$

(h) Symmetrical mechanism

$L_1 = 45.8$
$L_2 = 19.8$
$L_3 = 19.4$
$L_4 = 38.3$
$L_5 = 13.3$
$L_7 = 13.3$
$L_8 = 19.8$
$L_9 = 19.4$

all dimensions in mm

FIGURE P7-8

Problems 7-21 to 7-38. *Adapted from P. H. Hill and W. P. Rule. (1960). Mechanisms: Analysis and Design, with permission.*

†7-35   At $t = 0$, the non-Grashof linkage in Figure P7-8g has the local xy axis at $-119°$ and
        O_2A at $29°$ in the global XY coordinate system and $\omega_2 = 0$. Write a computer program
        or use an equation solver to find and plot ω_4, α_4, $\mathbf{V}_A$, $\mathbf{A}_A$, $\mathbf{V}_B$, and $\mathbf{A}_B$ in the local coor-
        dinate system for the maximum range of motion that this linkage allows if $\alpha_2 = 15$ rad/
        sec CCW, constant.

7-36   The 3-cylinder radial compressor in Figure P7-8c has its cylinders equispaced at
        $120°$. Find the piston accelerations $\mathbf{A}_6$, $\mathbf{A}_7$, $\mathbf{A}_8$ with the crank at $-53°$ using a graphical
        method if $\omega_2 = 15$ rad/sec CW, constant. (Print the figure's PDF file from the DVD and
        draw on it.)

†7-37   The 3-cylinder radial compressor in Figure P7-8c has its cylinders equispaced at $120°$.
        Find the piston accelerations $\mathbf{A}_6$, $\mathbf{A}_7$, $\mathbf{A}_8$ with the crank at $-53°$ using an analytical
        method if $\omega_2 = 15$ rad/sec CW, constant.

†7-38   For the 3-cylinder radial compressor in Figure P7-8f, write a program or use an equa-
        tion solver to find and plot the piston accelerations $\mathbf{A}_6$, $\mathbf{A}_7$, $\mathbf{A}_8$ for one revolution of the
        crank.

*†7-39 Figure P7-9 shows a linkage in one position. Find the instantaneous accelerations of
        points A, B, and P if link O_2A is rotating CW at 40 rad/sec.

*†7-40 Figure P7-10 shows a linkage and its coupler curve. Write a computer program or use
        an equation solver to calculate and plot the magnitude and direction of the acceleration
        of the coupler point P at $2°$ increments of crank angle for $\omega_2 = 100$ rpm. Check your
        result with program LINKAGES.

*†7-41 Figure P7-11 shows a linkage that operates at 500 crank rpm. Write a computer
        program or use an equation solver to calculate and plot the magnitude and direction
        of the acceleration of point B at $2°$ increments of crank angle. Check your result with
        program LINKAGES.

*†7-42 Figure P7-12 shows a linkage and its coupler curve. Write a computer program or use
        an equation solver to calculate and plot the magnitude and direction of the accelera-
        tion of the coupler point P at $2°$ increments of crank angle for $\omega_2 = 20$ rpm over the
        maximum range of motion possible. Check your result with program LINKAGES.

†7-43   Figure P7-13 (p. 388) shows a linkage and its coupler curve. Write a computer program
        or use an equation solver to calculate and plot the magnitude and direction of the ac-
        celeration of the coupler point P at $2°$ increments of crank angle for $\omega_2 = 80$ rpm over
        the maximum range of motion possible. Check your result with program LINKAGES.

FIGURE P7-9

Problem 7-39

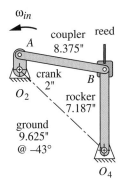

FIGURE P7-11

Problem 7-41 Loom
laybar drive

7

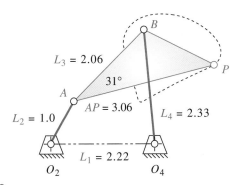

FIGURE P7-10

Problem 7-40

* Answers in Appendix F.

† These problems are suited
to solution using *Mathcad*,
Matlab, or *TKSolver* equation solver programs.

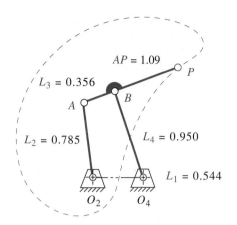

FIGURE P7-12

Problem 7-42

*†7-44   Figure P7-14 shows a linkage and its coupler curve. Write a computer program or use an equation solver to calculate and plot the magnitude and direction of the acceleration of the coupler point P at $2°$ increments of crank angle for $\omega_2 = 80$ rpm over the maximum range of motion possible. Check your result with program LINKAGES.

†7-45   Figure P7-15 shows a power hacksaw, used to cut metal. Link 5 pivots at O_5 and its weight forces the sawblade against the workpiece while the linkage moves the blade (link 4) back and forth on link 5 to cut the part. It is an offset crank-slider mechanism with the dimensions shown in the figure. Draw an equivalent linkage diagram, and then calculate and plot the acceleration of the sawblade with respect to the piece being cut over one revolution of the crank at 50 rpm.

†7-46   Figure P7-16 (p. 390) shows a walking-beam indexing and pick-and-place mechanism that can be analyzed as two fourbar linkages driven by a common crank. The link lengths are given in the figure. The phase angle between the two crankpins on links 4 and 5 is indicated. The product cylinders being pushed have 60-mm diameters. The point of contact between the left vertical finger and the leftmost cylinder in the position shown is 58 mm at $80°$ versus the left end of the parallelogram's coupler (point D). Calculate and plot the relative acceleration between points E and P for one revolution of gear 2.

FIGURE P7-13

Problem 7-43

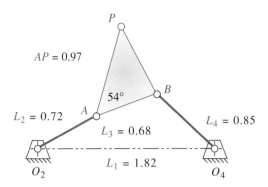

FIGURE P7-14

Problem 7-44

†7-47 Figure P7-17 (p. 390) shows a paper roll off-loading mechanism driven by an air cylinder. In the position shown O_4A is 0.3 m at 226° and $O_2O_4 = 0.93$ m at 163.2°. The V-links are rigidly attached to O_4A. The paper roll center is 0.707 m from O_4 at −181° with respect to O_4A. The air cylinder is retracted at a constant acceleration of 0.1 m/sec². Draw a kinematic diagram of the mechanism, write the necessary equations, and calculate and plot the angular acceleration of the paper roll and the linear acceleration of its center as it rotates through 90° CCW from the position shown.

†7-48 Figure P7-18 (p. 391) shows a mechanism and its dimensions. Find the accelerations of points A, B, and C for the position shown if $\omega_2 = 40$ rad/min and $\alpha_2 = -1500$ rad/min² as shown.

†7-49 Figure P7-19 (p. 391) shows a walking-beam mechanism. Calculate and plot the acceleration A_{out} for one revolution of the input crank 2 rotating at 100 rpm.

†7-50 Figure P7-20 (p. 392) shows a surface grinder. The workpiece is oscillated under the spinning 90-mm-diameter grinding wheel by the crank-slider linkage which has a 22-mm crank, a 157-mm connecting rod, and a 40-mm offset. The crank turns at 30 rpm, and the grinding wheel at 3450 rpm. Calculate and plot the acceleration of the grinding wheel contact point relative to the workpiece over one revolution of the crank.

† These problems are suited to solution using *Mathcad*, *Matlab*, or *TKSolver* equation solver programs.

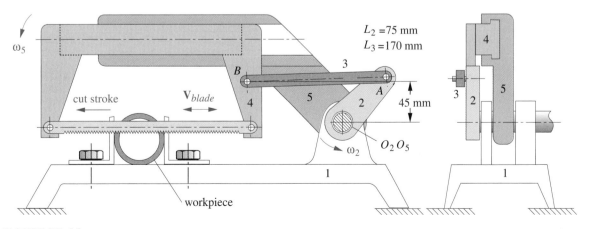

FIGURE P7-15

Problem 7-45 Power hacksaw. *Adapted from P. H. Hill and W. P Rule. (1960). Mechanisms: Analysis and Design, with permission.*

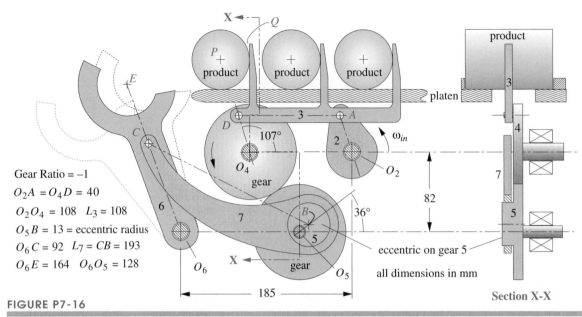

Gear Ratio = –1
$O_2A = O_4D = 40$
$O_2O_4 = 108$ $L_3 = 108$
$O_5B = 13$ = eccentric radius
$O_6C = 92$ $L_7 = CB = 193$
$O_6E = 164$ $O_6O_5 = 128$

FIGURE P7-16

Problem 7-46   Walking-beam indexer with pick-and-place mechanism   *Adapted from P. H. Hill and W. P. Rule. (1960). Mechanisms: Analysis and Design, with permission*

†7-51   Figure P7-21 (p. 392) shows a drag link mechanism with dimensions. Write the necessary equations and solve them to calculate the angular acceleration of link 4 for an input of $\omega_2 = 1$ rad/sec. Comment on uses for this mechanism.

7-52   Figure P7-22 (p. 393) shows a mechanism with dimensions. Use a graphical method to calculate the accelerations of points A, B, and C for the position shown. $\omega_2 = 20$ rad/sec.

7-53   Figure P7-23 (p. 393) shows a quick-return mechanism with dimensions. Use a graphical method to calculate the accelerations of points A, B, and C for the position shown. $\omega_2 = 10$ rad/sec.

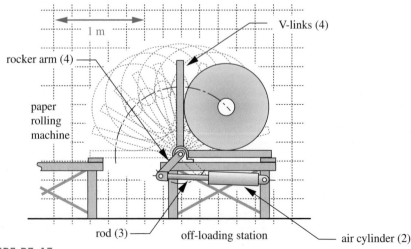

† These problems are suited to solution using *Mathcad*, *Matlab*, or *TKSolver* equation solver programs.

FIGURE P7-17

Problem 7-47

FIGURE P7-18

Problem 7-48 *From P. H. Hill and W. P. Rule. (1960). Mechanisms: Analysis and Design*

$L_2 = 0.80$ in
$L_4 = 2.97$
$L_5 = 2.61$
$\theta_2 = 241°$
$O_2O_4 = 1.85$ @ $278.5°$

3.25 in

†7-54   Figure P7-23 (p. 393) shows a quick-return mechanism with dimensions. Use an analytical method to calculate the accelerations of points *A*, *B*, and *C* for one revolution of the input link. $\omega_2 = 10$ rad/sec.

†7-55   Figure P7-24 (p. 393) shows a drum-pedal mechanism. $O_2A = 100$ mm at $162°$ and rotates to $171°$ at A'. $O_2O_4 = 56$ mm, $AB = 28$ mm, $AP = 124$ mm, and $O_4B = 64$ mm. The distance from O_4 to F_{in} is 48 mm. If the input velocity V_{in} is a constant magnitude of 3 m/sec, find the output acceleration over the range of motion.

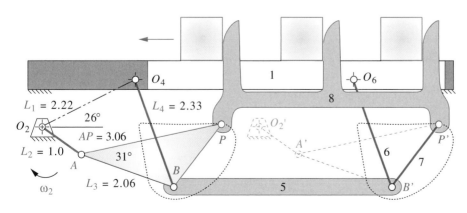

$L_1 = 2.22$
$26°$
$AP = 3.06$
$L_2 = 1.0$
$31°$
$L_3 = 2.06$
$L_4 = 2.33$

FIGURE P7-19

Problem 7-49 Straight-line walking-beam eightbar transport mechanism

† These problems are suited to solution using *Mathcad*, *Matlab*, or *TKSolver* equation solver programs.

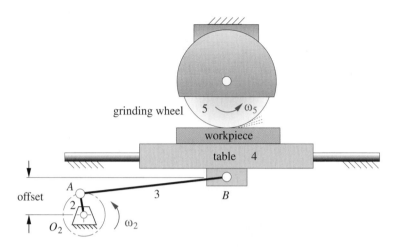

FIGURE P7-20

Problem 7-50 A surface grinder

*†7-56   A tractor-trailer tipped over while negotiating an on-ramp to the New York Thruway. The road has a 50-ft radius at that point and tilts 3° toward the outside of the curve. The 45-ft-long by 8-ft-wide by 8.5-ft-high trailer box (13 ft from ground to top) was loaded with 44 415 lb of paper rolls in two rows by two layers as shown in Figure P7-25 (p. 394). The rolls are 40 in diameter by 38 in long, and weigh about 900 lb each. They are wedged against backward rolling but not against sideward sliding. The empty trailer weighed 14 000 lb. The driver claims that he was traveling at less than 15 mph and that the load of paper shifted inside the trailer, struck the trailer sidewall, and tipped the truck. The paper company that loaded the truck claims the load was properly stowed and would not shift at that speed. Independent tests of the coefficient of friction between similar paper rolls and a similar trailer floor give a value of 0.43 ± 0.08. The composite center of gravity of the loaded trailer is estimated to be 7.5 ft above the road. Determine the truck speed that would cause the truck to just begin to tip and the speed at which the rolls will just begin to slide sideways. What do you think caused the accident?

$L_1 = 0.68$ in
$L_2 = 1.38$ in
$L_3 = 1.22$ in
$L_4 = 1.62$ in

* Answers in Appendix F.

† These problems are suited to solution using *Mathcad*, *Matlab*, or *TKSolver* equation solver programs.

FIGURE P7-21

Problem 7-51 *From P. H. Hill and W. P. Rule. (1960). Mechanisms: Analysis and Design*

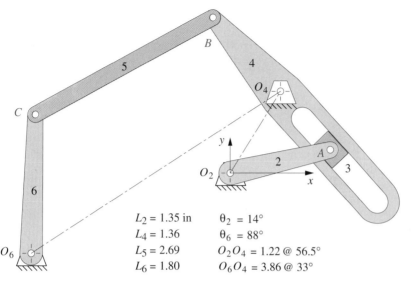

$L_2 = 1.35$ in $\qquad \theta_2 = 14°$
$L_4 = 1.36 \qquad\quad \theta_6 = 88°$
$L_5 = 2.69 \qquad\quad O_2O_4 = 1.22 @ 56.5°$
$L_6 = 1.80 \qquad\quad O_6O_4 = 3.86 @ 33°$

FIGURE P7-22

Problem 7-52 *From P. H. Hill and W. P. Rule. (1960). Mechanisms: Analysis and Design*

†7-57 Figure P7-26 (p. 394) shows a V-belt drive. The sheaves have pitch diameters of 150 and 300 mm, respectively. The smaller sheave is driven at a constant 1750 rpm. For a cross-sectional differential element of the belt, write the equations of its acceleration for one complete trip around both sheaves including its travel between the sheaves. Compute and plot the acceleration of the differential element versus time for one circuit around the belt path. What does your analysis tell about the dynamic behavior of the

FIGURE P7-24

Problem 7-55

$L_2 = 1.00$ in
$L_4 = 4.76$
$L_5 = 4.55$
$\theta_2 = 99°$
$O_4O_2 = 1.69 @ 15.5°$

2.86 in

FIGURE P7-23

Problem 7-53 to 7-54 *From P. H. Hill and W. P. Rule. (1960). Mechanisms: Analysis and Design*

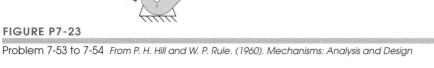

† These problems are suited to solution using *Mathcad*, *Matlab*, or *TKSolver* equation solver programs.

FIGURE P7-25

Problem 7-56

belt? Relate your findings to your personal observation of a belt of this type in operation. (Look in your school's machine shop or under the hood of an automobile—but mind your fingers!)

†7-58   Write a program using an equation solver or any computer language to solve for the displacements, velocities, and accelerations in an offset crank-slider linkage as shown in Figure P7-2 (p. 381). Plot the variation in all links' angular and all pins' linear positions, velocities, and accelerations with a constant angular velocity input to the crank over one revolution for both open and crossed configurations of the linkage. To test the program, use data from row *a* of Table P7-2 (p. 381). Check your results with program LINKAGES.

†7-59   Write a computer program or use an equation solver such as *Mathcad, Matlab*, or *TKSolver* to solve for the displacements, velocities, and accelerations in an inverted crank-slider linkage as shown in Figure P7-3 (p. 382). Plot the variation in all links' angular and all pins' linear positions, velocities, and accelerations with a constant angular velocity input to the crank over one revolution for both open and crossed configurations of the linkage. To test the program, use data from row *e* of Table P7-3 (p. 382) except for the value of α_2 which will be set to zero for this exercise.

---

† These problems are suited to solution using *Mathcad, Matlab*, or *TKSolver* equation solver programs.

FIGURE P7-26

Problem 7-57 A two-groove V-belt drive *Courtesy of T. B. Wood's Sons Co., Chambersburg, PA*

†7-60 Write a computer program or use an equation solver such as *Mathcad, Matlab,* or *TK-Solver* to solve for the displacements, velocities, and accelerations in a geared fivebar linkage as shown in Figure P7-4 (p. 383). Plot the variation in all links' angular and all pins' linear positions, velocities, and accelerations with a constant angular velocity input to the crank over one revolution for both open and crossed configurations of the linkage. To test the program, use data from row *a* of Table P7-4 (p. 383). Check your results with program LINKAGES.

7-61 Find the acceleration of the slider in Figure 3-33 (p. 152) for the position shown if $\theta_2 = 110°$ with respect to the global *X* axis assuming a constant $\omega_2 = 1$ rad/sec CW.

    a.   Using a graphical method.   †b.   Using an analytical method.

†7-62 Write a computer program or use an equation solver such as *Mathcad, Matlab,* or *TK-Solver* to calculate and plot the angular acceleration of link 4 and the linear acceleration of slider 6 in the sixbar crank-slider linkage of Figure 3-33 (p. 152) as a function of the angle of input link 2 for a constant $\omega_2 = 1$ rad/sec CW. Plot A_c both as a function of θ_2 and separately as a function of slider position as shown in the figure.

7-63 Find the angular acceleration of link 6 of the linkage in Figure 3-34 part (*b*) (p. 153) for the position shown ($\theta_6 = 90°$ with respect to the *x* axis) assuming constant $\omega_2 = 10$ rad/sec CW.

    a.   Using a graphical method.   †b.   Using an analytical method.

†7-64 Write a computer program or use an equation solver such as *Mathcad, Matlab,* or *TKSolver* to calculate and plot the angular acceleration of link 6 in the sixbar linkage of Figure 3-34 (p. 153) as a function of θ_2 for a constant $\omega_2 = 1$ rad/sec CW.

7-65 Use a compass and straightedge (ruler) to draw the linkage in Figure 3-35 (p. 153) with link 2 at 90° and find the angular acceleration of link 6 of the linkage assuming constant $\omega_2 = 10$ rad/sec CCW when $\theta_2 = 90°$.

    a.   Using a graphical method.   †b.   Using an analytical method.

†7-66 Write a computer program or use an equation solver such as *Mathcad, Matlab,* or *TKSolver* to calculate and plot the angular acceleration of link 6 in the sixbar linkage of Figure 3-35 (p. 153) as a function of θ_2 for a constant $\omega_2 = 1$ rad/sec CCW.

†7-67 Write a computer program or use an equation solver such as *Mathcad, Matlab,* or *TK-Solver* to calculate and plot the angular acceleration of link 8 in the linkage of Figure 3-36 (p. 154) as a function of θ_2 for a constant $\omega_2 = 1$ rad/sec CCW.

†7-68 Write a computer program or use an equation solver such as *Mathcad, Matlab,* or *TKSolver* to calculate and plot magnitude and direction of the acceleration of point *P* in Figure 3-37a (p. 154) as a function of θ_2. Also calculate and plot the acceleration of point *P* versus point *A*.

†7-69 Repeat problem 7-68 for the linkage in Figure 3-37b (p. 154).

7-70 Find the angular accelerations of links 3 and 4 and the linear accelerations of points *A, B,* and P_1 in the *XY* coordinate system for the linkage in Figure P7-27 (p. 396) in the position shown. Assume that $\theta_2 = 45°$ in the *XY* coordinate system and $\omega_2 = 10$ rad/sec, constant. The coordinates of the point P_1 on link 4 are (114.68, 33.19) with respect to the *xy* coordinate system.

    a.   Using a graphical method.   †b.   Using an analytical method.

†7-71 Using the data from Problem 7-70, write a computer program or use an equation solver such as *Mathcad, Matlab,* or *TKSolver* to calculate and plot magnitude and direction of the absolute acceleration of point P_1 in Figure P7-27 (p. 396) as a function of θ_2.

7-72 Find the angular accelerations of links 3 and 4, and the linear acceleration of point *P* in the *XY* coordinate system for the linkage in Figure P7-28 (p. 397) in the position shown. Assume that $\theta_2 = -94.121°$ in the *XY* coordinate system, $\omega_2 = 1$ rad/sec, and $\alpha_2 = 10$ rad/

7

† These problems are suited to solution using *Mathcad, Matlab,* or *TKSolver* equation solver programs.

FIGURE P7-27

Problems 7-70 to 7-71 An oil field pump - dimensions in inches

sec^2. The position of the coupler point P on link 3 with respect to point A is: $p = 15.00$, δ_3 = 0°.

    a.   Using a graphical method.   †b   Using an analytical method.

†7-73   For the linkage in Figure P7-28, write a computer program or use an equation solver such as *Mathcad, Matlab*, or *TKSolver* to calculate and plot the angular velocity and acceleration of links 2 and 4, and the magnitude and direction of the velocity and acceleration of point P as a function of θ_2 through its possible range of motion starting at the position shown. The position of the coupler point P on link 3 with respect to point A is: $p = 15.00$, $\delta_3 = 0°$. Assume that, @ $t = 0$, $\theta_2 = -94.121°$ in the XY coordinate system, $\omega_2 = 0$, and $\alpha_2 = 10$ rad/sec^2, constant.

7-74   Derive analytical expressions for the accelerations of points A and B in Figure P7-29 as a function of θ_3, ω_3, α_3, and the length AB of link 3. Use a vector loop equation. Code them in an equation solver or a programming language and plot them.

7-75   The linkage in Figure P7-30a (p. 398) has link 2 at 120° in the global XY coordinate system. Find α_6 and A_D in the global coordinate system for the position shown if ω_2 = 10 rad/sec CCW and $\alpha_2 = 50$ rad/sec^2 CW. Use the acceleration difference graphical method. (Print the figure from its PDF file on the DVD and draw on it.)

*7-76   The linkage in Figure P7-30a (p. 398) has link 2 at 120° in the global XY coordinate system. Find α_6 and A_D in the global coordinate system for the position shown if ω_2 = 10 rad/sec CCW and $\alpha_2 = 50$ rad/sec^2 CW. Use an analytical method.

7-77   The linkage in Figure P7-30b (p. 398) has link 3 perpendicular to the X axis and links 2 and 4 are parallel to each other. Find α_4, A_A, A_B, and A_P if $\omega_2 = 15$ rad/sec CW and $\alpha_2 = 100$ rad/ sec^2 CW. Use the acceleration difference graphical method. (Print the figure from its PDF file on the DVD and draw on it.)

*7-78   The linkage in Figure P7-30b (p. 398) has link 3 perpendicular to the X axis and links 2 and 4 are parallel to each other. Find α_4, A_A, A_B, and A_P if $\omega_2 = 15$ rad/sec CW and $\alpha_2 = 100$ rad/sec^2 CW. Use an analytical method.

* Answers in Appendix F.

† These problems are suited to solution using *Mathcad, Matlab*, or *TKSolver* equation solver programs.

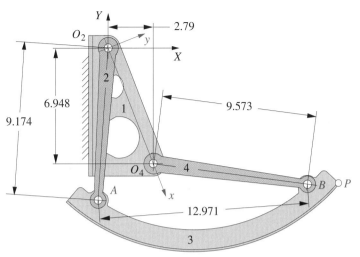

FIGURE P7-28

Problems 7-72 and 7-73 An aircraft overhead bin mechanism - dimensions in inches

FIGURE P7-29

Elliptical trammel - Problem 7-74

7-79   The crosshead linkage shown in Figure P7-30c (p. 398) has 2 *DOF* with inputs at crossheads 2 and 5. Find A_B, A_{P_3}, and A_{P_4} if the crossheads are each moving toward the origin of the *XY* coordinate system with a speed of 20 in/sec and are decelerating at 75 in/sec². Use the acceleration difference method. (Print the figure from its PDF file on the DVD and draw on it.)

[†]7-80   The crosshead linkage shown in Figure P7-30c (p. 398) has 2 *DOF* with inputs at crossheads 2 and 5. Find A_B, A_{P_3}, and A_{P_4} if the crossheads are each moving toward the origin of the *XY* coordinate system with a speed of 20 in/sec and are decelerating at 75 in/sec². Use an analytical method.

[†§]7-81   The crosshead linkage shown in Figure P7-30c (p. 398) has 2 *DOF* with inputs at crossheads 2 and 5. At $t = 0$, crosshead 2 is at rest at the origin of the global *XY* coordinate system and crosshead 5 is at rest at (70, 0). Write a computer program to find and plot A_{P_3} and A_{P_4} for the first 5 sec of motion if $A_2 = 0.5$ in/sec² upward and $A_5 = 0.5$ in/sec² to the left.

7-82   The linkage in Figure P7-30d (p. 398) has the path of slider 6 perpendicular to the global *X* axis and link 2 aligned with the global *X* axis. Find α_2 and A_A in the position shown if the velocity of the slider is constant at 20 in/sec downward. Use the acceleration difference graphical method. Print the figure's PDF file (on DVD) and draw on it.

[†]7-83   The linkage in Figure P7-30d (p. 398) has the path of slider 6 perpendicular to the global *X* axis and link 2 aligned with the global *X* axis. Find α_2 and A_A in the position shown if the velocity of the slider is constant at 20 in/sec downward. Use an analytical method.

[†]7-84   The linkage in Figure P7-30d (p. 398) has the path of slider 6 perpendicular to the global *X* axis and link 2 aligned with the global *X* axis at $t = 0$. Write a computer program or use an equation solver to find and plot A_D as a function of θ_2 over the possible range of motion of link 2 in the global *XY* coordinate system.

[†§]7-85   For the linkage of Figure P7-30e (p. 398), write a computer program or use an equation solver to find and plot A_D in the global coordinate system for one revolution of link 2 if ω_2 is constant at 10 rad/sec CW.

7

[†] These problems are suited to solution using *Mathcad, Matlab,* or *TKSolver* equation solver programs.

[§] Note that these can be long problems to solve and may be more appropriate for a project assignment than an overnight problem. In most cases, the solution can be checked with program LINKAGES.

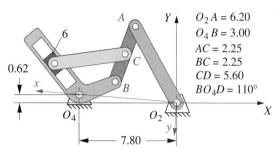

$O_2A = 6.20$
$O_4B = 3.00$
$AC = 2.25$
$BC = 2.25$
$CD = 5.60$
$BO_4D = 110°$

(a) Sixbar linkage

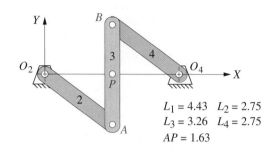

$L_1 = 4.43$ $L_2 = 2.75$
$L_3 = 3.26$ $L_4 = 2.75$
$AP = 1.63$

(b) Fourbar linkage

$AB = 34.32$
$BC = 50.4$
$AP_3 = 31.5$
$BP_3 = 22.2$
$BP_4 = 41.52$
$CP_4 = 27$

(c) Dual crosshead mechanism

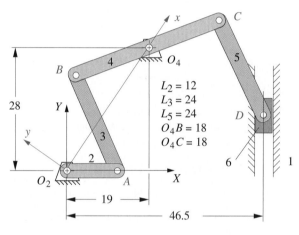

$L_2 = 12$
$L_3 = 24$
$L_5 = 24$
$O_4B = 18$
$O_4C = 18$

(d) Sixbar linkage

$L_2 = 5$ $L_3 = 5$ $L_5 = 15$ $BC = 8$
$O_2O_4 = 2.5$ $O_4B = 6$ $O_4C = 6$

(e) Drag link slider-crank

$L_2 = 5.0$
$L_3 = 8.4$
$L_4 = 2.5$
$L_5 = 8.9$
$L_6 = 3.2$
$L_7 = 6.4$
$O_2O_4 = 12.5$
$AC = 2.4$
$CD = 5.9$

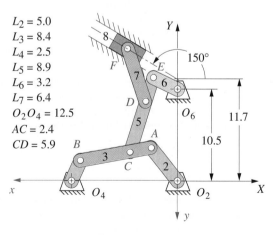

(f) Eightbar mechanism

FIGURE P7-30

Problems 7-75 to 7-86

7-86   The linkage of Figure P7-30f has link 2 at 130° in the global XY coordinate system. Find $\mathbf{A}_D$ in the global coordinate system for the position shown if $\omega_2 = 15$ rad/sec CW and $\alpha_2 = 50$ rad/sec² CW. Use the acceleration difference graphical method. (Print the figure from its PDF file on the DVD and draw on it.)

*7-87   Figure 3-14 (p. 122) shows a crank-shaper quick-return mechanism with the dimensions: $L_2 = 4.80$ in, $L_4 = 24.00$ in, $L_5 = 19.50$ in. The distance from link 4's pivot (O_4) to link 2's pivot (O_2) is 16.50 in. The vertical distance from O_2 to point C on link 6 is 6.465 in. Use a graphical method to find the acceleration of point C on link 6 when the linkage is near the rightmost position shown with $\theta_2 = 45°$ measured from an axis running from an origin at O_2 through O_4. Assume that link 2 has a constant angular velocity of 2 rad/sec CW.

§7-88   Use the data in Problem 7-87 and an analytical method to calculate and plot the acceleration of point C on link 6 of that mechanism for one revolution of input crank 2.

7-89   Figure P7-22 (p. 393) shows a mechanism with dimensions. Use a graphical method to determine the acceleration of points A and B for the position shown for $\omega_2 = 24$ rad/s CW. Ignore links 5 and 6.

7-90   Figure P7-22 (p. 393) shows a mechanism with dimensions. Use an analytical method to calculate the accelerations of points A and B for the position shown for $\omega_2 = 24$ rad/s CW. Ignore links 5 and 6.

7-91   Figure P7-23 (p. 393) shows a quick-return mechanism with dimensions. Use a graphical method to determine the accelerations of points A and B for the position shown for $\omega_2 = 16$ rad/s CCW. Ignore links 5 and 6.

7-92   Figure P7-23 (p. 393) shows a quick-return mechanism with dimensions. Use an analytical method to calculate the accelerations of points A and B for the position shown for $\omega_2 = 16$ rad/s CCW. Ignore links 5 and 6.

7-93   The general linkage configuration and terminology for an offset fourbar slider-crank linkage are shown in Fig P7-2 (p. 381). The link lengths and the values of d, $\dot{d}$, and $\ddot{d}$ are defined in Table P7-5. For the row(s) assigned, find the acceleration of the pin joint A and the angular acceleration of the crank using a graphical method.

7-94   The general linkage configuration and terminology for an offset fourbar slider-crank linkage are shown in Fig P7-2 (p. 381). The link lengths and the values of d, $\dot{d}$, and $\ddot{d}$ are defined in Table P7-5. For the rows assigned, find the acceleration of pin joint A and the angular acceleration of the crank using the analytic method. Draw the linkage to scale and label it before setting up the equations.

TABLE P7-5   Data for Problems 7-93 to 7-94‡

Row	Link 2	Link 3	Offset	d	$\dot{d}$	$\ddot{d}$
a	1.4	4	1	2.5	10	0
b	2	6	−3	5	−12	5
c	3	8	2	8	−15	−10
d	3.5	10	1	−8	24	−4
e	5	20	−5	15	−50	10
f	3	13	0	−12	−45	50
g	7	25	10	25	100	18

‡ Drawings of these linkages are in the *PDF Problem Workbook* folder on the DVD.

* Answers in Appendix F.

§ Note that these can be long problems to solve and may be more appropriate for a project assignment than an overnight problem. In most cases, the solution can be checked with program LINKAGES.

7.11   VIRTUAL LABORATORY

L7-1   View the video *Fourbar Linkage Virtual Laboratory* that is on the DVD. Open the file *Virtual Fourbar Linkage Lab 7-1.doc* from the DVD and follow the instructions as directed by your professor.

7

Chapter **8**

CAM DESIGN

It is much easier to design than to perform
SAMUEL JOHNSON

8.0 INTRODUCTION

Cam-follower systems are frequently used in all kinds of machines. The valves in your automobile engine are opened by cams. Machines used in the manufacture of many consumer goods are full of cams.[*] Compared to linkages, cams are easier to design to give a specific output function, but they are much more difficult and expensive to make than a linkage. Cams are a form of degenerate fourbar linkage in which the coupler link has been replaced by a half joint as shown in Figure 8-1 (p. 402). This topic was discussed in Section 2.10 (p. 48) on linkage transformation (see also Figure 2-12, p. 51). For any one instantaneous position of cam and follower, we can substitute an effective linkage that will, for that instantaneous position, have the same motion as the original. In effect, the cam-follower is a fourbar linkage with variable-length (effective) links. It is this conceptual difference that makes the cam-follower such a versatile and useful **function generator**. We can specify virtually any output function we desire and quite likely create a curved surface on the cam to generate that function in the motion of the follower. We are not limited to fixed-length links as we were in linkage synthesis. The cam-follower is an extremely useful mechanical device, without which the machine designer's tasks would be more difficult to accomplish. But, as with everything else in engineering, there are trade-offs. These will be discussed in later sections. A list of the variables used in this chapter is provided in Table 8-1 (p. 402).

This chapter will present the proper approach to designing a cam-follower system, and in the process also present some less than proper designs as examples of the problems that inexperienced cam designers often get into. Theoretical considerations of the mathematical functions commonly used for cam curves will be discussed. Methods for the derivation of custom polynomial functions, to suit any set of boundary conditions, will be presented. The task of sizing the cam with considerations of pressure angle and radius of curvature will be addressed, and manufacturing processes and their limitations

[*] View the video *Pick and Place Mechanism* on the book's DVD to see an example of a cam driven mechanism from an actual production machine.

TABLE 8-1	Notation Used in This Chapter

t = time, seconds

θ = camshaft angle, degrees or radians (rad)

ω = camshaft angular velocity, rad/sec

β = total angle of any segment, rise, fall, or dwell, degrees or rad

h = total lift (rise or fall) of any one segment, length units

s or S = follower displacement, length units

$v = ds/d\theta$ = follower velocity, length/rad

$V = dS/dt$ = follower velocity, length/sec

$a = dv/d\theta$ = follower acceleration, length/rad^2

$A = dV/dt$ = follower acceleration, length/sec^2

$j = da/d\theta$ = follower jerk, length/rad^3

$J = dA/dt$ = follower jerk, length/sec^3

$s \, v \, a \, j$ refers to the group of diagrams, length units versus radians

$S \, V \, A \, J$ refers to the group of diagrams, length units versus time

R_b = base circle radius, length units

R_p = prime circle radius, length units

R_f = roller follower radius, length units

ε = eccentricity of cam-follower, length units

ϕ = pressure angle, degrees or radians

ρ = radius of curvature of cam surface, length units

ρ_{pitch} = radius of curvature of pitch curve, length units

ρ_{min} = minimum radius of curvature of pitch curve or cam surface, length units

discussed. The computer program DYNACAM will be used throughout the chapter as a tool to present and illustrate design concepts and solutions. Information about this program is in Appendix A.

8.1   CAM TERMINOLOGY

Cam-follower systems can be classified in several ways: by *type of follower motion*, either **translating** or **rotating** (oscillating); by *type of cam*, radial, cylindrical, three-dimensional; by *type of joint closure*, either **force-** or **form**-closed; by *type of follower*, **curved** or **flat**, **rolling** or **sliding**; by *type of motion constraints*, **critical extreme position** (CEP), **critical path motion** (CPM); by *type of motion program*, **rise-fall** (RF), **rise-fall-dwell** (RFD), **rise-dwell-fall-dwell** (RDFD). We will now discuss each of these classification schemes in greater detail.

Type of Follower Motion

Figure 8-1a shows a system with an oscillating, or **rotating, follower**. Figure 8-1b shows a **translating follower**. These are analogous to the crank-rocker fourbar and the slider-crank fourbar linkages, respectively. An effective fourbar linkage can be substituted for the cam-follower system for any instantaneous position. The lengths of the effective links are determined by the instantaneous locations of the centers of curvature of cam and follower as shown in Figure 8-1. The velocities and accelerations of the cam-follower system can be found by analyzing the behavior of the effective linkage for any position. A proof of this can be found in reference [1]. Of course, the effective links change length as the cam-follower moves, giving it an advantage over a pure linkage as this allows greater flexibility in meeting the desired motion constraints.

The choice between these two forms of the cam-follower is usually dictated by the type of output motion desired. If true rectilinear translation is required, then the translating follower is dictated. If pure rotation output is needed, then the oscillator is the obvious choice. There are advantages to each of these approaches, separate from their motion characteristics, depending on the type of follower chosen. These will be discussed in a later section.

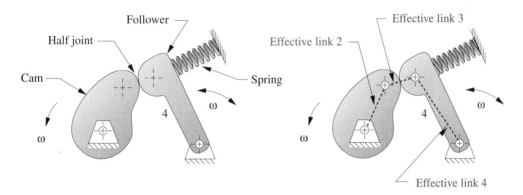

(a) An oscillating cam-follower has an effective pin-jointed fourbar equivalent

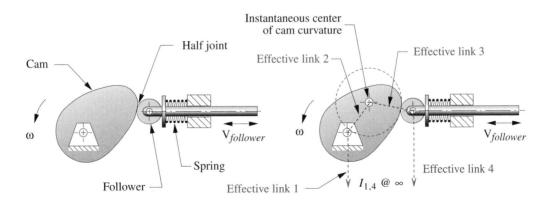

(b) A translating cam-follower has an effective fourbar slider-crank equivalent

FIGURE 8-1

Effective linkages in the cam-follower mechanism

Type of Joint Closure

Force and form closure were discussed in Section 2.3 (p. 32) on the subject of joints and have the same meaning here. **Force closure**, as shown in Figure 8-1 (p. 403), *requires an external force be applied to the joint* in order to keep the two links, cam and follower, physically in contact. This force is usually provided by a spring. This force, defined as positive in a direction that closes the joint, cannot be allowed to become negative. If it does, the links have lost contact because a *force-closed joint can only push, not pull.* **Form closure**, as shown in Figure 8-2, *closes the joint by geometry.* No external force is required. There are really two cam surfaces in this arrangement, one surface on each side of the follower. Each surface pushes, in its turn, to drive the follower in both directions.

Figure 8-2a and b shows track or groove cams that capture a single follower in the groove and both push and pull on the follower. Figure 8-2c shows another variety of

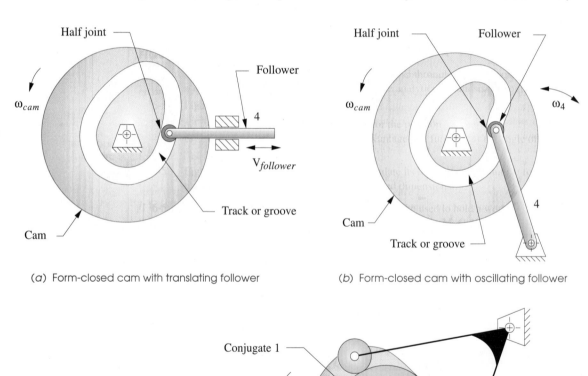

(a) Form-closed cam with translating follower   (b) Form-closed cam with oscillating follower

(c) Conjugate cams on common shaft

FIGURE 8-2

Form-closed cam-follower systems

form-closed cam-follower arrangement, called **conjugate cams**. There are two cams fixed on a common shaft that are mathematical conjugates of one another. Two roller followers, attached to a common arm, are each pushed in opposite directions by the conjugate cams. When form-closed cams are used in automobile or motorcycle engine valve trains, they are called **desmodromic**[*] cams. There are advantages and disadvantages to both force- and form-closed arrangements that are discussed in Section 8-7 (p. 467).

Type of Follower

Follower, in this context, refers only to that part of the follower link that contacts the cam. Figure 8-3 shows three common arrangements, **flat-faced**, **mushroom** (curved), and **roller**. The roller follower has the advantage of lower (rolling) friction than the sliding contact of the other two but can be more expensive. **Flat-faced followers** can package smaller than roller followers for some cam designs and are often favored for that reason as well as cost for automotive valve trains. **Roller followers** are most frequently used in production machinery where their ease of replacement and availability from bearing manufacturers' stock in any quantities are advantages. Grooved or track cams require roller followers. Roller followers are essentially ball or roller bearings with customized mounting details. Figure 8-5a (p. 407) shows two common types of commercial roller followers. Flat-faced or **mushroom followers** are usually custom-designed and manufactured for each application. For high-volume applications such as automobile engines, the quantities are high enough to warrant a custom-designed follower.

Type of Cam

The direction of the follower's motion relative to the axis of rotation of the cam determines whether it is a **radial** or **axial** cam. All cams shown in Figures 8-1 to 8-3 are radial

[*] More information on desmodromic cam-follower mechanisms can be found at http://members.chello. nl/~wgj.jansen/ where a number of models of their commercial implementations can be viewed in operation as movies.

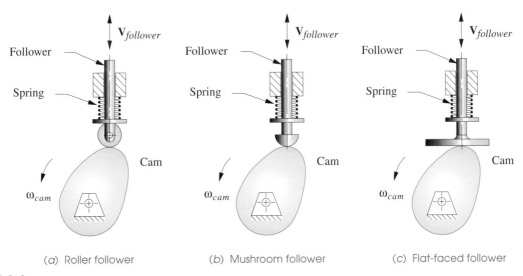

| (a) Roller follower | (b) Mushroom follower | (c) Flat-faced follower |

FIGURE 8-3

Three common types of cam followers

cams because the follower motion is generally in a radial direction. Open **radial cams** are also called **plate cams**.

Figure 8-4 shows an **axial cam** whose follower moves parallel to the axis of cam rotation. This arrangement is also called a **face** cam if open (force-closed) and a **cylindrical** or **barrel** cam if grooved or ribbed (form-closed).

Figure 8-5b shows a selection of cams of various types. Clockwise from the lower left, they are: an open (force-closed) axial or face cam; an axial grooved (track) cam (form-closed) with external gear; an open radial, or plate cam (force-closed); a ribbed axial cam (form-closed); an axial grooved (barrel) cam.

Three-dimensional cams (Figure 8-5c) are a combination of radial and axial cams. The input rotation of the cam drives a follower train having both radial and axial motion. The follower motion has two coupled degrees of freedom.

Type of Motion Constraints

There are two general categories of motion constraint, **critical extreme position** (CEP; also called endpoint specification) and **critical path motion** (CPM). **Critical extreme position** refers to the case in which the design specifications define the start and finish positions of the follower (i.e., extreme positions) but do not specify any constraints on the path motion between the extreme positions. This case is discussed in Sections 8.3 (p. 409) and 8.4 (p. 435) and is the easier of the two to design as the designer has great freedom to choose the cam functions that control the motion between extremes. **Critical path motion** is a more constrained problem than CEP because the path motion and/or one or more of its derivatives are defined over all or part of the interval of motion. This is analogous to **function generation** in the linkage design case except that with a cam we can achieve a continuous output function for the follower. Section 8.5 (p. 443) discusses this CPM case. It may only be possible to create an approximation of the specified function and still maintain suitable dynamic behavior.

FIGURE 8-4

Axial, cylindrical, or barrel cam with form-closed, translating follower

(a) Commercial roller followers
Courtesy of McGill Manufacturing Co.
South Bend, IN

(b) Commercial cams of various types
Courtesy of The Ferguson Co.
St. Louis, MO

(c) Three-dimensional cams
Courtesy of The Gillette Co.
Boston, MA

FIGURE 8-5

Cams and roller followers

Type of Motion Program

The motion programs **rise-fall** (RF), **rise-fall-dwell** (RFD), and **rise-dwell-fall-dwell** (RDFD) all refer mainly to the CEP case of motion constraint and in effect define how many dwells are present in the full cycle of motion, none (RF), one (RFD), or more than one (RDFD). **Dwells,** defined as *no output motion for a specified period of input motion,* are an important feature of cam-follower systems because it is very easy to create exact dwells in these mechanisms. The cam-follower is the design type of choice whenever a dwell is required. We saw in Section 3.9 (p. 146) how to design dwell linkages and found that at best we could obtain only an approximate dwell. The resulting single- or double-dwell linkages tend to be quite large for their output motion and are somewhat difficult to design. (See program SIXBAR for some built-in examples of these dwell linkages.) Cam-follower systems tend to be more compact than linkages for the same output motion.

If your need is for a **rise-fall** (RF) CEP motion, with no dwell, then you should really be considering a crank-rocker linkage rather than a cam-follower to obtain all the linkage's advantages over cams of reliability, ease of construction, and lower cost that were discussed in Section 2.18 (p. 69). If your needs for compactness outweigh those considerations, then the choice of a cam-follower in the RF case may be justified. Also, if you have a CPM design specification, and the motion or its derivatives are defined over the interval, then a cam-follower system is the logical choice in the RF case.

The **rise-fall-dwell** (RFD) and **rise-dwell-fall-dwell** (RDFD) cases are obvious choices for cam-followers for the reasons discussed above. However, each of these two cases has its own set of constraints on the behavior of the cam functions at the interfaces between the segments that control the rise, the fall, and the dwells. In general, we must match the **boundary conditions** (BCs) of the functions and their derivatives at all interfaces between the segments of the cam. This topic will be thoroughly discussed in the following sections.

8.2 $S\,V\,A\,J$ DIAGRAMS

The first task faced by the cam designer is to select the mathematical functions to be used to define the motion of the follower. The easiest approach to this process is to "linearize" the cam, i.e., "unwrap it" from its circular shape and consider it as a function plotted on cartesian axes. We plot the displacement function s, its first derivative velocity v, its second derivative acceleration a, and its third derivative jerk j, all on aligned axes as a function of camshaft angle θ as shown in Figure 8-6. Note that we can consider the independent variable in these plots to be either time t or shaft angle θ, as we know the constant angular velocity ω of the camshaft and can easily convert from angle to time and vice versa.

$$\theta = \omega t \qquad\qquad\qquad (8.1)$$

Figure 8-6a shows the specifications for a four-dwell cam that has eight segments, RDFDRDFDRDFD. Figure 8-6b shows the $s\,v\,a\,j$ curves for the whole cam over 360 degrees of camshaft rotation. A cam design begins with a definition of the required cam functions and their $s\,v\,a\,j$ diagrams. Functions for the nondwell cam segments should be chosen based on their velocity, acceleration, and jerk characteristics and the relationships at the interfaces between adjacent segments including the dwells. These function characteristics can be conveniently and quickly investigated with program DYNACAM which generated the data and plots shown in Figure 8-6.

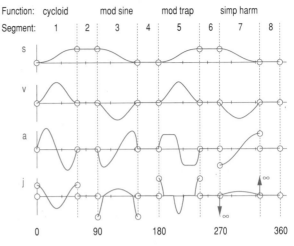

Segment Number	Function Used	Start Angle	End Angle	Delta Angle
1	Cycloid rise	0	60	60
2	Dwell	60	90	30
3	ModSine fall	90	150	60
4	Dwell	150	180	30
5	ModTrap rise	180	240	60
6	Dwell	240	270	30
7	SimpHarm fall	270	330	60
8	Dwell	330	360	30

(a) Cam program specifications

(b) Plots of cam-follower's s v a j diagrams

FIGURE 8-6

Cycloidal, modified sine, modified trapezoid, and simple harmonic motion functions on a four-dwell cam

8.3 DOUBLE-DWELL CAM DESIGN—CHOOSING S V A J FUNCTIONS

Many cam design applications require multiple dwells. The double-dwell case is quite common. Perhaps a **double-dwell** cam is driving a part feeding station on a production machine that makes toothpaste. This hypothetical cam's follower is fed an empty tooth-paste tube (during the low dwell), then moves the empty tube into a loading station (during the rise), holds the tube absolutely still in a **critical extreme position** (CEP) while tooth-paste is squirted into the open bottom of the tube (during the high dwell), and then retracts the filled tube back to the starting (zero) position and holds it in this other critical extreme position. At this point, another mechanism (during the low dwell) picks the tube up and carries it to the next operation, which might be to seal the bottom of the tube. A similar cam could be used to feed, align, and retract the tube at the bottom-sealing station as well.

Cam specifications such as this are often depicted on a timing diagram as shown in Figure 8-7 (p. 410) which is a graphical representation of the specified events in the ma-chine cycle. A **machine's cycle** is defined as *one revolution of its master driveshaft*. In a complicated machine, such as our toothpaste maker, there will be a **timing diagram** for each subassembly in the machine. The time relationships among all subassemblies are de-fined by their timing diagrams which are all drawn on a common time axis. Obviously all these operations must be kept in precise synchrony and time phase for the machine to work.

This simple example in Figure 8-7 is a critical extreme position (CEP) case, because nothing is specified about the functions to be used to get from the low dwell position (one extreme) to the high dwell position (other extreme). The designer is free to choose any function that will do the job. Note that these specifications contain only information about the displacement function. The higher derivatives are not specifically constrained in this example. We will now use this problem to investigate several different ways to meet the specifications.

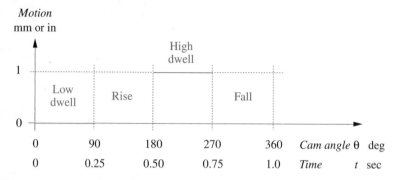

Motion
mm or in

FIGURE 8-7

Cam timing diagram

EXAMPLE 8-1

Naive Cam Design—A Bad Cam.

Problem: Consider the following cam design CEP specification:

dwell	at zero displacement for 90 degrees (low dwell)
rise	1 in (25 mm) in 90 degrees
dwell	at 1 in (25 mm) for 90 degrees (high dwell)
fall	1 in (25 mm) in 90 degrees
cam ω	2π rad/sec = 1 rev/sec

Solution:

1   The naive or inexperienced cam designer might proceed with a design as shown in Figure 8-8a. Taking the given specifications literally, it is tempting to merely "connect the dots" on the timing diagram to create the displacement (s) diagram. (After all, when we wrap this s diagram around a circle to create the actual cam, it will look quite smooth despite the sharp corners on the s diagram.) The mistake our beginning designer is making here is to ignore the effect on the higher derivatives of the displacement function that results from this simplistic approach.

2   Figure 8-8b, c, and d shows the problem. Note that we have to treat each segment of the cam (rise, fall, dwell) as a separate entity in developing mathematical functions for the cam. Taking the rise segment (#2) first, the displacement function in Figure 8-8a during this portion is a straight line, or first-degree polynomial. The general equation for a straight line is:

$$y = mx + b \qquad (8.2)$$

where m is the slope of the line and b is the y intercept. Substituting variables appropriate to this example in equation 8.2, angle θ replaces the independent variable x, and the displacement s replaces the dependent variable y. By definition, the constant slope m of the displacement is the velocity constant K_v.

3   For the rise segment, the y intercept b is zero because the low dwell position typically is taken as zero displacement by convention. Equation 8.2 then becomes:

$$s = K_v \theta \qquad (8.3)$$

4   Differentiating with respect to θ gives a function for velocity during the rise.

$$v = K_v = \text{constant} \qquad\qquad (8.4)$$

5   Differentiating again with respect to θ gives a function for acceleration during the rise.

$$a = 0 \qquad\qquad (8.5)$$

This seems too good to be true (and it is). Zero acceleration means zero dynamic force. This cam appears to have no dynamic forces or stresses in it!

Figure 8-8 shows what is really happening here. If we return to the displacement function and graphically differentiate it twice, we will observe that, from the definition of the derivative as the instantaneous slope of the function, the acceleration is in fact zero **during the interval**. *But, at the boundaries of the interval*, where rise meets low dwell on one side and high dwell on the other, note that *the velocity function is multivalued. There are discontinuities at these boundaries*. The effect of these discontinuities is to create a portion of the velocity curve that has **infinite slope** and zero duration. This results in the *infinite spikes of acceleration* shown at those points.

8

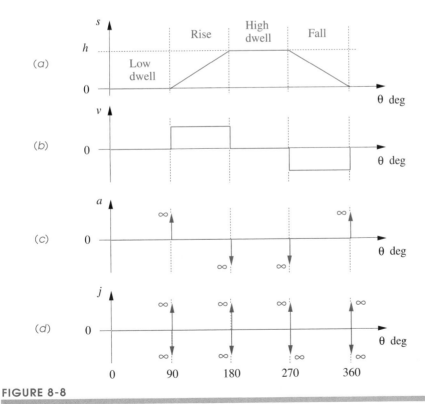

FIGURE 8-8

The *s v a j* diagrams of a "bad" cam design

These spikes are more properly called **Dirac delta functions**. Infinite acceleration cannot really be obtained, as it requires infinite force. Clearly the dynamic forces will be very large at these boundaries and will create high stresses and rapid wear. In fact, if this cam were built and run at any significant speeds, the sharp corners on the displacement diagram that are creating these theoretical infinite accelerations would be quickly worn to a smoother contour by the unsustainable stresses generated in the materials. *This is an unacceptable design.*

The unacceptability of this design is reinforced by the **jerk** diagram which shows theoretical values of **±infinity** at the discontinuities (the **doublet** function). The problem has been engendered by an inappropriate choice of displacement function. In fact, the cam designer should not be as concerned with the displacement function as with its higher derivatives.

The Fundamental Law of Cam Design

Any cam designed for operation at other than very low speeds must be designed with the following constraints:

The cam function must be continuous through the first and second derivatives of displacement across the entire interval (360 degrees).

Corollary:

The jerk function must be finite across the entire interval (360 degrees).

In any but the simplest of cams, the cam motion program cannot be defined by a single mathematical expression, but rather must be defined by several separate functions, each of which defines the follower behavior over one segment, or piece, of the cam. These expressions are sometimes called *piecewise functions*. These functions must have **third-order continuity** (the function plus two derivatives) at all boundaries. **The displacement, velocity and acceleration functions must have no discontinuities in them.**[*]

If any discontinuities exist in the acceleration function, then there will be infinite spikes, or Dirac delta functions, appearing in the derivative of acceleration, jerk. Thus the corollary merely restates the fundamental law of cam design. Our naive designer failed to recognize that by starting with a low-degree (linear) polynomial as the displacement function, discontinuities would appear in the upper derivatives.

Polynomial functions are one of the best choices for cams as we shall shortly see, but they do have one fault that can lead to trouble in this application. Each time they are differentiated, they reduce by one degree. Eventually, after enough differentiations, polynomials degenerate to zero degree (a constant value) as the velocity function in Figure 8-8b (p. 411) shows. Thus, by starting with a first-degree polynomial as a displacement function, it was inevitable that discontinuities would soon appear in its derivatives.

In order to obey the fundamental law of cam design, one must start with at least a fifth-degree polynomial (quintic) as the displacement function for a double-dwell cam. This will degenerate to a cubic function in the acceleration. The parabolic jerk function will have discontinuities, and the (unnamed) derivative of jerk will have infinite spikes in it. This is acceptable, as the jerk is still finite.

* This rule is stated by Neklutin[2] but is disputed by some other authors.[3],[4] Nevertheless, this author believes that it is a good (and simple) rule to follow in order to get acceptable dynamic results with high-speed cams. There are clear simulation data and experimental evidence that smooth jerk functions reduce residual vibrations in cam-follower systems.[10]

Simple Harmonic Motion (SHM)

Our naive cam designer recognized his mistake in choosing a straight-line function for the displacement. He also remembered a family of functions he had met in a calculus course that have the property of remaining continuous throughout any number of differentiations. These are the harmonic functions. On repeated differentiation, sine becomes cosine, which becomes negative sine, which becomes negative cosine, etc., ad infinitum. One never runs out of derivatives with the harmonic family of curves. In fact, differentiation of a harmonic function really only amounts to a 90° phase shift of the function. It is as though, when you differentiated it, you cut out, with a scissors, a different portion of the same continuous sine wave function, which is defined from minus infinity to plus infinity. The equations of simple harmonic motion (SHM) for a rise motion are:

$$s = \frac{h}{2}\left[1 - \cos\left(\pi\frac{\theta}{\beta}\right)\right] \tag{8.6a}$$

$$v = \frac{\pi}{\beta}\frac{h}{2}\sin\left(\pi\frac{\theta}{\beta}\right) \tag{8.6b}$$

$$a = \frac{\pi^2}{\beta^2}\frac{h}{2}\cos\left(\pi\frac{\theta}{\beta}\right) \tag{8.6c}$$

$$j = -\frac{\pi^3}{\beta^3}\frac{h}{2}\sin\left(\pi\frac{\theta}{\beta}\right) \tag{8.6d}$$

where h is the total rise, or lift, θ is the camshaft angle, and β is the total angle of the rise interval.

We have here introduced a notation to simplify the expressions. The independent variable in our cam functions is θ, the camshaft angle. The period of any one segment is defined as the angle β. Its value can, of course, be different for each segment. We normalize the independent variable θ by dividing it by the period of the segment β. Both θ and β are measured in radians (or both in degrees). The value of θ/β will then vary from 0 to 1 over any segment. It is a dimensionless ratio. Equations 8.6 define simple harmonic motion and its derivatives for this rise segment in terms of θ/β.

This family of harmonic functions appears, at first glance, to be well suited to the cam design problem of Figure 8-7 (p. 410). If we define the displacement function to be one of the harmonic functions, we should not "run out of derivatives" before reaching the acceleration function.

EXAMPLE 8-2

Sophomoric* Cam Design—Simple Harmonic Motion—Still a Bad Cam.

Problem: Consider the same cam design CEP specification as in Example 8-1 (p. 410):

dwell	at zero displacement for 90 degrees (low dwell)
rise	1 in (25 mm) in 90 degrees
dwell	at 1 in (25 mm) for 90 degrees (high dwell)
fall	1 in (25 mm) in 90 degrees
cam ω	2π rad/sec = 1 rev/sec

* **Sophomoric**, from sophomore, *def. wise fool*, from the Greek, *sophos = wisdom, moros = fool*.

8

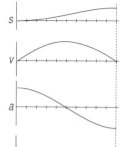

FIGURE 8-9

Simple harmonic motion with dwells has discontinuous acceleration.

Solution:

1. Figure 8-9 shows a full-rise simple harmonic function[†] applied to the rise segment of our cam design problem.

2. Note that the velocity function is continuous, as it matches the zero velocity of the dwells at each end. The peak value is 6.28 in/sec (160 mm/sec) at the midpoint of the rise.

3. The acceleration function, however, is **not** continuous. It is a half-period cosine curve and has nonzero values at start and finish that are ± 78.8 in/sec^2 (2.0 m/sec^2).

4. Unfortunately, the dwell functions, which adjoin this rise on each side, have zero acceleration as can be seen in Figure 8-6 (p. 409). Thus there are **discontinuities in the acceleration at each end of the interval** that uses this simple harmonic displacement function.

5. This violates the fundamental law of cam design and creates **infinite spikes of jerk** at the ends of this fall interval. **This is also an unacceptable design**.

What went wrong? While it is true that harmonic functions are differentiable ad infinitum, we are not dealing here with single harmonic functions. Our cam function over the entire interval is a **piecewise function** (Figure 8-6, p. 409) made up of several segments, some of which may be dwell portions or other functions. A dwell will always have zero velocity and zero acceleration. Thus we must match the dwells' zero values at the ends of those derivatives of any nondwell segments that adjoin them. The simple harmonic displacement function, when used with dwells, does **not** satisfy the fundamental law of cam design. Its second derivative, acceleration, is nonzero at its ends and thus does not match the dwells required in this example.

The only case in which the simple harmonic displacement function will satisfy the fundamental law is the non-quick-return RF case, i.e., rise in 180° and fall in 180° with no dwells. Then the cam profile, if run against a flat-faced follower, becomes an eccentric as shown in Figure 8-10. As a single continuous (not piecewise) function, its derivatives are continuous also. Figure 8-11 shows the displacement (in inches) and acceleration functions (in *g*'s) of an eccentric cam as actually measured on the follower. The noise, or "ripple," on the acceleration curve is due to small, unavoidable, manufacturing errors. Manufacturing limitations will be discussed in a later section.

Cycloidal Displacement

The two bad examples of cam design described above should lead the cam designer to the conclusion that consideration only of the displacement function when designing a cam is erroneous. The better approach is to start with consideration of the higher derivatives, especially acceleration. The acceleration function, and to a lesser extent the jerk function, should be the principal concern of the designer. In some cases, especially when the mass of the follower train is large, or when there is a specification on velocity, that function must be carefully designed as well.

[†] Though this is actually a half-period cosine wave, we will call it a *full-rise* (or *full-fall*) simple harmonic function to differentiate it from the *half-rise* (and *half-fall*) simple harmonic function which is actually a quarter-period cosine.

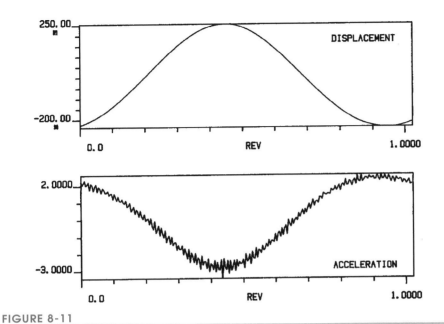

REV   1.0000

FIGURE 8-11

Displacement and acceleration as measured on the follower of an eccentric cam

$a\cos\omega t$

FIGURE 8-10

A flat-faced follower on an eccentric cam has simple harmonic motion.[*]

[*] If a roller follower is used instead of a flat-faced follower, then the trace of the roller follower center will still be a true eccentric, but the cam surface will not. This is due to the lead-lag error of the contact point of the roller with the cam surface. When going "uphill," the contact point leads the follower center and when going "downhill," it lags the center. This distorts the cam surface shape from that of a true eccentric circle. However, the motion of the follower will be simple harmonic motion as defined in Figure 8-10 regardless of follower type.

With this in mind, we will redesign the cam for the same example specifications as above. This time we will start with the acceleration function. The harmonic family of functions still has advantages that make them attractive for these applications. Figure 8-12 (p. 416) shows a full-period sinusoid applied as the acceleration function. It meets the constraint of zero magnitude at each end to match the dwell segments that adjoin it. The equation for a sine wave is:

$$a = C\sin\left(2\pi\frac{\theta}{\beta}\right) \tag{8.7}$$

We have again normalized the independent variable θ by dividing it by the period of the segment β with both θ and β measured in radians. The value of θ/β ranges from 0 to 1 over any segment and is a dimensionless ratio. Since we want a full-cycle sine wave, we must multiply the argument by 2π. The argument of the sine function will then vary between 0 and 2π regardless of the value of β. The constant C defines the amplitude of the sine wave.

Integrate to obtain velocity,

$$a = \frac{dv}{d\theta} = C\sin\left(2\pi\frac{\theta}{\beta}\right)$$

$$\int dv = \int C\sin\left(2\pi\frac{\theta}{\beta}\right)d\theta \tag{8.8}$$

$$v = -C\frac{\beta}{2\pi}\cos\left(2\pi\frac{\theta}{\beta}\right) + k_1$$

FIGURE 8-12

Sinusoidal
acceleration gives
cycloidal
displacement.

8

where k_1 is the constant of integration. To evaluate k_1, substitute the boundary condition $v = 0$ at $\theta = 0$, since we must match the zero velocity of the dwell at that point. The constant of integration is then:

and:

$$k_1 = C\frac{\beta}{2\pi}$$

$$v = C\frac{\beta}{2\pi}\left[1-\cos\left(2\pi\frac{\theta}{\beta}\right)\right]$$

(8.9)

Note that substituting the boundary values at the other end of the interval, $v = 0$, $\theta = \beta$, will give the same result for k_1. Integrate again to obtain displacement:

$$v = \frac{ds}{d\theta} = C\frac{\beta}{2\pi}\left[1-\cos\left(2\pi\frac{\theta}{\beta}\right)\right]$$

$$\int ds = \int\left\{C\frac{\beta}{2\pi}\left[1-\cos\left(2\pi\frac{\theta}{\beta}\right)\right]\right\}d\theta$$

(8.10)

$$s = C\frac{\beta}{2\pi}\theta - C\frac{\beta^2}{4\pi^2}\sin\left(2\pi\frac{\theta}{\beta}\right)+k_2$$

To evaluate k_2, substitute the boundary condition $s = 0$ at $\theta = 0$, since we must match the zero displacement of the dwell at that point. To evaluate the amplitude constant C, substitute the boundary condition $s = h$ at $\theta = \beta$, where h is the maximum follower rise (or lift) required over the interval and is a constant for any one cam specification.

$$k_2 = 0$$
$$C = 2\pi\frac{h}{\beta^2}$$

(8.11)

Substituting the value of the constant C in equation 8.7 (p. 415) for acceleration gives:

$$a = 2\pi\frac{h}{\beta^2}\sin\left(2\pi\frac{\theta}{\beta}\right)$$

(8.12a)

Differentiating with respect to θ gives the expression for jerk.

$$j = 4\pi^2\frac{h}{\beta^3}\cos\left(2\pi\frac{\theta}{\beta}\right)$$

(8.12b)

Substituting the values of the constants C and k_1 in equation 8.9 for velocity gives:

$$v = \frac{h}{\beta}\left[1-\cos\left(2\pi\frac{\theta}{\beta}\right)\right]$$

(8.12c)

This velocity function is the sum of a negative cosine term and a constant term. The coefficient of the cosine term is equal to the constant term. This results in a velocity curve that starts and ends at zero and reaches a maximum magnitude at $\beta/2$ as can be seen in Figure 8-12. Substituting the values of the constants C, k_1, and k_2 in equation 8.10 (p. 416) for displacement gives:

$$s = h\left[\frac{\theta}{\beta} - \frac{1}{2\pi}\sin\left(2\pi\frac{\theta}{\beta}\right)\right]$$ (8.12d)

Note that this displacement expression is the sum of a straight line of slope h and a negative sine wave. The sine wave is, in effect, "wrapped around" the straight line as can be seen in Figure 8-12. Equation 8.12d is the expression for a cycloid. This cam function is referred to either as **cycloidal displacement** or **sinusoidal acceleration**.

In the form presented, with θ (in radians) as the independent variable, the units of equation 8.12d are length, of equation 8.12c length/rad, of equation 8.12a length/rad^2, and of equation 8.12b length/rad^3. To convert these equations to a time base, multiply velocity v by the camshaft angular velocity ω (in rad/sec), multiply acceleration a by ω^2, and jerk j by ω^3.

EXAMPLE 8-3

Junior Cam Design—Cycloidal Displacement—An Acceptable Cam.

Problem: Consider the same cam design CEP specification as in Examples 8-1 and 8-2:

dwell	at zero displacement for 90 degrees (low dwell)
rise	1 in (25 mm) in 90 degrees
dwell	at 1 in (25 mm) for 90 degrees (high dwell)
fall	1 in (25 mm) in 90 degrees
cam ω	2π rad/sec = 1 rev/sec

Solution:

1   The cycloidal displacement function is an acceptable one for this double-dwell cam specification. Its derivatives are continuous through the acceleration function as seen in Figure 8-12. The peak acceleration is 100.4 in/sec^2 (2.55 m/sec^2).

2   The jerk curve in Figure 8-12 is discontinuous at its boundaries but is of finite magnitude, and this is acceptable. Its peak value is 2523 in/sec^3 (64 m/sec^3).

3   The velocity is smooth and matches the zeros of the dwell at each end. Its peak value is 8 in/sec (0.2 m/sec).

4   The only drawback to this function is that it has relatively large magnitudes of peak acceleration and peak velocity compared to some other possible functions for the double-dwell case.

The reader may open the file E08-03.cam in program DYNACAM to investigate this example in more detail.

Combined Functions

Dynamic force is proportional to acceleration. We generally would like to minimize dynamic forces, and thus should be looking to minimize the magnitude of the acceleration function as well as to keep it continuous. Kinetic energy is proportional to velocity squared. We also would like to minimize stored kinetic energy, especially with large mass follower trains, and so are concerned with the magnitude of the velocity function as well.

CONSTANT ACCELERATION   If we wish to minimize the peak value of the magnitude of the acceleration function for a given problem, the function that would best satisfy this constraint is the square wave as shown in Figure 8-13. This function is also called **constant acceleration**. The square wave has the property of minimum peak value for a given area in a given interval. However, this function is not continuous. It has discontinuities at the beginning, middle, and end of the interval, so, by itself, **this is unacceptable as a cam acceleration function**.

TRAPEZOIDAL ACCELERATION   The square wave's discontinuities can be removed by simply "knocking the corners off" the square wave function and creating the **trapezoidal acceleration** function shown in Figure 8-14a. The area lost from the "knocked off corners" must be replaced by increasing the peak magnitude above that of the original square wave in order to maintain the required specifications on lift and duration. But, this increase in peak magnitude is small, and the theoretical maximum acceleration can be significantly less than the theoretical peak value of the sinusoidal acceleration (cycloidal displacement) function. One disadvantage of this trapezoidal function is its discontinuous jerk function, as shown in Figure 8-14b. Ragged jerk functions such as this tend to excite vibratory behavior in the follower train due to their high harmonic content. The cycloidal's sinusoidal acceleration has a relatively smoother cosine jerk function with only two discontinuities in the interval and is preferable to the trapezoid's square waves of jerk. But the cycloidal's theoretical peak acceleration will be larger, which is not desirable. So, trade-offs must be made in selecting the cam functions.

MODIFIED TRAPEZOIDAL ACCELERATION   An improvement can be made to the trapezoidal acceleration function by substituting pieces of sine waves for the sloped sides

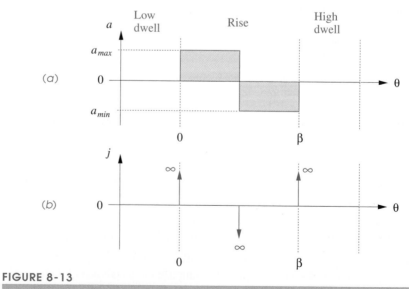

FIGURE 8-13

Constant acceleration gives infinite jerk.

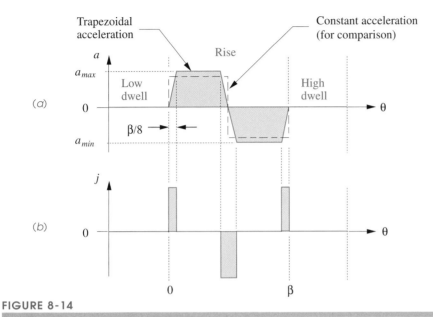

FIGURE 8-14

Trapezoidal acceleration gives finite jerk.

8

of the trapezoids as shown in Figure 8-15 (p. 420). This function is called the **modified trapezoidal acceleration** curve.[*] This function is a marriage of the sine acceleration and constant acceleration curves. Conceptually, a full period sine wave is cut into fourths and "pasted into" the square wave to provide a smooth transition from the zeros at the endpoints to the maximum and minimum peak values, and to make the transition from maximum to minimum in the center of the interval. The portions of the total segment period (β) used for the sinusoidal parts of the function can be varied. The most common arrangement is to cut the square wave at $\beta/8$, $3\beta/8$, $5\beta/8$, and $7\beta/8$ to insert the pieces of sine wave as shown in Figure 8-15.

The modified trapezoidal function defined above is one of many combined functions created for cams by piecing together various functions, while being careful to match the values of the s, v, and a curves at all the interfaces between the joined functions. It has the advantage of relatively low theoretical peak acceleration, and reasonably rapid, smooth transitions at the beginning and end of the interval. The modified trapezoidal cam function has been a popular and often used program for double-dwell cams.

MODIFIED SINUSOIDAL ACCELERATION[†]   The sine acceleration curve (cycloidal displacement) has the advantage of smoothness (less ragged jerk curve) compared to the modified trapezoid but has higher theoretical peak acceleration. By combining two harmonic (sinusoid) curves of different frequencies, we can retain some of the smoothness characteristics of the cycloid and also reduce the peak acceleration compared to the cycloid. As an added bonus we will find that the peak velocity is also lower than in either the cycloidal or modified trapezoid. Figure 8-16 (p. 422) shows how the modified sine acceleration curve is made up of pieces of two sinusoid functions, one of higher frequency than the other. The first and last quarters of the high-frequency (short period, $\beta/2$) sine

[*] Developed by C. N. Neklutin of Universal Match Corp. See ref.[2].

[†] Developed by E. H. Schmidt of DuPont.

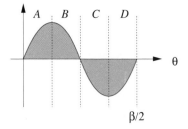

(a) Take a sine wave

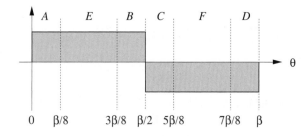

(b) Split the sine
wave apart

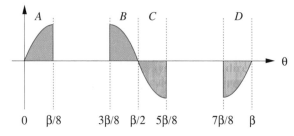

(c) Take a constant
acceleration
square wave

(d) Combine the two

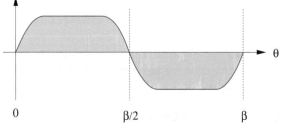

(e) Modified trapezoidal
acceleration

FIGURE 8-15

Creating the modified trapezoidal acceleration function

curve are used for the first and last eighths of the combined function. The center half of the low-frequency (long period, $3\beta/2$) sine wave is used to fill in the center three-fourths of the combined curve. Obviously, the magnitudes of the two curves and their derivatives must be matched at their interfaces in order to avoid discontinuities.

The SCCA Family of Double-Dwell Functions

SCCA stands for *Sine-Constant-Cosine-Acceleration* and refers to a family of acceleration functions that includes constant acceleration, simple harmonic, modified trapezoid, modified sine, and cycloidal curves.[11] These very different looking curves can all be defined by the same equation with only a change of numeric parameters. In like fashion, the equations for displacement, velocity, and jerk for all these SCCA functions differ only by their parametric values.

To reveal this similitude, it is first necessary to normalize the variables in the equations. We have already normalized the independent variable, cam angle θ, dividing it by the interval period β. We will further simplify the notation by defining

$$x = \frac{\theta}{\beta} \tag{8.13a}$$

The normalized variable x then runs from 0 to 1 over any interval. The normalized follower displacement is

$$y = \frac{s}{h} \tag{8.13b}$$

where s is the instantaneous follower displacement and h is the total lift. The normalized variable y then runs from 0 to 1 over any follower displacement.

The general shapes of the s v a j functions of the SCCA family are shown in Figure 8-17 (p. 423). The interval β is divided into five zones, numbered 1 through 5. Zones 0 and 6 represent the dwells on either side of the rise (or fall). The widths of zones 1 - 5 are defined in terms of β and one of three parameters, b, c, d. The values of these three parameters define the shape of the curve and define its identity within the family of functions. The normalized velocity, acceleration, and jerk are denoted, respectively, as:

$$y' = \frac{dy}{dx} \qquad\qquad y'' = \frac{d^2 y}{dx^2} \qquad\qquad y''' = \frac{d^3 y}{dx^3} \tag{8.14}$$

In zone 0, all functions are zero. The expressions for the functions within each other zone of Figure 8-17 are as follows:

Zone 1: $0 \leq x \leq \dfrac{b}{2}:$ $b \neq 0$

$$y = C_a\left[\frac{b}{\pi}x - \left(\frac{b}{\pi}\right)^2 \sin\left(\frac{\pi}{b}x\right)\right] \tag{8.15a}$$

$$y' = C_a\left[\frac{b}{\pi} - \frac{b}{\pi}\cos\left(\frac{\pi}{b}x\right)\right] \tag{8.15b}$$

(a) Sine wave #1
    of period β/2

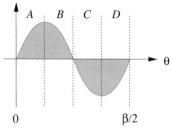

(b) Sine wave #2
    of period 3β/2

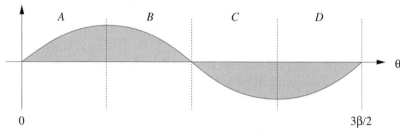

(c) Take 1st and 4th
    quarters of #1

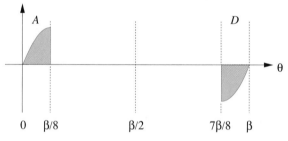

(d) Take 2nd and 3rd
    quarters of #2

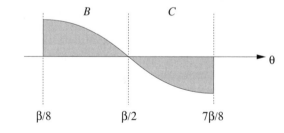

(e) Combine to get
    modified sine

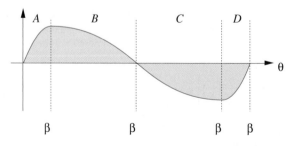

FIGURE 8-16

Creating the modified sine acceleration function

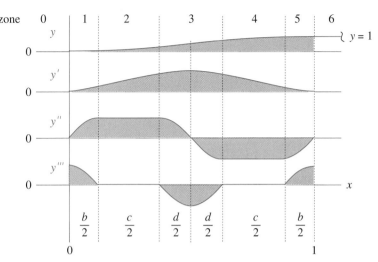

FIGURE 8-17

Parameters for the normalized SCCA family of curves

8

$$y'' = C_a \sin\left(\frac{\pi}{b}x\right) \tag{8.15c}$$

$$y''' = C_a \frac{\pi}{b} \cos\left(\frac{\pi}{b}x\right) \tag{8.15d}$$

Zone 2: $\frac{b}{2} \le x \le \frac{1-d}{2}$

$$y = C_a \left[\frac{x^2}{2} + b\left(\frac{1}{\pi} - \frac{1}{2}\right)x + b^2\left(\frac{1}{8} - \frac{1}{\pi^2}\right) \right] \tag{8.16a}$$

$$y' = C_a \left[x + b\left(\frac{1}{\pi} - \frac{1}{2}\right) \right] \tag{8.16b}$$

$$y'' = C_a \tag{8.16c}$$

$$y''' = 0 \tag{8.16d}$$

Zone 3: $\frac{1-d}{2} \le x \le \frac{1+d}{2} : \quad d \ne 0$

$$y = C_a \left\{ \left(\frac{b}{\pi} + \frac{c}{2}\right)x + \left(\frac{d}{\pi}\right)^2 + b^2\left(\frac{1}{8} - \frac{1}{\pi^2}\right) - \frac{(1-d)^2}{8} - \left(\frac{d}{\pi}\right)^2 \cos\left[\frac{\pi}{d}\left(x - \frac{1-d}{2}\right)\right] \right\} \tag{8.17a}$$

$$y' = C_a \left\{ \frac{b}{\pi} + \frac{c}{2} + \frac{d}{\pi} \sin\left[\frac{\pi}{d}\left(x - \frac{1-d}{2}\right)\right] \right\} \tag{8.17b}$$

$$y'' = C_a \cos\left[\frac{\pi}{d}\left(x - \frac{1-d}{2}\right)\right]$$

(8.17c)

$$y''' = -C_a \frac{\pi}{d} \sin\left[\frac{\pi}{d}\left(x - \frac{1-d}{2}\right)\right]$$

(8.17d)

Zone 4: $\dfrac{1+d}{2} \le x \le 1 - \dfrac{b}{2}$

$$y = C_a\left[-\frac{x^2}{2} + \left(\frac{b}{\pi} + 1 - \frac{b}{2}\right)x + \left(2d^2 - b^2\right)\left(\frac{1}{\pi^2} - \frac{1}{8}\right) - \frac{1}{4}\right]$$

(8.18a)

$$y' = C_a\left(-x + \frac{b}{\pi} + 1 - \frac{b}{2}\right)$$

(8.18b)

$$y'' = -C_a$$

(8.18c)

$$y''' = 0$$

(8.18d)

Zone 5: $1 - \dfrac{b}{2} \le x \le 1:$ $b \ne 0$

$$y = C_a\left\{\frac{b}{\pi}x + \frac{2\left(d^2 - b^2\right)}{\pi^2} + \frac{\left(1-b\right)^2 - d^2}{4} - \left(\frac{b}{\pi}\right)^2 \sin\left[\frac{\pi}{b}(x-1)\right]\right\}$$

(8.19a)

$$y' = C_a\left\{\frac{b}{\pi} - \frac{b}{\pi}\cos\left[\frac{\pi}{b}(x-1)\right]\right\}$$

(8.19b)

$$y'' = C_a \sin\left[\frac{\pi}{b}(x-1)\right]$$

(8.19c)

$$y''' = C_a \frac{\pi}{b}\cos\left[\frac{\pi}{b}(x-1)\right]$$

(8.19d)

Zone 6: $x > 1$
     $y = 1,$ $y' = y'' = y''' = 0$

(8.20)

The coefficient C_a is a dimensionless peak acceleration factor. It can be evaluated from the fact that, at the end of the rise in zone 5 when $x = 1$, the expression for displacement (equation 8.19a) must have $y = 1$ to match the dwell in zone 6. Setting the right side of equation 8.19a equal to 1 gives:

$$C_a = \frac{4\pi^2}{\left(\pi^2 - 8\right)\left(b^2 - d^2\right) - 2\pi(\pi - 2)b + \pi^2}$$

(8.21a)

We can also define dimensionless peak factors (coefficients) for velocity (C_v) and jerk (C_j) in terms of C_a. The velocity is a maximum at $x = 0.5$. Thus C_v will equal the right side of equation 8.17b (p. 423) when $x = 0.5$.

TABLE 8-2   Parameters and Coefficients for the SCCA Family of Functions

Function	b	c	d	C_v	C_a	C_j
Constant acceleration	0.00	1.00	0.00	2.0000	4.0000	infinite
Modified trapezoid	0.25	0.50	0.25	2.0000	4.8881	61.426
Simple harmonic	0.00	0.00	1.00	1.5708	4.9348	infinite
Modified sine	0.25	0.00	0.75	1.7596	5.5280	69.466
Cycloidal displacement	0.50	0.00	0.50	2.0000	6.2832	39.478

$$C_v = C_a \left(\frac{b+d}{\pi} + \frac{c}{2} \right) \qquad (8.21b)$$

The jerk is a maximum at $x = 0$. Setting the right side of equation 8.15d (p. 423) to zero gives:

$$C_j = C_a \frac{\pi}{b} \qquad b \neq 0 \qquad (8.21c)$$

Table 8-2 shows the values of b, c, d and the resulting factors C_v, C_a, and C_j for the five standard members of the SCCA family. There is an infinity of related functions with values of these parameters between those shown. Figure 8-18 shows these five members of the "acceleration family" superposed with their design parameters noted. Note that all the functions shown in Figure 8-18 were generated with the same set of equations (8.15 through 8.21, pp. 421- 425) with only changes to the values of the parameters b, c, and d. A *TKSolver* file (SCCA.tk) that is provided on the DVD calculates and plots any of the SCCA family of normalized functions, along with their coefficients C_v, C_a, and C_j, in response to the input of values for b, c, and d. Note also that there is an infinity of family members as b, c, and d can take on any set of values that add to 1.

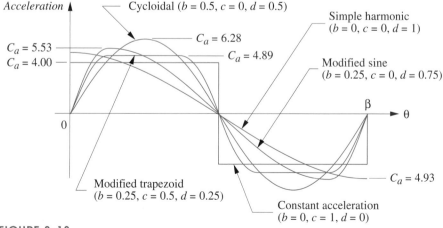

FIGURE 8-18

Comparison of five acceleration functions in the SCCA family

To apply the SCCA functions to an actual cam design problem only requires that they be multiplied or divided by factors appropriate to the particular problem, namely the actual rise h, the actual duration β (rad), and cam velocity ω (rad/sec).

$$
\begin{array}{llll}
s = hy & \text{length} & S = s & \text{length} \\[2mm]
v = \dfrac{h}{\beta} y' & \text{length/rad} & V = v\omega & \text{length/sec} \\[2mm]
a = \dfrac{h}{\beta^2} y'' & \text{length/rad}^2 & A = a\omega^2 & \text{length/sec}^2 \\[2mm]
j = \dfrac{h}{\beta^3} y''' & \text{length/rad}^3 & J = j\omega^3 & \text{length/sec}^3
\end{array}
\tag{8.22}
$$

Figure 8-19 shows a comparison of the shapes and relative magnitudes of five acceptable cam acceleration programs including the cycloidal, modified trapezoid, and modified sine acceleration curves.[*] The cycloidal curve has a theoretical peak acceleration that is approximately 1.3 times that of the modified trapezoid's peak value for the same cam specification. The peak value of acceleration for the modified sine is between those of the cycloidal and modified trapezoids. Table 8-3 lists the peak values of acceleration, velocity, and jerk for these functions in terms of the total rise h and period β.

Figure 8-20 compares the jerk curves for the same functions. The modified sine jerk is somewhat less ragged than the modified trapezoid jerk but not as smooth as that of the cycloid, which is a full-period cosine. Figure 8-21 (p. 428) compares their velocity curves. The peak velocities of the cycloidal and modified trapezoid functions are the same, so each will store the same peak kinetic energy in the follower train. The peak velocity of the modified sine is the lowest of the five functions shown. This is the principal advantage of the modified sine acceleration curve and the reason it is often chosen for applications in which the follower mass is very large.

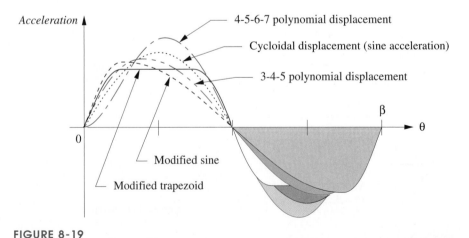

* The 3-4-5 and 4-5-6-7 polynomial functions also shown in the figure will be discussed in a later section.

FIGURE 8-19

Comparison of five acceptable double-dwell cam acceleration functions

TABLE 8-3   Factors for Peak Velocity and Acceleration of Some Cam Functions

Function	Max. Veloc.	Max. Accel.	Max. Jerk	Comments
Constant accel.	$2.000\ h/\beta$	$4.000\ h/\beta^2$	Infinite	∞ jerk—not acceptable
Harmonic disp.	$1.571\ h/\beta$	$4.945\ h/\beta^2$	Infinite	∞ jerk—not acceptable
Trapezoid accel.	$2.000\ h/\beta$	$5.300\ h/\beta^2$	$44\ h/\beta^3$	Not as good as mod. trap.
Mod. trap. accel.	$2.000\ h/\beta$	$4.888\ h/\beta^2$	$61\ h/\beta^3$	Low accel. but rough jerk
Mod. sine accel.	$1.760\ h/\beta$	$5.528\ h/\beta^2$	$69\ h/\beta^3$	Low veloc., good accel.
3-4-5 poly. disp.	$1.875\ h/\beta$	$5.777\ h/\beta^2$	$60\ h/\beta^3$	Good compromise
Cycloidal disp.	$2.000\ h/\beta$	$6.283\ h/\beta^2$	$40\ h/\beta^3$	Smooth accel. and jerk
4-5-6-7 poly. disp.	$2.188\ h/\beta$	$7.526\ h/\beta^2$	$52\ h/\beta^3$	Smooth jerk, high accel.

An example of such an application is shown in Figure 8-22 (p. 429) which is an indexing table drive used for automated assembly lines. The round indexing table is mounted on a tapered vertical spindle and driven as part of the follower train by a form-closed barrel cam that moves it through some angular displacement, and then holds the table still in a dwell (called a "stop") while an assembly operation is performed on the workpiece carried on the table. These indexers may have three or more stops, each corresponding to an index position. The table is solid steel and may be several feet in diameter; thus its mass is large. To minimize the stored kinetic energy, which must be dissipated each time the table is brought to a stop, the manufacturers often use the modified sine program on these multidwell cams, because of its lower peak velocity.

Let us again try to improve the double-dwell cam example using the SCCA combined functions of modified trapezoid and modified sine acceleration.

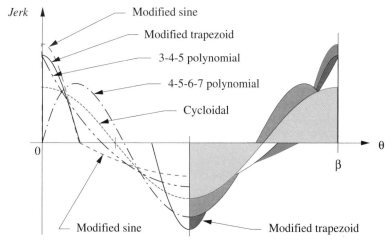

FIGURE 8-20

Comparison of five double-dwell cam jerk functions

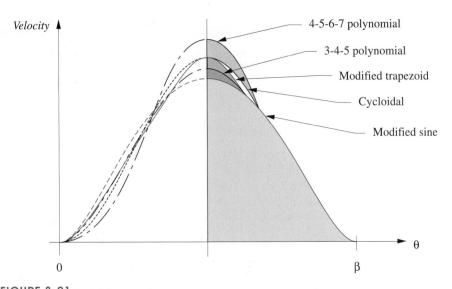

FIGURE 8-21

Comparison of five double-dwell cam velocity functions

✐ EXAMPLE 8-4

Senior Cam Design—Combined Functions—Better Cams.

Problem: Consider the same cam design CEP specification as in Examples 8-1 to 8-3:

dwell	at zero displacement for 90 degrees (low dwell)
rise	1 in (25 mm) in 90 degrees
dwell	at 1 in (25 mm) for 90 degrees (high dwell)
fall	1 in (25 mm) in 90 degrees
cam ω	2π rad/sec = 1 rev/sec

Solution:

1   The modified trapezoidal function is an acceptable one for this double-dwell cam specification. Its derivatives are continuous through the acceleration function as shown in Figure 8-19 (p. 426). The peak acceleration is 78.1 in/sec^2 (1.98 m/sec^2).

2   The modified trapezoidal jerk curve in Figure 8-20 (p. 427) is discontinuous at its boundaries but has finite magnitude of 3925 in/sec^3 (100 m/sec^3), and this is acceptable.

3   The modified trapezoidal velocity in Figure 8-21 is smooth and matches the zeros of the dwell at each end. Its peak magnitude is 8 in/sec (0.2 m/sec).

4   The advantage of this modified trapezoidal function is that it has smaller theoretical peak acceleration than the cycloidal but its peak velocity is identical to that of the cycloidal.

5   The modified sinusoid function is also an acceptable one for this double-dwell cam specification. Its derivatives are also continuous through the acceleration function as shown in Figure 8-19 (p. 426). Its peak acceleration is 88.3 in/sec^2 (2.24 m/sec^2).

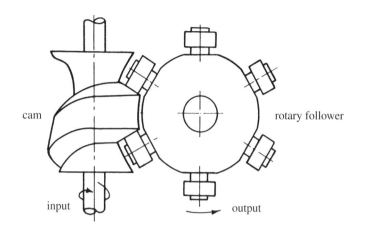

cam   rotary follower

input   output

LARGE DIA. DIAL MT'G.
SHAFT

PRELOADED
TIMKIN DIAL
SUPPORT
BEARING

RIBBED
CAST IRON
DIAL

HARDENED &
GROUND TOOL
STEEL CAM

STATIONARY CTR.
TOOL MT'G. POST

OPEN CENTER

RIBBED
CAST IRON
TOOL MT'G.
PLATE

CAGED ROLLER
BR'G. CAM FOLLOWERS

ACCESS COVER

FIGURE 8-22

Multistop rotary indexer   (*Courtesy of The Ferguson Co., St. Louis, MO*)

6   The modified sine jerk curve in Figure 8-20 (p. 427) is discontinuous at its boundaries but is of finite magnitude and is larger in magnitude at 4439 in/sec^3 (113 m/sec^3) but smoother than that of the modified trapezoid.

7   The modified sine velocity (Figure 8-21, p. 428) is smooth, matches the zeros of the dwell at each end, and is lower in peak magnitude than either the cycloidal or modified trapezoidal at 7 in/sec (0.178 m/sec). This is an advantage for high-mass follower systems as it reduces stored kinetic energy. This, coupled with a peak acceleration lower than the cycloidal (but higher than the modified trapezoidal), is its chief advantage.

Figure 8-23 shows the displacement curves for these three cam programs. (Open the file E08-04.cam in program DYNACAM to plot these also.) Note how little difference there is between the displacement curves despite the large differences in their acceleration waveforms in Figure 8-19 (p. 426). This is evidence of the smoothing effect of the integration process. Differentiating any two functions will exaggerate their differences. Integration tends to mask their differences. It is nearly impossible to recognize these very differently behaving cam functions by looking only at their displacement curves. This is further evidence of the folly of our earlier naive approach to cam design that dealt exclusively with the displacement function. The cam designer must be concerned with the higher derivatives of displacement. The displacement function is primarily of value to the manufacturer of the cam who needs its coordinate information in order to cut the cam.

FALL FUNCTIONS   We have used only the rise portion of the cam for these examples. The fall is handled similarly. The rise functions presented here are applicable to the fall with slight modification. To convert rise equations to fall equations, it is only necessary to subtract the rise displacement function s from the maximum lift h and to negate the higher derivatives, v, a, and j.

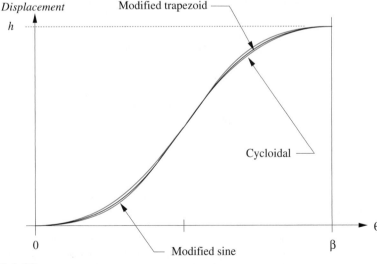

FIGURE 8-23

Comparison of three SCCA double-dwell cam displacement functions

Polynomial Functions

The class of polynomial functions is one of the more versatile types that can be used for cam design. They are not limited to single- or double-dwell applications and can be tailored to many design specifications. The general form of a polynomial function is:

$$s = C_0 + C_1 x + C_2 x^2 + C_3 x^3 + C_4 x^4 + C_5 x^5 + C_6 x^6 + \cdots + C_n x^n \qquad (8.23)$$

where s is the follower displacement; x is the independent variable, which in our case will be replaced by either θ/β or time t. The constant coefficients C_n are the unknowns to be determined in our development of the particular polynomial equation to suit a design specification. The degree of a polynomial is defined as the highest power present in any term. Note that a polynomial of degree n will have $n + 1$ terms because there is an x^0 or constant term with coefficient C_0, as well as coefficients through and including C_n.

We structure a polynomial cam design problem by deciding how many boundary conditions (BCs) we want to specify on the $s\ v\ a\ j$ diagrams. The number of BCs then determines the degree of the resulting polynomial. We can write an independent equation for each BC by substituting it into equation 8.16 (p. 423) or one of its derivatives. We will then have a system of linear equations that can be solved for the unknown coefficients $C_0, \ldots, C_n$. If k represents the number of chosen boundary conditions, there will be k equations in k unknowns $C_0, \ldots, C_n$ and the **degree** of the polynomial will be $n = k - 1$. The **order** of the n-degree polynomial is equal to the number of terms, k.

Double-Dwell Applications of Polynomials

THE 3-4-5 POLYNOMIAL   Reconsider the double-dwell problem of the previous three examples and solve it with polynomial functions. Many different polynomial solutions are possible. We will start with the simplest one possible for the double-dwell case.

EXAMPLE 8-5

The 3-4-5 Polynomial for the Double-Dwell Case.

Problem: Consider the same cam design CEP specification as in Examples 8-1 to 8-4:

dwell	at zero displacement for 90 degrees (low dwell)
rise	1 in (25 mm) in 90 degrees
dwell	at 1 in (25 mm) for 90 degrees (high dwell)
fall	1 in (25 mm) in 90 degrees
cam ω	2π rad/sec = 1 rev/sec

Solution:

1   To satisfy the fundamental law of cam design the values of the rise (and fall) functions at their boundaries with the dwells must match with no discontinuities in, at a minimum, s, v, and a.

2   Figure 8-24 (p. 433) shows the axes for the $s\ v\ a\ j$ diagrams on which the known data have been drawn. The dwells are the only fully defined segments at this stage. The requirement for continuity through the acceleration defines a minimum of **six boundary conditions** for the rise segment and six more for the fall in this problem. They are shown as filled circles on the

plots. For generality, we will let the specified total rise be represented by the variable h. The minimum set of required BCs for this example is then:

for the rise:

when	$\theta = 0$;	then	$s = 0$,	$v = 0$,	$a = 0$
when	$\theta = \beta_1$;	then	$s = h$,	$v = 0$,	$a = 0$

(a)

for the fall:

when	$\theta = 0$;	then	$s = h$,	$v = 0$,	$a = 0$
when	$\theta = \beta_2$;	then	$s = 0$,	$v = 0$,	$a = 0$

(b)

3　We will use the rise for an example solution. (The fall is a similar derivation.) We have six BCs on the rise. This requires six terms in the equation. The highest term will be fifth degree. We will use the normalized angle θ/β as our independent variable, as before. Because our boundary conditions involve velocity and acceleration as well as displacement, we need to differentiate equation 8.23 (p. 431) versus θ to obtain expressions into which we can substitute those BCs. Rewriting equation 8.23 to fit these constraints and differentiating twice, we get:

$$s = C_0 + C_1\left(\frac{\theta}{\beta}\right) + C_2\left(\frac{\theta}{\beta}\right)^2 + C_3\left(\frac{\theta}{\beta}\right)^3 + C_4\left(\frac{\theta}{\beta}\right)^4 + C_5\left(\frac{\theta}{\beta}\right)^5$$

(c)

$$v = \frac{1}{\beta}\left[C_1 + 2C_2\left(\frac{\theta}{\beta}\right) + 3C_3\left(\frac{\theta}{\beta}\right)^2 + 4C_4\left(\frac{\theta}{\beta}\right)^3 + 5C_5\left(\frac{\theta}{\beta}\right)^4\right]$$

(d)

$$a = \frac{1}{\beta^2}\left[2C_2 + 6C_3\left(\frac{\theta}{\beta}\right) + 12C_4\left(\frac{\theta}{\beta}\right)^2 + 20C_5\left(\frac{\theta}{\beta}\right)^3\right]$$

(e)

4　Substitute the boundary conditions $\theta = 0$, $s = 0$ into equation (c):

$$0 = C_0 + 0 + 0 + \cdots$$
$$C_0 = 0$$

(f)

5　Substitute $\theta = 0$, $v = 0$ into equation (d):

$$0 = \frac{1}{\beta}\left(C_1 + 0 + 0 + \cdots\right)$$
$$C_1 = 0$$

(g)

6　Substitute $\theta = 0$, $a = 0$ into equation (e):

$$0 = \frac{1}{\beta^2}\left(C_2 + 0 + 0 + \cdots\right)$$
$$C_2 = 0$$

(h)

7　Substitute $\theta = \beta$, $s = h$ into equation (c):

$$h = C_3 + C_4 + C_5$$

(i)

8　Substitute $\theta = \beta$, $v = 0$ into equation (d):

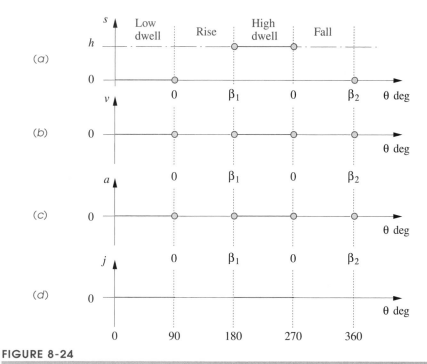

FIGURE 8-24

Minimum boundary conditions for the double-dwell case

8

$$0 = \frac{1}{\beta}\left(3C_3 + 4C_4 + 5C_5\right) \tag{j}$$

9   Substitute $\theta = \beta$, $a = 0$ into equation (e):

$$0 = \frac{1}{\beta^2}\left(6C_3 + 12C_4 + 20C_5\right) \tag{k}$$

10   Three of our unknowns are found to be zero, leaving three unknowns to be solved for, C_3, C_4, C_5. Equations (i), (j), and (k) can be solved simultaneously to get:

$$C_3 = 10h; \qquad C_4 = -15h; \qquad C_5 = 6h \tag{l}$$

11   The equation for this cam design's displacement is then:

$$s = h\left[10\left(\frac{\theta}{\beta}\right)^3 - 15\left(\frac{\theta}{\beta}\right)^4 + 6\left(\frac{\theta}{\beta}\right)^5\right] \tag{8.24}$$

12   The expressions for velocity and acceleration can be obtained by substituting the values of C_3, C_4, and C_5 into equations 8.18b and c (p. 424). This function is referred to as the **3-4-5 polynomial**, after its exponents. Open the file E08-07.cam in program DYNACAM to investigate this example in more detail.

FIGURE 8-25

3-4-5 polynomial rise. Its acceleration is very similar to the sinusoid of cycloidal motion

Figure 8-25 shows the resulting *s v a j* diagrams for a **3-4-5 polynomial rise** function with its boundary conditions circled. Note that the acceleration is continuous but the jerk is not, because we did not place any constraints on the boundary values of the jerk function. It is also interesting to note that the acceleration waveform looks very similar to the sinusoidal acceleration of the cycloidal function in Figure 8-12 (p. 416). Figure 8-19 (p. 426) shows the relative peak accelerations of this 3-4-5 polynomial compared to four other functions with the same *h* and β. Table 8-3 (p. 427) lists factors for the maximum velocity, acceleration, and jerk of these functions.

THE 4-5-6-7 POLYNOMIAL   We left the jerk unconstrained in the previous example. We will now redesign the cam for the same specifications but will also constrain the jerk function to be zero at both ends of the rise. It will then match the dwells in the jerk function with no discontinuities. This gives eight boundary conditions and yields a seventh-degree polynomial. The solution procedure to find the eight unknown coefficients is identical to that used in the previous example. Write the polynomial with the appropriate number of terms. Differentiate it to get expressions for all orders of boundary conditions. Substitute the boundary conditions and solve the resulting set of simultaneous equations.[*] This problem reduces to four equations in four unknowns, as the coefficients C_0, C_1, C_2, and C_3 turn out to be zero. For this set of boundary conditions the displacement equation for the rise is:

$$s = h\left[35\left(\frac{\theta}{\beta}\right)^4 - 84\left(\frac{\theta}{\beta}\right)^5 + 70\left(\frac{\theta}{\beta}\right)^6 - 20\left(\frac{\theta}{\beta}\right)^7 \right] \qquad (8.25)$$

This is known as the **4-5-6-7 polynomial**, after its exponents. Figure 8-26 shows the *s v a j* diagrams for this function with its boundary conditions circled. Compare these functions to the 3-4-5 polynomial functions shown in Figure 8-25. Note that the acceleration of the 4-5-6-7 starts off slowly, with zero slope (as we demanded with our zero jerk BC), and as a result goes to a larger peak value of acceleration in order to replace the missing area in the leading edge.

This **4-5-6-7 polynomial** function has the advantage of smoother jerk for better vibration control, compared to the **3-4-5 polynomial**, the **cycloidal**, and all other functions so far discussed, but it pays a price in the form of higher peak theoretical acceleration than all those functions. See also Table 8-3 (p. 427).

SUMMARY   The previous two sections have attempted to present an approach to the selection of appropriate double-dwell cam functions, using the common rise-dwell-fall-dwell cam as the example, and to point out some of the pitfalls awaiting the cam designer. The particular functions described are only a few of the ones that have been developed for this double-dwell case over many years, by many designers, but they are probably the most used and most popular among cam designers. Most of them are also included in program DYNACAM. There are many trade-offs to be considered in selecting a cam program for any application, some of which have already been mentioned, such as function continuity, peak values of velocity and acceleration, and smoothness of jerk. There are many other trade-offs still to be discussed in later sections of this chapter, involving the sizing and the manufacturability of the cam.

---

* Any matrix solving calculator, equation solver such as *Matlab*, *Mathcad*, or *TKSolver*, or programs MATRIX and DYNACAM (supplied with this text) will do the simultaneous equation solution for you. Programs MATRIX and DYNACAM are discussed in Appendix A. You need only to supply the desired boundary conditions to DYNACAM and the coefficients will be computed. The reader is encouraged to do so and examine the example problems presented here with the DYNACAM program.

8.4   SINGLE-DWELL CAM DESIGN—CHOOSING *S V A J* FUNCTIONS

Many applications in machinery require a **single-dwell** cam program, **rise-fall-dwell** (RFD). Perhaps a single-dwell cam is needed to lift and lower a roller that carries a moving paper web on a production machine that makes envelopes. This cam's follower lifts the paper up to one critical extreme position at the right time to contact a roller that applies a layer of glue to the envelope flap. Without dwelling in the up position, it immediately retracts the web back to the starting (zero) position and holds it in this other critical extreme position (low dwell) while the rest of the envelope passes by. It repeats the cycle for the next envelope as it comes by. Another common example of a single-dwell application is the cam that opens the valves in your automobile engine. This lifts the valve open on the rise, immediately closes it on the fall, and then keeps the valve closed in a dwell while the compression and combustion take place.

If we attempt to use the same type of cam programs as were defined for the double-dwell case for a single-dwell application, we will achieve a solution that may work but is not optimal. We will nevertheless do so here as an example in order to point out the problems that result. Then we will redesign the cam to eliminate those problems.

FIGURE 8-26

4-5-6-7 polynomial rise. Its jerk is piecewise continuous with the dwells.

8

✎ EXAMPLE 8-6

Using Cycloidal Motion for a Symmetrical Rise-Fall Single-Dwell Case.

Problem: Consider the following single-dwell cam specification:

rise	1 in (25 mm) in 90 degrees
fall	1 in (25 mm) in 90 degrees
dwell	at zero displacement for 180 degrees (low dwell)
cam ω	15 rad/sec

Solution:

1   Figure 8-27 shows a cycloidal displacement rise and separate cycloidal displacement fall applied to this single-dwell example. Note that the displacement (*s*) diagram looks acceptable in that it moves the follower from the low to the high position and back in the required intervals.

2   The velocity (*v*) also looks acceptable in shape in that it takes the follower from zero velocity at the low dwell to a peak value of 19.1 in/sec (0.49 m/sec) to zero again at the maximum displacement, where the glue is applied.

3   Figure 8-27 also shows the acceleration function for this solution. Its maximum absolute value is about 573 in/sec^2.

4   The problem is that this acceleration curve has an **unnecessary return to zero** at the end of the rise. It is unnecessary because the acceleration during the first part of the fall is also negative. It would be better to keep it in the negative region at the end of the rise.

5   This unnecessary oscillation to zero in the acceleration causes the jerk to have more abrupt changes and discontinuities. The only real justification for taking the acceleration to zero is the need to change its sign (as is the case halfway through the rise or fall) or to match an adjacent segment that has zero acceleration.

The reader may open the file E08-06.cam in program DYNACAM to investigate this example in more detail.

For the single-dwell case we would like a function for the rise that does not return its acceleration to zero at the end of the interval. The function for the fall should begin with the same nonzero acceleration value as ended the rise and then be zero at its terminus to match the dwell. One function that meets those criteria is the **double harmonic** which gets its name from its two cosine terms, one of which is a half-period harmonic and the other a full-period wave. The equations for the double harmonic functions are:

for the rise:

$$s = \frac{h}{2}\left\{\left[1-\cos\left(\pi\frac{\theta}{\beta}\right)\right]-\frac{1}{4}\left[1-\cos\left(2\pi\frac{\theta}{\beta}\right)\right]\right\}$$

$$v = \frac{\pi}{\beta}\frac{h}{2}\left[\sin\left(\pi\frac{\theta}{\beta}\right)-\frac{1}{2}\sin\left(2\pi\frac{\theta}{\beta}\right)\right]$$

$$a = \frac{\pi^2}{\beta^2}\frac{h}{2}\left[\cos\left(\pi\frac{\theta}{\beta}\right)-\cos\left(2\pi\frac{\theta}{\beta}\right)\right] \qquad (8.26a)$$

$$j = -\frac{\pi^3}{\beta^3}\frac{h}{2}\left[\sin\left(\pi\frac{\theta}{\beta}\right)-2\sin\left(2\pi\frac{\theta}{\beta}\right)\right]$$

for the fall:

$$s = \frac{h}{2}\left\{\left[1+\cos\left(\pi\frac{\theta}{\beta}\right)\right]-\frac{1}{4}\left[1-\cos\left(2\pi\frac{\theta}{\beta}\right)\right]\right\}$$

$$v = -\frac{\pi}{\beta}\frac{h}{2}\left[\sin\left(\pi\frac{\theta}{\beta}\right)+\frac{1}{2}\sin\left(2\pi\frac{\theta}{\beta}\right)\right]$$

$$a = -\frac{\pi^2}{\beta^2}\frac{h}{2}\left[\cos\left(\pi\frac{\theta}{\beta}\right)+\cos\left(2\pi\frac{\theta}{\beta}\right)\right] \qquad (8.26b)$$

$$j = \frac{\pi^3}{\beta^3}\frac{h}{2}\left[\sin\left(\pi\frac{\theta}{\beta}\right)+2\sin\left(2\pi\frac{\theta}{\beta}\right)\right]$$

Note that these double harmonic functions should **never** be used for the double-dwell case because their acceleration is nonzero at one end of the interval.

✒️ EXAMPLE 8-7

Double Harmonic Motion for Symmetrical Rise-Fall Single-Dwell Case.

Problem: Consider the same single-dwell cam specification as in Example 8-5 (p. 431):

rise	1 in (25 mm) in 90 degrees
fall	1 in (25 mm) in 90 degrees
dwell	at zero displacement for 180 degrees (low dwell)
cam ω	15 rad/sec

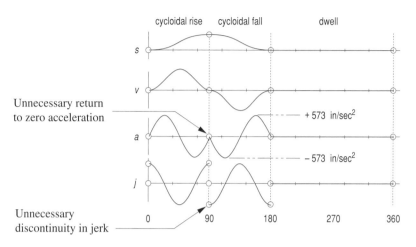

cycloidal rise   cycloidal fall   dwell

Unnecessary return
to zero acceleration

+ 573 in/sec²

− 573 in/sec²

Unnecessary
discontinuity in jerk

0   90   180   270   360

FIGURE 8-27

Cycloidal motion (or any double-dwell program) is a poor choice for the single-dwell case.

Solution:

1   Figure 8-28 (p. 438) shows a double harmonic rise and a double harmonic fall. The peak velocity is 19.5 in/sec (0.50 m/sec) which is similar to that of the cycloidal solution of Example 8-6.

2   Note that the acceleration of this double harmonic function does not return to zero at the end of the rise. This makes it more suitable for a single-dwell case in that respect.

3   The double harmonic jerk function peaks at 36 931 in/sec³ (938 m/sec³) and is quite smooth compared to the cycloidal solution.

4   Unfortunately, the peak negative acceleration is 900 in/sec², nearly twice that of the cycloidal solution. This is a smoother function but will develop higher dynamic forces. Open the file E08-07.cam in program DYNACAM to see this example in greater detail.

5   Another limitation of this function is that it may only be used for the case of an equal time (symmetrical) rise and fall. If the rise and fall times are different, the acceleration will be discontinuous at the juncture of rise and fall, violating the fundamental law of cam design.

Neither of the solutions in Examples 8-6 (p. 435) and 8-7 is optimal. We will now apply polynomial functions and redesign it to both improve its smoothness and reduce its peak acceleration.

Single-Dwell Applications of Polynomials

To solve the problem of Example 8-7 with a polynomial we must decide on a suitable set of boundary conditions. But first, we must decide how many segments to divide the cam cycle into. The problem statement seems to imply three segments, a rise, a fall, and a dwell. We could use those three segments to create the functions as we did in the two

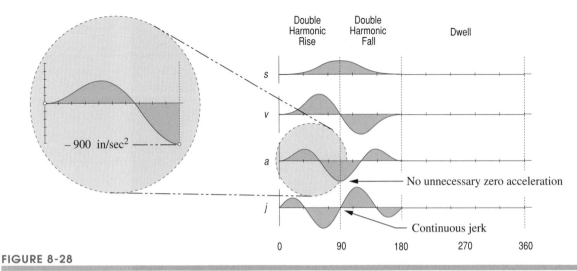

FIGURE 8-28
Double harmonic motion can be used for the single-dwell case if rise and fall durations are equal.

previous examples, but a better approach is to use only **two segments**, one for the rise-fall combined and one for the dwell. *As a general rule we would like to minimize the number of segments in our polynomial cam function*s. Any dwell requires its own segment. So, the minimum number possible in this case is two segments.

Another rule of thumb is that *we would like to minimize the number of boundary conditions specified*, because the degree of the polynomial is tied to the number of BCs. As the degree of the function increases, so will the number of its **inflection points** and its number of **minima and maxima**. The polynomial derivation process will guarantee that the function will pass through all specified BCs but says nothing about the function's behavior between the BCs. *A high-degree function may have undesirable oscillations between its BCs.*

With these assumptions we can select a set of boundary conditions for a trial solution. First we will restate the problem to reflect our two-segment configuration.

✎EXAMPLE 8-8

Designing a Polynomial for the Symmetrical Rise-Fall Single-Dwell Case.

Problem: Redefine the CEP specification from Examples 8-5 and 8-6 (pp. 431 and 435).

  rise-fall   1 in (25.4 mm) in 90° and fall 1 in (25.4 mm) in 90° over 180°
  dwell   at zero displacement for 180° (low dwell)
  cam ω 15 rad/sec

Solution:

  1 Figure 8-29 shows the minimum set of seven BCs for this symmetrical problem, which will

Segment number	Function used	Start angle	End angle	Delta angle
1	Poly 6	0	180	180

Boundary Conditions Imposed				Equation Resulting	
Function	Theta	% Beta	Boundary Cond.	Exponent	Coefficient
Displ	0	0	0	0	0
Veloc	0	0	0	1	0
Accel	0	0	0	2	0
Displ	180	1	0	3	64
Veloc	180	1	0	4	−192
Accel	180	1	0	5	192
Displ	90	0.5	1	6	−64

FIGURE 8-29

Boundary conditions and coefficients for a single-dwell polynomial application

give a sixth-degree polynomial. The dwell on either side of the combined rise-fall segment has zero values of $s, v, a,$ and j. The fundamental law of cam design requires that we match these zero values, through the acceleration function, at each end of the rise-fall segment.

2 These then account for six BCs; $s, v, a = 0$ at each end of the rise-fall segment.

3 We also must specify a value of displacement at the 1-in peak of the rise that occurs at $\theta = 90°$. This is the seventh BC. Note that due to symmetry, it is not necessary to specify the velocity to be zero at the peak. It will be anyway.

4 Figure 8-29 also shows the coefficients of the displacement polynomial that result from the simultaneous solution of the equations for the chosen BCs. For generality we have substituted the variable h for the specified 1-in rise. The function turns out to be a 3-4-5-6 polynomial whose equation is:

$$s = h\left[64\left(\frac{\theta}{\beta}\right)^3 - 192\left(\frac{\theta}{\beta}\right)^4 + 192\left(\frac{\theta}{\beta}\right)^5 - 64\left(\frac{\theta}{\beta}\right)^6\right] \qquad (a)$$

Figure 8-30 (p. 440) shows the $s\,v\,a\,j$ diagrams for this solution with its maximum values noted. Compare these acceleration and $s\,v\,a\,j$ curves to the double harmonic and cycloidal solutions to the same problem in Figures 8-27 (p. 437) and 8-28. Note that this sixth-degree polynomial function is as smooth as the double harmonic functions (Figure 8-28) and does not unnecessarily return the acceleration to zero at the top of the rise as does the cycloidal (Figure 8-27, p. 437). The polynomial has a peak acceleration of 547 in/sec^2, which is less than that of either the cycloidal or double harmonic solution. This 3-4-5-6 polynomial is a superior solution to either of those presented for the symmetrical rise-fall case and is an example of how polynomial functions can be easily tailored to particular design specifications. The reader may open the file E08-08.cam in program DYNACAM to investigate this example in greater detail.

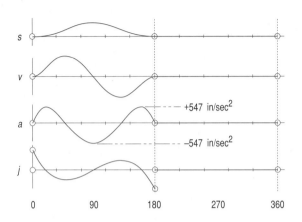

FIGURE 8-30

3-4-5-6 polynomial function for two-segment symmetrical rise-fall, single-dwell cam

Effect of Asymmetry on the Rise-Fall Polynomial Solution

The examples so far presented in this section all had equal time for rise and fall, referred to as a symmetrical rise-fall curve. What will happen if we need an asymmetric program and attempt to use a single polynomial as was done in the previous example?

EXAMPLE 8-9

Designing a Polynomial for an Asymmetrical Rise-Fall Single-Dwell Case.

Problem: Redefine the specification from Example 8-8 (p. 438) as:

         rise-fall   rise 1 in (25.4 mm) in 45° and fall 1 in (25.4 mm) in 135° over 180°
         dwell   at zero displacement for 180° (low dwell)
         cam ω   15 rad/sec

Solution:

1   Figure 8-31 shows the minimum set of seven BCs for this problem that will give a sixth-degree polynomial. The dwell on either side of the combined rise-fall segment has zero values for $S, V, A,$ and J. The fundamental law of cam design requires that we match these zero values, through the acceleration function, at each end of the rise-fall segment.

2   The endpoints account for six BCs; $S = V = A = 0$ at each end of the rise-fall segment.

3   We also must specify a value of displacement at the 1-in peak of the rise that occurs at $\theta = 45°$. This is the seventh BC.

4   Simultaneous solution of this equation set gives a 3-4-5-6 polynomial whose equation is:

$$s = h\left[151.704\left(\frac{\theta}{\beta}\right)^3 - 455.111\left(\frac{\theta}{\beta}\right)^4 + 455.111\left(\frac{\theta}{\beta}\right)^5 - 151.704\left(\frac{\theta}{\beta}\right)^6 \right] \qquad (a)$$

For generality we have substituted the variable h for the specified 1-in rise.

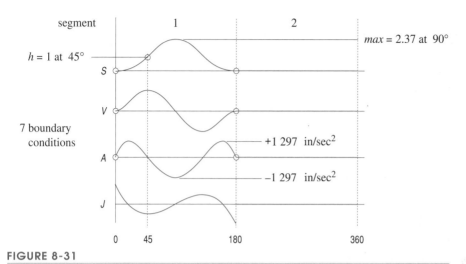

segment   1   2

max = 2.37 at 90°

h = 1 at 45°

S

V

7 boundary
conditions

A

+1 297 in/sec²

−1 297 in/sec²

J

0   45   180   360

FIGURE 8-31

Unacceptable polynomial for a two-segment asymmetrical rise-fall, single-dwell cam

5   Figure 8-31 shows the *S V A J* diagrams for this solution with its maximum values noted. Observe that the derived 6th-degree polynomial has obeyed the stated boundary conditions and does in fact pass through a displacement of 1 unit at 45°. But note also that it overshoots that point and reaches a height of 2.37 units at its peak. The acceleration peak is also 2.37 times that of the symmetrical case of Example 8-8 (p. 438). Without any additional boundary conditions applied, the function seeks symmetry. Note that the zero velocity point is still at 90° when we would like it to be at 45°. We can try to force the velocity to zero with an additional boundary condition of $V = 0$ when $\theta = 45°$.

6   Figure 8-32 (p. 442) shows the *S V A J* diagrams for a 7th-degree polynomial having 8 BCs, $S = V = A = 0$ at $\theta = 0°$, $S = V = A = 0$ at $\theta = 180°$, $S = 1$, $V = 0$ at $\theta = 45°$. Note that the resulting function has obeyed our commands and goes through those points, but "does its own thing" elsewhere. It now plunges to a negative displacement of −3.93, and the peak acceleration is much larger. This points out an inherent problem in polynomial functions, namely that their behavior between boundary conditions is not controllable and may create undesirable deviations in the follower motion. This problem is exacerbated as the degree of the function increases since it then has more roots and inflection points, thus allowing more oscillations between the boundary conditions.

7   Open the files Ex_08-09a and b in program DYNACAM to see this example in greater detail.

In this case, the rule of thumb to minimize the number of segments is in conflict with the rule of thumb to minimize the degree of the polynomial. One alternative solution to this asymmetrical problem is to use three segments, one for the rise, one for the fall, and one for the dwell. Adding segments will reduce the order of the functions and bring them under control.

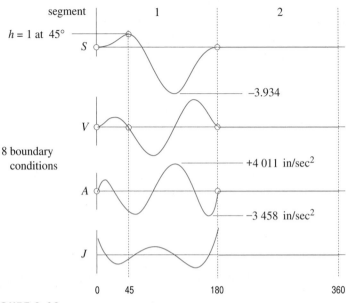

FIGURE 8-32

Unacceptable polynomial for a two-segment asymmetrical rise-fall, single-dwell cam

✏️ EXAMPLE 8-10

Designing a Three-Segment Polynomial for an Asymmetrical Rise-Fall Single-Dwell Case Using Minimum Boundary Conditions.

Problem: Redefine the specification from Example 8-9 (p. 440) as:

rise	1 in (25.4 mm) in 45°
fall	1 in (25.4 mm) in 135°
dwell	at zero displacement for 180° (low dwell)
cam ω	15 rad/sec

Solution:

1   The first attempt at this solution specifies 5 BCs; $S = V = A = 0$ at the start of the rise (to match the dwell), $S = 1$ and $V = 0$ at the end of the rise. Note that the rise segment BCs leave the acceleration at its end unspecified, but the fall segment BCs must include the value of the acceleration at the end of the rise that results from the calculation of its acceleration. Thus, the fall requires one more BC than the rise.

2   This results in the following 4th degree equation for the rise segment:

$$s = h\left[4\left(\frac{\theta}{\beta}\right)^3 - 3\left(\frac{\theta}{\beta}\right)^4\right] \qquad (a)$$

3   Evaluating the acceleration at the end of rise gives -4377.11 in/sec². This value becomes a

BC for the fall segment. The set of 6 BCs for the fall is then: $S = 1, V = 0, A = -4377.11$ at the start of the fall (to match the rise) and $S = V = A = 0$ at the end of the fall to match the dwell. The 5th-degree equation for the fall is then:

$$s = h\left[1 - 54\left(\frac{\theta}{\beta}\right)^2 + 152\left(\frac{\theta}{\beta}\right)^3 - 147\left(\frac{\theta}{\beta}\right)^4 + 48\left(\frac{\theta}{\beta}\right)^5\right] \qquad (b)$$

4 Figure 8-33 (p. 444) shows the $S V A J$ diagrams for this solution with its extreme values noted. Observe that this polynomial on the fall also has a problem—the displacement still goes negative.

5 The trick in this case (and in general) is to first calculate the segment with the smaller acceleration (here the second segment) because of its larger duration angle β. Then use its smaller acceleration value as a boundary condition on the first segment. The 5 BCs for segment 2 are then $S = 1$ and $V = 0$ at the start of the fall and $S = V = A = 0$ at the end of the fall (to match the dwell). These give the following 4th-degree polynomial for the fall.

$$s = h\left[1 - 6\left(\frac{\theta}{\beta}\right)^2 + 8\left(\frac{\theta}{\beta}\right)^3 - 3\left(\frac{\theta}{\beta}\right)^4\right] \qquad (c)$$

6 Evaluating the acceleration at the start of the fall gives -486.4 in/sec^2. This value becomes a BC for the rise segment. The set of 6 BCs for the rise is then: $S = V = A = 0$ at the start of the rise to match the dwells, and $S = 1, V = 0, A = -486.4$ at the end of the rise (to match the fall). The 5th-degree equation for the rise is then:

$$s = h\left[9.333\left(\frac{\theta}{\beta}\right)^3 - 13.667\left(\frac{\theta}{\beta}\right)^4 + 5.333\left(\frac{\theta}{\beta}\right)^5\right] \qquad (d)$$

7 The resulting cam design is shown in Figure 8-34 (p. 444). The displacement is now under control and the peak acceleration is much less than the previous design at about 2 024 in/sec^2.

8 The design of Figure 8-34 is acceptable (though not optimum)* for this example. Open the files Ex_08-10a and b in program DYNACAM to see this example in greater detail.

8.5 CRITICAL PATH MOTION (CPM)

Probably the most common application of **critical path motion** (CPM) specifications in production machinery design is the need for **constant velocity motion**. There are two general types of automated production machinery in common use, **intermittent motion** assembly machines and **continuous motion** assembly machines.

Intermittent motion assembly machines carry the manufactured goods from work-station to workstation, stopping the workpiece or subassembly at each station while another operation is performed upon it. The throughput speed of this type of automated production machine is typically limited by the dynamic forces that are due to accelerations

* An optimum solution to this generic problem can be found in reference [5].

FIGURE 8-33

Uacceptable polynomials for a three-segment asymmetrical rise-fall, single-dwell cam

and decelerations of the mass of the moving parts of the machine and its workpieces. The workpiece motion may be either in a straight line as on a conveyor or in a circle as on a rotary table as shown in Figure 8-22 (p. 429).

FIGURE 8-34

Acceptable polynomials for a three-segment asymmetrical rise-fall, single-dwell cam

Continuous motion assembly machines never allow the workpiece to stop and thus are capable of higher throughput speeds. All operations are performed on a moving target. Any tools that operate on the product have to "chase" the moving assembly line to do their job. Since the assembly line (often a conveyor belt or chain, or a rotary table) is moving at some constant velocity, there is a need for mechanisms to provide constant velocity motion, matched exactly to the conveyor, in order to carry the tools alongside for a long enough time to do their job. These cam driven "chaser" mechanisms must then return the tool quickly to its start position in time to meet the next part or subassembly on the conveyor (quick-return). There is a motivation in manufacturing to convert from intermittent motion machines to continuous motion in order to increase production rates. Thus there is some demand for this type of constant velocity mechanism. Though we met some linkages in Chapter 6 that give approximate constant velocity output, the cam-follower system is well suited to this problem, allowing theoretically exact constant follower velocity, and the polynomial cam function is particularly adaptable to the task.

Polynomials Used for Critical Path Motion

EXAMPLE 8-11

Designing a Polynomial for Constant Velocity Critical Path Motion.

Problem: Consider the following statement of a critical path motion (CPM) problem:

Accelerate	the follower from zero to 10 in/sec
Maintain	a constant velocity of 10 in/sec for 0.5 sec
Decelerate	the follower to zero velocity
Return	the follower to start position
Cycle time	exactly 1 sec

Solution:

1   This unstructured problem statement is typical of real design problems as was discussed in Chapter 1. No information is given as to the means to be used to accelerate or decelerate the follower or even as to the portions of the available time to be used for those tasks. A little reflection will cause the engineer to recognize that the specification on total cycle time in effect defines the camshaft velocity to be its reciprocal or **one revolution per second**. Converted to appropriate units, this is an angular velocity of 2π rad/sec.

2   The constant velocity portion uses half of the total period of 1 sec in this example. The designer must next decide how much of the remaining 0.5 sec to devote to each other phase of the required motion.

3   The problem statement seems to imply that four segments are needed. Note that the designer has to somewhat arbitrarily select the lengths of the individual segments (except the constant velocity one). Some iteration may be required to optimize the result. Program DYNACAM makes the iteration process quick and easy, however.

4   Assuming four segments, the timing diagram in Figure 8-35 (p. 446) shows an acceleration phase, a constant velocity phase, a deceleration phase, and a return phase, labeled as segments 1 through 4.

FIGURE 8-35

Constant velocity cam timing diagram

5   The segment angles (β's) are assumed, for a first approximation, to be 30° for segment 1, 180° for segment 2, 30° for segment 3, and 120° for segment 4 as shown in Figure 8-36. These angles may need to be adjusted in later iterations, except for segment 2 which is rigidly constrained in the specifications.

6   Figure 8-36 shows a tentative set of boundary conditions for the s v a j diagram. The solid circles indicate a set of boundary conditions that will constrain the continuous function to these specifications. These are for segment 1:

$$\text{when} \quad \theta = 0°; \qquad s = 0, \qquad v = 0, \qquad none$$
$$\text{when} \quad \theta = 30°; \qquad none, \qquad v = 10, \qquad a = 0 \qquad\qquad (a)$$

7   Note that the displacement at θ = 30° is left unspecified. The resulting polynomial function will provide us with the values of displacement at that point, which can then be used as a boundary condition for the next segment, in order to make the overall functions continuous as required. The acceleration at θ = 30° must be zero in order to match that of the constant velocity segment 2. The acceleration at θ = 0 is left unspecified. The resulting value will be used later to match the end of the last segment's acceleration.

8   Putting these four BCs for segment 1 into program DYNACAM yields a cubic function whose s v a j plots are shown in Figure 8-37. Its equation is:

$$s = 0.83376\left(\frac{\theta}{\beta}\right)^2 - 0.27792\left(\frac{\theta}{\beta}\right)^3 \qquad\qquad (8.27a)$$

The maximum displacement occurs at θ = 30°. This will be used as one BC for segment 2. The entire set for segment 2 is:

$$\text{when} \quad \theta = 30°; \qquad s = 0.556, \qquad v = 10$$
$$\text{when} \quad \theta = 210°; \qquad none, \qquad none \qquad\qquad (b)$$

9   Note that in the derivations and in the DYNACAM program each segment's local angles run from zero to the β for that segment. Thus, segment 2's local angles are 0° to 180°, which correspond to 30° to 210° globally in this example. We have left the displacement, velocity, and acceleration at the end of segment 2 unspecified. They will be determined by the computation.

FIGURE 8-36

A possible set of boundary conditions for the four-segment constant velocity solution

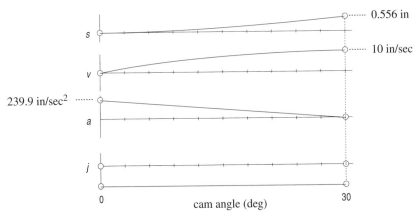

FIGURE 8-37

Segment one for the four-segment solution to the constant velocity problem (Example 8-11)

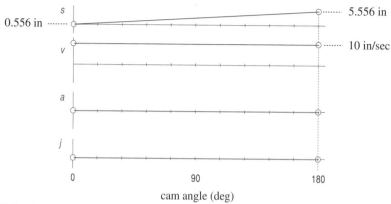

FIGURE 8-38

Segment two for the four-segment solution to the constant velocity problem (Example 8-11)

10 Since this is a constant velocity segment, its integral, the displacement function, must be a polynomial of degree one, i.e., a straight line. If we specify more than two BCs we will get a function of higher degree than one that will pass through the specified endpoints but may also oscillate between them and deviate from the desired constant velocity. Thus we can *only* provide two BCs, a slope and an intercept, as defined in equation 8.2 (p. 410). But, we must provide at least one displacement boundary condition in order to compute the coefficient C_0 from equation 8.23 (p. 431). Specifying the two BCs at only one end of the interval is perfectly acceptable. The equation for segment 2 is:

$$s = 5\left(\frac{\theta}{\beta}\right) + 0.556 \qquad (8.27\text{b})$$

11 Figure 8-38 shows the displacement and velocity plots of segment 2. The acceleration and jerk are both zero. The resulting displacement at $\theta = 210°$ is 5.556.

12 The displacement at the end of segment 2 is now known from its equation. The four boundary conditions for segment 3 are then:

when $\theta = 210°$; $s = 5.556,$ $v = 10,$ $a = 0$

when $\theta = 240°$;   *none,* $v = 0,$ *none*

$$(c)$$

13 This generates a cubic displacement function for segment 3 as in Figure 8-39. Its equation is:

$$s = -0.27792\left(\frac{\theta}{\beta}\right)^3 + 0.83376\left(\frac{\theta}{\beta}\right) + 5.556 \qquad (8.27\text{c})$$

14 The boundary conditions for the last segment 4 are now defined, as they must match those of the end of segment 3 and the beginning of segment 1. The displacement at the end of segment 3 is found from the computation in DYNACAM to be $s = 6.112$ at $\theta = 240°$ and the acceleration at that point is –239.9. We left the acceleration at the beginning of segment 1 unspecified.

FIGURE 8-39

Four-segment solution to the constant velocity problem of Example 8-11

From the second derivative of the equation for displacement in that segment we find that the acceleration is 239.9 at $\theta = 0°$. The BCs for segment 4 are then:

when $\theta = 240°$; $s = 6.112$, $v = 0$, $a = -239.9$

when $\theta = 360°$; $s = 0$, $v = 0$, $a = 239.9$ (d)

15   The equation for segment 4 is then:

$$s = -9.9894\left(\frac{\theta}{\beta}\right)^5 + 24.9735\left(\frac{\theta}{\beta}\right)^4 - 7.7548\left(\frac{\theta}{\beta}\right)^3 - 13.3413\left(\frac{\theta}{\beta}\right)^2 + 6.112 \qquad (8.27d)$$

16   Figure 8-39 shows the $s\ v\ a\ j$ plots for the complete cam. It obeys the fundamental law of cam design because the piecewise functions are continuous through the acceleration. The maximum value of acceleration is 257 in/sec^2. The maximum negative velocity is -29.4 in/sec. We now have four piecewise-continuous functions, equations 8.27 (pp. 446 - 449), which will meet the performance specifications for this problem.

The reader may open the file E08-11.cam in program DYNACAM to investigate this example in greater detail.

While this design is acceptable, it can be improved. One useful strategy in designing polynomial cams is to minimize the number of segments, provided that this does not result in functions of such high degree that they misbehave between boundary conditions. Another strategy is to always start with the segment for which you have the most information. In this example, the constant velocity portion is the most constrained and must be a separate segment, just as a dwell must be a separate segment. The rest of the cam motion exists only to return the follower to the constant velocity segment for the next cycle. If we start by designing the constant velocity segment, it may be possible to complete the cam with only one additional segment. We will now redesign this cam, to the same specifications but with only two segments as shown in Figure 8-40 (p. 450).

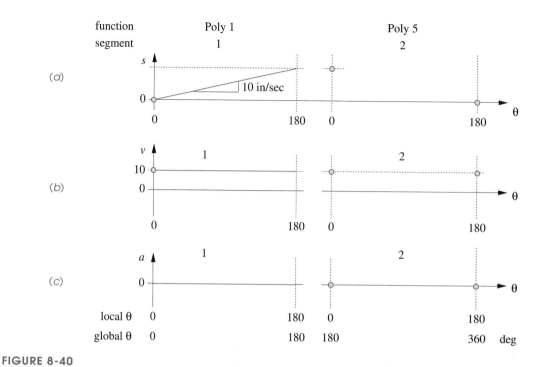

FIGURE 8-40

Boundary conditions for the two-segment constant velocity solution

✍️ **EXAMPLE 8-12**

Designing an Optimum Polynomial for Constant Velocity Critical Path Motion.

Problem: Redefine the problem statement of Example 8-11 (p. 445) to have only two segments.

Maintain	a constant velocity of 10 in/sec for 0.5 sec
Decelerate	and **accelerate** follower to constant velocity
Cycle time	exactly 1 sec

Solution: See Figures 8-40 and 8-41.

1   The BCs for the first, constant velocity, segment will be similar to our previous solution except for the global values of its angles and the fact that we will start at zero displacement rather than at 0.556 in. They are:

$$\text{when} \quad \theta = 0°; \qquad\qquad s = 0, \qquad\qquad v = 10$$

$$\text{when} \quad \theta = 180°; \qquad\qquad none, \qquad\qquad none \tag{a}$$

2   The displacement and velocity plots for this segment are identical to those in Figure 8-38 (p. 448) except that the displacement starts at zero. The equation for segment 1 is:

$$s = 5\left(\frac{\theta}{\beta}\right) \tag{8.28a}$$

3 The program calculates the displacement at the end of segment 1 to be 5.00 in. This defines that BC for segment 2. The set of BCs for segment 2 is then:

$$\text{when} \quad \theta = 180°; \qquad s = 5.00, \qquad v = 10, \qquad a = 0$$
$$\text{when} \quad \theta = 360°; \qquad s = 0, \qquad v = 10, \qquad a = 0 \tag{b}$$

The equation for segment 2 is:

$$s = -60\left(\frac{\theta}{\beta}\right)^5 + 150\left(\frac{\theta}{\beta}\right)^4 - 100\left(\frac{\theta}{\beta}\right)^3 + 5\left(\frac{\theta}{\beta}\right)^1 + 5 \tag{8.28b}$$

4 The *s v a j* diagrams for this design are shown in Figure 8-41. Note that they are much smoother than the four-segment design. The maximum acceleration in this example is now 230 in/sec², and the maximum negative velocity is –27.5 in/sec. These are both less than in the previous design of Example 8-11 (p. 445).

5 The fact that our displacement in this design contains negative values as shown in the *s* diagram of Figure 8-41 is of no concern. This is due to our starting with the beginning of the constant velocity portion as zero displacement. The follower has to go to a negative position in order to have distance to accelerate up to speed again. We will simply shift the displacement coordinates by that negative amount to make the cam. To do this, simply calculate the displacement coordinates for the cam. Note the value of the largest negative displacement. Add this value to the displacement boundary conditions for all segments and recalculate the cam functions with DYNACAM. (Do not change the BCs for the higher derivatives.) The finished cam's displacement profile will be shifted up such that its minimum value will now be zero.

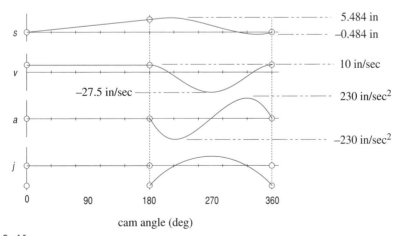

FIGURE 8-41
Two-segment solution to the constant velocity problem of Example 8-12

So, not only do we now have a smoother cam but the dynamic forces and stored kinetic energy are both lower. Note that we did not have to make any assumptions about the portions of the available nonconstant velocity time to be devoted to speeding up or slowing down. This all happened automatically from our choice of only two segments and the specification of the minimum set of necessary boundary conditions. This is clearly a superior design to the previous attempt and is in fact an optimal polynomial solution to the given specifications. The reader is encouraged to open the file E08-12.cam in program DYNACAM to investigate this example in more detail.

SUMMARY These sections have presented polynomial functions as the most versatile approach (of those shown here) to virtually any cam design problem. It is only since the development and general availability of computers that polynomial functions have become practical to use, as the computation to solve the simultaneous equations is often beyond hand calculation abilities. With the availability of a design aid to solve the equations such as program DYNACAM, polynomials have become a practical and preferable way to solve many, but not all, cam design problems. **Spline functions**, of which polynomials are a subset, offer even more flexibility in meeting boundary constraints and other cam performance criteria. Space does not permit a detailed exposition of spline functions as applied to cam systems here. See reference [6] for more information.

8.6   SIZING THE CAM—PRESSURE ANGLE AND RADIUS OF CURVATURE

Once the *s v a j* functions have been defined, the next step is to size the cam. There are two major factors that affect cam size, the **pressure angle** and the **radius of curvature**. Both of these involve either the **base circle radius** on the cam (R_b) when using flat-faced followers, or the **prime circle radius** on the cam (R_p) when using roller or curved followers.

The base circle's and prime circle's centers are at the center of rotation of the cam. The base circle is defined as *the smallest circle that can be drawn tangent to the physical cam surface* as shown in Figure 8-42. All radial cams will have a base circle, regardless of the follower type used.

The prime circle is only applicable to cams with roller followers or radiused (mushroom) followers and is measured to the center of the follower. The **prime circle** is defined as *the smallest circle that can be drawn tangent to the locus of the centerline of the follower* as shown in Figure 8-42. *The locus of the centerline of the follower* is called the **pitch curve**. Cams with roller followers are in fact defined for manufacture with respect to the pitch curve rather than with respect to the cam's physical surface. Cams with flat-faced followers must be defined for manufacture with respect to their physical surface, as there is no pitch curve.

The process of creating the physical cam from the *s* diagram can be visualized conceptually by imagining the *s* diagram to be cut out of a flexible material such as rubber. The *x* axis of the *s* diagram represents the circumference of a circle, which could be either the **base circle**, or the **prime circle**, around which we will "wrap" our "rubber" *s* diagram. We are free to choose the initial length of our *s* diagram's *x* axis, though the height of the displacement curve is fixed by the cam displacement function we have chosen. In effect we will choose the base or prime circle radius as a design parameter and stretch the length of the *s* diagram's axis to fit the circumference of the chosen circle.

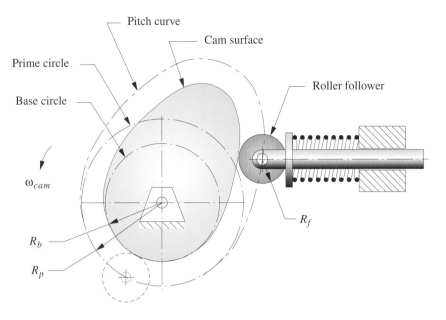

Pitch curve

Cam surface

Roller follower

ω_{cam}

R_f

R_b

R_p

FIGURE 8-42

Base circle R_b, prime circle R_p, and pitch curve of a radial cam with roller follower

We will present equations for pressure angle and radius of curvature only for radial cams with translating followers here. For related information on oscillating followers and axial (barrel) cams, see Chapter 7 of reference [5].

Pressure Angle—Translating Roller Followers

The **pressure angle** is defined as shown in Figure 8-43 (p. 454). It is the complement of the transmission angle that was defined for linkages in previous chapters and has a similar meaning with respect to cam-follower operation. By convention, the pressure angle is used for cams, rather than the transmission angle. Force can only be transmitted from cam to follower or vice versa along the **axis of transmission** which is perpendicular to the **axis of slip**, or common tangent.

PRESSURE ANGLE   The **pressure angle** ϕ is *the angle between the direction of motion (velocity) of the follower and the direction of the axis of transmission.*[*] When $\phi = 0$, all the transmitted force goes into motion of the follower and none into slip velocity. When ϕ becomes 90° there will be no motion of the follower. As a rule of thumb, we would like the pressure angle to be between zero and about 30° for translating followers to avoid excessive side load on the sliding follower. If the follower is oscillating on a pivoted arm, a pressure angle up to about 35° is acceptable. Values of ϕ greater than this can increase the follower sliding or pivot friction to undesirable levels and may tend to jam a translating follower in its guides.

ECCENTRICITY   Figure 8-44 (p. 455) shows the geometry of a cam and translating roller follower in an arbitrary position. This shows the general case in that the axis of

[*] Dresner and Buffington[7] point out that this definition is only valid for single-degree-of-freedom systems. For multi-input systems, a more complicated definition and calculation of pressure angle (or transmission angle) are needed.

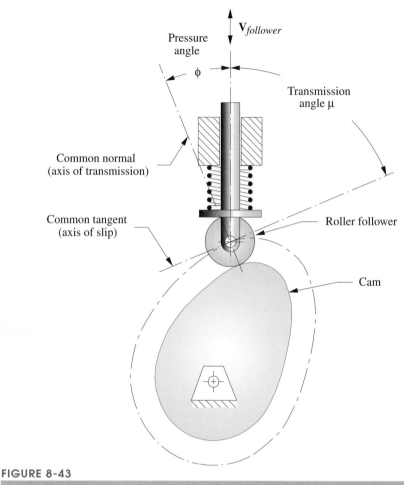

FIGURE 8-43

Cam pressure angle

motion of the follower does not intersect the center of the cam. There is an **eccentricity** ε defined as *the perpendicular distance between the follower's axis of motion and the center of the cam.* Often this eccentricity ε will be zero, making it an **aligned follower**, which is the special case.

In Figure 8-44, the axis of transmission is extended to intersect effective link 1, which is the ground link. (See Section 8.0, p. 401 and Figure 8-1, p. 403 for a discussion of effective links in cam systems.) This intersection is instant center $I_{2,4}$ (labeled B) which, by definition, has the same velocity in link 2 (the cam) and in link 4 (the follower). Because link 4 is in pure translation, all points on it have identical velocities $V_{follower}$, which are equal to the velocity of $I_{2,4}$ in link 2. We can write an expression for the velocity of $I_{2,4}$ in terms of cam angular velocity and the radius b from cam center to $I_{2,4}$,

$$V_{I_{2,4}} = b\omega = \dot{s} \qquad\qquad (8.29)$$

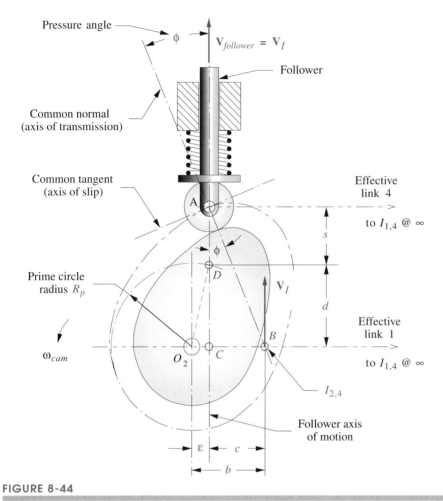

FIGURE 8-44

Geometry for the derivation of the equation for pressure angle

where s is the instantaneous displacement of the follower from the s diagram and $\dot{s}$ is its time derivative in units of length/sec. (Note that capital $S\ V\ A\ J$ denote time-based variables rather than functions of cam angle.)

But
$$\dot{s} = \frac{ds}{dt}$$

and
$$\frac{ds}{dt}\frac{d\theta}{d\theta} = \frac{ds}{d\theta}\frac{d\theta}{dt} = \frac{ds}{d\theta}\omega = v\omega$$

so
$$b\omega = v\omega$$

then
$$b = v \qquad\qquad\qquad (8.30)$$

This is an interesting relationship which says that the **distance b to the instant center $I_{2,4}$ is equal to the velocity of the follower** v in units of length per radian as derived in previous sections. We have reduced this expression to pure geometry, independent of the angular velocity ω of the cam.

Note that we can express the distance b in terms of the prime circle radius R_p and the eccentricity ε, by the construction shown in Figure 8-44. Swing the arc of radius R_p until it intersects the axis of motion of the follower at point D. This defines the length of line d from effective link 1 to this intersection. This is constant for any chosen prime circle radius R_p. Points A, C, and $I_{2,4}$ form a right triangle whose upper angle is the pressure angle ϕ and whose vertical leg is $(s + d)$, where s is the instantaneous displacement of the follower. From this triangle:

$$c = b - \varepsilon = (s+d)\tan\phi$$

and

$$b = (s+d)\tan\phi + \varepsilon$$

(8.31a)

Then from equation 8.30 (p. 455),

$$v = (s+d)\tan\phi + \varepsilon$$

(8.31b)

and from triangle CDO_2,

$$d = \sqrt{R_P^2 - \varepsilon^2}$$

(8.31c)

Substituting equation 8.31c into equation 8.31b and solving for ϕ give an expression for pressure angle in terms of displacement s, velocity v, eccentricity ε, and the prime circle radius R_p.

$$\phi = \arctan\frac{v - \varepsilon}{s + \sqrt{R_P^2 - \varepsilon^2}}$$

(8.31d)

The velocity v in this expression is in units of length/rad, and all other quantities are in compatible length units. We have typically defined s and v by this stage of the cam design process and wish to manipulate R_p and ε to get an acceptable maximum pressure angle ϕ. As R_p is increased, ϕ will be reduced. The only constraints against large values of R_p are the practical ones of package size and cost. Often there will be some upper limit on the size of the cam-follower package dictated by its surroundings. There will always be a cost constraint and bigger = heavier = more expensive.

Choosing a Prime Circle Radius

Both R_p and ε are within a transcendental expression in equation 8.31d, so they cannot be conveniently solved for directly. The simplest approach is to assume a trial value for R_p and an initial eccentricity of zero, and use program DYNACAM; your own program; or an equation solver such as *Matlab*, *TKSolver* or *Mathcad* to quickly calculate the values of ϕ for the entire cam, and then adjust R_p and repeat the calculation until an acceptable arrangement is found. Figure 8-45 shows the calculated pressure angles for a four-dwell cam. Note the similarity in shape to the velocity functions for the same cam in Figure 8-6 (p. 409), as that term is dominant in equation 8.31d.

USING ECCENTRICITY   If a suitably small cam cannot be obtained with acceptable pressure angle, then eccentricity can be introduced to change the pressure angle. Using eccentricity to control the pressure angle has its limitations. For a positive ω, a positive value of eccentricity will *decrease the pressure angle on the rise* but will *increase it on the fall*. Negative eccentricity does the reverse.

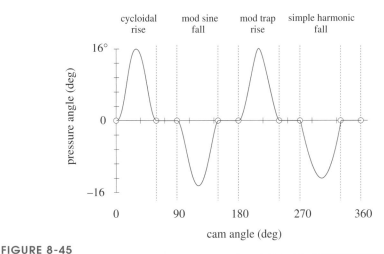

FIGURE 8-45

Pressure angle functions are similar in shape to velocity functions.

This is of little value with a form-closed (groove or track) cam, as it is driving the follower in both directions. For a force-closed cam with spring return, you can sometimes afford to have a larger pressure angle on the fall than on the rise because the stored energy in the spring is attempting to speed up the camshaft on the fall, whereas the cam is storing that energy in the spring on the rise. The limit of this technique can be the degree of overspeed attained with a larger pressure angle on the fall. The resulting variations in cam angular velocity may be unacceptable.

The most value gained from adding eccentricity to a follower comes in situations where the cam program is asymmetrical and significant differences exist (with no eccentricity) between maximum pressure angles on rise and fall. Introducing eccentricity can balance the pressure angles in this situation and create a smoother running cam.

If adjustments to R_p or ε do not yield acceptable pressure angles, the only recourse is to return to an earlier stage in the design process and redefine the problem. Less lift or more time to rise or fall will reduce the causes of the large pressure angle. Design is, after all, an iterative process.

Overturning Moment—Translating Flat-Faced Follower

Figure 8-46 (p. 458) shows a translating, flat-faced follower running against a radial cam. The pressure angle can be seen to be zero for all positions of cam and follower. This seems to be giving us something for nothing, which can't be true. As the contact point moves left and right, the point of application of the force between cam and follower moves with it. There is an overturning moment on the follower associated with this off-center force which tends to jam the follower in its guides, just as did too large a pressure angle in the roller follower case. In this case, we would like to keep the cam as small as possible in order to minimize the moment arm of the force. Eccentricity will affect the average value of the moment, but the peak-to-peak variation of the moment about that average is unaffected by eccentricity. Considerations of too-large pressure angle do not limit the

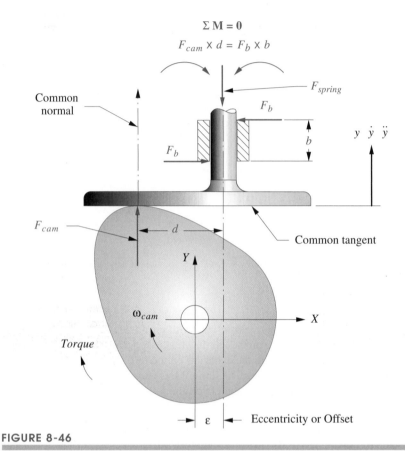

$$\Sigma\, M = 0$$
$$F_{cam} \times d = F_b \times b$$

FIGURE 8-46

Overturning moment on a flat-faced follower

size of this cam, but other factors do. The minimum radius of curvature (see below) of the cam surface must be kept large enough to avoid undercutting. This is true regardless of the type of follower used.

Radius of Curvature—Translating Roller Follower

The **radius of curvature** is a *mathematical property of a function*. Its value and use is not limited to cams but has great significance in their design. The concept is simple. No matter how complicated a curve's shape may be, nor how high the degree of the describing function, it will have an instantaneous radius of curvature at every point on the curve. These radii of curvature will have instantaneous centers (which may be at infinity), and the radius of curvature of any function is itself a function that can be computed and plotted. For example, the radius of curvature of a straight line is infinity everywhere; that of a circle is a constant value. A parabola has a constantly changing radius of curvature that approaches infinity. A cubic curve will have radii of curvature that are sometimes positive (convex) and sometimes negative (concave). The higher the degree of a function, in general, the more potential variety in its radius of curvature.

Cam contours are usually functions of high degree. When they are wrapped around their base or prime circles, they may have portions that are concave, convex, or flat. Infinitesimally short flats of infinite radius will occur at all inflection points on the cam surface where it changes from concave to convex or vice versa.

The radius of curvature of the finished cam is of concern regardless of the follower type, but the concerns are different for different followers. Figure 8-47 shows an obvious (and exaggerated) problem with a roller follower whose own (constant) radius of curvature R_f is too large to follow the locally smaller concave (negative) radius $-\rho$ on the cam. (Note that, normally, one would not use that large a roller compared to the cam.)

A more subtle problem occurs when the roller follower radius R_f is larger than the smallest positive (convex) local radius $+\rho$ on the cam. This problem is called **undercutting** and is depicted in Figure 8-48 (p. 460). Recall that for a roller follower cam, the cam contour is actually defined as the locus of the center of the roller follower, or the **pitch curve**. The machinist is given these x,y coordinate data (on computer tape or disk) and also told the radius of the follower R_f. The machinist will then cut the cam with a cutter of the same effective radius as the follower, following the pitch curve coordinates with the center of the cutter.

Figure 8-48a shows the situation in which the follower (cutter) radius R_f is at one point exactly equal to the minimum convex radius of curvature of the cam $(+\rho_{min})$. The cutter creates a perfect sharp point, or **cusp**, on the cam surface. This cam will not run very well at speed! Figure 8-48b shows the situation in which the follower (cutter) radius is greater than the minimum convex radius of curvature of the cam. The cutter now undercuts or removes material needed for cam contours in different locations and also creates a sharp point or cusp on the cam surface. This cam no longer has the same displacement function you so carefully designed.

The rule of thumb is to keep the absolute value of the minimum radius of curvature ρ_{min} of the cam pitch curve preferably at least 2 to 3 times as large as the radius of the roller follower R_f.

$$\left|\rho_{min}\right| >> R_f \tag{8.32}$$

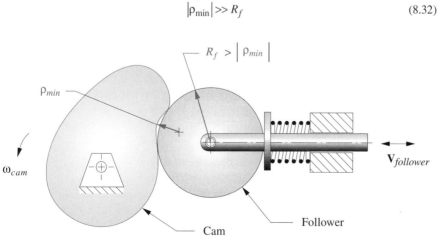

FIGURE 8-47

The result of using a roller follower larger than the one for which the cam was designed

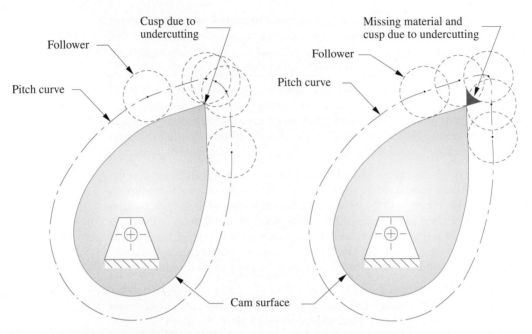

(a) Radius of curvature of pitch curve equals the radius of the roller follower

(b) Radius of curvature of pitch curve is less than the radius of the roller follower

FIGURE 8-48

Small positive radius of curvature can cause undercutting.

A derivation for radius of curvature can be found in any calculus text. For our case of a roller follower, we can write the equation for the radius of curvature of the pitch curve of the cam as:

$$\rho_{pitch} = \frac{\left[\left(R_P + s \right)^2 + v^2 \right]^{3/2}}{\left(R_P + s \right)^2 + 2v^2 - a\left(R_P + s \right)} \qquad (8.33)$$

In this expression, s, v, and a are the displacement, velocity, and acceleration of the cam program as defined in a previous section. Their units are length, length/rad, and length/rad^2, respectively. R_p is the prime circle radius. **Do not confuse** this *prime circle radius R_p* with the *radius of curvature*, ρ_{pitch}. R_p is a **constant value** which you choose as a design parameter and ρ_{pitch} is the constantly changing radius of curvature that results from your design choices.

Also do not confuse R_p, the *prime circle radius* with R_f, the *radius of the roller follower*. See Figure 8-43 (p. 454) for definitions. You can choose the value of R_f to suit the problem, so you might think that it is simple to satisfy equation 8.32 (p. 459) by just selecting a roller follower with a small value of R_f. Unfortunately it is more complicated than that, as a small roller follower may not be strong enough to withstand the dynamic forces from the cam. The radius of the pin on which the roller follower pivots is substantially smaller than R_f because of the space needed for roller or ball bearings within the follower. Dynamic forces will be addressed in later chapters where we will revisit this problem.

We can solve equation 8.33 for ρ_{pitch} since we know s, v, and a for all values of θ and can choose a trial R_p. If the pressure angle has already been calculated, the R_p found for its acceptable values should be used to calculate ρ_{pitch} as well. If a suitable follower radius cannot be found which satisfies equation 8.32 (p. 459) for the minimum values of ρ_{pitch} calculated from equation 8.33, then further iteration will be needed, possibly including a redefinition of the cam specifications.

Program DYNACAM calculates ρ_{pitch} for all values of θ for a user supplied prime circle radius R_p. Figure 8-49 shows ρ_{pitch} for the four-dwell cam of Figure 8-6 (p. 409). Note that this cam has both positive and negative radii of curvature. The large values of radius of curvature are truncated at arbitrary levels on the plot as they are heading to infinity at the inflection points between convex and concave portions. Note that the radii of curvature go out to positive infinity and return from negative infinity or vice versa at these inflection points (perhaps after a round trip through the universe?).

Once an acceptable prime circle radius and roller follower radius are determined based on pressure angle and radius of curvature considerations, the cam can be drawn in finished form and subsequently manufactured. Figure 8-50 shows the profile of the four-dwell cam from Figure 8-6. The cam surface contour is swept out by the envelope of follower positions just as the cutter will create the cam in metal. The sidebar shows the parameters for the design, which is an acceptable one. The ρ_{min} is 1.7 times R_f and the pressure angles are less than 30°. The contours on the cam surface appear smooth, with no sharp corners. Figure 8-51 (p. 463) shows the same cam with only one change. The radius of follower R_f has been made the same as the minimum radius of curvature, ρ_{min}. The sharp corners or cusps in several places indicate that undercutting has occurred. This has now become an **unacceptable cam**, *simply because of a roller follower that is too large.*

The coordinates for the cam contour, measured to the locus of the center of the roller follower, or the **pitch curve** as shown in Figure 8-50 (p. 462), are defined by the follow-

FIGURE 8-49

Radius of curvature of a four-dwell cam

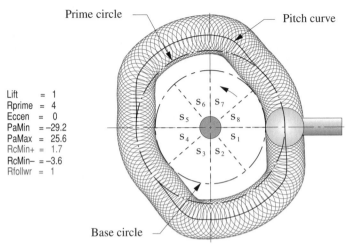

Prime circle
Pitch curve

Lift = 1
Rprime = 4
Eccen = 0
PaMin = –29.2
PaMax = 25.6
RcMin+ = 1.7
RcMin– = –3.6
Rfollwr = 1

s_6 s_7
s_5 s_8
s_4 s_1
s_3 s_2

Base circle

FIGURE 8-50

Radial plate cam profile is generated by the locus of the roller follower (or cutter)

ing expressions, referenced to the center of rotation of the cam. See Figure 8-44 (p. 455) for nomenclature. The subtraction of the cam input angle θ from 2π is necessary because the relative motion of the follower versus the cam is opposite to that of the cam versus the follower. In other words, to define the contour of the centerline of the follower's path around a stationary cam, we must move the follower (and also the cutter to make the cam) in the opposite direction of cam rotation.

$$x = \cos\lambda\sqrt{(d+s)^2 + \varepsilon^2}$$
$$y = \sin\lambda\sqrt{(d+s)^2 + \varepsilon^2}$$

(8.34)

where:

$$\lambda = (2\pi - \theta) - \arctan\left(\frac{\varepsilon}{d+s}\right)$$

Radius of Curvature—Translating Flat-Faced Follower

The situation with a flat-faced follower is different from that of a roller follower. A negative radius of curvature on the cam cannot be accommodated with a flat-faced follower. The flat follower obviously cannot follow a concave cam. Undercutting will occur when the radius of curvature becomes negative if a cam with that condition is made.

Figure 8-52 shows a cam and translating flat-faced follower in an arbitrary position. The origin of the global XY coordinate system is placed at the cam's center of rotation, and the X axis is defined parallel to the common tangent, which is the surface of the flat follower. The vector $\mathbf{r}$ is attached to the cam, rotates with it, and serves as the reference line to which the cam angle θ is measured from the X axis. The point of contact A is defined by the position vector $\mathbf{R}_A$. The instantaneous center of curvature is at C and the radius of curvature is ρ. R_b is the radius of the base circle and s is the displacement of the follower for angle θ. The eccentricity is ε.

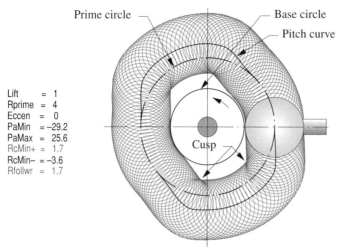

Lift   = 1
Rprime = 4
Eccen   = 0
PaMin   = −29.2
PaMax   = 25.6
RcMin+ = 1.7
RcMin− = −3.6
Rfollwr = 1.7

FIGURE 8-51

Cusps formed by undercutting due to radius of follower $R_f \geq$ cam radius of curvature ρ

We can define the location of contact point A from two vector loops (in complex notation).

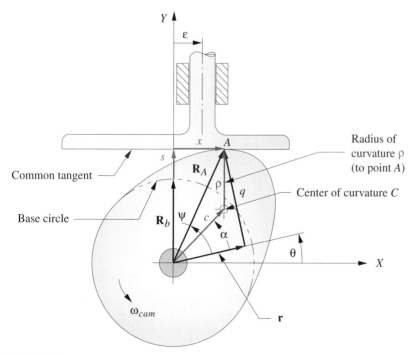

FIGURE 8-52

Geometry for derivation of radius of curvature and cam contour with flat-faced follower

$$\mathbf{R}_A = x + j\left(R_b + s\right)$$

and

$$\mathbf{R}_A = ce^{j(\theta+\alpha)} + j\rho$$

so:

$$ce^{j(\theta+\alpha)} + j\rho = x + j\left(R_b + s\right) \tag{8.35a}$$

Substitute the Euler equivalent (equation 4.4a, p. 185) in equation 8.35a and separate the real and imaginary parts.

real:

$$c\cos\left(\theta+\alpha\right) = x \tag{8.35b}$$

imaginary:

$$c\sin\left(\theta+\alpha\right) + \rho = R_b + s \tag{8.35c}$$

The center of curvature C is **stationary** on the cam meaning that the magnitudes of c and ρ, and angle α do not change for small changes in cam angle θ. (These values are not constant but are at stationary values. Their first derivatives with respect to θ are zero, but their higher derivatives are not zero.)

Differentiating equation 8.35a with respect to θ then gives:

$$jce^{j(\theta+\alpha)} = \frac{dx}{d\theta} + j\frac{ds}{d\theta} \tag{8.36}$$

Substitute the Euler equivalent (equation 4.4a, p. 185) in equation 8.36 and separate the real and imaginary parts.

real:

$$-c\sin\left(\theta+\alpha\right) = \frac{dx}{d\theta} \tag{8.37}$$

imaginary:

$$c\cos\left(\theta+\alpha\right) = \frac{ds}{d\theta} = v \tag{8.38}$$

Inspection of equations 8.35b and 8.36 shows that:

$$x = v \tag{8.39}$$

This is an interesting relationship that says the x position of the contact point between cam and follower is equal to the velocity of the follower in length/rad. This means that the v diagram gives a direct measure of the necessary minimum face width of the flat follower.

$$facewidth > v_{max} - v_{min} \tag{8.40}$$

If the velocity function is asymmetric, then a minimum-width follower will have to be asymmetric also, in order not to fall off the cam.

Differentiating equation 8.39 with respect to θ gives:

$$\frac{dx}{d\theta} = \frac{dv}{d\theta} = a \tag{8.41}$$

Equations 8.35c and 8.37 can be solved simultaneously and equation 8.41 substituted in the result to yield:

$$\rho = R_b + s + a \tag{8.42a}$$

and the minimum value of radius of curvature is

$$\rho_{min} = R_b + (s+a)_{min} \tag{8.42b}$$

BASE CIRCLE   Note that equations 8.42 define the radius of curvature in terms of the base circle radius and the displacement and acceleration functions from the *s v a j* diagrams only. Because ρ cannot be allowed to become negative with a flat-faced follower, we can formulate a relationship from equation 8.42b that will predict the minimum base circle radius R_b needed to avoid undercutting. The only factor on the right side of equations 8.42 that can be negative is the acceleration, a. We have defined s to be always positive, as is R_b. Therefore, the worst case for undercutting will occur when a is near its **largest negative value**, a_{min}, whose value we know from the a diagram. The minimum base circle radius can then be defined as:

$$R_{b_{min}} > \rho_{min} - (s+a)_{min} \tag{8.43}$$

Because the value of a_{min} is negative and it is also negated in equation 8.43, it dominates the expression. To use this relationship, we must choose some minimum radius of curvature ρ_{min} for the cam surface as a design parameter. Since the hertzian contact stresses at the contact point are a function of local radius of curvature, that criterion can be used to select ρ_{min}. That topic is beyond the scope of this text and will not be further explored here. See reference [1] for further information on contact stresses.

CAM CONTOUR   For a flat-faced follower cam, the coordinates of the physical cam surface must be provided to the machinist as there is no pitch curve to work to. Figure 8-52 (p. 463) shows two orthogonal vectors, **r** and **q**, which define the cartesian coordinates of contact point A between cam and follower with respect to a rotating axis coordinate system embedded in the cam. Vector **r** is the rotating "*x*" axis of this embedded coordinate system. Angle ψ defines the position of vector $\mathbf{R}_A$ in this system. Two vector loop equations can be written and equated to define the coordinates of all points on the cam surface as a function of cam angle θ.

$$\mathbf{R}_A = x + j(R_b + s)$$

and

$$\mathbf{R}_A = re^{j\theta} + qe^{j\left(\theta + \frac{\pi}{2}\right)}$$

so:

$$re^{j\theta} + qe^{j\left(\theta + \frac{\pi}{2}\right)} = x + j(R_b + s) \tag{8.44}$$

Divide both sides by $e^{j\theta}$:

$$r + jq = xe^{-j\theta} + j\left(R_b + s\right)e^{-j\theta} \tag{8.45}$$

Separate into real and imaginary components and substitute v for x from equation 8.39 p. 464):

real (x component):

$$r = \left(R_b + s\right)\sin\theta + v\cos\theta \tag{8.46a}$$

imaginary (y component):

$$q = \left(R_b + s\right)\cos\theta - v\sin\theta \tag{8.46b}$$

Equations 8.46 can be used to machine the cam for a flat-faced follower. These x, y components are in the rotating coordinate system that is embedded in the cam.

Note that none of the equations developed above for this case involve the **eccentricity**, ε. It is only a factor in cam size when a roller follower is used. It does not affect the geometry of a flat follower cam.

Figure 8-53 shows the result of trying to use a flat-faced follower on a cam whose theoretical path of follower point P has negative radius of curvature due to a base circle radius that is too small. If the follower tracked the path of P as is required to create the motion function defined in the s diagram, the cam surface would actually be as developed by the envelope of straight lines shown. But, these loci of the follower face are cutting into cam contours that are needed for other cam angles. The line running through the forest of follower loci is the path of point P needed for this design. The undercutting can be clearly seen as the crescent-shaped missing pieces at four places between the path of P and the follower face loci. Note that if the follower were zero width (at point P), it would work kinematically, but the stress at the knife edge would be infinite.

SUMMARY The task of sizing a cam is an excellent example of the need for and value of iteration in design. Rapid recalculation of the relevant equations with a tool such as

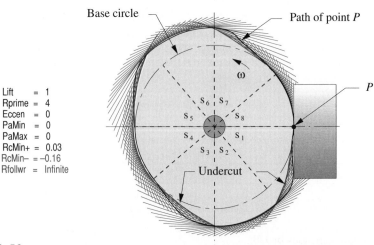

Lift	=	1
Rprime	=	4
Eccen	=	0
PaMin	=	0
PaMax	=	0
RcMin+	=	0.03
RcMin–	=	–0.16
Rfollwr	=	Infinite

FIGURE 8-53

Undercutting due to negative radius of curvature used with flat-faced follower program DYNACAM makes it possible to quickly and painlessly arrive at an acceptable solution while balancing the often conflicting requirements of pressure angle and radius of curvature constraints. In any cam, either the pressure angle or radius of curvature considerations will dictate the minimum size of the cam. Both factors must be checked. The choice of follower type, either roller or flat-faced, makes a big difference in the cam geometry. Cam programs that generate negative radii of curvature are unsuited to the flat-faced type of follower unless very large base circles are used to force ρ to be positive everywhere.

8.7   PRACTICAL DESIGN CONSIDERATIONS

The cam designer is often faced with many confusing decisions, especially at an early stage of the design process. Many early decisions, often made somewhat arbitrarily and without much thought, can have significant and costly consequences later in the design. The following is a discussion of some of the trade-offs involved with such decisions in the hope that it will provide the cam designer with some guidance in making these decisions.

Translating or Oscillating Follower?

There are many cases, especially early in a design, when either translating or rotating motion could be accommodated as output from the cam, though in other situations, the follower motion and geometry is dictated to the designer. If some design freedom is allowed, and straight-line motion is specified, the designer should consider the possibility of using an approximate straight-line motion, which is often adequate and can be obtained from a large-radius rocker follower. The rocker or oscillating follower has advantages over the translating follower when a roller is used. A round-cross-section translating follower slide is free to rotate about its axis of translation and needs to have some antirotation guiding provided (such as a keyway or second slide) to prevent z axis misalignment of the roller follower with the cam. Many commercial, nonrotating slide assemblies are now available, often fitted with ball bearings, and these provide a good way to deal with this issue. However, an oscillating follower arm will keep the roller follower aligned in the same plane as the cam with no guiding other than its own pivot.

Also, the pivot friction in an oscillating follower typically has a small moment arm compared to the moment of the force from the cam on the follower arm. But, the friction force on a translating follower has a one-to-one geometric relationship with the cam force. This can have a larger parasitic effect on the system.

Translating flat-faced followers are often deliberately arranged with their axis slightly out of the plane of the cam in order to create a rotation about their own axis due to the frictional moment resulting from the offset. The flat follower will then *precess* around its own axis and distribute the wear over its entire face surface. This is common practice with automotive valve cams that use flat-faced followers or "tappets."

Force- or Form-Closed?

A form-closed (track or groove) cam or conjugate cams are more expensive to make than a force-closed (open) cam simply because there are two surfaces to machine and grind. Also, heat treating will often distort the track of a form-closed cam, narrowing or widening it such that the roller follower will not fit properly. This virtually requires post heat-treat grinding for track cams in order to resize the slot. An open (force-closed) cam will also distort on heat-treating, but can still be usable without grinding.

FOLLOWER JUMP   The principal advantage of a form-closed (track) or conjugate-pair cam is that it does not need a return spring, and thus can be run at higher speeds than a force-closed cam whose spring and follower mass will go into resonance at some speed, causing potentially destructive follower jump. This phenomenon will be investigated in Chapter 15 on cam dynamics. High-speed automobile and motorcycle racing engines often use form-closed (desmodromic)* valve cam trains to allow higher engine rpm without incurring valve "float," or **follower jump**.

CROSSOVER SHOCK   Though the lack of a return spring can be an advantage, it comes, as usual, with a trade-off. In a form-closed (track) cam there will be **crossover shock** each time the acceleration changes sign. Crossover shock describes the impact force that occurs when the follower suddenly jumps from one side of the track to the other as the dynamic force (*ma*) reverses sign. There is no flexible spring in this system to absorb the force reversal as in the force-closed case. The high impact forces at crossover cause noise, high stresses, and local wear. Also, the roller follower has to reverse direction at each crossover, which causes sliding and accelerates follower wear. Studies have shown that roller followers running against a well-lubricated open radial cam have slip rates of less than 1%.[9]

Radial or Axial Cam?

This choice is largely dictated by the overall geometry of the machine for which the cam is being designed. If the follower must move parallel to the camshaft axis, then an axial cam is dictated. If there is no such constraint, a radial cam is probably a better choice simply because it is a less complicated, thus less expensive, cam to manufacture.

Roller or Flat-Faced Follower?

The roller follower is a better choice from a cam design standpoint simply because it accepts negative radius of curvature on the cam. This allows more variety in the cam program. Also, for any production quantities, the roller follower has the advantage of being available from several manufacturers in any quantity from one to a million. For low quantities it is not usually economical to design and build your own custom follower. In addition, replacement roller followers can be obtained from suppliers on short notice when repairs are needed. Also, they are not particularly expensive even in small quantities.

Perhaps the largest users of flat-faced followers are automobile engine makers. Their quantities are high enough to allow any custom design they desire. It can be made or purchased economically in large quantity and can be less expensive than a roller follower in that case. Also with engine valve cams, a flat follower can save space over a roller. Nevertheless, many manufacturers have switched to roller followers in automobile engine valve trains to reduce friction and improve fuel economy. Most new automotive IC engines designed in the United States in recent years have used roller followers for those reasons. Diesel engines have long used roller followers (tappets) as have racers who "hop-up" engines for high performance.

---

* More information on desmodromic cam-follower mechanisms can be found at http://members.chello.nl/~wgj.jansen/ where a number of models of their commercial implementations can be viewed in operation as movies.

Cams used in automated production line machinery use stock roller followers almost exclusively. The ability to quickly change a worn follower for a new one taken from the stockroom without losing much production time on the "line" is a strong argument in this environment. Roller followers come in several varieties (see Figure 8-5a, p. 407). They are based on roller or ball bearings. Plain bearing versions are also available for low-noise requirements. The outer surface, which rolls against the cam, can be either cylindrical or spherical in shape. The "crown" on the spherical follower is slight, but it guarantees that the follower will ride near the center of a flat cam even with some inaccuracy of alignment of the axes of rotation of cam and follower. If a cylindrical follower is chosen and care is not taken to align the axes of cam and roller follower, or if it deflects under load, the follower will ride on one edge and wear rapidly.

Commercial roller followers are typically made of high carbon alloy steel such as AISI 52100 and hardened to Rockwell HRC 60 - 62. The 52100 alloy is well suited to thin sections that must be heat-treated to a uniform hardness. Because the roller makes many revolutions for each cam rotation, its wear rate may be higher than that of the cam. Chrome plating the follower can markedly improve its life. Chrome is harder than steel at about HRC 70. Steel cams are typically hardened to a range of HRC 50 - 55.

To Dwell or Not to Dwell?

The need for a dwell is usually clear from the problem specifications. If the follower must be held stationary for any time, then a dwell is required. Some cam designers tend to insert dwells in situations where they are not specifically needed for follower stasis, in a mistaken belief that this is preferable to providing a rise-return motion when that is what is really needed. If the designer is attempting to use a double-dwell program in what really needs only to be a single-dwell case, with the motivation to "let the vibrations settle out" by providing a "short dwell" at the end of the motion, he or she is misguided. Instead, the designer probably should be using a different cam program, perhaps a polynomial or a B-spline tailored to the specifications. Taking the follower acceleration to zero, whether for an instant or for a "short dwell," is generally undesirable unless absolutely required for machine function. (See Examples 8-6, p. 435 , 8-7, p. 436, and 8-8, p. 438.) A dwell should be used only when the follower is required to be stationary for some measurable time. Moreover, if you do not need any dwell at all, consider using a linkage instead. They are a lot easier and less expensive to manufacture.

To Grind or Not to Grind?

Some production machinery cams are used as-milled, and not ground. Automotive valve cams are ground. The reasons are largely due to cost and quantity considerations as well as the high speeds of automotive cams. There is no question that a ground cam is superior to a milled cam. The question in each case is whether the advantage gained is worth the cost. In small quantities, as are typical of production machinery, grinding about doubles the cost of a cam. The advantages in terms of smoothness and quietness of operation, and of wear, are not in the same ratio as the cost difference. A well-machined cam can perform nearly as well as a well-ground cam and better than a poorly ground cam.[9, 10]

Automotive cams are made in large quantity, run at very high speed, and are expected to last for a very long time with minimal maintenance. This is a very challenging specification. It is a great credit to the engineering of these cams that they very seldom fail in 150,000 miles or more of operation. These cams are made on specialized equipment which keeps the cost of their grinding to a minimum.

Industrial production machine cams also see very long lives, often 10 to 20 years, running into billions of cycles at typical machine speeds. Unlike the typical automotive application, industrial cams often run around the clock, 7 days a week, 50+ weeks a year.

To Lubricate or Not to Lubricate?

Cams like lots of lubrication. Automotive cams are literally drowned in a flow of filtered and sometimes cooled engine oil. Many production machine cams run immersed in an oil bath. These are reasonably happy cams. Others are not so fortunate. Cams that operate in close proximity to the product on an assembly machine in which oil would cause contamination of the product (food products, personal products) often are run dry. Camera mechanisms, which are full of linkages and cams, are often run dry. Lubricant would eventually find its way to the film.

Unless there is some good reason to eschew lubrication, a cam-follower should be provided with a generous supply of clean lubricant, preferably a hypoid-type oil containing additives for boundary lubrication conditions. The geometry of a cam-follower joint (half-joint) is among the worst possible from a lubrication standpoint. Unlike a journal bearing, which tends to trap a film of lubricant within the joint, the half joint is continually trying to squeeze the lubricant out of itself. This can result in a boundary, or mixed boundary/EHD[*] lubrication state in which some metal-to-metal contact will occur. Lubricant must be continually resupplied to the joint. Another purpose of the liquid lubricant is to remove the heat of friction from the joint. If run dry, significantly higher material temperatures will result, with accelerated wear and possible early failure.

8.8   REFERENCES

1   **McPhate, A. J., and L. R. Daniel**. (1962). "A Kinematic Analysis of Fourbar Equivalent Mechanisms for Plane Motion Direct Contact Mechanisms." *Proc. of Seventh Conference on Mechanisms*, Purdue University, pp. 61-65.

2   **Neklutin, C. N.** (1954). "Vibration Analysis of Cams." *Machine Design*, **26**, pp. 190-198.

3   **Wiederrich, J. L., and B. Roth**. (1978). "Design of Low Vibration Cam Profiles." *Cams and Cam Mechanisms*, Jones, J. R., ed. Institution of Mechanical Engineers: London, pp. 3-8.

4   **Chew, M., and C. H. Chuang**. (1995). "Minimizing Residual Vibrations in High Speed Cam-Follower Systems Over a Range of Speeds." *Journal of Mechanical Design*, **117**(1), p. 166.

5   **Norton, R. L.** (2002). *Cam Design and Manufacturing Handbook*. Industrial Press: New York., pp. 108-115

6   **Ibid.**, pp. 69-126.

7   **Dresner, T. L., and K. W. Buffington**. (1991). "Definition of Pressure and Transmission Angles Applicable to Multi-Input Mechanisms." *Journal of Mechanical Design*, **113**(4), p. 495.

[*] Elasto-Hydrodynamic Lubrication—see ref. 5, p. 458.

8   **Casseres, M. G.** (1994). "An Experimental Investigation of the Effect of Manufacturing Methods and Displacement Functions on the Dynamic Performance of Quadruple Dwell Plate Cams." *M. S. Thesis*, Worcester Polytechnic Institute.

9   **Norton, R. L.** (1988). "Effect of Manufacturing Method on Dynamic Performance of Cams. " *Mechanism and Machine Theory*, **23**(3), pp. 191-208.

10   **Norton, R. L., et al.** (1988). "Analysis of the Effect of Manufacturing Methods and Heat Treatment on the Performance of Double Dwell Cams." *Mechanism and Machine Theory*, **23**(6), pp. 461-473.

11   **Jones, J. R., and J. E. Reeve.** (1978). "Dynamic Response of Cam Curves Based on Sinusoidal Segments." *Cams and Cam Mechanisms*, Jones, J. R., ed., Institution of Mechanical Engineers: London, pp. 14-24.

8.9   PROBLEMS[‡]

Programs DYNACAM *and* MATRIX *may be used to solve these problems or to check your solution where appropriate.*

*8-1   Figure P8-1 shows the cam and follower from Problem 6-65. Using graphical methods, find and sketch the equivalent fourbar linkage for this position of the cam and follower.

*8-2   Figure P8-1 shows the cam and follower from Problem 6-65. Using graphical methods, find the pressure angle at the position shown.

8-3   Figure P8-2 (p. 472) shows a cam and follower. Using graphical methods, find and sketch the equivalent fourbar linkage for this position of the cam and follower.

*8-4   Figure P8-2 shows a cam and follower. Using graphical methods, find the pressure angle at the position shown.

8-5   Figure P8-3 (p. 472) shows a cam and follower. Using graphical methods, find and sketch the equivalent fourbar linkage for this position of the cam and follower.

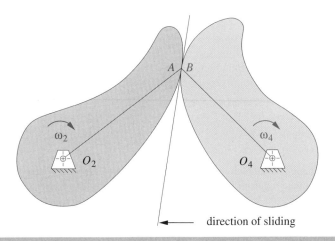

FIGURE P8-1

Problems 8-1 to 8-2 *Adapted from P. H. Hill and W. P. Rule. (1960). Mechanisms: Analysis and Design*

TABLE P8-0
Topic/Problem Matrix

8.1 Cam Terminology
8-1, 8-3, 8-5

8.3 Double-Dwell Cam Design
Simple Harmonic Motion (SHM)
8-26
Cycloidal Displacement
8-23
Modified Trapezoidal
8-7, 8-11, 8-21, 8-44
Modified Sinusoidal
8-8. 8-10, 8-22, 8-45
Polynomial
8-24, 8-25, 8-33, 8-46, 8-59, 8-60

8.4 Single-Dwell Cam Design
8-9, 8-41, 8-42, 8-47, 8-53

8.5 Critical Path Motion
8-17, 8-43, 8-48, 8-54

8.6 Sizing the Cam
Pressure Angle
8-2, 8-4, 8-6, 8-34, 8-56, 8-57, 8-58
Radius of Curvature - Roller Followers
8-18, 8-19, 8-20, 8-27, 8-28, 8-29, 8-30, 8-31, 8-32, 8-35, 8-36, 8-37, 8-38, 8-39, 8-40
Radius of Curvature - Flat-Faced Followers
8-49, 8-50, 8-51, 8-52
Roller & Flat-Faced Followers
8-12, 8-13, 8-14, 8-15

---

* Answers in Appendix F.

‡ Problem figures are provided as PDF file on the DVD. PDF filenames are the same as the figure number.

follower arm

4

3

ω_2

cam

2

FIGURE P8-2

Problems 8-3 to 8-4

4

3

2

ω_{cam}

FIGURE P8-3

Problems 8-5 to 8-6

---

‡ These problems are suited to solution using program DYNACAM, which is on the attached DVD.

*8-6   Figure P8-3 shows a cam and follower. Using graphical methods, find the pressure angle at the position shown.

‡8-7   Design a double-dwell cam to move a follower from 0 to 2.5" in 60°, dwell for 120°, fall 2.5" in 30°, and dwell for the remainder. The total cycle must take 4 sec. Choose suitable functions for rise and fall to minimize accelerations. Plot the $s\ v\ a\ j$ diagrams.

‡8-8   Design a double-dwell cam to move a follower from 0 to 1.5" in 45°, dwell for 150°, fall 1.5" in 90°, and dwell for the remainder. The total cycle must take 6 sec. Choose suitable functions for rise and fall to minimize velocities. Plot the $s\ v\ a\ j$ diagrams.

‡8-9   Design a single-dwell cam to move a follower from 0 to 2" in 60°, fall 2" in 90°, and dwell for the remainder. The total cycle must take 2 sec. Choose suitable functions for rise and fall to minimize accelerations. Plot the $s\ v\ a\ j$ diagrams.

‡8-10   Design a three-dwell cam to move a follower from 0 to 2.5" in 40°, dwell for 100°, fall 1.5" in 90°, dwell for 20°, fall 1" in 30°, and dwell for the remainder. The total cycle must take 10 sec. Choose suitable functions for rise and fall to minimize velocities. Plot the $s\ v\ a\ j$ diagrams.

‡8-11   Design a four-dwell cam to move a follower from 0 to 2.5" in 40°, dwell for 100°, fall 1.5" in 90°, dwell for 20°, fall 0.5" in 30°, dwell for 40°, fall 0.5" in 30°, and dwell for the remainder. The total cycle must take 15 sec. Choose suitable functions for rise and fall to minimize accelerations. Plot the $s\ v\ a\ j$ diagrams.

‡8-12   Size the cam from Problem 8-7 for a 1" radius roller follower considering pressure angle and radius of curvature. Use eccentricity only if necessary to balance those functions. Plot both those functions. Draw the cam profile. Repeat for a flat-faced follower. Which would you use?

‡8-13   Size the cam from Problem 8-8 for a 1.5" radius roller follower considering pressure angle and radius of curvature. Use eccentricity only if necessary to balance those functions. Plot both those functions. Draw the cam profile. Repeat for a flat-faced follower. Which would you use?

‡8-14   Size the cam from Problem 8-9 for a 0.5" radius roller follower considering pressure angle and radius of curvature. Use eccentricity only if necessary to balance those functions. Plot both those functions. Draw the cam profile. Repeat for a flat-faced follower. Which would you use?

‡8-15   Size the cam from Problem 8-10 for a 2" radius roller follower considering pressure angle and radius of curvature. Use eccentricity only if necessary to balance those functions. Plot both those functions. Draw the cam profile. Repeat for a flat-faced follower. Which would you use?

‡8-16   Size the cam from Problem 8-11 for a 0.5" radius roller follower considering pressure angle and radius of curvature. Use eccentricity only if necessary to balance those functions. Plot both those functions. Draw the cam profile. Repeat for a flat-faced follower. Which would you use?

‡8-17   A high friction, high inertia load is to be driven. We wish to keep peak velocity low. Combine segments of polynomial displacements with a constant velocity segment on both rise and fall to reduce the maximum velocity below that obtainable with a full period modified sine acceleration alone (i.e., one with no constant velocity portion). Rise

1" in 90°, dwell for 60°, fall in 50°, dwell for remainder. Compare the two designs and comment. Use an ω of one for comparison.

‡8-18   A constant velocity of 0.4 in/sec must be matched for 1.5 sec. Then the follower must return to your choice of start point and dwell for 2 sec. The total cycle time is 6 sec. Design a cam for a follower radius of 0.75" and a maximum pressure angle of 30° absolute value.

‡8-19   A constant velocity of 0.25 in/sec must be matched for 3 sec. Then the follower must return to your choice of start point and dwell for 3 sec. The total cycle time is 12 sec. Design a cam for a follower radius of 1.25" and a maximum pressure angle of 35° absolute value.

‡8-20   A constant velocity of 2 in/sec must be matched for 1 sec. Then the follower must return to your choice of start point. The total cycle time is 2.75 sec. Design a cam for a follower radius of 0.5" and a maximum pressure angle of 25° absolute value.

†8-21   Write a computer program or use an equation solver to calculate and plot the *s v a j* diagrams for a modified trapezoidal acceleration cam function for any specified values of lift and duration. Test it using a lift of 20 mm over 60° at 1 rad/sec.

†8-22   Write a computer program or use an equation solver to calculate and plot the *s v a j* diagrams for a modified sine acceleration cam function for any specified values of lift and duration. Test it using a lift of 20 mm over 60° at 1 rad/sec.

†8-23   Write a computer program or use an equation solver to calculate and plot the *s v a j* diagrams for a cycloidal displacement cam function for any specified values of lift and duration. Test it using a lift of 20 mm over 60° at 1 rad/sec.

†8-24   Write a computer program or use an equation solver to calculate and plot the *s v a j* diagrams for a 3-4-5 polynomial displacement cam function for any specified values of lift and duration. Test it using a lift of 20 mm over 60° at 1 rad/sec.

†8-25   Write a computer program or use an equation solver to calculate and plot the *s v a j* diagrams for a 4-5-6-7 polynomial displacement cam function for any specified values of lift and duration. Test it using a lift of 20 mm over 60° at 1 rad/sec.

†8-26   Write a computer program or use an equation solver to calculate and plot the *s v a j* diagrams for a simple harmonic displacement cam function for any specified values of lift and duration. Test it using a lift of 20 mm over 60° at 1 rad/sec.

†8-27   Write a computer program or use an equation solver to calculate and plot the pressure angle and radius of curvature for a modified trapezoidal acceleration cam function for any specified values of lift, duration, eccentricity, and prime circle radius. Test it using a lift of 20 mm over 60° at 1 rad/sec, and determine the prime circle radius needed to obtain a maximum pressure angle of 20°. What is the minimum diameter of roller follower needed to avoid undercutting with these data?

†8-28   Write a computer program or use an equation solver to calculate and plot the pressure angle and radius of curvature for a modified sine acceleration cam function for any specified values of lift, duration, eccentricity, and prime circle radius. Test it using a lift of 20 mm over 60° at 1 rad/sec, and determine the prime circle radius needed to obtain a maximum pressure angle of 20°. What is the minimum diameter of roller follower needed to avoid undercutting with these data?

† These problems are suited to solution using *Mathcad, Matlab,* or *TKSolver* equation solver programs.

‡ These problems are suited to solution using program DYNACAM, which is on the attached DVD.

†8-29   Write a computer program or use an equation solver to calculate and plot the pressure angle and radius of curvature for a cycloidal displacement cam function for any specified values of lift, duration, eccentricity, and prime circle radius. Test it using a lift of 20 mm over 60° at 1 rad/sec, and determine the prime circle radius needed to obtain a maximum pressure angle of 20°. What is the minimum diameter of roller follower needed to avoid undercutting with these data?

†8-30   Write a computer program or use an equation solver to calculate and plot the pressure angle and radius of curvature for a 3-4-5 polynomial displacement cam function for any specified values of lift, duration, eccentricity, and prime circle radius. Test it using a lift of 20 mm over 60° at 1 rad/sec, and determine the prime circle radius needed to obtain a maximum pressure angle of 20°. What is the minimum diameter of roller follower needed to avoid undercutting with these data?

†8-31   Write a computer program or use an equation solver to calculate and plot the pressure angle and radius of curvature for a 4-5-6-7 polynomial displacement cam function for any specified values of lift, duration, eccentricity, and prime circle radius. Test it using a lift of 20 mm over 60° at 1 rad/sec, and determine the prime circle radius needed to obtain a maximum pressure angle of 20°. What is the minimum diameter of roller follower needed to avoid undercutting with these data?

†8-32   Write a computer program or use an equation solver to calculate and plot the pressure angle and radius of curvature for a simple harmonic displacement cam function for any specified values of lift, duration, eccentricity, and prime circle radius. Test it using a lift of 20 mm over 60° at 1 rad/sec, and determine the prime circle radius needed to obtain a maximum pressure angle of 20°. What is the minimum diameter of roller follower needed to avoid undercutting with these data?

8-33   Derive equation 8.25 (p. 434) for the 4-5-6-7 polynomial function.

8-34   Derive an expression for the pressure angle of a barrel cam with zero eccentricity.

‡8-35   Design a radial plate cam to move a translating roller follower through 30 mm in 30°, dwell for 100°, fall 10 mm in 10°, dwell for 20°, fall 20 mm in 20°, and dwell for the remainder. Camshaft ω = 200 rpm. Minimize the follower's peak velocity and determine the minimum prime circle radius that will give a maximum 25° pressure angle. Determine the minimum radii of curvature on the pitch curve.

‡8-36   Repeat Problem 8-35, but minimize the follower's peak acceleration instead.

‡8-37   Repeat Problem 8-35, but minimize the follower's peak jerk instead.

‡8-38   Design a radial plate cam to lift a translating roller follower through 10 mm in 65°, return to 0 in 65° and dwell for the remainder. Camshaft ω = 3500 rpm. Minimize the cam size while not exceeding a 25° pressure angle. What size roller follower is needed?

† These problems are suited to solution using *Mathcad, Matlab,* or *TKSolver* equation solver programs.

‡8-39   Design a cam-driven quick-return mechanism for a 3:1 time ratio. The translating roller follower should move forward and back 50 mm and dwell in the back position for 80°. It should take one-third the time to return as to move forward. Camshaft ω = 100 rpm. Minimize the package size while maintaining a 25° maximum pressure angle. Draw a sketch of your design and provide s v a j, φ, and ρ diagrams.

‡ These problems are suited to solution using program DYNACAM, which is on the attached DVD.

‡8-40   Design a cam-follower system to drive a linear translating piston at constant velocity for 200° through a stroke of 100 mm at 60 rpm. Minimize the package size while maintaining a 25° maximum pressure angle. Draw a sketch of your design and provide s v a j, φ, and ρ diagrams.

‡8-41   Design a cam-follower system to rise 20 mm in 80°, fall 10 mm in 100°, dwell at 10 mm for 100°, fall 10 mm in 50°, and dwell at 0 for 30°. Total cycle time is 4 sec. Avoid unnecessary returns to zero acceleration. Minimize the package size and maximize the roller follower diameter while maintaining a 25° maximum pressure angle. Draw a sketch of your design and provide s v a j, ϕ, and ρ diagrams.

‡8-42   Design a single-dwell cam to move a follower from 0 to 35 mm in 75°, fall 35 mm in 120°, and dwell for the remainder. The total cycle time is 3 sec. Choose suitable functions to minimize acceleration and plot the s v a j diagrams for the rise/fall.

‡8-43   Design a cam to move a follower at a constant velocity of 100 mm/sec for 2 sec then return to its starting position with a total cycle time of 3 sec.

‡8-44   Design a double-dwell cam to move a follower from 0 to 50 mm in 75°, dwell for 75°, fall 50 mm in 75°, and dwell for the remainder. The total cycle must take 5 sec. Use a modified trapezoidal function for rise and fall and plot the s v a j diagrams.

‡8-45   Design a double-dwell cam to move a follower from 0 to 50 mm in 75°, dwell for 75°, fall 50 mm in 75°, and dwell for the remainder. The total cycle must take 5 sec. Use a modified sinusoidal function for rise and fall and plot the s v a j diagrams.

‡8-46   Design a double-dwell cam to move a follower from 0 to 50 mm in 75°, dwell for 75°, fall 50 mm in 75°, and dwell for the remainder. The total cycle must take 5 sec. Use a 4-5-6-7 polynomial function for rise and fall and plot the s v a j diagrams.

‡8-47   Design a single-dwell cam to move a follower from 0 to 65 mm in 90°, fall 65 mm in 180°, and dwell for the remainder. The total cycle time is 2 sec. Choose suitable functions to minimize acceleration and plot the s v a j diagrams for the rise/fall.

‡8-48   Design a cam to move a follower at a constant velocity of 200 mm/sec for 3 sec then return to its starting position with a total cycle time of 6 sec.

‡8-49   Size the cam from Problem 8-42 for a translating flat-faced follower considering follower face width and radius of curvature. Plot radius of curvature and cam profile.

‡8-50   Size the cam from Problem 8-44 for a translating flat-faced follower considering follower face width and radius of curvature. Plot radius of curvature and cam profile.

‡8-51   Size the cam from Problem 8-45 for a translating flat-faced follower considering follower face width and radius of curvature. Plot radius of curvature and cam profile.

‡8-52   Size the cam from Problem 8-46 for a translating flat-faced follower considering follower face width and radius of curvature. Plot radius of curvature and cam profile.

‡8-53   Design a single-dwell cam to move a follower from 0 to 50 mm in 100°, fall 50 mm in 120°, and dwell for the remainder. The total cycle time is 1 sec. Choose suitable functions to minimize acceleration and plot the s v a j diagrams for the rise/fall.

‡8-54   Design a cam to move a follower at a constant velocity of 300 mm/sec for 2 sec then return to its starting position with a total cycle time of 4 sec.

†8-55   Write a computer program or use an equation solver to calculate and plot the s v a j diagrams for the family of SCCA cam functions for any specified values of lift and duration. It should allow changing values of the parameters b, c, d, and C_a to plot any member of the family. Test all functions with 100 mm rise in 100°, dwell 80°, fall in 120°, dwell for remainder. Shaft turns at 1 rad/sec.

† These problems are suited to solution using *Mathcad*, *Matlab*, or *TKSolver* equation solver programs.

‡ These problems are suited to solution using program DYNACAM, which is on the attached DVD.

8

TABLE P8-1   Data for Problem 8-59

Event	S	V	A	J
First fall (60°)				
Beginning	20.5	0	0	0
Ending	15.0	0	0	
Second fall (90°)				
Beginning	15.0	0	0	
Ending	0.0	0	A_1	0
Rise (110°)				
Beginning	0.0	0	match A_1	0
Ending	20.5	0	0	0

[†]8-56   Write a computer program or use an equation solver such as *Mathcad* or *TKSolver* to calculate and plot the pressure angle for the cam of Problem 8-42 for any given prime circle radius and follower eccentricity. Test it using $R_p = 45$ mm and $e = 10$ mm.

[†]8-57   Write a computer program or use an equation solver such as *Mathcad* or *TKSolver* to calculate and plot the pressure angle for the cam of Problem 8-43 for any given prime circle radius and follower eccentricity. Test it using $R_p = 100$ mm and $e = -15$ mm.

[†]8-58   Write a computer program or use an equation solver such as *Mathcad* or *TKSolver* to calculate and plot the pressure angle for the rise segment of the cam of Problem 8-46 for any given prime circle radius and follower eccentricity. Test it using $R_p = 75$ mm and $e = 20$ mm.

[‡]8-59   Design a cam to move a follower from 20.5 to 15 mm in 60°, fall an additional 15 mm in 90°, rise 20.5 mm in 110°, and dwell for the remainder. Use polynomial functions for the rise and falls. Some of the boundary conditions are given in Table P8-1; however, in order to make the polynomials piecewise continuous, other boundary conditions will have to be determined. The shaft speed is 250 rpm. Plot the *s v a j* diagrams.

[‡]8-60   Design a cam to move a follower from 32 to 12 mm in 60°, fall an additional 12 mm in 50°, dwell 35°, rise 12 mm in 45°, rise an additional 20 mm in 65°, and dwell for the remainder. Use polynomial functions for the rises and falls. Velocity and acceleration are zero at the beginning and end of each event and jerk is zero at $\theta = 0°$, 110°, 145°, and 255°. The shaft speed is 37.5 rpm. Plot the *s v a j* diagrams.

[†] These problems are suited to solution using *Mathcad, Matlab,* or *TKSolver* equation solver programs.

[‡] These problems are suited to solution using program DYNACAM, which is on the attached DVD.

8.10   VIRTUAL LABORATORY

L8-1   View the video *Cam Machine Virtual Laboratory* that is on the DVD. Open the file *Virtual Cam Machine Lab.doc* from the DVD and follow the instructions as directed by your professor.

8.11   PROJECTS

These larger-scale project statements deliberately lack detail and structure and are loose-
ly defined. Thus, they are similar to the kind of "identification of need" or problem state-
ment commonly encountered in engineering practice. It is left to the student to structure
the problem through background research and to create a clear goal statement and set of
task specifications before attempting to design a solution. This design process is spelled
out in Chapter 1 and should be followed in all of these examples. Document all results in
a professional engineering report. (See Section 1.9 on p. 17 and Chap. 1 bibliography
on p. 29 for information on report writing.)

‡P8-1   A timing diagram for a halogen headlight filament insertion device is shown in Figure
P8-4. Four points are specified. Point *A* is the start of rise. At *B* the grippers close
to grab the filament from its holder. The filament enters its socket at *C* and is fully
inserted at *D*. The high dwell from *D* to *E* holds the filament stationary while it is
soldered in place. The follower returns to its start position from *E* to *F*. From *F* to *A*
the follower is stationary while the next bulb is indexed into position. It is desirable
to have low to zero velocity at point *B* where the grippers close on the fragile filament.
The velocity at *C* should not be so high as to "bend the filament in the breeze." Design
and size a complete cam-follower system to do this job.

‡P8-2   A cam-driven pump to simulate human aortic pressure is needed to serve as a consistent, repeatable pseudo-human input to a hospital's operating room computer monitor-
ing equipment, in order to test it daily. Figure P8-5 (p. 478) shows a typical aortic
pressure curve and a pump pressure-volume characteristic. Design a cam to drive the
piston and give as close an approximation to the aortic pressure curve shown as can be
obtained without violating the fundamental law of cam design. Simulate the dicrotic
notch as best you can.

‡P8-3   A fluorescent light bulb production machine moves 5500 lamps per hour through a
550°C oven on a chain conveyor which is in constant motion. The lamps are on 2-in
centerlines. The bulbs must be sprayed internally with a tin oxide coating as they leave
the oven, still hot. This requires a cam-driven device to track the bulbs at constant
velocity for the 0.5 sec required to spray them. The spray guns will fit on a 6 x 10 in
table. The spray creates hydrochloric acid, so all exposed parts must be resistant to that
environment. The spray head transport device will be driven from the conveyor chain
by a shaft having a 28-tooth sprocket in mesh with the chain. Design a complete spray
gun transport assembly to these specifications.

Timing Diagram

Cam angle, °	Point	s
120	A	0
140	B	2
150	C	3
180	D	3.5
300	E	3.5
360	F	0

Displacement Table

FIGURE P8-4

Data for cam design Project P8-1

† These problems are suited
to solution using *Mathcad,
Matlab,* or *TKSolver* equation solver programs.

‡ These problems are suited
to solution using program
DYNACAM, which is on the
attached DVD.

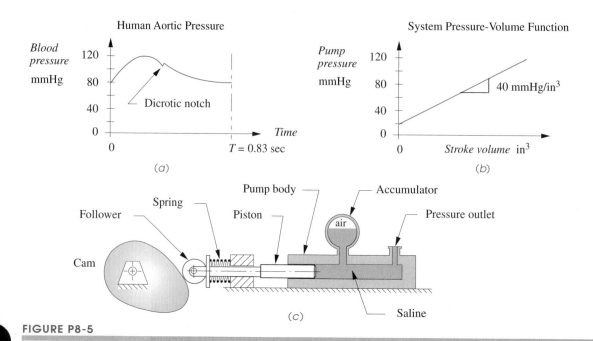

Human Aortic Pressure

System Pressure-Volume Function

(a)

(b)

(c)

FIGURE P8-5

Data for cam design Project P8-2

‡P8-4   A 30-ft-tall drop tower is being used to study the shape of water droplets as they fall through air. A camera is to be carried by a cam-operated linkage which will track the droplet's motion from the 8-ft to the 10-ft point in its fall (measured from release point at the top of the tower). The drops are released every 1/2 sec. Every drop is to be filmed. Design a cam and linkage which will track these droplets, matching their velocities and accelerations in the 1-ft filming window.

‡P8-5   A device is needed to accelerate a 3000-lb vehicle into a barrier with constant velocity, to test its 5 mph bumpers. The vehicle will start at rest, move forward, and have constant velocity for the last part of its motion before striking the barrier with the specified velocity. Design a cam-follower system to do this. The vehicle will leave contact with your follower just prior to the crash.

‡P8-6   An athletic footwear manufacturer wants a device to test rubber heels for their ability to withstand millions of cycles of force similar to that which a walking human's foot applies to the ground. Figure P8-6 shows a typical walker's force-time function and a pressure-volume curve for a piston-accumulator. Design a cam-follower system to drive the piston in a way that will create a force-time function on the heel similar to the one shown. Choose suitable piston diameters at each end.

‡ These problems are suited to solution using program DYNACAM, which is on the attached DVD.

‡P8-7   Figure P8-7 shows a timing diagram for a machine cam to drive a translating roller follower. Design suitable functions for all motions and size the cam for acceptable pressure angles and roller follower diameter. Note points of required zero velocity at particular displacements. Cam speed is 30 rpm. Hint: Segment 8 should be solved with polynomial functions, the fewer the better.

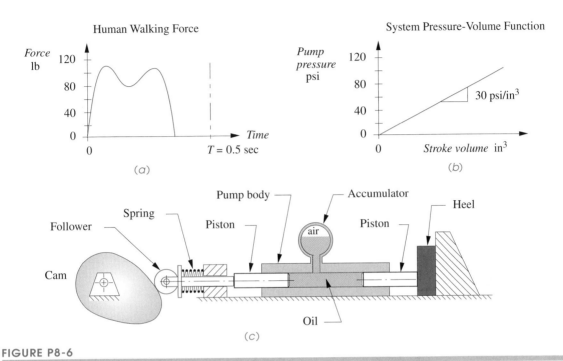

Human Walking Force

System Pressure-Volume Function

(a)

(b)

(c)

FIGURE P8-6

Data for cam design Project P8-6

8

‡P8-8   Design an engine exhaust-valve cam with 10-mm lift over 132 camshaft deg. The rest of the cycle is a dwell. The valve-open duration is measured between cam-follower displacements of 0.5 mm above the dwell position, where valve clearance is taken up and the valve begins to move as shown in Figure P8-8 (p. 480). Engine crankshaft speed ranges from 1000 to 10 000 rpm. The cam should take up the clearance with minimal impact, then continue to lift to 10-mm at 66° as rapidly as possible, close to the 0.5 mm point by 132° and then return it to zero at a controlled velocity. See Figure 8-3a (p. 406). Select a spring from the Appendix to prevent valve float (follower jump)

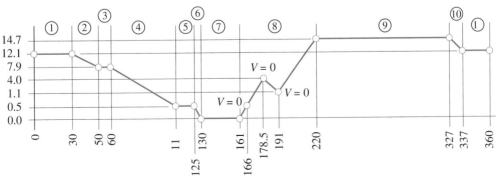

FIGURE P8-7

Timing diagram for Project P8-7. Displacements in mm (not to scale)

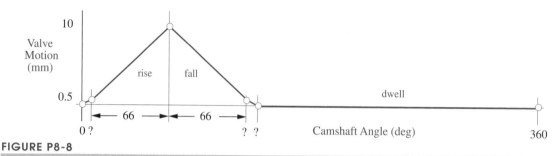

FIGURE P8-8

Timing diagram for Project P8-8—exhaust-valve cam. Determine suitable values for ? from problem statement.

assuming an effective follower train mass of 200 grams. The camshaft turns at half the crank speed.

‡P8-9   Design a cam-driven peanut-butter (PB) pump for a 600/min cookie assembly line. The cookies are spaced at 40-mm centers on a constant-velocity conveyor. A square, 1-mm thick patch containing 0.4 cc of peanut butter is applied to the cookie as it passes by a nozzle. Entrained air in the PB makes it compressible. Figure P8-5 (p. 477) shows a similar setup with a cam driving a follower attached to a piston pump. The peanut butter flows from the "pressure outlet." The accumulator represents entrained air in the PB. If pumped at constant rate with a piston pump, there is a lag at the start as the entrained air is compressed. Once compressed, it flows uniformly when the piston moves at constant velocity. At the end of the stroke, the stored energy in the entrained air causes "peanut-butter drool," making a messy cookie. To get a sharp-edged start to the "patch" of peanut butter we need an extra "kick" at the beginning of the pumping cycle to wind up the "air spring," followed by a period of constant velocity motion to lay down a uniform thickness of PB. At the end of the patch, we need a "sniff" to rapidly retract the piston slightly and prevent drool. The piston then returns to the start point at constant velocity to refill the pump and repeat the cycle. The velocity of the "kick" should be about 3 times the steady-state velocity and of as short a duration as practical. The velocity of the "sniff" is optimal at about –4 times the steady-state velocity with as short a duration as practical. Figure P8-9 shows the displacement timing diagram. Size the piston and design the piston-driver cam for good dynamic operation with reason-

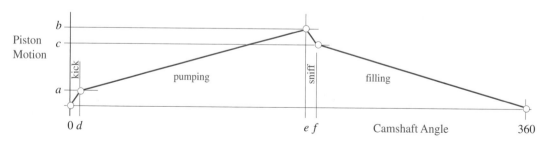

FIGURE P8-9

Timing diagram for Project P8-9—peanut butter pump. Determine suitable values for a—f from problem statement.

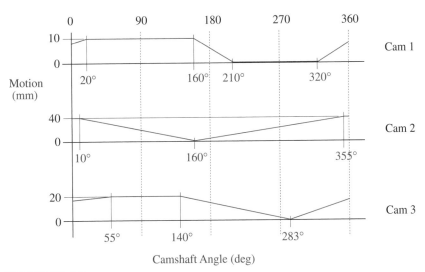

FIGURE P8-10

Timing diagram for Project P8-10

able accelerations and size it in a reasonable package. Select a return spring for a moving follower mass of 0.5 kg.

‡P8-10 Figure P8-10 shows timing diagrams for 3 cams used in a production machine. Design suitable *SVAJ* functions to run at 250 rpm with 10-kg effective mass on each follower. Size the cams for suitable pressure angles and radii of curvature using a 20-mm diameter roller follower. Select a suitable spring for each follower from the Appendix, specify its preload and sketch the assembly, showing all three cams on a common camshaft driving the three follower trains along the *X* axis.

8

Chapter **9**

GEAR TRAINS

Cycle and epicycle,
orb in orb
JOHN MILTON, PARADISE LOST

9.0 INTRODUCTION

The earliest known reference to gear trains is in a treatise by Hero of Alexandria (c. 100 B.C.). Gear trains are widely used in all kinds of mechanisms and machines, from can openers to aircraft carriers. Whenever a change in the speed or torque of a rotating device is needed, a gear train or one of its cousins, the belt or chain drive mechanism, will usually be used. This chapter will explore the theory of gear tooth action and the design of these ubiquitous devices for motion control. The calculations involved are trivial compared to those for cams or linkages. The shape of gear teeth has become quite standardized for good kinematic reasons that we will explore.

Gears of various sizes and styles are readily available from many manufacturers. Assembled gearboxes for particular ratios are also stock items. The kinematic design of gear trains is principally involved with the selection of appropriate ratios and gear diameters. A complete gear train design will necessarily involve considerations of strength of materials and the complicated stress states to which gear teeth are subjected. This text will not deal with the stress analysis aspects of gear design. There are many texts that do. Some are listed in the bibliography at the end of this chapter. This chapter will discuss the kinematics of gear tooth theory, gear types, and the kinematic design of gearsets and gear trains of simple, compound, reverted, and epicyclic types. Chain and belt drives will also be discussed. Examples of the use of these devices will be presented as well.

9.1   ROLLING CYLINDERS

The simplest means of transferring rotary motion from one shaft to another is a pair of rolling cylinders. They may be an external set of rolling cylinders as shown in Figure 9-1a or an internal set as in Figure 9-1b. Provided that sufficient friction is available at the rolling interface, this mechanism will work quite well. There will be no slip between the cylinders until the maximum available frictional force at the joint is exceeded by the demands of torque transfer.

A variation on this mechanism is what causes your car or bicycle to move along the road. Your tire is one rolling cylinder and the road the other (very large radius) one. Friction is all that prevents slip between the two, and it works well unless the friction coefficient is reduced by the presence of ice or other slippery substances. In fact, some early automobiles had rolling cylinder drives inside the transmission, as do some present-day snowblowers and garden tractors that use a rubber-coated wheel rolling against a steel disk to transmit power from the engine to the wheels.

A variant on the rolling cylinder drive is the flat or vee belt as shown in Figure 9-2. This mechanism also transfers power through friction and is capable of quite large power levels, provided enough belt cross section is provided. Friction belts are used in a wide variety of applications from small sewing machines to the alternator drive on your car, to multihorsepower generators and pumps. Whenever absolute phasing is not required and power levels are moderate, a friction belt drive may be the best choice. They are relatively quiet running, require no lubrication, and are inexpensive compared to gears and chain drives. A constant velocity transmission (CVT) as used in a number of automobiles is also a vee belt and pulley device in which the pulleys are adjusted in width to change the ratio. As one pulley widens, the other narrows to change the relative radii of the belt within their respective vees. The belt circumference, of course, remains the same.

Both rolling cylinders and belt (or chain) drives have effective linkage equivalents as shown in Figure 9-3 (p. 484). These effective linkages are valid only for one instantaneous position but nevertheless show that these devices are just another variation of the fourbar linkage in disguise.

(a) External set

(b) Internal set

FIGURE 9-1

Rolling cylinders

FIGURE 9-2

A two-groove vee belt drive   *Courtesy of T. B. Wood's Sons Co., Chambersburg, PA*

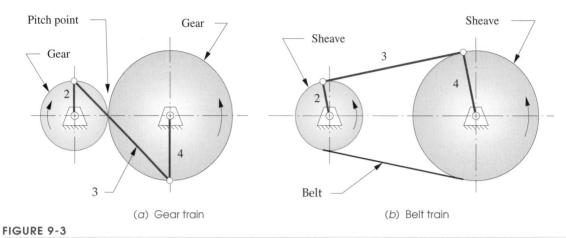

(a) Gear train   (b) Belt train

FIGURE 9-3

Gear and belt trains each have an equivalent fourbar linkage for any instantaneous position.

The principal drawbacks to the rolling cylinder drive (or smooth belt) mechanism are its relatively low torque capability and the possibility of slip. Some drives require absolute phasing of the input and output shafts for timing purposes. A common example is the valve train drive in an automobile engine. The valve cams must be kept in phase with the piston motion or the engine will not run properly. A smooth belt or rolling cylinder drive from crankshaft to camshaft would not guarantee correct phasing. In this case some means of preventing slip is needed.

This usually means adding some meshing teeth to the rolling cylinders. They then become gears as shown in Figure 9-4 and are together called a *gearset*. When two gears are placed in mesh to form a gearset such as this one, it is conventional to refer to the smaller of the two gears as the *pinion* and to the other as the *gear*.

9.2   THE FUNDAMENTAL LAW OF GEARING

Conceptually, teeth of any shape will prevent gross slip. Old water-powered mills and windmills used wooden gears whose teeth were merely round wooden pegs stuck into the rims of the cylinders. Even ignoring the crudity of construction of these early examples of gearsets, there was no possibility of smooth velocity transmission because the geometry of the tooth "pegs" violated the **fundamental law of gearing** which, if followed, provides that *the angular velocity ratio between the gears of a gearset remains constant throughout the mesh*. A more complete and formal definition of this law is given on p. 486. The angular velocity ratio (m_V) referred to in this law is the same one that we derived for the fourbar linkage in Section 6.4 (p. 299) and equation 6.10 (p. 301). It is equal to the ratio of the radius of the input gear to that of the output gear.

FIGURE 9-4

An external gearset

$$m_V = \frac{\omega_{out}}{\omega_{in}} = \pm\frac{r_{in}}{r_{out}} = \pm\frac{d_{in}}{d_{out}} \qquad (9.1a)$$

$$m_T = \frac{\omega_{in}}{\omega_{out}} = \pm\frac{r_{out}}{r_{in}} = \pm\frac{d_{out}}{d_{in}} \qquad (9.1b)$$

The **torque ratio** (m_T) was shown on p. 304 to be the reciprocal of the velocity ratio (m_V); thus a gearset is essentially a device to exchange torque for velocity or vice versa. Since there are no applied forces as in a linkage, but only applied torques on the gears, the **mechanical advantage** m_A of a gearset is equal to its torque ratio m_T. The most common application is to reduce velocity and increase torque to drive heavy loads as in your automobile transmission. Other applications require an increase in velocity, for which a reduction in torque must be accepted. In either case, it is usually desirable to maintain a constant ratio between the gears as they rotate. Any variation in ratio will show up as oscillation in the output velocity and torque even if the input is constant with time.

The radii in equation 9.1 are those of the rolling cylinders to which we are adding the teeth. The positive or negative sign accounts for internal or external cylinder sets as defined in Figure 9-1 (p. 483). An external set reverses the direction of rotation between the cylinders and requires the negative sign. An internal gearset or a belt or chain drive will have the same direction of rotation on input and output shafts and require the positive sign in equation 9.1. The surfaces of the rolling cylinders will become the **pitch circles,** and their diameters the **pitch diameters** of the gears. The contact point between the cylinders lies on the line of centers as shown in Figure 9-3a, and this point is called the **pitch point**.

In order for the fundamental law of gearing to be true, the gear tooth contours on mating teeth must be conjugates of one another. There is an infinite number of possible conjugate pairs that could be used, but only a few curves have seen practical application as gear teeth. The **cycloid** still is used as a tooth form in watches and clocks, but most other gears use the **involute** curve for their shape.

The Involute Tooth Form

The involute is a curve that can be generated by unwrapping a taut string from a cylinder (called the evolute) as shown in Figure 9-5. Note the following about this involute curve:

The string is always tangent to the cylinder.

The center of curvature of the involute is always at the point of tangency of the string with the cylinder.

A tangent to the involute is then always normal to the string, the length of which is the instantaneous radius of curvature of the involute curve.

Figure 9-6 (p. 486) shows two involutes on separate cylinders in contact or "in mesh." These represent gear teeth. The cylinders from which the strings are unwrapped are called the **base circles** of the respective gears. Note that the base circles are necessarily smaller than the pitch circles, which are at the radii of the original rolling cylinders, r_p and r_g. The gear tooth must project both below and above the rolling cylinder surface (pitch circle) and the *involute only exists outside of the base circle*. The amount of tooth that sticks out above the pitch circle is the **addendum**, shown as a_p and a_g for pinion and gear, respectively. These are equal for standard, full-depth gear teeth.

The geometry at this tooth-tooth interface is similar to that of a cam-follower joint as was defined in Figure 8-44 (p. 455). There is a **common tangent** to both curves at the contact point, and a **common normal**, perpendicular to the common tangent. Note that the common normal is, in fact, the "strings" of both involutes, which are colinear. Thus

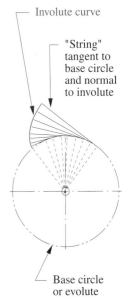

Involute curve

"String" tangent to base circle and normal to involute

Base circle or evolute

FIGURE 9-5

Development of the involute of a circle

9

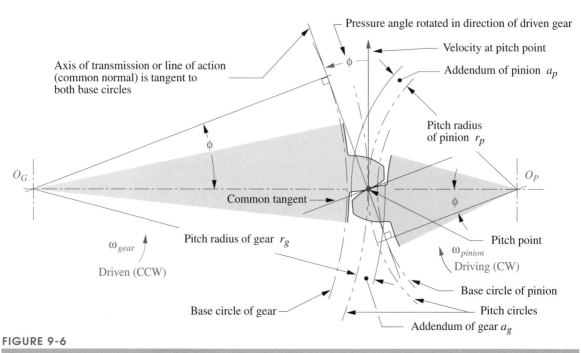

FIGURE 9-6

Contact geometry and pressure angle of involute gear teeth

the common normal, which is also the **axis of transmission**, always passes through the pitch point regardless of where in the mesh the two teeth are contacting.

Figure 9-7 shows a pair of involute tooth forms in two positions, just beginning contact and about to leave contact. The common normals of both these contact points still pass through the same pitch point. It is this property of the involute that causes it to obey the fundamental law of gearing. The ratio of the driving gear radius to the driven gear radius remains constant as the teeth move into and out of mesh.

From this observation of the behavior of the involute we can restate the **fundamental law of gearing** in a more kinematically formal way as: *the common normal of the tooth profiles, at all contact points within the mesh, must always pass through a fixed point on the line of centers, called the pitch point.* The gearset's velocity ratio will then be a constant defined by the ratio of the respective radii of the gears to the pitch point.

The points of beginning and leaving contact define the **mesh** of the pinion and gear. The distance along the line of action between these points within the mesh is called the **length of action**, Z, defined by the intersections of the respective addendum circles with the line of action, as shown in Figure 9-7. Variables are defined in Figures 9-6 and 9-7.

$$Z = \sqrt{\left(r_p + a_p\right)^2 - \left(r_p \cos\phi\right)^2} + \sqrt{\left(r_g + a_g\right)^2 - \left(r_g \cos\phi\right)^2} - C\sin\phi \qquad (9.2)$$

The distance along the pitch circle within the mesh is the **arc of action**, and the angles subtended by these points and the line of centers are the **angle of approach** and **angle of recess**. These are shown only on the gear in Figure 9-7 for clarity, but similar angles

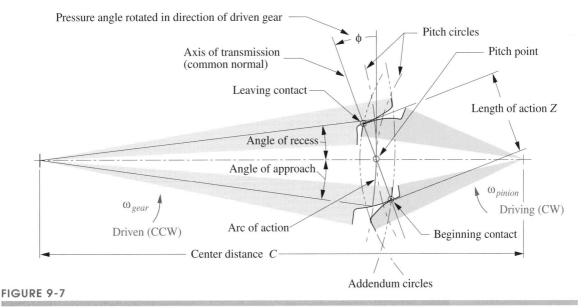

Pressure angle rotated in direction of driven gear

φ

Pitch circles

Axis of transmission
(common normal)

Pitch point

Leaving contact

Length of action Z

Angle of recess

Angle of approach

ω_{gear}

ω_{pinion}

Driving (CW)

Driven (CCW)

Arc of action

Beginning contact

Center distance C

Addendum circles

FIGURE 9-7

Pitch point, pitch circles, pressure angle, length of action, arc of action, and angles of approach and recess during the meshing of a gear and pinion

exist for the pinion. The arc of action on both pinion and gear pitch circles must be the same length for zero slip between the theoretical rolling cylinders.

Pressure Angle

The **pressure angle** in a gearset is similar to that of the cam and follower and is defined as the angle between the axis of transmission or line of action (common normal) and the direction of velocity at the pitch point as shown in Figures 9-6 and 9-7. Pressure angles of gearsets are standardized at a few values by the gear manufacturers. These are defined at the nominal center distance for the gearset as cut. The standard values are 14.5°, 20°, and 25° with 20° being the most commonly used and 14.5° now being considered obsolete. Any custom pressure angle can be made, but its expense over the available stock gears with standard pressure angles would be hard to justify. Special cutters would have to be made. Gears to be run together must be cut to the same nominal pressure angle.

Changing Center Distance

When involute teeth (or any teeth) have been cut into a cylinder, with respect to a particular base circle, to create a single gear, we do not yet have a pitch circle. The pitch circle only comes into being when we mate this gear with another to create a pair of gears, or gearset. There will be some range of center-to-center distances over which we can achieve a mesh between the gears. There will also be an ideal center distance (*CD*) that will give us the nominal pitch diameters for which the gears were designed. However, limitations of manufacturing processes give a low probability that we will be able to exactly achieve this ideal center distance in every case. More likely, there will be some error in the center distance, even if small.

What will happen to the adherence to the fundamental law of gearing if there is error in the location of the gear centers? If the gear tooth form is **not** an involute, then an error in center distance will violate the fundamental law, and there will be variation, or "ripple," in the output velocity. The output angular velocity will not be constant for a constant input velocity. However, **with an involute tooth form**, *center distance errors do not affect the velocity ratio*. This is the principal advantage of the involute over all other possible tooth forms and the reason why it is nearly universally used for gear teeth. Figure 9-8 shows what happens when the center distance is varied on an involute gearset. Note that the common normal still goes through a pitch point, common to all contact points within the mesh. But the pressure angle is affected by the change in center distance.

Figure 9-8 also shows the pressure angles for two different center distances. As the center distance increases, so will the pressure angle and vice versa. This is one result of a change, or error, in center distance when using involute teeth. Note that the fundamental law of gearing still holds in the modified center distance case. The common normal is

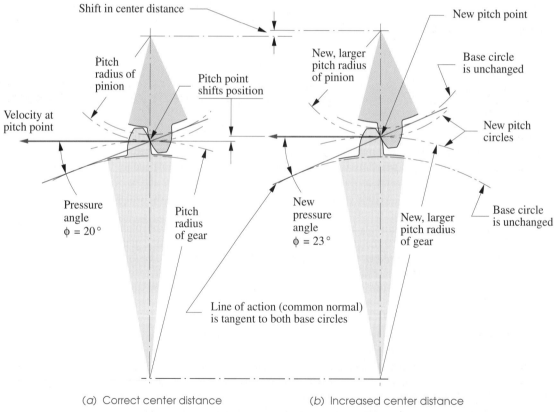

(a) Correct center distance   (b) Increased center distance

FIGURE 9-8

Changing center distance of involute gears changes the pressure angle and pitch diameters still tangent to the two base circles and still goes through the pitch point. The pitch point has moved, but in proportion to the move of the center distance and the gear radii. The velocity ratio is unchanged despite the shift in center distance. In fact, the velocity ratio of involute gears is fixed by the ratio of the base circle diameters, which are unchanging once the gear is cut.

Backlash

Another factor affected by changing center distance is backlash. Increasing the *CD* will increase the backlash and vice versa. **Backlash** is defined as *the clearance between mating teeth measured at the pitch circle.* Manufacturing tolerances preclude a zero clearance, as all teeth cannot be exactly the same dimensions, and all must mesh. So, there must be some small difference between the tooth thickness and the space width (see Figure 9-9, p. 490). As long as the gearset is run with a nonreversing torque, backlash should not be a problem. But, whenever torque changes sign, the teeth will move from contact on one side to the other. The backlash gap will be traversed and the teeth will impact with noticeable noise. This is the same phenomenon as crossover shock in the form-closed cam. As well as increasing stresses and wear, backlash can cause undesirable positional error in some applications. If the center distance is set exactly to match the theoretical value for the gearset, the tooth-to-tooth composite backlash tolerance is in the range of 0.0001 to 0.0007 in for precision gears. The increase in angular backlash as a function of error in center distance is approximately

$$\theta_B = 43\,200\,(\Delta C)\frac{\tan\phi}{\pi d} \quad \text{minutes of arc} \tag{9.3}$$

where ϕ = pressure angle, ΔC = error in center distance, and d = pitch diameter of the gear on the shaft where the backlash is measured.

In servomechanisms, where motors are driving, for example, the control surfaces on an aircraft, backlash can cause potentially destructive "hunting" in which the control system tries in vain to correct positional errors due to backlash "slop" in the mechanical drive system. Such applications need **antibacklash gears** which are really two gears back to back on the same shaft that can be rotated slightly at assembly with respect to one another, and then fixed so as to take up the backlash. In less critical applications, such as the propeller drive on a boat, backlash on reversal will not even be noticed.

The *American Gear Manufacturers Association* (**AGMA**) defines standards for gear design and manufacture. They define a spectrum of quality numbers and tolerances ranging from the lowest (3) to the highest precision (16). Obviously the cost of a gear will be a function of this quality index.

9.3 GEAR TOOTH NOMENCLATURE

Figure 9-9 shows two teeth of a gear with the standard nomenclature defined. **Pitch circle** and **base circle** have been defined above. The tooth height is defined by the **addendum** (*added on*) and the **dedendum** (*subtracted from*) that are referenced to the nominal pitch circle. The dedendum is slightly larger than the addendum to provide a small amount of **clearance** between the tip of one mating tooth (**addendum circle**) and the bottom of the

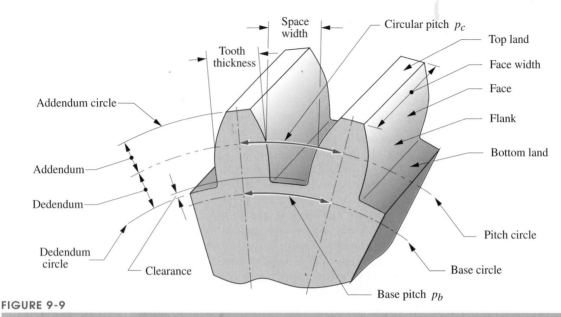

FIGURE 9-9

Gear tooth nomenclature

tooth space of the other (**dedendum circle**). The **tooth thickness** is measured at the pitch circle, and the tooth **space width** is slightly larger than the tooth thickness. The difference between these two dimensions is the **backlash**. The **face width** of the tooth is measured along the axis of the gear. The **circular pitch** is the arc length along the pitch circle circumference measured from a point on one tooth to the same point on the next. The circular pitch defines the tooth size. The other tooth dimensions are standardized based on that dimension as shown in Table 9-1. The definition of **circular pitch** p_c is:

$$p_c = \frac{\pi d}{N} \tag{9.4a}$$

where d = pitch diameter and N = number of teeth. The tooth pitch can also be measured along the base circle circumference and then is called the **base pitch** p_b.

$$p_b = p_c \cos\phi \tag{9.4b}$$

The units of p_c are inches or millimeters. A more convenient and common way to define tooth size is to relate it to the diameter of the pitch circle rather than its circumference. The **diametral pitch** p_d is:

$$p_d = \frac{N}{d} \tag{9.4c}$$

The units of p_d are reciprocal inches, or number of teeth per inch. This measure is only used in U.S. specification gears. Combining equations 9.4a and 9.4c gives the following relationship between circular pitch and diametral pitch.

$$p_d = \frac{\pi}{p_c} \tag{9.4d}$$

TABLE 9-1   AGMA Full-Depth Gear Tooth Specifications

Parameter	Coarse Pitch ($p_d < 20$)	Fine Pitch ($p_d \geq 20$)
Pressure angle ϕ	20° or 25°	20°
Addendum a	$1.000 / p_d$	$1.000 / p_d$
Dedendum b	$1.250 / p_d$	$1.250 / p_d$
Working depth	$2.000 / p_d$	$2.000 / p_d$
Whole depth	$2.250 / p_d$	$2.200 / p_d + 0.002$ in
Circular tooth thickness	$1.571 / p_d$	$1.571 / p_d$
Fillet radius—basic rack	$0.300 / p_d$	Not standardized
Minimum basic clearance	$0.250 / p_d$	$0.200 / p_d + 0.002$ in
Minimum width of top land	$0.250 / p_d$	Not standardized
Clearance (shaved or ground teeth)	$0.350 / p_d$	$0.350 / p_d + 0.002$ in

The SI system, used for metric gears, defines a parameter called the **module,** which is *the reciprocal of diametral pitch* with pitch diameter measured in millimeters.

$$m = \frac{d}{N} \qquad (9.4e)$$

The units of the module are millimeters. Unfortunately, metric gears are not interchangeable with U.S. gears, despite both being involute tooth forms, as their standards for tooth sizes are different. In the United States, gear tooth sizes are specified by diametral pitch, elsewhere by module. The conversion from one standard to the other is

$$m = \frac{25.4}{p_d} \qquad (9.4f)$$

where m is in mm and p_d is in inches.

The **velocity ratio** m_V and the **torque ratio** m_T of the gearset can be put into a more convenient form by substituting equation 9.4c into equations 9.1 (p. 484), noting that the diametral pitch of meshing gears must be the same.

$$m_V = \pm \frac{d_{in}}{d_{out}} = \pm \frac{N_{in}}{N_{out}} \qquad (9.5a)$$

$$m_T = \pm \frac{d_{out}}{d_{in}} = \pm \frac{N_{out}}{N_{in}} \qquad (9.5b)$$

Thus the velocity ratio and torque ratio can be computed from the number of teeth on the meshing gears, which are integers. Note that a minus sign implies an external gearset and a positive sign an internal gearset as shown in Figure 9-1 (p. 483). The gear ratio m_G is always > 1 and can be expressed in terms of either the velocity ratio or torque ratio depending on which is larger than 1. Thus m_G expresses the gear train's overall ratio independent of change in direction of rotation or of the direction of power flow through the train when operated as either a speed reducer or a speed increaser.

$$m_G = |m_V| \quad \text{or} \quad m_G = |m_T|, \quad \text{for} \quad m_G \geq 1 \qquad (9.5c)$$

STANDARD GEAR TEETH   Standard, full-depth gear teeth have equal addenda on pinion and gear, with the dedendum being slightly larger for clearance. The standard tooth dimensions are defined in terms of the diametral pitch. Table 9-1 (p. 491) shows the definitions of dimensions of standard, full-depth gear teeth as defined by the AGMA, and Figure 9-10 shows their shapes for three standard pressure angles. Figure 9-11 shows the actual sizes of 20° pressure angle, standard, full-depth teeth from $p_d = 4$ to 80. Note the inverse relationship between p_d and tooth size. While there are no theoretical restrictions on the possible values of diametral pitch, a set of standard values is defined based on available gear cutting tools. These standard tooth sizes are shown in Table 9-2 in terms of diametral pitch and in Table 9-3 (p. 494) in terms of metric module.

9.4   INTERFERENCE AND UNDERCUTTING

The involute tooth form is only defined outside of the base circle. In some cases, the dedendum will be large enough to extend below the base circle. If so, then the portion of tooth below the base circle will not be an involute and will interfere with the tip of the tooth on the mating gear, which is an involute. If the gear is cut with a standard gear shaper or a "hob," the cutting tool will also interfere with the portion of tooth below the base circle and will cut away the interfering material. This results in an undercut tooth as shown in Figure 9-12. This undercutting weakens the tooth by removing material at its root. The maximum moment and maximum shear from the tooth loaded as a cantilever beam both occur in this region. Severe undercutting will promote early tooth failure.

Interference (and undercutting caused by manufacturing tools) can be prevented simply by avoiding gears with too few teeth. If a gear has a large number of teeth, they will be small compared to its diameter. As the number of teeth is reduced for a fixed diameter gear, the teeth must become larger. At some point, the dedendum will exceed the radial distance between the base circle and the pitch circle, and interference will occur.

Table 9-4a (p. 494) shows the minimum number of pinion teeth that can mesh with a rack without interference as a function of pressure angle. Gears with this few teeth can be generated without undercutting only by a pinion cutter or by milling. Gears that are cut with a hob, which has the same action as a rack with respect to the gear being cut, must have more teeth to avoid undercutting the involute tooth form during manufacture. The minimum number of teeth that can be cut by a hob without undercutting as a function of pressure angle is shown in Table 9-4b (p. 495). Table 9-5a (p. 495) shows the maximum number of 20-degree pressure angle full-depth gear teeth that can mesh with a given number of pinion teeth without interference and Table 9-5b shows the same information for 25-degree pressure angle full-depth gear teeth. Note that the pinion tooth numbers in this table are all fewer than the minimum number of teeth that can be generated by a hob. As the mating gear gets smaller, the pinion can have fewer teeth and still avoid interference.

Unequal-Addendum Tooth Forms

In order to avoid interference and undercutting on small pinions, the tooth form can be changed from the standard, full-depth shapes of Figure 9-10 that have equal addenda on both pinion and gear to an involute shape with a longer addendum on the pinion and a

(a) $\phi = 14.5°$

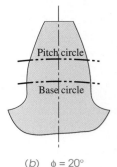

(b) $\phi = 20°$

(c) $\phi = 25°$

FIGURE 9-10

AGMA full-depth tooth profiles for three pressure angles

TABLE 9-2
Standard Diametral
Pitches

Coarse ($p_d < 20$)	Fine ($p_d \geq 20$)
1	20
1.25	24
1.5	32
1.75	48
2	64
2.5	72
3	80
4	96
5	120
6	
8	
10	
12	
14	
16	
18	

FIGURE 9-11

Actual tooth sizes for various diametral pitches *Courtesy of Barber-Colman Co., Loves Park, IL*

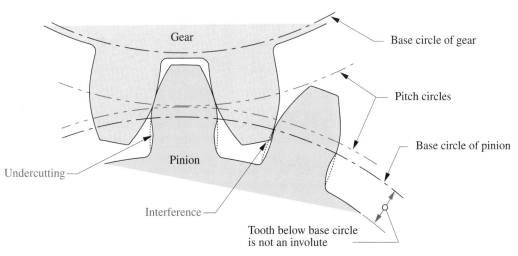

FIGURE 9-12

Interference and undercutting of teeth below the base circle

TABLE 9-3
Standard Metric Modules

Metric Module (mm)	Equivalent p_d (in^{-1})
0.3	84.67
0.4	63.50
0.5	50.80
0.8	31.75
1	25.40
1.25	20.32
1.5	16.93
2	12.70
3	8.47
4	6.35
5	5.08
6	4.23
8	3.18
10	2.54
12	2.12
16	1.59
20	1.27
25	1.02

TABLE 9-4a
Minimum Number of Pinion Teeth
To Avoid Interference Between a Full-Depth Pinion and a Full-Depth Rack

Pressure Angle (deg)	Minimum Number of Teeth
14.5	32
20	18
25	12

shorter one on the gear called **profile-shifted gears**. The AGMA defines addendum modification coefficients, x_1 and x_2, which always sum to zero, being equal in magnitude and opposite in sign. The positive coefficient x_1 is applied to increase the pinion addendum, and the negative x_2 decreases the gear addendum by the same amount. The total tooth depth remains the same. This shifts the pinion dedendum circle outside its base circle and eliminates that noninvolute portion of pinion tooth below the base circle. The standard coefficients are ±0.25 and ±0.50, which add or subtract 25% or 50% of the standard addendum. The limit of this approach occurs when the pinion tooth becomes pointed.

There are secondary benefits to this technique. The pinion tooth becomes thicker at its base and thus stronger. The gear tooth is correspondingly weakened, but since a full-depth gear tooth is stronger than a full-depth pinion tooth, this shift brings them closer to equal strength. A disadvantage of unequal-addendum tooth forms is an increase in sliding velocity at the tooth tip. The percent sliding between the teeth is greater than with equal addendum teeth which increases tooth-surface stresses. Friction losses in the gear mesh are also increased by higher sliding velocities. Figure 9-13 shows the contours of profile-shifted involute teeth. Compare these to standard tooth shapes in Figure 9-10 (p. 492).

9.5 CONTACT RATIO

The contact ratio m_p defines the average number of teeth in contact at any one time as:

$$m_p = \frac{Z}{p_b} \qquad (9.6a)$$

where Z is the length of action from equation 9.2 (p. 486) and p_b is the base pitch from equation 9.4b (p. 490). Substituting equations 9.4b and 9.4d into 9.6a defines m_p in terms of p_d:

$$m_p = \frac{p_d Z}{\pi \cos \phi} \qquad (9.6b)$$

The contact ratio m_p can also be expressed as a function only of pressure angle ϕ, number of pinion teeth, N_p, and the gear ratio m_G.

$$m_p = \frac{\sqrt{\left(\frac{N_p}{2}+1\right)^2 - \left(\frac{N_p}{2}\cos\phi\right)^2} + \sqrt{\left(\frac{m_G N_p}{2}+1\right)^2 - \left(\frac{m_G N_p}{2}\cos\phi\right)^2} - \frac{N_p}{2}\left(1+m_G\right)\sin\phi}{\pi \cos \phi} \qquad (9.6c)$$

If the contact ratio is 1, then one tooth is leaving contact just as the next is beginning contact. This is undesirable because slight errors in the tooth spacing will cause oscillations in the velocity, vibration, and noise. In addition, the load will be applied at the tip of the tooth, creating the largest possible bending moment. At larger contact ratios than 1, there is the possibility of load sharing among the teeth. For contact ratios between 1 and 2, which are common for spur gears, there will still be times during the mesh when one pair of teeth will be taking the entire load. However, these will occur toward the center of the mesh region where the load is applied at a lower position on the tooth, rather than at its tip. This point is called the **highest point of single-tooth contact** (HPSTC). The minimum acceptable contact ratio for smooth operation is 1.2. A minimum contact ratio of 1.4 is preferred and larger is better. Most spur gearsets will have contact ratios between 1.4 and 2.

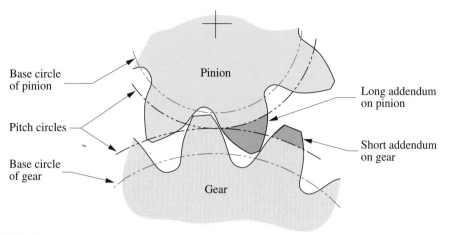

Base circle of pinion

Pinion

Pitch circles

Base circle of gear

Gear

Long addendum on pinion

Short addendum on gear

FIGURE 9-13

Profile-shifted teeth with long and short addenda to avoid interference and undercutting

TABLE 9-4b
Minimum Number of Pinion Teeth
To Avoid Undercutting When Cut With a Hob

Pressure Angle (deg)	Minimum Number of Teeth
14.5	37
20	21
25	14

TABLE 9-5a
Maximum Number of Gear Teeth
To Avoid Interference Between a 20° Full-Depth Pinion and Full-Depth Gears of Various Sizes

Number of Pinion Teeth	Maximum Gear Teeth
17	1309
16	101
15	45
14	26
13	16

TABLE 9-5b
Maximum Number of Gear Teeth
To Avoid Interference Between a 25° Full-Depth Pinion and Full-Depth Gears of Various Sizes

Number of Pinion Teeth	Maximum Gear Teeth
11	249
10	32
9	13

✎ **EXAMPLE 9-1**

Determining Gear Tooth and Gear Mesh Parameters.

Problem: Find the gear ratio, circular pitch, base pitch, pitch diameters, pitch radii, center distance, addendum, dedendum, whole depth, clearance, outside diameters, and contact ratio of a gearset with the given parameters. If the center distance is increased 2% what is the new pressure angle and increase in backlash?

Given: A $6p_d$, 20° pressure angle, 19-tooth pinion is meshed with a 37-tooth gear.

Assume: The tooth forms are standard AGMA full-depth involute profiles.

Solution:

1 The gear ratio is found from the tooth numbers on pinion and gear using equations 9.5a and 9.5c, (pp. 491, 492).

$$m_G = \frac{N_g}{N_p} = \frac{37}{19} = 1.947 \qquad (a)$$

2 The circular pitch can be found either from equation 9.4a or 9.4d (p. 490).

$$p_c = \frac{\pi}{p_d} = \frac{\pi}{6} = 0.524 \text{ in} \qquad (b)$$

3 The base pitch measured on the base circle is (from equation 9.4b):

$$p_b = p_c \cos\phi = 0.524\cos(20°) = 0.492 \text{ in} \qquad (c)$$

4 The pitch diameters and pitch radii of pinion and gear are found from equation 9.4c (p. 490).

$$d_p = \frac{N_p}{P_d} = \frac{19}{6} = 3.167 \text{ in,} \qquad r_p = \frac{d_p}{2} = 1.583 \text{ in} \qquad (d)$$

$$d_g = \frac{N_g}{P_d} = \frac{37}{6} = 6.167 \text{ in,} \qquad r_g = \frac{d_g}{2} = 3.083 \text{ in} \qquad (e)$$

5 The nominal center distance C is the sum of the pitch radii:

$$C = r_p + r_g = 4.667 \text{ in} \qquad (f)$$

6 The addendum and dedendum are found from the equations in Table 9-1 (p. 491):

$$a = \frac{1.0}{P_d} = 0.167 \text{ in,} \qquad b = \frac{1.25}{P_d} = 0.208 \text{ in} \qquad (g)$$

7 The whole depth h_t is the sum of the addendum and dedendum.

$$h_t = a + b = 0.167 + 0.208 = 0.375 \text{ in} \qquad (h)$$

8 The clearance is the difference between dedendum and addendum.

$$c = b - a = 0.208 - 0.167 = 0.042 \text{ in} \qquad (i)$$

9 The outside diameter of each gear is the pitch diameter plus two addenda:

$$D_{o_p} = d_p + 2a = 3.500 \text{ in,} \qquad D_{o_g} = d_g + 2a = 6.500 \text{ in} \qquad (j)$$

10 The contact ratio is found from equations 9.2 (p. 486) and 9.6a (p. 494).

$$Z = \sqrt{(r_p + a_p)^2 - (r_p \cos\phi)^2} + \sqrt{(r_g + a_g)^2 - (r_g \cos\phi)^2} - C \sin\phi$$

$$= \sqrt{(1.583 + 0.167)^2 - (1.583 \cos 20°)^2}$$

$$+ \sqrt{(3.083 + 0.167)^2 - (3.083 \cos 20°)^2} - 4.667 \sin 20° = 0.798 \text{ in}$$

$$m_p = \frac{Z}{p_b} = \frac{0.798}{0.492} = 1.62 \qquad (k)$$

11 If the center distance is increased from the nominal value due to assembly errors or other factors, the effective pitch radii will change by the same percentage. The gears' base radii will remain the same. The new pressure angle can be found from the changed geometry. For a 2% increase in center distance (1.02x):

$$\phi_{new} = \cos^{-1}\left(\frac{r_{\text{base circle}_p}}{1.02 r_p}\right) = \cos^{-1}\left(\frac{r_p \cos\phi}{1.02 r_p}\right) = \cos^{-1}\left(\frac{\cos 20°}{1.02}\right) = 22.89° \qquad (l)$$

12 The change in backlash as measured at the pinion is found from equation 9.3 (p. 489).

$$\theta_B = 43\,200 (\Delta C) \frac{\tan\phi}{\pi d} = 43\,200 (0.02)(4.667) \frac{\tan(22.89°)}{\pi(3.167)} = 171 \text{ minutes of arc} \qquad (m)$$

9.6   GEAR TYPES

Gears are made in many configurations for particular applications. This section describes some of the more common types.

Spur, Helical, and Herringbone Gears

SPUR GEARS   are ones in which the *teeth are parallel to the axis of the gear*. This is the simplest and least expensive form of gear to make. Spur gears can only be meshed if their axes are parallel. Figure 9-14 shows a spur gear.

HELICAL GEARS   are ones in which the teeth are at a helix angle ψ with respect to the axis of the gear as shown in Figure 9-15a. Figure 9-16 shows a pair of opposite-hand[*] **helical gears** in mesh. Their axes are parallel. Two **crossed helical gears** of the same hand can be meshed with their axes at an angle as shown in Figure 9-17 (p. 498). The helix angles can be designed to accommodate any skew angle between the nonintersecting shafts.

Helical gears are more expensive than spur gears but offer some advantages. They run quieter than spur gears because of the smoother and more gradual contact between their angled surfaces as the teeth come into mesh. Spur gear teeth mesh along their entire face width at once. The sudden impact of tooth on tooth causes vibrations that are heard as a "whine" which is characteristic of spur gears but is absent with helical gears. Also, for the same gear diameter and diametral pitch, a helical gear is stronger due to the slightly thicker tooth form in a plane perpendicular to the axis of rotation.

HERRINGBONE GEARS   are formed by joining two helical gears of identical pitch and diameter but of opposite hand on the same shaft. These two sets of teeth are often cut on the same gear blank. The advantage compared to a helical gear is the internal cancellation of its axial thrust loads since each "hand" half of the herringbone gear has an oppositely directed thrust load. Thus no thrust bearings are needed other than to locate the shaft axially. This type of gear is much more expensive than a helical gear and tends to be used in large, high-power applications such as ship drives, where the frictional losses from axial loads would be prohibitive. A herringbone gear is shown in Figure 9-15b. Its face view is the same as the helical gear's.

FIGURE 9-14

A spur gear
Courtesy of Martin Sprocket and Gear Co., Arlington, TX

* Helical gears are either right- or left-handed. Note that the gear of Figure 9-15a is left-handed because, if either face of the gear were placed on a horizontal surface, its teeth would slope up to the left.

(a) Helical gear   (b) Herringbone gear

FIGURE 9-15

A helical gear and a herringbone gear

FIGURE 9-16

Parallel axis helical gears
Courtesy of Martin Sprocket and Gear Co., Arlington, TX

FIGURE 9-17

Crossed axis helical
gears
*Courtesy of the Boston
Gear Division of IMO
Industries, Quincy, MA*

FIGURE 9-18

A worm and worm
gear (or worm wheel)
*Courtesy of Martin
Sprocket and Gear Co.,
Arlington, TX*

EFFICIENCY   The general definition of efficiency is *output power/input power* expressed as a percentage. A spur gearset can be 98 to 99% efficient. The helical gearset is less efficient than the spur gearset due to sliding friction along the helix angle. They also present a reaction force along the axis of the gear, which the spur gear does not. Thus helical gearsets must have thrust bearings as well as radial bearings on their shafts to prevent them from pulling apart along the axis. Some friction losses occur in the thrust bearings as well. A parallel helical gearset will be about 96 to 98% efficient, and a crossed helical set only 50 to 90% efficient. The parallel helical set (opposite hand but same helix angle) has line contact between the teeth and can handle high loads at high speeds. The crossed helical set has point contact and a large sliding component that limit its application to light load situations.

If the gearsets have to be shifted in and out of mesh while in motion, spur gears are a better choice than helical, as the helix angle interferes with the axial shifting motion. (Herringbone gears of course cannot be axially disengaged.) Truck transmissions often use spur gears for this reason, whereas automobile (standard) transmissions use helical, constant mesh gears for quiet running and have a synchromesh mechanism to allow shifting. These transmission applications will be described in a later section.

Worms and Worm Gears

If the helix angle is increased sufficiently, the result will be a **worm**, which has only one tooth wrapped continuously around its circumference a number of times, analogous to a screw thread. This worm can be meshed with a special **worm gear** (or **worm wheel**), whose axis is perpendicular to that of the worm as shown in Figure 9-18. Because the driving worm typically has only one tooth, the ratio of the gearset is equal to one over the number of teeth on the worm gear (see equations 9.5, p. 491). These teeth are not involutes over their entire face, which means that the center distance must be maintained accurately to guarantee conjugate action.

Worms and wheels are made and replaced as matched sets. These worm gearsets have the advantage of very high gear ratios in a small package and can carry very high loads especially in their single or double enveloping forms. **Single enveloping** means that the worm gear teeth are wrapped around the worm. **Double enveloping** sets also wrap the worm around the gear, resulting in an hourglass-shaped worm. Both of these techniques increase the surface area of contact between worm and wheel, increasing load carrying capacity and also cost. One trade-off in any wormset is very high sliding and thrust loads that make the wormset rather inefficient at 40 to 85% efficiency.

Perhaps the major advantage of the wormset is that it can be designed to be impossible to **backdrive**. A spur or helical gearset can be driven from either shaft, as a velocity step-up or step-down device. While this may be desirable in many cases, if the load being driven must be held in place after the power is shut off, the spur or helical gearset will not do. They will "backdrive." This makes them unsuitable for such applications as a jack to raise a car unless a brake is added to the design to hold the load. The wormset, on the other hand, can only be driven from the worm. The friction can be large enough to prevent it being backdriven from the worm wheel. Thus it can be used without a brake in load-holding applications such as jacks and hoists.

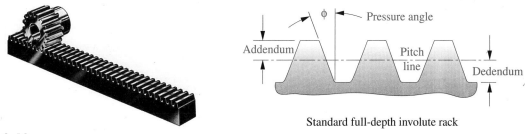

Standard full-depth involute rack

FIGURE 9-19

A rack and pinion   *Photo courtesy of Martin Sprocket and Gear Co., Austin, TX*

Rack and Pinion

If the diameter of the base circle of a gear is increased without limit, the base circle will become a straight line. If the "string" wrapped around this base circle to generate the involute were still in place after the base circle's enlargement to an infinite radius, the string would be pivoted at infinity and would generate an involute that is a straight line. This linear gear is called a **rack**. Its teeth are trapezoids, yet are true involutes. This fact makes it easy to create a cutting tool to generate involute teeth on circular gears, by accurately machining a rack and hardening it to cut teeth in other gears. Rotating the gear blank with respect to the rack cutter while moving the cutter axially back and forth across the gear blank will shape, or develop, a true involute tooth on the circular gear.

Figure 9-19 shows a **rack and pinion**. The most common application of this device is in rotary to linear motion conversion or vice versa. It can be backdriven, so it requires a brake if used to hold a load. An example of its use is in **rack-and-pinion steering** in automobiles. The pinion is attached to the bottom end of the steering column and turns with the steering wheel. The rack meshes with the pinion and is free to move left and right in response to your angular input at the steering wheel. The rack is also one link in a multibar linkage that converts the linear translation of the rack to the proper amount of angular motion of a rocker link attached to the front wheel assembly to steer the car.

Bevel and Hypoid Gears

BEVEL GEARS   For right-angle drives, crossed helical gears or a wormset can be used. For any angle between the shafts, including 90°, bevel gears may be the solution. Just as spur gears are based on rolling cylinders, **bevel gears** are based on rolling cones as shown in Figure 9-20 (p. 500). The angle between the axes of the cones and the included angles of the cones can be any compatible values as long as the apices of the cones intersect. If they did not intersect, there would be a mismatch of velocity at the interface. The apex of each cone has zero radius, thus zero velocity. All other points on the cone surface will have nonzero velocity. The velocity ratio of the bevel gears is defined by equation 9.1a (p. 484) using the pitch diameters at any common point of intersection of cone diameters.

SPIRAL BEVEL GEARS   If the teeth are parallel to the axis of the gear, it will be a straight bevel gear as shown in Figure 9-21, p. 500. If the teeth are angled with respect to the axis, it will be a **spiral bevel gear** (Figure 9-22, p. 500), analogous to a helical gear. The cone axes and apices must intersect in both cases. The advantages and disad-

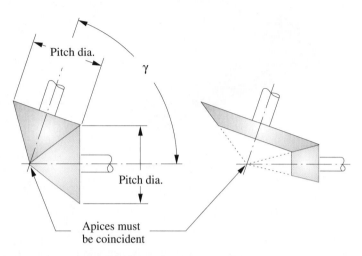

(a) Incorrect arrangement   (b) Correct arrangements

FIGURE 9-20

Bevel gears are based on rolling cones.

FIGURE 9-21

Straight bevel gears
*Courtesy of Martin
Sprocket and Gear,
Arlington, TX*

vantages of straight bevel and spiral bevel gears are similar to those of the spur gear and helical gear, respectively, regarding strength, quietness, and cost. Bevel gear teeth are not involutes but are based on an "octoid" tooth curve. They must be replaced in pairs (gearsets) as they are not universally interchangeable, and their center distances must be accurately maintained.

HYPOID GEARS   If the axes between the gears are nonparallel and also nonintersecting, bevel gears cannot be used. **Hypoid gears** will accommodate this geometry. Hypoid gears are based on rolling hyperboloids of revolution as shown in Figure 9-23. (The term *hypoid* is a contraction of *hyperboloid*.) The tooth form is not an involute. These hypoid gears are used in the final drive of front-engine, rear-wheel-drive automobiles, in order to lower the axis of the driveshaft below the center of the rear axle to reduce the "driveshaft hump" in the back seat.

Noncircular Gears

FIGURE 9-22

Spiral bevel gears
*Courtesy of the Boston
Gear Division of IMO
Industries, Quincy, MA*

Noncircular gears are based on the rolling centrodes of a Grashof double-crank fourbar linkage. Centrodes are the loci of the instant center I_{24} of the linkage and were described in Section 6.5 (p. 307). Figure 6-15b (p. 310) shows a pair of centrodes that could be used for noncircular gears. Teeth would be added to their circumferences in the same way that we add teeth to rolling cylinders for circular gears. The teeth then act to guarantee no slip. Figure 9-24 shows a pair of noncircular gears based on a different set of centrodes than those of Figure 6-15b. (The gears of Figure 9-24 really do make complete revolutions in mesh!) Of course, the velocity ratio of noncircular gears is not constant. That is their purpose, to provide a time-varying output function in response to a constant velocity input. Their instantaneous velocity ratio is defined by equation 6.11f (p. 302).

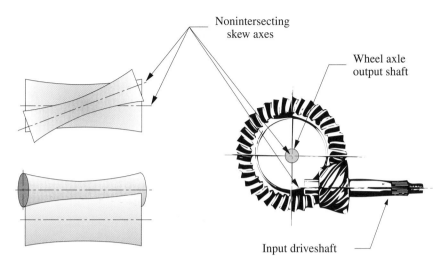

Nonintersecting
skew axes

Wheel axle
output shaft

Input driveshaft

(*a*) Rolling hyperboloids of revolution   (*b*) Automotive hypoid final drive gears
Courtesy of General Motors Co., Detroit, MI

FIGURE 9-23

Hypoid gears are based on hyperboloids of revolution.

These devices are used in a variety of rotating machinery such as printing presses where variation in the angular velocity of rollers is required on a cyclical basis.

Belt and Chain Drives

VEE BELTS   A **vee belt** drive is shown in Figure 9-2 (p. 483). Vee belts are made of elastomers (synthetic rubber) reinforced with synthetic or metallic cords for strength. The pulleys, or *sheaves*, have a matching vee groove that helps to grip the belt as belt tension jams the belt into the vee. Vee belts have a transmission efficiency of 95 to 98% when first installed. This will decrease to about 93% as the belt wears and slippage increases. Because of slip, the velocity ratio is neither exact nor constant. Flat belts running on flat and crowned pulleys are still used in some applications as well. As discussed above, slip is possible with untoothed belts and phasing cannot be guaranteed.

SYNCHRONOUS (TIMING) BELTS   The **synchronous belt** solves the phasing problem by preventing slip while retaining some of the advantages of vee belts and can cost less than gears or chains. Figure 9-25a (p. 502) shows a synchronous (or toothed) belt and its special gearlike pulleys or sheaves. These belts are made of a rubberlike material but are reinforced with steel or synthetic cords for higher strength and have molded-in teeth that fit in the grooves of the pulleys for positive drive. They are capable of fairly high torque and power transmission levels and are frequently used to drive automotive engine camshafts as shown in Figure 9-25b. They are more expensive than conventional vee belts and are noisier, but run cooler and last longer. Their transmission efficiency is 98% and stays at that level with use. Manufacturers' catalogs provide detailed information on sizing both vee and synchronous belts for various applications. See Bibliography (p. 532).

FIGURE 9-24

Noncircular gears

9

(a) Standard synchronous belt
*Courtesy of T. B. Wood's Sons Co.,
Chambersburg, PA*

(b) Engine valve camshaft drive
*Courtesy of Chevrolet Division,
General Motors Co., Detroit, MI*

FIGURE 9-25

Toothed synchronous belts and sprockets

CHAIN DRIVES are often used for applications where positive drive (phasing) is needed and large torque requirements or high temperature levels preclude the use of timing belts. When the input and output shafts are far apart, a chain drive may be the most economical choice. Conveyor systems often use chain drives to carry the work along the assembly line. Steel chain can be run through many (but not all) hostile chemical or temperature environments. Many types and styles of chain have been designed for various applications ranging from the common roller chain (Figure 9-26a) as used on your bicycle or motorcycle, to more expensive inverted tooth or "silent chain" designs (Figure 9-26b) used for camshaft drives in expensive automobile engines. Figure 9-27 shows a typical sprocket for roller chain. Note that the sprocket teeth are not the same shape as gear teeth and are not involutes. The sprocket tooth shape is dictated by the need to match

(a) Roller chain

(b) Inverted-tooth or silent chain

FIGURE 9-26

Chain types for power transmission *From Phelan, R. M. (1970). Fundamentals of Mechanical Design, 3rd ed., McGraw-Hill. NY.*

the contour of the portion of chain that nestles in the grooves. In this case the roller chain has cylindrical pins that engage the sprocket.

One unique limitation of chain drive is something called "**chordal action**." The links of the chain form a set of chords when wrapped around the circumference of the sprocket. As these links enter and leave the sprocket, they impart a "jerky" motion to the driven shaft that causes some variation, or ripple, on the output velocity. Chain drives do not exactly obey the fundamental law of gearing. If very accurate, constant output velocity is required, a chain drive may not be the best choice.

VIBRATION IN BELTS AND CHAINS You may have noticed when watching the operation of, for example, a vee belt such as your car engine's fan belt, that the belt span between pulleys vibrates laterally, even when the belt's linear velocity is constant. If you consider the acceleration of a belt particle as it travels around the belt path, you will realize that its acceleration is theoretically zero while traversing the unsupported spans between sheaves at constant velocity; but when it enters the wrap of a sheave, it suddenly acquires a nonzero centripetal acceleration that remains essentially constant in magnitude while the belt particle is on the sheave. Thus the acceleration of a belt particle has sudden jumps from zero to some constant magnitude or vice versa, four times per traverse for a simple two-sheave system such as that of Figure 9-2 (p. 483), and more if there are multiple sheaves. This provides theoretically infinite pulses of jerk to the belt particles at these transitions and this excites the lateral natural frequencies of the belt's unsupported span between sheaves. The result is lateral vibration of the belt span that creates dynamic variation in belt tension and noise.

FIGURE 9-27

A roller chain sprocket
Courtesy of Martin Sprocket and Gear Co., Arlington, TX

9.7   SIMPLE GEAR TRAINS

A gear train is any collection of two or more meshing gears. A simple gear train is one in which each shaft carries only one gear, the most basic, two-gear example of which is shown in Figure 9-4 (p. 484). The *velocity ratio* m_V (sometimes called *train ratio*) of this gearset is found by expanding equation 9.5a (p. 491). Figure 9-28 (p. 504) shows a simple gear train with five gears in series. The expression for this simple train's velocity ratio is:

$$m_V = \left(-\frac{N_2}{N_3}\right)\left(-\frac{N_3}{N_4}\right)\left(-\frac{N_4}{N_5}\right)\left(-\frac{N_5}{N_6}\right) = +\frac{N_2}{N_6}$$

or in general terms:

$$m_V = \pm\frac{N_{in}}{N_{out}} \qquad (9.7)$$

which is the same as equation 9.5a (p. 491) for a single gearset.

Each gearset potentially contributes to the overall train ratio, but in any case of a simple (series) train, the numerical effects of all gears except the first and last cancel out. The train ratio of a simple train is always just the ratio of the first gear over the last. Only the sign of the overall ratio is affected by the intermediate gears which are called *idlers* because typically no power is taken from their shafts. If all gears in the train are external and there is an even number of gears in the train, the output direction will be opposite that of the input. If there is an odd number of external gears in the train, the output will be in the same direction as the input. Thus a single, external idler gear *of any diameter* can be used to change the direction of the output gear without affecting its velocity.

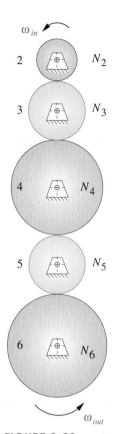

FIGURE 9-28

A simple gear train

A single gearset of spur, helical, or bevel gears *is usually limited to a ratio of about 10:1* simply because the gearset will become very large, expensive, and hard to package above that ratio if the pinion is kept above the minimum numbers of teeth shown in Table 9-4a or b (pp. 494-495). If the need is to get a larger train ratio than can be obtained with a single gearset, it is clear from equations 9.6 (p. 494) that the simple train will be of no help.

It is common practice to insert a single idler gear to change direction, but more than one idler is superfluous. There is little justification for designing a gear train as is shown in Figure 9-28. If the need is to connect two shafts that are far apart, a simple train of many gears could be used but will be more expensive than a chain or belt drive for the same application. Most gears are not cheap.

9.8   COMPOUND GEAR TRAINS

To get a train ratio of greater than about 10:1 with spur, helical, or bevel gears (or any combination thereof) it is necessary to **compound the train** (unless an epicyclic train is used—see Section 9.9, p. 513). A **compound train** *is one in which at least one shaft carries more than one gear.* This will be a parallel or series-parallel arrangement, rather than the pure series connections of the simple gear train. Figure 9-29 shows a compound train of four gears, two of which, gears 3 and 4, are fixed on the same shaft and thus have the same angular velocity.

The train ratio is now:

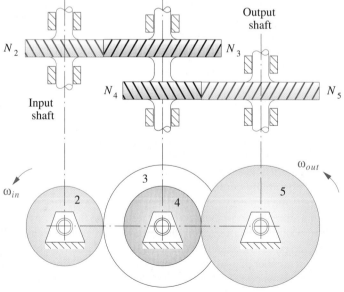

FIGURE 9-29

A compound gear train

$$m_V = \left(-\frac{N_2}{N_3}\right)\left(-\frac{N_4}{N_5}\right) \tag{9.8a}$$

This can be generalized for any number of gears in the train as:

$$m_V = \pm\frac{\text{product of number of teeth on driver gears}}{\text{product of number of teeth on driven gears}} \tag{9.8b}$$

Note that these intermediate ratios do not cancel and the overall train ratio is the product of the ratios of parallel gearsets. Thus a larger ratio can be obtained in a compound gear train despite the approximately 10:1 limitation on individual gearset ratios. The plus or minus sign in equation 9.8b depends on the number and type of meshes in the train, whether external or internal. Writing the expression in the form of equation 9.8a and carefully noting the sign of each mesh ratio in the expression will result in the correct algebraic sign for the overall train ratio.

Design of Compound Trains

If one is presented with a completed design of a compound gear train such as that in Figure 9-29, it is a trivial task to apply equation 9.8 and determine the train ratio. It is not so simple to do the inverse, namely, design a compound train for a specified train ratio.

EXAMPLE 9-2

Compound Gear Train Design.

Problem: Design a compound train for an exact train ratio of 180:1. Find a combination of gears that will give that ratio.

Solution:

1   The first step is to determine how many stages, or gearsets, are necessary. Simplicity is the mark of good design, so try the smallest possibility first. Take the square root of 180, which is 13.416. So, two stages each of that ratio will give approximately 180:1. However, this is larger than our design limit of 10:1 for each stage, so try three stages. The cube root of 180 is 5.646, well within 10, so three stages will do.

2   If we can find some integer ratio of gear teeth that will yield 5.646:1, we can simply use three of them to design our gearbox. Using a lower limit of 12 teeth for the pinion and trying several possibilities we get the gearsets shown in Table 9-6 as possibilities.

3   The number of gear teeth obviously must be an integer. The closest to an integer in Table 9-6 is the 79.05 result. Thus a 79:14 gearset comes closest to the desired ratio. Applying this ratio to all three stages will yield a train ratio of $(79/14)^3 = 179.68:1$, which is within 0.2% of 180:1. This may be an acceptable solution provided that the gearbox is not being used in a timing application. If the purpose of this gearbox is to step down the motor speed for a crane hoist, for example, an approximate ratio will be adequate.

4   Many gearboxes are used in production machinery to drive camshafts or linkages from a master driveshaft and must have the exact ratio needed or else the driven device will eventually get out of phase with the rest of the machine. If that were the case in this example, then the

TABLE 9-6
Example 9-2
Possible Gearsets for 180:1 Three-Stage Compound Train

Gearset Ratio	Pinion Teeth	Gear Teeth
5.646 x 12	=	67.75
5.646 x 13	=	73.40
5.646 x 14	=	79.05
5.646 x 15	=	84.69

FIGURE 9-30

Three-stage compound gear train for train ratio $m_V = 1:180$ (gear ratio $m_G = 180:1$)

TABLE 9-7

Example 9-2

Exact Solution for 180:1 Three-Stage Compound Train

Gearset Ratio	Pinion Teeth		Gear Teeth
6	x	14 =	84
6	x	14 =	84
5	x	14 =	70

solution found in step 3 would not be good enough. We will need to redesign it for exactly 180:1. Since our overall train ratio is an integer, it will be simplest to look for integer gearset ratios. Thus we need three integer factors of 180. The first solution above gives us a reasonable starting point in the cube root of 180, which is 5.646. If we round this up (or down) to an integer, we may be able to find a suitable combination.

5 Two compounded stages of 6:1 together give 36:1. Dividing 180 by 36 gives 5. Thus the stages shown in Table 9-7 provide one possible exact solution.

This solution, shown in Figure 9-30, meets our design criteria. It has the correct, exact ratio; the stages are all less than 10:1; and no pinion has less than 14 teeth, which avoids undercutting if 25° pressure angle gears are used (Table 9-4b, p. 495).

Design of Reverted Compound Trains

In the preceding example the input and output shaft locations are in different locations. This may well be acceptable or even desirable in some cases, depending on other packaging constraints in the overall machine design. Such a gearbox, whose *input and output shafts are not coincident*, is called a **nonreverted compound train**. In some cases, such as automobile transmissions, it is desirable or even necessary to have the *output shaft con-*

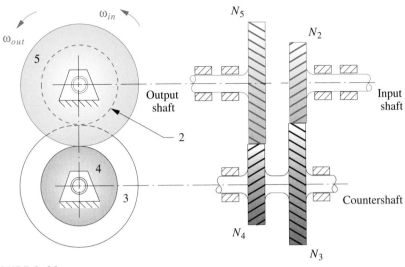

FIGURE 9-31

A reverted compound gear train

centric with the input shaft. This is referred to as "reverting the train" or "bringing it back onto itself." The design of a **reverted compound train** is more complicated because of the additional constraint that the center distances of the stages must be equal. Referring to Figure 9-31, this constraint can be expressed in terms of their pitch radii, pitch diameters, or numbers of teeth (provided that all gears have the same diametral pitch).

$$r_2 + r_3 = r_4 + r_5 \tag{9.9a}$$

or
$$d_2 + d_3 = d_4 + d_5 \tag{9.9b}$$

If p_d is the same for all gears, equation 9.4c (p. 490) can be substituted in equation 9.9b and the diametral pitch terms will cancel giving

$$N_2 + N_3 = N_4 + N_5 \tag{9.9c}$$

EXAMPLE 9-3

Reverted Gear Train Design.

Problem: Design a reverted compound train for an exact train ratio of 18:1.

Solution:

1   Though it is not at all necessary to have integer gearset ratios in a compound train (only integer tooth numbers), if the train ratio is an integer, it is easier to design with integer ratio gearsets.

2   The square root of 18 is 4.2426, well within our 10:1 limitation. So two stages will suffice in this gearbox.

TABLE 9-8
Example 9-3
Possible Gearsets for 18:1
Two-Stage Reverted
Compound Train

Gearset Ratio	Pinion Teeth	Gear Teeth
4.2426 x 12 =		50.91
4.2426 x 13 =		55.15
4.2426 x 14 =		59.40
4.2426 x 15 =		63.64
4.2426 x 16 =		67.88
4.2426 x 17 =		72.12
4.2426 x 18 =		76.37
4.2426 x 19 =		80.61
4.2426 x 20 =		84.85

3   If we could form two identical stages, each with a ratio equal to the square root of the overall train ratio, the train would be reverted by default. Table 9-8 (p. 507) shows that there are no reasonable combinations of tooth ratios that will give the exact square root needed. Moreover, this square root is not a rational number, so we cannot get an exact solution by this approach.

4   Instead, let's factor the train ratio. All numbers in the factors 9 × 2 and 6 × 3 are less than 10, so they are acceptable on that basis. It is probably better to have the ratios of the two stages closer in value to one another for packaging reasons, so the 6 × 3 choice will be tried.

5   Figure 9-31 (p. 507) shows a two-stage reverted train. Note that, unlike the nonreverted train in Figure 9-29 (p. 504), the input and output shafts are now in-line and cantilevered; thus each must have double bearings on one end for moment support and a good bearing ratio as was defined in Section 2.18 (p. 69).

6   Equation 9.8 (p. 505) states the relationship for its compound train ratio. In addition, we have the constraint that the center distances of the stages must be equal. Use equation 9.9c (p. 507) and set it equal to an arbitrary constant K to be determined.

$$N_2 + N_3 = N_4 + N_5 = K \qquad (a)$$

7   We wish to solve equations 9.8 and 9.9c simultaneously. We can separate the terms in equation 9.8 and set them each equal to one of the stage ratios chosen for this design.

$$\frac{N_2}{N_3} = \frac{1}{6}$$
$$N_3 = 6N_2 \qquad (b)$$

$$\frac{N_4}{N_5} = \frac{1}{3}$$
$$N_5 = 3N_4 \qquad (c)$$

8   Separating the terms in equation (a):

$$N_2 + N_3 = K \qquad (d)$$
$$N_4 + N_5 = K \qquad (e)$$

9   Substituting equation (b) in (d) and equation (c) in (e) yields:

$$N_2 + 6N_2 = K = 7N_2 \qquad (f)$$
$$N_4 + 3N_4 = K = 4N_4 \qquad (g)$$

10 To make equations (f) and (g) compatible, K must be set to at least the lowest common multiple of 7 and 4, which is 28. This yields values of $N_2 = 4$ teeth and $N_4 = 7$ teeth.

11 Since a four-tooth gear will have unacceptable undercutting, we need to increase our value of K sufficiently to make the smallest pinion large enough.

12 A new value of $K = 28 \times 4 = 112$ will increase the four-tooth gear to a 16-tooth gear, which is acceptable for a 25° pressure angle (Table 9-4b, p. 490). With this assumption of $K = 112$, equations (b), (c), (f), and (g) can be solved simultaneously to give:

$$N_2 = 16 \qquad\qquad N_3 = 96$$
$$N_4 = 28 \qquad\qquad N_5 = 84 \qquad (h)$$

which is a viable solution for this reverted train.

FIGURE 9-32

A commercial, three-stage reverted compound gearbox
Courtesy of Boston Gear Division of IMO Industries, Quincy, MA

The same procedure outlined here can be applied to the design of reverted trains involving several stages such as the helical gearbox in Figure 9-32.

An Algorithm for the Design of Compound Gear Trains

The examples of compound gear train design presented above used integer train ratios. If the required train ratio is noninteger, it is more difficult to find a combination of integer tooth numbers that will give the exact train ratio. Sometimes an irrational gear ratio may be needed for such tasks as conversion of English to metric measure within a machine tool gear train or when π is a factor in the ratio. Then the closest approximation to the desired irrational train ratio that can be contained in a reasonable package is needed.

DilPare [1] and Selfridge and Riddle [2] have devised algorithms to solve this problem. Both require a computer for their solution. The Selfridge and Riddle approach will be described here. It is applicable to two- or three-stage compound trains. A low limit N_{min} and a high limit N_{max} on the acceptable number of teeth for any gear must be specified. An error tolerance ε expressed as a percentage of the desired train ratio R (made always > 1) is also selected. For a two-stage compound train the ratio will be as shown in equation 9.5c (p. 492) expanded according to equation 9.8b (p. 505) with the signs neglected for this analysis.

$$R = m_G = \frac{N_3 N_5}{N_2 N_4} \qquad (9.10a)$$

The range of acceptable ratios is determined by the choice of error tolerance ε.

$$R_{low} = R - \varepsilon$$
$$R_{high} = R + \varepsilon \qquad (9.10b)$$

$$R_{low} \leq \frac{N_3 N_5}{N_2 N_4} \leq R_{high} \tag{9.10c}$$

Then, since the tooth numbers must be integers:

$$N_3 N_5 \leq INT\left(N_2 N_4 R_{high}\right) \tag{9.10d}$$

Let:
$$P = INT\left(N_2 N_4 R_{high}\right) \tag{9.10e}$$

Also from equation 9.10c,

$$N_3 N_5 \geq INT\left(N_2 N_4 R_{low}\right) \tag{9.10f}$$

Let:
$$Q = INT\left(N_2 N_4 R_{low}\right) + 1 \tag{9.10g}$$

rounding up to the next integer.

A search is done on all values of a temporary parameter K defined as $Q \leq K \leq P$ to see if a usable product pair can be found. Because of multiplicative symmetry, the largest value of N_3 that need be considered is

$$N_3 \leq \sqrt{P} \tag{9.11a}$$

Let:
$$N_p = \sqrt{P} \tag{9.11b}$$

The smallest value of N_3 that need be considered occurs when K is at its smallest value Q and N_5 takes its largest value N_{high}. (N_3 is also constrained by N_{low}.)

$$N_3 \geq \frac{Q}{N_{high}} \tag{9.11c}$$

Let:
$$N_m = INT\left(\frac{Q + N_{high} - 1}{N_{high}}\right) \tag{9.11d}$$

which also rounds up to the next integer.

The search finds those values of N_3 that meet $N_m \leq N_3 \leq N_p$ and $N_5 = K / N_3$. The computer code for this algorithm is shown in Table 9-9. The complete program COM-POUND.TK is on the DVD included with this book, encoded for use with the *TKSolver* program. The code can be easily rewritten for other equation solvers or compilers.

This algorithm is extendable to three-stage compound gear trains, and the two-stage version can be modified to force reversion of the train by adding a center distance calculation for each gearset and a comparison to a selected tolerance on center distance. These files are also on the DVD as TRIPLE.TK and REVERT.TK, respectively. These programs each generate a table of all solutions that meet the stated error criteria within the tooth limits specified.

TABLE 9-9   Algorithm for Design of Two-Stage Compound Gear Trains
From *TKSolver* file Compound.tk. Based on Reference (2)

" *Ratio* is the desired gear train ratio and must be > 1. *Nmin* is the minimum number of teeth acceptable on any pinion.
" *Nmax* is the maximum number of teeth acceptable on any gear. *eps1* is initial estimate of the error tolerance on *Ratio*.
" *eps* is the tolerance used in the computation, initialized to *eps1* but modified (doubled) until solutions are found.
" *counter* indicates how many times the initial tolerance was doubled. Note that a large initial value on *eps1* will cause long
" computation times whereas a too-small value (that gives no solutions) will quickly be increased and lead to a faster solution.
" *pinion1*, *pinion2*, *gear1*, and *gear2* are tooth numbers for solution.

$eps = eps1$	" initialize error bound
$counter = 0$	" initialize counter
redo:	" reentry point for additional tries at solution
$S = 1$	" initialize the array pointer
$R\_high = Ratio + eps$	" initialize tolerance bands around ratio
$R\_low = Ratio - eps$	" initialize tolerance bands around ratio
$Nh3 = INT(Nmax^2 / R\_high)$	" intermediate value for computation
$Nh4 = INT(Nmax / SQRT (R\_high))$	" intermediate value for computation
For $pinion1 = Nmin$ to $Nh4$	" loop for first pinion
$Nhh = MIN (Nmax, INT (Nh3 / pinion1))$	" intermediate value for computation
For $pinion2 = pinion1$ to Nhh	" loop for 2nd pinion
$P = INT(pinion1 * pinion2 * R\_high)$	" intermediate value for computation
$Q = INT(pinion1 * pinion2 * R\_low) + 1$	" intermediate value for computation
IF ($P < Q$) THEN GOTO np2	" skip to next pinion2 if true
$Nm = MAX (Nmin, INT ((Q + Nmax - 1) / Nmax))$	" intermediate value for computation
$Np = SQRT(P)$	" intermediate value for computation
For $K = Q$ to P	" loop for parameter K
For $gear1 = Nm$ to Np	" loop for first gear
IF (MOD(K, gear1) <> 0) Then GOTO ng1	" not a match - skip to next gear1
$gear2 = K / gear1$	" find second gear tooth number
$error = (Ratio - K / (pinion1 * pinion2))$	" find error in ratio
"check to see if is within current tolerance	
IF error > eps THEN GOTO ng1	" is out of bounds - skip to next gear1
" else load solution into arrays	
$pin1[S] = pinion1$	
$pin2[S] = pinion2$	
$gear1[S] = gear1$	
$gear2[S] = gear2$	
$error[S] = ABS(error)$	
$ratio1[S] = gear1 / pinion1$	
$ratio2[S] = gear2 / pinion2$	
$ratio[S] = ratio1[S] * ratio2[S]$	
$S = S + 1$	" increment array pointer
ng1: Next gear1	
Next K	
np2: Next pinion2	
Next pinion1	
" test to see if any solution occurred with current eps value	
IF (Length (pin1) = 0) Then GOTO again ELSE Return	" have a solution
again:	" double eps value and try again
$eps = eps * 2$	
$counter = counter + 1$	
GOTO redo	

9

EXAMPLE 9-4

Compound Gear Train Design to Approximate an Irrational Ratio.

Problem: Find a pair of gearsets which when compounded will give a train ratio of 3.14159:1 with an error of < 0.0005%. Limit gears to tooth numbers between 15 and 100. Also determine the tooth numbers for the smallest error possible if the two gearsets must be reverted.

Solution:

1   Input data to the algorithm are $R = 3.141\ 59$, $N_{low} = 15$, $N_{high} = 100$, initial $\varepsilon = 3.141\ 59$ E-5.

2   The *TKSolver* file COMPOUND.TK (see Table 9-9, p. 511) was used to generate the nonreverted solutions shown in Table 9-10.

3   The best nonreverted solution (7th row in Table 9-10) has an error in ratio of 7.849 9 E-06 (0.000 249 87%) giving a ratio of 3.141 582 with gearsets of 29:88 and 85:88 teeth.

4   The *TKSolver* file REVERT.TK (see DVD) was used to generate the reverted solutions shown in Table 9-11.

5   The best reverted solution has an error in ratio of –9.619 8 E-04 (–0.030 62%) giving a ratio of 3.142 562 with gearsets of 22:39 and 22:39 teeth.

6   Note that imposing the additional constraint of reversion has reduced the number of possible solutions effectively to one (the two solutions in Table 9-11 differ by a factor of 2 in tooth

* Note that this gear train combination gives an approximation for π that is exact to 4 decimal places. But, this example asks for an approximation to 5 decimal places within a tolerance of 5 ten-thousandths of one percent. This ratio is off by one thousandth of a percent of the desired 5-place value.

TABLE 9-10   Nonreverted Gearsets and Errors in Ratio for Example 9-4

N_2	N_3	Ratio1	N_4	N_5	Ratio2	m_V	Error
17	54	3.176	91	90	0.989	3.141 564	2.568 2 E-05
17	60	3.529	91	81	0.890	3.141 564	2.568 2 E-05
22	62	2.818	61	68	1.115	3.141 580	1.026 8 E-05
23	75	3.261	82	79	0.963	3.141 569	2.054 1 E-05
25	51	2.040	50	77	1.540	3.141 600*	1.000 0 E-05
28	85	3.036	86	89	1.035	3.141 611	2.129 6 E-05
29	88	3.034	85	88	1.035	3.141 582†	7.849 9 E-06
33	68	2.061	61	93	1.525	3.141 580	1.026 8 E-05
41	75	1.829	46	79	1.717	3.141 569	2.054 1 E-05
43	85	1.977	56	89	1.589	3.141 611	2.129 6 E-05
43	77	1.791	57	100	1.754	3.141 575	1.513 3 E-05

† This is the closest possible approximation to a 5-place value for π in a nonreverted gear train within the given limitations on gear sizes.

TABLE 9-11   Reverted Gearsets and Errors in Ratio for Example 9-4

N_2	N_3	Ratio1	N_4	N_5	Ratio2	m_V	Error
22	39	1.773	22	39	1.773	3.142 562	-9.619 8 E-04
44	78	1.773	44	78	1.773	3.142 562	-9.619 8 E-04

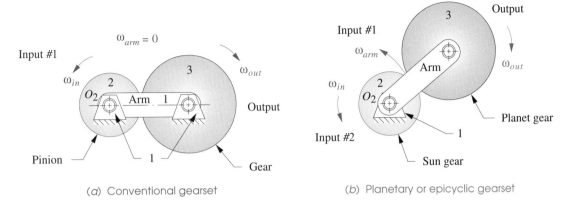

$\omega_{arm} = 0$

Input #1

ω_{in}

O_2

2

Arm 1

3

ω_{out}

Output

Pinion

1

Gear

(a) Conventional gearset

Input #1

ω_{arm}

ω_{in}

O_2

2

Arm

3

Output

ω_{out}

Planet gear

Input #2

1

Sun gear

(b) Planetary or epicyclic gearset

FIGURE 9-33

Conventional gearsets are special cases of planetary or epicyclic gearsets

numbers but have the same error) and the error is much greater than that of even the worst of the 11 nonreverted solutions in Table 9-10.

9.9 EPICYCLIC OR PLANETARY GEAR TRAINS

The conventional gear trains described in the previous sections are all one-degree-of-freedom (*DOF*) devices. Another class of gear train has wide application, the **epicyclic or planetary train**. This is a two-*DOF* device. Two inputs are needed to obtain a predictable output. In some cases, such as the automotive differential, one input is provided (the driveshaft) and two frictionally coupled outputs are obtained (the two driving wheels). In other applications such as automatic transmissions, aircraft engine to propeller reductions, and in-hub bicycle transmissions, two inputs are provided (one usually being a zero velocity, i.e., a fixed gear), and one controlled output results.

Figure 9-33a shows a conventional, one-*DOF* gearset in which link 1 is immobilized as the ground link. Figure 9-33b shows the same gearset with link 1 now free to rotate as an **arm** that connects the two gears. Now only the joint O_2 is grounded and the system *DOF* = 2. This has become an **epicyclic** train with a **sun gear** and a **planet gear** orbiting around the sun, held in orbit by the **arm**. Two inputs are required. Typically, the arm and the sun gear will each be driven in some direction at some velocity. In many cases, one of these inputs will be zero velocity, i.e., a brake applied to either the arm or the sun gear. Note that a zero velocity input to the arm merely makes a conventional train out of the epicyclic train as shown in Figure 9-33a. Thus the conventional gear train is simply a special case of the more complex epicyclic train, in which its arm is held stationary.

In this simple example of an epicyclic train, the only gear left to take an output from, after putting inputs to sun and arm, is the planet. It is a bit difficult to get a usable output from this orbiting gear as its pivot is moving. A more useful configuration is shown in Figure 9-34 to which a ring gear has been added. This **ring gear** meshes with the planet

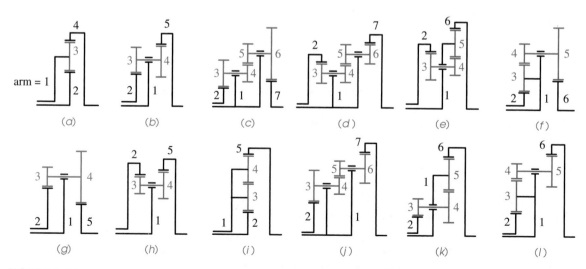

FIGURE 9-34

Planetary gearset with ring gear used as output

and pivots at O_2, so it can be easily tapped as the output member. Most planetary trains will be arranged with ring gears to bring the planetary motion back to a grounded pivot. Note how the sun gear, ring gear, and arm are all brought out as concentric hollow shafts

FIGURE 9-35

Levai's 12 possible epicyclic trains (3)

so that each can be accessed to tap its angular velocity and torque as either an input or an output.

Epicyclic trains come in many varieties. Levai[3] cataloged 12 possible types of basic epicyclic trains as shown in Figure 9-35. These basic trains can be connected together to create a larger number of trains having more degrees of freedom. This is done in automotive automatic transmissions as described in a later section.

While it is relatively easy to visualize the power flow through a conventional gear train and observe the directions of motion for its member gears, it is very difficult to determine the behavior of a planetary train by observation. We must do the necessary calculations to determine its behavior and may be surprised at the often counterintuitive results. Since the gears are rotating with respect to the arm and the arm itself has motion, we have a velocity difference problem here that requires equation 6.5b (p. 287) be applied to this problem. Rewriting the velocity difference equation in terms of angular velocities specific to this system, we get:

$$\omega_{gear} = \omega_{arm} + \omega_{gear/arm} \qquad (9.12)$$

Equations 9.12 and 9.5a (p. 491) are all that is needed to solve for the velocities in an epicyclic train, provided that the tooth numbers and two input conditions are known.

The Tabular Method

One approach to the analysis of velocities in an epicyclic train is to create a table which represents equation 9.12 for each gear in the train.

EXAMPLE 9-5

Epicyclic Gear Train Analysis by the Tabular Method.

Problem: Consider the train in Figure 9-34, with the tooth numbers and initial conditions:

Sun gear	N_2 = 40-tooth external gear
Planet gear	N_3 = 20-tooth external gear
Ring gear	N_4 = 80-tooth internal gear
Input to arm	200 rpm clockwise
Input to sun	100 rpm clockwise

We wish to find the absolute output angular velocity of the ring gear.

	1	2	3	
Gear #	ω_{gear} =	ω_{arm} +	$\omega_{gear/arm}$	*Gear ratio*

FIGURE 9-36

Table for the solution of planetary gear trains

Gear #	1 $\omega_{gear} =$	2 W $_{arm}$ +	3 $\omega_{gear/arm}$	Gear Ratio
2	−100	−200		
				−40/20
3		−200		
				+20/80
4		−200		

FIGURE 9-37

Given data for planetary gear train from Example 9-5 placed in solution table

Solution:

1   The solution table is set up with a column for each term in equation 9.12 (p. 515) and a row for each gear in the train. It will be most convenient if we can arrange the table so that meshing gears occupy adjacent rows. The table for this method, prior to data entry, is shown in Figure 9-36 (p. 515).

2   Note that the gear ratios are shown straddling the rows of gears to which they apply. The gear ratio column is placed next to the column containing the velocity differences $\omega_{gear/arm}$ because the gear ratios only apply to the velocity difference. The gear ratios **cannot be directly applied to the absolute velocities** in the ω_{gear} column.

3   The solution strategy is simple but is fraught with opportunities for careless errors. Note that we are solving a vector equation with scalar algebra and the signs of the terms denote the sense of the ω vectors which are all directed along the Z axis. Great care must be taken to get the signs of the input velocities and of the gear ratios correct in the table, or the answer will be wrong. Some gear ratios may be negative if they involve external gearsets, and others will be positive if they involve an internal gear. We have both types in this example.

4   The first step is to enter the known data as shown in Figure 9-37 which in this case are the arm velocity (in all rows) and the absolute velocity of gear 2 in column 1. The gear ratios can also be calculated and placed in their respective locations. Note that these ratios should be calculated for each gearset in a consistent manner, following the power flow through the train. That is, starting at gear 2 as the driver, it drives gear 3 directly. This makes its ratio $-N_2/N_3$, or input over output, not the reciprocal. *This ratio is negative because the gearset is external.*

Gear #	1 $\omega_{gear} =$	2 ω_{arm} +	3 $\omega_{gear/arm}$	Gear Ratio
2	−100	−200	+100	
				−40/20
3	−400	−200	−200	
				+20/80
4	−250	−200	−50	

FIGURE 9-38

Solution for planetary gear train from Example 9-5

Gear 3 in turn drives gear 4 so its ratio is +N_3/N_4. *This is a positive ratio because of the internal gear.*

5 Once any one row has two entries, the value for its remaining column can be calculated from equation 9.12 (p. 515). Once any one value in the velocity difference column (column 3) is found, the gear ratios can be applied to calculate all other values in that column. Finally, the remaining rows can be calculated from equation 9.12 to yield the absolute velocities of all gears in column 1. These computations are shown in Figure 9-38 which completes the solution.

6 The overall train value for this example can be calculated from the table and is from arm to ring gear +1.25:1 and from sun gear to ring gear +2.5:1.

In this example, the arm velocity was given. If it is to be found as the output, then it must be entered in the table as an unknown, x, and the equations solved for that unknown.

FERGUSON'S PARADOX   Epicyclic trains have several advantages over conventional trains including higher train ratios in smaller packages, reversion by default, and simultaneous, concentric, bidirectional outputs available from a single unidirectional input. These features make planetary trains popular as automatic transmissions in automobiles and trucks, etc.

The so-called **Ferguson paradox** of Figure 9-39 illustrates all these features of the planetary train. It is a **compound epicyclic train** with one 20-tooth planet gear (gear 5) carried on the arm and meshing simultaneously with three sun gears. These sun gears have 100 teeth (gear 2), 99 teeth (gear 3), and 101 teeth (gear 4), respectively. The center distances between all sun gears and the planet are the same despite the slightly different pitch diameters of each sun gear. This is possible because of the properties of the involute tooth form as described in Section 9.2 (p. 484). Each sun gear will run smoothly with the planet gear. Each gearset will merely have a slightly different pressure angle.

FIGURE 9-39

Ferguson's paradox compound planetary gear train

		1	2	3	
Gear #		$\omega_{gear} =$	$\omega_{arm} +$	$\omega_{gear/arm}$	Gear Ratio
2		0	+100		
					−100/20
5			+100		
					−20/99
3			+100		
5			+100		
					−20/101
4			+100		

FIGURE 9-40

Given data for Ferguson's paradox planetary gear train from Example 9-6

EXAMPLE 9-6

Analyzing Ferguson's Paradox by the Tabular Method.

Problem: Consider Ferguson's paradox train in Figure 9-39 (p. 517) which has the following tooth numbers and initial conditions:

Sun gear # 2	$N_2 = 100$-tooth external gear
Sun gear # 3	$N_3 = 99$-tooth external gear
Sun gear # 4	$N_4 = 101$-tooth external gear
Planet gear	$N_5 = 20$-tooth external gear
Input to sun # 2	0 rpm
Input to arm	100 rpm counterclockwise

Sun gear 2 is fixed to the frame, thus providing one input (zero velocity) to the system. The arm is driven at 100 rpm counterclockwise as the second input. Find the angular velocities of the two outputs that are available from this compound train, one from gear 3 and one from gear 4, both of which are free to rotate on the main shaft.

Solution:

1   The tabular solution for this train is set up in Figure 9-40 which shows the given data. Note that the row for gear 5 is repeated for clarity in applying the gear ratio between gears 5 and 4.

2   The known input values of velocity are the arm angular velocity and the zero absolute velocity of gear 2.

3   The gear ratios in this case are all negative because of the external gear sets, and their values reflect the direction of power flow from gear 2 to 5, then 5 to 3, and 5 to 4 in the second branch.

4   Figure 9-41 shows the calculated values added to the table. Note that for a **counterclockwise** 100 rpm input to the arm, we get a **counterclockwise** 1 rpm output from gear 4 and a **clockwise** 1 rpm output from gear 3, simultaneously.

Gear #	1 $\omega_{gear} =$	2 $\omega_{arm} +$	3 $\omega_{gear/arm}$	Gear Ratio
2	0	+100	−100	
				−100/20
5	+600	+100	+500	
				−20/99
3	−1.01	+100	−101.01	
5	+600	+100	+500	
				−20/101
4	+0.99	+100	−99.01	

FIGURE 9-41

Solution to Ferguson's paradox planetary gear train from Example 9-6

This result accounts for the use of the word **paradox** to describe this train. Not only do we get a much larger ratio (100:1) than we could from a conventional train with gears of 100 and 20 teeth, but we have our choice of output directions!

Automotive automatic transmissions use compound planetary trains, which are always in mesh, and which give different ratio forward speeds, plus reverse, by simply engaging and disengaging brakes on different members of the train. The brake provides zero velocity input to one train member. The other input is from the engine. The output is thus modified by the application of these internal brakes in the transmission according to the selection of the operator (**P**ark, **R**everse, **N**eutral, **D**rive, etc.).

The Formula Method

It is not necessary to tabulate the solution to an epicyclic train. The velocity difference formula can be solved directly for the train ratio. We can rearrange equation 9.12 (p. 515) to solve for the velocity difference term. Then, let ω_F represent the angular velocity of the first gear in the train (chosen at either end), and ω_L represent the angular velocity of the last gear in the train (at the other end).

For the first gear in the system:

$$\omega_{F/arm} = \omega_F - \omega_{arm} \qquad (9.13a)$$

For the last gear in the system:

$$\omega_{L/arm} = \omega_L - \omega_{arm} \qquad (9.13b)$$

Dividing the last by the first:

$$\frac{\omega_{L/arm}}{\omega_{F/arm}} = \frac{\omega_L - \omega_{arm}}{\omega_F - \omega_{arm}} = R \qquad (9.13c)$$

This gives an expression for the fundamental train value R which defines a velocity ratio for the train with the arm held stationary. The leftmost side of equation 9.13c involves only the velocity difference terms that are relative to the arm. This fraction is equal to the ratio of the products of tooth numbers of the gears from first to last in the train as defined in equation 9.8b (p. 505) which can be substituted for the leftmost side of equation 9.13c.

$$R = \pm \frac{\text{product of number of teeth on driver gears}}{\text{product of number of teeth on driven gears}} = \frac{\omega_L - \omega_{arm}}{\omega_F - \omega_{arm}} \qquad (9.14)$$

This equation can be solved for any one of the variables on the right side provided that the other two are defined as the two inputs to this two-*DOF* train. Either the velocities of the arm plus one gear must be known or the velocities of two gears, the first and last, as so designated, must be known. Another limitation of this method is that both the first and last gears chosen must be pivoted to ground (not orbiting), and there must be a path of meshes connecting them, which may include orbiting planet gears. Let us use this method to again solve the Ferguson paradox of Example 9-6.

✎ EXAMPLE 9-7

Analyzing Ferguson's Paradox by the Formula Method.

Problem: Consider the same Ferguson paradox train as in Example 9-6 which has the following tooth numbers and initial conditions (see Figure 9-37, p. 516):

Sun gear #2	N_2 = 100-tooth external gear
Sun gear #3	N_3 = 99-tooth external gear
Sun gear #4	N_4 = 101-tooth external gear
Planet gear	N_5 = 20-tooth external gear
Input to sun #2	0 rpm
Input to arm	100 rpm counterclockwise

Sun gear 2 is fixed to the frame, providing one input (zero velocity) to the system. The arm is driven at 100 rpm CCW as the second input. Find the angular velocities of the two outputs that are available from this compound train, one from gear 3 and one from gear 4, both of which are free to rotate on the main shaft.

Solution:

1 We will have to apply equation 9.14 twice, once for each output gear. Taking gear 3 as the last gear in the train with gear 2 as the first, we have:

$$N_2 = 100 \qquad N_3 = 99 \qquad N_5 = 20s$$
$$\omega_{arm} = +100 \qquad \omega_F = 0 \qquad \omega_L = ? \qquad (a)$$

2 Substituting in equation 9.14 we get:

$$\left(-\frac{N_2}{N_5}\right)\left(-\frac{N_5}{N_3}\right) = \frac{\omega_L - \omega_{arm}}{\omega_F - \omega_{arm}}$$
$$\left(-\frac{100}{20}\right)\left(-\frac{20}{99}\right) = \frac{\omega_3 - 100}{0 - 100} \qquad (b)$$
$$\omega_3 = -1.01$$

3 Now taking gear 4 as the last gear in the train with gear 2 as the first, we have:

$$N_2 = 100 \qquad N_4 = 101 \qquad N_5 = 20$$

$$\omega_{arm} = +100 \qquad \omega_F = 0 \qquad \omega_L = ? \tag{c}$$

4 Substituting in equation 9.14, we get:

$$\left(-\frac{N_2}{N_5}\right)\left(-\frac{N_5}{N_4}\right) = \frac{\omega_L - \omega_{arm}}{\omega_F - \omega_{arm}}$$

$$\left(-\frac{100}{20}\right)\left(-\frac{20}{101}\right) = \frac{\omega_4 - 100}{0 - 100} \tag{d}$$

$$\omega_4 = +0.99$$

These are the same results as were obtained with the tabular method.

9.10   EFFICIENCY OF GEAR TRAINS

The general definition of efficiency is *output power/input power.* It is expressed as a fraction (decimal %) or as a percentage. The efficiency of a conventional gear train (simple or compound) is very high. The power loss per gearset is only about 1 to 2% depending on such factors as tooth finish and lubrication. A gearset's basic efficiency is termed E_0. An external gearset will have an E_0 of about 0.98 or better and an external-internal gearset about 0.99 or better. When multiple gearsets are used in a conventional simple or compound train, the overall efficiency of the train will be the product of the efficiencies of all its stages. For example, a two-stage train with both gearset efficiencies of $E_0 = 0.98$ will have an overall efficiency of $\eta = 0.98^2 = 0.96$.

Epicyclic trains, if properly designed, can have even higher overall efficiencies than conventional trains. But, if the epicyclic train is poorly designed, its efficiency can be so low that it will generate excessive heat and may even be unable to operate at all. This strange result can come about if the orbiting elements (planets) in the train have high losses that absorb a large amount of "circulating power" within the train. It is possible for this circulating power to be much larger than the throughput power for which the train was designed, resulting in excessive heating or stalling. The computation of the overall efficiency of an epicyclic train is much more complicated than the simple multiplication indicated above that works for conventional trains. Molian[4] presents a concise derivation.

To calculate the overall efficiency η of an epicyclic train we need to define a basic ratio ρ which is related to the fundamental train value R defined in equation 9.13c (p. 519):

$$\text{if } |R| \geq 1, \text{ then } \rho = R \text{ else } \rho = 1/R \tag{9.15}$$

This constrains ρ to represent a speed increase rather than a decrease regardless of which way the gear train is intended to operate.

For the purpose of calculating torque and power in an epicyclic gear train, we can consider it to be a "black box" with three concentric shafts as shown in Figure 9-42. These shafts are labeled 1, 2, and arm and connect to either "end" of the gear train and to its arm, respectively. Two of these shafts can serve as inputs and the third as output in any combination. The details of the gear train's internal configuration are not needed if we

FIGURE 9-42

Generic epicyclic gear train

know its basic ratio ρ and the basic efficiency E_0 of its gearsets. All the analysis is done relative to the arm of the train since the internal power flow and losses are only affected by rotation of shafts 1 and 2 with respect to the arm, not by rotation of the entire unit. We also model it as having a single planet gear for the purpose of determining E_0 on the assumption that the power and the losses are equally divided among all gears actually in the train. Counterclockwise torques and angular velocities are considered positive. Power is the product of torque and angular velocity, so a positive power is an input (torque and velocity in same direction) and negative power is an output.

If the gear train is running at constant speed or is changing speed too slowly to significantly affect its internal kinetic energy, then we can assume static equilibrium and the torques will sum to zero.

$$T_1 + T_2 + T_{arm} = 0 \qquad (9.16)$$

The sum of power in and out must also be zero, but the direction of power flow affects the computation. If the power flows from shaft 1 to shaft 2, then:

$$E_0 T_1 (\omega_1 - \omega_{arm}) + T_2 (\omega_2 - \omega_{arm}) = 0 \qquad (9.17a)$$

If the power flows from shaft 2 to shaft 1, then:

$$T_1 (\omega_1 - \omega_{arm}) + E_0 T_2 (\omega_2 - \omega_{arm}) = 0 \qquad (9.17b)$$

If the power flows from shaft 1 to 2, equations 9.16 and 9.17a are solved simultaneously to obtain the system torques. If the power flows in the other direction, then equations 9.16 and 9.17b are used instead. Substitution of equation 9.13c (p. 519) in combination with equation 9.15 (p. 521) introduces the basic ratio ρ and after simultaneous solution yields:

power flow from 1 to 2 $\qquad\qquad T_1 = \dfrac{T_{arm}}{\rho E_0 - 1} \qquad (9.18a)$

$$T_2 = -\dfrac{\rho E_0 T_{arm}}{\rho E_0 - 1} \qquad (9.18b)$$

power flow from 2 to 1 $\qquad\qquad T_1 = \dfrac{E_0 T_{arm}}{\rho - E_0} \qquad (9.19a)$

$$T_2 = -\dfrac{\rho T_{arm}}{\rho - E_0} \qquad (9.19b)$$

Once the torques are found, the input and output power can be calculated using the known input and output velocities (from a kinematic analysis as described above) and the efficiency then determined from *output power/input power*.

There are eight possible cases depending on which shaft is fixed, which shaft is input, and whether the basic ratio ρ is positive or negative. These cases are shown in Table 9-12[4] which includes expressions for the train efficiency as well as for the torques. Note that the torque on one shaft is always known from the load required to be driven or the power available from the driver, and this is needed to calculate the other two torques.

TABLE 9-12   Torques and Efficiencies in an Epicyclic Train [4]

Case	ρ	Fixed Shaft	Input Shaft	Train Ratio	T_1	T_2	T_{arm}	Efficiency (η)
1	$> +1$	2	1	$1-\rho$	$-\dfrac{T_{arm}}{1-\rho E_0}$	$\dfrac{\rho E_0 T_{arm}}{1-\rho E_0}$	T_{arm}	$\dfrac{\rho E_0 - 1}{\rho - 1}$
2	$> +1$	2	arm	$\dfrac{1}{1-\rho}$	T_1	$-\rho \dfrac{T_1}{E_0}$	$\left(\dfrac{\rho - E_0}{E_0}\right) T_1$	$\dfrac{E_0(\rho - 1)}{\rho - E_0}$
3	$> +1$	1	2	$\dfrac{\rho - 1}{\rho}$	$\dfrac{T_{arm}}{\rho E_0 - 1}$	$-\dfrac{\rho E_0 T_{arm}}{\rho E_0 - 1}$	T_{arm}	$\dfrac{\rho E_0 - 1}{E_0(\rho - 1)}$
4	$> +1$	1	arm	$\dfrac{\rho}{\rho - 1}$	$-\dfrac{E_0}{\rho} T_2$	T_2	$-\left(\dfrac{\rho - E_0}{\rho}\right) T_2$	$\dfrac{\rho - 1}{\rho - E_0}$
5	≤ -1	2	1	$1-\rho$	$-\dfrac{T_{arm}}{1-\rho E_0}$	$\dfrac{\rho E_0 T_{arm}}{1-\rho E_0}$	T_{arm}	$\dfrac{\rho E_0 - 1}{\rho - 1}$
6	≤ -1	2	arm	$\dfrac{1}{1-\rho}$	T_1	$-\rho \dfrac{T_1}{E_0}$	$\left(\dfrac{\rho - E_0}{E_0}\right) T_1$	$\dfrac{E_0(\rho - 1)}{\rho - E_0}$
7	≤ -1	1	2	$\dfrac{\rho - 1}{\rho}$	$\dfrac{E_0 T_{arm}}{\rho - E_0}$	$-\dfrac{\rho T_{arm}}{\rho - E_0}$	T_{arm}	$\dfrac{\rho - E_0}{\rho - 1}$
8	≤ -1	1	arm	$\dfrac{\rho}{\rho - 1}$	$-\dfrac{T_2}{\rho E_0}$	T_2	$-\left(\dfrac{\rho E_0 - 1}{\rho E_0}\right) T_2$	$\dfrac{E_0(\rho - 1)}{\rho E_0 - 1}$

9

✎ **EXAMPLE 9-8**

Determining the Efficiency of an Epicyclic Gear Train.*

Problem: Find the overall efficiency of the epicyclic train shown in Figure 9-43.[5] The basic efficiency E_0 is 0.9928 and the gear tooth numbers are: $N_A = 82t$, $N_B = 84t$,

FIGURE 9-43

Epicyclic train for Example 9-8[5]

* This example is adapted from reference [5].

$N_C = 86t$, $N_D = 82t$, $N_E = 82t$, and $N_F = 84t$. Gear A *(shaft 2)* is fixed to the frame, providing a zero velocity input. The arm is driven as the second input.

Solution:

1 Find the basic ratio ρ for the gear train using equations 9.14 (p. 520) and 9.15 (p. 521). Note that gears B and C have the same velocity as do gears D and E, so their ratios are 1 and thus are omitted.

$$\rho = \frac{N_F N_D N_B}{N_E N_C N_A} = \frac{84(82)(84)}{82(86)(82)} = \frac{1764}{1763} \cong 1.000567 \qquad (a)$$

2 The combination of $\rho > 1$, shaft 2 fixed and input to the arm corresponds to Case 2 in Table 9-12, (p. 523) giving an efficiency of:

$$\eta = \frac{E_0(\rho - 1)}{\rho - E_0} = \frac{0.9928(1.000567 - 1)}{1.000567 - 0.9928} = 0.073 = 7.3\% \qquad (b)$$

3 This is a very low efficiency which makes this gearbox essentially useless. About 93% of the input power is being circulated within the gear train and wasted as heat.

The above example points out a problem with epicyclic gear trains that have basic ratios near unity. They have low efficiency and are useless for transmission of power. Large speed ratios with high efficiency can only be obtained with trains having large basic ratios.[5]

FIGURE 9-44

Four-speed manual synchromesh automobile transmission *From Crouse, W. H. (1980). Automotive Mechanics, 8th ed., McGraw-Hill, New York, NY, p. 480. Reprinted with permission.*

9.11   TRANSMISSIONS

COMPOUND REVERTED GEAR TRAINS   are commonly used in manual (nonautomatic) automotive transmissions to provide user-selectable ratios between the engine and the drive wheels for torque multiplication (mechanical advantage). These gearboxes usually have from three to six forward speeds and one reverse. Most modern transmissions of this type use helical gears for quiet operation. These gears are **not** moved into and out of engagement when shifting from one speed to another except for reverse. Rather, the desired ratio gears are selectively locked to the output shaft by synchromesh mechanisms as in Figure 9-44 which shows a four-speed, manually shifted, synchromesh automotive transmission.

The input shaft is at top left. The input gear is always in mesh with the leftmost gear on the countershaft at the bottom. This countershaft has several gears integral with it, each of which meshes with a different output gear that is freewheeling on the output shaft. The output shaft is concentric with the input shaft, making this a reverted train, but the input and output shafts only connect through the gears on the countershaft except in "top gear" (fourth speed), for which the input and output shafts are directly coupled together with a synchromesh clutch for a 1:1 ratio.

The synchromesh clutches are beside each gear on the output shaft and are partially hidden by the shifting collars that move them left and right in response to the driver's hand on the shift lever. These clutches act to lock one gear to the output shaft at a time to provide a power path from input to output of a particular ratio. The arrows on the figure show the power path for third-speed forward, which is engaged. Reverse gear, on

FIGURE 9-45

Four-speed automatic automobile transmission   *Courtesy of Mercedes Benz of North America Inc.*

the lower right, engages an idler gear which is physically shifted into and out of mesh at standstill.

PLANETARY OR EPICYCLIC TRAINS   are commonly used in automatic shifting automotive transmissions as shown in Figure 9-45 (p. 525). The input shaft, which couples to the engine's crankshaft, is one input to the multi-*DOF* transmission that consists of several stages of epicyclic trains. Automatic transmissions can have any number of ratios. Automotive examples typically have had from two to six forward speeds. Truck and bus automatic transmissions may have more.

Three epicyclic gearsets can be seen near the center of the four-speed transmission in Figure 9-45. They are controlled by hydraulically operated multidisk clutches and brakes within the transmission that impart zero velocity (second) inputs to various elements of the train to create one of four forward velocity ratios plus reverse in this particular example. The clutches force zero relative velocity between the two elements engaged, and the brakes force zero absolute velocity on the element. Since all gears are in constant mesh, the transmission can be shifted under load by switching the internal brakes and clutches on and off. They are controlled by a combination of inputs that include driver selection (PRND), road speed, throttle position, engine load and speed, and other factors that are automatically monitored and computer controlled. Some modern transmission controllers use artificial intelligence techniques to learn and adapt to the operator's style of driving by automatically resetting the shift points for gentle or aggressive performance based on driving habits.

FIGURE 9-46

Torque converter   *Courtesy of Mannesmann Sachs AG, Schweinfurt, Germany*

At the left side of Figure 9-45 (p. 525) is a turbine-like fluid coupling between engine and transmission, called a **torque converter,** a cutaway of which is shown in Figure 9-46. This device allows sufficient slip in the coupling fluid to let the engine idle with the transmission engaged and the vehicle's wheels stopped. The engine-driven *impeller blades*, running in oil, transmit torque by pumping oil past a set of stationary *stator blades* and against the *turbine blades* attached to the transmission input shaft. The stator blades, which do not move, serve to redirect the flow of oil exiting the impeller blades to a more favorable angle relative to the turbine blades. This redirection of flow is responsible for the torque multiplication that gives the device its name, torque converter. Without the stator blades, it is just a *fluid coupling* that will transmit, but not multiply, the torque. In a torque converter, the maximum torque increase of about 2x occurs at stall when the transmission's turbine is stopped and the engine-driven impeller is turning, creating maximum slip between the two. This torque boost aids in accelerating the vehicle from rest when its inertia must be overcome. The torque multiplication decreases to one at zero slip between impeller and turbine. However, the device cannot reach a zero slip condition on its own. It will always operate with a few percent of slip. This wastes energy in steady-state operation, as when the vehicle is traveling at constant speed on level ground. To conserve this energy, most torque converters are now equipped with an electromechanical lockup clutch that engages above about 30 mph in top gear and locks the stator to the impeller, making the transmission efficiency then 100%. When speed drops below a set speed, or when the transmission downshifts, the clutch is disengaged, allowing the torque converter to again perform its function.

Figure 9-47a shows a schematic of the same transmission as in Figure 9-45. Its three epicyclic stages, two clutches (C_1, C_2), and three band brakes (B_1, B_2, B_3) are depicted. Figure 9-47b shows an activation table of the brake-clutch combinations for each speed ratio of this transmission.[6]

An historically interesting example of an epicyclic train used in a manually shifted gearbox is the Ford Model T transmission shown and described in Figure 9-48 (p. 528). Over 9 million were produced from 1909 to 1927, before the invention of the synchro-mesh mechanism shown in Figure 9-44 (p. 524). Conventional (compound-reverted) transmissions as used in most other automobiles of that era (and into the 1930s) were unaffectionately known as "crashboxes," the name being descriptive of the noise made

* Frederick Lanchester, a major automotive pioneer, actually invented the compound epicyclic manual transmission and patented it in England in 1898, well before Ford (from 1909 to 1927) made the Model T (and money) by the millions. Lanchester died poor.

9

Range	Clutch/Brake Activation				
	C_1	C_2	B_1	B_2	B_3
First	x			x	
Second	x				x
Third	x				x
Fourth	x	x			
Reverse		x	x		

(a) Schematic of 4-speed automatic transmission   (b) Clutch / brake activation table

FIGURE 9-47

Schematic of automatic transmission from Figure 9-45 *Adapted from reference (6)*

The input from the engine is to arm 2. Gear 6 is rigidly attached to the output shaft which drives the wheels. Gears 3, 4, and 5 rotate at the same speed.

There are two forward speeds. Low (1:2.75) is selected by engaging band brake B_2 to lock gear 7 to the frame. Clutch C is disengaged.

High (1:1) is selected by engaging clutch C which locks the input shaft directly to the output shaft.

Reverse (1:–4) is obtained by engaging brake band B_1 to lock gear 8 to the frame. Clutch C is disengaged.

$N_3 = 27$, $N_6 = 27$
$N_4 = 33$, $N_7 = 21$
$N_5 = 24$, $N_8 = 30$

FIGURE 9-48

Ford Model T epicyclic transmission *From Phelan, R. M. (1970). Fundamentals of Mechanical Design, 3rd ed., McGraw-Hill. NY.*

when shifting unsynchronized gears into and out of mesh while in motion. Henry Ford had a better idea. His Model T gears were in constant mesh. The two forward speeds and one reverse were achieved by engaging/disengaging a clutch and band brakes in various combinations via foot pedals. These provided second inputs to the epicyclic train which, like Ferguson's paradox, gave bidirectional outputs, all without any "crashing" of gear teeth. This Model T transmission is the precursor to all modern automatic transmissions which replace the T's foot pedals with automated hydraulic operation of the clutches and brakes.[*]

* Non-full-time 4WD is common in trucks and differs from AWD in that it lacks the center differential, making it usable only when the road is slippery. Any required differences in rotational velocity between rear and front driven wheels is then accommodated by tire slip. On dry pavement, a non-full-time 4WD vehicle will not handle well and can be dangerous. Unlike vehicles with AWD, which is always engaged, non-full-time 4WD vehicles normally operate in 2WD and require driver action to obtain 4WD. Manufacturers caution against shifting these vehicles into 4WD unless traction is poor.

FIGURE 9-49

Continuously Variable Transmission (CVT) *Courtesy of ZF Getriebe GmbH, Saabrucken, Germany*

CONTINUOUSLY VARIABLE TRANSMISSION (CVT)   A transmission that has no gears, the CVT uses two sheaves or pulleys that adjust their axial widths simultaneously in opposite directions to change the ratio of the belt drive that runs in the sheaves. This concept has been around for over a century and in fact was used on some very early automobiles as the final drive and transmission combined. It is finding renewed application in the 21st century in the quest for higher-efficiency vehicle drives. Figure 9-49 shows a commercial automobile CVT that uses a steel, segmented "belt" of vee cross section that runs on adjustable width sheaves. To change the transmission ratio, one sheave's width is opened and the other closed in concert such that the effective pitch radii deliver the desired ratio. It thus has an infinity of possible ratios, varying continuously between two limits. The ratio is adjustable while running under load. The CVT shown is designed and computer controlled to keep the vehicle's engine running at essentially constant speed at an rpm that delivers the best fuel economy, regardless of vehicle speed. Similar designs of CVTs that use conventional rubber vee belts have long been used in low-power machinery such as snow blowers and lawn tractors.

9.12   DIFFERENTIALS

A differential is a device that allows a difference in velocity (and displacement) between two elements. This requires a 2-*DOF* mechanism such as an epicyclic gear train. Perhaps the most common application of differentials is in the final drive mechanisms of wheeled land vehicles as shown in Figure P9-3 (p. 535). When a four-wheeled vehicle turns, the wheels on the outside of the turn must travel farther than the inside wheels due to their different turning radii as shown in Figure 9-50. Without a differential mechanism between the inner and outer driving wheels, the tires must slip on the road surface for the vehicle to turn. If the tires have good traction, a nondifferentiated drive train will attempt to go in a straight line at all times and will fight the driver in turns. In a "full-time" four-wheel -drive[*] (4WD) vehicle (sometimes called "all wheel drive" or AWD) an additional differential is needed between the front and rear wheels to allow the wheel velocities at each end

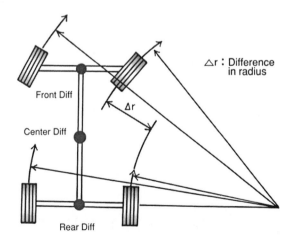

FIGURE 9-50

Turning behavior of a four-wheel vehicle *Courtesy of Tochigi Fuji Sangyo, Japan*

FRONT

Center differential ●

Front differential ●

Rear differential ●

REAR

FIGURE 9-51

An all-wheel-drive (AWD) chassis and drive train *Courtesy of Tochigi Fuji Sangyo, Japan*

(a)

(b)

(c)

FIGURE 9-52

Differentials
*Courtesy of Tochigi Fuji
Sangyo, Japan*

9

of the vehicle to vary in proportion to the traction developed at either end of the vehicle under slippery conditions. Figure 9-51 shows an AWD automotive chassis with its three differentials. In this example, the center differential is packaged with the transmission and front differential but effectively is in the driveshaft between the front and rear wheels as shown in Figure 9-50 (p. 529). Differentials are made with various gear types. For rear axle applications, a bevel gear epicyclic is commonly used as shown in Figure 9-52a and in Figure P9-3. For center and front differentials, helical or spur gear arrangements are often used as in Figure 9-52b and c.

An epicyclic train used as a differential has one input and two outputs. Taking the rear differential in an automobile as an example, its input is from the driveshaft and its outputs are to the right and left wheels. The two outputs are coupled through the road via the traction (friction) forces between tires and pavement. The relative velocity between each wheel can vary from zero when both tires have equal traction and the car is not turning, to twice the epicyclic train's input speed when one wheel is on ice and the other has traction. Front or rear differentials split the torque equally between their two wheel outputs. Since power is the product of torque and angular velocity, and power out cannot exceed power in, the power is split between the wheels according to their velocities. When traveling straight ahead (both wheels having traction), half the power goes to each wheel. As the car turns, the faster wheel gets more power and the slower one less. When one wheel loses traction (as on ice), it gets *all* the power (50% torque x 200% speed), and the wheel with traction gets zero power (50% torque x 0% speed). This is why 4WD or AWD is needed in slippery conditions. In AWD, the center differential splits the torque between front and rear in some proportion. If one end of the car loses traction, the other may still be able to control it provided it still has traction.

(a) TORSEN® Type 1 differential   (b) TORSEN® Type 2 differential

FIGURE 9-53

TORSEN® differentials *Courtesy of Zexel Torsen Inc., Rochester, NY*

LIMITED SLIP DIFFERENTIALS   Because of their behavior when one wheel loses traction, various differential designs have been created to limit the slip between the two outputs under those conditions. These are called limited slip differentials and typically provide some type of friction device between the two output gears to transmit some torque but still allow slip for turning. Some use a fluid coupling between the gears, and others use spring-loaded friction disks or cones as can be seen in Figure 9-52a. Some use an electrically controlled clutch within the epicyclic train to lock it up on demand for off-road applications as shown in Figure 9-52b. The TORSEN® (from TORque SENsing) differential of Figure 9-53, invented by V. Gleasman, uses wormsets whose resistance to backdriving provides torque coupling between the outputs. The lead angle of the worm determines the percent of torque transmitted across the differential. These differentials are used in many AWD vehicles including the Army's High Mobility Multipurpose Wheeled Vehicle (HMMWV) known as the "Humvee" or "Hummer."

9.13   REFERENCES

1   **DilPare, A. L.** (1970). "A Computer Algorithm to Design Compound Gear Trains for Arbitrary Ratio." *J. of Eng. for Industry,* **93B**(1), pp. 196-200.

2   **Selfridge, R. G., and D. L. Riddle**. (1978). "Design Algorithms for Compound Gear Train Ratios." ASME Paper: 78-DET-62.

3   **Levai, Z.** (1968). "Structure and Analysis of Planetary Gear Trains." *Journal of Mechanisms,* **3**, pp. 131-148.

4   **Molian, S.** (1982). *Mechanism Design: An Introductory Text.* Cambridge University Press: Cambridge, p. 148.

5   **Auksmann, B., and D. A. Morelli**. (1963). "Simple Planetary-Gear System." ASME Paper: 63-WA-204.

6   **Pennestri, E., et al.** (1993). "A Catalog of Automotive Transmissions with Kinematic and Power Flow Analyses." *Proc. of 3rd Applied Mechanisms and Robotics Conference,* Cincinnati, p. 57-1.

---

* Answers in Appendix F.

---

† These problems are suited
to solution using *Mathcad,
Matlab,* or *TKSolver* equation solver programs.

9.14 BIBLIOGRAPHY

Useful websites for information on gear, belt, or chain drives

http://www.howstuffworks.com/gears.htm

http://www.efunda.com/DesignStandards/gears/gears_introduction.cfm

http://www.oit.doe.gov/bestpractices/pdfs/motor3.pdf

http://www.gates.com/index.cfm

http://www.bostongear.com/

http://www.martinsprocket.com/

9.15 PROBLEMS‡

*†9-1   A 24-tooth gear has AGMA standard full-depth involute teeth with diametral pitch of 5. Calculate the pitch diameter, circular pitch, addendum, dedendum, tooth thickness, and clearance.

†9-2   A 40-tooth, 10 p_d gear has AGMA standard full-depth involute teeth. Calculate pitch diameter, circular pitch, addendum, dedendum, tooth thickness, and clearance.

†9-3   A 30-tooth, 12 p_d gear has AGMA standard full-depth involute teeth. Calculate the pitch diameter, circular pitch, addendum, dedendum, tooth thickness, and clearance.

9-4   Using any available string, some tape, a pencil, and a drinking glass or tin can, generate and draw an involute curve on a piece of paper. With your protractor, show that all normals to the curve are tangent to the base circle.

*9-5   A spur gearset must have pitch diameters of 2.5 and 8 in. What is the largest standard tooth size, in terms of diametral pitch p_d, that can be used without having any interference or undercutting? Find the number of teeth on the hob-cut gear and pinion for this p_d.

     a.   For a 20° pressure angle.
     b.   For a 25° pressure angle. (Note that diametral pitch need not be an integer.)

*†9-6   Design a simple, spur gear train for a ratio of –7:1 and diametral pitch of 10. Specify pitch diameters and numbers of teeth. Calculate the contact ratio.

*†9-7   Design a simple, spur gear train for a ratio of +6:1 and diametral pitch of 5. Specify pitch diameters and numbers of teeth. Calculate the contact ratio.

†9-8   Design a simple, spur gear train for a ratio of –7:1 and diametral pitch of 8. Specify pitch diameters and numbers of teeth. Calculate the contact ratio.

†9-9   Design a simple, spur gear train for a ratio of +6.5:1 and diametral pitch of 5. Specify pitch diameters and numbers of teeth. Calculate the contact ratio.

*†9-10   Design a compound, spur gear train for a ratio of –80:1 and diametral pitch of 12. Specify pitch diameters and numbers of teeth. Sketch the train to scale.

†9-11   Design a compound, spur gear train for a ratio of 50:1 and diametral pitch of 8. Specify pitch diameters and numbers of teeth. Sketch the train to scale.

---

‡ All problem figures are provided as PDF files, and some are also provided as animated Working Model files; all are on the DVD. PDF filenames are the same as the figure number. Run the file *Animations.html* to access and run the animations.

*†9-12 Design a compound, spur gear train for a ratio of 120:1 and diametral pitch of 5. Specify pitch diameters and numbers of teeth. Sketch the train to scale.

†9-13 Design a compound, spur gear train for a ratio of –250:1 and diametral pitch of 9. Specify pitch diameters and numbers of teeth. Sketch the train to scale.

*†9-14 Design a compound, reverted, spur gear train for a ratio of 28:1 and diametral pitch of 8. Specify pitch diameters and numbers of teeth. Sketch the train to scale.

†9-15 Design a compound, reverted, spur gear train for a ratio of 40:1 and diametral pitch of 8. Specify pitch diameters and numbers of teeth. Sketch the train to scale.

*†9-16 Design a compound, reverted, spur gear train for a ratio of 65:1 and diametral pitch of 8. Specify pitch diameters and numbers of teeth. Sketch the train to scale.

†9-17 Design a compound, reverted, spur gear train for a ratio of 7:1 and diametral pitch of 4. Specify pitch diameters and numbers of teeth. Sketch the train to scale.

†9-18 Design a compound, reverted, spur gear train for a ratio of 12:1 and diametral pitch of 6. Specify pitch diameters and numbers of teeth. Sketch the train to scale.

*†9-19 Design a compound, reverted, spur gear transmission that will give two shiftable ratios of +3:1 forward and –4.5:1 reverse with diametral pitch of 6. Specify pitch diameters and numbers of teeth. Sketch the train to scale.

†9-20 Design a compound, reverted, spur gear transmission that will give two shiftable ratios of +5:1 forward and –3.5:1 reverse with diametral pitch of 6. Specify pitch diameters and numbers of teeth. Sketch the train to scale.

*†9-21 Design a compound, reverted, spur gear transmission that will give three shiftable ratios of +6:1, +3.5:1 forward, and –4:1 reverse with diametral pitch of 8. Specify pitch diameters and numbers of teeth. Sketch the train to scale.

†9-22 Design a compound, reverted, spur gear transmission that will give three shiftable ratios of +4.5:1, +2.5:1 forward, and –3.5:1 reverse with diametral pitch of 5. Specify pitch diameters and numbers of teeth. Sketch the train to scale.

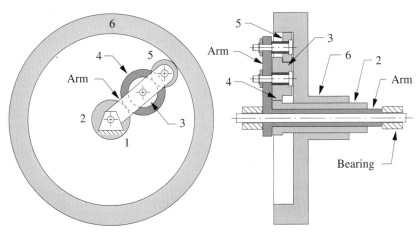

FIGURE P9-1

Planetary gearset for Problem 9-25 (p. 535)

* Answers in Appendix F.

† These problems are suited to solution using *Mathcad*, *Matlab*, or *TKSolver* equation solver programs.

TABLE P9-1　Data for Problem 9-25

Row	N_2	N_3	N_4	N_5	N_6	ω_2	ω_6	ω_{arm}
a	30	25	45	50	200	?	20	– 50
b	30	25	45	50	200	30	?	– 90
c	30	25	45	50	200	50	0	?
d	30	25	45	30	160	?	40	– 50
e	30	25	45	30	160	50	?	– 75
f	30	25	45	30	160	50	0	?

FIGURE P9-2

Compound planetary gear train for Problem 9-26

TABLE P9-2　Data for Problem 9-26

Row	N_2	N_3	N_4	N_5	N_6	ω_2	ω_6	ω_{arm}
a	50	25	45	30	40	?	20	– 50
b	30	35	55	40	50	30	?	– 90
c	40	20	45	30	35	50	0	?
d	25	45	35	30	50	?	40	– 50
e	35	25	55	35	45	30	?	– 75
f	30	30	45	40	35	40	0	?

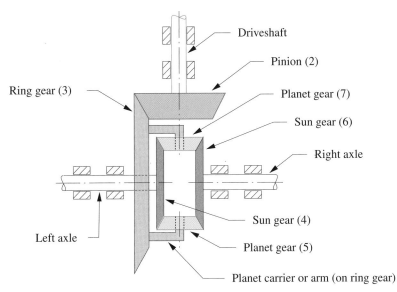

FIGURE P9-3

Automotive differential planetary gear train for Problem 9-27

†9-23 Design the rolling cones for a –3:1 ratio and a 60° included angle between the shafts. Sketch the train to scale.

†9-24 Design the rolling cones for a –4.5:1 ratio and a 40° included angle between the shafts. Sketch the train to scale.

*†9-25 Figure P9-1 (p. 533) shows a compound planetary gear train (not to scale). Table P9-1 gives data for gear numbers of teeth and input velocities. For the row(s) assigned, find the variable represented by a question mark.

*†9-26 Figure P9-2 shows a compound planetary gear train (not to scale). Table P9-2 gives data for gear numbers of teeth and input velocities. For the row(s) assigned, find the variable represented by a question mark.

*†9-27 Figure P9-3 shows a planetary gear train used in an automotive rear-end differential (not to scale). The car has wheels with a 16-in rolling radius and is moving forward in a straight line at 55 mph. The engine is turning 2500 rpm. The transmission is in direct drive (1:1) with the driveshaft.

    a. What are the rear wheels' rpm and the gear ratio between ring and pinion?

    b. As the car hits a patch of ice, the right wheel speeds up to 800 rpm. What is the speed of the left wheel? Hint: The average of both wheels' rpm is a constant.

    c. Calculate the fundamental train value of the epicyclic stage.

†9-28 Design a speed-reducing planetary gearbox to be used to lift a 5-ton load 50 ft with a motor that develops 20 lb-ft of torque at its operating speed of 1750 rpm. The available winch drum has no more than a 16-in diameter when full of its steel cable. The speed reducer should be no larger in diameter than the winch drum. Gears of no more than about 75 teeth are desired, and diametral pitch needs to be no smaller than 6 to stand the stresses. Make multiview sketches of your design and show all calculations. How long will it take to raise the load with your design?

(a) *(b)*

FIGURE P9-4

Problems 9-33 to 9-34 *From P. H. Hill and W. P. Rule. (1960). Mechanisms: Analysis and Design, with permission*

9

*†9-29   Determine all possible two-stage compound gear combinations that will give an approximation to the Naperian base 2.71828. Limit tooth numbers to between 18 and 80. Determine the arrangement that gives the smallest error.

†9-30   Determine all possible two-stage compound gear combinations that will give an approximation to 2π. Limit tooth numbers to between 15 and 90. Determine the arrangement that gives the smallest error.

†9-31   Determine all possible two-stage compound gear combinations that will give an approximation to $\pi/2$. Limit tooth numbers to between 20 and 100. Determine the arrangement that gives the smallest error.

† These problems are suited to solution using *Mathcad, Matlab,* or *TKSolver* equation solver programs.

(a) *(b)*

FIGURE P9-5

Problems 9-35 to 9-36 *From P. H. Hill and W. P. Rule. (1960). Mechanisms: Analysis and Design, with permission*

†9-32 Determine all possible two-stage compound gear combinations that will give an approximation to $3\pi/2$. Limit tooth numbers to between 20 and 100. Determine the arrangement that gives the smallest error.

†9-33 Figure P9-4a shows a reverted clock train. Design it using 25° nominal pressure angle gears of 24 p_d having between 12 and 150 teeth. Determine the tooth numbers and nominal center distance. If the center distance has a manufacturing tolerance of ± 0.006 in, what will the pressure angle and backlash at the minute hand be at each extreme of the tolerance?

†9-34 Figure P9-4b shows a three-speed shiftable transmission. Shaft F, with the cluster of gears E, G, and H, is capable of sliding left and right to engage and disengage the three gearsets in turn. Design the three reverted stages to give output speeds at shaft F of 150, 350, and 550 rpm for an input speed of 450 rpm to shaft D.

*†9-35 Figure P9-5a shows a compound epicyclic train used to drive a winch drum. Gear A is driven at 18 rpm CW and gear D is fixed to ground. Tooth numbers are in the figure. Find speed and direction of the drum. What is train efficiency for gearsets $E_0 = 0.97$?

†9-36 Figure P9-5b shows a compound epicyclic train with its tooth numbers. The arm is driven CCW at 20 rpm. Gear A is driven CW at 40 rpm. Find speed of ring gear D.

*†9-37 Figure P9-6a shows an epicyclic train with its tooth numbers. The arm is driven CCW at 50 rpm and gear A on shaft 1 is fixed to ground. Find speed of gear D on shaft 2. What is the efficiency of this train if the basic gearsets have $E_0 = 0.96$?

†9-38 Figure P9-6b shows a differential with its tooth numbers. Gear A is driven CCW at 10 rpm and gear B is driven CW at 24 rpm. Find the speed of gear D.

*†9-39 Figure P9-7a (p. 538) shows a gear train containing both compound-reverted and epicyclic stages. Tooth numbers are in the figure. The motor is driven CW at 1500 rpm. Find the speeds of shafts 1 and 2.

* Answers in Appendix F.

† These problems are suited to solution using *Mathcad*, *Matlab*, or *TKSolver* equation solver programs.

(a)

(b)

FIGURE P9-6

Problems 9-37 to 9-38 *From P. H. Hill and W. P. Rule. (1960). Mechanisms: Analysis and Design, with permission*

FIGURE P9-7

Problems 9-39 to 9-40 *From P. H. Hill and W. P. Rule. (1960). Mechanisms: Analysis and Design, with permission*

†9-40   Figure P9-7b shows an epicyclic train used to drive a winch drum. The arm is driven at 250 rpm CCW and gear A, on shaft 2, is fixed to ground. Find speed and direction of the drum on shaft 1. What is train efficiency if the basic gearsets have $E_0 = 0.98$?

*†9-41   Figure P9-8 shows an epicyclic train with its tooth numbers. Gear 2 is driven at 600 rpm CW and gear D is fixed to ground. Find speed and direction of gears 1 and 3.

†9-42   Figure P9-9 shows a compound epicyclic train. Shaft 1 is driven at 300 rpm CCW and gear A is fixed to ground. The tooth numbers are indicated in the figure. Determine the speed and direction of shaft 2.

*†9-43   Figure P9-10 shows a compound epicyclic train. Shaft 1 is driven at 60 rpm. Tooth numbers are in the figure. Find speed and direction of gears G and M.

* Answers in Appendix F.

† These problems are suited to solution using *Mathcad*, *Matlab*, or *TKSolver* equation solver programs.

FIGURE P9-8

Problem 9-41 *From P. H. Hill and W. P. Rule. (1960). Mechanisms: Analysis and Design, with permission*

FIGURE P9-9

Problem 9-42 *From P. H. Hill and W. P. Rule. (1960). Mechanisms: Analysis and Design, with permission*

†9-44   Calculate the ratios in the Model T transmission shown in Figure 9-48 (p. 528) and
         prove that the values shown in the figure's sidebar are correct.

†9-45   Do Problem 7-57 on p. 393.

9-46   Figure P9-11 shows an involute generated from a base circle of radius r_b. Point *A* is
         simultaneously on the base circle and the involute. Point *B* is any point on the involute
         curve and point *C* is on the base circle where a line drawn from point *B* is tangent to
         the base circle. Point *O* is the center of the base circle. The angle ϕ_B (angle *BOC*) is
         known as the *involute pressure angle* corresponding to point *B* (not to be confused with

† These problems are suited
to solution using *Mathcad,
Matlab,* or *TKSolver* equation solver programs.

FIGURE P9-10

Problem 9-43 *From P. H. Hill and W. P. Rule. (1960). Mechanisms: Analysis and Design, with permission*

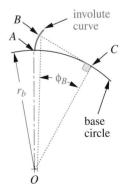

FIGURE P9-11

Problem 9-46

the *pressure angle of two gears in mesh*, which is defined on page 487). The angle *AOB* is known as the *involute of* ϕ_B and is often designated as *inv* ϕ_B. Using the definition of the involute tooth form and Figure 9-5 (p. 485), derive an equation for *inv* ϕ_B as a function of ϕ_B alone.

9-47   Using data and definitions from Problem 9-46, show that when point *B* is at the pitch circle the *involute pressure angle* is equal to the *pressure angle of two gears in mesh*.

9-48   Using data and definitions from Problem 9-46, and with point *B* at the pitch circle where the involute pressure angle ϕ_B is equal to the pressure angle ϕ of two gears in mesh, derive equation 9.4b (p. 490).

9-49   Using Figures 9-6 and 9-7 (pp. 486-487), derive equation 9.2 (p. 486), which is used to calculate the length of action of a pair of meshing gears.

†9-50   Backlash of 0.03 mm measured on the pitch circle of a 40-mm-diameter pinion in mesh with a 100-mm-diameter gear is desired. If the gears are standard, full-depth, with 25° pressure angles, by how much should the center distance be increased?

†9-51   Backlash of 0.0012 in measured on the pitch circle of a 2.000-in-diameter pinion in mesh with a 5.000-in-diameter gear is desired. If the gears are standard, full-depth, with 25° pressure angles, by how much should the center distance be increased?

†9-52   A 22-tooth gear has standard full-depth involute teeth with a module of 6. Calculate the pitch diameter, circular pitch, addendum, dedendum, tooth thickness, and clearance using the AGMA specifications in Table 9-1 (p. 491) substituting m for $1/p_d$.

†9-53   A 40-tooth gear has standard full-depth involute teeth with a module of 3. Calculate the pitch diameter, circular pitch, addendum, dedendum, tooth thickness, and clearance using the AGMA specifications in Table 9-1 (p. 491) substituting m for $1/p_d$.

†9-54   A 30-tooth gear has standard full-depth involute teeth with a module of 2. Calculate the pitch diameter, circular pitch, addendum, dedendum, tooth thickness, and clearance using the AGMA specifications in Table 9-1 (p. 491) substituting m for $1/p_d$.

†9-55   Determine the minimum number of teeth on a pinion with a 20° pressure angle for the following gear-to-pinion ratios such that there will be no tooth-to-tooth interference: 1:1, 2:1, 3:1, 4:1, 5:1.

†9-56   Determine the minimum number of teeth on a pinion with a 25° pressure angle for the following gear-to-pinion ratios such that there will be no tooth-to-tooth interference: 1:1, 2:1, 3:1, 4:1, 5:1.

†9-57   A pinion with a 3.000-in pitch diameter is to mesh with a rack. What is the largest standard tooth size, in terms of diametral pitch, that can be used without having any interference?   a.   For a 20° pressure angle   b.   For a 25° pressure angle

†9-58   A pinion with a 75-mm pitch diameter is to mesh with a rack. What is the largest standard tooth size, in terms of metric module, that can be used without having any interference?   a.   For a 20° pressure angle   b.   For a 25° pressure angle

†9-59   In order to have a smooth-running gearset it is desired to have a contact ratio of at least 1.5. If the gears have a pressure angle of 25° and gear ratio of 4, what is the minimum number of teeth on the pinion that will yield the required minimum contact ratio?

†9-60   In order to have a smooth-running gearset it is desired to have a contact ratio of at least 1.5. If the gears have a pressure angle of 25° and a 20-tooth pinion, what is the minimum gear ratio that will yield the required minimum contact ratio?

† These problems are suited to solution using *Mathcad, Matlab,* or *TKSolver* equation solver programs.

†9-61   Calculate and plot the train ratio of a noncircular gearset, as a function of input angle, that is based on the centrodes of Figure 6-15b (p. 310). The link length ratios are $L_1/L_2 = 1.60$, $L_3/L_2 = 1.60$, and $L_4/L_2 = 1.00$.

†9-62   Repeat problem 9-61 for a fourbar linkage with link ratios of $L_1/L_2 = 1.80$, $L_3/L_2 = 1.80$, and $L_4/L_2 = 1.00$.

†9-63   Figure 9-35b (p. 514) shows (schematically) a compound epicyclic train. The tooth numbers are 50, 25, 35, and 90 for gears 2, 3, 4, and 5, respectively. The arm is driven at 180 rpm CW and gear 5 is fixed to ground. Determine the speed and direction of gear 2. What is the efficiency of this train if the basic gearsets have $E_0 = 0.98$?

†9-64   Figure 9-35h (p. 514) shows (schematically) a compound epicyclic train. The tooth numbers are 80, 20, 25, and 85 for gears 2, 3, 4, and 5, respectively. Gear 2 is driven at 200 rpm CCW. Determine the speed and direction of the arm if gear 5 is fixed to ground. What is the efficiency of this train if the basic gearsets have $E_0 = 0.98$?

†9-65   Figure 9-35i (p. 514) shows (schematically) a compound epicyclic train. The tooth numbers are 24, 18, 20, and 90 for gears 2, 3, 4, and 5, respectively. The arm is driven at 100 rpm CCW and gear 2 is fixed to ground. Determine the speed and direction of gear 5. What is the efficiency of this train if the basic gearsets have $E_0 = 0.98$?

9-66   Using Figure 9-8 (p. 488), derive an equation for the operating pressure angle of two gears in mesh as a function of their base circle radii, the standard center distance, and the change in center distance.

**†9-67   A pinion and gear in mesh have base circle radii of 1.8126 and 3.6252 in, respectively. If they were cut with a standard pressure angle of 25°, determine their operating pressure angle if the standard center distance is increased by 0.062 in.

†9-68   A pinion and gear in mesh have base circle radii of 1.35946 and 2.26577 in, respectively. If they have a standard center distance of 4.000 in, determine the standard pressure angle and the operating pressure angle if the standard center distance is increased by 0.050 in.

**†9-69   A 25-tooth pinion meshes with a 60-tooth gear. They have a diametral pitch of 4, a pressure angle of 20°, and AGMA full-depth involute profiles. Find the gear ratio, circular pitch, base pitch, pitch diameters, standard center distance, addendum, dedendum, whole depth, clearance, outside diameters and contact ratio of the gearset.

†9-70   A 15-tooth pinion meshes with a 45-tooth gear. They have a diametral pitch of 5, a pressure angle of 25°, and AGMA full-depth involute profiles. Find the gear ratio, circular pitch, base pitch, pitch diameters, standard center distance, addendum, dedendum, whole depth, clearance, outside diameters and contact ratio of the gearset.

**†9-71   Design a compound, spur gear train that will reduce the speed of a 900-rpm synchronous AC motor to exactly 72 revolutions per hour with the output rotating in the same direction as the motor. Use gears with a pressure angle of 25° and minimize the package size.

†9-72   A gearset with a contact ratio of at least 1.5 is desired. If the gears have standard AGMA full-depth teeth with a pressure angle of 25°, and the pinion has 21 teeth, what is the minimum gear ratio that will give the required minimum contact ratio?

†9-73   Provide a preliminary design (pitch diameters and numbers of teeth) for a gear set with a gear ratio of $m_G = 4$, a diametral pitch $p_d = 8$, and a contact ratio of at least 1.5.

---

† These problems are suited to solution using *Mathcad*, *Matlab*, or *TKSolver* equation solver programs.

9

The entire world of machinery ... is inspired by the play of organs of reproduction. The designer animates artificial objects by simulating the movements of animals engaged in propagating the species. Our machines are Romeos of steel and Juliets of cast iron.

J. COHEN. (1966). *Human Robots in Myth and Science,* Allen & Unwin, London, p. 67.

DYNAMICS OF MACHINERY

Chapter 10

DYNAMICS FUNDAMENTALS

*He has half the deed done
who has made a beginning*
HORACE, 65-8 B.C.

10.0   INTRODUCTION

Part I of this text has dealt with the **kinematics** of mechanisms while temporarily ignoring the forces present in those mechanisms. This second part will address the problem of determining the forces present in moving mechanisms and machinery. This topic is called **kinetics** or **dynamic force analysis**. We will start with a brief review of some fundamentals needed for dynamic analysis. It is assumed that the reader has had an introductory course in dynamics. If that topic is rusty, one can review it by referring to reference [1] or to any other text on the subject.

10.1   NEWTON'S LAWS OF MOTION

Dynamic force analysis involves the application of **Newton's** three **laws of motion** which are:

1   *A body at rest tends to remain at rest and a body in motion at constant velocity will tend to maintain that velocity unless acted upon by an external force.*

2   *The time rate of change of momentum of a body is equal to the magnitude of the applied force and acts in the direction of the force.*

3   *For every action force there is an equal and opposite reaction force.*

The second law is expressed in terms of rate of change of *momentum*, $\mathbf{M} = m\mathbf{v}$, where m is mass and $\mathbf{v}$ is velocity. Mass m is assumed to be constant in this analysis. The time rate of change of $m\mathbf{v}$ is $m\mathbf{a}$, where $\mathbf{a}$ is the acceleration of the mass center.

$$\mathbf{F} = m\mathbf{a} \tag{10.1}$$

$\mathbf{F}$ is the resultant of all forces on the system acting at the mass center.

We can differentiate between two subclasses of dynamics problems depending upon which quantities are known and which are to be found. The "**forward dynamics problem** is the one in which we know everything about the external loads (forces and/or torques) being exerted on the system, and we wish to determine the accelerations, velocities, and displacements which result from the application of those forces and torques. This subclass is typical of the problems you probably encountered in an introductory dynamics course, such as determining the acceleration of a block sliding down a plane, acted upon by gravity. Given $\mathbf{F}$ and m, solve for $\mathbf{a}$.

The second subclass of dynamics problem, called the **inverse dynamics problem**, is one in which we know the (desired) accelerations, velocities, and displacements to be imposed upon our system and wish to solve for the magnitudes and directions of the forces and torques which are necessary to provide the desired motions and which result from them. This inverse dynamics case is sometimes also called **kinetostatics**. Given $\mathbf{a}$ and m, solve for $\mathbf{F}$.

Whichever subclass of problem is addressed, it is important to realize that they are both dynamics problems. Each merely solves $\mathbf{F} = m\mathbf{a}$ for a different variable. To do so we must first review some fundamental geometric and mass properties which are needed for the calculations.

10.2 DYNAMIC MODELS

It is often convenient in dynamic analysis to create a simplified model of a complicated part. These models are sometimes considered to be a collection of point masses connected by massless rods. For a model of a rigid body to be **dynamically equivalent** to the original body, three things must be true:

1 *The mass of the model must equal that of the original body.*

2 *The center of gravity must be in the same location as that of the original body.*

3 *The mass moment of inertia must equal that of the original body.*

10.3 MASS

Mass is not weight! Mass is an invariant property of a rigid body. The weight of the same body varies depending on the gravitational system in which it sits. See Section 1.10 (p. 17) for a discussion of the use of proper mass units in various measuring systems. We will assume the mass of our parts to be constant with time. For most earthbound machines, this is reasonable. The rate at which an automobile loses mass due to fuel consumption, for example, is slow enough to be ignored when calculating dynamic forces over short time spans. However, this would not be a safe assumption for a vehicle such as the Space Shuttle, whose mass changes rapidly and drastically during lift-off.

When designing machinery, we must first do a complete kinematic analysis of our design, as described in Part I of this text, in order to obtain information about the accelerations of the moving parts. We next want to use Newton's second law to calculate the dynamic forces. But to do so we need to know the masses of all the moving parts that have these known accelerations. These parts do not exist yet! As with any design problem, we lack sufficient information at this stage of the design to accurately determine the best sizes and shapes of the parts. We must estimate the masses of the links and other parts of the design in order to make a first pass at the calculation. We will then have to iterate to better and better solutions as we generate more information. See Section 1.5 (p. 7) on the design process to review the use of iteration in design.

A first estimate of your parts' masses can be obtained by assuming some reasonable shapes and sizes for all the parts and choosing appropriate materials. Then calculate the volume of each part and multiply its volume by the material's **mass density** (not weight density) to obtain a first approximation of its mass. These mass values can then be used in Newton's equation. The densities of some common engineering materials can be found in Appendix B.

How will we know whether our chosen sizes and shapes of links are even acceptable, let alone optimal? Unfortunately, we will not know until we have carried the computations all the way through a complete stress and deflection analysis of the parts. It is often the case, especially with long, thin elements such as shafts or slender links, that the deflections of the parts under their dynamic loads will limit the design even at low stress levels. In some cases the stresses will be excessive.

We will probably discover that the parts fail under the dynamic forces. Then we will have to go back to our original assumptions about the shapes, sizes, and materials of these parts; redesign them; and repeat the force, stress, and deflection analyses. Design is, unavoidably, an **iterative process**.

The topic of stress and deflection analysis is beyond the scope of this text and will not be further discussed here. (See reference 2.). It is mentioned only to put our discussion of dynamic force analysis into context. We are analyzing these dynamic forces primarily to provide the information needed to do the stress and deflection analyses on our parts! It is also worth noting that, unlike a static force situation in which a failed design might be fixed by adding more mass to the part to strengthen it, to do so in a dynamic force situation can have a deleterious effect. More mass with the same acceleration will generate even higher forces and thus higher stresses! The machine designer often needs to remove mass (in the right places) from parts in order to reduce the stresses and deflections due to $\mathbf{F} = m\mathbf{a}$. Thus the designer needs to have a good understanding of both material properties and stress and deflection analysis to properly shape and size parts for minimum mass while maximizing the strength and stiffness needed to withstand the dynamic forces.

10.4 MASS MOMENT AND CENTER OF GRAVITY

When the mass of an object is distributed over some dimensions, it will possess a moment with respect to any axis of choice. Figure 10-1 (p. 548) shows a mass of general shape in an *xyz* axis system. A differential element of mass is also shown. The **mass moment (first moment of mass)** of the differential element is equal to the **product of its mass and its distance** from the axis of interest. With respect to the *x, y,* and *z* axes these are:

$$dM_x = x\,dm \qquad\qquad (10.2a)$$

$$dM_y = y\,dm \qquad\qquad (10.2b)$$

$$dM_z = z\,dm \qquad\qquad (10.2c)$$

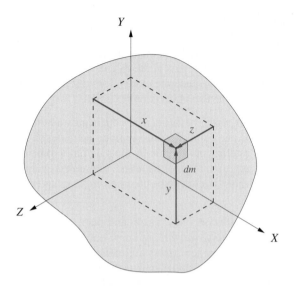

FIGURE 10-1

A generalized mass element in a 3-D coordinate system

To obtain the mass moments of the body we integrate each of these expressions.

$$M_x = \int x \, dm \tag{10.3a}$$

$$M_y = \int y \, dm \tag{10.3b}$$

$$M_z = \int z \, dm \tag{10.3c}$$

If the mass moment with respect to a particular axis is numerically zero, then that axis passes through the **center of mass (CM)** of the object, which for earthbound systems is coincident with its **center of gravity (CG)**. By definition the summation of first moments about all axes through the center of gravity is zero. We will need to locate the *CG* of all moving bodies in our designs because the linear acceleration component of each body is calculated acting at that point.

It is often convenient to model a complicated shape as several interconnected simple shapes whose individual geometries allow easy computation of their masses and the locations of their local *CGs*. The global *CG* can then be found from the summation of the first moments of these simple shapes set equal to zero. Appendix C contains formulas for the volumes and locations of centers of gravity of some common shapes.

Figure 10-2 shows a simple model of a mallet broken into two cylindrical parts, the handle and the head, which have masses m_h and m_d, respectively. The individual centers of gravity of the two parts are at l_d and $l_h/2$, respectively, with respect to the axis ZZ. We want to find the location of the composite center of gravity of the mallet with respect to ZZ. Summing the first moments of the individual components about ZZ and setting them equal to the moment of the entire mass about ZZ:

$$\sum M_{ZZ} = m_h \frac{l_h}{2} + m_d l_d = (m_h + m_d) d \tag{10.3d}$$

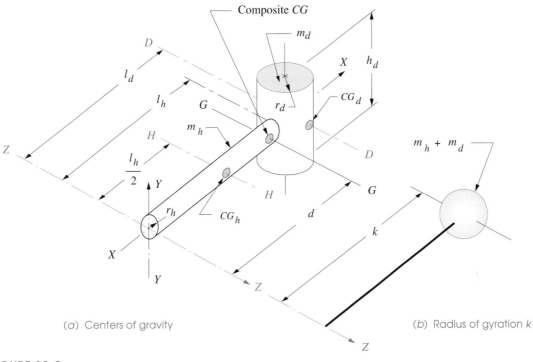

(a) Centers of gravity (b) Radius of gyration k

FIGURE 10-2

Dynamic models, composite center of gravity, and radius of gyration of a mallet

This equation can be solved for the distance d along the X axis, which, in this symmetrical example, is the only dimension of the composite CG not discernible by inspection. The y and z components of the composite CG are both zero.

$$d = \frac{m_h \dfrac{l_h}{2} + m_d l_d}{m_h + m_d} \qquad (10.3e)$$

10.5 MASS MOMENT OF INERTIA (SECOND MOMENT OF MASS)

Newton's law applies to systems in rotation as well as to those in translation. The rotational form of Newton's second law is:

$$\mathbf{T} = I\alpha \qquad (10.4)$$

where $\mathbf{T}$ is resultant torque about the mass center, α is angular acceleration, and I is mass moment of inertia about an axis through the mass center.

Mass moment of inertia is referred to some axis of rotation, usually one through the CG. Refer again to Figure 10-1 that shows a mass of general shape and an XYZ axis system. A differential element of mass is also shown. The **mass moment of inertia** of the differential element is equal to the **product of its mass and the square of its distance** from the axis of interest. With respect to the X, Y, and Z axes they are:

$$dI_x = r_x^2 dm = (y^2 + z^2)dm \qquad\qquad (10.5a)$$

$$dI_y = r_y^2 dm = (x^2 + z^2)dm \qquad\qquad (10.5b)$$

$$dI_z = r_z^2 dm = (x^2 + y^2)dm \qquad\qquad (10.5c)$$

The exponent of 2 on the radius term gives this property its other name of **second moment of mass**. To obtain the mass moments of inertia of the entire body we integrate each of these expressions.

$$I_x = \int (y^2 + z^2)dm \qquad\qquad (10.6a)$$

$$I_y = \int (x^2 + z^2)dm \qquad\qquad (10.6b)$$

$$I_z = \int (x^2 + y^2)dm \qquad\qquad (10.6c)$$

While it is fairly intuitive to appreciate the physical significance of the first moment of mass, it is more difficult to do the same for the second moment, or moment of inertia.

Consider equation 10.4 (p. 549). It says that torque is proportional to angular acceleration, and the constant of proportionality is this moment of inertia, I. Picture a common hammer or mallet as depicted in Figure 10-2 (p. 549). The head, made of steel, has large mass compared to the light wooden handle. When gripped properly, at the end of the handle, the radius to the mass of the head is large. Its contribution to the total I of the mallet is proportional to the square of the radius from the axis of rotation (your wrist at axis ZZ) to the head. Thus it takes considerably more torque to swing (and thus angularly accelerate) the mallet when it is held properly than if held near the head. As a child you probably chose to hold a hammer close to its head because you lacked the strength to provide the larger torque it needed when held properly. You also found it ineffective in driving nails when held close to the head because you were unable to store very much **kinetic energy** in it. In a translating system kinetic energy is:

$$KE = \frac{1}{2}mv^2 \qquad\qquad (10.7a)$$

and in a rotating system kinetic energy is:

$$KE = \frac{1}{2}I\omega^2 \qquad\qquad (10.7b)$$

Thus the kinetic energy stored in the mallet is also proportional to its moment of inertia I and to ω^2. So, you can see that holding the mallet close to its head reduces the I and lowers the energy available for driving the nail.

Moment of inertia then is one indicator of the ability of the body to store rotational kinetic energy and is also an indicator of the amount of torque that will be needed to rotationally accelerate the body. Unless you are designing a device intended for the storage and transfer of large amounts of energy (punch press, drop hammer, rock crusher etc.) you will probably be trying to minimize the moments of inertia of your rotating parts. Just as mass is a measure of resistance to linear acceleration, moment of inertia is a measure of resistance to angular acceleration. A large I will require a large driving torque and thus a larger and more powerful motor to obtain the same acceleration. Later we will see how to make moment of inertia work for us in rotating machinery by using flywheels with large I. The units of moment of inertia can be determined by doing a unit balance on either equation 10.4 (p. 549) or equation 10.7 and are shown in Table 1-4 (p. 19). In the **ips** system they are lb-in-sec^2 or blob-in^2. In the **SI** system, they are N-m-s^2 or kg-m^2.

10.6 PARALLEL AXIS THEOREM (TRANSFER THEOREM)

The moment of inertia of a body with respect to any axis (ZZ) can be expressed as the sum of its moment of inertia about an axis (GG) parallel to ZZ through its CG and the product of the mass and the square of the perpendicular distance between those parallel axes.

$$I_{ZZ} = I_{GG} + md^2 \qquad (10.8)$$

where ZZ and GG are parallel axes, GG goes through the CG of the body or assembly, m is the mass of the body or assembly, and d is the perpendicular distance between the parallel axes. This property is most useful when computing the moment of inertia of a complex shape which has been broken into a collection of simple shapes as shown in Figure 10-2a (p. 549), which represents a simplistic model of a mallet. The mallet is broken into two cylindrical parts, the handle and the head, which have masses m_h and m_d, and radii r_h and r_d, respectively. The expressions for the mass moments of inertia of a cylinder with respect to axes through its CG can be found in Appendix C and are for the handle about its CG axis HH:

$$I_{HH} = \frac{m_h \left(3r_h^2 + l_h^2 \right)}{12} \qquad (10.9a)$$

and for the head about its CG axis DD:

$$I_{DD} = \frac{m_d \left(3r_d^2 + h_d^2 \right)}{12} \qquad (10.9b)$$

Using the parallel axis theorem to transfer the moment of inertia to the axis ZZ at the end of the handle:

$$I_{ZZ} = \left[I_{HH} + m_h \left(\frac{l_h}{2} \right)^2 \right] + \left[I_{DD} + m_d l_d^2 \right] \qquad (10.9c)$$

10.7 DETERMINING MASS MOMENT OF INERTIA

There are several ways to determine the mass moment of inertia of a part. If the part is in the process of being designed, then an analytical method is needed. If the part exists, then either an analytical or experimental method can be used.

Analytical Methods

While it is possible to integrate equations 10.6 numerically for a part of any arbitrary shape, the work involved to do this by hand is usually prohibitively tedious and time consuming. If a part of complicated shape can be broken down into subparts that have simple geometry such as cylinders, rectangular prisms, spheres, etc., as was done with the mallet of Figure 10-2 (p. 549), then the mass moments of inertia of each subpart about its own CG can be calculated. These values must each be referred to the desired axis of rotation using the transfer theorem (equation 10.8) then summed to obtain an approximate value of the complete part's moment of inertia about the desired axis. Formulas for the mass moments of inertia of some simple geometric solids are shown in Appendix C.

(a)

(b)

FIGURE 10-3

Measuring moment of
inertia

If a CAD solids modeling package is used to design the part's geometry, then the task of determining all its mass properties is greatly simplified. Most CAD packages will calculate the mass and mass moments of inertia of a solid 3-D part about any set of selected axes to good accuracy. This is, by far, the preferred method and is only one of the many advantages of using a solids modeling CAD package for mechanical design work.

Experimental Methods

If the part has been designed and built, its mass moment of inertia can be determined approximately by a simple experiment. This requires that the part be swung about any axis (other than one that passes through its CG) parallel to that about which the moment is sought and its period of pendular oscillation measured. Figure 10-3a shows a part (a connecting rod) suspended on a knife-edge pivot at ZZ and rotated through a small angle θ as shown in Figure 10-3b. Its weight force W acts at its CG and has a component $W \sin \theta$ perpendicular to the radius r from the pivot to the CG. From equation 10.4 (p. 549):

$$\mathbf{T}_{ZZ} = I_{ZZ}\,\alpha \tag{10.10a}$$

substituting equivalent expressions for $\mathbf{T}_{ZZ}$ and α:

$$-(W \sin \theta)r = I_{ZZ}\frac{d^2\theta}{dt^2} \tag{10.10b}$$

where the negative sign is used because the torque is in the opposite direction to angle θ.

For small values of θ, $\sin \theta = \theta$, approximately, so:

$$-W\theta r = I_{ZZ}\frac{d^2\theta}{dt^2}$$

$$\frac{d^2\theta}{dt^2} = -\frac{W\,r}{I_{ZZ}}\theta \tag{10.10c}$$

Equation 10.10c is a second-order differential equation with constant coefficients that has the well-known solution:

$$\theta = C\left(\sin\sqrt{\frac{Wr}{I_{ZZ}}}t\right) + D\left(\cos\sqrt{\frac{Wr}{I_{ZZ}}}t\right) \tag{10.10d}$$

The constants of integration C and D can be found from the initial conditions defined at the instant the part is released and allowed to swing.

$$\text{at: } t = 0, \qquad \theta = \theta_{max}, \qquad \omega = \frac{d\theta}{dt} = 0; \qquad \text{then:} \qquad C = 0, \qquad D = \theta_{max}$$

and:

$$\theta = \theta_{max}\left(\cos\sqrt{\frac{Wr}{I_{ZZ}}}t\right) \tag{10.10e}$$

Equation 10.10e defines the part's motion as a cosine wave that completes a full cycle of period τ sec when

$$\sqrt{\frac{Wr}{I_{ZZ}}}\,\tau = 2\pi \tag{10.10f}$$

The weight of the part is easily measured. The *CG* location can be found by balancing the part on a knife edge or suspending it from two different locations, either approach giving the distance r. The period of oscillation τ can be measured with a stopwatch, preferably over a number of cycles to reduce experimental error. With these data, equation 10.10f can be solved for the mass moment of inertia I_{ZZ} about the pivot ZZ as:

$$I_{ZZ} = Wr \left(\frac{\tau}{2\pi} \right)^2 \qquad (10.10g)$$

and the moment of inertia I_{GG} about the *CG* can then be found using the transfer theorem (equation 10.8, p. 551).

$$I_{ZZ} = I_{GG} + mr^2$$

$$I_{GG} = Wr \left(\frac{\tau}{2\pi} \right)^2 - \frac{W}{g} r^2 \qquad (10.10h)$$

10.8 RADIUS OF GYRATION

The **radius of gyration** of a body is defined as the radius at which the entire mass of the body could be concentrated such that the resulting model will have the same moment of inertia as the original body. The mass of this model must be the same as that of the original body. Let I_{ZZ} represent the mass moment of inertia about ZZ from equation 10.9c (p. 551) and m the mass of the original body. From the parallel axis theorem, a concentrated mass m at a radius k will have a moment of inertia:

$$I_{ZZ} = mk^2 \qquad (10.11a)$$

Since we want I_{ZZ} to be equal to the original moment of inertia, the required **radius of gyration** at which we will concentrate the mass m is then:

$$k = \sqrt{\frac{I_{ZZ}}{m}} \qquad (10.11b)$$

Note that this property of radius of gyration allows the construction of an even simpler dynamic model of the system in which all the system mass is concentrated in a "point mass" at the end of a massless rod of length k. Figure 10-2b (p. 549) shows such a model of the mallet in Figure 10-2a.

By comparing equation 10.11a with equation 10.8 (p. 551), it can be seen that the radius of gyration k will always be larger than the radius to the composite *CG* of the original body.

$$I_{GG} + md^2 = I_{ZZ} = mk^2 \qquad \therefore k > d \qquad (10.11c)$$

Appendix C contains formulas for the moments of inertia and radii of gyration of some common shapes.

10

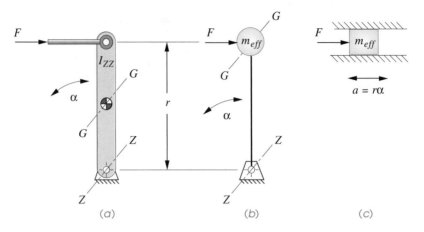

FIGURE 10-4

Modeling a rotating link as a translating mass

10.9　MODELING ROTATING LINKS

Many mechanisms contain links that oscillate in pure rotation. As a first approximation, it is possible to model these links as lumped masses in translation. The error in so doing will be acceptably small if the angular rotation of the link is small. Then the difference between the length of the arc over a small angle and its chord is small.

The goal is to model the distributed mass of the rotating link as a lumped, point mass placed at the point of attachment to its adjacent link, connected to its pivot by a rigid but massless rod. Figure 10-4 shows a link, rotating about an axis ZZ, and its lumped dynamic model. The mass of the lump placed at the link radius r must have the same moment of inertia about the pivot ZZ as the original link. The mass moment of inertia I_{ZZ} of the original link must be known or estimated. The mass moment of inertia of a point mass at a radius is found from the transfer theorem. Since a point mass, by definition, has no dimension, its moment of inertia I_{GG} about its center of mass is zero and equation 10.8 (p. 551) reduces to

$$I_{ZZ} = mr^2 \qquad (10.12a)$$

The effective mass m_{eff} to be placed at the radius r is then

$$m_{eff} = \frac{I_{ZZ}}{r^2} \qquad (10.12b)$$

For small angles of rotation, the rotating link can then be modeled as a mass m_{eff} in pure rectilinear translation for inclusion in a model such as that shown in Figure 10-11 (p. 566).

10.10　CENTER OF PERCUSSION

The **center of percussion** is a point on a body which, when struck with a force, will have associated with it another point called the **center of rotation** at which there will be a zero reaction force. You have probably experienced the result of "missing the center of percussion" when you hit a baseball or softball with the wrong spot on the bat. The "right place on the bat" to hit the ball is the center of percussion associated with the point where your hands grip the bat (the center of rotation). Hitting the ball at other than the center of percussion results in a stinging force being delivered to your hands. Hit the right spot and you feel no force (or pain). The center of percussion is sometimes called the "sweet spot" on a bat, tennis racquet, or golf club. In the case of our mallet example, a center of percussion at the head corresponds to a center of rotation near the end of the handle, and the handle is usually contoured to encourage gripping it there.

The explanation of this phenomenon is quite simple. To make the example two-dimensional and eliminate the effects of friction, consider a hockey stick of mass m lying on the ice as shown in Figure 10-5a. Strike it a sharp blow at point P with a force $\mathbf{F}$ perpendicular to the stick axis. The stick will begin to travel across the ice in complex planar motion, both rotating and translating. Its complex motion at any instant can be considered as the superposition of two components: pure translation of its center of gravity G in the direction of $\mathbf{F}$ and pure rotation about that point G. Set up an embedded coordinate system centered at G with the X axis along the stick in its initial position as shown. The translating component of acceleration of the CG resulting from the force $\mathbf{F}$ is (from Newton's law)

$$A_{G_y} = \frac{F}{m} \qquad (10.13a)$$

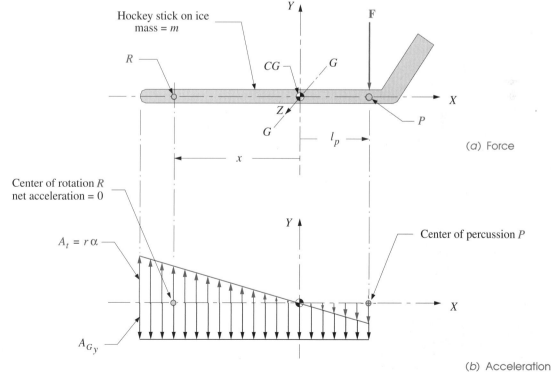

(a) Force

(b) Acceleration

FIGURE 10-5

Center of percussion and center of rotation

and the angular acceleration is:

$$\alpha = \frac{T}{I_{GG}} \tag{10.13b}$$

where I_{GG} is its mass moment of inertia about the line GG through the CG (out of the page along the Z axis). But torque is also:

$$T = F l_P \tag{10.13c}$$

where l_p is the distance along the X axis from point G to point P so:

$$\alpha = \frac{F l_P}{I_{GG}} \tag{10.13d}$$

The total linear acceleration at any point along the stick will be the sum of the linear acceleration A_{G_y} of the CG and the tangential component $(r\alpha)$ of the angular acceleration as shown in Figure 10-4b (p. 554).

$$A_{ytotal} = A_{G_y} + r\alpha$$

$$= \frac{F}{m} + x\left(\frac{F l_P}{I_{GG}}\right) \tag{10.14}$$

where x is the distance to any point along the stick. Equation 10.14 can be set equal to zero and solved for the value of x for which the $r\alpha$ component exactly cancels the A_{G_y}-component. This will be the **center of rotation** at which there is no translating acceleration, and thus no linear dynamic force. The solution for x when $A_{ytotal} = 0$ is:

$$x = -\frac{I_{GG}}{m l_P} \tag{10.15a}$$

and substituting equation 10.11b (p. 553):

$$x = -\frac{k^2}{l_P} \tag{10.15b}$$

where the radius of gyration k is calculated with respect to the line GG through the CG.

Note that this relationship between the center of percussion and the center of rotation involves only geometry and mass properties. The magnitude of the applied force is irrelevant, but its location l_p completely determines x. Thus there is **not** just one center of percussion on a body. Rather there will be pairs of points. For every point (center of percussion) at which a force is applied there will be a corresponding center of rotation at which the reaction force felt will be zero. This center of rotation need not fall within the physical length of the body, however. Consider the value of x predicted by equation 10.15b if you strike the body at its CG.

10.11 LUMPED PARAMETER DYNAMIC MODELS

Figure 10-6a shows a simple plate or disk cam driving a spring-loaded, roller follower. This is a force-closed system which depends on the spring force to keep the cam and follower in contact at all times. Figure 10-6b shows a lumped parameter model of this

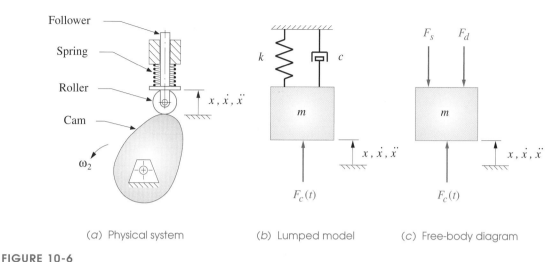

(a) Physical system   (b) Lumped model   (c) Free-body diagram

FIGURE 10-6

One-*DOF* lumped parameter model of a cam-follower system

system in which all the **mass** which moves with the follower train is lumped together as m, all the springiness in the system is lumped within the **spring constant** k, and all the **damping** or resistance to movement is lumped together as a damper with coefficient c. The sources of mass which contribute to m are fairly obvious. The mass of the follower stem, the roller, its pivot pin, and any other hardware attached to the moving assembly all add together to create m. Figure 10-6c shows the free-body diagram of the system acted upon by the cam force F_c, the spring force F_s, and the damping force F_d. There will of course also be the effects of mass times acceleration on the system.

Spring Constant

We have been assuming all links and parts to be rigid bodies in order to do the kinematic analyses, but to do a more accurate force analysis we need to recognize that these bodies are not truly rigid. The springiness in the system is assumed to be linear, thus describable by a spring constant k. A spring constant is defined as the force per unit deflection.

$$k = \frac{F_s}{x} \qquad (10.16)$$

The total spring constant k of the system is a combination of the spring constants of the actual coil spring, plus the spring constants of all other parts which are deflected by the forces. The roller, its pin, and the follower stem are all springs themselves as they are made of elastic materials. The spring constant for any part can be obtained from the equation for its deflection under the applied loading. Any deflection equation relates force to displacement and can be algebraically rearranged to express a spring constant. An individual part may have more than one k if it is loaded in several modes as, for example, a camshaft with a spring constant in bending and also one in torsion. We will discuss the procedures for combining these various spring constants in the system together into a combined, effective spring constant k in the next section. For now let us just assume that we can so combine them for our analysis and create an overall k for our lumped parameter model.

Damping

The friction, more generally called **damping**, is the most difficult parameter of the three to model. It needs to be a combination of all the damping effects in the system. These may be of many forms. **Coulomb friction** results from two dry or lubricated surfaces rubbing together. The contact surfaces between cam and follower and between the follower and its sliding joint can experience coulomb friction. It is generally considered to be independent of velocity magnitude but has a different, larger value when velocity is zero (static friction force F_{st} or *stiction*) than when there is relative motion between the parts (dynamic friction F_d). Figure 10-7a shows a plot of coulomb friction force versus relative velocity v at the contact surfaces. Note that friction always opposes motion, so the friction force abruptly changes sign at $v = 0$. The stiction F_{st} shows up as a larger spike at zero v than the dynamic friction value F_d. Thus, this is a **nonlinear** friction function. It is multivalued at zero. In fact, at zero velocity, the friction force can be any value between $-F_{st}$ and $+F_{st}$. It will be whatever force is needed to balance the system forces and create equilibrium. When the applied force exceeds F_{st}, motion begins and the friction force suddenly drops to F_d. This nonlinear damping creates difficulties in our simple model since we want to describe our system with linear differential equations having known solutions.

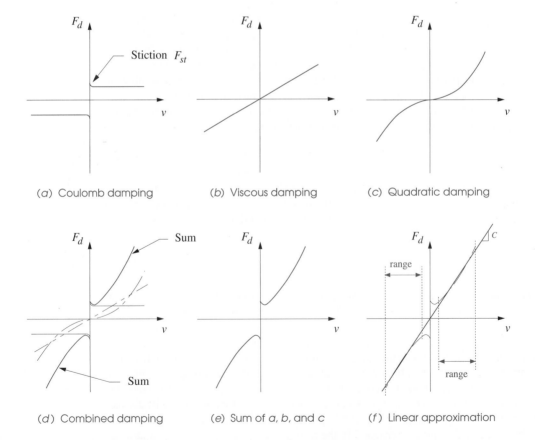

(a) Coulomb damping   (b) Viscous damping   (c) Quadratic damping

(d) Combined damping   (e) Sum of a, b, and c   (f) Linear approximation

FIGURE 10-7

Modeling damping

Other sources of damping may be present besides coulomb friction. **Viscous damping** results from the shearing of a fluid (lubricant) in the gap between the moving parts and is considered to be a linear function of relative velocity as shown in Figure 10-7b. **Quadratic damping** results from the movement of an object through a fluid medium as with an automobile pushing through the air or a boat through the water. This factor is a fairly negligible contributor to a cam-follower's overall damping unless the speeds are very high or the fluid medium is very dense. Quadratic damping is a function of the square of the relative velocity as shown in Figure 10-7c. The relationship of the dynamic damping force F_d as a function of relative velocity for all these cases can be expressed as:

$$F_d = cv|v|^{r-1} \qquad (10.17a)$$

where c is the constant damping coefficient, v is the relative velocity, and r is a constant which defines the type of damping.

For coulomb damping, $r = 0$ and:

$$F_d = \pm c \qquad (10.17b)$$

For viscous damping, $r = 1$ and:

$$F_d = cv \qquad (10.17c)$$

For quadratic damping, $r = 2$ and:

$$F_d = \pm cv^2 \qquad (10.17d)$$

If we combine these three forms of damping, their sum will look like Figure 10-7d and e. This is obviously a nonlinear function. But we can approximate it over a reasonably small range of velocity as a linear function with a slope c which is then a *pseudo-viscous damping coefficient*. This is shown in Figure 10-7f. While not an exact method to account for the true damping, this approach has been found to be acceptably accurate for a first approximation during the design process. The damping in these kinds of mechanical systems can vary quite widely from one design to the next due to different geometries, pressure or transmission angles, types of bearings, lubricants or their absence, etc. It is very difficult to accurately predict the level of damping (i.e., the value of c) in advance of the construction and testing of a prototype, which is the best way to determine the damping coefficient. If similar devices have been built and tested, their history can provide a good prediction. For the purpose of our dynamic modeling, we will assume *pseudo-viscous damping* and some value for c.

10.12 EQUIVALENT SYSTEMS

More complex systems than that shown in Figure 10-6 (p. 557) will have multiple masses, springs, and sources of damping connected together as shown in Figure 10-11 (p. 566). These models can be analyzed by writing dynamic equations for each subsystem and then solving the set of differential equations simultaneously. This allows a multi-degree-of-freedom analysis, with one-*DOF* for each subsystem included in the analysis. Koster[3] found in his extensive study of vibrations in cam mechanisms that a five-*DOF* model which included the effects of both torsional and bending deflection of the camshaft, backlash (see Section 10.2, p. 546) in the driving gears, squeeze effects of the lubricant, nonlinear coulomb damping, and motor speed variation gave a very good prediction of the actual, measured follower response. But he also found that a single-*DOF* model as shown in Figure 10-5 (p. 555) gave a reasonable simulation of the same system. We can then take the simpler approach and lump all the subsystems of Figure 10-11 (p. 566) together into a single-*DOF* **equivalent system** as shown in Figure 10-6 (p. 557). The combining of the various springs, dampers, and masses must be done carefully to properly approximate their dynamic interactions with each other.

There are only two types of variables active in any dynamic system. These are given the general names of *through variable* and *across variable*. These names are descriptive of their actions within the system. A **through variable** *passes through the system*. An **across variable** *exists across the system*. The power in the system is the product of the through and across variables. Table 10-1 lists the through and across variables for various types of dynamic systems.

We commonly speak of the voltage across a circuit and the current flowing through it. We also can speak of the velocity across a mechanical "circuit" or system and the force which flows through it. Just as we can connect electrical elements such as resistors, capacitors, and inductors together in series or parallel or a combination of both to make an electrical circuit, we can connect their mechanical analogs, dampers, springs, and masses together in series, parallel, or a combination thereof to make a mechanical system. Table 10-2 shows the analogs between three types of physical systems. The fundamental relationships between through and across variables in electrical, mechanical, and fluid systems are shown in Table 10-3.

TABLE 10-1   Through and Across Variables in Dynamic Systems

System Type	Through Variable	Across Variable	Power Units
Electrical	Current (i)	Voltage (e)	ei = watts
Mechanical	Force (F)	Velocity (v)	Fv = (in-lb)/sec
Fluid	Flow (Q)	Pressure (P)	PQ = (in-lb)/sec

TABLE 10-2   Physical Analogs in Dynamic Systems

System Type	Energy Dissipator	Energy Storage	Energy Storage
Electrical	Resistor (R)	Capacitor (C)	Inductor (L)
Mechanical	Damper (c)	Mass (m)	Spring (k)
Fluid	Fluid resistor (R_f)	Accumulator (C_f)	Fluid inductor (L_f)

TABLE 10-3   Relationships Between Variables in Dynamic Systems

System Type	Resistance	Capacitance	Inductance
Electrical	$i = \dfrac{1}{R} e$	$i = C \dfrac{de}{dt}$	$i = \dfrac{1}{L} \int e\, dt$
Mechanical	$F = cv$	$F = m \dfrac{dv}{dt}$	$F = k \int v\, dt$
Fluid	$Q = \dfrac{1}{R_f} P$	$Q = C_f \dfrac{dP}{dt}$	$Q = \dfrac{1}{L_f} \int P\, dt$

Recognizing a series or parallel connection between elements in an electrical circuit is fairly straightforward, as their interconnections are easily seen. Determining how mechanical elements in a system are interconnected is more difficult as their interconnections are sometimes hard to see. The test for series or parallel connection is best done by examining the forces and velocities (or the integral of velocity, displacement) that exist in the particular elements. If two elements have the same force passing through them, they are in series. If two elements have the same velocity or displacement, they are in parallel.

Combining Dampers

DAMPERS IN SERIES   Figure 10-8a shows three dampers in series. The force passing through each damper is the same, and their individual displacements and velocities are different.

$$F = c_1\left(x_1 - x_2\right) = c_2\left(x_2 - x_3\right) = c_3 x_3$$

or:

$$\frac{F}{c_1} = x_1 - x_2; \qquad \frac{F}{c_2} = x_2 - x_3; \qquad \frac{F}{c_3} = x_3$$

combining:

$$x_{total} = \left(x_1 - x_2\right) + \left(x_2 - x_3\right) + x_3 = \frac{F}{c_1} + \frac{F}{c_2} + \frac{F}{c_3}$$

then:

$$x_{total} = F\frac{1}{c_{eff}} = F\left(\frac{1}{c_1} + \frac{1}{c_2} + \frac{1}{c_3}\right)$$

$$\frac{1}{c_{eff}} = \frac{1}{c_1} + \frac{1}{c_2} + \frac{1}{c_3}$$

$$c_{eff} = \frac{1}{\dfrac{1}{c_1} + \dfrac{1}{c_2} + \dfrac{1}{c_3}} \qquad (10.18a)$$

The reciprocal of the effective damping of the dampers in series is the sum of the reciprocals of their individual damping coefficients.

DAMPERS IN PARALLEL   Figure 10-8b shows three dampers in parallel. The force passing through each damper is different, and their displacements and velocities are all the same.

$$F = F_1 + F_2 + F_3$$
$$F = c_1 x + c_2 x + c_3 x$$
$$F = \left(c_1 + c_2 + c_3\right)x$$
$$F = c_{eff}\, x$$

$$c_{eff} = c_1 + c_2 + c_3 \qquad (10.18b)$$

The effective damping of the three is the sum of their individual damping coefficients.

Combining Springs

Springs are the mechanical analog of electrical inductors. Figure 10-9a (p. 562) shows three springs in series. The force passing through each spring is the same, and their individual displacements are different. A force F applied to the system will create a total deflection which is the sum of the individual deflections. The spring force is defined from the relationship in equation 10.16 (p. 557):

(a) Series

(b) Parallel

FIGURE 10-8

Dampers in series and in parallel

10

(a) Series

$$F = k_{eff} x_{total}$$

where:

$$x_{total} = (x_1 - x_2) + (x_2 - x_3) + x_3 \qquad (10.19a)$$

$$(x_1 - x_2) = \frac{F}{k_1} \qquad (x_2 - x_3) = \frac{F}{k_2} \qquad x_3 = \frac{F}{k_3} \qquad (10.19b)$$

Substituting, we find that the reciprocal of the effective k of **springs in series** is the sum of the reciprocals of their individual spring constants.

$$\frac{F}{k_{eff}} = \frac{F}{k_1} + \frac{F}{k_2} + \frac{F}{k_3}$$

$$(10.19c)$$

$$k_{eff} = \frac{1}{\dfrac{1}{k_1} + \dfrac{1}{k_2} + \dfrac{1}{k_3}}$$

Figure 10-9b shows three springs in parallel. The force passing through each spring is different, and their displacements are all the same. The total force is the sum of the individual forces.

$$F_{total} = F_1 + F_2 + F_3 \qquad (10.20a)$$

Substituting equation 10.19b we find that the effective k of **springs in parallel** is the sum of the individual spring constants.

$$k_{eff} x = k_1 x + k_2 x + k_3 x$$

$$k_{eff} = k_1 + k_2 + k_3 \qquad (10.20b)$$

Combining Masses

Masses are the mechanical analog of electrical capacitors. The inertial forces associated with all moving masses are referenced to the ground plane of the system because the acceleration in $F = ma$ is absolute. Thus all masses are connected in parallel and combine in the same way as do capacitors in parallel with one terminal connected to a common ground.

$$m_{eff} = m_1 + m_2 + m_3 \qquad (10.21)$$

Lever and Gear Ratios

Whenever an element is separated from the point of application of a force or from another element by a **lever ratio** or **gear ratio**, its effective value will be modified by that ratio. Figure 10-10a shows a spring at one end (A) and a mass at the other end (B) of a lever. We wish to model this system as a single-*DOF* lumped parameter system. There are two possibilities in this case. We can either transfer an equivalent mass m_{eff} to point A and attach it to the existing spring k, as shown in Figure 10-10b, or we can transfer an equivalent spring k_{eff} to point B and attach it to the existing mass m as shown in Figure 10-10c. In either case, for the lumped model to be equivalent to the original system, it must have the same energy in it.

(b) Parallel

FIGURE 10-9

Springs in series and in parallel

10

(a) Physical system

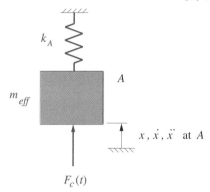

(b) Equivalent mass at point A

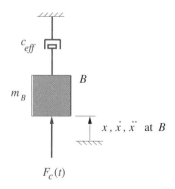

(c) Equivalent spring at point B

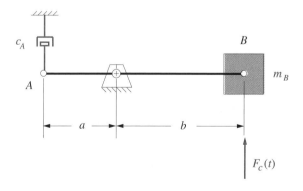

(d) Physical system

(e) Equivalent damper at point B

FIGURE 10-10

Lever or gear ratios affect the equivalent system

First find the effective mass that must be placed at point A to eliminate the lever. Equating the kinetic energies in the masses at points A and B:

$$\frac{1}{2}m_B v_B^2 = \frac{1}{2}m_{eff} v_A^2 \qquad (10.22a)$$

The velocities at each end of the lever can be related by the lever ratio:

$$v_A = \left(\frac{a}{b}\right) v_B$$

substituting:

$$m_B v_B^2 = m_{eff}\left(\frac{a}{b}\right)^2 v_B^2$$

$$m_{eff} = \left(\frac{b}{a}\right)^2 m_B \qquad (10.22b)$$

The effective mass varies from the original mass by the square of the lever ratio. Note that if the lever were instead a pair of gears of radii a and b, the result would be the same.

Now find the effective spring that would have to be placed at B to eliminate the lever. Equating the potential energies in the springs at points A and B:

$$\frac{1}{2}k_A x_A^2 = \frac{1}{2}k_{eff} x_B^2 \qquad (10.23a)$$

The deflection at B is related to the deflection at A by the lever ratio:

$$x_B = \left(\frac{b}{a}\right) x_A$$

substituting:

$$k_A x_A^2 = k_{eff}\left(\frac{b}{a}\right)^2 x_A^2$$

$$k_{eff} = \left(\frac{a}{b}\right)^2 k_A \qquad (10.23b)$$

The effective k varies from the original k by the square of the lever ratio. If the lever were instead a pair of gears of radii a and b, the result would be the same. So, gear or lever ratios can have a large effect on the lumped parameters' values in the simplified model.

Damping coefficients are also affected by the lever ratio. Figure 10-10d (p. 563) shows a damper and a mass at opposite ends of a lever. If the damper at A is to be replaced by a damper at B, then the two dampers must produce the same moment about the pivot, thus:

$$F_{d_A} a = F_{d_B} b \qquad (10.23c)$$

Substitute the product of the damping coefficient and velocity for force:

$$\left(c_A x_A\right) a = \left(c_{B_{eff}} x_B\right) b \qquad (10.23d)$$

The velocities at points A and B in Figure 10-10d can be related from kinematics:

$$\omega = \frac{x_A}{a} = \frac{x_B}{b}$$

$$x_A = x_B \frac{a}{b} \qquad (10.23e)$$

Substituting in equation 10.23d we get an expression for the effective damping coefficient at B resulting from a damper at A.

$$\left(c_A x_B \frac{a}{b} \right) a = \left(c_{B_{eff}} x_B \right) b$$

$$c_{B_{eff}} = c_A \left(\frac{a}{b} \right)^2 \qquad\qquad (10.23f)$$

Again, the square of the lever ratio determines the effective damping. The equivalent system is shown in Figure 10-10e (p. 563).

✏️ EXAMPLE 10-1

Creating a Single-*DOF* Equivalent System Model of a Multielement Dynamic System.

Given: An automotive valve cam with translating flat follower, long pushrod, rocker arm, valve, and valve spring is shown in Figure 10-11a (p. 566).

Problem: Create a suitable, approximate, single-*DOF*, lumped parameter model of the system. Define its effective mass, spring constant, and damping in terms of the individual elements' parameters.

Solution:

1   Break the system into individual elements as shown in Figure 10-11b. Each significant moving part is assigned a lumped mass element which has a connection to ground through a damper. There is also elasticity and damping within the individual elements, shown as connecting springs and dampers. The rocker arm is modeled as two lumped masses at its ends, connected with a rigid, massless rod for the crank and conrod of the slider-crank linkage. (See also Section 13.4, p. 675.) The breakdown shown represents a six-*DOF* model as there are six independent displacement coordinates, x_1 through x_6.

2   Define the individual spring constants of each element which represents the elasticity of a lumped mass from the elastic deflection formula for the particular part. For example, the pushrod is loaded in compression, so its relevant deflection formula and its k are:

$$x = \frac{Fl}{AE} \qquad \text{and} \qquad k_{pr} = \frac{F}{x} = \frac{AE}{l} \qquad\qquad (a)$$

where A is the cross-sectional area of the pushrod, l is its length, and E is Young's modulus for the material. The k of the tappet element will have the same expression. The expression for the k of a helical coil compression spring, as used for the valve spring, can be found in any spring design manual or machine design text and is:

$$k_{sp} = \frac{d^4 G}{8 D^3 N} \qquad\qquad (b)$$

where d is the wire diameter, D is the mean coil diameter, N is the number of coils, and G is the modulus of rupture of the material.

The rocker arm also acts as a spring, as it is a beam in bending. It can be modeled as a double cantilever beam with its deflection on each side of the pivot considered separately. These spring effects are shown in the model as if they were compression springs, but that is just schematic. They really represent the bending deflection of the rocker arms. From the deflection formula for a cantilever beam with concentrated load:

$$x = \frac{Fl^3}{3EI} \qquad \text{and} \qquad k_{ra} = \frac{3EI}{l^3} \qquad (c)$$

where I is the cross-sectional second moment of area of the beam, l is its length, and E is Young's modulus for the material. The spring constants of any other elements in a system can be obtained in similar fashion from their deflection formulas.

(a) Physical model

(b) Six-DOF model

(c) One-DOF model with lever arm

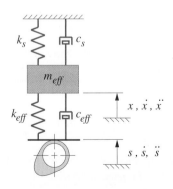

(d) One-DOF lumped model

FIGURE 10-11

Lumped parameter models of an overhead valve engine cam-follower system

3 The dampers shown connected to ground represent the friction or viscous damping at the interfaces between the elements and the ground plane. The dampers between the masses represent the internal damping in the parts, which typically is quite small. These values either will have to be estimated from experience or measured in prototype assemblies.

4 The rocker arm provides a lever ratio which must be taken into account. The strategy will be to combine all elements on each side of the lever separately into two lumped parameter models as shown in Figure 10-11c, and then transfer one of those across the lever pivot to create one, single-*DOF* model as shown in Figure 10-11d.

5 The next step is to determine the types of connections, either series or parallel, between the elements. The masses are all in parallel as they each communicate their inertial force directly to ground and have independent displacements. On the left and right sides, respectively, the effective masses are:

$$m_L = m_{tp} + m_{pr} + m_{ra} \qquad\qquad m_R = m_{rb} + m_v \qquad (d)$$

Note that m_v includes about one-third of the spring's mass to account for that portion of the spring that is moving. The two springs shown representing the bending deflection of the camshaft split the force between them, so they are in parallel and thus add directly.

$$k_{cs} = k_{cs_1} + k_{cs_2} \qquad (e)$$

Note that, for completeness, the torsional deflection of the camshaft should also be included but is omitted in this example to reduce complexity. The combined camshaft spring rate and all the other springs shown on the left side are in series as they each have independent deflections and the same force passes through them all. On the right side, the spring of the rocker arm is in series with that of the left side, but the physical valve spring is in parallel with the effective spring of the follower-train elements as it has a separate path from the effective mass at the valve to ground. (The follower-train elements all communicate back to ground through the cam pivots.) The effective spring rates of the follower-train elements for each side of the rocker arm are then:

$$k_L = \cfrac{1}{\cfrac{1}{k_{cs}} + \cfrac{1}{k_{tp}} + \cfrac{1}{k_{pr}} + \cfrac{1}{k_{ra}}} \qquad\qquad k_R = k_{rb} \qquad (f)$$

The dampers are in a combination of series and parallel. The dampers c_{cs1} and c_{cs2} supporting the camshaft represent the friction in the two camshaft bearings and are in parallel.

$$c_{cs} = c_{cs_1} + c_{cs_2} \qquad (g)$$

The ones representing internal damping are in series with one another and with the combined shaft damping.[*]

$$c_{in_L} = \cfrac{1}{\cfrac{1}{c_{tp}} + \cfrac{1}{c_{pr}} + \cfrac{1}{c_{ra}} + \cfrac{1}{c_{cs}}} \qquad\qquad c_{in_R} = c_{rb} \qquad (h)$$

where c_{in_L} is all internal damping on the left side and c_{in_R} is all internal damping on the right side of the rocker arm pivot. The combined internal damping c_{in_L} goes to ground through c_{rg} and the combined internal damping c_{in_R} goes to ground through the valve spring c_s. These two combinations are then in parallel with all the other dampers that go to ground. The combined dampings for each side of the system are then:

$$c_L = c_{tg} + c_{rg} + c_{in_L} \qquad\qquad c_R = c_{vg} + c_{in_R} \qquad (i)$$

* This analysis assumes that the internal damping values (c's) of the elements are very small and vary approximately proportionally to the stiffness (k's) of the respective elements to which they apply. Because damping is typically small in these systems, its effect on the equivalent spring rate is small, but the reverse is not true since high stiffness will affect damping levels. A very stiff element will deflect less under a given load than a less stiff one. If damping is proportional to velocity across the element, then a small deflection will have small velocity. Even if the damping coefficient of that element is large, it will have little effect on the system due to the element's relatively high stiffness. A more accurate way to estimate damping must take the interaction between the k's and c's into account. For n springs k_1, $k_2, \ldots, k_n$ in series, placed in parallel with n dampers $c_1, c_2, \ldots, c_n$ in series, the effective damping can be shown to be:

$$c_{eff} = k_{eff} \sum_{i=1}^{n} \frac{c_i}{k_i^2}$$

As a practical matter, however, it is usually quite difficult to determine the values of the individual damping elements that are needed to do a calculation such as shown above and in equation (h).

10

6 The system can now be reduced to a single-*DOF* model with masses and springs lumped on either end of the rocker arm as shown in Figure 10-11c (p. 566). We will bring the elements at point *B* across to point *A*. Note that we have reversed the sign convention across the pivot so that positive motion on one side results also in positive motion on the other. The damper, mass, and spring constant are affected by the square of the lever ratio as shown in equations 10.22 and 10.23 (pp. 564-565).

$$m_{eff} = m_L + \left(\frac{b}{a}\right)^2 m_R$$

$$k_{eff} = k_L + \left(\frac{b}{a}\right)^2 k_R \qquad (j)$$

$$c_{eff} = c_L + \left(\frac{b}{a}\right)^2 c_R$$

These are shown in Figure 10-11d (p. 566) on the final, one-*DOF* lumped model of the system. It shows all the elasticity of the follower-train elements lumped into the effective spring k_{eff} and the damping as c_{eff}. The cam's displacement input $s(t)$ acts against a massless but rigid shoe. The valve spring and the valve's damping against ground provide forces that keep the joint between cam and follower closed. If the cam and follower separate, the system changes dynamically from that shown.

Note that this one-*DOF* model provides only an approximation of this complex system's behavior. Even though it may be an oversimplification, it is nevertheless still useful as a first approximation and serves in this context as an example of the general method involved in modelling dynamic systems. A more complex model with multiple degrees of freedom will provide a better approximation of the dynamic system's behavior.

10.13 SOLUTION METHODS

Dynamic force analysis can be done by any of several methods. Two will be discussed here, **superposition** and **linear simultaneous equation solution**. Both methods require that the system be linear.

These dynamic force problems typically have a large number of unknowns and thus have multiple equations to solve. The method of superposition attacks the problem by solving for parts of the solution and then adding (superposing) the partial results together to get the complete result. For example, if there are two loads applied to the system, we solve independently for the effects of each load, and then add the results. In effect we solve an *N*-variable system by doing sequential calculations on parts of the problem. It can be thought of as a "serial processing" approach.

Another method writes all the relevant equations for the entire system as a set of linear simultaneous equations. These equations can then be solved simultaneously to obtain the results. This can be thought of as analogous to a "parallel processing" approach. A convenient approach to the solution of sets of simultaneous equations is to put them in a standard matrix form and use a numerical matrix solver to obtain the answers. Matrix solvers are built into most engineering and scientific pocket calculators. Some spreadsheet packages and equation solvers will also do a matrix solution. A brief introduction to matrix solution of simultaneous equations was presented in Section 5.6 (p. 240). Appendix A describes the use of the computer program MATRIX, included on DVD with this text. This program allows the rapid calculation of the solution to systems of up to 16 simultaneous equations. Please refer to the sections in Chapter 5 to review these calculation procedures. Reference [4] provides an introduction to matrix algebra.

We will use both superposition and simultaneous equation solution to solve various dynamic force analysis problems in the remaining chapters. Both have their place, and one can serve as a check on the results from the other. So it is useful to be familiar with more than one approach. Historically, superposition was the only practical method for systems involving large numbers of equations until computers became available to solve large sets of simultaneous equations. Now the simultaneous equation solution method is more popular.

10.14 THE PRINCIPLE OF D'ALEMBERT

Newton's second law (equations 10.1, p. 546 and 10.4, p. 549) is all that is needed to solve any dynamic force system by the newtonian method. Jean le Rond d'Alembert (1717-1783), a French mathematician, rearranged Newton's equations to create a "quasi-static" situation from a dynamic one. D'Alembert's versions of equations 10.1 and 10.4 are:

$$\sum \mathbf{F} - m\mathbf{a} = 0$$
$$\sum \mathbf{T} - I\alpha = 0$$

(10.24)

All d'Alembert did was to move the terms from the right side to the left, changing their algebraic signs in the process as required. These are obviously still the same equations as 10.1 and 10.4, algebraically rearranged. The motivation for this algebraic manipulation was to make the system look like a statics problem in which, for equilibrium, all forces and torques must sum to zero. Thus, this is sometimes called a quasi-static problem when expressed in this form. The premise is that by placing an "inertia force" equal to $-ma$ and an "inertia torque" equal to $-I\alpha$ on our free-body diagrams, the system will then be in a state of "dynamic equilibrium" and can be solved by the familiar methods of statics. These inertia forces and torques are equal in magnitude, opposite in sense, and along the same line of action as ma and $I\alpha$. This was a useful and popular approach which made the solution of dynamic force analysis problems somewhat easier when graphical vector solutions were the methods of choice.

With the availability of calculators and computers that can solve the simultaneous equations for these problems, there is now little motivation to labor through the complicated tedium of a graphical force analysis. It is for this reason that graphical force analysis methods are not presented in this text. However, d'Alembert's concept of "inertia forces and torques" still has, at a minimum, historical value and, in many instances, can prove useful in understanding what is going on in a dynamic system. Moreover, the concept of inertia force has entered the popular lexicon and is often used in a lay context when discussing motion. Thus we present a simple example of its use here and will use it again in our discussion of dynamic force analysis later in this text where it helps us to understand some topics such as balancing and superposition.

The popular term **centrifugal force**, used by laypersons everywhere to explain why a mass on a rope keeps the rope taut when swung in a circle, is in fact a d'Alembert inertial force. Figure 10-12a (p. 570) shows such a mass, being rotated at the end of a flexible but inextensible cord at a constant angular velocity ω and constant radius r. Figure 10-12b

10

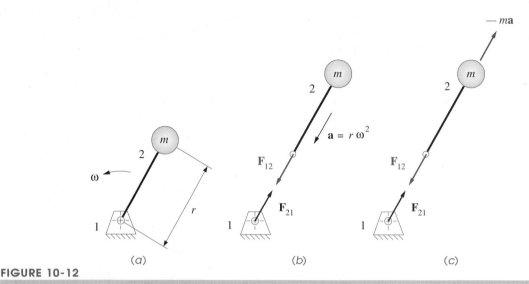

FIGURE 10-12

Centripetal and centrifugal forces

shows "pure" free-body diagrams of both members in this system, the ground link (1) and the rotating link (2). The only real force acting on link 2 is the force of link 1 on 2, F_{12}. Since angular acceleration is zero in this example, the acceleration acting on the link is only the $r\omega^2$ component, which is a **centripetal acceleration**, i.e., directed *toward the center*. The force at the pin from Newton's equation 10.1 (p. 546) is then:

$$F_{12} = mr\omega^2 \tag{10.25a}$$

Note that this force is directed toward the center, so it is a *centripetal* not a *centrifugal* (away from center) force. The force $\mathbf{F}_{21}$ which link 2 exerts on link 1 can be found from Newton's third law and is obviously equal and opposite to $\mathbf{F}_{12}$.

$$F_{21} = -F_{12} \tag{10.25b}$$

Thus it is the reaction force on link 1 which is centrifugal, not the force on link 2. Of course, it is this reaction force that your hand (link 1) feels, and this gives rise to the popular conception of something pulling centrifugally on the rotating weight. Now let us look at this through d'Alembert's eyes. Figure 10-12c shows another set of free-body diagrams done according to the principle of d'Alembert. Here we show a negative $m\mathbf{a}$ inertia force applied to the mass on link 2. The force at the pin from d'Alembert's equation is:

$$F_{12} - mr\omega^2 = 0$$
$$F_{12} = mr\omega^2 \tag{10.25c}$$

Not surprisingly, the result is the same as equation 10.25a, as it must be. The only difference is that the free-body diagram shows an inertia force applied to the rotating mass on link 2. This is the centrifugal force of popular repute which takes the credit (or blame) for keeping the cord taut.

Clearly, any problem can be solved for the right answer no matter how we may algebraically rearrange the correct equations. So, if it helps our understanding to think in terms of these inertia forces being applied to a dynamic system, we will do so. When dealing with the topic of balancing, this approach does, in fact, help to visualize the effects of the balance masses on the system.

10.15 ENERGY METHODS—VIRTUAL WORK

The newtonian methods of dynamic force analysis of Section 10.1 (p. 545) have the advantage of providing complete information about all interior forces at pin joints as well as about the external forces and torques on the system. One consequence of this fact is the relative complexity of their application which requires the simultaneous solution of large systems of equations. Other methods are available for the solution of these problems which are easier to implement but give less information. Energy methods of solution are of this type. Only the external, work-producing, forces and torques are found by these methods. The internal joint forces are not computed. One chief value of the energy approach is its use as a quick check on the correctness of the newtonian solution for input torque. Usually we are forced to use the more complete newtonian solution in order to obtain force information at pin joints so that pins and links can be analyzed for failure due to stress.

The *law of conservation of energy* states that energy can be neither created nor destroyed, only converted from one form to another. Most machines are designed specifically to convert energy from one form to another in some controlled fashion. Depending on the efficiency of the machine, some portion of the input energy will be converted to heat which cannot be completely recaptured. But large quantities of energy will typically be stored temporarily within the machine in both potential and kinetic form. It is not uncommon for the magnitude of this internally stored energy, on an instantaneous basis, to far exceed the magnitude of any useful external work being done by the machine.

Work is defined as *the dot product of force and displacement*. It can be positive, negative, or zero and is a scalar quantity.

$$W = \mathbf{F} \cdot \mathbf{R} \qquad (10.26a)$$

Since the forces at the pin joints between the links have no relative displacement associated with them, they do no work on the system, and thus will not appear in the work equation. The work done by the system plus losses is equal to the energy delivered to the system.

$$E = W + losses \qquad (10.26b)$$

Pin-jointed linkages with low-friction bearings at the pivots can have high efficiencies, above 95%. Thus it is not unreasonable, for a first approximation in designing such a mechanism, to assume the losses to be zero. **Power** is the time rate of change of energy:

$$P = \frac{dE}{dt} \qquad (10.26c)$$

Since we are assuming the machine member bodies to be rigid, only a change of position of the *CG*s of the members will alter the stored potential energy in the system. The gravitational forces of the members in moderate- to high-speed machinery often tend to be dwarfed by the dynamic forces from the accelerating masses. For these reasons we will ignore the weights and the gravitational potential energy and consider only the kinetic

10

energy in the system for this analysis. The time rate of change of the kinetic energy stored within the system for linear and angular motion, respectively, is then:

$$\frac{d\left(\frac{1}{2}m\mathbf{v}^2\right)}{dt} = m\mathbf{a} \cdot \mathbf{v} \tag{10.27a}$$

and:

$$\frac{d\left(\frac{1}{2}I\omega^2\right)}{dt} = I\alpha \cdot \omega \tag{10.27b}$$

These are, of course, expressions for power in the system, equivalent to:

$$P = \mathbf{F} \cdot \mathbf{v} \tag{10.27c}$$

and:

$$P = \mathbf{T} \cdot \omega \tag{10.27d}$$

The rate of change of energy in the system at any instant must balance between that which is externally supplied and that which is stored within the system (neglecting losses). Equations 10.27a and b represent change in the energy stored in the system, and equations 10.27c and d represent change in energy passing into or out of the system. In the absence of losses, these two must be equal in order to conserve energy. We can express this relationship as a summation of all the delta energies (or power) due to each moving element (or link) in the system.

$$\sum_{k=2}^{n} \mathbf{F}_k \cdot \mathbf{v}_k + \sum_{k=2}^{n} \mathbf{T}_k \cdot \omega_k = \sum_{k=2}^{n} m_k \mathbf{a}_k \cdot \mathbf{v}_k + \sum_{k=2}^{n} I_k \alpha_k \cdot \omega_k \tag{10.28a}$$

The subscript k represents each of the n links or moving elements in the system, starting with link 2 because link 1 is the stationary ground link. Note that all the angular and linear velocities and accelerations in this equation must have been calculated, for all positions of the mechanism of interest, from a prior kinematic analysis. Likewise, the masses and mass moments of inertia of all moving links must be known.

If we use the principle of d'Alembert to rearrange this equation, we can more easily "name" the terms for discussion purposes.

$$\sum_{k=2}^{n} \mathbf{F}_k \cdot \mathbf{v}_k + \sum_{k=2}^{n} \mathbf{T}_k \cdot \omega_k - \sum_{k=2}^{n} m_k \mathbf{a}_k \cdot \mathbf{v}_k - \sum_{k=2}^{n} I_k \alpha_k \cdot \omega_k = 0 \tag{10.28b}$$

The first two terms in equation 10.28b represent, respectively, the change in energy due to all **external forces** and all **external torques** applied to the system. These would include any forces or torques from other mechanisms which impinge upon any of these links and also includes the driving torque. The second two terms represent, respectively, the change in energy due to all **inertia forces** and all **inertia torques** present in the system. These last two terms define the change in stored kinetic energy in the system at each time step. The only unknown in this equation when properly set up is the **driving torque** (or driving force) to be supplied by the mechanism's motor or actuator. This driving torque (or force) is then the only variable which can be solved for with this approach. The internal joint forces are not present in the equation as they do no net work on the system.

Equation 10.28b is sometimes called the **virtual work equation**, which is something of a misnomer, as it is in fact a **power equation** as can be seen from its units. When this analysis approach is applied to a statics problem, there is no motion. The term **virtual work** comes from the concept of each force causing an infinitesimal, or virtual, displacement of the static system element to which it is applied over an infinitesimal delta time. The dot product of the force and the virtual displacement is the virtual work. In the limit, this becomes the instantaneous power in the system. We will present an example of the use of this method of virtual work in the next chapter along with examples of the newtonian solution applied to linkages in motion.

10.16 REFERENCES

1   **Beer, F. P., and E. R. Johnson.** (1984). *Vector Mechanics for Engineers, Statics and Dynamics*, McGraw-Hill Inc., New York.

2   **Norton, R. L.** (2011). *Machine Design: An Integrated Approach*, 4ed. Prentice-Hall, Upper Saddle River, NJ.

3   **Koster, M. P.** (1974). *Vibrations of Cam Mechanisms*. Phillips Technical Library Series, Macmillan: London.

4   **Jennings, A**. (1977). *Matrix Computation for Engineers and Scientists*. John Wiley and Sons, New York.

10.17 PROBLEMS[‡]

*[†]10-1   The mallet shown in Figure 10-2 (p. 549) has the following specifications: The steel head has a 1.25-in diameter and is 3.5 in tall; the wood handle is 1.5-in diameter and 10 in long and necks down to 3/4 in wide where it enters the head. Find the location of its composite *CG*, and its moment of inertia and radius of gyration about axis ZZ. Assume the wood has a density equal to 0.9 times that of water.

*[†]10-2   Repeat Problem 10-1 using a wooden mallet head of 2.5-in diameter. Assume the wood has a density equal to 0.85 times that of water.

[†]10-3   Calculate the location of the composite *CG*, the mass moment of inertia and the radius of gyration with respect to the specified axis, for whichever of the following commonly available items that are assigned. (Note these are not short problems.)

     a.   A good-quality writing pen, about the pivot point at which you grip it to write. (How does placing the cap on the upper end of the pen affect these parameters when you write?)

     b.   Two table knives, one metal and one plastic, about the pivot axis when held for cutting. Compare the calculated results and comment on what they tell you about the dynamic usability of the two knives (ignore sharpness considerations).

     c.   A ball-peen hammer (available in any university machine shop), about the center of rotation (after you calculate its location for the proper center of percussion).

     d.   A baseball bat (see the coach) about the center of rotation (after you calculate its location for the proper center of percussion).

     e.   A cylindrical coffee mug, about the handle hole.

*[†]10-4   Set up these equations in matrix form. Use program MATRIX, *Mathcad*, or a calculator which has matrix math capability to solve them.

     a. $-5x \quad -2y \quad +12z \quad -w \quad = -9$

          $x \quad +3y \quad -2z \quad +4w \quad = 10$

10

[‡] All problem figures are provided as PDF files, and some are also provided as animated Working Model files; all are on the DVD. PDF filenames are the same as the figure number. Run the file *Animations.html* to access and run the animations.

[*] Answers in Appendix F.

[†] These problems are suited to solution using *Mathcad*, *Matlab*, or *TKSolver* equation solver programs.

$$
\begin{array}{lllll}
-x & -y & +z & & = -7 \\
3x & -3y & +7z & +9w & = -6
\end{array}
$$

b.
$$
\begin{array}{lllll}
3x & -5y & +17z & -5w & = -5 \\
-2x & +9y & -14z & +6w & = 22 \\
-x & -y & & -2w & = 13 \\
4x & -7y & +8z & +4w & = -9
\end{array}
$$

[†]10-5   Figure P10-1 shows a bracket made of steel.

    a.   Find the location of its centroid referred to point B.
    b.   Find its mass moment of inertia I_{xx} about the X axis through point B.
    c.   Find its mass moment of inertia I_{yy} about the Y axis through point B.

[*†]10-6   Two springs are connected in series. One has a k of 34 and the other a k of 3.4. Calculate their effective spring constant. Which spring dominates? Repeat with the two springs in parallel. Which spring dominates? (Use any unit system.)

[†]10-7   Repeat Problem 10-6 with k_1 = 125 and k_2 = 25. (Use any unit system.)

[†]10-8   Repeat Problem 10-6 with k_1 = 125 and k_2 = 115. (Use any unit system.)

[*†]10-9   Two dampers are connected in series. One has a damping factor c_1 = 12.5 and the other, c_2 = 1.2. Calculate their effective damping constant. Which damper dominates? Repeat with the two dampers in parallel. Which damper dominates? (Use any units.)

[†]10-10   Repeat Problem 10-9 with c_1 = 12.5 and c_2 = 2.5. (Use any unit system.)

[†]10-11   Repeat Problem 10-9 with c_1 = 12.5 and c_2 = 10. (Use any unit system.)

[*†]10-12   A mass of m = 2.75 and a spring with k = 48 are attached to one end of a lever at a radius of 5. Calculate the effective mass and effective spring constant at a radius of 10 on the same lever. (Use any unit system.)

[†]10-13   A mass of m = 1.5 and a spring with k = 24 are attached to one end of a lever at a radius of 3. Calculate the effective mass and effective spring constant at a radius of 10 on the same lever. (Use any unit system.)

FIGURE P10-1

Problem 10-5

FIGURE P10-2

Problems 10-16, 10-17, 10-21, 10-26 and 10-29

*†10-14   A mass of m = 6.5 and a spring with k = 25 are attached to one end of a lever at a radius of 15. Calculate the effective mass and effective spring constant at a radius of 5 on the same lever. (Use any unit system.)

†10-15   Refer to Figure 10-11 (p. 566) and Example 10-1 (p. 565). The dimensions for the valve train are (in ips unit system): Tappet is a solid cylinder 0.75 diameter by 1.25 long. Pushrod is a hollow tube. 0.375 outside diameter by 0.25 inside diameter by 12 long. Rocker arm has an average cross section of 1 wide by 1.5 high. Length a = 2, b = 3. Camshaft is 1 diameter by 3 between bearing supports, cam in center. Valve spring k = 200. All parts are steel. Find the effective spring constant and effective mass of a single-*DOF* equivalent system placed on the cam side of the rocker arm.

†10-16   Figure P10-2 shows a cam-follower system. The dimensions of the solid, rectangular 2 x 2.5 in cross-section aluminum arm are given. The cutout for the 2-in dia by 1.5-in wide steel roller follower is 3 in long. Find the arm's mass, center of gravity location, and mass moment of inertia about both its *CG* and the arm pivot. Create a linear, one-*DOF* lumped mass model of the dynamic system referenced to the cam-follower. Ignore damping.

†10-17   The cam in Figure P10-2 is a pure eccentric with eccentricity = 0.5 in and turns at 500 rpm. The spring has a rate of 123 lb/in and a preload of 173 lb. Use the method of virtual work to find the torque required to rotate the cam through one revolution. Use the data from the solution to Problem 10-16.

†10-18   The cam in Figure P10-3 is a pure eccentric with eccentricity = 20 mm and turns at 200 rpm. The mass of the follower is 1 kg. The spring has a rate of 10 N/m and a preload of 0.2 N. Use the method of virtual work to find the torque required to rotate the cam through one revolution.

†10-19   Repeat Problem 10-18 using a cam with a 20-mm, symmetric, double harmonic rise in 180° and double harmonic fall in 180°. See Chapter 8 for cam formulas.

*†10-20   A 3000-lb automobile has a final drive ratio of 1:3 and transmission gear ratios of 1:4, 1:3, 1:2, and 1:1 in first through fourth speeds, respectively. What is the effective mass of the vehicle as felt at the engine flywheel in each gear?

$a \cos \omega t$

FIGURE P10-3

Problems 10-18 to 10-19

* Answers in Appendix F.

† These problems are suited to solution using *Mathcad, Matlab,* or *TKSolver* equation solver programs.

10

FIGURE P10-4

Problem 10-27

*†10-21   Determine the effective spring constant and effective preload of the spring in Figure P10-2 (p. 575) as reflected back to the cam-follower. See Problem 10-17 for additional data.

†10-22   What is the effective inertia of a load applied at the drum of Figure P9-5a (p. 536) as reflected back to gear A?

†10-23   What is the effective inertia of a load W applied at the drum of Figure P9-7b (p. 538) as reflected back to the arm?

†10-24   Refer to Figure 10-10 (p. 563). Given $a = 100$ mm, $b = 150$ mm, $k_A = 2000$ N/m, and $m_B = 2$ kg, find the equivalent mass at point A and the equivalent spring at point B.

*†10-25   Repeat Problem 10-24 with $a = 50$ mm, $b = 150$ mm, $k_A = 1000$ N/m, and $m_B = 3$ kg.

*†10-26   For the cam-follower arm in Figure P10-2, determine a new location for its fixed pivot that will have zero reaction force when the cam applies its force to the follower.

†10-27   Figure P10-4 shows a fourbar mechanism. The crank is 1.00 in wide by 0.5 in thick. The coupler and rocker are both 0.75 in wide by 0.5 in thick. All links are made from steel. The ends of the links have a full radius equal to one half of the link width. The pivot pins all have a diameter of 0.25 in. Find the moment of inertia of the crank and the rocker about their fixed pivots and the moment of inertia of the coupler about its CG.

†10-28   The rocker in Figure 10-11a (p. 566) has the following dimensions: $a = 50.8$ mm, $b = 76.2$ mm. Its total weight is 10.1 N and, when supported on knife-edges at A and B, the weights at the supports were found to be 4.3 N and 5.8 N, respectively. The rocker was supported at its pivot point with a low-friction ball bearing and the period of oscillation was found to be 0.75 sec. What is the approximate moment of inertia of the rocker with respect to its pivot axis?

†10-29   The arm in Problem 10-16 and Figure P10-2 has been redesigned such that the cross-section is no longer uniform and the material changed from aluminum to steel. However, the dimensions shown in the figure remain unchanged. The new arm has a total weight of 15.3 lb and, when supported on knife-edges at points 9.5 in to the left of the pivot and 17.5 in to the right of the pivot, the weights at the supports were found to be 7.1 lb and 8.2 lb, respectively. The arm was supported at its pivot point with a low-friction ball bearing and the period of oscillation was found to be 2.0 sec. What is the approximate moment of inertia of the arm with respect to its pivot axis?

†10-30   Figure P10-5 shows a cam-follower system that drives slider 6 through an external output arm 3. Arms 2 and 3 are both rigidly attached to the 0.75-in-dia shaft X-X, which rotates in bearings that are supported by the housing. The pin-to-pin dimensions of the links are shown. The cross-sections of arms 2, 3, and 5 are solid, rectangular 1.5 × 0.75 in steel. The ends of these links have a full radius equal to one-half of the link width. Link 4 is 1-in-dia × 0.125 wall round steel tubing. Link 6 is a 2-in-dia × 6-in-long solid steel cylinder. Find the effective mass and effective spring constant of the follower train referenced to the cam-follower roller if the spring at A has a rate of 150 lb/in.

†10-31   The spring in Figure P10-5 has a rate of 150 lb/in with a preload of 60 lb. Determine the effective spring constant and preload of the spring as reflected back to the cam-follower. See Problem 10-30 for detail about the system needed to solve the problem.

†10-32   A company wants to manufacture chimes that are made from hollow tubes of various lengths. Regardless of length they will be hung from a hole that is 25 mm from one end of the tube. Develop an equation that will give the distance from this hole to the point where the chime should be struck such that there will be zero reaction force at the hole where the chime is hung. The distance should be a function of the length (L),

---

* Answers in Appendix F.

† These problems are suited to solution using *Mathcad, Matlab,* or *TKSolver* equation solver programs.

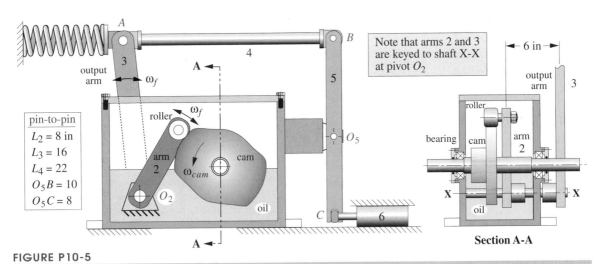

FIGURE P10-5

Problems 10-30 to 10-31 and 10-33

outside diameter (*OD*), and inside diameter (*ID*) of the chime as well as the distance from the end to the hanging hole (25 mm) only. Solve your equation for the following dimensions: $L = 300$ mm, $OD = 35$ mm, $ID = 30$ mm.

†10-33   What is the amount by which the roller arm of Problem 10-30 must be extended on the opposite side of the pivot axis O_2 in order to make the pivot axis a *center of rotation* if the point where the cam-follower is mounted is a *center of percussion*?

**†10-34   Figure P7-30d (p. 398) shows a sixbar mechanism with link lengths given in centimeters. The crank (2) is 30 mm wide by 10 mm thick. The couplers (3 and 5) are both 24 mm wide by 8 mm thick. The rocker (4) is 40 mm wide by 12 mm thick. All links are made from steel. The ends of the links have a full radius equal to one-half of the link width. The pivot pins all have a diameter of 8 mm. Find the moment of inertia of the crank and rocker about their fixed pivots and the moment of inertia of the couplers about their *CG*s.

**†10-35   A certain baseball bat has a mass of 1 kg and a mass moment of inertia about its *CG* of 0.08 kg-m². Its *CG* is located 630 mm from the end closest to the grip. If the center of a batter's grip is located 75 mm from the same end of the bat, at what point on the bat (measured from the end closest to the grip) should the batter hit the ball to produce no reaction at the grip?

†10-36   The cam of Example 8-8 (p. 438) drives an aligned translating roller follower. The effective mass of the follower and the mechanism that it actuates is 0.45 kg. The follower spring has a rate of 8 N/m with a preload at zero displacement of 0.3 N. Use the method of virtual work to find and plot the torque required to rotate the cam through one rise-fall segment.

10-37   Figure P10-6a shows a typical binary link with full-radius ends. Figure P10-6b shows a full-radius end and gives moments of inertia about its *CG* and about an axis through the center point of the radius R. Table P10-1 gives data for the length L between holes of diameter d; the end radius R; the thickness t; and the material of the link. For the row(s) assigned, find the moment of inertia of the link about one fixed pivot and about its *CG*.

†10-38   Using the definition of the binary link in Problem 10-37, write a computer program or use an equation solver to calculate the moment of inertia of the link about either of its fixed pivots and about its *CG*. Use the data in row a of Table P10-1 to test your program.

---

* Answers in Appendix F.

---

† These problems are suited to solution using *Mathcad*, *Matlab*, or *TKSolver* equation solver programs.

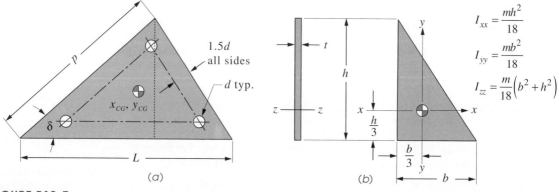

FIGURE P10-6

Problems 10-37 and 10-38

10-39   Figure P10-7a shows a simplified ternary link (the vertices would normally be rounded). Table P10-1 gives data for the length L, of the base of the link; the angle δ between the base and one side; the distance p along that side; the diameter d of the three holes; the thickness t; and the material of the link. Figure P10-7b shows a (right) triangular plate and gives the moment of inertia about its CG. Break the ternary link of (a) into two right triangles and for the row(s) assigned, find the moment of inertia of the link about its CG by using the parallel axis theorem to combine the two triangles. As a first approximation, ignore the holes.

† These problems are suited to solution using *Mathcad, Matlab,* or *TKSolver* equation solver programs.

†10-40   Using the definition of the ternary link in Problem 10-39, write a computer program or use an equation solver to calculate the moment of inertia of the link in (a) about its CG. Use the data in row a of Table P10-1 to test your program. Include the effect of the holes in your calculation.

TABLE P10-1   Data for Problems 10-37 to 10-40   Lengths in mm, angles in deg.

Row	L	R	d	t	p	δ	Material
a	225	13	8	12	185	20	Steel
b	50	10	6	12	37	45	Steel
c	175	15	10	15	125	60	Aluminum
d	75	12	8	8	50	50	Titanium
e	187	18	12	12	150	30	Aluminum
f	138	12	12	10	75	70	Steel

FIGURE P10-7

Problems 10-39 and 10-40

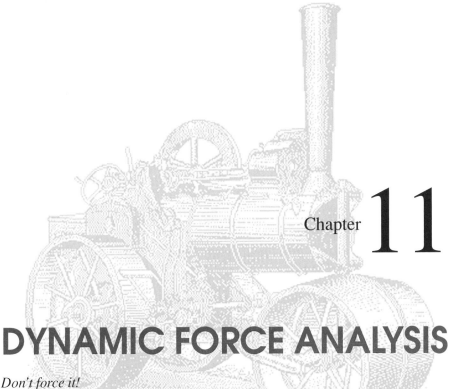

Chapter **11**

DYNAMIC FORCE ANALYSIS

Don't force it!
Use a bigger hammer
ANONYMOUS

11.0 INTRODUCTION

When kinematic synthesis and analysis have been used to define a geometry and set of motions for a particular design task, it is logical and convenient to then use a **kineto-static,** or **inverse dynamics,** solution to determine the forces and torques in the system. We will take that approach in this chapter and concentrate on solving for the forces and torques that result from, and are required to drive, our kinematic system in such a way as to provide the designed accelerations. Numerical examples are presented throughout this chapter. These examples are also provided as disk files for input to either program MATRIX or LINKAGES. These programs are described in Appendix A. The reader is encouraged to open the referenced files in these programs and investigate the examples in more detail. The file names are noted in the discussion of each example.

11.1 NEWTONIAN SOLUTION METHOD

Dynamic force analysis can be done by any of several methods. The one which gives the most information about forces internal to the mechanism requires only the use of Newton's law as defined in equations 10.1 (p. 546) and 10.4 (p. 549). These can be written as a summation of all forces and torques in the system.

$$\sum \mathbf{F} = m\mathbf{a} \qquad\qquad \sum \mathbf{T} = I_G \alpha \qquad\qquad (11.1a)$$

It is also convenient to separately sum force components in X and Y directions, with the coordinate system chosen for convenience. The torques in our two-dimensional system are all in the Z direction. This lets us break the two vector equations into three scalar equations:

$$\sum F_x = ma_x \qquad\qquad \sum F_y = ma_y \qquad\qquad \sum T = I_G \alpha \qquad\qquad (11.1b)$$

These three equations must be written for each moving body in the system which will lead to a set of linear simultaneous equations for any system. The set of simultaneous equations can most conveniently be solved by a matrix method as was shown in Chapter 5. These equations do not account for the gravitational force (weight) on a link. If the kinematic accelerations are large compared to gravity, which is often the case, then the weight forces can be ignored in the dynamic analysis. If the machine members are very massive or moving slowly with small kinematic accelerations, or both, the weight of the members may need to be included in the analysis. The weight can be treated as an external force acting on the CG of the member at a constant angle.

11.2   SINGLE LINK IN PURE ROTATION

As a simple example of this solution procedure, consider the single link in pure rotation shown in Figure 11-1a. In any of these kinetostatic dynamic force analysis problems, the kinematics of the problem must first be fully defined. That is, the angular accelerations of all rotating members and the linear accelerations of the CGs of all moving members must be found for all positions of interest. The mass of each member and the mass moment of inertia I_G with respect to each member's CG must also be known. In addition there may be external forces or torques applied to any member of the system. These are all shown in the figure.

While this analysis can be approached in many ways, it is useful for the sake of consistency to adopt a particular arrangement of coordinate systems and stick with it. We present such an approach here which, if carefully followed, will tend to minimize the chances of error. The reader may wish to develop his or her own approach once the principles are understood. The underlying mathematics is invariant, and one can choose coordinate systems for convenience. The vectors which are acting on the dynamic system in any loading situation are the same at a particular time regardless of how we may decide to resolve them into components for the sake of computation. The solution result will be the same.

We will first set up a nonrotating, local coordinate system on each moving member, located at its CG. (In this simple example we have only one moving member.) All externally applied forces, whether due to other connected members or to other systems must then have their points of application located in this local coordinate system. Figure 11-1b shows a free-body diagram of the moving link 2. The pin joint at O_2 on link 2 has a force $\mathbf{F}_{12}$ due to the mating link 1, the x and y components of which are F_{12x} and F_{12y}. These subscripts are read "force of link 1 on 2" in the x or y direction. This subscript notation scheme will be used consistently to indicate which of the "action-reaction" pair of forces at each joint is being solved for.

Note: x, y is a local, nonrotating coordinate system (LNCS), attached to the link

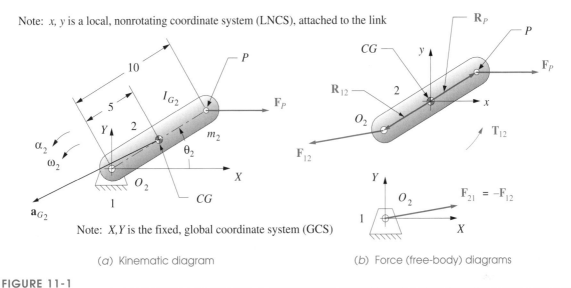

Note: X, Y is the fixed, global coordinate system (GCS)

(a) Kinematic diagram   (b) Force (free-body) diagrams

FIGURE 11-1

Dynamic force analysis of a single link in pure rotation

There is also an externally applied force $\mathbf{F}_P$ shown at point P, with components F_{P_x} and F_{P_y}. The points of application of these forces are defined by position vectors $\mathbf{R}_{12}$ and $\mathbf{R}_P$, respectively. These position vectors are defined with respect to the local coordinate system at the CG of the member. We will need to resolve them into x and y components. There will have to be a source torque available on the link to drive it at the kinematically defined accelerations. This is one of the unknowns to be solved for. The source torque is the torque delivered *from the ground to the driver link 2* and so is labeled $\mathbf{T}_{12}$. The other two unknowns in this example are the force components at the pin joint F_{12x} and F_{12y}.

We have three unknowns and three equations, so the system can be solved. Equations 11.1 can now be written for the moving link 2. Any applied forces or torques whose directions are known must retain the proper signs on their components. We will assume all unknown forces and torques to be positive. Their true signs will "come out in the wash."

$$\sum \mathbf{F} = \mathbf{F}_P + \mathbf{F}_{12} = m_2 \mathbf{a}_G$$
$$\sum \mathbf{T} = \mathbf{T}_{12} + \left(\mathbf{R}_{12} \times \mathbf{F}_{12}\right) + \left(\mathbf{R}_P \times \mathbf{F}_P\right) = I_G \alpha \qquad (11.2)$$

The force equation can be broken into its two components. The torque equation contains two cross product terms which represent torques due to the forces applied at a distance from the CG. When these cross products are expanded, the system of equations becomes:

$$F_{P_x} + F_{12_x} = m_2 a_{G_x}$$
$$F_{P_y} + F_{12_y} = m_2 a_{G_y} \qquad (11.3)$$
$$T_{12} + \left(R_{12_x} F_{12_y} - R_{12_y} F_{12_x}\right) + \left(R_{P_x} F_{P_y} - R_{P_y} F_{P_x}\right) = I_G \alpha$$

This can be put in matrix form with the coefficients of the unknown variables forming the **A** matrix, the unknown variables the **B** vector, and the constant terms the **C** vector and then solved for **B**.

$$[\mathbf{A}] \qquad \times \quad [\mathbf{B}] = \qquad [\mathbf{C}]$$

$$\begin{bmatrix} 1 & 0 & 0 \\ 0 & 1 & 0 \\ -R_{12_y} & R_{12_x} & 1 \end{bmatrix} \times \begin{bmatrix} F_{12_x} \\ F_{12_y} \\ T_{12} \end{bmatrix} = \begin{bmatrix} m_2 a_{G_x} - F_{P_x} \\ m_2 a_{G_y} - F_{P_y} \\ I_G \alpha - \left(R_{P_x} F_{P_y} - R_{P_y} F_{P_x} \right) \end{bmatrix} \qquad (11.4)$$

Note that the **A** matrix contains all the geometric information and the **C** matrix contains all the dynamic information about the system. The **B** matrix contains all the unknown forces and torques. We will now present a numerical example to reinforce your understanding of this method.

EXAMPLE 11-1

Dynamic Force Analysis of a Single Link in Pure Rotation. (See Figure 11-1, p. 581)

Given: The 10-in-long link shown weighs 4 lb. Its *CG* is on the line of centers at the 5-in point. Its mass moment of inertia about its *CG* is 0.08 lb-in-sec^2. Its kinematic data are:

θ_2 deg	ω_2 rad/sec	α_2 rad/sec^2	a_{G_2} in/sec^2
30	20	15	2001 @ 208°

An external force of 40 lb at 0° is applied at point *P*.

Find: The **force** F_{12} at pin joint O_2 and the driving **torque** T_{12} needed to maintain motion with the given acceleration for this instantaneous position of the link.

Solution:

1   Convert the given weight to proper mass units, in this case blobs:

$$mass = \frac{weight}{g} = \frac{4 \text{ lb}}{386 \text{ in/sec}^2} = 0.0104 \text{ blob} \qquad (a)$$

2   Set up a local coordinate system at the *CG* of the link and draw all applicable vectors acting on the system as shown in the figure. Draw a free-body diagram as shown.

3   Calculate the *x* and *y* components of the position vectors $\mathbf{R}_{12}$ and $\mathbf{R}_P$ in this coordinate system:

$$\mathbf{R}_{12} = 5 \text{ in @ } \angle 210°; \qquad R_{12_x} = -4.33, \qquad R_{12_y} = -2.50$$
$$\mathbf{R}_P = 5 \text{ in @ } \angle 30°; \qquad R_{P_x} = +4.33, \qquad R_{P_y} = +2.50 \qquad (b)$$

4   Calculate the *x* and *y* components of the acceleration of the *CG* in this coordinate system:

$$\mathbf{a}_G = 2001 @ \angle 208°; \qquad a_{G_x} = -1766.78, \qquad a_{G_y} = -939.41 \qquad (c)$$

5   Calculate the x and y components of the external force at P in this coordinate system:

$$\mathbf{F}_P = 40 \ @ \ \angle 0°; \qquad\qquad F_{P_x} = 40, \qquad\qquad F_{P_y} = 0 \qquad\qquad (d)$$

6   Substitute these given and calculated values into the matrix equation 11.4:

$$\begin{bmatrix} 1 & 0 & 0 \\ 0 & 1 & 0 \\ 2.50 & -4.33 & 1 \end{bmatrix} \times \begin{bmatrix} F_{12_x} \\ F_{12_y} \\ T_{12} \end{bmatrix} = \begin{bmatrix} (0.01)(-1766.78) - 40 \\ (0.01)(-939.41) - 0 \\ (0.08)(15) - \{(4.33)(0) - (2.5)(40)\} \end{bmatrix}$$

$$(e)$$

$$\begin{bmatrix} 1 & 0 & 0 \\ 0 & 1 & 0 \\ 2.50 & -4.33 & 1 \end{bmatrix} \times \begin{bmatrix} F_{12_x} \\ F_{12_y} \\ T_{12} \end{bmatrix} = \begin{bmatrix} -57.67 \\ -9.39 \\ 101.2 \end{bmatrix}$$

7   Solve this system either by inverting matrix **A** and premultiplying that inverse times matrix **C** using a pocket calculator with matrix capability, with *Mathcad, Matlab*; or by putting the values for matrices **A** and **C** into program MATRIX provided with this text.

Program MATRIX gives the following solution:

$$F_{12_x} = -57.67 \text{ lb}, \qquad F_{12_y} = -9.39 \text{ lb}, \qquad T_{12} = 204.72 \text{ lb-in} \qquad (f)$$

Converting the force to polar coordinates:

$$\mathbf{F}_{12} = 58.43 \ @ \ \angle 189.25° \qquad\qquad (g)$$

Open the disk file E11-01.mtr in program MATRIX to exercise this example.

11.3   FORCE ANALYSIS OF A THREEBAR CRANK-SLIDE LINKAGE

11

When there is more than one link in the assembly, the solution simply requires that the three equations 11.1b (p. 580) be written for each link and then solved simultaneously. Figure 11-2a (p. 584) shows a threebar crank-slide linkage. This linkage has been simplified from the fourbar slider-crank (see Figure 11-4, p. 597) by replacing the kinematically redundant slider block (link 4) with a half joint as shown. This linkage transformation reduces the number of links to three with no change in degree of freedom (see Section 2.10, p. 48). Only links 2 and 3 are moving. Link 1 is ground. Thus we should expect to have six equations in six unknowns (three per moving link).

Figure 11-2b shows the linkage "exploded" into its three separate links, drawn as free bodies. A kinematic analysis must have been done in advance of this dynamic force analysis in order to determine, for each moving link, its angular acceleration and the linear acceleration of its *CG*. For the kinematic analysis, only the link lengths from pin to pin were required. For a dynamic analysis the mass (m) of each link, the location of its *CG*, and its mass moment of inertia (I_G) about that *CG* are also needed.

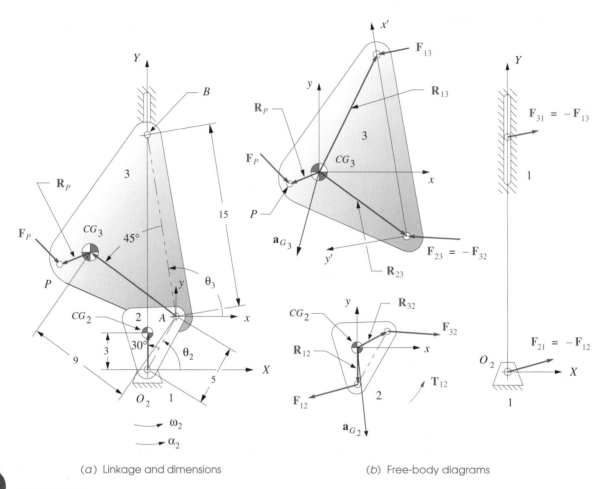

(a) Linkage and dimensions　　　　　　　　　　(b) Free-body diagrams

FIGURE 11-2

Dynamic force analysis of a crank-slide linkage

The *CG* of each link is initially defined by a position vector rooted at one pin joint whose angle is measured with respect to the line of centers of the link in the local, rotating coordinate system (LRCS) x', y'. This is the most convenient way to establish the *CG* location since the link line of centers is the kinematic definition of the link. However, we will need to define the link's dynamic parameters and force locations with respect to a local, nonrotating coordinate system (LNCS) x, y located at its *CG* and which is always parallel to the global coordinate system (GCS) *XY*. The position vector locations of all attachment points of other links and points of application of external forces must be defined with respect to the link's LNCS. Note that these kinematic and applied force data must be available for all positions of the linkage for which a force analysis is desired. In the following discussion and examples, only one linkage position will be addressed. The process is identical for each succeeding position and only the calculations must be repeated. Obviously, a computer will be a valuable aid in accomplishing the task.

Link 2 in Figure 11-2b shows forces acting on it at each pin joint, designated $\mathbf{F}_{12}$ and $\mathbf{F}_{32}$. By convention their subscripts denote the force that the adjoining link is exerting *on* the link being analyzed; that is, $\mathbf{F}_{12}$ is the force of 1 *on* 2 and $\mathbf{F}_{32}$ is the force of 3 *on* 2. Obviously there is also an equal and opposite force at each of these pins which would be designated as $\mathbf{F}_{21}$ and $\mathbf{F}_{23}$, respectively. The choice of which of the members of these pairs of forces to be solved for is arbitrary. As long as proper bookkeeping is done, their identities will be maintained.

When we move to link 3, we maintain the same convention of showing forces acting *on* the link in its free-body diagram. Thus at instant center I_{23} we show $\mathbf{F}_{23}$ acting on link 3. However, because we showed force $\mathbf{F}_{32}$ acting at the same point on link 2, this introduces an additional unknown to the problem for which we need an additional equation. The equation is available from Newton's third law:

$$\mathbf{F}_{23} = -\mathbf{F}_{32} \tag{11.5}$$

Thus we are free to substitute the negative reaction force for any action force at any joint. This has been done on link 3 in the figure in order to reduce the unknown forces at that joint to one, namely $\mathbf{F}_{32}$. The same procedure is followed at each joint with one of the action-reaction forces arbitrarily chosen to be solved for and its negative reaction applied to the mating link.

The naming convention used for the position vectors ($\mathbf{R}_{ap}$) which locate the pin joints with respect to the *CG* in the link's nonrotating local coordinate system is as follows. The first subscript (a) denotes the adjoining link to which the position vector points. The second subscript (p) denotes the parent link to which the position vector belongs. Thus in the case of link 2 in Figure 11-2b, vector $\mathbf{R}_{12}$ locates the attachment point of link 1 to link 2, and $\mathbf{R}_{32}$ the attachment point of link 3 to link 2. Note that in some cases these subscripts will match those of the pin forces shown acting at those points, but where the negative reaction force has been substituted as described above, the subscript order of the force and its position vector will not agree. This can lead to confusion and must be carefully watched for typographical errors when setting up the problem.

Any external forces acting on the links are located in similar fashion with a position vector to a point on the line of application of the force. This point is given the same letter subscript as that of the external force. Link 3 in the figure shows such an external force $\mathbf{F}_P$ acting on it at point P. The position vector $\mathbf{R}_P$ locates that point with respect to the *CG*. It is important to note that the *CG* of each link is consistently taken as the point of reference for all forces acting on that link. Left to its own devices, an unconstrained body in complex motion will spin about its own *CG*; thus we analyze its linear acceleration at that point and apply the angular acceleration about the *CG* as a center.

Equations 11.1 (p. 580) are now written for each moving link. For link 2, with the cross products expanded:

$$F_{12_x} + F_{32_x} = m_2 a_{G_{2x}}$$
$$F_{12_y} + F_{32_y} = m_2 a_{G_{2y}} \tag{11.6a}$$
$$T_{12} + \left(R_{12_x} F_{12_y} - R_{12_y} F_{12_x} \right) + \left(R_{32_x} F_{32_y} - R_{32_y} F_{32_x} \right) = I_{G_2} \alpha_2$$

For link 3, with the cross products expanded, note the substitution of the reaction force $-\mathbf{F}_{32}$ for $\mathbf{F}_{23}$:

$$F_{13_x} - F_{32_x} + F_{P_x} = m_3 a_{G_{3x}}$$
$$F_{13_y} - F_{32_y} + F_{P_y} = m_3 a_{G_{3y}} \qquad (11.6b)$$
$$\left(R_{13_x} F_{13_y} - R_{13_y} F_{13_x}\right) - \left(R_{23_x} F_{32_y} - R_{23_y} F_{32_x}\right) + \left(R_{P_x} F_{P_y} - R_{P_y} F_{P_x}\right) = I_{G_3}\alpha_3$$

Note also that T_{12}, the source torque, only appears in the equation for link 2 as that is the driver crank to which the motor is attached. Link 3 has no externally applied torque but does have an external force F_P which might be due to whatever link 3 is pushing on to do its external work.

There are seven unknowns present in these six equations, $F_{12x}, F_{12y}, F_{32x}, F_{32y}, F_{13x}, F_{13y}$, and T_{12}. But, F_{13y} is due only to friction at the joint between link 3 and link 1. We can write a relation for the friction force f at that interface such as $f = \pm\mu N$, where $\pm\mu$ is a known coefficient of coulomb friction. The friction force always opposes motion. The kinematic analysis will provide the velocity of the link at the sliding joint. The direction of f will always be the opposite of this velocity. Note that μ is a nonlinear function which has a discontinuity at zero velocity; thus at the linkage positions where velocity is zero, the inclusion of μ in these linear equations is not valid. (See Figure 10-5a, p. 555.) In this example, the normal force N is equal to F_{13x} and the friction force f is equal to F_{13y}. For linkage positions with nonzero velocity, we can eliminate F_{13y} by substituting into equation 11.6b,

$$F_{13_y} = \mu F_{13_x} \qquad (11.6c)$$

where the sign of F_{13y} is taken as the opposite of the sign of the velocity at that point. We are then left with six unknowns in equations 11.6 and can solve them simultaneously. We also rearrange equations 11.6a and 11.6b to put all known terms on the right side.

$$F_{12_x} + F_{32_x} = m_2 a_{G_{2x}}$$
$$F_{12_y} + F_{32_y} = m_2 a_{G_{2y}}$$
$$T_{12} + R_{12_x} F_{12_y} - R_{12_y} F_{12_x} + R_{32_x} F_{32_y} - R_{32_y} F_{32_x} = I_{G_2}\alpha_2$$

$$(11.6d)$$

$$F_{13_x} - F_{32_x} = m_3 a_{G_{3x}} - F_{P_x}$$
$$\pm \mu F_{13_x} - F_{32_y} = m_3 a_{G_{3y}} - F_{P_y}$$
$$\left(\pm\mu R_{13_x} - R_{13_y}\right)F_{13_x} - R_{23_x} F_{32_y} + R_{23_y} F_{32_x} = I_{G_3}\alpha_3 - R_{P_x} F_{P_y} + R_{P_y} F_{P_x}$$

Putting these six equations in matrix form we get:

$$
\begin{bmatrix}
1 & 0 & 1 & 0 & 0 & 0 \\
0 & 1 & 0 & 1 & 0 & 0 \\
-R_{12_y} & R_{12_x} & -R_{32_y} & R_{32_x} & 0 & 1 \\
0 & 0 & -1 & 0 & 1 & 0 \\
0 & 0 & 0 & -1 & \mu & 0 \\
0 & 0 & R_{23_y} & -R_{23_x} & \left(\mu R_{13_x} - R_{13_y}\right) & 0
\end{bmatrix}
\times
\begin{bmatrix}
F_{12_x} \\
F_{12_y} \\
F_{32_x} \\
F_{32_y} \\
F_{13_x} \\
T_{12}
\end{bmatrix}
=
\begin{bmatrix}
m_2 a_{G_{2x}} \\
m_2 a_{G_{2y}} \\
I_{G_2}\alpha_2 \\
m_3 a_{G_{3x}} - F_{P_x} \\
m_3 a_{G_{3y}} - F_{P_y} \\
I_{G_3}\alpha_3 - R_{P_x} F_{P_y} + R_{P_y} F_{P_x}
\end{bmatrix}
\qquad (11.7)
$$

This system can be solved by using program MATRIX or any other matrix solving calculator. As an example of this solution consider the following linkage data.

✍EXAMPLE 11-2

Dynamic Force Analysis of a Threebar Crank-Slide Linkage with Half Joint. (See Figure 11-2, p. 584.)

Given: The 5-in long crank (link 2) shown weighs 2 lb. Its *CG* is at 3 in and 30° from the line of centers. Its mass moment of inertia about its *CG* is 0.05 lb-in-sec². Its acceleration is defined in its LNCS, *x,y*. Its kinematic data are:

θ_2 deg	ω_2 rad/sec	α_2 rad/sec²	a_{G_2} in/sec²
60	30	−10	2700.17 @ −89.4°

The coupler (link 3) is 15 in long and weighs 4 lb. Its *CG* is at 9 in and 45° from the line of centers. Its mass moment of inertia about its *CG* is 0.10 lb-in-sec². Its acceleration is defined in its LNCS, *x,y*. Its kinematic data are:

θ_3 deg	ω_3 rad/sec	α_3 rad/sec²	a_{G_3} in/sec²
99.59	−8.78	−136.16	3453.35 @ 254.4°

The sliding joint on link 3 has a velocity of 96.95 in/sec in the +*Y* direction.

There is an external force of 50 lb at − 45°, applied at point *P* which is located at 2.7 in and 101° from the *CG* of link 3, measured in the link's embedded, rotating coordinate system or LRCS *x′, y′* (origin at *A* and *x* axis from *A* to *B*). The coefficient of friction μ is 0.2.

Find: The **forces** F_{12}, F_{32}, F_{13} at the joints and the driving **torque** T_{12} needed to maintain motion with the given acceleration for this instantaneous position of the link.

Solution:

1   Convert the given weights to proper mass units, in this case blobs:

$$mass_{link2} = \frac{weight}{g} = \frac{2\ lb}{386\ in/sec^2} = 0.0052\ \ blob \qquad (a)$$

$$mass_{link3} = \frac{weight}{g} = \frac{4\ lb}{386\ in/sec^2} = 0.0104\ \ blob \qquad (b)$$

2   Set up a local, nonrotating *xy* coordinate system (LNCS) at the *CG* of each link, and draw all applicable position and force vectors acting within or on that system as shown in Figure 11-2 (p. 584). Draw a free-body diagram of each moving link as shown.

3   Calculate the *x* and *y* components of the position vectors R_{12}, R_{32}, R_{23}, R_{13}, and R_P in the LNCS coordinate system:

$$R_{12} = 3.00 \ @ \ \angle \ 270.0°; \qquad R_{12_x} = 0.000, \qquad R_{12_y} = -3.0$$

$$R_{32} = 2.83 \ @ \ \angle \ 28.0°; \qquad R_{32_x} = 2.500, \qquad R_{32_y} = 1.333$$

$$R_{23} = 9.00 \ @ \ \angle \ 324.5°; \qquad R_{23_x} = 7.329, \qquad R_{23_y} = -5.224 \qquad (c)$$

$$R_{13} = 10.72 \ @ \ \angle \ 63.14°; \qquad R_{13_x} = 4.843, \qquad R_{13_y} = 9.563$$

$$R_P = 2.70 \ @ \ \angle \ 201.0°; \qquad R_{P_x} = -2.521, \qquad R_{P_y} = -0.968$$

These position vector angles are measured with respect to the LNCS which is always parallel to the global coordinate system (GCS), making the angles the same in both systems.

4   Calculate the x and y components of the acceleration of the CGs of all moving links in the global coordinate system:

$$\mathbf{a}_{G_2} = 2700.17 \ @ \ \angle \ -89.4°; \qquad a_{G_{2_x}} = 28.28, \qquad a_{G_{2_y}} = -2700$$

$$\mathbf{a}_{G_3} = 3453.35 \ @ \ \angle \ 254.4°; \qquad a_{G_{3_x}} = -930.82, \qquad a_{G_{3_y}} = -3325.54 \qquad (d)$$

5   Calculate the x and y components of the external force at P in the global coordinate system:

$$\mathbf{F}_P = 50@\angle -45°; \qquad F_{P_x} = 35.36, \qquad F_{P_y} = -35.36 \qquad (e)$$

6   Substitute these given and calculated values into the matrix equation 11.7 (p. 586).

$$
\begin{bmatrix}
1 & 0 & 1 & 0 & 0 & 0 \\
0 & 1 & 0 & 1 & 0 & 0 \\
3 & 0 & -1.333 & 2.5 & 0 & 1 \\
0 & 0 & -1 & 0 & 1 & 0 \\
0 & 0 & 0 & -1 & 0.2 & 0 \\
0 & 0 & -5.224 & -7.329 & [(0.2)4.843-(9.563)] & 0
\end{bmatrix}
\times
\begin{bmatrix}
F_{12_x} \\
F_{12_y} \\
F_{32_x} \\
F_{32_y} \\
F_{13_x} \\
T_{12}
\end{bmatrix}
=
$$

$$
\begin{bmatrix}
(0.005)(28.28) \\
(0.005)(-2700) \\
(0.05)(-10) \\
(0.01)(-930.82) - 35.36 \\
(0.01)(-3325.54) - (-35.36)
\end{bmatrix}
=
\begin{bmatrix}
0.141 \\
-13.500 \\
-0.500 \\
-44.668 \\
2.105
\end{bmatrix}
\qquad (f)
$$

7   Solve this system either by inverting matrix **A** and premultiplying that inverse times matrix **C** using a pocket calculator with matrix capability; or by inputting the values for matrices **A** and **C** to program MATRIX provided with this text which gives the following solution:

$$
\begin{bmatrix}
F_{12_x} \\
F_{12_y} \\
F_{32_x} \\
F_{32_y} \\
F_{13_x} \\
T_{12}
\end{bmatrix}
=
\begin{bmatrix}
-39.232 \\
-10.336 \\
39.373 \\
-3.164 \\
-5.295 \\
177.590
\end{bmatrix}
\qquad (g)
$$

Converting the forces to polar coordinates:

$$\mathbf{F}_{12} = 40.57 \text{ lb @ } \angle\, 194.76°$$
$$\mathbf{F}_{32} = 39.50 \text{ lb @ } \angle\, {-}4.60° \tag{h}$$
$$\mathbf{F}_{13} = 5.40 \text{ lb @ } \angle\, 191.31°$$

Open the disk file E11-02.mtr in program MATRIX to exercise this example.

11.4 FORCE ANALYSIS OF A FOURBAR LINKAGE

Figure 11-3a (p. 590) shows a fourbar linkage. All dimensions of link lengths, link positions, locations of the links' CGs, linear accelerations of those CGs, and link angular accelerations and velocities have been previously determined from a kinematic analysis. We now wish to find the forces acting at all the pin joints of the linkage for one or more positions. The procedure is exactly the same as that used in the previous two examples. This linkage has three moving links. Equation 11.1 (p. 580) provides three equations for any link or rigid body in motion. We should expect to have nine equations in nine unknowns for this problem.

Figure 11-3b shows the free-body diagrams for all links, with all forces shown. Note that an external force $\mathbf{F}_P$ is shown acting on link 3 at point P. Also an external torque $\mathbf{T}_4$ is shown acting on link 4. These external loads are due to some other mechanism (device, person, thing, etc.) pushing or twisting against the motion of the linkage. Any link can have any number of external loads and torques acting on it. Only one external torque and one external force are shown here to serve as examples of how they are handled in the computation. (Note that a more complicated force system, if present, could also be reduced to the combination of a single force and torque on each link.)

To solve for the pin forces it is necessary that these applied external forces and torques be defined for all positions of interest. We will solve for one member of the pair of action-reaction forces at each joint, and also for the driving torque $\mathbf{T}_{12}$ needed to be supplied at link 2 in order to maintain the kinematic state as defined. The force subscript convention is the same as that defined in the previous example. For example, $\mathbf{F}_{12}$ is the force of 1 *on* 2 and $\mathbf{F}_{32}$ is the force of 3 *on* 2. The equal and opposite forces at each of these pins are designated $\mathbf{F}_{21}$ and $\mathbf{F}_{23}$, respectively. All the unknown forces in the figure are shown at arbitrary angles and lengths as their true values are still to be determined.

The linkage kinematic parameters are defined with respect to a global XY system (GCS) whose origin is at the driver pivot O_2 and whose X axis goes through link 4's fixed pivot O_4. The mass (m) of each link, the location of its CG, and its mass moment of inertia (I_G) about that CG are also needed. The CG of each link is initially defined within each link with respect to a local moving and rotating axis system (LRCS) embedded in the link because the CG is an unchanging physical property of the link. The origin of this x', y' axis system is at one pin joint and the x' axis is the line of centers of the link. The CG position within the link is defined by a position vector in this LRCS. The instantaneous location of the CG can easily be determined for each dynamic link position by adding the angle of the internal CG position vector to the current GCS angle of the link.

11

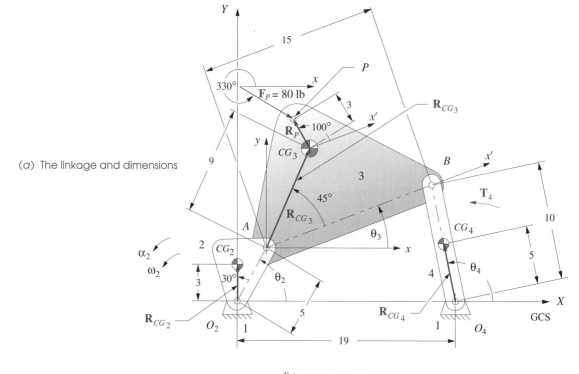

(a) The linkage and dimensions

(b) Free-body diagrams

FIGURE 11-3

Dynamic force analysis of a fourbar linkage. (See also Figure P11-2, p. 584)

We need to define each link's dynamic parameters and force locations with respect to a local, moving, but nonrotating axis system (LNCS) x,y located at its CG as shown on each free-body diagram in Figure 11-3b. The position vector locations of all attachment points of other links and points of application of external forces must be defined with respect to this LNCS axis system. These kinematic and applied force data differ for each position of the linkage. In the following discussion and examples, only one linkage position will be addressed. The process is identical for each succeeding position.

Equations 11.1 (p. 580) are written for each moving link. For link 2, the result is identical to that done for the crank-slider example in equation 11.6a (p. 585).

$$F_{12_x} + F_{32_x} = m_2 a_{G_{2x}}$$
$$F_{12_y} + F_{32_y} = m_2 a_{G_{2y}} \qquad (11.8a)$$
$$T_{12} + \left(R_{12_x} F_{12_y} - R_{12_y} F_{12_x}\right) + \left(R_{32_x} F_{32_y} - R_{32_y} F_{32_x}\right) = I_{G_2} \alpha_2$$

For link 3, with substitution of the reaction force $-\mathbf{F}_{32}$ for $\mathbf{F}_{23}$, the result is similar to equation 11.6b with some subscript changes to reflect the presence of link 4.

$$F_{43_x} - F_{32_x} + F_{P_x} = m_3 a_{G_{3x}}$$
$$F_{43_y} - F_{32_y} + F_{P_y} = m_3 a_{G_{3y}} \qquad (11.8b)$$
$$\left(R_{43_x} F_{43_y} - R_{43_y} F_{43_x}\right) - \left(R_{23_x} F_{32_y} - R_{23_y} F_{32_x}\right) + \left(R_{P_x} F_{P_y} - R_{P_y} F_{P_x}\right) = I_{G_3} \alpha_3$$

For link 4, substituting the reaction force $-\mathbf{F}_{43}$ for $\mathbf{F}_{34}$, a similar set of equations 11.1 (p. 580) can be written:

$$F_{14_x} - F_{43_x} = m_4 a_{G_{4x}}$$
$$F_{14_y} - F_{43_y} = m_4 a_{G_{4y}} \qquad (11.8c)$$
$$\left(R_{14_x} F_{14_y} - R_{14_y} F_{14_x}\right) - \left(R_{34_x} F_{43_y} - R_{34_y} F_{43_x}\right) + T_4 = I_{G_4} \alpha_4$$

Note that $\mathbf{T}_{12}$, the source torque, only appears in the equation for link 2, the motor-driven crank. Link 3, in this example, has no externally applied torque (though it could have) but does have an external force $\mathbf{F}_P$. Link 4, in this example, has no external force acting on it (though it could have) but does have an external torque $\mathbf{T}_4$. (The driving link 2 could also have an externally applied force on it though it usually does not.) There are nine unknowns present in these nine equations, F_{12x}, F_{12y}, F_{32x}, F_{32y}, F_{43x}, F_{43y}, F_{14x}, F_{14y}, and T_{12}, so we can solve them simultaneously. We rearrange terms in equations 11.8 to put all known constant terms on the right side and then put them in matrix form.

$$
\begin{bmatrix}
1 & 0 & 1 & 0 & 0 & 0 & 0 & 0 & 0 \\
0 & 1 & 0 & 1 & 0 & 0 & 0 & 0 & 0 \\
-R_{12_y} & R_{12_x} & -R_{32_y} & R_{32_x} & 0 & 0 & 0 & 0 & 1 \\
0 & 0 & -1 & 0 & 1 & 0 & 0 & 0 & 0 \\
0 & 0 & 0 & -1 & 0 & 1 & 0 & 0 & 0 \\
0 & 0 & R_{23_y} & -R_{23_x} & -R_{43_y} & R_{43_x} & 0 & 0 & 0 \\
0 & 0 & 0 & 0 & -1 & 0 & 1 & 0 & 0 \\
0 & 0 & 0 & 0 & 0 & -1 & 0 & 1 & 0 \\
0 & 0 & 0 & 0 & R_{34_y} & -R_{34_x} & -R_{14_y} & R_{14_x} & 0
\end{bmatrix}
\times
\begin{bmatrix}
F_{12_x} \\
F_{12_y} \\
F_{32_x} \\
F_{32_y} \\
F_{43_x} \\
F_{43_y} \\
F_{14_x} \\
F_{14_y} \\
T_{12}
\end{bmatrix}
=
\begin{bmatrix}
m_2 a_{G_{2x}} \\
m_2 a_{G_{2y}} \\
I_{G_2} \alpha_2 \\
m_3 a_{G_{3x}} - F_{P_x} \\
m_3 a_{G_{3y}} - F_{P_y} \\
I_{G_3} \alpha_3 - R_{P_x} F_{P_y} + R_{P_y} F_{P_x} \\
m_4 a_{G_{4x}} \\
m_4 a_{G_{4y}} \\
I_{G_4} \alpha_4 - T_4
\end{bmatrix}
\qquad (11.9)
$$

11

This system can be solved by using program MATRIX or any matrix solving calculator. As an example of this solution consider the following linkage data.

✍️ EXAMPLE 11-3

Dynamic Force Analysis of a Fourbar Linkage. (See Figure 11-3, p. 590)

Given: The 5-in-long crank (link 2) shown weighs 1.5 lb. Its *CG* is at 3 in @ +30° from the line of centers (LRCS). Its mass moment of inertia about its *CG* is 0.4 lb-in-sec². Its kinematic data are:

θ_2 deg	ω_2 rad/sec	α_2 rad/sec²	a_{G_2} in/sec²
60	25	−40	1878.84 @ 273.66°

The coupler (link 3) is 15 in long and weighs 7.7 lb. Its *CG* is at 9 in @ 45° off the line of centers (LRCS). Its mass moment of inertia about its *CG* is 1.5 lb-in-sec². Its kinematic data are:

θ_3 deg	ω_3 rad/sec	α_3 rad/sec²	a_{G_3} in/sec²
20.92	−5.87	120.9	3646.1 @ 226.5°

The ground link is 19 in long. The rocker (link 4) is 10 in long and weighs 5.8 lb. Its *CG* is at 5 in @ 0° on the line of centers (LRCS). Its mass moment of inertia about its *CG* is 0.8 lb-in-sec². There is an external torque on link 4 of 120 lb-in (GCS). An external force of 80 lb @ 330° acts on link 3 in the GCS, applied at point *P* at 3 in @ 100° from the *CG* of link 3 (LRCS). The kinematic data are:

θ_4 deg	ω_4 rad/sec	α_4 rad/sec²	a_{G_4} in/sec²
104.41	7.93	276.29	1416.8 @ 207.2°

Find: Forces $\mathbf{F}_{12}$, $\mathbf{F}_{32}$, $\mathbf{F}_{43}$, and $\mathbf{F}_{14}$ at the joints and the driving torque $\mathbf{T}_{12}$ needed to maintain motion with the given acceleration for this instantaneous position of the link.

Solution:

1   Convert the given weight to proper mass units, in this case blobs:

$$mass_{link2} = \frac{weight}{g} = \frac{1.5 \text{ lb}}{386 \text{ in/sec}^2} = 0.004 \text{ blob} \qquad (a)$$

$$mass_{link3} = \frac{weight}{g} = \frac{7.7 \text{ lb}}{386 \text{ in/sec}^2} = 0.020 \text{ blob} \qquad (b)$$

$$mass_{link4} = \frac{weight}{g} = \frac{5.8 \text{ lb}}{386 \text{ in/sec}^2} = 0.015 \text{ blob} \qquad (c)$$

2   Set up an LNCS *xy* coordinate system at the *CG* of each link, and draw all applicable vectors acting on that system as shown in the figure. Draw a free-body diagram of each moving link.

3   Calculate the *x* and *y* components of the position vectors $\mathbf{R}_{12}$, $\mathbf{R}_{32}$, $\mathbf{R}_{23}$, $\mathbf{R}_{43}$, $\mathbf{R}_{34}$, $\mathbf{R}_{14}$, and $\mathbf{R}_P$ in the link's LNCS. $\mathbf{R}_{43}$, $\mathbf{R}_{34}$, and $\mathbf{R}_{14}$ will have to be calculated from the given link geometry data using the law of cosines and law of sines. Note that the current value of link 3's position angle (θ_3) in the GCS must be added to the angles of all position vectors before creating their

x,y components in the LNCS if their angles were originally measured with respect to the link's embedded, local rotating coordinate system (LRCS).

$$\mathbf{R}_{12} = 3.00 \ @ \ \angle \ 270.00°; \qquad R_{12_x} = 0.000, \qquad R_{12_y} = -3$$

$$\mathbf{R}_{32} = 2.83 \ @ \ \angle \ 28.00°; \qquad R_{32_x} = 2.500, \qquad R_{32_y} = 1.333$$

$$\mathbf{R}_{23} = 9.00 \ @ \ \angle \ 245.92°; \qquad R_{23_x} = -3.672, \qquad R_{23_y} = -8.217$$

$$\mathbf{R}_{43} = 10.72 \ @ \ \angle \ -15.46°; \qquad R_{43_x} = 10.332, \qquad R_{43_y} = -2.858 \qquad (d)$$

$$\mathbf{R}_{34} = 5.00 \ @ \ \angle \ 104.41°; \qquad R_{34_x} = -1.244, \qquad R_{34_y} = 4.843$$

$$\mathbf{R}_{14} = 5.00 \ @ \ \angle \ 284.41°; \qquad R_{14_x} = 1.244, \qquad R_{14_y} = -4.843$$

$$\mathbf{R}_P = 3.00 \ @ \ \angle \ 120.92°; \qquad R_{P_x} = -1.542, \qquad R_{P_y} = 2.574$$

4 Calculate the *x* and *y* components of the acceleration of the *CG*s of all moving links in the global coordinate system (GCS):

$$\mathbf{a}_{G_2} = 1878.84 \ @ \ \angle 273.66°; \qquad a_{G_{2x}} = 119.94, \qquad a_{G_{2y}} = -1875.01$$

$$\mathbf{a}_{G_3} = 3646.10 \ @ \ \angle 226.51°; \qquad a_{G_{3x}} = -2509.35, \qquad a_{G_{3y}} = -2645.23 \qquad (e)$$

$$\mathbf{a}_{G_4} = 1416.80 \ @ \ \angle 207.24°; \qquad a_{G_{4x}} = -1259.67, \qquad a_{G_{4y}} = -648.50$$

5 Calculate the *x* and *y* components of the external force at *P* in the GCS:

$$\mathbf{F}_{P3} = 80 \ @ \ \angle \ 330°; \qquad F_{P3_x} = 69.28, \qquad F_{P3_y} = -40.00 \qquad (f)$$

6 Substitute these given and calculated values into the matrix equation 11.9 (p. 591).

$$\begin{bmatrix} 1 & 0 & 1 & 0 & 0 & 0 & 0 & 0 & 0 \\ 0 & 1 & 0 & 1 & 0 & 0 & 0 & 0 & 0 \\ 3 & 0 & -1.330 & 2.5 & 0 & 0 & 0 & 0 & 1 \\ 0 & 0 & -1 & 0 & 1 & 0 & 0 & 0 & 0 \\ 0 & 0 & 0 & -1 & 0 & 1 & 0 & 0 & 0 \\ 0 & 0 & -8.217 & 3.673 & 2.861 & 10.339 & 0 & 0 & 0 \\ 0 & 0 & 0 & 0 & -1 & 0 & 1 & 0 & 0 \\ 0 & 0 & 0 & 0 & 0 & -1 & 0 & 1 & 0 \\ 0 & 0 & 0 & 0 & 4.843 & 1.244 & 4.843 & 1.244 & 0 \end{bmatrix} \times \begin{bmatrix} F_{12_x} \\ F_{12_y} \\ F_{32_x} \\ F_{32_y} \\ F_{43_x} \\ F_{43_y} \\ F_{14_x} \\ F_{14_y} \\ T_{12} \end{bmatrix} =$$

(g)

$$\begin{bmatrix} (0.004)(119.94) \\ (0.004)(-1875.01) \\ (0.4)(-40) \\ (0.02)(-2509.35)-(69.28) \\ (0.02)(-2645.23)-(-40) \\ (1.5)(120.9)-[(-1.542)(-40)-(2.574)(69.28)] \\ (0.015)(-1259.67) \\ (0.015)(-648.50) \\ (0.8)(276.29)-(120) \end{bmatrix} = \begin{bmatrix} 0.480 \\ -7.500 \\ -16.000 \\ -119.465 \\ -12.908 \\ 298.003 \\ -18.896 \\ -9.727 \\ 101.031 \end{bmatrix}$$

11

7 Solve this system either by inverting matrix **A** and premultiplying that inverse times matrix **C** using a pocket calculator with matrix capability, or by inputting the values for matrices **A** and **C** to program MATRIX provided with this text, which gives the following solution:

$$
\begin{bmatrix}
F_{12_x} \\
F_{12_y} \\
F_{32_x} \\
F_{32_y} \\
F_{43_x} \\
F_{43_y} \\
F_{14_x} \\
F_{14_y} \\
T_{12}
\end{bmatrix}
=
\begin{bmatrix}
-117.65 \\
-107.84 \\
118.13 \\
100.34 \\
-1.34 \\
87.43 \\
-20.23 \\
77.71 \\
243.23
\end{bmatrix}
\qquad (h)
$$

Converting the forces to polar coordinates:

$$
\begin{aligned}
\mathbf{F}_{12} &= \ 159.60 \text{ lb} \ @ \angle \ \ 222.52° \\
\mathbf{F}_{32} &= \ 154.99 \text{ lb} \ @ \angle \ \ \ \ 40.35° \\
\mathbf{F}_{43} &= \ \ \ 87.44 \text{ lb} \ @ \angle \ \ \ \ 90.88° \\
\mathbf{F}_{14} &= \ \ \ 80.30 \text{ lb} \ @ \angle \ \ 104.59°
\end{aligned}
\qquad (i)
$$

8 The pin-force magnitudes in (i) are needed to size the pivot pins and links against failure and to select pivot bearings that will last for the required life of the assembly. The driving torque T_{12} defined in (h) is needed to select a motor or other device capable of supplying the power to drive the system. See Section 2.19 (p. 74) for a brief discussion of motor selection. Issues of stress calculation and failure prevention are beyond the scope of this text, but note that those calculations cannot be done until a good estimate of the dynamic forces and torques on the system has been made by methods such as those shown in this example.

This solves the linkage for one position. A new set of values can be put into the **A** and **C** matrices for each position of interest at which a force analysis is needed. Open the disk file E11-03.mtr in program MATRIX to exercise this example. The disk file E11-03.4br can also be opened in program LINKAGES and will run the linkage through a series of positions starting with the stated parameters as initial conditions. The linkage will slow to a stop and then run in reverse due to the negative acceleration. The matrix for equation (g) can be seen within LINKAGES using *Dynamics/Solve/Show Matrix*.

It is worth noting some general observations about this method at this point. The solution is done using cartesian coordinates of all forces and position vectors. Before being placed in the matrices, these vector components must be defined in the global coordinate system (GCS) or in nonrotating, local coordinate systems, parallel to the global coordinate system, with their origins at the links' *CG*s (LNCS). Some of the linkage parameters are normally expressed in such coordinate systems, but others are not, and so must be transformed to the proper coordinate system. The kinematic data should all be computed in the global system or in parallel, **nonrotating**, local systems placed at the *CG*s of individual links. Any external forces on the links must also be defined in the global system.

However, the position vectors that define intralink locations, such as the pin joints versus the *CG*, or which locate points of application of external forces versus the *CG* are defined in local, **rotating** coordinate systems embedded in the links (LRCS). Thus these position vectors must be redefined in a **nonrotating**, parallel system before being used in the matrix. An example of this is vector $\mathbf{R}_p$, which was initially defined as 3 in at 100° in link 3's embedded, **rotating** coordinate system. Note in Example 11-3 (p. 592) that its cartesian coordinates for use in the equations were calculated after adding the current value of θ_3 to its angle. This redefined $\mathbf{R}_p$ as 3 in at 120.92° in the **nonrotating** local system. The same was done for position vectors $\mathbf{R}_{12}$, $\mathbf{R}_{32}$, $\mathbf{R}_{23}$, $\mathbf{R}_{43}$, $\mathbf{R}_{34}$, and $\mathbf{R}_{14}$. In each case the **intralink angle** of these vectors (which is independent of linkage position) was added to the current link angle to obtain its position in the xy system at the link's *CG*. The proper definition of these position vector components is critical to the solution, and it is very easy to make errors in defining them.

To further confuse things, even though the position vector $\mathbf{R}_p$ is initially measured in the link's embedded, rotating coordinate system, the force $\mathbf{F}_p$, which it locates, is not. The force $\mathbf{F}_p$ is not part of the link, as is $\mathbf{R}_p$, but rather is part of the external world, so it is defined in the global system.

11.5   FORCE ANALYSIS OF A FOURBAR SLIDER-CRANK LINKAGE

The approach taken for the pin-jointed fourbar is equally valid for a fourbar slider-crank linkage. The principal difference will be that the slider block will have no angular acceleration. Figure 11-4 (p. 597) shows a fourbar slider-crank with an external force on the slider block, link 4. This is representative of the mechanism used extensively in piston pumps and internal combustion engines. We wish to determine the forces at the joints and the driving torque needed on the crank to provide the specified accelerations. A kinematic analysis must have previously been done in order to determine all position, velocity, and acceleration information for the positions being analyzed. Equations 11.1 (p. 580) are written for each link. For link 2:

$$F_{12_x} + F_{32_x} = m_2 a_{G2_x}$$
$$F_{12_y} + F_{32_y} = m_2 a_{G2_y} \tag{11.10a}$$
$$T_{12} + \left(R_{12_x} F_{12_y} - R_{12_y} F_{12_x}\right) + \left(R_{32_x} F_{32_y} - R_{32_y} F_{32_x}\right) = I_{G2} \alpha_2$$

This is identical to equation 11.8a (p. 591) for the "pure" fourbar linkage. For link 3:

$$F_{43_x} - F_{32_x} = m_3 a_{G3_x}$$
$$F_{43_y} - F_{32_y} = m_3 a_{G3_y} \tag{11.10b}$$
$$\left(R_{43_x} F_{43_y} - R_{43_y} F_{43_x}\right) - \left(R_{23_x} F_{32_y} - R_{23_y} F_{32_x}\right) = I_{G3} \alpha_3$$

This is similar to equation 11.8b, lacking only the terms involving $\mathbf{F}_p$ since there is no external force shown acting on link 3 of our example slider-crank. For link 4:

$$F_{14_x} - F_{43_x} + F_{P_x} = m_4 a_{G4_x}$$
$$F_{14_y} - F_{43_y} + F_{P_y} = m_4 a_{G4_y} \tag{11.10c}$$
$$\left(R_{14_x} F_{14_y} - R_{14_y} F_{14_x}\right) - \left(R_{34_x} F_{43_y} - R_{34_y} F_{43_x}\right) + \left(R_{P_x} F_{P_y} - R_{P_y} F_{P_x}\right) = I_{G4} \alpha_4$$

11

These contain the external force $\mathbf{F}_p$ shown acting on link 4.

For the inversion of the slider-crank shown, the slider block, or piston, is in pure translation against the stationary ground plane; thus it can have no angular acceleration or angular velocity. Also, the position vectors in the torque equation (equation 11.10c, p. 595) are all zero as the force $\mathbf{F}_p$ acts at the CG. Thus the torque equation for link 4 (third expression in equation 11.10c) is zero for this inversion of the slider-crank linkage. Its linear acceleration also has no y component.

$$\alpha_4 = 0, \qquad\qquad a_{G_{4_y}} = 0 \qquad\qquad (11.10d)$$

The only x directed force that can exist at the interface between links 4 and 1 is friction. Assuming coulomb friction, the x component can be expressed in terms of the y component of force at this interface. We can write a relation for the friction force f at that interface such as $f = \pm\mu N$, where $\pm\mu$ is a known coefficient of friction. The plus and minus signs on the coefficient of friction are to recognize the fact that the friction force always opposes motion. The kinematic analysis will provide the velocity of the link at the sliding joint. The sign of μ will always be the opposite of the sign of this velocity.

$$F_{14_x} = \pm\mu F_{14_y} \qquad\qquad (11.10e)$$

Substituting equations 11.10d and 11.10e into the reduced equation 11.10c (p. 595) yields:

$$\pm\mu F_{14_y} - F_{43_x} + F_{P_x} = m_4 a_{G_4}$$
$$F_{14_y} - F_{43_y} + F_{P_y} = 0 \qquad\qquad (11.10f)$$

This last substitution has reduced the unknowns to eight, $F_{12x}, F_{12y}, F_{32x}, F_{32y}, F_{43x}, F_{43y}, F_{14y},$ and $T_{12};$ thus we need only eight equations. We can now use the eight equations in 11.10a, b, and f to assemble the matrices for solution.

$$\begin{bmatrix} 1 & 0 & 1 & 0 & 0 & 0 & 0 & 0 \\ 0 & 1 & 0 & 1 & 0 & 0 & 0 & 0 \\ -R_{12_y} & R_{12_x} & -R_{32_y} & R_{32_x} & 0 & 0 & 0 & 1 \\ 0 & 0 & -1 & 0 & 1 & 0 & 0 & 0 \\ 0 & 0 & 0 & -1 & 0 & 1 & 0 & 0 \\ 0 & 0 & R_{23_y} & -R_{23_x} & -R_{43_y} & R_{43_x} & 0 & 0 \\ 0 & 0 & 0 & 0 & -1 & 0 & \pm\mu & 0 \\ 0 & 0 & 0 & 0 & 0 & -1 & 1 & 0 \end{bmatrix} \times \begin{bmatrix} F_{12_x} \\ F_{12_y} \\ F_{32_x} \\ F_{32_y} \\ F_{43_x} \\ F_{43_y} \\ F_{14_y} \\ T_{12} \end{bmatrix}$$

$$= \begin{bmatrix} m_2 a_{G_{2_x}} \\ m_2 a_{G_{2_y}} \\ I_{G_2}\alpha_2 \\ m_3 a_{G_{3_x}} \\ m_3 a_{G_{3_y}} \\ I_{G_3}\alpha_3 \\ m_4 a_{G_4} - F_{P_x} \\ -F_{P_y} \end{bmatrix} \qquad\qquad (11.10g)$$

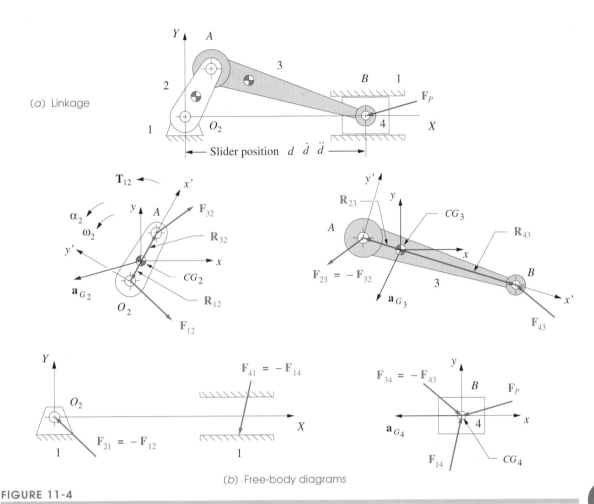

(a) Linkage

Slider position d $\dot{d}$ $\ddot{d}$

(b) Free-body diagrams

FIGURE 11-4

Dynamic force analysis of the fourbar slider-crank linkage

11

Solution of this matrix equation 11.10g plus equation 11.10e will yield complete dynamic force information for the fourbar slider-crank linkage.

11.6   FORCE ANALYSIS OF THE INVERTED CRANK-SLIDER

Another inversion of the fourbar crank-slider was also analyzed kinematically in Part I. It is shown in Figure 11-5 (p. 598). Link 4 does have an angular acceleration in this inversion. In fact, it must have the same angle, angular velocity, and angular acceleration as link 3 because they are rotationally coupled by the sliding joint. We wish to determine the forces at all pin joints and at the sliding joint as well as the driving torque needed to create the desired accelerations. Each link's joints are located by position vectors referenced to nonrotating local xy coordinate systems at each link's CG as before. The sliding joint is located by the position vector $\mathbf{R}_{43}$ to the center of the slider, point B. The instantaneous position of point B was determined from the kinematic analysis as length b referenced

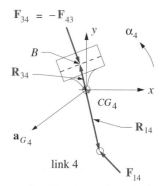

FIGURE 11-5

Dynamic forces in the inverted slider-crank fourbar linkage

to instant center I_{23} (point A). See Sections 4.8 (p. 197), 6.7 (p. 315), and 7.3 (p. 358) to review the position, velocity, and acceleration analysis of this mechanism. Recall that this mechanism has a nonzero Coriolis component of acceleration. The force between link 3 and link 4 within the sliding joint is distributed along the unspecified length of the slider block. For this analysis the distributed force can be modeled as a force concentrated at point B within the sliding joint. We will neglect friction in this example.

The equations for links 2 and 3 are identical to those for the noninverted crank-slider (equations 11.10a and b, p. 595). The equations for link 4 are the same as equations 11.10c (p. 595) except for the absence of the terms involving $\mathbf{F}_p$ since no external force is shown acting on link 4 in this example. The slider joint can only transmit force from link 3 to link 4 or vice versa along a line perpendicular to the axis of slip. This line is called the axis of transmission. In order to guarantee that the force $\mathbf{F}_{34}$ or $\mathbf{F}_{43}$ is always perpendicular to the axis of slip, we can write the following relation:

$$\hat{\mathbf{u}} \cdot \mathbf{F}_{43} = 0 \tag{11.11a}$$

which expands to:

$$u_x F_{43_x} + u_y F_{43_y} = 0 \tag{11.11b}$$

The dot product of two vectors will be zero when the vectors are mutually perpendicular. The unit vector $\hat{\mathbf{u}}$ is in the direction of link 3 which is defined from the kinematic analysis as θ_3.

$$u_x = \cos\theta_3, \qquad u_y = \sin\theta_3 \tag{11.11c}$$

Equation 11.11b provides a tenth equation, but we have only nine unknowns, F_{12x}, F_{12y}, F_{32x}, F_{32y}, F_{43x}, F_{43y}, F_{14x}, F_{14y}, and T_{12}, so one of our equations is redundant. Since we must include equation 11.11, we will combine the torque equations for links 3 and 4 rewritten here in vector form and without the external force $\mathbf{F}_p$.

$$\left(\mathbf{R}_{43} \times \mathbf{F}_{43}\right) - \left(\mathbf{R}_{23} \times \mathbf{F}_{32}\right) = I_{G_3}\alpha_3 = I_{G_3}\alpha_4$$

$$\left(\mathbf{R}_{14} \times \mathbf{F}_{14}\right) - \left(\mathbf{R}_{34} \times \mathbf{F}_{43}\right) = I_{G_4}\alpha_4 \tag{11.12a}$$

Note that the angular acceleration of link 3 is the same as that of link 4 in this linkage. Adding these equations gives:

$$\left(\mathbf{R}_{43} \times \mathbf{F}_{43}\right) - \left(\mathbf{R}_{23} \times \mathbf{F}_{32}\right) + \left(\mathbf{R}_{14} \times \mathbf{F}_{14}\right) - \left(\mathbf{R}_{34} \times \mathbf{F}_{43}\right) = \left(I_{G_3} + I_{G_4}\right)\alpha_4 \tag{11.12b}$$

Expanding and collecting terms:

$$\left(R_{43x} - R_{34_x}\right)F_{43_y} + \left(R_{34_y} - R_{43_y}\right)F_{43_x} - R_{23_x}F_{32_y}$$

$$+ R_{23_y}F_{32_x} + R_{14_x}F_{14_y} - R_{14_y}F_{14_x} = \left(I_{G_3} + I_{G_4}\right)\alpha_4 \tag{11.12c}$$

Equations 11.10a, 11.11b, 11.12c, and the four force equations from equations 11.10b and 11.10c (p. 595) (excluding the external force $\mathbf{F}_P$) give us nine equations in the nine unknowns which we can put in matrix form for solution.

11

$$
\begin{bmatrix}
1 & 0 & 1 & 0 & 0 & 0 & 0 & 0 & 0 \\
0 & 1 & 0 & 1 & 0 & 0 & 0 & 0 & 0 \\
-R_{12_y} & R_{12_x} & -R_{32_y} & R_{32_x} & 0 & 0 & 0 & 0 & 1 \\
0 & 0 & -1 & 0 & 1 & 0 & 0 & 0 & 0 \\
0 & 0 & 0 & -1 & 0 & 1 & 0 & 0 & 0 \\
0 & 0 & R_{23_y} & -R_{23_x} & \left(R_{34}-R_{43_y}\right) & \left(R_{43}-R_{34_x}\right) & -R_{14_y} & R_{14_x} & 0 \\
0 & 0 & 0 & 0 & -1 & 0 & 1 & 0 & 0 \\
0 & 0 & 0 & 0 & 0 & -1 & 0 & 1 & 0 \\
0 & 0 & 0 & 0 & u_x & u_y & 0 & 0 & 0
\end{bmatrix}
$$

$$
\times \begin{bmatrix}
F_{12_x} \\
F_{12_y} \\
F_{32_x} \\
F_{32_y} \\
F_{43_x} \\
F_{43_y} \\
F_{14_x} \\
F_{14_y} \\
T_{12}
\end{bmatrix}
=
\begin{bmatrix}
m_2 a_{G_{2_x}} \\
m_2 a_{G_{2_y}} \\
I_{G_2}\alpha_2 \\
m_3 a_{G_{3_x}} \\
m_3 a_{G_{3_y}} \\
\left(I_{G_3}+I_{G_4}\right)\alpha_4 \\
m_4 a_{G_{4_x}} \\
m_4 a_{G_{4_y}} \\
0
\end{bmatrix}
\tag{11.13}
$$

11.7   FORCE ANALYSIS—LINKAGES WITH MORE THAN FOUR BARS

This matrix method of force analysis can easily be extended to more complex assemblages of links. The equations for each link are of the same form. We can create a more general notation for equations 11.1 (p. 580) to apply them to any assembly of n pin-connected links. Let j represent any link in the assembly. Let $i = j - 1$ be the previous link in the chain and $k = j + 1$ be the next link in the chain; then, using the vector form of equations 11.1:

$$
\mathbf{F}_{ij} + \mathbf{F}_{jk} + \sum \mathbf{F}_{ext_j} = m_j\,\mathbf{a}_{G_j}
\tag{11.14a}
$$

$$
\left(\mathbf{R}_{ij}\times\mathbf{F}_{ij}\right)+\left(\mathbf{R}_{jk}\times\mathbf{F}_{jk}\right)+\sum\mathbf{T}_j+\left(\mathbf{R}_{ext_j}\times\sum\mathbf{F}_{ext_j}\right)=I_{G_j}\alpha_j
\tag{11.14b}
$$

where:

$$
j = 2, 3, \ldots, n; \qquad i = j - 1; \qquad k = j + 1,\ j \neq n; \qquad \text{if } j = n, k = 1
$$

and

$$
\mathbf{F}_{ji} = -\mathbf{F}_{ij}; \qquad \mathbf{F}_{kj} = -\mathbf{F}_{jk}
\tag{11.14c}
$$

The sum of forces vector equation 11.14a can be broken into its two x and y component equations and then applied, along with the sum of torques equation 11.14b, to each of the links in the chain to create the set of simultaneous equations for solution. Any link may have external forces and/or external torques applied to it. All will have pin forces.

Since the nth link in a closed chain connects to the first link, the value of k for the nth link is set to 1. In order to reduce the number of variables to a tractable quantity, substitute the negative reaction forces from equation 11.14c where necessary as was done in the examples in this chapter. When sliding joints are present, it will be necessary to add constraints on the allowable directions of forces at those joints as was done in the inverted crank-slider derivation above.

11.8   SHAKING FORCE AND SHAKING MOMENT

It is usually of interest to know the net effect of the dynamic forces as felt on the ground plane as this can set up vibrations in the structure that supports the machine. For our simple examples of three- and fourbar linkages, there are only two points at which the dynamic forces can be delivered to link 1, the ground plane. More complicated mechanisms will have more joints with the ground plane. The forces delivered by the moving links to the ground at the fixed pivots O_2 and O_4 are designated $\mathbf{F}_{21}$ and $\mathbf{F}_{41}$ by our subscript convention as defined in Section 11.1 (p. 579). Since we chose to solve for $\mathbf{F}_{12}$ and $\mathbf{F}_{14}$ in our solutions, we simply negate those forces to obtain their equal and opposite counterparts (see also equation 11.5, p. 585).

$$\mathbf{F}_{21} = -\mathbf{F}_{12} \qquad\qquad \mathbf{F}_{41} = -\mathbf{F}_{14} \qquad\qquad (11.15a)$$

The *sum of all the forces acting on the ground plane* is called the **shaking force ($\mathbf{F}_s$)** as shown in Figure 11-6.[*] In these simple examples it is equal to:

$$\mathbf{F}_s = \mathbf{F}_{21} + \mathbf{F}_{41} \qquad\qquad (11.15b)$$

The *reaction moment felt by the ground plane* is called the **shaking moment ($\mathbf{M}_s$)** as shown in Figure 11-7 (p. 602).[*] This is the negative of the source torque ($\mathbf{T}_{21} = -\mathbf{T}_{12}$) plus the cross products of the ground-pin forces and their distances from the reference point. The shaking moment about the crank pivot O_2 is:

* The LINKAGES files (F11-06.4br & F11-07.4br) that generated the plots in Figures 11-6 and 11-7 may be opened in that program to see more details on the linkage's dynamics.

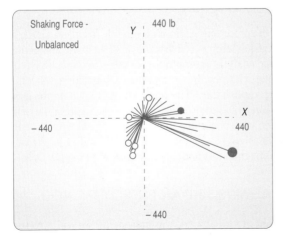

Link No.	Length in	Mass Units	Inertia Units	CG Posit	at Deg	Ext. Force lb	at Deg
1	5.5						
2	2.0	.002	.004	1.0	0		
3	6.0	.030	.060	2.5	30	12	270
4	3.0	.010	.020	1.5	0	60	−45

Coupler pt. = 3 in @ 45°
Open/Crossed = open

Ext. Force 3 acts at 5 in @ 30° vs. CG of Link 3
Ext. Force 4 acts at 5 in @ 90° vs. CG of Link 4
Ext. Torque 3 = − 20 lb-in
Ext. Torque 4 = 25 lb-in

Start Alpha2 = 0 rad/sec²
Start Omega2 = 50 rad/sec
Start Theta2 = 0°
Final Theta2 = 360°
Delta Theta2 = 10°

FIGURE 11-6

Linkage data and polar plot of shaking force for an unbalanced crank-rocker fourbar linkage from program LINKAGES

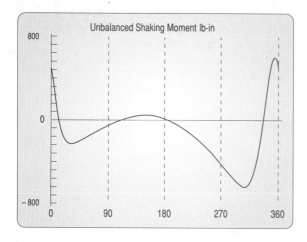

Link	Length	Mass	Inertia	CG	at	Ext. Force	at
No.	in	Units	Units	Posit	Deg	lb	Deg
1	5.5						
2	2.0	.002	.004	1.0	0		
3	6.0	.030	.060	2.5	30	12	270
4	3.0	.010	.020	1.5	0	60	−45

Coupler pt. = 3 in @ 45°
Open/Crossed = open

Ext. Force 3 acts at 5 in @ 30° vs. CG of Link 3
Ext. Force 4 acts at 5 in @ 90° vs. CG of Link 4
Ext. Torque 3 = − 20 lb-in
Ext. Torque 4 = 25 lb-in

Start Alpha2 = 0 rad/sec^2
Start Omega2 = 50 rad/sec
Start Theta2 = 0°
Final Theta2 = 360°
Delta Theta2 = 2°

FIGURE 11-7

Linkage data and shaking moment curve for an unbalanced crank-rocker fourbar linkage from program LINKAGES

$$\mathbf{M}_s = \mathbf{T}_{21} + \left(\mathbf{R}_1 \times \mathbf{F}_{41} \right) \tag{11.15c}$$

The shaking force will tend to move the ground plane back and forth, and the shaking moment will tend to rock the ground plane about the driveline axis. Both will cause vibrations. We are usually looking to minimize the effects of the shaking force and shaking moment on the frame. This can sometimes be done by balancing, sometimes by the addition of a flywheel to the system, and sometimes by shock mounting the frame to isolate the vibrations from the rest of the assembly. Most often we will use a combination of all three approaches. We will investigate some of these techniques in Chapter 12.

11.9 PROGRAM LINKAGES

The matrix methods introduced in the preceding sections all provide force and torque information for one position of the linkage assembly as defined by its kinematic and geometric parameters. To do a complete force analysis for multiple positions of a machine requires that these computations be repeated with new input data for each position. A computer program is the obvious way to accomplish this. The program LINKAGES on the attached DVD computes the kinematic parameters for those linkages over changes in time or driver (crank) angle plus the forces and torques concomitant with the linkage kinematics and link geometry. Examples of its output are shown in Figures 11-6 (p. 601) and 11-7. Please refer to Appendix A for more information about this and other programs on the DVD.

11.10 LINKAGE TORQUE ANALYSIS BY AN ENERGY METHOD

In Section 10.15 (p. 571) the method of virtual work was presented. We will now use that approach to solve the linkage from Example 11-3 (p. 592) as a check on its solution by the newtonian method used in that example. The kinematic data given in Example 11-3 did not include information on the angular velocities of all the links, the linear velocities of the centers of gravities of the links, and the linear velocity of the point P of application of the external force on link 3. Velocity data were not needed for the newtonian solution but are needed for the virtual work approach and are detailed below. Equation 10.28a is repeated here and renumbered.

$$\sum_{k=2}^{n} \mathbf{F}_k \cdot \mathbf{v}_k + \sum_{k=2}^{n} \mathbf{T}_k \cdot \omega_k = \sum_{k=2}^{n} m_k \mathbf{a}_k \cdot \mathbf{v}_k + \sum_{k=2}^{n} I_k \alpha_k \cdot \omega_k \qquad (11.16a)$$

Expanding the summations, still in vector form:

$$\left(\mathbf{F}_{P_3} \cdot \mathbf{v}_{P_3} + \mathbf{F}_{P_4} \cdot \mathbf{v}_{P_4} \right) + \left(\mathbf{T}_{12} \cdot \omega_2 + \mathbf{T}_3 \cdot \omega_3 + \mathbf{T}_4 \cdot \omega_4 \right)$$
$$= \left(m_2 \mathbf{a}_{G_2} \cdot \mathbf{v}_{G_2} + m_3 \mathbf{a}_{G_3} \cdot \mathbf{v}_{G_3} + m_4 \mathbf{a}_{G_4} \cdot \mathbf{v}_{G_4} \right) \qquad (11.16b)$$
$$+ \left(I_{G_2} \alpha_2 \cdot \omega_2 + I_{G_3} \alpha_3 \cdot \omega_3 + I_{G_4} \alpha_4 \cdot \omega_4 \right)$$

Expanding the dot products to create a scalar equation:

$$\left(F_{P_{3x}} V_{P_{3x}} + F_{P_{3y}} V_{P_{3y}} \right) + \left(F_{P_{4x}} V_{P_{4x}} + F_{P_{4y}} V_{P_{4y}} \right) + \left(T_{12} \omega_2 + T_3 \omega_3 + T_4 \omega_4 \right)$$
$$= m_2 \left(a_{G_{2x}} V_{G_{2x}} + a_{G_{2y}} V_{G_{2y}} \right) + m_3 \left(a_{G_{3x}} V_{G_{3x}} + a_{G_{3y}} V_{G_{3y}} \right) \qquad (11.16c)$$
$$+ m_4 \left(a_{G_{4x}} V_{G_{4x}} + a_{G_{4y}} V_{G_{4y}} \right) + \left(I_{G_2} \alpha_2 \omega_2 + I_{G_3} \alpha_3 \omega_3 + I_{G_4} \alpha_4 \omega_4 \right)$$

✒️EXAMPLE 11-4

Analysis of a Fourbar Linkage by the Method of Virtual Work. (See Figure 11-3, p. 590.)

Given: The 5-in-long crank (link 2) shown weighs 1.5 lb. Its CG is at 3 in at +30° from the line of centers. Its mass moment of inertia about its CG is 0.4 lb-in-sec². Its kinematic data are:

θ_2 deg	ω_2 rad/sec	α_2 rad/sec²	V_{G_2} in/sec
60	25	−40	75 @ 180°

The coupler (link 3) is 15 in long and weighs 7.7 lb. Its CG is at 9 in at 45° off the line of centers. Its mass moment of inertia about its CG is 1.5 lb-in-sec². Its kinematic data are:

θ_3 deg	ω_3 rad/sec	α_3 rad/sec²	V_{G_3} in/sec
20.92	−5.87	120.9	72.66 @ 145.7°

There is an external force on link 3 of 80 lb at 330°, applied at point P which is located 3 in @ 100° from the CG of link 3. The linear velocity of that point is 67.2 in/sec at 131.94°.

The rocker (link 4) is 10 in long and weighs 5.8 lb. Its CG is at 5 in at 0° off the line of centers. Its mass moment of inertia about its CG is 0.8 lb-in-sec². Its kinematic data are:

θ_4 deg	ω_4 rad/sec	α_4 rad/sec²	V_{G_4} in/sec
104.41	7.93	276.29	39.66 @ 194.41°

11

There is an external torque on link 4 of 120 lb-in. The ground link is 19 in long.

Find: The driving torque T_{12} needed to maintain motion with the given acceleration for this instantaneous position of the link.

Solution:

1   The torque, angular velocity, and angular acceleration vectors in this two-dimensional problem are all directed along the Z axis, so their dot products each have only one term. Note that in this particular example there is no force $\mathbf{F}_{P_4}$ and no torque $\mathbf{T}_3$.

2   The cartesian coordinates of the acceleration data were calculated in Example 11-3 (p. 592).

$$\mathbf{a}_{G_2} = 1878.84 \ @ \ \angle \ 273.66°; \qquad a_{G_{2_x}} = 119.94, \qquad a_{G_{2_y}} = -1875.01$$

$$\mathbf{a}_{G_3} = 3646.10 \ @ \ \angle \ 226.51°; \qquad a_{G_{3_x}} = -2509.35, \qquad a_{G_{3_y}} = -2645.23 \qquad (a)$$

$$\mathbf{a}_{G_4} = 1416.80 \ @ \ \angle \ 207.24°; \qquad a_{G_{4_x}} = -1259.67, \qquad a_{G_{4_y}} = -648.50$$

3   The x and y components of the external force at P in the global coordinate system were also calculated in Example 11-3:

$$\mathbf{F}_{P_3} = 80 \ @ \ \angle 330°; \qquad F_{P_{3_x}} = 69.28, \qquad F_{P_{3_y}} = -40.00 \qquad (b)$$

4   Converting the velocity data for this example to cartesian coordinates:

$$\mathbf{V}_{G_2} = 75.00 \ @ \ \angle \ 180.00°; \qquad V_{G_{2_x}} = -75.00, \qquad V_{G_{2_y}} = 0$$

$$\mathbf{V}_{G_3} = 72.66 \ @ \ \angle \ 145.70°; \qquad V_{G_{3_x}} = -60.02, \qquad V_{G_{3_y}} = 40.95$$

$$\mathbf{V}_{G_4} = 39.66 \ @ \ \angle \ 194.41°; \qquad V_{G_{4_x}} = -38.41, \qquad V_{G_{4_y}} = -9.87 \qquad (c)$$

$$\mathbf{V}_{P_2} = 67.20 \ @ \ \angle \ 131.94°; \qquad V_{P_2} = -44.91, \qquad V_{P_2} = 49.99$$

5   Substituting the example data into equation 11.16c (p. 603):

$$\left[(69.28)(-44.91)+(-40)(49.99) \right]+[0]+\left[25T_{12}+(0)+(120)(7.93) \right]$$

$$= \frac{1.5}{386} \left[(119.94)(-75)+(-1875.01)(0) \right]$$

$$+ \frac{7.7}{386} \left[(-2509.35)(-60.02)+(-2645.23)(40.95) \right] \qquad (d)$$

$$+ \frac{5.8}{386} \left[(-1259.67)(-38.41)+(-648.50)(-9.87) \right]$$

$$+ \left[(0.4)(-40)(25)+(1.5)(120.9)(-5.87)+(0.8)(276.29)(7.93) \right]$$

6   The only unknown in this equation is the input torque T_{12} which calculates to:

$$\mathbf{T}_{12} = 243.2 \ \hat{\mathbf{k}} \qquad (e)$$

which is the same as the answer obtained in Example 11-3.

This method of virtual work is useful if a quick answer is needed for the input torque, but it does not give any information about the joint forces.

11.11 CONTROLLING INPUT TORQUE—FLYWHEELS

The typically large variation in accelerations within a mechanism can cause significant oscillations in the torque required to drive it at a constant or near constant speed. The peak torques needed may be so high as to require an overly large motor to deliver them. However, the average torque over the cycle, due mainly to losses and external work done, may often be much smaller than the peak torque. We would like to provide some means to smooth out these oscillations in torque during the cycle. This will allow us to size the motor to deliver the average torque rather than the peak torque. One convenient and relatively inexpensive means to this end is the addition of a **flywheel** to the system.

TORQUE VARIATION   Figure 11-8 shows the variation in the input torque for a crank-rocker fourbar linkage over one full revolution of the drive crank. It is running at a constant angular velocity of 50 rad/sec. The torque varies a great deal within one cycle of the mechanism, going from a positive peak of 341.7 lb-in to a negative peak of –166.4 lb-in. The **average value of this torque** over the cycle is only 70.2 lb-in, being due to the *external work done plus losses*. This linkage has only a 12-lb external force applied to link 3 at the *CG* and a 25 lb-in external torque applied to link 4. These small external loads cannot account for the large variation in input torque required to maintain constant crank speed. What then is the explanation? The large variations in torque are evidence of the kinetic energy that is stored in the links as they move. We can think of the positive pulses of torque as representing energy delivered by the driver (motor) and stored temporarily in the moving links, and the negative pulses of torque as energy attempting to return from the links to the driver. Unfortunately most motors are designed to deliver energy but not to take it back. Thus the "returned energy" has no place to go.

Figure 11-9 (p. 606) shows the speed-torque characteristic of a permanent magnet (PM) DC electric motor. Other types of motors will have differently shaped functions that relate motor speed to torque as shown in Figures 2-41 and 2-42 (pp. 77-78), but all drivers (sources) will have some such characteristic curve. As the torque demands on the motor

11

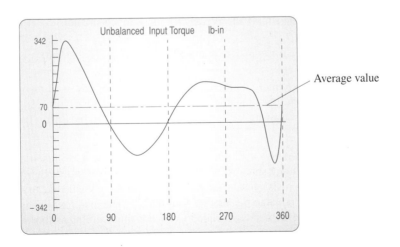

FIGURE 11-8

Input torque curve for an unbalanced crank-rocker fourbar linkage

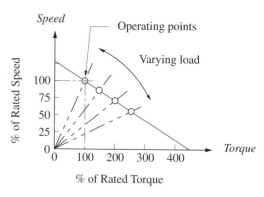

(a) Speed–torque characteristic of a PM electric motor   (b) Load lines superposed on speed–torque curve

FIGURE 11-9

DC permanent magnet (PM) electric motor's typical speed-torque characteristic

change, the motor's speed must also change according to its inherent characteristic. This means that the torque curve being demanded in Figure 11-8 (p. 605) will be very difficult for a standard motor to deliver without drastic changes in its speed.

The computation of the torque curve in Figure 11-8 was made on the assumption that the crank (thus the motor) speed was a constant value. All the kinematic data used in the force and torque calculation were generated on that basis. With the torque variation shown we would have to use a large-horsepower motor to provide the power required to reach that peak torque at the design speed:

$$Power = torque \times angular\ velocity$$

$$Peak\ power = 341.7\ \text{lb-in} \times 50\ \frac{\text{rad}}{\text{sec}} = 17\ 085\ \frac{\text{in-lb}}{\text{sec}} = 2.59\ \text{hp}$$

The power needed to supply the average torque is much smaller.

$$Average\ power = 70.2\ \text{lb-in} \times 50\ \frac{\text{rad}}{\text{sec}} = 3510\ \frac{\text{in-lb}}{\text{sec}} = 0.53\ \text{hp}$$

It would be extremely inefficient to specify a motor based on the peak demand of the system, as most of the time it will be underutilized. We need something in the system which is capable of storing kinetic energy. One such kinetic energy storage device is called a **flywheel**.

FLYWHEEL ENERGY   Figure 11-10 shows a **flywheel**, designed as a flat circular disk, attached to a motor shaft which might also be the driveshaft for the crank of our linkage. The motor supplies a torque magnitude T_M which we would like to be as constant as possible, i.e., to be equal to the average torque T_{avg}. The load (our linkage), on the other side of the flywheel, demands a torque T_L which is time varying as shown in Figure 11-8. The kinetic energy in a rotating system is:

$$E = \frac{1}{2}I\omega^2 \qquad\qquad (11.17)$$

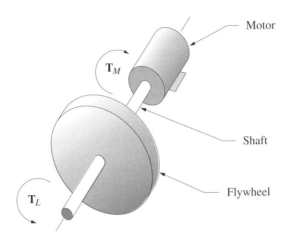

FIGURE 11-10

Flywheel on a driveshaft

where I is the moment of inertia of all rotating mass on the shaft. This includes the I of the motor rotor and of the linkage crank plus that of the flywheel. We want to determine how much I we need to add in the form of a flywheel to reduce the speed variation of the shaft to an acceptable level. We begin by writing Newton's law for the free-body diagram in Figure 11-10.

$$\sum T = I\alpha$$

$$T_L - T_M = I\alpha$$

but we want:

$$T_M = T_{avg}$$

so:

$$T_L - T_{avg} = I\alpha \qquad (11.18a)$$

substituting:

$$\alpha = \frac{d\omega}{dt} = \frac{d\omega}{dt}\left(\frac{d\theta}{d\theta}\right) = \omega\frac{d\omega}{d\theta}$$

gives:

$$T_L - T_{avg} = I\omega\frac{d\omega}{d\theta}$$

$$\left(T_L - T_{avg}\right)d\theta = I\omega\,d\omega \qquad (11.18b)$$

and integrating:

$$\int_{\theta@\,\omega_{min}}^{\theta@\,\omega_{max}} \left(T_L - T_{avg}\right)d\theta = \int_{\omega_{min}}^{\omega_{max}} I\omega\,d\omega$$

$$(11.18c)$$

$$\int_{\theta@\,\omega_{min}}^{\theta@\,\omega_{max}} \left(T_L - T_{avg}\right)d\theta = \frac{1}{2}I\left(\omega_{max}^2 - \omega_{min}^2\right)$$

The left side of this expression represents the change in energy E between the maximum and minimum shaft ω's and is equal to the *area under the torque-time diagram*[*] (Figures 11-8, p. 605, and 11-11, p. 609) between those extreme values of ω. The right side of equation 11.18c is the change in energy stored in the flywheel. The only way we

11

[*] There is often confusion between torque and energy because they appear to have the same units of *lb-in (in-lb)* or *N-m (m-N)*. This leads some students to think that they are the same quantity, but they are not. Torque ≠ energy. The **integral** of torque with respect to angle, measured in radians, **is** equal to energy. This integral has the units of *in-lb-rad*. The radian term is usually omitted since it is in fact unity. Power in a rotating system is equal to torque × angular velocity (measured in *rad/sec*), and the power units are then *(in-lb-rad)/sec*. When power is integrated versus time to get energy, the resulting units are *in-lb-rad*, the same as the integral of torque versus angle. The radians are again usually dropped, contributing to the confusion.

can extract energy from the flywheel is to slow it down as shown in equation 11.17 (p. 606). Adding energy will speed it up. Thus it is impossible to obtain exactly constant shaft velocity in the face of changing energy demands by the load. The best we can do is to minimize the speed variation ($\omega_{max} - \omega_{min}$) by providing a flywheel with sufficiently large I.

✐ EXAMPLE 11-5

Determining the Energy Variation in a Torque-Time Function.

Given: An input torque-time function which varies over its cycle. Figure 11-11 shows the input torque curve from Figure 11-8 (p. 605). The torque is varying during the 360° cycle about its average value.

Find: The total energy variation over one cycle.

Solution:

1   Calculate the average value of the torque-time function over one cycle, which in this case is 70.2 lb-in. (Note that in some cases the average value may be zero.)

2   Note that the *integration on the left side of equation 11.18c (p. 607) is done with respect to the average line of the torque function, not with respect to the θ axis.* (From the definition of the average, the sum of positive area above an average line is equal to the sum of negative area below that line.) The integration limits in equation 11.18 are from the shaft angle θ at which the shaft ω is a minimum to the shaft angle θ at which ω is a maximum.

3   The minimum ω will occur after the maximum positive energy has been delivered from the motor to the load, i.e., at a point (θ) where the summation of positive energy (area) in the torque pulses is at its largest positive value.

4   The maximum ω will occur after the maximum negative energy has been returned to the load, i.e., at a point (θ) where the summation of energy (area) in the torque pulses is at its largest negative value.

5   To find these locations in θ corresponding to the maximum and minimum ω's and thus find the amount of energy needed to be stored in the flywheel, we need to numerically integrate each pulse of this function from crossover to crossover with the average line. The crossover points in Figure 11-11 have been labeled A, B, C, and D. (Program LINKAGES does this integration for you numerically, using a trapezoidal rule.)

6   The LINKAGES program prints the table of areas shown in Figure 11-11. The positive and negative pulses are separately integrated as described above. Reference to the plot of the torque function will indicate whether a positive or negative pulse is the first encountered in a particular case. The first pulse in this example is a positive one.

7   The remaining task is to accumulate these pulse areas beginning at an arbitrary crossover (in this case point A) and proceeding pulse by pulse across the cycle. Table 11-1 shows this process and the result.

8   Note in Table 11-1 that the minimum shaft speed occurs after the largest accumulated positive energy pulse (+200.73 in-lb) has been delivered from the driveshaft to the system. Delivery

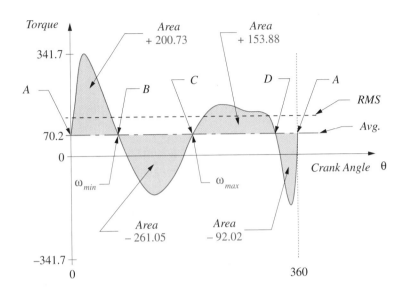

FIGURE 11-11

Integrating the pulses above and below the average value in the input torque function

of energy slows the motor down. Maximum shaft speed occurs after the largest accumulated negative energy pulse (–60.32 in-lb) has been returned from the system by the driveshaft. This return of stored energy will speed up the motor. The total energy variation is the algebraic difference between these two extreme values, which in this example is –261.05 in-lb. This negative energy coming out of the system needs to be absorbed by the flywheel and then returned to the system *during each cycle* to smooth the variations in shaft speed.

SIZING THE FLYWHEEL   We now must determine how large a flywheel is needed to absorb this energy with an acceptable change in speed. The change in shaft speed during a cycle is called its *fluctuation (Fl)* and is equal to:

$$Fl = \omega_{max} - \omega_{min} \qquad\qquad (11.19a)$$

TABLE 11-1   Integrating the Torque Function

From	Δ Area = ΔE	Accum. Sum = E	
A to B	+200.73	+200.73	ω_{min} @ B
B to C	–261.05	–60.32	ω_{max} @ C
C to D	+153.88	+93.56	
D to A	–92.02	+1.54	
	Total Δ Energy	= E @ ω_{max} – E @ ω_{min}	
		= $(-60.32) - (+200.73) = -261.05$ in-lb	

We can normalize this to a dimensionless ratio by dividing it by the average shaft speed. This ratio is called the *coefficient of fluctuation (k)*.

$$k = \frac{\omega_{max} - \omega_{min}}{\omega_{avg}} \qquad (11.19b)$$

This *coefficient of fluctuation* is a design parameter to be chosen by the designer. It typically is set to a value between 0.01 and 0.05, which corresponds to a 1 to 5% fluctuation in shaft speed. The smaller this chosen value, the larger the flywheel will have to be. This presents a design trade-off. A larger flywheel will add more cost and weight to the system, which factors have to be weighed against the smoothness of operation desired.

We found the required change in energy E by integrating the torque curve

$$\int_{\theta@\,\omega_{min}}^{\theta@\,\omega_{max}} \left(T_L - T_{avg} \right) d\theta = E \qquad (11.20a)$$

and can now set it equal to the right side of equation 11.18c (p. 607):

$$E = \frac{1}{2} I \left(\omega_{max}^2 - \omega_{min}^2 \right) \qquad (11.20b)$$

Factoring this expression:

$$E = \frac{1}{2} I \left(\omega_{max} + \omega_{min} \right) \left(\omega_{max} - \omega_{min} \right) \qquad (11.20c)$$

If the torque-time function were a pure harmonic, then its average value could be expressed exactly as:

$$\omega_{avg} = \frac{\omega_{max} + \omega_{min}}{2} \qquad (11.21)$$

Our torque functions will seldom be pure harmonics, but the error introduced by using this expression as an approximation of the average is acceptably small. We can now substitute equations 11.19b and 11.21 into equation 11.20c to get an expression for the mass moment of inertia I_s of the system flywheel needed.

$$E = \frac{1}{2} I_s \left(2\omega_{avg} \right) \left(k\,\omega_{avg} \right)$$

$$I_s = \frac{E}{k\,\omega_{avg}^2} \qquad (11.22)$$

Equation 11.22 can be used to design the physical flywheel by choosing a desired coefficient of fluctuation k, and using the value of E from the numerical integration of the torque curve (see Table 11-1) and the average shaft ω to compute the needed system I_s. The physical flywheel's mass moment of inertia I_f is then set equal to the required system I_s. But if the moments of inertia of the other rotating elements on the same driveshaft (such as the motor) are known, the physical flywheel's required I_f can be reduced by those amounts.

FIGURE 11-12

Input torque curve for the linkage in Figure 11-8 (p. 605) after smoothing with a flywheel

The most efficient flywheel design in terms of maximizing I_f for minimum material used is one in which the mass is concentrated in its rim and its hub is supported on spokes, like a carriage wheel. This puts the majority of the mass at the largest radius possible and minimizes the weight for a given I_f. Even if a flat, solid circular disk flywheel design is chosen, either for simplicity of manufacture or to obtain a flat surface for other functions (such as an automobile clutch), the design should be done with an eye to reducing weight and thus cost. Since in general $I = mr^2$, a thin disk of large diameter will need fewer pounds of material to obtain a given I than will a thicker disk of smaller diameter. Dense materials such as cast iron and steel are the obvious choices for a flywheel. Aluminum is seldom used. Though many metals (lead, gold, silver, platinum) are more dense than iron and steel, one can seldom get the accounting department's approval to use them in a flywheel.

Figure 11-12 shows the change in the input torque $\mathbf{T}_{12}$ for the linkage in Figure 11-8 (p. 605) after the addition of a flywheel sized to provide a coefficient of fluctuation of 0.05. The oscillation in torque about the unchanged average value is now 5%, much less than what it was without the flywheel. A much smaller-horsepower motor can now be used because the flywheel is available to absorb the energy returned from the linkage during its cycle.

11.12 A LINKAGE FORCE TRANSMISSION INDEX

The transmission angle was introduced in Chapter 2 and used in subsequent chapters as an index of merit to predict the kinematic behavior of a linkage. A too-small transmission angle predicts problems with motion and force transmission in a fourbar linkage. Unfortunately, the transmission angle has limited application. It is only useful for fourbar linkages and then only when the input and output torques are applied to links that are pivoted to ground (i.e., the crank and rocker). When external forces are applied to the coupler link, the transmission angle tells nothing about the linkage's behavior.

Holte and Chase [1] define a joint-force index (JFI) which is useful as an indicator of any linkage's ability to smoothly transmit energy regardless of where the loads are applied on the linkage. It is applicable to higher-order linkages as well as to the fourbar linkage. The JFI at any instantaneous position is defined as the ratio of the maximum static force in any joint of the mechanism to the applied external load. If the external load is a force, then it is:

$$\text{JFI} = MAX \left| \frac{F_{ij}}{F_{ext}} \right| \qquad \text{for all pairs } i, j \qquad (11.23a)$$

If the external load is a torque, then it is:

$$\text{JFI} = MAX \left| \frac{F_{ij}}{T_{ext}} \right| \qquad \text{for all pairs } i, j \qquad (11.23b)$$

where, in both cases, F_{ij} is the force in the linkage joint connecting links i and j.

The F_{ij} are calculated from a static force analysis of the linkage. Dynamic forces can be much greater than static forces if speeds are high. However, if this static force transmission index indicates a problem in the absence of any dynamic forces, then the situation will obviously be worse at speed. The largest joint force at each position is used rather than a composite or average value on the assumption that high friction in any one joint is sufficient to hamper linkage performance regardless of the forces at other joints.

Equation 11.23a is dimensionless and so can be used to compare linkages of different design and geometry. Equation 11.23b has dimensions of reciprocal length, so caution must be exercised when comparing designs when the external load is a torque. Then the units used in any comparison must be the same, and the compared linkages should be similar in size.

Equations 11.23 apply to any one instantaneous position of the linkage. As with the transmission angle, this index must be evaluated for all positions of the linkage over its expected range of motion and the largest value of that set found. The peak force may move from pin to pin as the linkage rotates. If the external loads vary with linkage position, they can be accounted for in the calculation.

Holte and Chase suggest that the JFI be kept below a value of about 2 for linkages whose output is a force. Larger values may be tolerable especially if the joints are designed with good bearings that are able to handle the higher loads.

There are some linkage positions in which the JFI can become infinite or indeterminate as when the linkage reaches an immovable position, defined as the input link or input joint being inactive. This is equivalent to a stationary configuration as described in earlier chapters provided that the input joint is inactive in the particular stationary configuration. These positions need to be identified and avoided in any event, independent of the determination of any index of merit. In some cases the mechanism may be immovable but still capable of supporting a load. See reference [1] for more detailed information on these special cases.

11.13 PRACTICAL CONSIDERATIONS

This chapter has presented some approaches to the computation of dynamic forces in moving machinery. The newtonian approach gives the most information and is necessary in order to obtain the forces at all pin joints so that stress analyses of the members can be done. Its application is really quite straightforward, requiring only the creation of correct free-body diagrams for each member and the application of the two simple vector equations which express Newton's second law to each free body. Once these equations are expanded for each member in the system and placed in standard matrix form, their solution (with a computer) is a trivial task.

The real work in designing these mechanisms comes in the determination of the shapes and sizes of the members. In addition to the kinematic data, the force computation requires only the masses, *CG* locations, and mass moments of inertia versus those *CG*s for its completion. These three geometric parameters completely characterize the member for dynamic modeling purposes. Even if the link shapes and materials are completely defined at the outset of the force analysis process (as with the redesign of an existing system), it is a tedious exercise to calculate the dynamic properties of complicated shapes. Current solids modeling CAD systems make this step easy by computing these parameters automatically for any part designed within them.

If, however, you are starting from scratch with your design, the *blank-paper syndrome* will inevitably rear its ugly head. A first approximation of link shapes and selection of materials must be made in order to create the dynamic parameters needed for a "first pass" force analysis. A stress analysis of those parts, based on the calculated dynamic forces, will invariably find problems that require changes to the part shapes, thus requiring recalculation of the dynamic properties and recomputation of the dynamic forces and stresses. This process will have to be repeated in circular fashion (*iteration*—see Chapter 1, p. 8) until an acceptable design is reached. The advantage of using a computer to do these repetitive calculations is obvious and cannot be overstressed. An equation solver program such as *Mathcad, Matlab, or TKSolver* will be a useful aid in this process by reducing the amount of computer programming necessary.

Students with no design experience are often not sure how to approach this process of designing parts for dynamic applications. The following suggestions are offered to get you started. As you gain experience, you will develop your own approach.

It is often useful to create complex shapes from a combination of simple shapes, at least for first approximation dynamic models. For example, a link could be considered to be made up of a hollow cylinder at each pivot end, connected by a rectangular prism along the line of centers. It is easy to calculate the dynamic parameters for each of these simple shapes and then combine them. The steps would be as follows (repeated for each link):

1   Calculate the volume, mass, *CG* location, and mass moments of inertia with respect to the local *CG* of each separate part of your built-up link. In our example link these parts would be the two hollow cylinders and the rectangular prism.

2   Find the location of the composite *CG* of the assembly of the parts into the link by the method shown in Section 10.4 (p. 547) and equations 10.3 (pp. 548-549). See also Figure 10-2 (p. 549).

3   Use the *parallel axis theorem* (equation 10.8, p. 551) to transfer the mass moments of inertia of each part to the common, composite *CG* for the link. Then add the individual, transferred *I*'s of the parts to get the total *I* of the link about its composite *CG*. See Section 10.6 (p. 551).

11

11

§ All problem figures are
provided as PDF files, and
some are also provided as
animated Working Model
files; all are on the DVD.
PDF filenames are the same
as the figure number. Run
the file *Animations.html* to
access and run the anima-
tions.

* See footnote opp. page.

† See footnote opp. page.

‡ See footnote opp. page.

Steps 1 to 3 will create the link geometry data for each link needed for the dynamic force analysis as derived in this chapter.

4 Do the dynamic force analysis.

5 Do a dynamic stress and deflection analysis of all parts.

6 Redesign the parts and repeat steps 1 to 5 until a satisfactory result is achieved.

Remember that lighter (lower-mass) links will have smaller inertial forces on them and thus could have lower stresses despite their smaller cross sections. Also, smaller mass moments of inertia of the links can reduce the driving torque requirements, especially at higher speeds. But be cautious about the dynamic deflections of thin, light links becoming too large. We are assuming rigid bodies in these analyses. That assumption will not be valid if the links are too flexible. Always check the deflections as well as the stresses in your designs.

11.14 REFERENCE

1   **Holte, J. E., and T. R. Chase**. (1994). "A Force Transmission Index for Planar Link-age Mechanisms." *Proc. of 23rd Biennial Mechanisms Conference*, Minneapolis, MN, p. 377.

11.15 PROBLEMS§

11-1   Draw free-body diagrams of the links in the geared fivebar linkage shown in Figure 4-11 (p. 194) and write the dynamic equations to solve for all forces plus the driving torque. Assemble the symbolic equations in matrix form for solution.

11-2   Draw free-body diagrams of the links in the sixbar linkage shown in Figure 4-12 (p. 197) and write the dynamic equations to solve for all forces plus the driving torque. Assemble the symbolic equations in matrix form for solution.

*†‡11-3   Table P11-1 shows kinematic and geometric data for several crank-slider linkages of the type and orientation shown in Figure P11-1 (p. 616). The point locations are defined as described in the text. For the row(s) in the table assigned, use the matrix method of Section 11.5 (p. 595) and program MATRIX, *Mathcad, Matlab, TKSolver*, or a matrix solving calculator to solve for forces and torques at the position shown. Also compute the shaking force and shaking torque. Consider the coefficient of friction μ between slider and ground to be zero. You may check your solution by opening the solution files (located in the Solutions folder on the DVD) named P11-03x (where x is the row letter) into program LINKAGES.

*†11-4   Repeat Problem 11-3 using the method of virtual work to solve for the input torque on link 2. Additional data for corresponding rows are given in Table P11-2.

*†11-5   Table P11-3 (p. 618) shows kinematic and geometric data for several pin-jointed fourbar linkages of the type and orientation shown in Figure P11-2 (p. 619). All have $\theta_1 = 0$. The point locations are defined as described in the text. For the row(s) in the table assigned, use the matrix method of Section 11.4 (p. 589) and program MATRIX or a matrix solving calculator to solve for forces and torques at the position shown. You may check your solution by opening the solution files from the DVD named P11-05x (where x is the row letter) into program LINKAGES.

TABLE P11-1 Data for Problem 11-3 (See Figure P11-1 for Nomenclature, p. 616)

Part 1   Lengths in inches, angles in degrees, mass in blobs, angular velocity in rad/sec

Row	link 2	link 3	offset	θ_2	ω_2	α_2	m_2	m_3	m_4
a	4	12	0	45	10	20	0.002	0.020	0.060
b	3	10	1	30	15	− 5	0.050	0.100	0.200
c	5	15	−1	260	20	15	0.010	0.020	0.030
d	6	20	1	− 75	−10	− 10	0.006	0.150	0.050
e	2	8	0	135	25	25	0.001	0.004	0.014
f	10	35	2	120	5	− 20	0.150	0.300	0.050
g	7	25	−2	− 45	30	− 15	0.080	0.200	0.100

Part 2   Angular acceleration in rad/sec^2, moments of Inertia in blob-in^2, torque in lb-in

Row	I_2	I_3	Rg_2 mag	δ_2 ang	Rg_3 mag	δ_3 ang	F_{P_3} mag	δF_{P_3} ang	R_{P_3} mag	δR_P ang	T_3
a	0.10	0.2	2	0	5	0	0	0	0	0	20
b	0.20	0.4	1	20	4	− 30	10	45	4	30	− 35
c	0.05	0.1	3	− 40	9	50	32	270	0	0	− 65
d	0.12	0.3	3	120	12	60	15	180	2	60	− 12
e	0.30	0.8	0.5	30	3	75	6	− 60	2	75	40
f	0.24	0.6	6	45	15	135	25	270	0	0	− 75
g	0.45	0.9	4	− 45	10	225	9	120	5	45	− 90

Part 3   Forces in lb, linear accelerations in in/sec^2

Row	θ_3	α_3	ag_2 mag	ag_2 ang	ag_3 mag	ag_3 ang	ag_4 mag	ag_4 ang
a	166.40	− 2.40	203.96	213.69	371.08	200.84	357.17	180
b	177.13	34.33	225.06	231.27	589.43	200.05	711.97	180
c	195.17	−134.76	1 200.84	37.85	2 088.04	43.43	929.12	0
d	199.86	− 29.74	301.50	230.71	511.74	74.52	23.97	180
e	169.82	113.12	312.75	−17.29	976.79	−58.13	849.76	0
f	169.03	3.29	192.09	23.66	302.50	−29.93	301.92	0
g	186.78	− 172.20	3 600.50	90.95	8 052.35	134.66	4 909.27	180

TABLE P11-2 Data for Problem 11-4

See also Table P11-1. Unit system is the same as in that table.

Row	ω_3	Vg_2 mag	Vg_2 ang	Vg_3 mag	Vg_3 ang	Vg_4 mag	Vg_4 ang	V_{P_3} mag	V_{P_3} ang
a.	− 2.43	20.0	135	35.24	152.09	35.14	180	35.24	152.09
b.	− 3.90	15.0	140	40.35	140.14	24.45	180	26.69	153.35
c.	1.20	60.0	310	89.61	−8.23	93.77	0	89.61	− 8.23
d.	0.83	30.0	315	69.10	191.15	63.57	180	70.63	191.01
e.	4.49	12.5	255	56.02	211.93	29.01	180	61.36	204.87
f.	0.73	30.0	255	60.89	210.72	38.46	180	60.89	210.72
g.	−5.98	120.0	0	211.46	61.31	166.14	0	208.60	53.19

11

* Answers in Appendix F.

† These problems are suited to solution using *Mathcad*, *Matlab*, or *TKSolver* equation solver programs.

‡ These problems are suited to solution using program LINKAGES which is on the attached DVD.

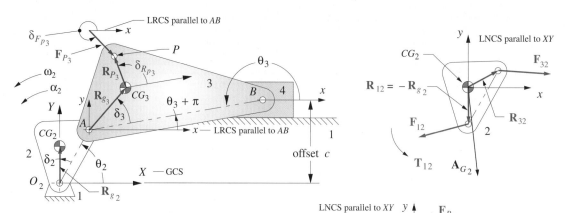

Generic linkage and free-body diagrams

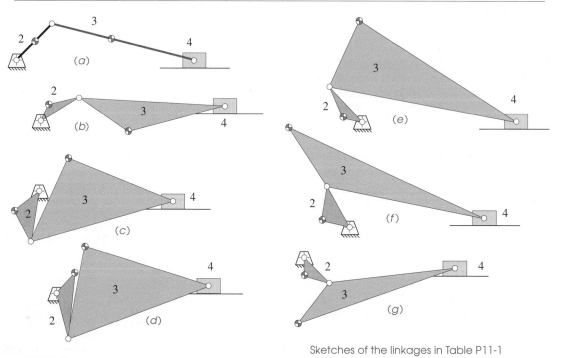

Sketches of the linkages in Table P11-1

FIGURE P11-1

Linkage geometry, notation, and free-body diagrams for problems 11-3 to 11-4 (p. 614)

*†11-6   Repeat Problem 11-5 using the method of virtual work to solve for the input torque on link 2. Additional data for corresponding rows are given in Table P11-4 (p. 620).

*‡11-7   For the row(s) assigned in Table P11-3 (*a-f*) (p. 618), input the associated disk file to program LINKAGES, calculate the linkage parameters for crank angles from zero to 360° by 5° increments with $\alpha_2 = 0$, and design a steel disk flywheel to smooth the input torque using a coefficient of fluctuation of 0.05. Minimize the flywheel weight.

‡11-8   Figure P11-3 (p. 620) shows a fourbar linkage and its dimensions. The steel crank and rocker have uniform cross sections 1 in wide by 0.5 in thick. The aluminum coupler is 0.75 in thick. In the instantaneous position shown, the crank O_2A has $\omega = 40$ rad/sec and $\alpha = -20$ rad/sec². There is a horizontal force at P of $F = 50$ lb. Find all pin forces and the torque needed to drive the crank at this instant.

‡11-9   Figure P11-4a (p. 620) shows a fourbar linkage and its dimensions in meters. The steel crank and rocker have uniform cross sections of 50 mm wide by 25 mm thick. The aluminum coupler is 25 mm thick. In the instantaneous position shown, the crank O_2A has $\omega = 10$ rad/sec and $\alpha = 5$ rad/sec². There is a vertical force at P of $F = 100$ N. Find all pin forces and the torque needed to drive the crank at this instant.

‡11-10   Figure P11-4b shows a fourbar linkage and its dimensions in meters. The steel crank and rocker have uniform cross sections of 50 mm wide by 25 mm thick. The aluminum coupler is 25 mm thick. In the instantaneous position shown, the crank O_2A has $\omega = 15$ rad/sec and $\alpha = -10$ rad/sec². The horizontal force applied at point P is $F = 500$ N. Find all pin forces and the torque needed to drive the crank at this instant.

‡11-11   Figure P11-5a (p. 621) shows a fourbar linkage and its dimensions in meters. The steel crank, coupler, and rocker have uniform cross sections of 50 mm wide by 25 mm thick. In the instantaneous position shown, the crank O_2A has $\omega = 15$ rad/sec and $\alpha = -10$ rad/sec². There is a vertical force at P of $F = 500$ N. Find all pin forces and the torque needed to drive the crank at this instant.

*†‡11-12 Figure P11-5b shows a fourbar linkage and its dimensions in meters. The steel crank, coupler, and rocker have uniform cross sections of 60-mm diameter. In the instantaneous position shown, the crank O_2A has $\omega = -10$ rad/sec and $\alpha = 10$ rad/sec². There is a horizontal force at P of $F = 500$ N. Find all pin forces and the torque needed to drive the crank at this instant.

*†‡11-13 Figure P11-6 (p. 621) shows a water jet loom laybar drive mechanism driven by a pair of Grashof crank-rocker fourbar linkages. The crank rotates at 500 rpm. The laybar is carried between the coupler-rocker joints of the two linkages at their respective instant centers $I_{3,4}$. The combined weight of the reed and laybar is 29 lb. A 540-lb beat-up force from the cloth is applied to the reed as shown. The steel links have a 2 × 1-in uniform cross section. Find the forces on the pins for one revolution of the crank. Find the torque-time function required to drive the system.

*†11-14   Figure P11-7 (p. 622) shows a crimping tool. Find the force F_{hand} needed to generate a 2000-lb F_{crimp}. Find the pin forces. What is this linkage's joint force transmission index (JFI) in this position?

†11-15   Figure P11-8 (p. 622) shows a walking-beam conveyor mechanism that operates at slow speed (25 rpm). The boxes being pushed each weigh 50 lb. Determine the pin forces in the linkage and the torque required to drive the mechanism through one revolution. Neglect the masses of the links.

* Answers in Appendix F.

† These problems are suited to solution using *Mathcad, Matlab,* or *TKSolver* equation solver programs.

‡ These problems are suited to solution using program LINKAGES which is on the attached DVD.

11

TABLE P11-3   Data for Problems 11-5 and 11-7 (See Figure P11-2 for Nomenclature)

Part 1 — Lengths in inches, angles in degrees, angular acceleration in rad/sec^2

Row	link 2	link 3	link 4	link 1	θ_2	θ_3	θ_4	α_2	α_3	α_4
a	4	12	8	15	45	24.97	99.30	20	75.29	244.43
b	3	10	12	6	30	90.15	106.60	-5	140.96	161.75
c	5	15	14	2	260	128.70	151.03	15	78.78	53.37
d	6	19	16	10	-75	91.82	124.44	-10	-214.84	-251.82
e	2	8	7	9	135	34.02	122.71	25	71.54	-14.19
f	17	35	23	4	120	348.08	19.01	-20	-101.63	-150.86
g	7	25	10	19	100	4.42	61.90	-15	-17.38	-168.99

Part 2 — Angular velocity in rad/sec, mass in blobs, moment of Inertia in blob-in^2, torque in lb-in

Row	ω_2	ω_3	ω_4	m_2	m_3	m_4	I_2	I_3	I_4	T_3	T_4
a	20	-5.62	3.56	0.002	0.02	0.10	0.10	0.20	0.50	-15	25
b	10	-10.31	-7.66	0.050	0.10	0.20	0.20	0.40	0.40	12	0
c	20	16.60	14.13	0.010	0.02	0.05	0.05	0.10	0.13	-10	20
d	20	3.90	-3.17	0.006	0.15	0.07	0.12	0.30	0.15	0	30
e	20	1.06	5.61	0.001	0.04	0.09	0.30	0.80	0.30	25	40
f	20	18.55	21.40	0.150	0.30	0.25	0.24	0.60	0.92	0	-25
g	20	4.10	16.53	0.080	0.20	0.12	0.45	0.90	0.54	0	0

Part 3 — Lengths in inches, angles in degrees, linear accelerations in in/sec^2

Row	R_{g2} mag	R_{g2} ang	R_{g3} mag	R_{g3} ang	R_{g4} mag	R_{g4} ang	a_{g2} mag	a_{g2} ang	a_{g3} mag	a_{g3} ang
a	2	0	5	0	4	30	801.00	222.14	1 691.49	208.24
b	1	20	4	-30	6	40	100.12	232.86	985.27	194.75
c	3	-40	9	50	7	0	1 200.84	37.85	3 120.71	22.45
d	3	120	12	60	6	-30	1 200.87	226.43	4 543.06	81.15
e	0.5	30	3	75	2	-40	200.39	341.42	749.97	295.98
f	6	45	15	135	10	25	2 403.00	347.86	12 064.20	310.22
g	4	-45	10	225	4	45	1 601.12	237.15	2 562.10	-77.22

Part 4 — Linear accelerations in in/sec^2, forces in lb, lengths in inches, angles in degrees

Row	a_{g4} mag	a_{g4} ang	F_{P3} mag	δ_{FP3} ang	R_{P3} mag	δ_{RP3} ang	F_{P4} mag	δ_{FP4} ang	R_{P4} mag	δ_{RP4} ang
a	979.02	222.27	0	0	0	0	40	-30	8	0
b	1 032.32	256.52	4	30	10	45	15	-55	12	0
c	1 446.58	316.06	0	0	0	0	75	45	14	0
d	1 510.34	2.15	2	45	15	180	20	270	16	0
e	69.07	286.97	9	0	6	-60	16	60	7	0
f	4 820.72	242.25	0	0	0	0	23	0	23	0
g	1 284.55	-41.35	12	-60	9	120	32	20	10	0

11

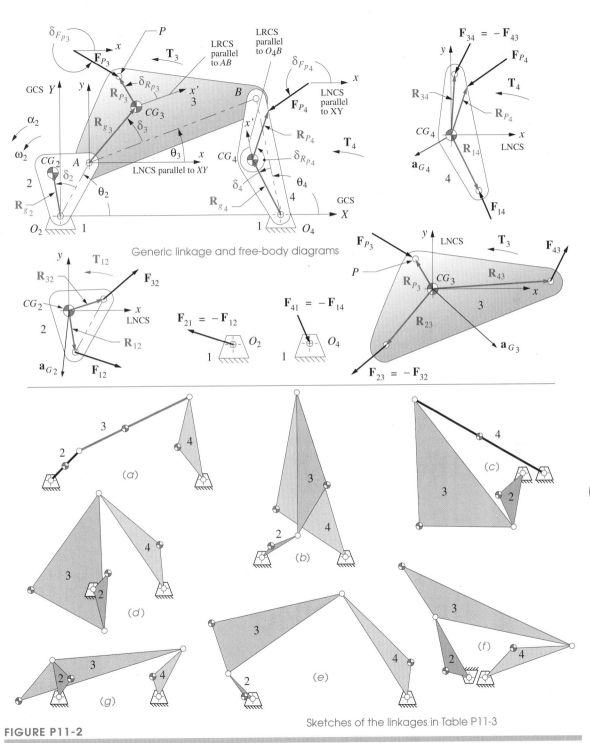

Generic linkage and free-body diagrams

Sketches of the linkages in Table P11-3

FIGURE P11-2

Linkage geometry, notation, and free-body diagrams for Problems 11-5 to11-7

11

TABLE P11-4 Data for Problem 11-6 (p. 617)

Row	Vg_2 mag	Vg_2 ang	Vg_3 mag	Vg_3 ang	Vg_4 mag	Vg_4 ang	VP_3 mag	VP_3 ang	VP_4 mag	VP_4 ang
a	40.00	135.00	54.44	145.19	14.23	219.30	54.44	145.19	41.39	−160.80
b	10.00	140.00	21.46	14.74	45.94	56.60	122.10	40.04	130.51	29.68
c	60.00	−50.00	191.94	299.70	98.91	241.03	191.94	−60.30	296.73	−118.97
d	60.00	135.00	94.36	353.80	19.03	4.44	152.51	−3.13	67.86	26.38
e	10.00	255.00	42.89	223.13	11.22	172.71	37.01	−140.37	48.41	−155.86
f	120.00	255.00	618.05	211.39	213.98	134.01	618.03	−148.61	692.08	116.52
g	80.00	145.00	118.29	205.52	66.10	196.90	154.85	−152.36	217.15	164.33

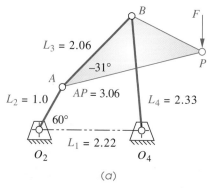

Dimensions in inches

FIGURE P11-3

Problem 11-8 (p. 617)

†11-16 Figure P11-9 (p. 622) shows a surface grinder table crank-slider drive that operates at 120 rpm. The crank radius is 22 mm, the coupler is 157 mm, and its offset is 40 mm. The mass of table and workpiece combined is 50 kg. Find the pin forces, slider side loads, and driving torque over one revolution. Neglect the mass of the crank and connecting rod.

†11-17 Figure P11-10 (p. 623) shows a crank-slider power hacksaw that operates at 50 rpm. The balanced crank is 75 mm; the uniform cross section coupler is 170 mm long, weighs 2 kg, and its offset is 45 mm. Link 4 weighs 15 kg. Find the pin forces, slider side loads, and driving torque over one revolution for a cutting force of 250 N in the forward direction and 50 N during the return stroke.

†11-18 Figure P11-11 (p. 623) shows a slider-crank paper roll off-loading station. The paper rolls have a 0.9-m *OD* and 0.22-m *ID*, are 3.23 m long, and have a density of 984 kg/m³. The forks that support the roll are 1.2 m long. The motion is slow so inertial loading can be neglected. Find the force required of the air cylinder to rotate the roll through 90°.

†11-19 Derive an expression for the relationship between flywheel mass and the dimensionless parameter radius/thickness (*r/t*) for a solid disk flywheel of moment of inertia *I*. Plot this function for an arbitrary value of *I* and determine the optimum *r/t* ratio to minimize flywheel weight for that *I*.

† These problems are suited to solution using *Mathcad*, *Matlab*, or *TKSolver* equation solver programs.

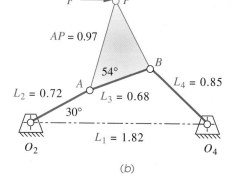

FIGURE P11-4

Problems 11-9 to 11-10 (p. 617)

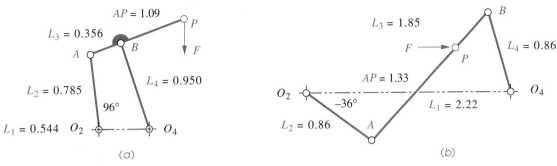

FIGURE P11-5

Problems 11-11 to 11-12 (p. 617)

FIGURE P11-6

Problem 11-13 (p. 617) - Fourbar linkage for laybar drive, showing forces and accelerations

all lengths in inches

$AB = 0.80$, $BC = 1.23$, $CD = 1.55$, $AD = 2.4$

FIGURE P11-7

Problem 11-14 (p. 617)

All lengths in inches

FIGURE P11-8

Problem 11-15 (p. 617)

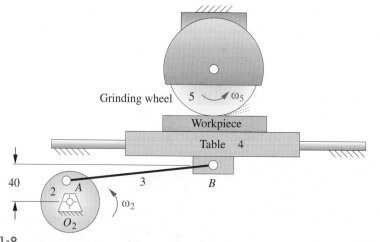

FIGURE P11-9

Problem 11-16 (p. 620)

$L_2 = 75$ mm
$L_3 = 170$ mm   2 kg

FIGURE P11-10

Problem 11-17 (p. 620) Power hacksaw *Adapted from P. H. Hill and W. P. Rule. (1960). Mechanisms: Analysis and Design*

11-20   Figure P11-12 (p. 624) shows an oil field pump mechanism. The head of the rocker
arm is shaped such that the lower end of a flexible cable attached to it will always be
directly over the well head regardless of the position of the rocker arm 4. The pump
rod, which connects to the pump in the well casing, is connected to the lower end of the
cable. The force in the pump rod on the up stroke is 2970 lb and the force on the down
stroke is 2300 lb. Link 2 weighs 598.3 lb and has a mass moment of inertia of 11.8
lb-in-sec^2 (blob-in^2); both include the counterweight. Its *CG* is on the link centerline,
13.2 in from O_2. Link 3 weighs 108 lb and its *CG* is on the link centerline, 40 in from
A. It has a mass moment of inertia of 150 lb-in-sec^2 (blob-in^2). Link 4 weighs 2706
lb and has a mass moment of inertia of 10 700 lb-in-sec^2 (blob-in^2); both include the
counterweight. Its *CG* is on the link centerline where shown. The crank turns at a con-

FIGURE P11-11

Problem 11-18 (p. 620)

| $B\text{-}CG_4 = 32.00$ |
| $P\text{-}CG_4 = 124.44$ |
| $O_4\text{-}CG_4 = 79.22$ |

FIGURE P11-12

Problems 11-20 (p. 623) to 11-23   An oil field pump - dimensions in inches

stant speed of 4 rpm CCW. At the instant shown in the figure the crank angle is at 45° with respect to the global coordinate system. Find all pin forces and the torque needed to drive the crank for the position shown. Include gravity forces because the links are heavy and the speed is low.

†11-21   Use the information in Problem 11-20 to find and plot all pin forces and the torque needed to drive the crank for one revolution of the crank.

†11-22   Use the information in Problem 11-20 to find the torque needed to drive the crank for the position shown using the method of virtual work.

†11-23   Use the information in Problem 11-20 to find and plot the torque needed to drive the crank for one revolution of the crank using the method of virtual work.

†11-24   In Figure P11-13, links 2 and 4 each weigh 2 lb and there are 2 of each (another set behind). Their *CGs* are at their midpoints. Link 3 weighs 10 lb. The mass moments of inertia of links 2, 3, and 4 are 0.071, 0.430, and 0.077 lb-in-sec² (blob-in²), respectively. Find the torque needed to begin a slow CCW rotation of link 2 from the position shown using the method of virtual work. Include gravity forces because the links are heavy and the speed is low.

†*11-25   The linkage in Figure P11-14 has $L_1 = 9.5$, $L_2 = 5.0$, $L_3 = 7.4$, $L_4 = 8.0$, and $AP = 8.9$ in. The steel crank and rocker have uniform cross sections 1 in wide by 0.5 in thick. The aluminum coupler is 0.75 in thick. In the instantaneous position shown, the crank O_2A has $\omega = 40$ rad/sec and $\alpha = -20$ rad/sec². There is a horizontal force at P of $F = 50$ lb. Find the torque needed to drive the crank at the position shown using the method of virtual work.

11-26   For the linkage defined in Problem 11-25 use program LINKAGES to find and plot all pin forces and the torque needed to drive the crank at a constant speed of 40 rad/sec for one revolution of the crank.

* Answers in Appendix F.

† These problems are suited
to solution using *Mathcad*,
Matlab, or *TKSolver* equation solver programs.

FIGURE P11-13

Problem 11-24   An aircraft overhead bin mechanism - dimensions in inches

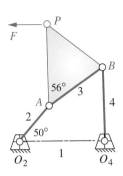

FIGURE P11-14

Problems 11-25 to 11-27

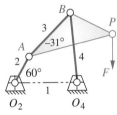

FIGURE P11-15

Problems 11-28 to 11-30

†11-27   For the linkage defined in Problem 11-25 find and plot the torque needed to drive the crank at a constant speed of 40 rad/sec for one revolution of the crank using the method of virtual work.

†11-28   The linkage in Figure P11-15 has $L_1 = 2.22$, $L_2 = 1.0$, $L_3 = 2.06$, $L_4 = 2.33$, and $AP = 3.06$ m. The steel crank and rocker have uniform cross sections of 50 mm wide by 25 mm thick. The aluminum coupler is 25 mm thick. In the instantaneous position shown, the crank O_2A has $\omega = 10$ rad/sec and $\alpha = 5$ rad/sec². There is a vertical force at P of $F = 100$ N. Find the torque needed to drive the crank at the position shown using the method of virtual work.

11-29   For the linkage defined in Problem 11-28 use program LINKAGES to find and plot all pin forces and the torque needed to drive the crank at a constant speed of 10 rad/sec for one revolution of the crank.

†11-30   For the linkage defined in Problem 11-28 find and plot the torque needed to drive the crank at a constant speed of 10 rad/sec for one revolution of the crank using the method of virtual work.

†11-31   The linkage in Figure P11-16 has $L_1 = 1.82$, $L_2 = 0.72$, $L_3 = 1.43$, $L_4 = 1.60$, and $AP = 0.97$ m. The steel crank and rocker have uniform cross sections 50 mm wide by 25 mm thick. The aluminum coupler is 25 mm thick. In the instantaneous position shown, the crank O_2A has $\omega = 15$ rad/sec and $\alpha = -10$ rad/sec². There is a horizontal force at P of $F = 200$ N. Find the torque needed to drive the crank at the position shown using the method of virtual work.

11-32   For the linkage defined in Problem 11-31 use program LINKAGES to find and plot all pin forces and the torque needed to drive the crank at a constant speed of 15 rad/sec for one revolution of the crank using the method of virtual work.

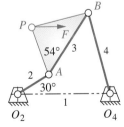

FIGURE P11-16

Problems 11-31 to 11-33

11

† These problems are suited to solution using *Mathcad*, *Matlab*, or *TKSolver* equation solver programs.

†11-33   For the linkage defined in Problem 11-31 find and plot the torque needed to drive the crank at a constant speed of 15 rad/sec for one revolution of the crank using the method of virtual work.

†11-34   The linkage in Figure P11-17 has $L_1 = 1.0$, $L_2 = 0.356$, $L_3 = 0.785$, $L_4 = 0.95$, and $AP = 1.09$ m. The steel crank, coupler, and rocker have uniform cross sections of 50 mm wide by 25 mm thick. In the instantaneous position shown, the crank O_2A has $\omega = 15$ rad/sec and $\alpha = -10$ rad/sec^2. The vertical force at P is $F = 500$ N. Find the torque needed to drive the crank at the position shown using the method of virtual work.

11-35   For the linkage defined in Problem 11-34 use program LINKAGES to find and plot all pin forces and the torque needed to drive the crank at a constant speed of 15 rad/sec for one revolution of the crank using the method of virtual work.

†11-36   For the linkage defined in Problem 11-34 find and plot the torque needed to drive the crank at a constant speed of 15 rad/sec for one revolution of the crank using the method of virtual work.

†11-37   The linkage in Figure P11-18 has $L_1 = 2.22$, $L_2 = 0.86$, $L_3 = 1.85$, $L_4 = 1.86$, and $AP = 1.33$ m. The steel crank, coupler, and rocker have uniform cross sections of 50-mm diameter. In the instantaneous position shown, the crank O_2A has $\omega = -10$ rad/sec and $\alpha = 10$ rad/sec^2. There is a horizontal force at P of $F = 300$ N. Find the torque needed to drive the crank at the position shown using the method of virtual work.

11-38   For the linkage defined in Problem 11-37 use program LINKAGES to find and plot all pin forces and the torque needed to drive the crank at a constant speed of 10 rad/sec for one revolution of the crank.

†11-39   For the linkage defined in Problem 11-37 find and plot the torque needed to drive the crank at a constant speed of 10 rad/sec for one revolution of the crank using the method of virtual work.

†*11-40   Design a steel disk flywheel to smooth the input torque for the crank of Problem 11-26 using a coefficient of fluctuation of 0.05 while minimizing flywheel weight.

†11-41   Design a steel disk flywheel to smooth the input torque for the crank of Problem 11-29 using a coefficient of fluctuation of 0.05 while minimizing flywheel weight.

†11-42   Design a steel disk flywheel to smooth the input torque for the crank of Problem 11-32 using a coefficient of fluctuation of 0.07 while minimizing flywheel weight.

†11-43   Design a steel disk flywheel to smooth the input torque for the crank of Problem 11-35 using a coefficient of fluctuation of 0.05 while minimizing flywheel weight.

†11-44   Design a steel disk flywheel to smooth the input torque for the crank of Problem 11-38 using a coefficient of fluctuation of 0.06 while minimizing flywheel weight.

11-45   Table P11-5 gives kinematic and geometric data for a slider-crank linkage of the type and orientation shown in Figure 11-4 (p. 597). For the row(s) in the table assigned, solve for the three pin forces and the torque available at the crank for the position shown.

11-46   Table P11-5 gives kinematic and geometric data for a slider-crank linkage of the type and orientation shown in Figure 11-4 (p. 597). For the row(s) assigned in the table, solve for the torque available at the crank using the method of virtual work for the position shown, assuming no friction losses.

---

* Answers in Appendix F.

---

† These problems are suited to solution using *Mathcad, Matlab,* or, *TKSolver* equation solver programs.

TABLE P11-5 Data for Problems 11-45 to 11-46 (See Figure 11-4 for Nomenclature)

Part 1   Lengths (inches), velocity (in/sec), acceleration (in/sec²)

Row	link 2	link 3	d	$\dot{d}$	$\ddot{d}$	R_{12}	R_{23}	μ
a	4	12	14	400	−22 760	1.3	3.0	0.15
b	3	10	8	375	67 350	1.0	2.5	0.00
c	5	15	12	390	36 400	1.7	3.8	0.10
d	6	20	18	700	45 430	2.0	5.0	0.18
e	2	8	8	225	3 010	0.7	2.0	0.08
f	10	35	35	−900	69 750	3.3	8.8	0.12
g	7	25	25	−935	209 900	2.3	6.2	0.14

Part 2   force (lbf, deg), mass (blobs), moments of inertia (blob-in²)

Row	F_P mag	F_P ang	m_2	m_3	m_4	I_{G_2}	I_{G_3}
a	60	180	0.002	0.020	0.060	0.10	0.2
b	45	180	0.050	0.100	0.200	0.20	0.4
c	75	180	0.010	0.020	0.030	0.05	0.1
d	90	180	0.006	0.150	0.050	0.12	0.3
e	30	180	0.001	0.004	0.014	0.30	0.8
f	150	180	0.150	0.300	0.050	0.24	0.6
g	110	180	0.080	0.200	0.100	0.45	0.9

11.16 VIRTUAL LABORATORY

L11-1   View the video *Fourbar Linkage Virtual Laboratory* that is on the DVD. Open the file
Virtual Fourbar Linkage Lab 11-1.doc from the DVD and follow the instructions as
directed by your professor. For this lab it is suggested that you analyze only the data
for the unbalanced conditions of the linkage.

11.17 PROJECTS

The following problem statement applies to all the projects listed.

*These larger-scale project statements deliberately lack detail and structure and are loosely defined.
Thus, they are similar to the kind of "identification of need" or problem statement commonly
encountered in engineering practice. It is left to the student to structure the problem through*
background research *and to create a* **clear goal statement** *and set of* **performance specifications**
*before attempting to design a solution. This design process is spelled out in Chapter 1 and should
be followed in all of these examples. All results should be documented in a professional engineering
report. See the Bibliography in Chapter 1 for references on report writing.*

   *Some of these project problems are based on the kinematic design projects in Chapter 3. Those
kinematic devices can now be designed more realistically with consideration of the dynamic forces
that they generate. The strategy in most of the following project problems is to keep the dynamic
pin forces and thus the shaking forces to a minimum and also keep the input torque-time curve as
smooth as possible to minimize power requirements.* **All these problems can be solved with a**

pin-jointed fourbar linkage. *This fact will allow you to use program* LINKAGES *to do the kinematic and dynamic computations on a large number and variety of designs in a short time. There are infinities of viable solutions to these problems.* **Iterate to find the best one!** *All links must be designed in detail as to their geometry (mass, moment of inertia, etc.). An equation solver such as* Mathcad, Matlab, *or* TKSolver *will be useful here. Determine all pin forces, shaking force, shaking torque, and input horsepower required for your designs.*

P11-1   The tennis coach needs a better tennis ball server for practice. This device must fire a sequence of standard tennis balls from one side of a standard tennis court over the net such that they land and bounce within each of the three court areas defined by the court's white lines. The order and frequency of a ball's landing in any one of the three court areas must be random. The device should operate automatically and unattended except for the refill of balls. It should be capable of firing 50 balls between reloads. The timing of ball releases should vary. For simplicity, a motor-driven pin-jointed linkage design is preferred. This project asks you to design such a device to be mounted upon a tripod stand of 5-foot height. Design it, and the stand, for stability against tipover due to the shaking forces and shaking torques which should also be minimized in the design of your linkage. Minimize the input torque.

P11-2   The "Save the Skeet" foundation has requested a more humane skeet launcher be designed. While they have not yet succeeded in passing legislation to prevent the wholesale slaughter of these little devils, they are concerned about the inhumane aspects of the large accelerations imparted to the skeet as it is launched into the sky for the sportsperson to shoot down. The need is for a skeet launcher that will smoothly accelerate the clay pigeon onto its desired trajectory. Design a skeet launcher to be mounted upon a child's "little red wagon." Control your design parameters so as to minimize the shaking forces and torques so that the wagon will remain as nearly stationary as possible during the launch of the clay pigeon.

P11-3   The coin-operated "kid bouncer" machines found outside supermarkets typically provide a very unimaginative rocking motion to the occupant. There is a need for a superior "bouncer" which will give more interesting motions while remaining safe for small children. Design it for mounting in the bed of a pickup truck. Keep the shaking forces to a minimum and the input torque-time curve as smooth as possible.

P11-4   NASA wants a zero-*g* machine for astronaut training to carry one person and provide a negative 1-*g* acceleration for as long as possible. Design this device and mount it to the ground plane so as to minimize dynamic forces and driving torque.

P11-5   The Amusement Machine Co. Inc. wants a portable "WHIP" ride which will give two or four passengers a thrilling but safe ride and which can be trailed behind a pickup truck from one location to another. Design this device and its mounting hardware to the truck bed minimizing the dynamic forces and driving torque.

P11-6   The Air Force has requested a pilot training simulator that will give potential pilots exposure to *g* forces similar to those they will experience in dogfight maneuvers. Design this device and mount it to the ground plane so as to minimize dynamic forces and driving torque.

P11-7   Cheers needs a better "mechanical bull" simulator for their "yuppie" bar in Boston. It must give a thrilling "bucking bronco" ride but be safe. Design this device and mount it to the ground plane so as to minimize dynamic forces and driving torque.

P11-8   Gargantuan Motors Inc. is designing a new light military transport vehicle. Their current windshield wiper linkage mechanism develops such high shaking forces when run at its highest speed that the engines are falling out! Design a superior windshield wiper mechanism to sweep the 20-lb armored wiper blade through a 90° arc while minimizing both input torque and shaking forces. The wind load on the blade, perpendicular to the windshield, is 50 lb. The coefficient of friction of the wiper blade on glass is 0.9.

P11-9   The Army's latest helicopter gunship is to be fitted with the Gatling gun, which fires 50-mm-diameter, 2-cm-long spent uranium slugs at a rate of 10 rounds per second. The reaction (recoil) force may upset the helicopter's stability. A mechanism is needed that can be mounted to the frame of the helicopter and which will provide a synchronous shaking force, 180° out of phase with the recoil force pulses, to counteract the recoil of the gun. Design such a linkage and minimize its torque and power drawn from the aircraft's engine. Total weight of your device should also be minimized.

P11-10   Steel pilings are universally used as foundations for large buildings. These are often driven into the ground by hammer blows from a "pile driver." In certain soils (sandy, muddy) the piles can be "shaken" into the ground by attaching a "vibratory driver" that imparts a vertical, dynamic shaking force at or near the natural frequency of the pile-earth system. The pile can literally be made to "fall into the ground" under optimal conditions. Design a fourbar linkage-based pile shaker mechanism which, when its ground link is firmly attached to the top of a piling (supported from a crane hook), will impart a dynamic shaking force that is predominantly directed along the piling's long, vertical axis. Operating speed should be in the vicinity of the natural frequency of the pile-earth system.

P11-11   Paint can shaker mechanisms are common in paint and hardware stores. While they do a good job of mixing the paint, they are also noisy and transmit their vibrations to the shelves and counters. A better design of the paint can shaker is possible using a balanced fourbar linkage. Design such a portable device to sit on the floor (not bolted down) and minimize the shaking forces and vibrations while still effectively mixing the paint.

P11-12   Convertible automobiles are once again popular. While offering the pleasure of open-air motoring, they offer little protection to the occupants in a rollover accident. Permanent roll bars are ugly and detract from the open feeling of a true convertible. An automatically deployable roll bar mechanism is needed that will be out of sight until needed. In the event that sensors in the vehicle detect an imminent rollover, the mechanism should deploy within 250 ms. Design a collapsible/deployable roll bar mechanism to retrofit to the convertible of your choice.

11

Chapter 12

BALANCING

Moderation is best,
and to avoid all extremes
PLUTARCH

12.0   INTRODUCTION

Any link or member that is in pure rotation can, theoretically, be perfectly balanced to eliminate all shaking forces and shaking moments. It is accepted design practice to balance all rotating members in a machine unless shaking forces are desired (as in a vibrating shaker mechanism, for example). A rotating member can be balanced either statically or dynamically. Static balance is a subset of dynamic balance. To achieve complete balance requires that dynamic balancing be done. In some cases, static balancing can be an acceptable substitute for dynamic balancing and is generally easier to do.

Rotating parts can, and generally should, be designed to be inherently balanced by their geometry. However, the vagaries of production tolerances guarantee that there will still be some small unbalance in each part. Thus a balancing procedure will have to be applied to each part after manufacture. The amount and location of any imbalance can be measured quite accurately and compensated for by adding or removing material in the correct locations.

In this chapter we will investigate the mathematics of determining and designing a state of static and dynamic balance in rotating elements and also in mechanisms having complex motion, such as the fourbar linkage. The methods and equipment used to measure and correct imbalance in manufactured assemblies will also be discussed. It is quite

12

convenient to use the method of d'Alembert (see Section 10.14, p. 569) when discussing rotating imbalance, applying inertia forces to the rotating elements, so we will do that.

12.1 STATIC BALANCE

Despite its name, **static balance** *does* apply to things in motion. The unbalanced forces of concern are due to the accelerations of masses in the system. The requirement for **static balance** is simply that *the sum of all forces on the moving system (including d'Alembert inertial forces) must be zero.*

$$\sum \mathbf{F} - m\mathbf{a} = 0 \qquad (12.1)$$

This is simply a restatement of Newton's law as discussed in Section 10.14 (p. 569).

Another name for static balance is **single-plane balance**, which means that *the masses which are generating the inertia forces are in, or nearly in, the same plane.* It is essentially a two-dimensional problem. Some examples of common devices which meet this criterion, and thus can successfully be statically balanced, are a single gear or pulley on a shaft, a bicycle or motorcycle tire and wheel, a thin flywheel, an airplane propeller, an individual turbine blade-wheel (but not the entire turbine). The common denominator among these devices is that they are all short in the axial direction compared to the radial direction, and thus can be considered to exist in a single plane. An automobile tire and wheel is only marginally suited to static balancing as it is reasonably thick in the axial direction compared to its diameter. Despite this fact, auto tires are sometimes statically balanced. More often they are dynamically balanced and will be discussed under that topic.

Figure 12-1a (p. 633) shows a link in the shape of a vee which is part of a linkage. We want to statically balance it. We can model this link dynamically as two point masses m_1 and m_2 concentrated at the local *CG*s of each "leg" of the link as shown in Figure 12-1b. These point masses each have a mass equal to that of the "leg" they replace and are supported on massless rods at the position ($\mathbf{R}_1$ or $\mathbf{R}_2$) of that leg's *CG*. We can solve for the required amount and location of a third "balance mass" m_b to be added to the system at some location $\mathbf{R}_b$ in order to satisfy equation 12.1.

Assume that the system is rotating at some constant angular velocity ω. The accelerations of the masses will then be strictly centripetal (toward the center), and the inertia forces will be centrifugal (away from the center) as shown in Figure 12-1. Since the system is rotating, the figure shows a "freeze-frame" image of it. The position at which we "stop the action" for the purpose of drawing the picture and doing the calculations is both arbitrary and irrelevant to the computation. We will set up a coordinate system with its origin at the center of rotation and resolve the inertial forces into components in that system. Writing vector equation 12.1 for this system, we get:

$$-m_1\mathbf{R}_1\omega^2 - m_2\mathbf{R}_2\omega^2 - m_b\mathbf{R}_b\omega^2 = 0 \qquad (12.2a)$$

Note that the only forces acting on this system are the inertia forces. For balancing, it does not matter what external forces may be acting on the system. External forces cannot be balanced by making any changes to the system's internal geometry. Note that the ω^2 terms cancel. For balancing, it also does not matter how fast the system is rotating, only

12

that it *is* rotating. (*The ω will determine the magnitudes of these forces, but we are going to force their sum to be zero anyway.*)

Dividing out the ω^2 and rearranging, we get:

$$m_b \mathbf{R}_b = -m_1 \mathbf{R_1} - m_2 \mathbf{R_2} \tag{12.2b}$$

Breaking into x and y components:

$$m_b R_{b_x} = -\left(m_1 R_{1_x} + m_2 R_{2_x} \right)$$
$$m_b R_{b_y} = -\left(m_1 R_{1_y} + m_2 R_{2_y} \right) \tag{12.2c}$$

The terms on the right sides are known. We can readily solve for the mR_x and mR_y products needed to balance the system. It will be convenient to convert the results to polar coordinates.

$$\theta_b = \arctan \frac{m_b R_{b_y}}{m_b R_{b_x}}$$
$$= \arctan \frac{-\left(m_1 R_{1_y} + m_2 R_{2_y} \right)}{-\left(m_1 R_{1_x} + m_2 R_{2_x} \right)} \tag{12.2d}$$

$$R_b = \sqrt{R_{b_x}^2 + R_{b_y}^2}$$
$$m_b R_b = m_b \sqrt{R_{b_x}^2 + R_{b_y}^2}$$
$$= \sqrt{m_b^2 \left(R_{b_x}^2 + R_{b_y}^2 \right)}$$
$$= \sqrt{m_b^2 R_{b_x}^2 + m_b^2 R_{b_y}^2} \tag{12.2e}$$
$$= \sqrt{\left(m_b R_{b_x} \right)^2 + \left(m_b R_{b_y} \right)^2}$$

The angle at which the balance mass must be placed (with respect to our arbitrarily oriented freeze-frame coordinate system) is θ_b, found from equation 12.2d. Note that the signs of the numerator and denominator of equation 12.2d must be individually maintained and a two-argument arctangent computed in order to obtain θ_b in the correct quadrant. Most calculators and computers will give an arctangent result only between ±90°.

The $m_b R_b$ product is found from equation 12.2e. There is now an infinity of solutions available. We can either select a value for m_b and solve for the necessary radius R_b at which it should be placed, or choose a desired radius and solve for the mass that must be placed there. Packaging constraints may dictate the maximum radius possible in some cases. The balance mass is confined to the "single plane" of the unbalanced masses.

Once a combination of m_b and R_b is chosen, it remains to design the physical counterweight. The chosen radius R_b is the distance from the pivot to the *CG* of whatever shape we create for the counterweight mass. Our simple dynamic model, used to calculate the mR product, assumed a point mass and a massless rod. These ideal devices do not exist. A possible shape for this counterweight is shown in Figure 12-1c. Its mass must be m_b, distributed so as to place its *CG* at radius R_b at angle θ_b.

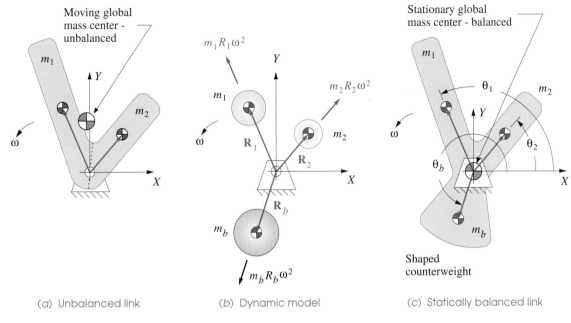

(a) Unbalanced link   (b) Dynamic model   (c) Statically balanced link

FIGURE 12-1

Static balancing a link in pure rotation

EXAMPLE 12-1

Static Balancing.

Given: The system shown in Figure 12-1 has the following data:

$m_1 = 1.2$ kg $R_1 = 1.135$ m @ $\angle 113.4°$

$m_2 = 1.8$ kg $R_2 = 0.822$ m @ $\angle 48.8°$

$\omega = 40$ rad/sec

Find: The mass-radius product and its angular location needed to statically balance the
              system.

Solution:

1   Resolve the position vectors into xy components in the arbitrary coordinate system associated
    with the freeze-frame position of the linkage chosen for analysis.

$R_1 = 1.135$ @ $\angle 113.4°$; $R_{1_x} = -0.451$, $R_{1_y} = 1.042$

                                                                                                    (a)

$R_2 = 0.822$ @ $\angle 48.8°$; $R_{2_x} = +0.541$, $R_{2_y} = 0.618$

2   Solve equations 12.2c.

$$m_b R_{b_x} = -m_1 R_{1_x} - m_2 R_{2_x} = -(1.2)(-0.451) - (1.8)(0.541) = -0.433$$

$$m_b R_{b_y} = -m_1 R_{1_y} - m_2 R_{2_y} = -(1.2)(1.042) - (1.8)(0.618) = -2.363$$

(b)

3   Solve equations 12.2d and 12.2e (p. 632).

$$\theta_b = \arctan \frac{-2.363}{-0.433} = 259.6°$$

(c)

$$m_b R_b = \sqrt{(-0.433)^2 + (-2.363)^2} = 2.402 \ \text{kg-m}$$

4   This mass-radius product of 2.402 kg-m can be obtained with a variety of shapes appended to the assembly. Figure 12-1c (p. 633) shows a particular shape whose CG is at a radius of $R_b = 0.806$ m at the required angle of 259.6°. The mass required for this counterweight design is then:

$$m_b = \frac{2.402 \ \text{kg-m}}{0.806 \ \text{m}} = 2.980 \ \text{kg}$$

(d)

at a chosen CG radius of:

$$R_b = 0.806 \ \text{m}$$

(e)

Many other shapes are possible. As long as they provide the required mass-radius product at the required angle, the system will be statically balanced. Note that the value of ω was not needed in the calculation.

12.2   DYNAMIC BALANCE

Dynamic balance is sometimes called **two-plane balance**. It requires that two criteria be met. The sum of the forces must be zero (static balance) plus the sum of the moments[*] must also be zero.

$$\sum \mathbf{F} = 0$$

$$\sum \mathbf{M} = 0$$

(12.3)

These moments act in planes that include the axis of rotation of the assembly such as planes XZ and YZ in Figure 12-2. The moment's vector direction, or axis, is perpendicular to the assembly's axis of rotation.

Any rotating object or assembly which is relatively long in the axial direction compared to the radial direction requires dynamic balancing for complete balance. It is possible for an object to be statically balanced but not be dynamically balanced. Consider the assembly in Figure 12-2. Two equal masses are at identical radii, 180° apart rotationally, but separated along the shaft length. A summation of $-m\mathbf{a}$ forces due to their rotation will be always zero. However, in the side view, their inertia forces form a couple which rotates with the masses about the shaft. This rocking couple causes a moment on the ground plane, alternately lifting and dropping the left and right ends of the shaft.

[*] We will use the term *moment* in this text to refer to "turning forces" whose vectors are perpendicular to an axis of rotation or "long axis" of an assembly, and the term *torque* to refer to "turning forces" whose vectors are parallel to an axis of rotation.

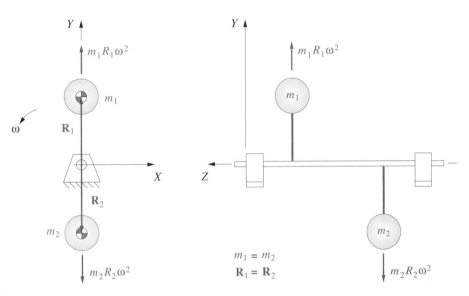

FIGURE 12-2

Balanced forces—unbalanced moment

Some examples of devices which require dynamic balancing are rollers, crankshafts, camshafts, axles, clusters of multiple gears, motor rotors, turbines, propeller shafts. The common denominator among these devices is that their mass may be unevenly distributed both rotationally around their axis and longitudinally along their axis.

To correct dynamic imbalance requires either adding or removing the right amount of mass at the proper angular locations *in two correction planes* separated by some distance along the shaft. This will create the necessary counterforces to statically balance the system and also provide a countercouple to cancel the unbalanced moment. When an automobile tire and wheel is dynamically balanced, the two correction planes are the inner and outer edges of the wheel rim. Correction weights are added at the proper locations in each of these correction planes based on a measurement of the dynamic forces generated by the unbalanced, spinning wheel.

It is always good practice to first statically balance all individual components that go into an assembly, if possible. This will reduce the amount of dynamic imbalance that must be corrected in the final assembly and also reduce the bending moment on the shaft. A common example of this situation is the aircraft turbine which consists of a number of circular turbine wheels arranged along a shaft. Since these spin at high speed, the inertia forces due to any imbalance can be very large. The individual wheels are statically balanced before being assembled to the shaft. The final assembly is then dynamically balanced.

Some devices do not lend themselves to this approach. An electric motor rotor is essentially a spool of copper wire wrapped in a complex pattern around the shaft. The mass of the wire is not uniformly distributed either rotationally or longitudinally, so it will not be balanced. It is not possible to modify the windings' local mass distribution after

12

the fact without compromising electrical integrity. Thus the entire rotor imbalance must be countered in the two correction planes after assembly.

Consider the system of three lumped masses arranged around and along the shaft in Figure 12-3. Assume that, for some reason, they cannot be individually statically balanced within their own planes. We then create two correction planes labeled A and B. In this design example, the unbalanced masses m_1, m_2, m_3 and their radii R_1, R_2, R_3 are known along with their angular locations θ_1, θ_2, and θ_3. We want to dynamically balance the system. A three-dimensional coordinate system is applied with the axis of rotation in the Z direction. Note that the system has again been stopped in an arbitrary freeze-frame position. Angular acceleration is assumed to be zero. The summation of forces is:

$$-m_1\mathbf{R}_1\omega^2 - m_2\mathbf{R}_2\omega^2 - m_3\mathbf{R}_3\omega^2 - m_A\mathbf{R}_A\omega^2 - m_B\mathbf{R}_B\omega^2 = 0 \qquad (12.4a)$$

Dividing out the ω^2 and rearranging we get:

$$m_A\mathbf{R}_A + m_B\mathbf{R}_B = -m_1\mathbf{R}_1 - m_2\mathbf{R}_2 - m_3\mathbf{R}_3 \qquad (12.4b)$$

Breaking into x and y components:

$$m_A R_{A_x} + m_B R_{B_x} = -m_1 R_{1_x} - m_2 R_{2_x} - m_3 R_{3_x}$$
$$m_A R_{A_y} + m_B R_{B_y} = -m_1 R_{1_y} - m_2 R_{2_y} - m_3 R_{3_y} \qquad (12.4c)$$

Equations 12.4c have four unknowns in the form of the $m\mathbf{R}$ products at plane A and the $m\mathbf{R}$ products at plane B. To solve we need the sum of the moments equation which we can take about a point in one of the correction planes such as point O. The moment arm z distances of each force measured from plane A are labeled l_1, l_2, l_3, l_B in the figure; thus

$$\left(m_B\mathbf{R}_B\omega^2\right)l_B = -\left(m_1\mathbf{R}_1\omega^2\right)l_1 - \left(m_2\mathbf{R}_2\omega^2\right)l_2 - \left(m_3\mathbf{R}_3\omega^2\right)l_3 \qquad (12.4d)$$

Dividing out the ω^2, breaking into x and y components, and rearranging:

The moment in the XZ plane (i.e., about the Y axis) is:

$$m_B R_{B_x} = \frac{-\left(m_1 R_{1_x}\right)l_1 - \left(m_2 R_{2_x}\right)l_2 - \left(m_3 R_{3_x}\right)l_3}{l_B} \qquad (12.4e)$$

The moment in the YZ plane (i.e., about the X axis) is:

$$m_B R_{B_y} = \frac{-\left(m_1 R_{1_y}\right)l_1 - \left(m_2 R_{2_y}\right)l_2 - \left(m_3 R_{3_y}\right)l_3}{l_B} \qquad (12.4f)$$

These can be solved for the $m\mathbf{R}$ products in x and y directions for correction plane B which can then be substituted into equation 12.4c to find the values needed in plane A. Equations 12.2d and 12.2e (p. 632) can then be applied to each correction plane to find the angles at which the balance masses must be placed and the $m\mathbf{R}$ products needed in each plane. The physical counterweights can then be designed consistent with the constraints outlined in Section 12.1 (p. 631) on static balance. Note that the radii R_A and R_B do not have to be the same value.

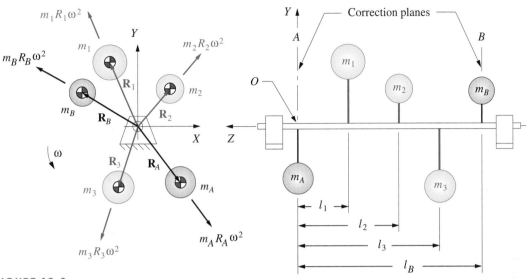

FIGURE 12-3

Two-plane dynamic balancing

EXAMPLE 12-2

Dynamic Balancing.

Given: The system shown in Figure 12-3 has the following data:

$$m_1 = 1.2 \text{ kg} \qquad\qquad R_1 = 1.135 \text{ m @} \angle 113.4°$$
$$m_2 = 1.8 \text{ kg} \qquad\qquad R_2 = 0.822 \text{ m @} \angle 48.8°$$
$$m_3 = 2.4 \text{ kg} \qquad\qquad R_3 = 1.04 \text{ m @} \angle 251.4°$$

The z distances in meters from plane A are:

$$l_1 = 0.854, \qquad l_2 = 1.701, \qquad l_3 = 2.396, \qquad l_B = 3.097$$

Find: The mass-radius products and their angular locations needed to dynamically balance the system using the correction planes A and B.

Solution:

1   Resolve the position vectors into xy components in the arbitrary coordinate system associated with the freeze-frame position of the linkage chosen for analysis.

$$R_1 = 1.135 \text{ @ } \angle 113.4°; \qquad R_{1_x} = -0.451, \qquad R_{1_y} = +1.042$$
$$R_2 = 0.822 \text{ @ } \angle 48.8°; \qquad R_{2_x} = +0.541, \qquad R_{2_y} = +0.618 \qquad (a)$$
$$R_3 = 1.040 \text{ @ } \angle 251.4°; \qquad R_{3_x} = -0.332, \qquad R_{3_y} = -0.986$$

2   Solve equation 12.4e (p. 636) for summation of moments about point O.

$$m_B R_{B_x} = \frac{-\left(m_1 R_{1_x}\right)l_1 - \left(m_2 R_{2_x}\right)l_2 - \left(m_3 R_{3_x}\right)l_3}{l_B}$$

$$= \frac{-1.2(-0.451)(0.854) - 1.8(0.541)(1.701) - 2.4(-0.332)(2.396)}{3.097} = 0.230$$

(b)

$$m_B R_{B_y} = \frac{-\left(m_1 R_{1_y}\right)l_1 - \left(m_2 R_{2_y}\right)l_2 - \left(m_3 R_{3_y}\right)l_3}{l_B}$$

$$= \frac{-1.2(1.042)(0.854) - 1.8(0.618)(1.701) - 2.4(-0.986)(2.396)}{3.097} = 0.874$$

(c)

3   Solve equations 12.2d and 12.2e (p. 632) for the mass radius product in plane B.

$$\theta_B = \arctan \frac{0.874}{0.230} = 75.27°$$

(d)

$$m_B R_B = \sqrt{(0.230)^2 + (0.874)^2} = 0.904 \text{ kg-m}$$

4   Solve equations 12.4c (p. 636) for forces in x and y directions.

$$m_A R_{A_x} = -m_1 R_{1_x} - m_2 R_{2_x} - m_3 R_{3_x} - m_B R_{B_x}$$

$$m_A R_{A_y} = -m_1 R_{1_y} - m_2 R_{2_y} - m_3 R_{3_y} - m_B R_{B_y}$$

(e)

$$m_A R_{A_x} = -1.2(-0.451) - 1.8(0.541) - 2.4(-0.332) - 0.230 = 0.134$$

$$m_A R_{A_y} = -1.2(1.042) - 1.8(0.618) - 2.4(-0.986) - 0.874 = -0.870$$

5   Solve equations 12.2d and 12.2e (p. 632) for the mass-radius product in plane A.

$$\theta_A = \arctan \frac{-0.870}{0.134} = -81.25°$$

(f)

$$m_A R_A = \sqrt{(0.134)^2 + (-0.870)^2} = 0.880 \text{ kg-m}$$

6   These mass-radius products can be obtained with a variety of shapes appended to the assembly in planes A and B. Many shapes are possible. As long as they provide the required mass-radius products at the required angles in each correction plane, the system will be dynamically balanced.

So, when the design is still on the drawing board, these simple analysis techniques can be used to determine the necessary sizes and locations of balance masses for any assembly in pure rotation for which the mass distribution is defined. This two-plane balance method can be used to dynamically balance any system in pure rotation, and all such systems should be balanced unless the purpose of the device is to create shaking forces or moments.

12.3   BALANCING LINKAGES

Many methods have been devised to balance linkages. Some achieve a complete balance of one dynamic factor, such as shaking force, at the expense of other factors such as shaking moment or driving torque. Others seek an optimum arrangement that collectively minimizes (but does not zero) shaking forces, moments, and torques for a best compromise. Lowen and Berkof, [1] and Lowen, Tepper, and Berkof [2] give comprehensive reviews of the literature on this subject up to 1983. Additional work has been done on the problem since that time, some of which is noted in the references at the end of this chapter. Kochev[15] presents a general theory for complete shaking moment balancing and a critical review of known methods.

Complete balance of any mechanism can be obtained by creating a second "mirror image" mechanism connected to it so as to cancel all dynamic forces and moments. Certain configurations of multicylinder internal combustion engines do this. The pistons and cranks of some cylinders cancel the inertial effects of others. We will explore these engine mechanisms in Chapter 14. However, this approach is expensive and is only justified if the added mechanism serves some second purpose such as increasing power, as in the case of additional cylinders in an engine. Adding a "dummy" mechanism whose only purpose is to cancel dynamic effects is seldom economically justifiable.

Most practical linkage balancing schemes seek to minimize or eliminate one or more of the dynamic effects (forces, moments, torques) by redistributing the mass of the existing links. This typically involves adding counterweights and/or changing the shapes of links to relocate their *CG*s. More elaborate schemes add geared counterweights to some links in addition to redistributing their mass. As with any design endeavor, there are trade-offs. For example, elimination of shaking forces usually increases the shaking moment and driving torque. We can only present a few approaches to this problem in the space available. The reader is directed to the literature for information on other methods.

Complete Force Balance of Linkages

The rotating links (cranks, rockers) of a linkage can be individually statically balanced by the rotating balance methods described in Section 12.1 (p. 631). The effects of the couplers, which are in complex motion, are more difficult to compensate for. Note that the process of statically balancing a rotating link, in effect, forces its mass center (*CG*) to be at its fixed pivot and thus stationary. In other words the condition of **static balance** can also be **defined as** one of *making the mass center stationary*. A coupler has no fixed pivot, and thus its mass center is, in general, always in motion.

Any mechanism, no matter how complex, will have, for every instantaneous position, a single, overall, *global mass center* located at some particular point. We can calculate its location knowing only the link masses and the locations of the *CG*s of the individual links at that instant. The global mass center normally will change position as the linkage moves. If we can somehow force this global mass center to be stationary, we will have a state of static balance for the overall linkage.

The Berkof-Lowen method of linearly independent vectors[3] provides a means to calculate the magnitude and location of counterweights to be placed on the rotating links which will make the global mass center stationary for all positions of the linkage. Place-

12

ment of the proper balance masses on the links will cause the dynamic forces on the fixed pivots to always be equal and opposite, i.e., a couple, thus creating static balance ($\Sigma F = 0$ but $\Sigma M \neq 0$) in the moving linkage.

This method works for any n-link planar linkage having a combination of revolute (pin) and prismatic (slider) joints, provided that there exists a path to the ground from every link which only contains revolute joints.[4] In other words, if all possible paths from any one link to the ground contain sliding joints, then the method fails. Any linkage of n links that meets the above criterion can be balanced by the addition of $n/2$ balance weights, each on a different link.[4] We will apply the method from reference [3] to a fourbar linkage. Unfortunately, doing so will increase the total mass of the original linkage by a factor of 2-3 for fourbar linkages and substantially more for complex mechanisms.[15]

Figure 12-4 shows a fourbar linkage with its overall global mass center located by the position vector $\mathbf{R}_t$. The individual CGs of the links are located *in the global system* by position vectors $\mathbf{R}_2$, $\mathbf{R}_3$, and $\mathbf{R}_4$ (magnitudes R_2, R_3, R_4), rooted at its origin, the crank pivot O_2. The link lengths are defined by position vectors labeled $\mathbf{L}_1, \mathbf{L}_2, \mathbf{L}_3, \mathbf{L}_4$ (magnitudes l_1, l_2, l_3, l_4), and the local position vectors which locate the CGs *within each link* are $\mathbf{B}_2, \mathbf{B}_3, \mathbf{B}_4$ (magnitudes b_2, b_3, b_4). The angles of the vectors $\mathbf{B}_2, \mathbf{B}_3, \mathbf{B}_4$ are ϕ_2, ϕ_3, ϕ_4 measured internal to the links with respect to the links' lines of centers $\mathbf{L}_2, \mathbf{L}_3, \mathbf{L}_4$. The instantaneous link angles which locate $\mathbf{L}_2, \mathbf{L}_3, \mathbf{L}_4$ in the global system are $\theta_2, \theta_3, \theta_4$. The total mass of the system is simply the sum of the individual link masses:

$$m_t = m_2 + m_3 + m_4 \tag{12.5a}$$

The total mass moment about the origin must be equal to the sum of the mass moments due to the individual links:

$$\sum M_{O_2} = m_t \mathbf{R}_t = m_2 \mathbf{R}_2 + m_3 \mathbf{R}_3 + m_4 \mathbf{R}_4 \tag{12.5b}$$

The position of the global mass center is then:

$$\mathbf{R}_t = \frac{m_2 \mathbf{R}_2 + m_3 \mathbf{R}_3 + m_4 \mathbf{R}_4}{m_t} \tag{12.5c}$$

and from the linkage geometry:

$$\mathbf{R}_2 = b_2\, e^{j(\theta_2 + \phi_2)} = b_2\, e^{j\theta_2} e^{j\phi_2}$$
$$\mathbf{R}_3 = l_2\, e^{j\theta_2} + b_3\, e^{j(\theta_3 + \phi_3)} = l_2\, e^{j\theta_2} + b_3\, e^{j\theta_3} e^{j\phi_3} \tag{12.5d}$$
$$\mathbf{R}_4 = l_1\, e^{j\theta_1} + b_4\, e^{j(\theta_4 + \phi_4)} = l_1\, e^{j\theta_1} + b_4\, e^{j\theta_4} e^{j\phi_4}$$

We can solve for the location of the global mass center for any link position for which we know the link angles $\theta_2, \theta_3, \theta_4$. We want to make this position vector $\mathbf{R}_t$ be a constant. The first step is to substitute equations 12.5d into 12.5b,

$$m_t \mathbf{R}_t = m_2 \left(b_2\, e^{j\theta_2} e^{j\phi_2} \right) + m_3 \left(l_2\, e^{j\theta_2} + b_3\, e^{j\theta_3} e^{j\phi_3} \right) + m_4 \left(l_1\, e^{j\theta_1} + b_4\, e^{j\theta_4} e^{j\phi_4} \right) \tag{12.5e}$$

and rearrange to group the constant terms as coefficients of the time-dependent terms:

$$m_t \mathbf{R}_t = \left(m_4 l_1\, e^{j\theta_1} \right) + \left(m_2 b_2 e^{j\phi_2} + m_3 l_2 \right) e^{j\theta_2} + \left(m_3 b_3\, e^{j\phi_3} \right) e^{j\theta_3} + \left(m_4 b_4\, e^{j\phi_4} \right) e^{j\theta_4} \tag{12.5f}$$

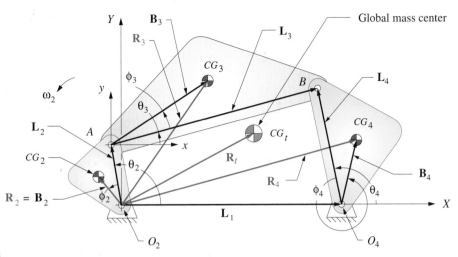

FIGURE 12-4

Static (force) balancing a fourbar linkage

Note that the terms in parentheses are all constant with time. The only time-dependent terms are the ones containing θ_2, θ_3, and θ_4.

We can also write the vector loop equation for the linkage,

$$l_2\, e^{j\theta_2} + l_3\, e^{j\theta_3} - l_4\, e^{j\theta_4} - l_1\, e^{j\theta_1} = 0 \qquad (12.6a)$$

and solve it for one of the unit vectors that define a link direction, say link 3:

$$e^{j\theta_3} = \frac{l_1\, e^{j\theta_1} - l_2\, e^{j\theta_2} + l_4\, e^{j\theta_4}}{l_3} \qquad (12.6b)$$

Substitute this into equation 12.5f to eliminate the θ_3 term and rearrange:

$$m_t\mathbf{R}_t = \left(m_2 b_2 e^{j\phi_2} + m_3 l_2\right) e^{j\theta_2} + \frac{1}{l_3}\left(m_3 b_3\, e^{j\phi_3}\right)\left(l_1\, e^{j\theta_1} - l_2\, e^{j\theta_2} + l_4\, e^{j\theta_4}\right)$$

$$+ \left(m_4 b_4\, e^{j\phi_4}\right) e^{j\theta_4} + \left(m_4 l_1\, e^{j\theta_1}\right) \qquad (12.7a)$$

and collect terms:

$$m_t\mathbf{R}_t = \left(m_2 b_2 e^{j\phi_2} + m_3 l_2 - m_3 b_3\, \frac{l_2}{l_3} e^{j\phi_3}\right) e^{j\theta_2} + \left(m_4 b_4\, e^{j\phi_4} + m_3 b_3\, \frac{l_4}{l_3} e^{j\phi_3}\right) e^{j\theta_4}$$

$$+ m_4 l_1\, e^{j\theta_1} + m_3 b_3\, \frac{l_1}{l_3} e^{j\phi_3}\, e^{j\theta_1} \qquad (12.7b)$$

This expression gives us the tool to force $\mathbf{R}_t$ to be a constant and make the linkage mass center stationary. For that to be so, the terms in parentheses which multiply the only two time-dependent variables, θ_2 and θ_4, must be forced to be zero. (The fixed link angle θ_1 is a constant.) Thus the requirement for linkage force balance is:

12

$$m_2 b_2 e^{j\phi_2} + m_3 l_2 - m_3 b_3 \frac{l_2}{l_3} e^{j\phi_3} = 0$$

$$m_4 b_4\, e^{j\phi_4} + m_3 b_3 \frac{l_4}{l_3} e^{j\phi_3} = 0$$

$$(12.8a)$$

Rearrange to isolate one link's terms (say link 3) on one side of each of these equations:

$$m_2 b_2 e^{j\phi_2} = m_3 \left(b_3 \frac{l_2}{l_3} e^{j\phi_3} - l_2 \right)$$

$$m_4 b_4\, e^{j\phi_4} = -m_3 b_3 \frac{l_4}{l_3} e^{j\phi_3}$$

$$(12.8b)$$

We now have two equations involving three links. The parameters for any one link can be assumed and the other two solved for. A linkage is typically first designed to satisfy the required motion and packaging constraints before this force-balancing procedure is attempted. In that event, the link geometry and masses are already defined, at least in a preliminary way. A useful strategy is to leave the link 3 mass and *CG* location as originally designed and calculate the necessary masses and *CG* locations of links 2 and 4 to satisfy these conditions for balanced forces. Links 2 and 4 are in pure rotation, so it is straightforward to add counterweights to them in order to move their *CG*s to the necessary locations. With this approach, the right sides of equations 12.8b are reducible to numbers for a designed linkage. We want to solve for the mass radius products $m_2 b_2$ and $m_4 b_4$ and also for the angular locations of the *CG*s within the links. Note that the angles ϕ_2 and ϕ_4 in equations 12.8 are measured with respect to the lines of centers of their respective links.

Equations 12.8b are vector equations. Substitute the Euler identity (equation 4.4a, p. 185) to separate into real and imaginary components, and solve for the x and y components of the mass-radius products.

$$\left(m_2 b_2 \right)_x = m_3 \left(b_3 \frac{l_2}{l_3} \cos\phi_3 - l_2 \right)$$

$$\left(m_2 b_2 \right)_y = m_3 \left(b_3 \frac{l_2}{l_3} \sin\phi_3 \right)$$

$$(12.8c)$$

$$\left(m_4 b_4 \right)_x = -m_3 b_3 \frac{l_4}{l_3} \cos\phi_3$$

$$\left(m_4 b_4 \right)_y = -m_3 b_3 \frac{l_4}{l_3} \sin\phi_3$$

$$(12.8d)$$

These components of the *mR* product needed to force balance the linkage represent the entire amount needed. If links 2 and 4 are already designed with some individual unbalance (the *CG* not at pivot), then the existing *mR* product of the unbalanced link must be subtracted from that found in equations 12.8c and 12.8d in order to determine the size and location of additional counterweights to be added to those links. As we did with the balance of rotating links, any combination of mass and radius that gives the desired product is acceptable. Use equations 12.2d and 12.2e (p. 632) to convert the cartesian *mR* products in equations 12.8c and 12.8d to polar coordinates in order to find the magnitude

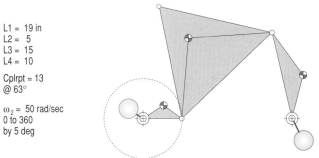

FIGURE 12-5

A balanced fourbar linkage showing balance masses applied to links 2 and 4

and angle of the counterweight's *mR* vector. Note that the angle of the *mR* vector for each link will be referenced to that link's line of centers. Design the shape of the physical counterweights to be put on the links as discussed in Section 12.1 (p. 631).

12.4 EFFECT OF BALANCING ON SHAKING AND PIN FORCES

Figure 12-5 shows a fourbar linkage* to which balance masses have been added in accord with equations 12.8. Note the counterweights placed on links 2 and 4 at the calculated locations for complete force balance. Figure 12-6a shows a polar plot of the shaking forces of this linkage without the balance masses. The maximum is 462 lb at 15°. Figure 12-6b shows the shaking forces after the balance masses are added. The shaking forces are reduced to essentially zero. The small residual forces seen in Figure 12-6b are due to computational round-off errors—the method gives theoretically exact results.

* Open the disk file F12-05.4br in program LINKAGES to see more details on this linkage and its balancing.

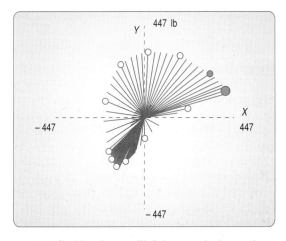

(*a*) Shaking force with linkage unbalanced

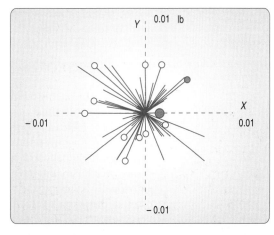

(*b*) Shaking force with linkage balanced

FIGURE 12-6

Polar plot of unbalanced shaking forces on ground plane of the fourbar linkage of Figure 12-5

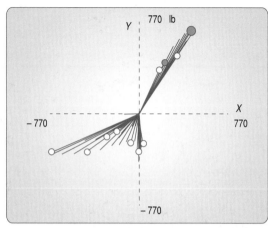

(a) Crank pivot force F21   (b) Rocker pivot force F41

FIGURE 12-7

Polar plots of forces F21 and F41 acting on the ground plane of the force-balanced fourbar linkage of Figure 12-5

The pin forces at the crank and rocker pivots have not disappeared as a result of adding the balance masses, however. Figures 12-7a and b, respectively, show the forces on crank and rocker pivots after balancing. These forces are now equal and opposite. After balancing, the pattern of forces at pivot O_2 is the mirror image of the pattern at pivot O_4. The **net shaking force** is the vector sum of these two sets of forces for each time step (Section 11.8, p. 601). The equal and opposite pairs of forces acting at the ground pivots at each time step create a time-varying shaking couple that rocks the ground plane. These pin forces can be larger due to the balance weights and if so will increase the shaking couple compared to its former value in the unbalanced linkage—one trade-off for reducing the shaking forces to zero. The stresses in the links and pins may also increase as a result of force balancing.

12.5   EFFECT OF BALANCING ON INPUT TORQUE

Individually balancing a link which is in pure rotation by the addition of a counterweight will have the side effect of increasing its mass moment of inertia. The "flywheel effect" of the link is increased by this increase in its moment of inertia. Thus the torque needed to accelerate that link will be greater. The input torque will be unaffected by any change in the I of the input crank when it is run at constant angular velocity. But, any rockers in the mechanism will have angular accelerations even when the crank does not. Thus, individually balancing the rockers will tend to increase the required input torque even at constant input crank velocity.

Adding counterweights to the rotating links, necessary to force balance the entire linkage, both increases the links' mass moments of inertia and also (individually) *unbalances* those rotating links in order to gain the global balance. Then the CGs of the rotating links will not be at their fixed pivots. Any angular acceleration of these links will add to the torque loading on the linkage. Balancing an entire linkage by this method then

FIGURE 12-8

Unbalanced and balanced input torque curves for the fourbar linkage of Figure 12-5

can have the side effect of increasing the variation in the required input torque. A larger flywheel may be needed on a balanced linkage in order to achieve the same coefficient of fluctuation as the unbalanced version of the linkage.

Figure 12-8 shows the input torque curve for the unbalanced linkage and for the same linkage after complete force balancing has been done. The peak value of the required input torque has increased as a result of force balancing.

Note, however, that the degree of increase in the input torque due to force balancing is dependent upon the choice of radii at which the balance masses are placed. The extra mass moment of inertia that the balance mass adds to a link is proportional to the square of the radius to the *CG* of the balance mass. The force balance algorithm only computes the required mass-radius product. Placing the balance mass at as small a radius as possible will minimize the increase in input torque. Weiss and Fenton [5] have shown that a circular counterweight placed tangent to the link's pivot center (Figure 12-9, p. 646) is a good compromise between added weight and increased moment of inertia. To reduce the torque penalty further, one could also choose to do less than a complete force balance and accept some shaking force in trade.

12.6 BALANCING THE SHAKING MOMENT IN LINKAGES

The shaking moment $\mathbf{M}_s$ about the crank pivot O_2 in a force-balanced linkage is the sum of the reaction torque $\mathbf{T}_{21}$ and the shaking couple (ignoring any externally applied loads)[6]*

$$\mathbf{M}_s = \mathbf{T}_{21} + (\mathbf{R}_1 \times \mathbf{F}_{41}) \qquad (12.9)$$

where $\mathbf{T}_{21}$ is the negative of the driving torque $\mathbf{T}_{12}$, $\mathbf{R}_1$ is the position vector from O_2 to O_4 (i.e., link 1), and $\mathbf{F}_{41}$ is the force of the rocker on the ground plane. In a general linkage, the magnitude of the shaking moment can be reduced but cannot be eliminated by means of mass redistribution within its links. Complete balancing of the shaking moment requires the addition of supplementary links and/or rotating counterweights.[7]

* Note that this statement is only true if the linkage is force-balanced which makes the moment of the shaking couple a free vector. Otherwise it is referenced to the chosen global coordinate system. See reference [6] for complete derivations of the shaking moment for both force-balanced and unbalanced linkages.

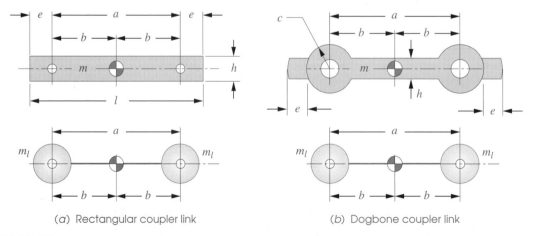

FIGURE 12-9

An inline fourbar
linkage [6], [7]
with optimally
located circular
counterweights. (5)

Many techniques have been developed that use optimization methods to find a link-age-mass configuration that will minimize the shaking moment alone or in combination with minimizing shaking force and/or input torque. Hockey [8], [9] shows that the fluctuation in kinetic energy and input torque of a mechanism may be reduced by proper distribution of mass within its links and that this approach is more weight efficient than adding a flywheel to the input shaft. Berkof [10] also describes a method to minimize the input torque by internal mass rearrangement. Lee and Cheng [11] and Qi and Pennestri [12] show methods to optimally balance the combined shaking force, shaking moment, and input torque in high-speed linkages by mass redistribution and addition of counterweights. Porter et al. [13] suggest using a genetic algorithm to optimize the same set of parameters. Bagci [14] describes several approaches to balancing shaking forces and shaking moments in the fourbar slider-crank linkage. Kochev[15] provides a general theory for complete force and moment balance. Esat and Bahai[16] describe a theory for complete force and moment balance that requires rotating counterweights on the coupler. Arakelian and Smith[17] derive a method for the complete force and moment balance of Watt's and Stephenson's sixbar linkages. Most of these methods require significant computing resources, and space does not permit a complete discussion of them all here. The reader is directed to the references for more information.

Berkof's method for complete moment balancing of the fourbar linkage [7] is simple and useful even though it is limited to "inline" linkages, i.e., those whose link *CG*s lie on their respective link centerlines as shown in Figure 12-9. This is not an overly restrictive constraint since many practical linkages are made with straight links. Even if a link must have a shape that deviates from its line of centers, its *CG* can still be placed on that line by adding mass to the link in the proper location, increased mass being the trade-off.

* This method of moment
balancing is "recognized
as a superior technique and
recommended when ap-
plicable."[15]

For complete moment balancing by Berkof's method, in addition to being an inline linkage, the coupler must be reconfigured to become a **physical pendulum*** such that it is dynamically equivalent to a lumped mass model as shown in Figure 12-10. The coupler is shown in Figure 12-10a as a uniform rectangular bar of mass *m*, length *a* and width *h* and in Figure 12-10b as a "dogbone." These are only two of many possibilities. We

(*a*) Rectangular coupler link   (*b*) Dogbone coupler link

FIGURE 12-10

Making the coupler link a physical pendulum

want the lumped masses to be at the pivot pins, connected by a "massless" rod. Then the coupler's lumped masses will be in pure rotation either as part of the crank or as part of the rocker. This can be accomplished by adding mass as indicated by dimension e at the coupler ends.[†]

The three requirements for dynamic equivalence were stated in Section 10.2 (p. 546) and are equal mass, same CG location, and same mass moment of inertia. The first and second of these are easily satisfied by placing $m_l = m/2$ at each pin. The third requirement can be stated in terms of radius of gyration k instead of moment of inertia using equation 10.11b (p. 553).

$$k = \sqrt{\frac{I}{m}} \qquad (12.10)$$

Taking each lump separately as if the massless rod were split at the CG into two rods each of length b, the moment of inertia I_l of each lump will be

$$I_l = \frac{I}{2} = m_l b^2$$

and
$$I = 2 m_l b^2 = m b^2 \qquad (12.11a)$$

then
$$k = \sqrt{\frac{m b^2}{m}} = b = \frac{a}{2} \qquad (12.11b)$$

For the link configuration in Figure 12-10a, this will be satisfied if the link dimensions have the following dimensionless ratio (assuming constant link thickness).

$$\frac{e}{h} = \frac{1}{2} \sqrt{3\left(\frac{a}{h}\right)^2 - 1} - \frac{a}{2h} \qquad (12.12)$$

where e defines the length of the material that must be added at each end to satisfy equation 12.11b.

For the link configuration in Figure 12-10b, the length e of the added material of width h needed to make it a physical pendulum can be found from

$$A\left(\frac{e}{h}\right)^3 + B\left(\frac{e}{h}\right)^2 + C\left(\frac{e}{h}\right) + D = 0$$

$$(12.13)$$

where:
$$A = 8$$
$$B = 12\left(\frac{a}{c}\right) + 24$$
$$C = 24\left(\frac{a}{c}\right) + 26$$
$$D = -2\left(\frac{a}{c}\right)^3 + 13\left(\frac{a}{c}\right) + 12\pi - 10$$

The second step is to force-balance the linkage with its modified coupler using the method of Section 12.3 (p. 639) and define the required counterweights on links 2 and 4. With the shaking forces eliminated, the shaking moment is a free vector, as is the input torque.

12

[†] Note that this arrangement also makes each pin joint the center of percussion for the other pin as the center of rotation. This means that a force applied at either pin will have a zero reaction force at the other pin, effectively decoupling them dynamically. See Section 10.10 (p. 554) and also Figure 13-10 (p. 678) for further discussion of this effect.

Then as the third step, the shaking moment can be counteracted by adding geared inertia counterweights to links 2 and 4 as shown in Figure 12-11. These must turn in the opposite direction to the links, so they require a gear ratio of -1. Such an inertia counterweight can balance any planar moment that is proportional to an angular acceleration and does not introduce any net inertia forces to upset the force balance of the linkage. Trade-offs include increased input torque and larger pin forces resulting from the torque required to accelerate the additional rotational inertia. There can also be large loads on the gear teeth and impact when torque reversals take up the gearsets' backlash, causing noise.

The shaking moment of an inline fourbar linkage is derived in reference [6] as

$$\mathbf{M}_s = \sum_{i=2}^{4} A_i \alpha_i \qquad (12.14)$$

where:
$$A_2 = -m_2 \left(k_2^2 + r_2^2 + a_2 r_2 \right)$$
$$A_3 = -m_3 \left(k_3^2 + r_3^2 - a_3 r_3 \right)$$
$$A_4 = -m_4 \left(k_4^2 + r_4^2 + a_4 r_4 \right)$$

α_i is the angular acceleration of link i. The other variables are defined in Figure 12-11.

Adding the effects of the two inertia counterweights gives

$$\mathbf{M}_s = \sum_{i=2}^{4} A_i \alpha_i + I_2 \alpha_2 + I_4 \alpha_4 \qquad (12.15)$$

The shaking moment can be forced to zero if

$$I_2 = -A_2$$
$$I_4 = -A_4 \qquad (12.16)$$
$$A_3 = 0, \quad \text{or} \quad k_3^2 = r_3 (a_3 - r_3)$$

This leads to a set of five design equations that must be satisfied for complete force and moment balancing of an inline fourbar linkage.[*]

$$m_2 r_2 = m_3 b_3 \left(\frac{a_2}{a_3} \right) \qquad (12.17a)$$

$$m_4 r_4 = m_3 r_3 \left(\frac{a_4}{a_3} \right) \qquad (12.17b)$$

$$k_3^2 = r_3 b_3 \qquad (12.17c)$$

$$I_2 = m_2 \left(k_2^2 + r_2^2 + a_2 r_2 \right) \qquad (12.17d)$$

$$I_4 = m_4 \left(k_4^2 + r_4^2 + a_4 r_4 \right) \qquad (12.17e)$$

Equations 12.17a and 12.17b are the force-balance criteria of equation 12.8 (p. 642) written for the inline linkage case. Equation 12.17c defines the coupler as a physical

[*] These components of the mR product needed to force-balance the linkage represent the entire amount needed. If links 2 and 4 are already designed with some individual unbalance (i.e., the CG not at pivot), then the existing mR product of the unbalanced link must be subtracted from that found in equations 12.17a and 12.17b in order to determine the size and location of additional counterweights to be added to those links.

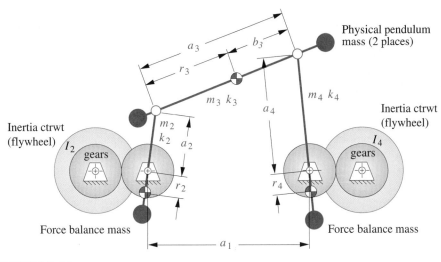

FIGURE 12-11

Completely force and moment balanced inline fourbar linkage with physical pendulum coupler and inertia counterweights on rotating links

pendulum. Equations 12.17d and 12.17e define the mass moments of inertia required for the two inertia counterweights. Note that if the linkage is run at constant angular velocity, α_2 will be zero in equation 12.14 (p. 648) and the inertia counterweight on link 2 can be omitted.

12.7 MEASURING AND CORRECTING IMBALANCE

While we can do a great deal to ensure balance when designing a machine, variations and tolerances in manufacturing will preclude even a well-balanced design from being in perfect balance when built. Thus there is need for a means to measure and correct the imbalance in rotating systems. Perhaps the best example assembly to discuss is that of the automobile tire and wheel, with which most readers will be familiar. Certainly the design of this device promotes balance, as it is essentially cylindrical and symmetrical. If manufactured to be perfectly uniform in geometry and homogeneous in material, it should be in perfect balance as is. But typically it is not. The wheel (or rim) is more likely to be close to balanced, as manufactured, than is the tire. The wheel is made of a homogeneous metal and has fairly uniform geometry and cross section. The tire, however, is a composite of synthetic rubber elastomer and fabric cord or metal wire. The whole is compressed in a mold and steam-cured at high temperature. The resulting material varies in density and distribution, and its geometry is often distorted in the process of removal from the mold and cooling.

STATIC BALANCING   After the tire is assembled to the wheel, the assembly must be balanced to reduce vibration at high speeds. The simplest approach is to statically balance it, though it is not really an ideal candidate for this approach as it is thick axially compared to its diameter. To do so it is typically suspended in a horizontal plane on a cone through its center hole. A bubble level is attached to the wheel, and weights are placed at positions around the rim of the wheel until it sits level. These weights are then attached to the rim at those points. This is a single-plane balance and thus can only cancel the unbalanced forces. It has no effect on any unbalanced moments due to uneven distribution of mass along the axis of rotation. It also is not very accurate.

DYNAMIC BALANCING   The better approach is to dynamically balance it. This requires a dynamic balancing machine be used. Figure 12-12 shows a schematic of such a device used for balancing wheels and tires or any other rotating assembly. The assembly to be balanced is mounted temporarily on an axle, called a mandrel, which is supported in bearings within the balancer. These two bearings are each mounted on a suspension which contains a transducer that measures dynamic force. A common type of force transducer contains a piezoelectric crystal which delivers a voltage proportional to the force applied. This voltage is amplified electronically and delivered to circuitry or software which can compute its peak magnitude and the phase angle of that peak with respect to some time reference signal. The reference signal is supplied by a shaft encoder on the mandrel which provides a short duration electrical pulse once per revolution in exactly the same angular location. This encoder pulse triggers the computer to begin processing the force signal. The encoder may also provide some large number of additional pulses equispaced around the shaft circumference (often 1024). These are used to trigger the recording of each data sample from the transducers in exactly the same location around the shaft and to provide a measure of shaft velocity via an electronic counter.

The assembly to be balanced is then "spun up" to some angular velocity, usually with a friction drive contacting its circumference. The drive torque is then removed and the drive motor stopped, allowing the assembly to "freewheel." (This is to avoid measuring any forces due to imbalances in the drive system.) The measuring sequence is begun, and the dynamic forces at each bearing are measured simultaneously and their waveforms stored. Many cycles can be measured and averaged to improve the quality of the measure-

FIGURE 12-12

A dynamic wheel balancer

ment. Because forces are being measured at two locations displaced along the axis, both summation of moment and summation of force data are computed.

The force signals are sent to a built-in computer for processing and computation of the needed balance masses and locations. The data needed from the measurements are the magnitudes of the peak forces and the angular locations of those peaks with respect to the shaft encoder's reference angle (which corresponds to a known point on the wheel). The axial locations of the wheel rim's inside and outside edges (the correction planes) with respect to the balance machine's transducer locations are provided to the machine's computer by operator measurement. From these data the net unbalanced force and net unbalanced moment can be calculated since the distance between the measured bearing forces is known. The mass-radius products needed in the correction planes on each side of the wheel can then be calculated from equations 12.3 (p. 634) in terms of the mR product of the balance weights. The correction radius is that of the wheel rim. The balance masses and angular locations are calculated for each correction plane to put the system in dynamic balance. Weights having the needed mass are clipped onto the inside and outside wheel rims (which are the correction planes in this case), at the proper angular locations. The result is a fairly accurately dynamically balanced tire and wheel.

12.8   REFERENCES

1   **Lowen, G. G., and R. S. Berkof.** (1968). "Survey of Investigations into the Balancing of Linkages." *J. Mechanisms*, **3**(4), pp. 221-231.

2   **Lowen, G. G., et al.** (1983). "Balancing of Linkages—An Update." *Journal of Mechanism and Machine Theory*, **18**(3), pp. 213-220.

3   **Berkof, R. S., and G. G. Lowen.** (1969). "A New Method for Completely Force Balancing Simple Linkages." *Trans. ASME J. of Eng. for Industry,* February, pp. 21-26.

4   **Tepper, F. R., and G. G. Lowen.** (1972). "General Theorems Concerning Full Force Balancing of Planar Linkages by Internal Mass Redistribution." *Trans ASME J. of Eng. for Industry*, **94 series B**(3), pp. 789-796.

5   **Weiss, K., and R. G. Fenton.** (1972). "Minimum Inertia Weight." *Mech. Chem. Engng. Trans. I.E. Aust.,* **MC8**(1), pp. 93-96.

6   **Berkof, R. S., and G. G. Lowen.** (1971). "Theory of Shaking Moment Optimization of Force-Balanced Four-Bar Linkages." *Trans. ASME J. of Eng. for Industry,* February, pp. 53-60.

7   **Berkof, R. S.** (1972). "Complete Force and Moment Balancing of Inline Four-Bar Linkages." *J. Mechanism and Machine Theory*, **8** (August)**,** pp. 397-410.

8   **Hockey, B. A.** (1971). "An Improved Technique for Reducing the Fluctuation of Kinetic Energy in Plane Mechanisms." *J. Mechanisms*, **6**, pp. 405-418.

9   **Hockey, B. A.** (1972). "The Minimization of the Fluctuation of Input Torque in Plane Mechanisms." *Mechanism and Machine Theory*, **7**, pp. 335-346.

10   **Berkof, R. S.** (1979). "The Input Torque in Linkages." *Mechanism and Machine Theory*, **14**, pp. 61-73.

11   **Lee, T. W., and C. Cheng.** (1984). "Optimum Balancing of Combined Shaking Force, Shaking Moment, and Torque Fluctuations in High Speed Linkages." *Trans. ASME J. Mechanisms, Transmission, Automation and Design*, **106**, pp. 242-251.

12   **Qi, N. M., and E. Pennestri.** (1991). "Optimum Balancing of Fourbar Linkages." *Mechanism and Machine Theory*, **26**(3), pp. 337-348.

13   **Porter, B., et al.** (1994). "Genetic Design of Dynamically Optimal Fourbar Linkages." *Proc. of 23rd Biennial Mechanisms Conference*, Minneapolis, MN, p. 413.

12

14   **Bagci, C.** (1975). "Shaking Force and Shaking Moment Balancing of the Plane Slider-Crank Mechanism." *Proc. of the 4th OSU Applied Mechanism Conference*, Stillwater, OK, p. 25-1.

15   **Kochev, I. S.** (2000). "General Theory of Complete Shaking Moment Balancing of Planar Linkages: A Critical Review." *Mechanism and Machine Theory*, **35**, pp. 1501-1514.

16   **Esat, I., and H. Bahai.** (1999). "A Theory of Complete Force and Moment Balancing of Planar Linkage Mechanisms." *Mechanism and Machine Theory*, **34**, pp. 903-922.

17   **Arakelian, V. H., and M. R. Smith.** (1999). "Complete Shaking Force and Shaking Moment Balancing of Linkages." *Mechanism and Machine Theory*, **34**, pp. 1141-1153.

12.9 PROBLEMS

*†12-1   A system of two coplanar arms on a common shaft, as shown in Figure 12-1 (p. 633), is to be designed. For the row(s) assigned in Table P12-1, find the shaking force of the linkage when run unbalanced at 10 rad/sec and design a counterweight to statically balance the system. Work in any consistent units system you prefer.

†12-2   The minute hand on Big Ben weighs 40 lb and is 10 ft long. Its *CG* is 4 ft from the pivot. Calculate the *mR* product and angular location needed to statically balance this link and design a physical counterweight, positioned close to the center. Select material and design the detailed shape of the counterweight which is of 2-in uniform thickness in the *Z* direction.

†12-3   A "V for victory" advertising sign is being designed to be oscillated about the apex of the V, on a billboard, as the rocker of a fourbar linkage. The angle between the legs of the V is 20°. Each leg is 8 ft long and 1.5 ft wide. Material is 0.25-in-thick aluminum. Design the V link for static balance.

†12-4   A three-bladed ceiling fan has 1.5-ft by 0.25-ft equispaced rectangular blades that nominally weigh 2 lb each. Manufacturing tolerances will cause the blade weight to vary up to plus or minus 5%. The mounting accuracy of the blades will vary the location of the *CG* versus the spin axis by plus or minus 10% of the blades' diameters. Calculate the weight of the largest steel counterweight needed at a 2-in radius to statically balance the worst-case blade assembly if the minimum blade radius is 6 in.

*†12-5   A system of three noncoplanar weights is arranged on a shaft generally as shown in Figure 12-3 (p. 637). For the dimensions from the row(s) assigned in Table P12-2, find the shaking forces and shaking moment when run unbalanced at 100 rpm and specify the *mR* product and angle of the counterweights in correction planes *A* and *B* needed to dynamically balance the system. The correction planes are 20 units apart. Work in any consistent units system you prefer.

* Answers in Appendix F.

† These problems are suited to solution using *Mathcad, Matlab*, or *TKSolver* equation solver programs.

TABLE P12-1   Data for Problem 12-1

Row	m_1	m_2	R_1	R_2
a	0.20	0.40	1.25 @ 30°	2.25 @ 120°
b	2.00	4.36	3.00 @ 45°	9.00 @ 320°
c	3.50	2.64	2.65 @ 100°	5.20 @ −60°
d	5.20	8.60	7.25 @ 150°	6.25 @ 220°
e	0.96	3.25	5.50 @ −30°	3.55 @ 120°

TABLE P12-2 Data for Problem 12-5

Row	m_1	m_2	m_3	l_1	l_2	l_3	R_1	R_2	R_3
a	0.20	0.40	1.24	2	8	17	1.25 @ 30°	2.25 @ 120°	5.50 @ – 30°
b	2.00	4.36	3.56	5	7	16	3.00 @ 45°	9.00 @ 320°	6.25 @ 220°
c	3.50	2.64	8.75	4	9	11	2.65 @ 100°	5.20 @ – 60°	1.25 @ 30°
d	5.20	8.60	4.77	7	12	16	7.25 @ 150°	6.25 @ 220°	9.00 @ 320°
e	0.96	3.25	0.92	1	3	18	5.50 @ 30°	3.55 @ 120°	2.65 @ 100°

*†12-6 A wheel and tire assembly has been run at 100 rpm on a dynamic balancing machine as shown in Figure 12-12 (p. 650). The force measured at the left bearing had a peak of 5 lb at a phase angle of 45° with respect to the zero reference angle on the tire. The force measured at the right bearing had a peak of 2 lb at a phase angle of –120° with respect to the reference zero on the tire. The center distance between the two bearings on the machine is 10 in. The left edge of the wheel rim is 4 in from the centerline of the closest bearing. The wheel is 7 in wide at the rim. Calculate the size and location, with respect to the tire's zero reference angle, of balance weights needed on each side of the rim to dynamically balance the tire assembly. The wheel rim diameter is 15 in.

*†12-7 Repeat Problem 12-6 for measured forces of 6 lb at a phase angle of –60° with respect to the reference zero on the tire, measured at the left bearing, and 4 lb at a phase angle of 150° with respect to the reference zero on the tire, measured at the right bearing. The wheel diameter is 16 in.

*†‡12-8 Table P11-3 (p. 618) shows geometric and kinematic data of some fourbar linkages.

   a. For the row(s) from Table P11-3 assigned in this problem, calculate the size and angular locations of the counterbalance mass-radius products needed on links 2 and 4 to completely force-balance the linkage by the method of Berkof and Lowen. Check your manual calculation with program LINKAGES.

   b. Calculate the input torque for the linkage both with and without the added balance weights and compare the results. Use program LINKAGES.

*†12-9 Link 2 in Figure P12-1 rotates at 500 rpm. The links are steel with cross sections of 1 × 2 in. Half of the 29-lb weight of the laybar and reed is supported by the linkage at point B. Design counterweights to force-balance the linkage and determine its change in peak torque versus the unbalanced condition. See Problem 11-13 (p. 617) for more information on the overall mechanism.

†‡12-10 Figure P12-2a (p. 654) shows a fourbar linkage and its dimensions in meters. The steel crank and rocker have uniform cross sections 50 mm wide by 25 mm thick. The aluminum coupler is 25 mm thick. The crank O_2A rotates at a constant speed of $\omega = 40$ rad/sec. Design counterweights to force-balance the linkage and determine its change in peak torque versus the unbalanced condition.

†‡12-11 Figure P12-2b shows a fourbar linkage and its dimensions in meters. The steel crank and rocker have uniform cross sections 50 mm wide by 25 mm thick. The aluminum coupler is 25 mm thick. The crank O_2A rotates at a constant speed of $\omega = 50$ rad/sec. Design counterweights to force-balance the linkage and determine its change in peak torque versus the unbalanced condition.

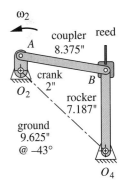

FIGURE P12-1

Problem 12-9

12

---

* Answers in Appendix F.

---

† These problems are suited to solution using *Mathcad*, *Matlab*, or *TKSolver* equation solver programs.

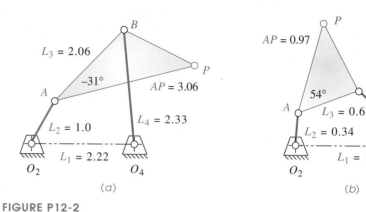

FIGURE P12-2

Problems 12-10 to 12-11

FIGURE P12-3

Problem 12-13

---

†12-12   Write a computer program or use an equation solver such as *Mathcad, Matlab,* or *TK-Solver* to solve for the mass-radius products that will force-balance any fourbar linkage for which the geometry and mass properties are known.

†12-13   Figure P12-3 shows a system with two weights on a rotating shaft. $W_1 = 15$ lb @ $0°$ at a 6-in radius and $W_2 = 20$ lb @ $270°$ at a 5-in radius. Determine the magnitudes and angles of the balance weights needed to dynamically balance the system. The balance weight in plane 3 is placed at a radius of 5 in and in plane 4 of 8 in.

**†12-14   Figure P12-4 shows a system with two weights on a rotating shaft. $W_1 = 20$ lb @ $45°$ at a 6-in radius and $W_2 = 15$ lb @ $300°$ at a 4-in radius. Determine the radii and angles of the balance weights needed to dynamically balance the system. The balance weight in plane 3 weighs 20 lb and in plane 4 weighs 40 lb.

†12-15   Figure P12-5 shows a system with two weights on a rotating shaft. $W_1 = 10$ lb @ $90°$ at a 3-in radius and $W_2 = 15$ lb @ $240°$ at a 3-in radius. Determine the magnitudes and angles of the balance weights needed to dynamically balance the system. The balance weights in planes 3 and 4 are placed at a 3-in radius.

**†12-16   Figure P12-6 shows a system with three weights on a rotating shaft. $W_1 = 6$ lb @ $120°$ at a 5-in radius, $W_2 = 12$ lb @ $240°$ at a 4-in radius, and $W_3 = 9$ lb @ $300°$ at a 8-in radius. Determine the magnitudes and angles of the balance weights needed to dynamically balance the system. The balance weights in planes 4 and 5 are placed at a 4s-in radius.

†12-17   Figure P12-7 (p. 656) shows a system with three weights on a rotating shaft. $W_2 = 10$ lb @ $90°$ at a 3-in radius, $W_3 = 10$ lb @ $180°$ at a 4-in radius, and $W_4 = 8$ lb @ $315°$ at a 4-in radius. Determine the magnitudes and angles of the balance weights needed to dynamically balance the system. The balance weight in plane 1 is placed at a radius of 4 in and in plane 5 of 3 in.

**†12-18   The 400-mm-dia steel roller in Figure P12-8 (p. 656) has been tested on a dynamic balancing machine at 100 rpm and shows an unbalanced force of $F_1 = 0.291$ N @ $\theta_1 = 45°$ in the *xy* plane at 1 and $F_4 = 0.514$ N @ $\theta_4 = 210°$ in the *xy* plane at 4. Determine the angular locations and required diameters of 25-mm-deep holes drilled radially inward from the surface in planes 2 and 3 to dynamically balance the system.

FIGURE P12-4

Problem 12-14

FIGURE P12-5

Problem 12-15

†12-19   The 500-mm-dia steel roller in Figure P12-8 (p. 656) has been tested on a dynamic balancing machine at 100 rpm and shows an unbalanced force of $F_1 = 0.23$ N @ $\theta_1 = 30°$ in the xy plane at 1 and $F_4 = 0.62$ N @ $\theta_4 = 135°$ in the x-y plane at 4. Determine the angular locations and required diameters of 25-mm-deep holes drilled radially inward from the surface in planes 2 and 3 to dynamically balance the system.

†‡12-20   The linkage in Figure P12-9a (p. 656) has rectangular steel links of 20 × 10 mm cross section similar to that shown in Figure 12-10a (p. 646). Design the necessary balance weights and other features necessary to completely eliminate the shaking force and shaking moment. State all assumptions.

†‡12-21   Repeat Problem 12-20 using links configured as in Figure 12-10b (p. 646) with the same cross section but having "dogbone" end diameters of 50 mm.

†‡12-22   The linkage in Figure P12-9b (p. 656) has rectangular steel links of 20 × 10 mm cross section similar to that shown in Figure 12-10a. Design the necessary balance weights and other features necessary to completely eliminate the shaking force and shaking moment. State all assumptions.

†‡12-23   Repeat Problem 12-22 using steel links configured as in Figure 12-10b (p. 646) with a 20 × 10 mm cross section and having "dogbone" end diameters of 50 mm.

†12-24   The device in Figure P12-10 (p. 656) is used to balance fan blade/hub assemblies running at 600 rpm. The center distance between the two bearings on the machine is 250 mm. The left edge of the fan hub (plane A) is 100 mm from the centerline of the closest bearing (at F_2). The hub is 75 mm wide along its axis and has a diameter of 200

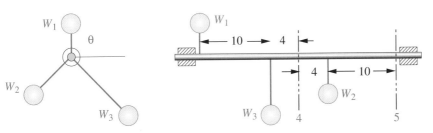

FIGURE P12-6

Problem 12-16

* Answers in Appendix F.

† These problems are suited to solution using *Mathcad*, *Matlab*, or *TKSolver* equation solver programs.

12

FIGURE P12-7

Problem 12-17 (p. 654)

FIGURE P12-8

Problems 12-18 and 12-19 (pp. 654-655)

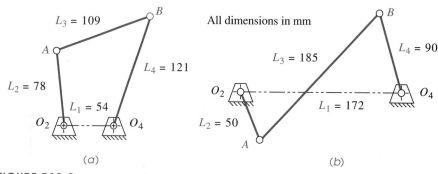

(a)   (b)

FIGURE P12-9

Problems 12-20 to 12-23 (p. 655)

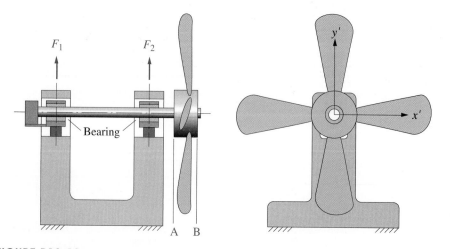

FIGURE P12-10

Problems 12-24 (p. 655) to 12-26

mm along the surfaces where balancing weights are fastened. The peak magnitude of force F_1 is 0.5 N at a phase angle of 30° with respect to the rotating x' axis. Force F_2 had a peak of 0.2 N at a phase angle of –130°. Calculate the magnitudes and locations with respect to the x' axis of balance weights placed in planes A and B of the hub to dynamically balance the fan assembly.

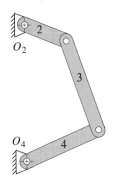

†12-25 Repeat Problem 12-24 using the following data. The hub is 55 mm wide and has a diameter of 150 mm along the surfaces where balancing weights are fastened. The force F_1 measured at the left bearing had a peak of 1.5 N at a phase angle of 60° with respect to the rotating x' axis. The force F_2 measured at the right bearing had a peak of 2.0 N at a phase angle of –180° with respect to the rotating x' axis.

†12-26 Repeat Problem 12-24 using the following data. The hub is 125 mm wide and has a diameter of 250 mm along the surfaces where balancing weights are fastened. The force F_1 measured at the left bearing had a peak of 1.1 N at a phase angle of 120° with respect to the rotating x' axis. The force F_2 measured at the right bearing had a peak of 1.8 N at a phase angle of –93° with respect to the rotating x' axis.

schematic- not to scale

FIGURE P12-11
Problems 12-27 to 12-30

†‡12-27 Figure P12-11 shows a fourbar linkage. $L_1 = 160$, $L_2 = 58$, $L_3 = 108$, and $L_4 = 110$ mm. All links are 4-mm-thick by 20-mm-wide steel. The square ends of link 3 extend 10 mm beyond the pivots. The other links' ends have 10-mm radii about the hole. Design counterweights to force-balance the linkage using the Berkof-Lowen method.

†‡12-28 Use the data of Problem 12-27 to design the necessary balance weights and other features to completely eliminate the shaking force and shaking moment the linkage exerts on the ground link.

†‡12-29 The linkage in Figure P12-11 has link lengths $L_1 = 3.26$, $L_2 = 2.75$, $L_3 = 3.26$, $L_4 = 2.95$ in. All links are 0.5-in-wide × 0.2-in-thick steel. The square ends of link 3 extend 0.25 in beyond the pivots. Links 2 and 4 have rounded ends that have a radius of 0.25 in. Design counterweights to force-balance the linkage using the Berkof-Lowen method.

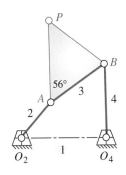

†‡12-30 Use the data of Problem 12-29 to design the necessary balance weights and other features to completely eliminate the shaking force and shaking moment the linkage exerts on the ground link.

FIGURE P12-12
Problems 12-33 to 12-34

†‡12-31 The linkage in Figure P12-11 has link lengths $L_1 = 8.88$, $L_2 = 3.44$, $L_3 = 7.40$, $L_4 = 5.44$ in. All links have a uniform 0.5-in-wide × 0.2-in-thick cross section and are made from aluminum. Link 3 has squared ends that extend 0.25 in from the pivot point centers. Links 2 and 4 have rounded ends that have a radius of 0.25 in. Design counterweights to force-balance the linkage using the method of Berkof and Lowen.

12

†‡12-32 Use the data of Problem 12-31 to design the necessary balance weights and other features to completely eliminate the shaking force and shaking moment the linkage exerts on the ground link.

†‡12-33 The linkage in Figure P12-12 has $L_1 = 9.5$, $L_2 = 5.0$, $L_3 = 7.4$, $L_4 = 8.0$, and $AP = 8.9$ in. Links 2 and 4 are rectangular steel with a 1-in wide × 0.12-in thick cross section and 0.5-in-radius ends. The coupler is 0.25-in-thick aluminum with 0.5-in radii at points A, B, and P. Design counterweights to force-balance the linkage using the Berkof-Lowen method.

† These problems are suited to solution using *Mathcad*, *Matlab*, or *TKSolver* equation solver programs.

†‡12-34 Use the data of Problem 12-33, changing link 3 to be steel with the same cross-section dimensions as links 2 and 4, to design the necessary balance weights and other features necessary to completely eliminate the shaking force and shaking moment the linkage exerts on the ground link.

‡ These problem solutions can be checked with program LINKAGES which is on the attached DVD.

$L_1 = 2.72$ in
$L_2 = 5.52$ in
$L_3 = 4.88$ in
$L_4 = 6.48$ in

FIGURE P12-13

Problems 12-35 and 12-36

i	W_i (lb)	r_i (in)	δ_i (°)
1	1.50	12.01	–0.25
2	1.48	11.97	0.75
3	1.54	11.95	0.25
4	1.55	12.03	–1.00
5	1.49	12.04	–0.50

†‡12-35   Figure P12-13 shows a fourbar linkage and its dimensions in inches. All links are 0.08-in-thick steel and have a uniform cross section 0.26 in wide × 0.12 in thick. Links 2 and 4 have rounded ends with a 0.13-in radius. Link 3 has squared ends that extend 0.13 in from the pivot point centers. Design counterweights to force-balance the linkage using the method of Berkof and Lowen.

†‡12-36   Use the data of Problem 12-35 to design the necessary balance weights and other features to completely eliminate the shaking force and shaking moment the linkage exerts on the ground link.

†12-37   A manufacturing company makes 5-blade ceiling fans. Before assembling the fan blades onto the hub, the blades are weighed and the location of the CG is determined as a distance from the center of rotation and an angular offset from the geometric center of the blade. At final assembly a technician is provided with the weight and CG data for the 5 blades. Write a computer program or use an equation solver such as *Mathcad* or *TKSolver* to calculate the required weight and angular position of a balance weight that is attached to the hub at a radius of 2.5 in. Use the geometric center of blade one as a reference axis. Test your program with the data given in Table P12-3.

†*12-38   The motor rotor shown in Figure P12-14 has been tested on a dynamic balance machine at 1800 rpm and shows unbalanced forces of $F_1 = 2.43$ lb @ $\theta_1 = 34.5°$ in the *xy* plane at 1 and $F_4 = 5.67$ lb @ $\theta_4 = 198°$ in the *xy* plane at 4. Balance weights consist of cylindrical disks whose center of rotation is a drilled hole located at a distance *e* from the center of the disk. The net weight of each disk is 0.50 lb and the disks are located on planes 2 and 3. Determine the angular locations of the line through the drilled hole and the center of the disk with respect to the *x* axis and the eccentric distances *e* to dynamically balance the system.

†12-39   The motor rotor shown in Figure P12-14 has been tested on a dynamic balance machine at 1450 rpm and shows unbalanced forces of $F_1 = 4.82$ lb @ $\theta_1 = 163°$ in the *xy* plane at 1 and $F_4 = 7.86$ lb @ $\theta_4 = 67.8°$ in the *xy* plane at 4. Balance weights consist of cylindrical disks whose center of rotation is a drilled hole located at a distance *e* from the center of the disk. The net weight of each disk is 0.375 lb and the disks are located on planes 2 and 3. Determine the angular locations of the line through the drilled hole and the center of the disk with respect to the *x* axis and the eccentric distances *e* to dynamically balance the system.

---

* Answers in Appendix F.

---

† These problems are suited to solution using *Mathcad, Matlab,* or *TKSolver* equation solver programs.

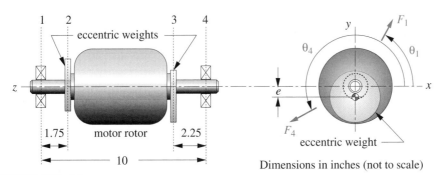

FIGURE P12-14

Problems 12-38 and 12-39

TABLE P12-4 Data for Problem 12-40   Lengths in mm.

Row	L_1	L_2	L_3	L_4	r	e	d	t	Material
a	375	100	300	200	13	13	6	4	Steel
b	150	75	250	300	12	15	6	4	Steel
c	50	125	375	350	15	15	8	6	Aluminum
d	250	150	475	400	20	20	10	3	Titanium
e	225	50	200	175	15	16	8	6	Aluminum
f	475	175	625	250	25	30	12	5	Steel

[†]12-40   Table P12-4 gives the geometry and kinematic data for several fourbar linkages similar to that shown in Figure P12-11 (p. 657). For the row(s) assigned in Table P12-4, design counterweights of the type shown in Figure P12-15 for links 2 and 4 to completely force-balance the linkage by the method of Berkof and Lowen. The square ends of link 3 extend a distance e from the hole center. The other links' ends are full round with a radius r about the hole center. All pin holes have the same diameter d, and all links have the same width, $2r$, and thickness t. The hole-to-hole link lengths are L_1, L_2, L_3, and L_4. The counterweight will be integrally machined with the link and will have the same thickness as the link.

$$\bar{r} = \frac{2}{3}\frac{\left(R^3 - r^3\right)}{\left(R^2 - r^2\right)}$$

$$A = \frac{\alpha}{2}\left(R^2 - r^2\right)$$

FIGURE P12-15

Problem 12-40

12.10 VIRTUAL LABORATORY

L12-1   View the video *Fourbar Linkage Virtual Laboratory* that is on the DVD. Open the file *Virtual Fourbar Linkage Lab 12-1.doc* from the DVD and follow the instructions as directed by your professor. For this lab it is suggested that you compare the data for the balanced and unbalanced conditions of the linkage.

[†] These problems are suited to solution using *Mathcad, Matlab,* or *TKSolver* equation solver programs.

Chapter 13

ENGINE DYNAMICS

*I have always thought that the substitution of the
internal combustion machine for the horse marked
a very gloomy milestone in the progress of mankind.*
WINSTON S. CHURCHILL

13.0 INTRODUCTION

The previous chapters have introduced analysis techniques for the determination of dynamic forces, moments, and torques in machinery. Shaking forces, moments, and their balancing have also been discussed. We will now attempt to integrate all these dynamic considerations into the design of a common device, the slider-crank linkage as used in the internal combustion engine.[*] This deceptively simple mechanism will be found to be actually quite complex in terms of the dynamic considerations necessary to its design for high-speed operation. Thus it will serve as an excellent example of the application of the dynamics concepts just presented. We will not address the thermodynamic aspects of the internal combustion engine beyond defining the combustion forces which are necessary to drive the device. Many other texts, such as those listed in the bibliography at the end of this chapter, deal with the very complex thermodynamic and fluid dynamic aspects of this ubiquitous device. We will concentrate only on its kinematics and mechanical dynamics aspects. It is not our intention to make an "engine designer" of the student so much as to apply dynamic principles to a realistic design problem of general interest and also to convey the complexity and fascination involved in the design of an apparently simple dynamic device.

Some students may have had the opportunity to disassemble and service an internal combustion engine, but many have never done so. Thus we will begin with very fundamental descriptions of engine design and operation. The program ENGINE, supplied with this text, is designed to reinforce and amplify the concepts presented. It will perform all the tedious computations necessary to provide the student with dynamic information for

[*] Basic information on the operation of engines with animations can be found at: http://www.Howstuffworks.com/engine.htm

Carburetor*
Pushrod
Rocker arm
Valve
Combustion chamber
Piston
Wrist pin
Connecting rod
Crankpin
Crank center
Crankshaft

Intake manifold
Head
Camshaft
Exhaust manifold
Engine block
Water jackets
Starter motor
Oil pickup
Oil sump
Oil pan

FIGURE 13-1

Cutaway cross section of a vee-eight engine
Adapted from a drawing by Lane Thomas, Western Carolina University, Dept. of Industrial Education, with permission

13

design choices and trade-offs. The student is encouraged to use this program concurrent with a reading of the text. Many examples and illustrations within the text will be generated with this program and reference will frequently be made to it. A user manual for program ENGINE is provided in Appendix A which can be read or referred to at any time in order to gain familiarity with the program's operation. Examples used in Chapters 13 and 14 which deal with engine dynamics are built into program ENGINE for student observation and exercise. They can be found on a dropdown menu in that program. Other example engine files for program ENGINE are on the DVD.

* Carburetors have been replaced by fuel injection systems on automotive and other engines that are required to meet increasingly stringent exhaust emission control regulations in the United States. Fuel injection gives better control over the fuel-air mixture than a carburetor.

FIGURE 13-2

Finite-element models of an engine piston (*a*), connecting rod (*b*), and crankshaft (*c*) *Courtesy of General Motors Co.*

13.1 ENGINE DESIGN

Figure 13-1 (p. 661) shows a detailed cross section of an internal combustion engine. The basic mechanism consists of a crank, a connecting rod (coupler), and piston (slider). Since this figure depicts a **multicylinder vee-eight** engine configuration, there are four cranks arranged on a crankshaft, and eight sets of connecting rods and pistons, four in the left bank of cylinders and four in the right bank. Only two piston-connecting rod assemblies are visible in this view, both on a common crank pin. The others are behind those shown. Figure 13-2 shows finite-element models of a piston, connecting rod, and crankshaft for a four-cylinder inline engine. The most usual arrangement is an inline engine with cylinders all in a common plane. Three-, four-, five-, and six-cylinder **inline engines** are in production the world over. **Vee engines** in four-, six-, eight-, ten-, and twelve-cylinder versions are also in production, with vee six and vee eight being the most popular vee configurations. The geometric arrangements of the crankshaft and cylinders have a significant effect on the dynamic condition of the engine. We will explore these effects of multicylinder arrangements in the next chapter. At this stage we wish to deal only with the design of a **single-cylinder** engine. After optimizing the geometry and dynamic condition of one cylinder, we will be ready to assemble combinations of cylinders into multicylinder configurations.

A schematic of the basic **one-cylinder slider-crank** mechanism and the terminology for its principal parts are shown in Figure 13-3. Note that it is "backdriven" compared to

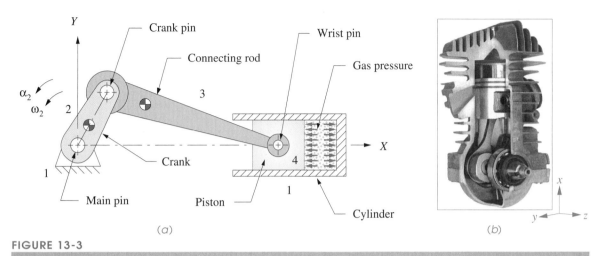

FIGURE 13-3

Fourbar slider-crank mechanism (*a*) for single-cylinder internal combustion engine (*b*) *Mahle Inc., Morristown, NJ*

the linkages we have been analyzing in previous chapters. That is, the explosion of the combustible mixture in the cylinder drives the piston to the left in Figure 13-3 or down in Figure 13-4 (p. 664), turning the crank. The crank torque that results is ultimately delivered to the drive wheels of the vehicle through a transmission (see Section 9.11, p. 524) to propel the car, motorcycle, or other device. The same slider-crank mechanism can also be used "forward-driven," by motor-driving the crank and taking the output energy from the piston end. It is then called a **piston pump** and is used to compress air and pump well water, gasoline, and other liquids.

In the internal combustion engine of Figure 13-3, it should be fairly obvious that at most we can only expect energy to be delivered from the exploding gases to the crank during the power stroke of the cycle. The piston must return from bottom dead center (BDC) to top dead center (TDC) on its own momentum before it can receive another push from the next explosion. In fact, some rotational kinetic energy must be stored in the crankshaft merely to carry it through the TDC and BDC points as the moment arm for the gas force at those points is zero. This is why an internal combustion engine must be "spun up" with a hand crank, pull rope, or starter motor to get it running.

There are two common combustion cycles in use in internal combustion engines, the **Clerk two-stroke cycle** and the **Otto four-stroke cycle**, named after their nineteenth century inventors. The four-stroke cycle is most common in automobile, truck, and stationary gasoline engines. The two-stroke cycle is used in motorcycles, outboard motors, chain saws, and other applications where its better power-to-weight ratio outweighs its drawbacks of higher pollution levels and poor fuel economy compared to the four-stroke.

FOUR-STROKE CYCLE   The **Otto four-stroke cycle** is shown in Figure 13-4 (p. 664). It takes four full strokes of the piston to complete one Otto cycle. A piston stroke is defined as its travel from TDC to BDC or the reverse. Thus there are two strokes per 360° crank revolution and it takes 720° of crankshaft rotation to complete one four-stroke cycle. This engine requires at least two valves per cylinder, one for intake and one for exhaust. For discussion, we can start the cycle at any point as it repeats every two crank

13

(*a*) Intake stroke   (*b*) Compression stroke   (*c*) Power stroke   (*d*) Exhaust stroke

(*e*) The gas pressure curve

13

FIGURE 13-4

The Otto four-stroke combustion cycle

revolutions. Figure 13-4a shows the **intake stroke** which starts with the piston at TDC. A mixture of fuel and air is drawn into the cylinder from the induction system (the fuel injectors, or the carburetor and intake manifold in Figure 13-1, p. 661) as the piston descends to BDC, increasing the volume of the cylinder and creating a slight negative pressure.

During the **compression stroke** in Figure 13-4b, all valves are closed and the gas is compressed as the piston travels from BDC to TDC. Slightly before TDC, a spark is ignited to explode the compressed gas. The pressure from this explosion builds very quickly and pushes the piston down from TDC to BDC during the **power stroke** shown in Figure 13-4c. The exhaust valve is opened and the piston's **exhaust stroke** from BDC to TDC (Figure 13-4d) pushes the spent gases out of the cylinder into the exhaust manifold (see also Figure 13-1, p. 661) and thence to the catalytic converter for cleaning before being dumped out the tailpipe. The cycle is then ready to repeat with another intake stroke. The valves are opened and closed at the right times in the cycle by a camshaft which is driven in synchrony with the crankshaft by gears, chain, or toothed belt drive. (See Figure 9-25, p. 502.) Figure 13-4e shows the gas pressure curve for one cycle. With a one-cylinder Otto cycle engine, power is delivered to the crankshaft, at most, 25% of the time as there is only 1 power stroke per 2 revolutions.

TWO-STROKE CYCLE   The **Clerk two-stroke cycle** is shown in Figure 13-5 (p. 666). This engine does not need any valves, though to increase its efficiency it is sometimes provided with a passive (pressure differential operated) one at the intake port. It does not have a camshaft or valve train or cam drive gears to add weight and bulk to the engine. As its name implies, it requires only two strokes, or 360°, to complete its cycle. There is a passageway, called a transfer port, between the combustion chamber above the piston and the crankcase below. There is also an exhaust port in the side of the cylinder. The piston acts to sequentially block or expose these ports as it moves up and down. The crankcase is sealed and mounts the carburetor on it, serving also as the intake manifold.

Starting at TDC (Figure 13-5a), the two-stroke cycle proceeds as follows: The spark plug ignites the fuel-air charge, compressed on the previous revolution. The expansion of the burning gases drives the piston down, delivering torque to the crankshaft. Partway down, the piston uncovers the exhaust port, allowing the burned (and also some unburned) gases to begin to escape to the exhaust system.

As the piston descends (Figure 13-5b), it compresses the charge of fuel-air mixture in the sealed crankcase. The piston blocks the intake port, preventing blowback through the carburetor. As the piston clears the transfer port in the cylinder wall, its downward motion pushes the new fuel-air charge up through the transfer port to the combustion chamber. The momentum of the exhaust gases leaving the chamber on the other side helps pull in the new charge as well.

The piston passes BDC (Figure 13-5c) and starts up, pushing out the remaining exhaust gases. The exhaust port is closed by the piston as it ascends, allowing compression of the new charge. As the piston approaches TDC, it exposes the intake port (Figure 13-5d), sucking a new charge of air and fuel into the expanded crankcase from the carburetor. Slightly before TDC, the spark is ignited and the cycle repeats as the piston passes TDC.

Clearly, this Clerk cycle is not as efficient as the Otto cycle in which each event is more cleanly separated from the others. Here there is much mixing of the various phases of the cycle. Unburned hydrocarbons are exhausted in larger quantities. This accounts for the poor fuel economy and dirty emissions of the Clerk engine.[*] It is nevertheless popular in applications where low weight is paramount.

Lubrication is also more difficult in the two-stroke engine than in the four-stroke as the crankcase is not available as an oil sump. Thus the lubricating oil must be mixed with the fuel. This further increases the emissions problem compared to the Otto cycle engine which burns raw gasoline and pumps its lubricating oil separately throughout the engine.

13

[*] Research and development is underway to clean up the emissions of the two-stroke engine by using fuel injection and compressed air scavenging of the cylinders. These efforts may yet bring this potentially more powerful engine design into compliance with air quality specifications.

Power

(a)

Down stroke

Exhaust

Transfer

(b)

Intake

(d)

Up stroke

Compression

(c)

FIGURE 13-5

The Clerk two-stroke combustion cycle

DIESEL CYCLE   The **diesel cycle** can be either two-stroke or four-stroke. It is a **compression-ignition** cycle. No spark is needed to ignite the air-fuel mixture. The air is compressed in the cylinder by a factor of about 14 to 15 (versus 8 to 10 in the spark engine), and a low volatility fuel is injected into the cylinder just before TDC. The heat of compression causes the explosion. Diesel engines are larger and heavier than spark ignition engines for the same power output because the higher pressures and forces at which they operate require stronger, heavier parts. Two-stroke cycle diesel engines are quite common. Diesel fuel is a better lubricant than gasoline.

GAS FORCE   In all the engines discussed here, the usable **output torque** is created from the explosive gas pressure generated within the cylinder either once or twice per two revolutions of the crank, depending on the cycle used.   The magnitude and shape of this explosion pressure curve will vary with the engine design, stroke cycle, fuel used, speed of operation, and other factors related to the thermodynamics of the system. For our purpose of analyzing the mechanical dynamics of the system, we need to keep the gas pressure function consistent while we vary other design parameters in order to compare the results of our mechanical design changes. For this purpose, program ENGINE has been provided with a built-in **gas pressure curve** whose peak value is about 600 psi and whose shape is similar to the curve from a real engine. Figure 13-6 (p. 668) shows the **gas force curve** that results from the built-in gas pressure function in program ENGINE applied to a piston of particular area, for both two- and four-stroke engines. Changes in piston area will obviously affect the gas force magnitude for this consistent pressure function, but no changes in engine design parameters input to this program will change its built-in gas pressure curve. To see this gas force curve, run program ENGINE and select any one of the example engines from the pulldown menu. Then calculate and plot *Gas Force*.

13.2   SLIDER-CRANK KINEMATICS

In Chapters 4, 6, 7, and 11 we developed general equations for the exact solution of the positions, velocities, accelerations, and forces in the pin-jointed fourbar linkage, and also for two inversions of the **slider-crank linkage**, using vector equations. We could again apply that method to the analysis of the "standard" slider-crank linkage, used in the majority of internal combustion engines, as shown in Figure 13-7 (p. 669). Note that its slider motion has been aligned with the X axis. This is a "nonoffset" slider-crank, because the slider axis extended passes through the crank pivot. It also has its slider block translating against the stationary ground plane; thus there will be no Coriolis component of acceleration (see Section 7.3, p. 358).

The simple geometry of this particular inversion of the slider-crank mechanism allows a very straightforward approach to the exact analysis of its slider's position, velocity, and acceleration, using only plane trigonometry and scalar equations. Because of this method's simplicity and to present an alternative solution approach we will analyze this device again.

Let the crank radius be r and the conrod length be l. The angle of the crank is θ, and the angle that the conrod makes with the X axis is ϕ. For any constant crank angular velocity ω, the crank angle $\theta = \omega t$. The instantaneous piston position is x. Two right triangles rqs and lqu are constructed. Then from geometry:

13

(a) Otto four-stroke cycle   (b) Clerk two-stroke cycle

1 Cylinder

Bore = 3.00 in
Stroke = 3.54
B/S   = 0.85
L/R   = 3.50
RPM   = 3400
P_{max} = 600 psi

FIGURE 13-6

Gas force functions in the two-stroke and four-stroke cycle engines

$$q = r\sin\theta = l\sin\phi$$
$$\theta = \omega t \tag{13.1a}$$

$$\sin\phi = \frac{r}{l}\sin\omega t$$

$$s = r\cos\omega t$$
$$u = l\cos\phi \tag{13.1b}$$
$$x = s + u = r\cos\omega t + l\cos\phi$$

$$\cos\phi = \sqrt{1 - \sin^2\phi} = \sqrt{1 - \left(\frac{r}{l}\sin\omega t\right)^2} \tag{13.1c}$$

$$x = r\cos\omega t + l\sqrt{1 - \left(\frac{r}{l}\sin\omega t\right)^2} \tag{13.1d}$$

Equation 13.1d is an exact expression for the piston position x as a function of r, l, and ωt. This can be differentiated versus time to obtain exact expressions for the velocity and acceleration of the piston. For a steady-state analysis we will assume ω to be constant.

$$\dot{x} = -r\omega\left[\sin\omega t + \frac{r}{2l}\frac{\sin 2\omega t}{\sqrt{1 - \left(\frac{r}{l}\sin\omega t\right)^2}}\right] \tag{13.1e}$$

$$\ddot{x} = -r\omega^2\left\{\cos\omega t - \frac{r\left[l^2\left(1 - 2\cos^2\omega t\right) - r^2\sin^4\omega t\right]}{\left[l^2 - (r\sin\omega t)^2\right]^{\frac{3}{2}}}\right\} \tag{13.1f}$$

(a) The linkage geometry

Note:
Link 3 can be considered as a
2-force member for this gas-force
analysis because the inertia forces
are being temporarily ignored.
They will be superposed later.

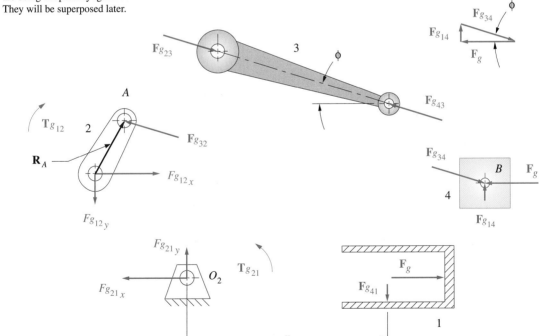

(b) Free-body diagrams

FIGURE 13-7

Fourbar slider-crank linkage position and gas force analysis (See Figure 13-12, p. 680 for the inertia force analysis.)

13

Equations 13.1 (p. 668) can easily be solved with a computer for all values of ωt needed. But, it is rather difficult to look at equation 13.1f and visualize the effects of changes in the design parameters r and l on the acceleration. It would be useful if we could derive a simpler expression, even if approximate, that would allow us to more easily predict the results of design decisions involving these variables. To do so, we will use the binomial theorem to expand the radical in equation 13.1d for piston position to put the equations for position, velocity, and acceleration in simpler, approximate forms which will shed some light on the dynamic behavior of the mechanism.

The general form of the binomial theorem is:

$$(a+b)^n = a^n + na^{n-1}b + \frac{n(n-1)}{2!}a^{n-2}b^2 + \frac{n(n-1)(n-2)}{3!}a^{n-3}b^3 + \cdots \qquad (13.2a)$$

The radical in equation 13.1d is:

$$\sqrt{1-\left(\frac{r}{l}\sin\omega t\right)^2} = \left[1-\left(\frac{r}{l}\sin\omega t\right)^2\right]^{\frac{1}{2}} \qquad (13.2b)$$

where, for the binomial expansion:

$$a = 1 \qquad\qquad b = -\left(\frac{r}{l}\sin\omega t\right)^2 \qquad\qquad n = \frac{1}{2} \qquad (13.2c)$$

It expands to:

$$1 - \frac{1}{2}\left(\frac{r}{l}\sin\omega t\right)^2 - \frac{1}{8}\left(\frac{r}{l}\sin\omega t\right)^4 - \frac{1}{16}\left(\frac{r}{l}\sin\omega t\right)^6 - \cdots \qquad (13.2d)$$

or:

$$1 - \left(\frac{r^2}{2l^2}\right)\sin^2\omega t - \left(\frac{r^4}{8l^4}\right)\sin^4\omega t - \left(\frac{r^6}{16l^6}\right)\sin^6\omega t - \cdots \qquad (13.2e)$$

Each nonconstant term contains the **crank-conrod ratio** r/l to some power. Applying some engineering common sense to the depiction of the slider-crank in Figure 13-7a (p. 669), we can see that if r/l were greater than 1 the crank could not make a complete revolution. In fact if r/l even gets close to 1, the piston will hit the fixed pivot O_2 before the crank completes its revolution. If r/l is as large as 1/2, the transmission angle $(\pi/2 - \phi)$ will be too small (see Sections 3.3, p. 100 and 4.11, p. 204) and the linkage will not run well. A practical upper limit on the value of r/l is about 1/3. Most slider-crank linkages will have this **crank-conrod ratio** somewhere between 1/3 and 1/5 for smooth operation. If we substitute this practicl upper limit of $r/l = 1/3$ into equation 13.2e, we get:

$$1 - \left(\frac{1}{18}\right)\sin^2\omega t - \left(\frac{1}{648}\right)\sin^4\omega t - \left(\frac{1}{11\,664}\right)\sin^6\omega t - \cdots$$

$$\qquad (13.2f)$$

or: $\qquad 1 - 0.055\,56\sin^2\omega t - 0.001\,54\sin^4\omega t - 0.000\,09\sin^6\omega t - \cdots$

Clearly we can drop all terms after the second with very small error. Substituting this approximate expression for the radical in equation 13.1d (p. 668) gives an approximate expression for piston displacement with only a fraction of one percent error.

$$x = r\cos\omega t + l\left[1 - \left(\frac{r^2}{2l^2}\right)\sin^2\omega t\right] \qquad (13.3a)$$

Substitute the trigonometric identity:

$$\sin^2\omega t = \frac{1 - \cos 2\omega t}{2} \qquad (13.3b)$$

and simplify:

$$x = l - \frac{r^2}{4l} + r\left(\cos\omega t + \frac{r}{4l}\cos 2\omega t\right) \qquad (13.3c)$$

Differentiate for velocity of the piston (with constant ω):

$$\dot{x} \cong -r\omega\left(\sin\omega t + \frac{r}{2l}\sin 2\omega t\right) \qquad (13.3d)$$

Differentiate again for acceleration (with constant ω):

$$\ddot{x} \cong -r\omega^2\left(\cos\omega t + \frac{r}{l}\cos 2\omega t\right) \qquad (13.3e)$$

The process of binomial expansion has, in this particular case, led us to Fourier series approximations of the exact expressions for the piston displacement, velocity, and acceleration. Fourier[*] showed that any periodic function can be approximated by a series of sine and cosine terms of integer multiples of the independent variable. Recall that we dropped the fourth, sixth, and subsequent power terms from the binomial expansion, which would have provided $\cos 4\omega t$, $\cos 6\omega t$, etc., terms in this expression. These multiple-angle functions are referred to as the **harmonics** of the fundamental $\cos \omega t$ term. The $\cos \omega t$ term repeats once per crank revolution and is called the fundamental frequency or the **primary component**. The second harmonic, $\cos 2\omega t$, repeats twice per crank revolution and is called the **secondary component**. The higher harmonics were dropped when we truncated the series. The constant term in the displacement function is the **DC component** or **average value**. The complete function is the sum of its harmonics. The Fourier series form of the expressions for displacement and its derivatives lets us see the relative contributions of the various harmonic components of the functions. This approach will prove to be quite valuable when we attempt to dynamically balance an engine design.

Program ENGINE calculates the position, velocity, and acceleration of the piston according to equations 13.3c, d, and e. Figure 13-8a, b, and c (p. 672) shows these functions for this example engine in the program as plotted for constant crank ω over two full revolutions. The acceleration curve shows the effects of the second harmonic term most clearly because that term's coefficient is larger than its correspondent in either of the other two functions. The fundamental ($-\cos \omega t$) term gives a pure harmonic function with a period of 360°. This fundamental term dominates the function as it has the largest coefficient in equation 13.3e. The flat top and slight dip in the positive peak acceleration

[*] Baron Jean Baptiste Joseph Fourier (1768-1830) published the description of the mathematical series which bears his name in *The Analytic Theory of Heat* in 1822. The Fourier series is widely used in harmonic analysis of all types of physical systems. Its general form is:

$$y = \frac{a_0}{2} + \left(a_1\cos x + b_1\sin x\right)$$
$$+ \left(a_2\cos 2x + b_2\sin 2x\right)$$
$$+ \cdots + \left(a_n\cos nx + b_n\sin nx\right)$$

13

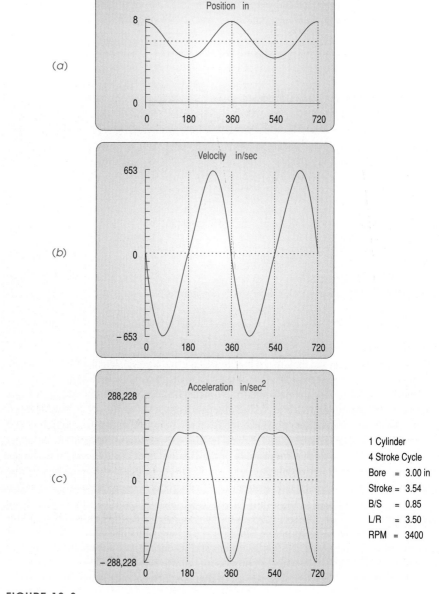

(a)

(b)

(c)

1 Cylinder
4 Stroke Cycle
Bore = 3.00 in
Stroke = 3.54
B/S = 0.85
L/R = 3.50
RPM = 3400

FIGURE 13-8

Position, velocity, and acceleration functions for a single-cylinder engine

* If you think that number is high, consider the typical Nascar pushrod V-8 that turns up to 9600 rpm and Formula 1, V-12, and V-8 racing engines that red-line at over 19 000 rpm. As an exercise, calculate their peak accelerations assuming the same dimensions as our example engine.

of Figure 13-8c are caused by the cos $2\omega t$ second harmonic adding or subtracting from the fundamental. Note the very high value of peak acceleration of the piston even at the midrange engine speed of 3400 rpm. It is 747 *g*'s! At 6000 rpm this increases to nearly 1300 *g*'s.* This is a moderately sized engine, of 3-in (76 mm) bore and 3.54-in (89 mm) stroke, with 25-in^3 (400-cc) displacement per cylinder (a 1.6-L 4-cylinder engine).

SUPERPOSITION   We will now analyze the dynamic behavior of the single-cylinder engine based on the approximate kinematic model developed in this section. Since we have several sources of dynamic excitation to deal with, we will use the method of superposition to separately analyze them and then combine their effects. We will first consider the **forces and torques** which are due to the presence of the **explosive gas forces** in the cylinder, which drive the engine. Then we will analyze the **inertia forces and torques** that result from the high-speed motion of the elements. The total force and torque state of the machine at any instant will be the sum of these components. Finally we will look at the **shaking forces and torques** on the ground plane and the **pin forces** within the linkage that result from the combination of applied and dynamic forces on the system.

13.3   GAS FORCE AND GAS TORQUE

The **gas force** is due to the gas pressure from the exploding fuel-air mixture impinging on the top of the piston surface as shown in Figure 13-3 (p. 663). Let F_g = gas force, P_g = gas pressure, A_p = area of piston, and B = bore of cylinder, which is also equal to the piston diameter. Then:

$$\mathbf{F}_g = -P_g A_p \,\hat{\mathbf{i}}; \qquad\qquad A_p = \frac{\pi}{4} B^2$$

(13.4)

$$\mathbf{F}_g = -\frac{\pi}{4} P_g B^2 \,\hat{\mathbf{i}}$$

The negative sign is due to the choice of engine orientation in the coordinate system of Figure 13-3. The **gas pressure** P_g in this expression is a function of crank angle ωt and is defined by the thermodynamics of the engine. A typical **gas pressure curve** for a four-stroke engine is shown in Figure 13-4 (p. 664). The **gas force curve** shape is identical to that of the gas pressure curve as they differ only by a constant multiplier, the piston area A_p. Figure 13-6 (p. 668) shows the approximation of the gas force curve used in program ENGINE for both four- and two-stroke engines.

The **gas torque** in Figure 13-9 (p. 674) is due to the gas force acting at a moment arm about the crank center O_2 in Figure 13-7 (p. 669). This moment arm varies from zero to a maximum as the crank rotates. The distributed gas force over the piston surface has been resolved to a single force acting through the mass center of link 4 in the free-body diagrams of Figure 13-7b. The concurrent force system at point B is resolved in the vector diagram showing that:

$$\mathbf{F}_{g_{14}} = F_g \tan\phi \,\hat{\mathbf{j}} \tag{13.5a}$$

$$\mathbf{F}_{g_{34}} = -F_g \,\hat{\mathbf{i}} - F_g \tan\phi \,\hat{\mathbf{j}} \tag{13.5b}$$

From the free-body diagrams in Figure 13-7 (p. 669) we can see that:

$$\mathbf{F}_{g_{41}} = -\mathbf{F}_{g_{14}}$$

$$\mathbf{F}_{g_{43}} = -\mathbf{F}_{g_{34}}$$

$$\mathbf{F}_{g_{23}} = -\mathbf{F}_{g_{43}}$$

$$\mathbf{F}_{g_{32}} = -\mathbf{F}_{g_{23}}$$

so: $$\mathbf{F}_{g_{32}} = -\mathbf{F}_{g_{34}} = F_g \,\hat{\mathbf{i}} + F_g \tan\phi \,\hat{\mathbf{j}} \tag{13.5c}$$

13

(a) Otto four-stroke cycle

(b) Clerk two-stroke cycle

FIGURE 13-9

Gas torque functions in the two-stroke and four-stroke cycle engines

The **driving torque** $\mathbf{T}_{g21}$ at link 2 due to the gas force can be found from the cross product of the position vector to point A and the force at point A.

$$\mathbf{T}_{g21} = \mathbf{R}_A \times \mathbf{F}_{g32} \tag{13.6a}$$

This expression can be expanded and will involve the crank length r and the angles θ and ϕ as well as the gas force $\mathbf{F}_g$. Note from the free-body diagram for link 1 that we can also express the torque in terms of the forces $\mathbf{F}_{g14}$ or $\mathbf{F}_{g41}$, which act always perpendicular to the motion of the slider (neglecting friction), and the distance x, which is their instantaneous moment arm about O_2. The reaction torque $\mathbf{T}_{g12}$ due to the gas force trying to rock the ground plane is:

$$\mathbf{T}_{g12} = F_{g41} \cdot x \; \hat{\mathbf{k}} \tag{13.6b}$$

If you have ever abruptly opened the throttle of a running automobile engine while working on it, you probably noticed the engine move to the side as it rocked in its mounts from the reaction torque. The driving torque $\mathbf{T}_{g21}$ is the negative of this reaction torque.

$$\mathbf{T}_{g21} = -\mathbf{T}_{g12}$$
$$\mathbf{T}_{g21} = -F_{g41} \cdot x \; \hat{\mathbf{k}} \tag{13.6c}$$

and:

$$F_{g14} = -F_{g41}$$

so:

$$\mathbf{T}_{g21} = F_{g14} \cdot x \; \hat{\mathbf{k}} \tag{13.6d}$$

Equation 13.6d gives us an expression for **gas torque** which involves the displacement of the piston x for which we have already derived equation 13.3a (p. 671). Substituting equation 13.3a for x and the magnitude of equation 13.5a (p. 673) for F_{g14}, we get:

$$\mathbf{T}_{g21} = \left(F_g \tan\phi\right)\left[l - \frac{r^2}{4l} + r\left(\cos\omega t + \frac{r}{4l}\cos 2\omega t\right)\right]\hat{\mathbf{k}} \tag{13.6e}$$

Equation 13.6e contains the conrod angle ϕ as well as the independent variable, crank angle ωt. We would like to have an expression which involves only ωt. We can substitute an expression for $\tan\phi$ generated from the geometry of Figure 13-7a (p. 669).

$$\tan\phi = \frac{q}{u} = \frac{r\sin\omega t}{l\cos\phi} \qquad (13.7a)$$

Substitute equation 13.1c (p. 668) for cos ϕ:

$$\tan\phi = \frac{r\sin\omega t}{l\sqrt{1-\left(\frac{r}{l}\sin\omega t\right)^2}} \qquad (13.7b)$$

The radical in the denominator can be expanded using the binomial theorem as was done in equations 13.2 (p. 670), and the first two terms retained for a good approximation to the exact expression,

$$\frac{1}{\sqrt{1-\left(\frac{r}{l}\sin\omega t\right)^2}} = 1 + \frac{r^2}{2l^2}\sin^2\omega t \qquad (13.7c)$$

giving:

$$\tan\phi = \frac{r}{l}\sin\omega t\left(1+\frac{r^2}{2l^2}\sin^2\omega t\right) \qquad (13.7d)$$

Substitute this into equation 13.6e for the gas torque:

$$\mathbf{T}_{g_{21}} \cong F_g\left[\frac{r}{l}\sin\omega t\left(1+\frac{r^2}{2l^2}\sin^2\omega t\right)\right]\left[1-\frac{r^2}{4l}+r\left(\cos\omega t+\frac{r}{4l}\cos 2\omega t\right)\right]\hat{\mathbf{k}} \qquad (13.8a)$$

Expand this expression and neglect any terms containing the conrod crank ratio r/l raised to any power greater than one since these will have very small coefficients as was seen in equation 13.2. This results in a simpler, but even more approximate expression for the gas torque:

$$\mathbf{T}_{g_{21}} \cong F_g\, r\sin\omega t\left(1+\frac{r}{l}\cos\omega t\right) \qquad (13.8b)$$

Note that the **exact value** of this **gas torque** can always be calculated from equations 13.1d (p. 668), 13.5a (p. 673) and 13.6d (p. 674) in combination, or from the expansion of equation 13.6a, if you require a more accurate answer. For design purposes the approximate equation 13.8b will usually be adequate. Program ENGINE calculates the gas torque using equation 13.8b and its built-in gas pressure curve to generate the gas force function. Plots of the gas torque for two- and four-stroke cycles are shown in Figure 13-9. Note the similarity in shape to that of the gas force curve in Figure 13-6 (p. 668). Note also that the two-stroke has theoretically twice the power available as the four-stroke, all other factors equal, because there are twice as many torque pulses per unit time. The poorer efficiency of the two-stroke significantly reduces this theoretical advantage, however.

13.4 EQUIVALENT MASSES

To do a complete dynamic force analysis on any mechanism we need to know the geometric properties (mass, center of gravity, mass moment of inertia) of the moving links as was discussed in previous chapters (see Sections 10.3 to 10.10, pp. 546-554, and Chapter 11).

13

This is easy to do if the link already is designed in detail and its dimensions are known. When designing the mechanism from scratch, we typically do not yet know that level of detail about the links' geometries. But we must nevertheless make some estimate of their geometric parameters in order to begin the iteration process which will eventually converge on a detailed design.

In the case of this slider-crank mechanism, the **crank** is in **pure rotation** and the **piston** is in **pure translation**. By assuming some reasonable geometries and materials we can make approximations of their dynamic parameters. Their kinematic motions are easily determined. We have already derived expressions for the piston motion in equations 13.3 (p. 671). Further, if we balance the rotating crank, as described and recommended in the previous chapter, then the *CG* of the crank will be motionless at its center O_2 and will not contribute to the dynamic forces. We will do this in a later section.

The conrod is in complex motion. To do an exact dynamic analysis as was derived in Section 11.5 (p. 595), we need to determine the linear acceleration of its *CG* for all positions. At the outset of the design, the conrod's *CG* location is not accurately defined. To "bootstrap" the design, we need a simplified model of this connecting rod which we can later refine, as more dynamic information is generated about our engine design. The requirements for a dynamically equivalent model were stated in Section 10.2 (p. 546) and are repeated here as Table 13-1 for your convenience.

If we could model our still-to-be-designed conrod as two, lumped, point masses, concentrated one at the crank pin (point *A* in Figure 13-7, p. 669), and one at the wrist pin (point *B* in Figure 13-7), we would at least know what the motions of these lumps are. The lump at *A* would be in pure rotation as part of the crank, and the lump at point *B* would be in pure translation as part of the piston. These lumped, point masses have no dimension and are assumed to be connected with a magical, massless but rigid rod.[*]

DYNAMICALLY EQUIVALENT MODEL   Figure 13-10a (p. 678) shows a typical conrod. Figure 13-10b shows a generic two-mass model of the conrod. One mass m_t is located at distance l_t from the *CG* of the original rod, and the second mass m_p at distance l_p from the *CG*. The mass of the original part is m_3, and its moment of inertia about its *CG* is I_{G_3}. Expressing the three requirements for dynamic equivalence from Table 13-1 mathematically in terms of these variables, we get:

$$m_p + m_t = m_3 \qquad (13.9a)$$

$$m_p l_p = m_t l_t \qquad (13.9b)$$

$$m_p l_p^2 + m_t l_t^2 = I_{G_3} \qquad (13.9c)$$

There are four unknowns in these three equations, m_p, l_p, m_t, l_t, which means we must choose a value for any one variable to solve the system. Let us choose the distance

[*]   These lumped mass models have to be made with *very* special materials. *Unobtainium 206* has the property of infinite mass density, thus occupies no space and can be used for "point masses." *Unobtainium 208* has infinite stiffness and zero mass and thus can be used for rigid but "massless rods."

TABLE 13-1   Requirements for Dynamic Equivalence

1   The mass of the model must equal that of the original body.

2   The center of gravity must be in the same location as that of the original body.

3   The mass moment of inertia must equal that of the original body.

l_t equal to the distance to the wrist pin, l_b, as shown in Figure 13-10c (p. 678). This will put one mass at a desired location. Solving equations 13.9a and 13.9b simultaneously with that substitution gives expressions for the two lumped masses:

$$m_p = m_3 \frac{l_b}{l_p + l_b}$$

$$m_b = m_3 \frac{l_p}{l_p + l_b}$$
(13.9d)

Substituting equation 13.9d into 13.9c gives a relation between l_p and l_b:

$$m_3 \frac{l_b}{l_p + l_b} l_p^2 + m_3 \frac{l_p}{l_p + l_b} l_b^2 = I_{G_3} = m_3 l_p l_b$$
(13.9e)

$$l_p = \frac{I_{G_3}}{m_3 l_b}$$

Please refer to Section 10.10 (p. 554) and equation 10.13 (p. 555) that define the *center of percussion* and its geometric relationship to a corresponding *center of rotation*. Equation 13.9e is the same as equation 11.13 (p. 600) (except for sign which is due to an arbitrary choice of the link's orientation in the coordinate system). The distance l_p is the location of the center of percussion corresponding to a center of rotation at l_b. Thus our second mass m_p must be placed at the link's **center of percussion** P (using point B as its center of rotation) to obtain exact dynamic equivalence. The masses must be as defined in equation 13.9d.

The geometry of the typical conrod, as shown in Figures 13-2 (p. 662) and 13-10a (p. 678), is large at the crank pin end (A) and small at the wrist pin end (B). This puts the CG close to the "big end." The center of percussion P will be even closer to the big end than is the CG. For this reason, we can place the second lumped mass, which belongs at P, at point A with relatively small error in our dynamic model's accuracy. This approximate model is adequate for our initial design calculations. Once a viable design geometry is established, we will have to do a complete and exact force analysis with the methods of Chapter 11 before considering the design complete.

Making this substitution of distance l_a for l_p and renaming the lumped masses at those distances m_{3a} and m_{3b}, to reflect both their identity with link 3 and with points A and B, we rewrite equations 13.9d.

Let
$$l_p = l_a$$

then:
$$m_{3a} = m_3 \frac{l_b}{l_a + l_b}$$
(13.10a)

and:
$$m_{3b} = m_3 \frac{l_a}{l_a + l_b}$$
(13.10b)

These define the amounts of the total conrod mass to be placed at each end, to approximately dynamically model that link. Figure 13-10d shows this dynamic model. In the absence of any data on the shape of the conrod at the outset of a design, preliminary dynamic force information can be obtained by using the rule of thumb of placing two-thirds of the conrod's mass at the crank pin and one-third at the wrist pin.

13

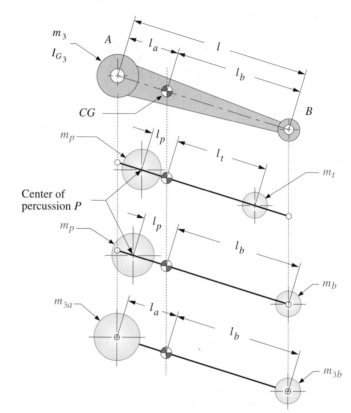

(a) Original connecting rod

(b) Generic two-mass model

(c) Exact dynamic model

(d) Approximate model

FIGURE 13-10

Lumped mass dynamic models of a connecting rod

STATICALLY EQUIVALENT MODEL   We can create a similar lumped mass model of the crank. Even though we intend to balance the crank before we are done, for generality we will initially model it *unbalanced* as shown in Figure 13-11. Its *CG* is located at some distance r_{G_2} from the pivot, O_2, on the line to the crank pin, A. We would like to model it as a lumped mass at A on a massless rod pivoted at O_2. If our principal concern is with a steady-state analysis, then the crank velocity ω will be held constant. An absence of angular acceleration on the crank allows a statically equivalent model to be used because the equation $\mathbf{T} = I\alpha$ will be zero regardless of the value of I. A **statically equivalent model** needs only to have equivalent mass and equivalent first moments as shown in Table 13-2. The moments of inertia need not match. We model it as two lumped masses, one at point A and one at the fixed pivot O_2. Writing the two requirements for static equivalence from Table 13-2:

$$m_2 = m_{2a} + m_{2O_2}$$

$$m_{2a}r = m_2 r_{G_2}$$

$$m_{2a} = m_2 \frac{r_{G_2}}{r}$$

(13.11)

TABLE 13-2	Requirements for Static Equivalence
1	The mass of the model must equal that of the original body.
2	The center of gravity must be in the same location as that of the original body.

The lumped mass m_{2a} can be placed at point A to represent the unbalanced crank. The second lumped mass, at the fixed pivot O_2, is not necessary to any calculations as that point is stationary.

These simplifications lead to the lumped parameter model of the slider-crank linkage shown in Figure 13-12 (p. 680). The crank pin, point A, has two masses concentrated at it, the equivalent mass of the crank m_{2a} and the portion of conrod m_{3a}. Their sum is m_A. At the wrist pin, point B, two masses are also concentrated, the piston mass m_4 and the remaining portion of the conrod mass m_{3b}. Their sum is m_B. This model has masses which are either in pure rotation (m_A) or in pure translation (m_B), so it is very easy to dynamically analyze.

$$m_A = m_{2a} + m_{3a}$$
$$m_B = m_{3b} + m_4 \qquad (13.12)$$

VALUE OF MODELS   *The value of constructing simple, lumped mass models of complex systems increases with the complexity of the system being designed.* It makes little sense to spend large amounts of time doing sophisticated, detailed analyses of designs which are so ill-defined at the outset that their conceptual viability is as yet unproven. It is better to get a reasonably approximate and rapid answer that tells you the concept needs to be rethought, than to spend a greater amount of time reaching the same conclusion to more decimal places.

FIGURE 13-11

Statically equivalent lumped mass model of a crank

13.5   INERTIA AND SHAKING FORCES

The simplified, lumped mass model of Figure 13-12 can be used to develop expressions for the forces and torques due to the accelerations of the masses in the system. The method of d'Alembert is of value in visualizing the effects of these moving masses on the system and on the ground plane. Accordingly, the free-body diagrams of Figure 13-12b show the d'Alembert inertia forces acting on the masses at points A and B. Friction is again ignored. The acceleration for point B is given in equation 13.3e (p. 671). The acceleration of point A in pure rotation is obtained by differentiating the position vector $\mathbf{R}_A$ twice, assuming a constant crankshaft ω, which gives:

$$\mathbf{R}_A = r \cos \omega t \, \hat{\mathbf{i}} + r \sin \omega t \, \hat{\mathbf{j}}$$
$$\mathbf{a}_A = -r \omega^2 \cos \omega t \, \hat{\mathbf{i}} - r \omega^2 \sin \omega t \, \hat{\mathbf{j}} \qquad (13.13)$$

The total inertia force $\mathbf{F}_i$ is the sum of the centrifugal (inertia) force at point A and the inertia force at point B.

$$\mathbf{F}_i = -m_A \, \mathbf{a}_A - m_B \, \mathbf{a}_B \qquad (13.14a)$$

13

(a) Dynamic model

(b) Free-body diagrams

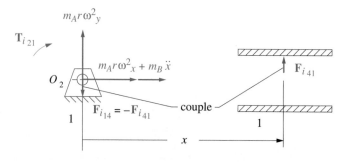

FIGURE 13-12

Lumped mass dynamic model of the slider-crank—arrows show vector direction and sense, labels show magnitude

Breaking it into x and y components:

$$F_{i_x} = -m_A \left(-r\omega^2 \cos\omega t \right) - m_B\, x \qquad (13.14b)$$

$$F_{i_y} = -m_A \left(-r\omega^2 \sin\omega t \right) \qquad (13.14c)$$

Note that only the x component is affected by the acceleration of the piston. Substituting equation 13.3e (p. 671) into equation 13.14b:

$$F_{i_x} \cong -m_A \left(-r\omega^2 \cos\omega t \right) - m_B \left[-r\omega^2 \left(\cos\omega t + \frac{r}{l} \cos 2\omega t \right) \right]$$
$$\qquad (13.14d)$$

$$F_{i_y} = -m_A \left(-r\omega^2 \sin\omega t \right)$$

Notice that the x-directed inertia forces have primary components at crank frequency and secondary (second harmonic) forces at twice crank frequency. There are also small-magnitude, higher, even harmonics which we truncated in the binomial expansion of the piston displacement function. The force due to the rotating mass at point A has only a primary component.

The **shaking force** was defined in Section 11.8 (p. 601) to be *the sum of all forces acting on the ground plane*. From the free-body diagram for link 1 in Figure 13-12:

$$\sum F_{s_x} \cong -m_A \left(r\omega^2 \cos\omega t \right) - m_B \left[r\omega^2 \left(\cos\omega t + \frac{r}{l} \cos 2\omega t \right) \right]$$
$$\qquad (13.14e)$$

$$\sum F_{s_y} = -m_A \left(r\omega^2 \sin\omega t \right) + F_{i_{41}} - F_{i_{41}}$$

Note that the side force of the piston on the cylinder wall $F_{i_{41}}$ is cancelled by an equal and opposite force $F_{i_{14}}$ passed through the connecting rod and crankshaft to the main pin at O_2. These two forces create a couple that provides the shaking torque. The shaking force $\mathbf{F}_s$ is equal to the negative of the inertia force.

$$\mathbf{F}_s = -\mathbf{F}_i \qquad (13.14f)$$

Note that the gas force from equation 13.4 (p. 673) does not contribute to the shaking force. Only inertia forces and external forces are felt as shaking forces. The gas force is an internal force which is cancelled within the mechanism. It acts equally and oppositely on both the piston top and the cylinder head as shown in Figure 13-7 (p. 669).

Program ENGINE calculates the shaking force at constant ω, for any combination of linkage parameters input to it. Figure 13-13 (p. 682) shows the shaking force plot for the same unbalanced built-in example engine as shown in the acceleration plot (Figure 13-8c, p. 672). The linkage orientation is the same as in Figure 13-12 with the x axis horizontal. The x component is larger than the y component due to the high acceleration of the piston. The forces are seen to be quite large despite this being a relatively small (0.4 liter per cylinder) engine running at moderate speed (3400 rpm). We will soon investigate techniques to reduce or eliminate this shaking force from the engine. It is an undesirable feature which creates noise and vibration.

13

FIGURE 13-13

Shaking force in an unbalanced slider-crank linkage

13.6　INERTIA AND SHAKING TORQUES

The **inertia torque** results from the action of the inertia forces at a moment arm. The inertia force at point A in Figure 13-12 (p. 680) has two components, radial and tangential. The radial component has no moment arm. The tangential component has a moment arm of crank radius r. If the crank ω is constant, the mass at A will not contribute to inertia torque. The inertia force at B has a nonzero component perpendicular to the cylinder wall except when the piston is at TDC or BDC. As we did for the gas torque, we can express the inertia torque in terms of the couple $-\mathbf{F}_{i_{41}}$, $\mathbf{F}_{i_{41}}$ whose forces act always perpendicular to the motion of the slider (neglecting friction), and the distance x, which is their instantaneous moment arm (see Figure 13-12, p. 680). The inertia torque is:

$$\mathbf{T}_{i_{21}} = \left(F_{i_{41}} \cdot x \right) \hat{\mathbf{k}} = \left(-F_{i_{14}} \cdot x \right) \hat{\mathbf{k}} \tag{13.15a}$$

Substituting for $F_{i_{14}}$ (see Figure 13-12b, p. 680) and for x (see equation 13.3a,. p. 671) we get:

$$\mathbf{T}_{i_{21}} \cong -\left(-m_B \ddot{x} \tan \phi \right) \left[l - \frac{r^2}{4l} + r \left(\cos \omega t + \frac{r}{4l} \cos 2\omega t \right) \right] \hat{\mathbf{k}} \tag{13.15b}$$

We previously developed expressions for $\ddot{x}$ (equation 13.3e, p. 671) and $\tan \phi$ (equation 13.7d, p. 675) which can now be substituted.

$$\mathbf{T}_{i_{21}} \cong m_B \left[-r\omega^2 \left(\cos \omega t + \frac{r}{l} \cos 2\omega t \right) \right]$$

$$\cdot \left[\frac{r}{l} \sin \omega t \left(1 + \frac{r^2}{2l^2} \sin^2 \omega t \right) \right]$$

$$\cdot \left[l - \frac{r^2}{4l} + r \left(\cos \omega t + \frac{r}{4l} \cos 2\omega t \right) \right] \hat{\mathbf{k}} \tag{13.15c}$$

Expanding this and then dropping all terms with coefficients containing r/l to powers higher than one gives the following approximate equation for inertia torque with constant shaft ω:

$$\mathbf{T}_{i_{21}} = -m_B r^2 \omega^2 \sin \omega t \left(\frac{r}{2l} + \cos \omega t + \frac{3r}{2l} \cos 2\omega t \right) \hat{\mathbf{k}} \qquad (13.15\text{d})$$

This contains products of sine and cosine terms. Putting it entirely in terms of harmonics will be instructive, so substitute the identities:

$$2 \sin \omega t \cos 2\omega t = \sin 3\omega t - \sin \omega t$$
$$2 \sin \omega t \cos \omega t = \sin 2\omega t$$

to get: $$\mathbf{T}_{i_{21}} = \frac{1}{2} m_B r^2 \omega^2 \left(\frac{r}{2l} \sin \omega t - \sin 2\omega t - \frac{3r}{2l} \sin 3\omega t \right) \hat{\mathbf{k}} \qquad (13.15\text{e})$$

This shows that the **inertia torque** has a third harmonic term as well as a first and second. The second harmonic is the dominant term as it has the largest coefficient because r/l is always less than 2/3.

The **shaking torque** is equal to the inertia torque.

$$\mathbf{T}_s = \mathbf{T}_{i_{21}} \qquad (13.15\text{f})$$

Program ENGINE calculates the inertia torque from equation 13.15e. Figure 13-14 (p. 684) shows a plot of the inertia torque for this built-in example engine. Note the dominance of the second harmonic. The ideal magnitude for the inertia torque is zero, as it is parasitic. Its average value is always zero, so *it contributes nothing to the net driving torque*. It merely creates large positive and negative oscillations in the total torque which increase vibration and roughness. We will soon investigate means to reduce or eliminate this inertia and shaking torque in our engine designs. It is possible to cancel their effects by proper arrangement of the cylinders in a multicylinder engine, as will be explored in the next chapter.

13.7 TOTAL ENGINE TORQUE

The total engine torque is the sum of the gas torque and the inertia torque.

$$\mathbf{T}_{total} = \mathbf{T}_g + \mathbf{T}_i \qquad (13.16)$$

The gas torque is less sensitive to engine speed than is the inertia torque, which is a function of ω^2. So the relative contributions of these two components to the total torque will vary with engine speed. Figure 13-15a (p. 686) shows the total torque for this example engine plotted by program ENGINE for an idle speed of 800 rpm. Compare this to the gas torque plot of the same engine in Figure 13-9a (p. 674). The inertia torque component is negligible at this slow speed compared to the gas torque component. Figure 13-15c shows the same engine run at 6000 rpm. Compare this to the plot of inertia torque in Figure 13-14 (p. 684). The inertia torque component is dominating at this high speed. At the midrange speed of 3400 rpm (Figure 13-15b, p. 686), a mix of both components is seen.

13

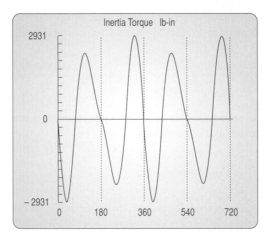

FIGURE 13-14

Inertia torque in the slider-crank linkage

13.8   FLYWHEELS

We saw in Section 11.11 (p. 605) that large oscillations in the torque-time function can be significantly reduced by the addition of a flywheel to the system. The single-cylinder engine is a prime candidate for the use of a flywheel. The intermittent nature of its power strokes makes one mandatory as it will store the kinetic energy needed to carry the piston through the Otto cycle's exhaust, intake, and compression strokes during which work must be done on the system. Even the two-stroke engine needs a flywheel to drive the piston up on the compression stroke.

The procedure for designing an engine flywheel is identical to that described in Section 11.11 for the fourbar linkage. The total torque function for one revolution of the crank is integrated, pulse by pulse, with respect to its average value. These integrals represent energy fluctuations in the system. The maximum change in energy under the torque curve during one cycle is the amount needed to be stored in the flywheel. Equation 11.20c (p. 610) expresses this relationship. Program ENGINE does the numerical integration of the total torque function and presents a table similar to the one shown in Figure 11-11 (p. 609). These data and the designer's choice of a coefficient of fluctuation k (see equation 11.19, p. 609) are all that is needed to solve equations 11.20 and 11.21 for the required moment of inertia of the flywheel.

The calculation must be done at some average crank ω. Since the typical engine operates over a wide range of speeds, some thought needs to be given to the most appropriate speed to use in the flywheel calculation. The flywheel's stored kinetic energy is proportional to ω^2 (see equation 11.17, p. 606). Thus at high speeds a flywheel can have a small moment of inertia and still be effective. The slowest operating speed will require the largest flywheel and should be the one used in the computation of required flywheel size.

Program ENGINE plots the flywheel-smoothed total torque for a user-supplied coefficient of fluctuation k. Figure 13-16 (p. 687) shows the smoothed torque functions for $k = 0.05$ corresponding to the unsmoothed ones in Figure 13-15 (p. 686). Note that the smoothed curves shown for each engine speed are what would result with the flywheel size necessary to obtain that coefficient of fluctuation at that speed. In other words, the flywheel applied to the 800-rpm engine is much larger than the one on the 6000-rpm engine, in these plots. Compare corresponding rows (speeds) between Figures 13-15 (p. 686) and 13-16 (p. 687) to see the effect of the addition of a flywheel. But do not directly compare parts a, b, and c within Figure 13-16 as to the amount of smoothing since the flywheel sizes used are different at each operating speed.

An engine flywheel is usually designed as a flat disk, bolted to one end of the crankshaft. One flywheel face is typically used for the clutch to run against. The clutch is a friction device which allows disconnection of the engine from the drive train (the wheels of a vehicle) when no output is desired. The engine can then remain running at idle speed with the vehicle or output device stopped. When the clutch is engaged, all engine torque is transmitted through it, by friction, to the output shaft.

13.9 PIN FORCES IN THE SINGLE-CYLINDER ENGINE

In addition to calculating the overall effects on the ground plane of the dynamic forces present in the engine, we also need to know the magnitudes of the forces at the pin joints. These forces will dictate the design of the pins and the bearings at the joints. Though we were able to lump the mass due to both conrod and piston, or conrod and crank, at points A and B for an overall analysis of the linkage's effects on the ground plane, we cannot do so for the pin force calculations. This is because the pins feel the effect of the conrod pulling on one "side" and the piston (or crank) pulling on the other "side" of the pin as shown in Figure 13-17 (p. 688). Thus we must separate the effects of the masses of the links joined by the pins.

We will calculate the effect of each component due to the various masses and the gas force and then superpose them to obtain the complete pin force at each joint. We need a bookkeeping system to keep track of all these components. We have already used some subscripts for these forces, so we will retain them and add others. The resultant bearing loads have the following components:

1   The gas force components, with the subscript g, as in F_g.

2   The inertia force due to the piston mass, with subscript ip, as in F_{ip}.

3   The inertia force due to the mass of the conrod at the wrist pin, with subscript iw, as in F_{iw}.

4   The inertia force due to the mass of the conrod at the crank pin, with subscript ic, as in F_{ic}.

5   The inertia force due to the mass of the crank at the crank pin, with subscript ir, as in F_{ir}.

The link number designations will be added to each subscript in the same manner as before, indicating the link from which the force is coming as the first number and the link being analyzed as the second. (See Section 11.2, p. 580 for further discussion of this notation.)

Figure 13-18 (p. 689) shows the free-body diagrams for the inertia force $\mathbf{F}_{ipB}$ due only to the acceleration of the mass of the piston, m_4. Those components are:

13

(a) 800 rpm

(b) 3400 rpm

1 Cylinder
4 Stroke Cycle
Bore = 3.00 in
Stroke = 3.54
B/S   = 0.85
L/R   = 3.50
RPM = 3400
m_A = 0.0284 bl
m_B = 0.0116 bl
P_{max} = 600 psi

(c) 6000 rpm

FIGURE 13-15

The total torque function's shape and magnitude vary with crankshaft speed

(a) 800 rpm

(b) 3400 rpm

1 Cylinder
4 Stroke Cycle
Bore   = 3.00 in
Stroke = 3.54
B/S   = 0.85
L/R   = 3.50
RPM   = 3400
m_A = 0.0284 bl
m_B = 0.0116 bl
P_{max} = 600 psi

(c) 6000 rpm

FIGURE 13-16

The total torque function's shape and magnitude vary with crankshaft speed

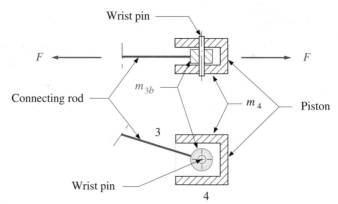

FIGURE 13-17

Forces on a pivot pin

$$\mathbf{F}_{ip_B} = -m_4 a_B \hat{\mathbf{i}} \tag{13.17a}$$

$$\mathbf{F}_{ip_{14}} = -\mathbf{F}_{ip_B} \tan\phi \, \hat{\mathbf{j}} = m_4 a_B \tan\phi \, \hat{\mathbf{j}} \tag{13.17b}$$

$$\mathbf{F}_{ip_{34}} = -\mathbf{F}_{ip_B} - \mathbf{F}_{ip_{14}} = m_4 a_B \, \hat{\mathbf{i}} - m_4 a_B \tan\phi \, \hat{\mathbf{j}} \tag{13.17c}$$

$$\mathbf{F}_{ip_{32}} = -\mathbf{F}_{ip_{34}} = -m_4 a_B \, \hat{\mathbf{i}} + m_4 a_B \tan\phi \, \hat{\mathbf{j}} \tag{13.17d}$$

$$\mathbf{F}_{ip_{12}} = -\mathbf{F}_{ip_{32}} = \mathbf{F}_{ip_{34}} \tag{13.17e}$$

Figure 13-19 (p. 690) shows the free-body diagrams for the forces due only to the acceleration of the mass of the conrod located at the wrist pin, m_{3b}. Those components are:

$$\mathbf{F}_{iw_B} = -m_{3b} a_B \hat{\mathbf{i}} \tag{13.18a}$$

$$\mathbf{F}_{iw_{34}} = \mathbf{F}_{iw_{41}} = F_{iw_B} \tan\phi \, \hat{\mathbf{j}} = -m_{3b} a_B \tan\phi \, \hat{\mathbf{j}} \tag{13.18b}$$

$$\mathbf{F}_{iw_{43}} = -\mathbf{F}_{iw_{34}} = m_{3b} a_B \tan\phi \, \hat{\mathbf{j}} \tag{13.18c}$$

$$\mathbf{F}_{iw_{23}} = -\mathbf{F}_{iw_B} - \mathbf{F}_{iw_{43}} = m_{3b} a_B \, \hat{\mathbf{i}} - m_{3b} a_B \tan\phi \, \hat{\mathbf{j}} \tag{13.18d}$$

$$\mathbf{F}_{iw_{12}} = -\mathbf{F}_{iw_{32}} = \mathbf{F}_{iw_{23}} \tag{13.18e}$$

Figure 13-20a (p. 690) shows the free-body diagrams for the forces due only to the acceleration of the mass of the conrod located at the crank pin, m_{3a}. That component is:

$$\mathbf{F}_{ic} = -\mathbf{F}_{ic_{12}} = \mathbf{F}_{ic_{21}} = -m_{3a}\mathbf{a}_A \tag{13.19a}$$

Substitute equation 13.13 (p. 679) for $\mathbf{a}_A$:

$$\mathbf{F}_{ic_{21}} = -\mathbf{F}_{ic_{12}} = m_{3a} r \omega^2 \left(\cos\omega t \, \hat{\mathbf{i}} + \sin\omega t \, \hat{\mathbf{j}} \right) \tag{13.19b}$$

Figure 13-20b (p. 690) shows the free-body diagrams for the forces due only to the acceleration of the lumped mass of the crank at the crank pin, m_{2a}. These affect only the main pin at O_2. That component is:

$$\mathbf{F}_{ir} = -\mathbf{F}_{in_{12}} = \mathbf{F}_{in_{21}} = -m_{2a}\mathbf{a}_A$$

$$\mathbf{F}_{in_{21}} = m_{2a} r \omega^2 \left(\cos\omega t \, \hat{\mathbf{i}} + \sin\omega t \, \hat{\mathbf{j}} \right) \tag{13.19c}$$

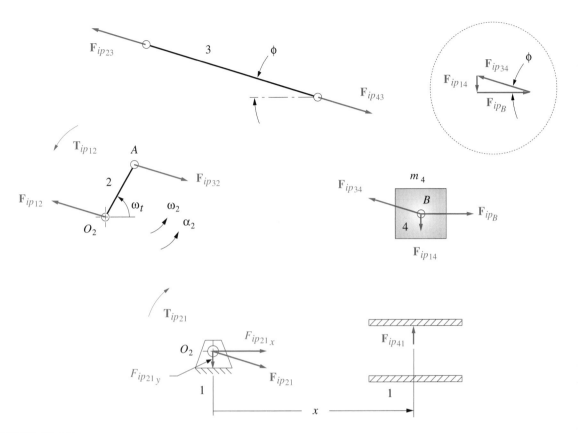

FIGURE 13-18

Free-body diagrams for forces due to the piston mass

The gas force components were shown in Figure 13-7 (p. 669) and defined in equations 13.5 (p. 673).

We can now sum the components of the forces at each pin joint. For the sidewall force $\mathbf{F}_{41}$ of the piston against the cylinder wall:

$$
\begin{aligned}
\mathbf{F}_{41} &= \mathbf{F}_{g41} + \mathbf{F}_{ip41} + \mathbf{F}_{iw41} \\
&= -F_g \tan\phi \,\hat{\mathbf{j}} - m_4 a_B \tan\phi \,\hat{\mathbf{j}} - m_{3b} a_B \tan\phi \,\hat{\mathbf{j}} \\
&= -\left[(m_4 + m_{3b}) a_B + F_g\right] \tan\phi \,\hat{\mathbf{j}}
\end{aligned} \tag{13.20}
$$

The total force $\mathbf{F}_{34}$ on the wrist pin is:

$$
\begin{aligned}
\mathbf{F}_{34} &= \mathbf{F}_{g34} + \mathbf{F}_{ip34} + \mathbf{F}_{iw34} \\
&= \left(-F_g \,\hat{\mathbf{i}} - F_g \tan\phi \,\hat{\mathbf{j}}\right) + \left(m_4 a_B \,\hat{\mathbf{i}} - m_4 a_B \tan\phi \,\hat{\mathbf{j}}\right) + \left(-m_{3b} a_B \tan\phi \,\hat{\mathbf{j}}\right) \\
&= \left(-F_g + m_4 a_B\right) \hat{\mathbf{i}} - \left[F_g + (m_4 + m_{3b}) a_B\right] \tan\phi \,\hat{\mathbf{j}}
\end{aligned} \tag{13.21}
$$

13

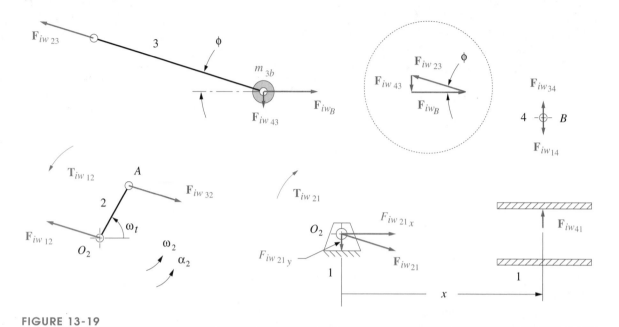

FIGURE 13-19

Free-body diagrams for forces due to the conrod mass concentrated at the wrist pin

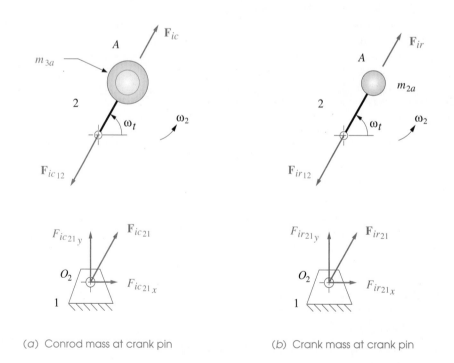

(a) Conrod mass at crank pin   (b) Crank mass at crank pin

FIGURE 13-20

Free-body diagrams for forces due to masses at the crank pin

The total force $\mathbf{F}_{32}$ on the crank pin is:

$$
\begin{aligned}
\mathbf{F}_{32} &= \mathbf{F}_{g32} + \mathbf{F}_{ip32} + \mathbf{F}_{iw32} + \mathbf{F}_{ic32} \\
&= \left(F_g\,\hat{\mathbf{i}} + F_g\,\tan\phi\,\hat{\mathbf{j}} \right) + \left(-m_4 a_B\,\hat{\mathbf{i}} + m_4 a_B\,\tan\phi\,\hat{\mathbf{j}} \right) \\
&\quad + \left(-m_{3b} a_B\,\hat{\mathbf{i}} + m_{3b} a_B\,\tan\phi\,\hat{\mathbf{j}} \right) + \left[m_{3a} r\omega^2 \left(\cos\omega t\,\hat{\mathbf{i}} + \sin\omega t\,\hat{\mathbf{j}} \right) \right] \\
&= \left[m_{3a} r\omega^2 \cos\omega t - \left(m_{3b} + m_4 \right) a_B + F_g \right]\hat{\mathbf{i}} \\
&\quad + \left\{ m_{3a} r\omega^2 \sin\omega t + \left[\left(m_{3b} + m_4 \right) a_B + F_g \right]\tan\phi \right\}\hat{\mathbf{j}}
\end{aligned}
\tag{13.22}
$$

The total force $\mathbf{F}_{21}$ on the main journal is:

$$
\mathbf{F}_{21} = \mathbf{F}_{32} + \mathbf{F}_{ir21} = \mathbf{F}_{32} + m_{2a} r\omega^2 \left(\cos\omega t\,\hat{\mathbf{i}} + \sin\omega t\,\hat{\mathbf{j}} \right)
\tag{13.23}
$$

Note that, unlike the inertia force in equation 13.14 (p. 681), which was unaffected by the gas force, these pin forces **are** a function of the gas force as well as of the inertia forces. Engines with larger piston diameters will experience greater pin forces as a result of the explosion pressure acting on their larger piston area.

Program ENGINE calculates the pin forces on all joints using equations 13.20 (p. 689) to 13.23. Figure 13-21 (p. 692) shows the wrist-pin force on the same unbalanced engine example as shown in previous figures, for three engine speeds. The "bow tie" loop is the inertia force and the "teardrop" loop is the gas force portion of the force curve. An interesting trade-off occurs between the gas force components and the inertia force components of the pin forces. At a low speed of 800 rpm (Figure 13-21a), the gas force dominates as the inertia forces are negligible at small ω. The peak wrist-pin force is then about 4200 lb. At high speed (6000 rpm), the inertia components dominate and the peak force is about 4500 lb (Figure 13-21c). But at a midrange speed (3400 rpm), the inertia force cancels some of the gas force and the peak force is only about 3200 lb (Figure 13-21b). These plots show that the pin forces can be quite large even in a moderately sized (0.4 liter/cylinder) engine. The pins, links, and bearings all have to be designed to withstand hundreds of millions of cycles of these reversing forces without failure.

Figure 13-22 (p. 693) shows further evidence of the interaction of the gas forces and inertia forces on the crank pin and the wrist pin. Figures 13-22a and 13-22c show the variation in the magnitude of the inertia force component on the crank pin and wrist pin, respectively, over one full revolution of the crank as the engine speed is increased from idle to redline. Figure 13-22b and d show the variation in the total force on the same respective pins with both the inertia and gas force components included. These two plots show only the first 90° of crank revolution where the gas force in a four-stroke cylinder occurs. Note that the gas force and inertia force components counteract one another resulting in one particular speed where the total pin force is a minimum during the power stroke. This is the same phenomenon as seen in Figure 13-21.

Figure 13-23 (p. 694) shows the forces on the main pin and crank pin at three engine speeds for the same **unbalanced** single-cylinder engine example as in previous figures. These forces are plotted as hodographs in a local rotating coordinate system (LRCS) $x'y'$ embedded in the crankshaft. Figure 13-23a shows that at 800 rpm (idle speed) the crank pin and main pin forces are essentially equal and opposite because the inertia force

13

(a) 800 rpm

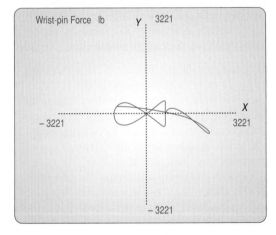

(b) 3400 rpm

1 Cylinder
4 Stroke Cycle
Bore = 3.00 in
Stroke = 3.54
B/S = 0.85
L/R = 3.50
RPM = 3400
m_{2a} = 0.0150 bl
m_{3a} = 0.0134 bl
m_{3b} = 0.0066 bl
m_4 = 0.0050 bl
P_{max} = 600 psi

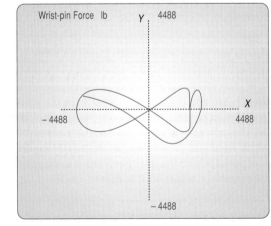

(c) 6000 rpm

FIGURE 13-21

Forces on the wrist pin of the single-cylinder engine at various speeds components are so small in comparison to the gas force components, which dominate at low speed. Only half the circumference of either pin sees any force. At 3400 rpm (Figure 13-23b, p. 694), the inertia force effects are evident and the angular portions of the main pin and crank pin that see any force are now only 39° and 72°, respectively. The effects of the gas force create asymmetry of the force hodographs about the x' axis. The differences between the main pin and crank pin forces are due to the different mass terms in their equations (e.g., compare equations 13.22 and 13.23, p. 691).

In Figure 13-23c the engine is at redline (6000 rpm) and the inertia force components are now dominant, raising the peak force levels and making the hodographs nearly symmetrical about the x' axis. The angular portions of main pin and crank pin that see any force are now reduced to 30° and 54°, respectively. This force distribution causes crank pins to wear only on a portion of their circumference. As shown in the next section, crank balancing affects the force distribution on the main pins.

Note that the numerical values of force and torque in the figures of this chapter are unique to the arbitrary choice of engine parameters used for their example engine and should not be extrapolated to any other engine design. Also, the gas force function used in program ENGINE to generate the figures is both approximate and invariant with engine speed, unlike in a real engine. Use the equations of this chapter to calculate forces and torques using mass, geometry, and gas force data appropriate to your particular engine design.

13.10 BALANCING THE SINGLE-CYLINDER ENGINE

The derivations and figures in the preceding sections have shown that significant forces are developed both on the pivot pins and on the ground plane due to the gas forces and the inertia and shaking forces. Balancing will not have any effect on the gas forces, which are internal, but it can have a dramatic effect on the inertia and shaking forces. The main pin force can be reduced, but the crank-pin and wrist-pin forces will be unaffected by any crankshaft balancing done. Figure 13-13 (p. 682) shows the unbalanced shaking force as felt on the ground plane of our 0.4-liter single-cylinder example engine from program ENGINE. It is about 9700 lb even at the moderate speed of 3400 rpm. At 6000 rpm it increases to over 30 000 lb. The methods of Chapter 12 can be applied to this mechanism to balance the members in pure rotation and reduce these large shaking forces.

Figure 13-24a (p. 695) shows the dynamic model of our slider-crank with the conrod mass lumped at both crank pin A and wrist pin B. We can consider this single-cylinder engine to be a single-plane device, thus suitable for static balancing (see Section 13.1, p. 662). It is straightforward to statically balance the crank. We need a balance mass at some radius, 180° from the lumped mass at point A whose mr product is equal to the product of the mass at A and its radius r. Applying equation 12.2 (p. 631) to this simple problem, we get:

$$m_{bal}\mathbf{R}_{bal} = -m_A\mathbf{R}_A \qquad (13.24)$$

Any combination of mass and radius that gives this product, placed at 180° from point A, will balance the crank. For simplicity of example, we will use a balance radius equal to r. Then a mass equal to m_A placed at A' will exactly balance the rotating masses. The CG of the crank will then be at the fixed pivot O_2 as shown in Figure 13-24a. In a

Force

Speed

0

Angle 2π

(a) Crank-pin inertia force

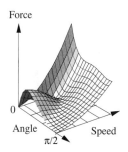

Force

0

Angle   Speed

$\pi/2$

(b) Crank-pin total force

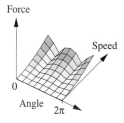

Force

Speed

0

Angle 2π

(c) Wrist-pin inertia force

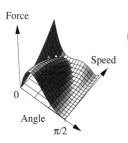

Force

Speed

0

Angle

$\pi/2$

(d) Wrist-pin total force

FIGURE 13-22

Pin force variation

13

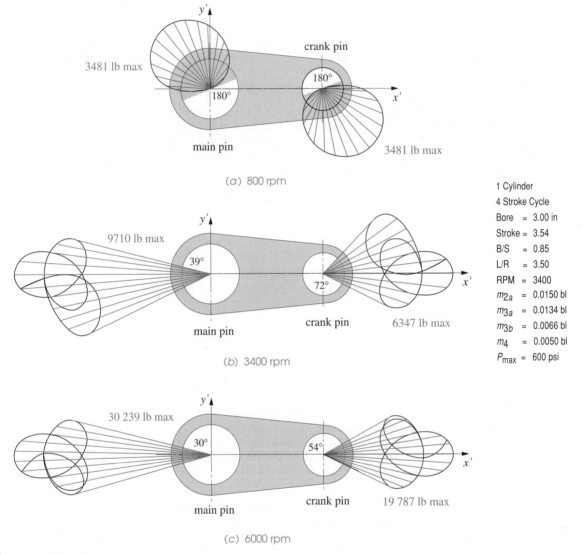

crank pin

3481 lb max

180°

180°

main pin

3481 lb max

x'

y'

(*a*) 800 rpm

9710 lb max

39°

72°

crank pin   6347 lb max

main pin

(*b*) 3400 rpm

30 239 lb max

30°

54°

crank pin   19 787 lb max

main pin

(*c*) 6000 rpm

1 Cylinder
4 Stroke Cycle
Bore   = 3.00 in
Stroke = 3.54
B/S   = 0.85
L/R   = 3.50
RPM   = 3400
m_{2a} = 0.0150 bl
m_{3a} = 0.0134 bl
m_{3b} = 0.0066 bl
m_4 = 0.0050 bl
P_{max} = 600 psi

FIGURE 13-23

Hodographs of dynamic forces on main pin and crank pin of an unbalanced, one-cylinder, four-stroke engine running at various speeds

real crankshaft, actually placing the counterweight *CG* at this large a radius will not work. The balance mass has to be kept close to the centerline to clear the piston at BDC. Figure 13-2c (p. 662) shows the shape of typical crankshaft counterweights.

Figure 13-25a shows the shaking force from the same engine as in Figure 13-13 (p. 682) after the crank has been exactly balanced in this manner. The *Y* component of the shaking force has been reduced to zero and the *X* component to 3343 lb at 3400 rpm. This is a factor of three reduction over the unbalanced engine. Note that the only source of *Y*-directed inertia force is the rotating mass at point *A* of Figure 13-24 (see equations

(a) Crank exactly balanced

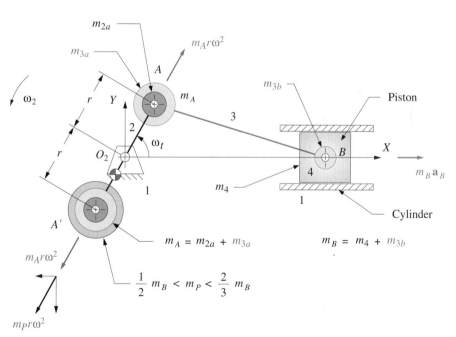

13

(b) Crank overbalanced

FIGURE 13-24

Balancing and overbalancing the single-cylinder engine

13.14, p. 679). What remains after balancing the rotating mass is the force due to the acceleration of the piston and conrod masses at point B of Figure 13-24 which are in linear translation along the X axis, as shown by the inertia force $-m_B a_B$ at point B in that figure.

To completely eliminate this reciprocating unbalanced shaking force would require the introduction of another reciprocating mass, which oscillates 180° out of phase with the piston. Adding a second piston and cylinder, properly arranged, can accomplish this. One of the principal advantages of multicylinder engines is their ability to reduce or eliminate the shaking forces. We will investigate this in the next chapter.

In the single-cylinder engine, there is no way to completely eliminate the reciprocating unbalance with a single, rotating counterweight, but we can reduce the shaking force still further. Figure 13-24b shows an additional amount of mass m_P added to the counterweight at point A'. (Note that the crank's CG has now moved away from the fixed pivot.) This extra balance mass creates an additional inertia force $(-m_P r \omega^2)$ as is shown, broken into X and Y components, in the figure. The Y component is not opposed by any other inertia forces present, but the X component will always be opposite to the reciprocating inertia force at point B. Thus this extra mass, m_P, which *overbalances the crank*, will reduce the X-directed shaking force at the expense of adding back some Y-directed shaking force. This is a useful trade-off as the direction of the shaking force is usually of less concern than is its magnitude. Shaking forces create vibrations in the supporting structure which are transmitted through it and modified by it. As an example, it is unlikely that you could define the direction of a motorcycle engine's shaking forces by feeling their resultant vibrations in the handlebars. But you **will** detect an increase in the magnitude of the shaking forces from the larger amplitude of vibration they cause in the cycle frame.

The correct amount of additional "overbalance" mass needed to minimize the peak shaking force, regardless of its direction, will vary with the particular engine design. It will usually be between one-half and two-thirds of the reciprocating mass at point B

| 1 Cylinder |
| 4 Stroke Cycle |
| Bore = 3.00 in |
| Stroke = 3.54 |
| B/S = 0.85 |
| L/R = 3.50 |
| RPM = 3400 |
| m_A = 0.0284 bl |
| m_B = 0.0116 bl |

(a) Crank exactly balanced   (b) Crank overbalanced

FIGURE 13-25

Effects of balancing and overbalancing on the shaking force in the slider-crank linkage (piston plus conrod at wrist pin), if placed at the crank radius r. Of course, once this mass-radius product is determined, it can be achieved with any combination of mass and radius. Figure 13-25b shows the minimum shaking force achieved for this engine with the addition of 65.5% of the mass at B acting at radius r. The shaking force has now been reduced to 1684 lb at 3400 rpm, which is **17% of its original unbalanced value** of 9710 lb. The benefits of balancing, and of overbalancing in the case of the single-cylinder engine, should now be obvious.

Effect of Crankshaft Balancing on Pin Forces

Of the pin forces, only the main pin force is affected by the addition of balance mass to the crankshaft. This is because its equation (13.23, p. 691) is the only one of the pin-force equations (13.20 to 13.23) that contains the mass of the crank. Table 13-3 shows the magnitudes of the shaking forces (in GCS) and main pin forces (in LCRS) for the single-cylinder example engine of Figure 13-23 (p. 694) at three engine speeds and under three conditions of balance: unbalanced, exactly balanced with a counterweight mass equal to the total mass m_A at the crank pin (Figure 13-25a), and overbalanced with the mass needed to minimize the single-cylinder shaking force (Figure 13-25b). Note that both balancing and overbalancing reduce the main pin force, though to a lesser degree than they reduce the shaking force in some cases. At idle speed the gas force far exceeds the inertia force and, since balancing can only affect the latter, the reduction in main pin force is less at idle speed than at higher engine speeds. The main pin forces in the overbalanced case most closely track the shaking forces at the redline speed where inertia force dominates gas force. Note that overbalancing the crank reduces the main pin force below that of the exact balancing case at all speeds.[*]

Figure 13-26 shows the effect of balancing and overbalancing on the main pin force magnitude and its distribution. Not only is the peak unbalanced main pin force (Figure 13-26a) nearly 3 times the magnitude of the exactly balanced case (Figure 13-26b) but the forces in the unbalanced case are concentrated over a small portion of the pin circumference. (See also Figure 13-23, p. 694.) The exactly balanced crankshaft has its main pin force distributed over more than half its circumference and the overbalanced crankshaft puts the force completely around the pin circumference as shown in Figure 13-26c.

13.11 DESIGN TRADE-OFFS AND RATIOS

In the design of any system or device, no matter how simple, there will always be conflicting demands, requirements, or desires which must be traded off to achieve the best design

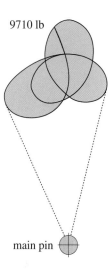

9710 lb

main pin

(a) Unbalanced

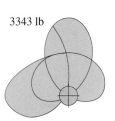

3343 lb

(b) Exact balance

2868 lb

(c) Overbalanced

13

FIGURE 13-26

Force on main pin at 3400 rpm with different crank balance states, shown at same scale in local rotating coordinates (LRCS)

* See footnote next page.

TABLE 13-3   Effect of Crank Balance Mass on Shaking Force and Mainpin Force

Balance Mode	Peak Shaking Force Mag. (lb)			Peak Mainpin Force Mag. (lb)		
	Idle	Midrange	Redline	Idle	Midrange	Redline
Unbalanced	538	9710	30 239	3481	9710	30 239
Exact balance	185	3343	10 412	4095	3343	10 412
Overbalanced	33	1684	5246	3675	2868	5886

compromise. This single-cylinder engine is no exception. There are two dimensionless design ratios which can be used to characterize an engine's dynamic behavior in a general way. The first is the crank/conrod ratio r/l, introduced in Section 13.2 (p. 667), or its inverse, the **conrod/crank ratio** l/r. The second is the **bore/stroke ratio** B/S.

Conrod/Crank Ratio

The crank/conrod ratio r/l appears in all the equations for acceleration, forces, and torques. In general, the smaller the r/l ratio, the smoother will be the acceleration function and thus all other factors which it influences. Program ENGINE uses the inverse of this ratio as an input parameter. The **conrod/crank ratio** l/r must be greater than about two to obtain acceptable transmission angles in the slider-crank linkage. The ideal value for l/r from a kinematic standpoint would be infinity as that would result in the piston acceleration function being a pure harmonic. The second and all subsequent harmonic terms in equations 13.3 (p. 671) would be zero in this event, and the peak value of acceleration would be a minimum. However, an engine this tall would not package very well, and package considerations often dictate the maximum value of the l/r ratio. Most engines will have an l/r ratio between three and five which values give acceptable smoothness in a reasonably short engine.

Bore/Stroke Ratio

The bore B of the cylinder is essentially equal to the diameter of the piston. (There is a small clearance.) The stroke S is defined as the distance travelled by the piston from TDC to BDC and is twice the crank radius, $S = 2r$. The bore appears in the equation for gas force (equation 13.4, p. 673) and thus also affects gas torque. The crank radius appears in every equation. An engine with a B/S ratio of 1 is referred to as a "square" engine. If B/S is larger than 1, it is "oversquare"; if less, "undersquare." The designer's choice of this ratio can have a significant effect on the dynamic behavior of the engine. Assuming that the displacement, or stroke volume V, of the engine has been chosen and is to remain constant, this displacement can be achieved with an infinity of combinations of bore and stroke ranging from a "pancake" piston with tiny stroke to a "pencil" piston with very long stroke.

$$V = \frac{\pi B^2}{4} S \qquad (13.25)$$

There is a classic design trade-off here between B and S for a constant stroke volume V. A large bore and small stroke will result in high gas forces which will affect pin forces adversely. A large stroke and small bore will result in high inertia forces which will affect pin forces (as well as other forces and torques) adversely. So there should be an optimum value for the B/S ratio in each case, which will minimize these adverse effects. Most production engines have B/S ratios in the range of about 0.75 to 1.5.

It is left as an exercise for the reader to investigate the effects of variation in the B/S and l/r ratios on forces and torques in the system. Program ENGINE will demonstrate the effects of changes made independently to each of these ratios, while all other design parameters are held constant. The reader is encouraged to experiment with the program to gain insight into the role of these ratios in the dynamic performance of the engine.

Materials

There will always be a strength/weight trade-off. The forces in this device can be quite high, due both to the explosion and to the inertia of the moving elements. We would like to keep the masses of the parts as low as possible as the accelerations are typically very high, as seen in Figure 13-8c (p. 672). But the parts must be strong enough to withstand the forces, so materials with good strength-to-weight ratios are needed. Pistons are usually made of an aluminum alloy, either cast or forged. Conrods are most often cast ductile iron or forged steel, except in very small engines (lawn mower, chain saw, motorcycle) where they may be aluminum alloy. Some high-performance engines (e.g. Acura NSX) have titanium connecting rods. Crankshafts are usually forged steel or cast ductile iron, and wrist pins are of hardened steel tubing or rod. Plain bearings of a special soft, nonferrous metal alloy called babbitt are usually used. In the four-stroke engine these are pressure lubricated with oil pumped through drilled passageways in the block, crankshaft, and connecting rods. In the two-stroke engine, the fuel carries the lubricant to these parts. Engine blocks are cast iron or cast aluminum alloy. The chrome-plated steel piston rings seal and wear well against gray cast iron cylinders. Most aluminum blocks are fitted with cast iron liners around the cylinder bores. Some are unlined and made of a high-silicon aluminum alloy which is specially cooled after casting to precipitate the hard silicon in the cylinder walls for wear resistance.

13.12 BIBLIOGRAPHY

1 **Heywood, J. B.** (1988). *Internal Combustion Engine Fundamentals.* McGraw-Hill: New York.

2 **Taylor, C. F.** (1966). *The Internal Combustion Engine in Theory and Practice.* MIT Press: Cambridge, MA.

3 **Heisler, H.** (1999). *Vehicle and Engine Technology, 2ed.* SAE: Warrendale, PA.

13.13 PROBLEMS‡

*†13-1 A slider-crank linkage has $r = 3$ and $l = 12$, $\omega = 200$ rad/sec at time $t = 0$. Its initial crank angle is zero. Calculate the piston acceleration at $t = 1$ sec. Use two methods, the exact solution, and the approximate Fourier series solution, and compare the results.

†13-2 Repeat Problem 13-1 for $r = 4$ and $l = 15$ and $t = 0.9$ sec.

*†13-3 A slider-crank linkage has $r = 3$ and $l = 12$, and a piston bore $B = 2$. The peak gas pressure in the cylinder occurs at a crank angle of $10°$ and is 1000 pressure units. Calculate the gas force and gas torque at this position.

†13-4 A slider-crank linkage has $r = 4$ and $l = 15$, and a piston bore $B = 3$. The peak gas pressure in the cylinder occurs at a crank angle of $5°$ and is 600 pressure units. Calculate the gas force and gas torque at this position.

*†13-5 Repeat Problem 13-3 using the exact method of calculation of gas torque and compare its result to that obtained by the approximate equation 13.8b (p. 675). What is the % error?

†13-6 Repeat Problem 13-4 using the exact method of calculation of gas torque and compare its result to that obtained by the approximate equation 13.8b (p. 675). What is the % error?

TABLE P13-0
Topic/Problem Matrix

13.2 Slider-Crank Kinematics

13-1, 13-2, 13-34, 13-35, 13-36, 13-37

13.3 Gas Force and Gas Torque

13-3, 13-4, 13-5, 13-6, 13-38, 13-39, 13-40, 13-41, 13-42

13.4 Equivalent Masses

13-7, 13-8, 13-9, 13-10, 13-43, 13-44, 13-45, 13-46

13.6 Inertia and Shaking Torques

13-11, 13-12, 13-13, 13-14, 13-47, 13-48, 13-49, 13-50

13.9 Pin Forces

13-15, 13-16, 13-17, 13-18, 13-23, 13-24, 13-25, 13-26, 13-27, 13-28, 13-33, 13-51, 13-52, 13-53, 13-54

13.10 Balancing the Single-Cylinder Engine

13-19, 13-20, 13-21, 13-22, 13-29, 13-30, 13-31, 13-32, 13-55, 13-56, 13-57, 13-58

---

‡ All problem figures are provided as PDF files, and some are also provided as animated Working Model files; all are on the DVD. PDF filenames are the same as the figure number. Run the file *Animations.html* to access and run the animations.

---

* Answers in Appendix F.

---

† These problems are suited to solution using *Mathcad, Matlab,* or *TKSolver* equation solver programs.

13

*†13-7   A connecting rod of length $l = 12$ has a mass $m_3 = 0.020$. Its mass moment of inertia is 0.620. Its CG is located at $0.4l$ from the crank pin, point A.

   a.   Calculate an exact dynamic model using two lumped masses, one at the wrist pin, point B, and one at whatever other point is required. Define the lumped masses and their locations.

   b.   Calculate an approximate dynamic model using two lumped masses, one at the wrist pin, point B, and one at the crank pin, point A. Define the lumped masses and their locations.

   c.   Calculate the error in the mass moment of inertia of the approximate model as a percentage of the original mass moment of inertia.

†13-8   Repeat Problem 13-7 for these data: $l = 15$, mass $m_3 = 0.025$, mass moment of inertia is 1.020. Its CG is located at $0.25l$ from the crank pin, point A.

*†13-9   A crank of length $r = 3.5$ has a mass $m_2 = 0.060$. Its mass moment of inertia about its pivot is 0.300. Its CG is at $0.30r$ from the main pin, O_2. Calculate a statically equivalent two-lumped mass dynamic model with the lumps placed at the main pin and crank pin. What is the percent error in the model's moment of inertia about the crank pivot?

†13-10   Repeat Problem 13-9 for a crank length $r = 4$, a mass $m_2 = 0.050$, a mass moment of inertia about its pivot of 0.400. Its CG is located at $0.40r$ from the main pin, point O_2.

*†13-11   Combine the data from problems 13-7 and 13-9. Run the linkage at a constant 2000 rpm. Calculate the inertia force and inertia torque at $\omega t = 45°$. Piston mass = 0.012.

†13-12   Combine the data from problems 13-7 and 13-10. Run the linkage at a constant 3000 rpm. Calculate the inertia force and inertia torque at $\omega t = 30°$. Piston mass = 0.019.

†13-13   Combine the data from problems 13-8 and 13-9. Run the linkage at a constant 2500 rpm. Calculate the inertia force and inertia torque at $\omega t = 24°$. Piston mass = 0.023.

*†13-14   Combine the data from problems 13-8 and 13-10. Run the linkage at a constant 4000 rpm. Calculate the inertia force and inertia torque at $\omega t = 18°$. Piston mass = 0.015.

†13-15   Combine the data from problems 13-7 and 13-9. Run the linkage at a constant 2000 rpm. Calculate the pin forces at $\omega t = 45°$. Piston mass = 0.022. $F_g = 300$.

†13-16   Combine the data from problems 13-7 and 13-10. Run the linkage at a constant 3000 rpm. Calculate the pin forces at $\omega t = 30°$. Piston mass = 0.019. $F_g = 600$.

†13-17   Combine the data from problems 13-8 and 13-9. Run the linkage at a constant 2500 rpm. Calculate the pin forces at $\omega t = 24°$. Piston mass = 0.032. $F_g = 900$.

†13-18   Combine the data from problems 13-8 and 13-10. Run the linkage at a constant 4000 rpm. Calculate the pin forces at $\omega t = 18°$. Piston mass = 0.014. $F_g = 1200$.

*†‡13-19   Using the data from Problem 13-11:

   a.   Exactly balance the crank and recalculate the inertia force.

   b.   Overbalance the crank with approximately two-thirds of the mass at the wrist pin placed at radius $-r$ on the crank and recalculate the inertia force.

   c.   Compare these results to those for the unbalanced crank.

†‡13-20   Repeat Problem 13-19 using the data from Problem 13-12.

* Answers in Appendix F.

† These problems are suited to solution using *Mathcad, Matlab,* or *TKSolver* equation solver programs.

‡ These problems are suited to solution using program ENGINE which is on the attached DVD.

†‡13-21   Repeat Problem 13-19 using the data from Problem 13-13.

†‡13-22   Repeat Problem 13-19 using the data from Problem 13-14.

†13-23   Combine the necessary equations to develop expressions that show how each of these dynamic parameters varies as a function of the crank/conrod ratio alone:

   a.   Piston acceleration   b.   Inertia force
   c.   Inertia torque   d.   Pin forces

   Plot the functions. Check your conclusions with program ENGINE.

   Hint: *Consider all other parameters to be temporarily constant. Set the crank angle to some value such that the gas force is nonzero.*

†13-24   Combine the necessary equations to develop expressions that show how each of these dynamic parameters varies as a function of the bore/stroke ratio alone:

   a.   Gas force   b.   Gas torque   c.   Inertia force
   d.   Inertia torque   e.   Pin forces

   Plot the functions. Check your conclusions with program ENGINE.

   Hint: *Consider all other parameters to be temporarily constant. Set the crank angle to some value such that the gas force is nonzero.*

†13-25   Develop an expression to determine the optimum bore/stroke ratio to minimize the wrist-pin force. Plot the function.

†‡13-26   Use program ENGINE, your own computer program, or an equation solver to calculate the maximum value and the polar-plot shape of the force on the main pin of a 1-in³ displacement, single-cylinder engine with bore = 1.12838 in for the following situations.

   a.   Piston, conrod, and crank masses = 0
   b.   Piston mass = 1 blob, conrod and crank masses = 0
   c.   Conrod mass = 1 blob, piston and crank masses = 0
   d.   Crank mass = 1 blob, conrod and piston masses = 0

   Place the *CG* of the crank at $0.5r$ and the conrod at $0.33l$. Compare and explain the differences in the main pin force under these different conditions with reference to the governing equations.

†13-27   Repeat Problem 13-26 for the crank pin.

†13-28   Repeat Problem 13-26 for the wrist pin.

†‡13-29   Use program ENGINE, your own computer program, or an equation solver to calculate the maximum value and the polar-plot shape of the force on the main pin of a 1-in³ single-cylinder engine with bore = 1.12838 in for the following situations.

   a.   Engine unbalanced.
   b.   Crank exactly balanced against mass at crank pin.
   c.   Crank optimally overbalanced against masses at crank pin and wrist pin.

   Piston, conrod, and crank masses = 1. Place the *CG* of the crank at $0.5r$ and the conrod at $0.33l$. Compare and explain the differences in the main pin force under these conditions with reference to the governing equations.

†‡13-30   Repeat Problem 13-29 for the crankpin force.

13

---

† These problems are suited to solution using *Mathcad, Matlab,* or *TKSolver* equation solver programs.

---

‡ These problems are suited to solution using program ENGINE which is on the attached DVD.

†‡13-31 Repeat Problem 13-29 for the wrist pin force.

†‡13-32 Repeat Problem 13-29 for the shaking force.

†13-33 Figure P13-1 shows a single-cylinder air compressor stopped at top dead center (TDC). There is a static pressure $P = 100$ psi trapped in the 3-in-bore cylinder. The entire assembly weighs 30 lb. Draw the necessary free-body diagrams to determine the forces at points A, B, C, and the supports R_1 and R_2, which are symmetrically located about the piston centerline. Assume that the piston remains stationary.

†13-34 Calculate and plot the position, velocity, and acceleration of a slider-crank linkage with $r = 3$, $l = 12$, and $\omega = 200$ rad/sec over one cycle using the exact solution and the approximate Fourier series solution. Also, calculate and plot the percent difference between the exact and approximate solutions for acceleration.

†13-35 Repeat Problem 13-34 for $r = 3$, $l = 15$, and $\omega = 100$ rad/sec.

†13-36 A slider-crank linkage has $r = 3$, $l = 9$. It has an angular velocity of 100 rad/sec at time $t = 0$. Its initial crank angle is zero. Calculate the piston acceleration at $t = 0.01$ sec. Compare the exact solution to the approximate Fourier series solution.

†13-37 Repeat Problem 13-36 with $r = 3$, $l = 15$, and $t = 0.02$.

†13-38 The following equation is an approximation of the gas force over 180° of crank angle.

$$F_g = \begin{cases} F_{g_{max}} \sin\left[(\omega t/\beta)(\pi/2)\right], & 0 \le \omega t \le \beta \\ F_{g_{max}} \left\{1 + \cos\left[\pi(\omega t - \beta)/(\pi - \beta)\right]\right\}\big/2, & \beta < \omega t \le \pi \end{cases}$$

Using this equation with $\beta = 15°$ and $F_{g_{max}} = 1200$ lb, calculate and plot the approximate gas torque for $r = 4$ in and $l = 12$ in. What is the total energy delivered by the gas force over the 180° of motion? What is the average power delivered if the crank rotates at a constant speed of 1500 rpm?

†13-39 A slider-crank linkage has $r = 3.75$, $l = 11$ and a piston bore of $B = 2.5$. The peak gas pressure in the cylinder occurs at a crank angle of 12° and is 1150 pressure units. Calculate the gas force and gas torque at this position.

†13-40 Repeat Problem 13-39 using the exact method of calculation of the gas torque and compare the result to that obtained by the approximate expression in equation 13.8b (p. 675). What is the percent error?

†13-41 A slider-crank linkage has $r = 4.12$, $l = 14.5$ and a piston bore of $B = 2.25$. The peak gas pressure in the cylinder occurs at a crank angle of 9° and is 1325 pressure units. Calculate the gas force and gas torque at this position.

†13-42 Repeat Problem 13-41 using the exact method of calculation of the gas torque and compare the result to that obtained by the approximate expression in equation 13.8b (p. 675). What is the percent error?

†13-43 A slider-crank linkage has crank $m_2 = 0.045$, crank $r_{G_2} = 0.4r$, conrod $m_3 = 0.12$, and piston $m_4 = 0.15$. Determine the approximately dynamically equivalent two-mass lumped parameter model for this linkage with the masses placed at the crank and wrist pins.

†13-44 If the conrod in Problem 13-43 has $l = 12.5$, $I_{G_3} = 0.15$ and its CG is located 4.5 units from point A; calculate the sizes of two dynamically equivalent masses and the location of one if the other is placed at point B (see Figure 13-10, p. 678).

† These problems are suited to solution using *Mathcad*, *Matlab*, or *TKSolver* equation solver programs.

‡ These problems are suited to solution using program ENGINE which is on the attached DVD.

†13-45 A slider-crank linkage has crank $m_2 = 0.060$, crank $r_{G_2} = 0.38r$, conrod $m_3 = 0.18$, and piston $m_4 = 0.16$. Determine the approximately dynamically equivalent two-mass lumped parameter model for this linkage with the masses placed at the crank and wrist pins.

†13-46 If the conrod in Problem 13-45 has $l = 10.4$, $I_{G_3} = 0.12$ and its CG is located 4.16 units from point A; calculate the sizes of two dynamically equivalent masses and the location of one if the other is placed at point B (see Figure 13-10, p. 678).

†13-47 A slider-crank linkage has $r = 3.13$, $l = 12.5$, crank $m_2 = 0.045$, crank $r_{G_2} = 0.4r$, conrod $m_3 = 0.12$, conrod $r_{G_3} = 0.36l$, and piston $m_4 = 0.15$. Crank $\omega = 1800$ rpm. Calculate the inertia force and inertia torque for a crank position of $\omega t = 30°$.

†13-48 A slider-crank linkage has $r = 2.6$, $l = 10.4$, crank $m_2 = 0.060$, crank $r_{G_2} = 0.38r$, conrod $m_3 = 0.18$, conrod $r_{G_3} = 0.4l$, and piston $m_4 = 0.16$. Crank $\omega = 1850$ rpm. Calculate the inertia force and inertia torque for a crank position of $\omega t = 20°$.

†13-49 A slider-crank linkage has $r = 2.6$, $l = 10.4$, crank $m_2 = 0.045$, crank $r_{G_2} = 0.4r$, conrod $m_3 = 0.12$, conrod $r_{G_3} = 0.36l$, and piston $m_4 = 0.15$. Crank $\omega = 2000$ rpm. Calculate the inertia force and inertia torque for a crank position of $\omega t = 25°$.

†13-50 A slider-crank linkage has $r = 3.13$, $l = 12.5$, crank $m_2 = 0.060$, crank $r_{G_2} = 0.38r$, conrod $m_3 = 0.18$, conrod $r_{G_3} = 0.4l$, and piston $m_4 = 0.15$. Crank $\omega = 1500$ rpm. Calculate the inertia force and inertia torque for a crank position of $\omega t = 22°$.

†13-51 A slider-crank linkage has $r = 3.13$, $l = 12.5$, crank $m_2 = 0.045$, crank $r_{G_2} = 0.4r$, conrod $m_3 = 0.12$, conrod $r_{G_3} = 0.36l$, and piston $m_4 = 0.15$. Crank $\omega = 1800$ rpm. Calculate the pin forces for a crank position of $\omega t = 30°$ and a gas force of $F_g = 450$.

†13-52 A slider-crank linkage has $r = 2.6$, $l = 10.4$, crank $m_2 = 0.060$, crank $r_{G_2} = 0.38r$, conrod $m_3 = 0.18$, conrod $r_{G_3} = 0.4l$, and piston $m_4 = 0.16$. Crank $\omega = 1850$ rpm. Calculate the pin forces for a crank position of $\omega t = 20°$ and a gas force of $F_g = 600$.

†13-53 A slider-crank linkage has $r = 2.6$, $l = 10.4$, crank $m_2 = 0.045$, crank $r_{G_2} = 0.4r$, conrod $m_3 = 0.12$, conrod $r_{G_3} = 0.36l$, and piston $m_4 = 0.15$. Crank $\omega = 2000$ rpm. Calculate the pin forces for a crank position of $\omega t = 25°$ and a gas force of $F_g = 350$.

†13-54 A slider-crank linkage has $r = 3.13$, $l = 12.5$, crank $m_2 = 0.060$, crank $r_{G_2} = 0.38r$, conrod $m_3 = 0.18$, conrod $r_{G_3} = 0.4l$, and piston $m_4 = 0.15$. Crank $\omega = 1500$ rpm. Calculate the pin forces for a crank position of $\omega t = 22°$ and a gas force of $F_g = 550$.

†‡13-55 Using the data from Problem 13-47:

    a.   Exactly balance the crank and recalculate the inertia force.
    b.   Overbalance the crank with approximately two-thirds of the mass at the wrist pin placed at radius $-r$ on the crank and recalculate the inertia force.
    c.   Compare these results to those for the unbalanced crank.

†‡13-56 Using the data from Problem 13-48:

    a.   Exactly balance the crank and recalculate the inertia force.
    b.   Overbalance the crank with approximately two-thirds of the mass at the wrist pin placed at radius $-r$ on the crank and recalculate the inertia force.
    c.   Compare these results to those for the unbalanced crank.

†‡13-57 Using the data from Problem 13-49:

    a.   Exactly balance the crank and recalculate the inertia force.
    b.   Overbalance the crank with approximately two-thirds of the mass at the wrist pin placed at radius $-r$ on the crank and recalculate the inertia force.
    c.   Compare these results to those for the unbalanced crank.

13

† These problems are suited to solution using *Mathcad, Matlab,* or *TKSolver* equation solver programs.

‡ These problems are suited to solution using program ENGINE which is on the attached DVD.

†‡13-58   Using the data from Problem 13-50 (p. 703):

    a.   Exactly balance the crank and recalculate the inertia force.

    b.   Overbalance the crank with approximately two-thirds of the mass at the wrist pin placed at radius $-r$ on the crank and recalculate the inertia force. Compare these results to those for the unbalanced crank.

13.14 PROJECTS

These are loosely structured design problems intended for solution using program ENGINE. All involve the design of a single-cylinder engine and differ only in the specific data for the engine. The general problem statement is:

Design a single-cylinder engine for a specified displacement and stroke cycle. Optimize the conrod/ crank ratio and bore/stroke ratio to minimize shaking forces, shaking torque, and pin forces, also considering package size. Design your link shapes and calculate realistic dynamic parameters (mass, CG location, moment of inertia) for those links using the methods shown in Chapter 10 and Section 11.13 (p. 612). Dynamically model the links as described in this chapter. Balance or overbalance the linkage as needed to achieve these results. Overall smoothness of total torque is desired. Design and size a minimum weight flywheel by the method of Section 11.11 (p. 605) to smooth the total torque. Write an engineering report on your design.

P13-1   Two-stroke cycle with a displacement of 0.125 liter.

P13-2   Four-stroke cycle with a displacement of 0.125 liter.

P13-3   Two-stroke cycle with a displacement of 0.25 liter.

P13-4   Four-stroke cycle with a displacement of 0.25 liter.

P13-5   Two-stroke cycle with a displacement of 0.50 liter.

P13-6   Four-stroke cycle with a displacement of 0.50 liter.

P13-7   Two-stroke cycle with a displacement of 0.75 liter.

P13-8   Four-stroke cycle with a displacement of 0.75 liter.

† These problems are suited to solution using *Mathcad, Matlab,* or *TKSolver* equation solver programs.

‡ These problems are suited to solution using program ENGINE which is on the attached DVD.

13

Chapter 14

MULTICYLINDER ENGINES

*Look long on an engine,
it is sweet to the eyes*
MACKNIGHT BLACK

14.0 INTRODUCTION

The previous chapter discussed the design of the slider-crank mechanism as used in the single-cylinder internal combustion engine and piston pumps. We will now extend the design to multicylinder configurations. Some of the problems with shaking forces and torques can be alleviated by proper combination of multiple slider-crank linkages on a common crankshaft. Program ENGINE, included with this text, will calculate the equations derived in this chapter and allow the student to exercise many variations of an engine design in a short time. Some examples are provided as disk files to be read into the program. These are noted in the text. The student is encouraged to investigate these examples with program ENGINE in order to develop an understanding of and insight to the subtleties of this topic. A user manual for program ENGINE is provided on the DVD and context-sensitive help is available within the program. See Appendix A for more information on using the programs that come with the book.

As with the single-cylinder case, we will not address the thermodynamic aspects of the internal combustion engine beyond the definition of the combustion forces necessary to drive the device presented in the previous chapter. We will concentrate on the engine's kinematic and mechanical dynamics aspects. It is not our intention to make an "engine designer" of the reader so much as to apply dynamic principles to a realistic design problem of general interest and also to convey the complexity and fascination involved in the design of a more complicated dynamic device than the single-cylinder engine.

14

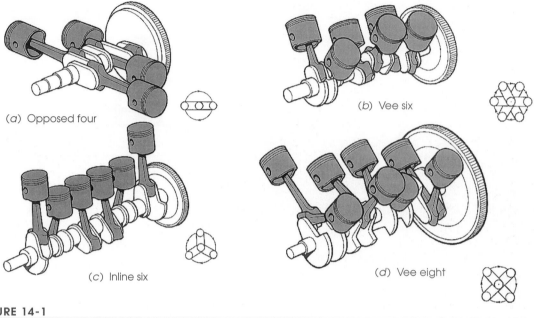

(a) Opposed four

(b) Vee six

(c) Inline six

(d) Vee eight

FIGURE 14-1

Various multicylinder engine configurations
Illustrations copyright Eaglemoss Publications/Car Care Magazine. Reprinted with permission.

FIGURE 14-2

Cutaway views of a four-stroke, four-cylinder inline engine *Courtesy of FIAT Corporation, Italy*

FIGURE 14-3

Crankshaft from an inline four-cylinder engine with pistons, connecting rods, and flywheel
Illustration copyright Eaglemoss Publications/Car Care Magazine. Reprinted with permission.

14.1 MULTICYLINDER ENGINE DESIGNS

Multicylinder engines are designed in a wide variety of configurations from the simple inline arrangement to vee, opposed, and radial arrangements some of which are shown in Figure 14-1. These arrangements may use any of the stroke cycles discussed in Chapter 13, Clerk, Otto, or diesel.

INLINE ENGINES   The most common and simplest arrangement is an inline engine with its cylinders all in a common plane as shown in Figure 14-2. Two-,[*] three-,[*] four-, five-, and six-cylinder **inline engines** are in common use. Each cylinder will have its individual slider-crank mechanism consisting of a crank, conrod, and piston. The cranks are formed together in a common **crankshaft** as shown in Figure 14-3. Each cylinder's crank on the crankshaft is referred to as a **crank throw**. These crank throws will be arranged with some **phase angle** relationship one to the other, in order to stagger the motions of the pistons in time. It should be apparent from the discussion of shaking forces and balancing in the previous chapter that we would like to have pistons moving in opposite directions to one another at the same time in order to cancel the reciprocating inertial forces. The optimum phase angle relationships between the crank throws will differ depending on the number of cylinders and the stroke cycle of the engine. There will usually be one or a small number of viable crank throw arrangements for a given engine configuration to accomplish this goal. The engine in Figure 14-2 is a four-stroke cycle, four-cylinder, inline engine with its crank throws at 0, 180, 180, and 0° phase angles which we will soon see are optimum for this engine. Figure 14-3 shows the crankshaft, connecting rods, and pistons for the same design of engine as in Figure 14-2.

14

_____

[*] Used mainly in motor-cycles and boats.

Engine power and torque curves

FIGURE 14-4

Cross section of a BMW 5-liter V-12 engine and its power and torque curves   *Courtesy of BMW of North America Inc.*

FIGURE 14-5

Cutaway view of a 5-liter BMW V-12 engine   *Courtesy of BMW of North America Inc.*

14

VEE ENGINES in two-,[*] four-,[*] six-, eight-, ten-,[†] and twelve-cylinder[‡] versions are produced, with vee-six and vee-eight being the most common configurations. Figure 14-4 shows a cross section and Figure 14-5 a cutaway of a 60° vee-twelve engine. **Vee engines** can be thought of as *two inline engines grafted together onto a common crankshaft*. The two "inline" portions, or **banks**, are arranged with some **vee angle** between them. Figure 14-1d (p. 706) shows a vee-eight engine. Its crank throws are at 0, 90, 270, and 180° respectively. A vee-eight's vee angle is 90°. The geometric arrangements of the crankshaft (phase angles) and cylinders (vee angle) have a significant effect on the dynamic condition of the engine. We will soon explore these relationships in detail.

OPPOSED ENGINES are essentially vee engines with a vee angle of 180°. The pistons in each bank are on opposite sides of the crankshaft as shown in Figure 14-6. This arrangement promotes cancellation of inertial forces and is popular in aircraft engines.[§] It has also been used in some automobile and motorcycle applications.[‖]

RADIAL ENGINES have their cylinders arranged radially around the crankshaft in nearly a common plane. These were common on World War II vintage aircraft as they allowed large displacements, and thus high power, in a compact form whose shape was well suited to that of an airplane. Typically air-cooled, the cylinder arrangement allowed good exposure of all cylinders to the airstream. Large versions had multiple rows of radial cylinders, rotationally staggered to allow cooling air to reach the back rows. The gas turbine jet engine has rendered these radial aircraft engines obsolete.

ROTARY ENGINES (Figure 14-7, p. 710) were an interesting variant on the aircraft radial engine and were used in World War I airplanes.[Δ] Although they were similar in appearance and cylinder arrangement to the radial engine, the anomaly was that the **crankshaft** was the stationary ground plane. The propeller was attached to the crankcase (engine block), which rotated around the stationary crankshaft! It is a kinematic inversion of the slider-crank. (See Figure 2-15b, p. 55.) One advantage is that the piston mass centers can be in pure rotation and so do not impart any vibration to the airframe. All seven connecting rods and pistons are in the same plane. One connecting rod (the "mother" rod) pivots on the crank pin and carries six "daughter" rods on it, as in the radial engine. At least it didn't need a flywheel.

[*] Mainly in motorcycles, and boats.

[†] Honda, Chrysler, Ford, Porsche.

[‡] BMW, Jaguar, Mercedes.

[§] Continental six-cylinder aircraft engine.

[‖] Original VW "Beetle" four cylinder, Subaru four and six, Honda motorcycle four and six, Ferrari twelve, Porsche six, the ill-fated Corvair six, and the short-lived Tucker (Continental) six, among others.

[Δ] Lubrication in the rotary engine was a problem. So-called "loss lubrication" was used, meaning that the lubricant (castor oil) passed through the engine and out the exhaust. This somewhat limited the time that the pilot in his open cockpit could endure inhaling the exhaust. The white silk scarf popular with these pilots was not used for warmth, but rather was placed over mouth and nose to reduce the ingestion of castor oil which is a laxative.

14

FIGURE 14-6

Chevrolet Corvair horizontally opposed six-cylinder engine
Courtesy of Chevrolet Division, General Motors Corp.

FIGURE 14-7

The Gnome rotary engine (circa 1915). Note the multiple connecting rods on the single stationary crank pin.[1]

Many other configurations of engines have been tried over the century of development of this ubiquitous device. The bibliography at the end of this chapter contains several references which describe other engine designs, the usual, unusual, and exotic. We will begin our detailed exploration of multicylinder engine design with the simplest configuration, the inline engine, and then progress to the vee and opposed versions.

14.2   THE CRANK PHASE DIAGRAM

Fundamental to the design of any multicylinder engine (or piston pump) is the arrangement of crank throws on the crankshaft. We will use the four-cylinder inline engine as an example. Many choices are possible for the crank phase angles in the four-cylinder engine. We will start, for example, with the one that seems most obvious from a commonsense standpoint. There are 360° in any crankshaft. We have four cylinders, so an arrangement of 0, 90, 180, and 270° seems appropriate. The **delta phase angle** $\Delta\phi$ between throws is then 90°. In general, for maximum cancellation of inertia forces, which have a period of one revolution, the optimum delta phase angle will be:

$$\Delta\phi_{\text{inertia}} = \frac{360°}{n} \tag{14.1}$$

where n is the number of cylinders.

We must establish some convention for the measurement of these phase angles which will be:

1   The first (front) cylinder will be number 1 and its phase angle will always be zero.
It is the reference cylinder for all others.

2   The phase angles of all other cylinders will be measured with respect to the crank throw for cylinder 1.

3   Phase angles are measured internal to the crankshaft, that is, with respect to a rotating coordinate system embedded in the first crank throw.

4   Cylinders will be numbered consecutively from front to back of the engine.

The phase angles are defined in a **crank phase diagram** as shown in Figure 14-8 (p. 712) for a four-cylinder, inline engine. Figure 14-8a shows the crankshaft with the throws numbered clockwise around the axis. The shaft is rotating counterclockwise. The pistons are oscillating horizontally in this diagram, along the x axis. Cylinder 1 is shown with its piston at top dead center (TDC). Taking that position as the starting point for the abscissas (thus time zero) in Figure 14-8b, we plot the velocity of each piston for two revolutions of the crank (to accommodate one complete four-stroke cycle). Piston 2 arrives at TDC 90° after piston 1 has left. Thus we say that cylinder 2 lags cylinder 1 by 90 degrees. By convention a *lagging event is defined as having a negative phase angle*, shown by the clockwise numbering of the crank throws. The velocity plots clearly show that each cylinder arrives at TDC (zero velocity) 90° later than the one before it. Negative velocity on the plots in Figure 14-8b indicates piston motion to the left (downstroke) in Figure 14-8a; positive velocity indicates motion to the right (up stroke).

For the discussion in this chapter we will assume counterclockwise rotation of all crankshafts, and all phase angles will thus be negative. We will, however, omit the negative signs on the listings of phase angles with the understanding that they follow this convention.

Figure 14-8 shows the timing of events in the cycle and is a necessary and useful aid in defining our crankshaft design. However, it is not necessary to go to the trouble of drawing the correct sinusoidal shapes of the velocity plots to obtain the needed information. All that is needed is a schematic indication of the relative positions within the cycle of the ups and downs of the various cylinders. This same information is conveyed by the simplified crank phase diagram shown in Figure 14-9 (p. 713). Here the piston motions are represented by rectangular blocks with a negative block arbitrarily used to denote a piston downstroke and a positive one a piston upstroke. It is strictly schematic. The positive and negative values of the blocks imply nothing more than that stated. Such a schematic crank phase diagram can (and should) be drawn for any proposed arrangement of crankshaft phase angles. To draw it, simply shift each cylinder's blocks to the right by its phase angle with respect to the first cylinder.

14.3   SHAKING FORCES IN INLINE ENGINES

We want to determine the overall shaking force which results from our chosen crankshaft phase angle arrangement. The individual cylinders will each contribute to the total shaking force. We can superpose their effects, taking their phase shifts into account. Equation 13.14e (p. 681) defined the shaking force for one cylinder with the crankshaft rotating at constant ω.

for $\alpha = 0$:

$$\mathbf{F_s} \cong \left[m_A r \omega^2 \cos \omega t + m_B r \omega^2 \left(\cos \omega t + \frac{r}{l} \cos 2\omega t \right) \right] \hat{\mathbf{i}} + \left[m_A r \omega^2 \sin \omega t \right] \hat{\mathbf{j}} \qquad (14.2a)$$

(a) Crankshaft phase angles

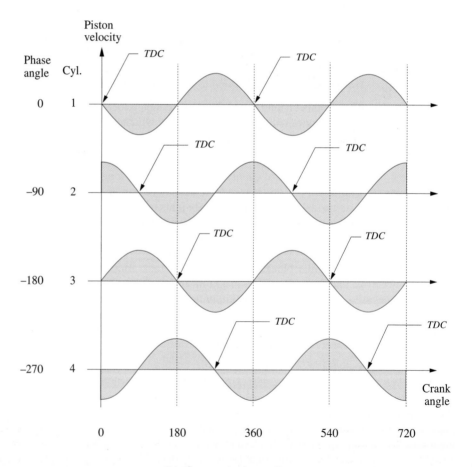

(b) The crank phase diagram

FIGURE 14-8

Crank phase angles and the phase diagram

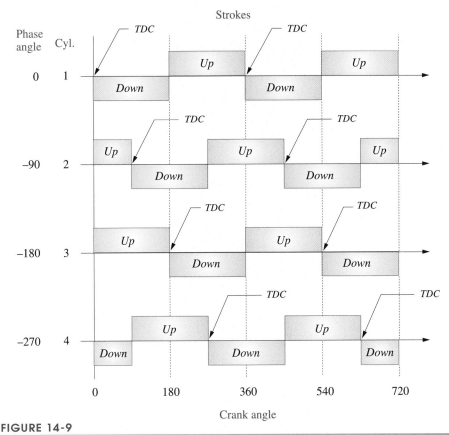

FIGURE 14-9

The schematic crank phase diagram

This expression is for an unbalanced crank. In multicylinder engines each crank throw on the crankshaft is at least counterweighted to eliminate the shaking force effects of the combined mass m_A of crank and conrod assumed concentrated at the crank pin. (See Section 13.10 and equation 13.24, p. 693.) The need for overbalancing is less if the crankshaft phase angles are arranged to cancel the effects of the reciprocating masses at the wrist pins. This inherent balance is possible in inline engines of three or more cylinders, but not in some two-cylinder inline engines. Sometimes the crank throws in an inherently balanced multicylinder engine are also overbalanced in order to reduce the main pin bearing forces as was described in Section 13.10 (p. 693).[*]

If we provide balance masses with an mr product equal to $m_A r_A$ on each crank throw as shown in Figure 14-3 (p. 707), the terms in equation 14.2a which include m_A will be eliminated, reducing it to:

$$\mathbf{F_s} \cong m_B r \omega^2 \left(\cos \omega t + \frac{r}{l} \cos 2\omega t \right) \hat{\mathbf{i}} \qquad (14.2b)$$

Recall that these are approximate expressions that exclude all harmonics above the second and also assume that each crank throw is exactly balanced, not under- or overbalanced.

[*] A 90° vee-eight engine typically has approximately $m_B/2$ of overbalance mass added per crank throw to reduce the main pin bearing forces.

14

We will assume that all cylinders in the engine are of equal displacement and that all pistons and all conrods are interchangeable. This is desirable both for dynamic balance and for lower production costs. If we let the crank angle ωt represent the instantaneous position of the reference crank throw for cylinder 1, the corresponding positions of the other cranks can be defined by their phase angles as shown in Figure 14-8 (p. 712). The total shaking force for a multicylinder inline engine is then:[*]

$$\mathbf{F_s} \cong m_B r\omega^2 \sum_{i=1}^{n}\left[\cos(\omega t - \phi_i) + \frac{r}{l}\cos 2(\omega t - \phi_i)\right]\hat{\mathbf{i}} \tag{14.2c}$$

where n = number of cylinders and $\phi_1 = 0$. Substitute the identity:

$$\cos(a-b) = \cos a \cos b + \sin a \sin b$$

and factor to:

$$\mathbf{F_s} \cong m_B r\omega^2 \left[\begin{array}{c} \cos\omega t \sum_{i=1}^{n}\cos\phi_i + \sin\omega t \sum_{i=1}^{n}\sin\phi_i \\ + \frac{r}{l}\left(\cos 2\omega t \sum_{i=1}^{n}\cos 2\phi_i + \sin 2\omega t \sum_{i=1}^{n}\sin 2\phi_i\right) \end{array}\right]\hat{\mathbf{i}} \tag{14.2d}$$

The ideal value for the shaking force is zero. This expression can only be zero for all values of ωt if:

$$\sum_{i=1}^{n}\cos\phi_i = 0 \qquad\qquad \sum_{i=1}^{n}\sin\phi_i = 0 \tag{14.3a}$$

$$\sum_{i=1}^{n}\cos 2\phi_i = 0 \qquad\qquad \sum_{i=1}^{n}\sin 2\phi_i = 0 \tag{14.3b}$$

Thus, there are some combinations of phase angles ϕ_i which will cause cancellation of the shaking force through the second harmonic. If we wish to cancel higher harmonics, we could reintroduce those harmonics' terms that were truncated from the Fourier series representation and find that the fourth and sixth harmonics will be cancelled if:

$$\sum_{i=1}^{n}\cos 4\phi_i = 0 \qquad\qquad \sum_{i=1}^{n}\sin 4\phi_i = 0 \tag{14.3c}$$

$$\sum_{i=1}^{n}\cos 6\phi_i = 0 \qquad\qquad \sum_{i=1}^{n}\sin 6\phi_i = 0 \tag{14.3d}$$

[*] The effect of overbalancing the crank throws is not included in equations 14.2c and 14.2d as shown. The crankshaft is assumed to be exactly balanced here. See Appendix G for the complete equations that include the effects of crank overbalance. Program ENGINE uses the equations from Appendix G to account for the effects of overbalancing in multicylinder engines.

Equations 14.3 provide us with a convenient predictor of the shaking force behavior of any proposed inline engine design. Program ENGINE calculates equations 14.3a and b and displays a table of their values. Note that **both the sine and cosine summations of any multiple of the phase angles must be zero for that harmonic of the shaking force to be zero.** The calculation for our example of a four-cylinder engine with phase angles of $\phi_1 = 0$, $\phi_2 = 90$, $\phi_3 = 180$, $\phi_4 = 270°$ in Table 14-1 shows that the shaking forces are zero for the first, second, and sixth harmonics and nonzero for the fourth. So, our commonsense choice in this instance has proven a good one as far as shaking forces are

TABLE 14-1   Force Balance State of a 4-Cylinder Inline Engine with a 0, 90, 180, 270° Crankshaft

Primary forces:	$\displaystyle\sum_{i=1}^{n}\sin\phi_i=0$	$\displaystyle\sum_{i=1}^{n}\cos\phi_i=0$
Secondary forces:	$\displaystyle\sum_{i=1}^{n}\sin 2\phi_i=0$	$\displaystyle\sum_{i=1}^{n}\cos 2\phi_i=0$
Fourth harmonic forces:	$\displaystyle\sum_{i=1}^{n}\sin 4\phi_i=0$	$\displaystyle\sum_{i=1}^{n}\cos 4\phi_i=4$
Sixth harmonic forces :	$\displaystyle\sum_{i=1}^{n}\sin 6\phi_i=0$	$\displaystyle\sum_{i=1}^{n}\cos 6\phi_i=0$

concerned. As was shown in equation 13.2f (p. 670), the coefficients of the fourth and sixth harmonic terms are minuscule, so their contributions, if any, can be ignored. The primary component is of most concern, because of its potential magnitude. The secondary (second harmonic) term is less critical than the primary as it is multiplied by r/l which is generally less than 1/3. An unbalanced secondary harmonic of shaking force is undesirable but can be lived with, especially if the engine is of small displacement (less than about 1/2 liter per cylinder).

To see more details on the results of this 0, 90, 180, 270° inline four-cylinder engine configuration, run program ENGINE, select the configuration from the *Example* pulldown menu, and then *Plot* the shaking force. See Appendix A for more detailed instructions on the use of program ENGINE.

14.4   INERTIA TORQUE IN INLINE ENGINES

The inertia torque for a single-cylinder engine was defined in Section 13.6 and equation 13.15e (p. 683). We are concerned with reducing this inertia torque, preferably to zero, because it combines with the gas torque to form the total torque. (See Section 13.7, p. 683.) Inertia torque adds nothing to the net driving torque as its average value is always zero, but it does create large oscillations in the total torque which detracts from its smoothness. Inertia torque oscillations can be masked to a degree with the addition of sufficient flywheel to the system, or their external, net effect can be cancelled by the proper choice of phase angles. However, the torque oscillations, even if hidden from the outside observer, or made to sum to zero, are still present within the crankshaft and can lead to torsional fatigue failure if the part is not properly designed. (See also Figure 14-23, p. 736.) The approximate one-cylinder inertia torque equation for three harmonics is:

$$\mathbf{T}_{i21} \cong \frac{1}{2}m_B r^2 \omega^2\left(\frac{r}{2l}\sin\omega t - \sin 2\omega t - \frac{3r}{2l}\sin 3\omega t\right)\hat{\mathbf{k}} \qquad (14.4a)$$

Summing for all cylinders and including their phase angles:

14

$$\mathbf{T}_{i21} \cong \frac{1}{2} m_B r^2 \omega^2 \sum_{i=1}^{n} \left[\frac{r}{2l} \sin\left(\omega t - \phi_i\right) - \sin 2\left(\omega t - \phi_i\right) - \frac{3r}{2l} \sin 3\left(\omega t - \phi_i\right) \right] \hat{\mathbf{k}} \qquad (14.4b)$$

Substitute the identity:

$$\sin(a-b) = \sin a \cos b - \cos a \sin b$$

and factor to:

$$\mathbf{T}_{i21} \cong \frac{1}{2} m_B r^2 \omega^2 \begin{bmatrix} \frac{r}{2l}\left(\sin\omega t \sum_{i=1}^{n}\cos\phi_i - \cos\omega t \sum_{i=1}^{n}\sin\phi_i \right) \\[2mm] -\left(\sin 2\omega t \sum_{i=1}^{n}\cos 2\phi_i - \cos 2\omega t \sum_{i=1}^{n}\sin 2\phi_i \right) \\[2mm] -\frac{3r}{2l}\left(\sin 3\omega t \sum_{i=1}^{n}\cos 3\phi_i - \cos 3\omega t \sum_{i=1}^{n}\sin 3\phi_i \right) \end{bmatrix} \hat{\mathbf{k}} \qquad (14.4c)$$

This can only be zero for all values of ωt if:

$$\sum_{i=1}^{n} \sin\phi_i = 0 \qquad\qquad \sum_{i=1}^{n} \cos\phi_i = 0 \qquad (14.5a)$$

$$\sum_{i=1}^{n} \sin 2\phi_i = 0 \qquad\qquad \sum_{i=1}^{n} \cos 2\phi_i = 0 \qquad (14.5b)$$

$$\sum_{i=1}^{n} \sin 3\phi_i = 0 \qquad\qquad \sum_{i=1}^{n} \cos 3\phi_i = 0 \qquad (14.5c)$$

Equations 14.5 provide us with a convenient predictor of the inertia torque behavior of any proposed inline engine design. Calculation for our example of a four-cylinder engine with phase angles of $\phi_1 = 0$, $\phi_2 = 90$, $\phi_3 = 180$, $\phi_4 = 270°$ shows that the inertia torque components are zero for the first, second, and third harmonics. So, our current example is a good one for inertia torques as well.

14.5 SHAKING MOMENT IN INLINE ENGINES

We were able to consider the single-cylinder engine to be a single-plane, or two-dimensional, device and thus could statically balance it. The multicylinder engine is three-dimensional. Its multiple cylinders are distributed along the axis of the crankshaft. Even though we may have cancellation of the shaking forces, there may still be unbalanced moments in the plane of the engine block. We need to apply the criteria for dynamic balance. (See Section 12.2 and equation 12.3, p. 634.) Figure 14-10 shows a schematic of an inline four-cylinder engine with crank phase angles of $\phi_1 = 0$, $\phi_2 = 90$, $\phi_3 = 180$, $\phi_4 = 270°$. The spacing between the cylinders is normally uniform. We can sum moments in the plane of the cylinders about any convenient point such as L at the centerline of the first cylinder:

14

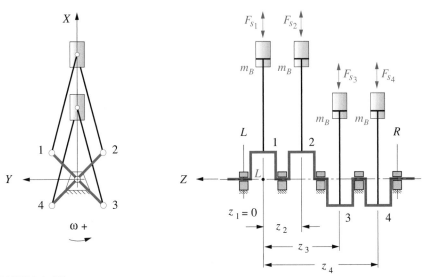

FIGURE 14-10

Moment arms of the shaking moment

$$\sum M_L = \sum_{i=1}^{n} z_i \mathbf{F}_{s_i} \; \hat{\mathbf{j}} \qquad (14.6a)$$

where $\mathbf{F}_{s_i}$ is the shaking force and z_i is the moment arm of the ith cylinder.* Substituting equation 14.2d (p. 714) for $\mathbf{F}_{s_i}$:

$$\sum M_L \doteq m_B r \omega^2 \left[\begin{array}{c} \cos \omega t \displaystyle\sum_{i=1}^{n} z_i \cos \phi_i + \sin \omega t \displaystyle\sum_{i=1}^{n} z_i \sin \phi_i \\[2em] + \dfrac{r}{l} \left(\cos 2\omega t \displaystyle\sum_{i=1}^{n} z_i \cos 2\phi_i + \sin 2\omega t \displaystyle\sum_{i=1}^{n} z_i \sin 2\phi_i \right) \end{array} \right] \hat{\mathbf{j}} \qquad (14.6b)$$

This expression can only be zero for all values of ωt if:

$$\sum_{i=1}^{n} z_i \cos \phi_i = 0 \qquad\qquad \sum_{i=1}^{n} z_i \sin \phi_i = 0 \qquad (14.7a)$$

$$\sum_{i=1}^{n} z_i \cos 2\phi_i = 0 \qquad\qquad \sum_{i=1}^{n} z_i \sin 2\phi_i = 0 \qquad (14.7b)$$

These will guarantee no shaking moments through the second harmonic. We can extend this to higher harmonics as we did for the shaking force.

$$\sum_{i=1}^{n} z_i \cos 4\phi_i = 0 \qquad\qquad \sum_{i=1}^{n} z_i \sin 4\phi_i = 0 \qquad (14.7c)$$

$$\sum_{i=1}^{n} z_i \cos 6\phi_i = 0 \qquad\qquad \sum_{i=1}^{n} z_i \sin 6\phi_i = 0 \qquad (14.7d)$$

* The effect of overbalancing the crank throws is not included in equation 14.6b as shown. The crankshaft is assumed to be exactly balanced here. See Appendix G for the complete equations that include the effects of crank overbalance. Program ENGINE uses the equations from Appendix G to account for the effects of overbalancing in multicylinder engines.

14

TABLE 14-2   Moment Balance State of a 4-Cylinder, Inline Engine with a 0, 90, 180, 270° Crankshaft, and $z_1 = 0$, $z_2 = 1$, $z_3 = 2$, $z_4 = 3$

Primary moments:	$\displaystyle\sum_{i=1}^{n} z_i \sin\phi_i = -2$	$\displaystyle\sum_{i=1}^{n} z_i \cos\phi_i = -2$
Secondary moments:	$\displaystyle\sum_{i=1}^{n} z_i \sin 2\phi_i = 0$	$\displaystyle\sum_{i=1}^{n} z_i \cos 2\phi_i = -2$

Note that both the sine and cosine summations of any multiple of the phase angles must be zero for that harmonic of the shaking moment to be zero. The calculation for our example of a four-cylinder engine with phase angles of $\phi_1 = 0$, $\phi_2 = 90$, $\phi_3 = 180$, $\phi_4 = 270°$, and an assumed cylinder spacing of one length unit ($z_2 = 1$, $z_3 = 2$, $z_3 = 4$) in Table 14-2, shows that the shaking moments are not zero for any of these harmonics. So, our choice of phase angles, which is a good one for shaking forces and torques, fails the test for zero shaking moments. The coefficients of the fourth and sixth harmonic terms in the moment equations are minuscule, so they will be ignored. The secondary (second harmonic) term is less critical than the primary as it is multiplied by r/l which is generally less than 1/3. An unbalanced secondary harmonic of shaking moment is undesirable but can be tolerated, especially if the engine is of small displacement (less than about 1/2 liter per cylinder). The primary component is of greatest concern, because of its magnitude. If we wish to use this crankshaft configuration, we will need to apply a balancing technique to the engine as described in a later section to at least eliminate the primary moment. A large shaking moment is undesirable as it will cause the engine to **pitch** forward and back (like a bucking bronco) as the moment oscillates from positive to negative in the plane of the cylinders. *Do not confuse this shaking moment with the shaking torque* which acts to **roll** the engine back and forth about the Z axis of the crankshaft.

Figure 14-11 shows the primary and secondary components of the shaking moment for this example engine for two revolutions of the crank. Each is a pure harmonic of zero average value. The total moment is the sum of these two components. This engine configuration is a built-in example in program ENGINE. See Appendix A for information about the program.

14.6   EVEN FIRING

The inertial forces, torques, and moments are only one set of criteria which need to be considered in the design of multicylinder engines. Gas force and gas torque considerations are equally important. In general, it is desirable to create a firing pattern among the cylinders that is evenly spaced in time. If the cylinders fire unevenly, vibrations will be created which may be unacceptable. Smoothness of the power pulses is desired. The power pulses depend on the stroke cycle. If the engine is a two-stroke, there will be one power pulse per revolution in each of its n cylinders. The optimum delta phase angle between the cylinders' crank throws for evenly spaced power pulses will then be:

$$\Delta\phi_{two\ stroke} = \frac{360°}{n} \tag{14.8a}$$

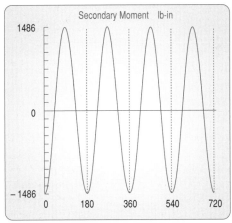

Inline 4-Cylinder
Bore = 3.00 in
Stroke = 3.54
B/S   = 0.85
L/R   = 3.50
m_A = 0.0284 bl
m_B = 0.0116 bl
P_{max} = 600 psi
RPM = 3400

Phase Angles:
0 90 180 270

Power Strokes:
0 90 180 270

FIGURE 14-11

Primary and secondary moments in the 0, 90, 180, 270° crankshaft four-cylinder engine

For a four-stroke engine there will be one power pulse in each cylinder every two revolutions. The optimum delta phase angle of the crank throws for evenly spaced power pulses will then be:

$$\Delta\phi_{four\ stroke} = \frac{720°}{n}$$

(14.8b)

Compare equations 14.8a and 14.8b to equation 14.1 (p. 710) which defined the optimum delta phase angle for cancellation of inertia forces. A two-stroke engine can have both even firing and inertia balance, but a four-stroke engine has a conflict between these two criteria. Thus some design trade-offs will be necessary to obtain the best compromise between these factors in the four-stroke case.

Two-Stroke Cycle Engine

To determine the firing pattern of an engine design we must return to the crank phase diagram. Figure 14-12 (p. 720) reproduces Figure 14-9 (p. 713) and adds new information to it. It shows the power pulses for a **two-stroke cycle, four-cylinder engine** with the $\phi_i = 0, 90, 180, 270°$ phase angle crank configuration. Note that each cylinder's negative block in Figure 14-12 is shifted to the right by its phase angle with respect to reference cylinder 1. In this schematic representation, only the negative blocks on the diagram are available for power pulses as they represent the downstroke of the piston. By convention, cylinder 1 fires first, so its negative block at 0° is labeled *Power*. The other cylinders may be fired in any order, but their power pulses should be as evenly spaced as possible across the interval.

The available power pulse spacings are dictated by the crank phase angles. There may be more than one firing order which will give even firing, especially with large numbers of cylinders. In this simple example the firing order 1, 2, 3, 4 will work as it will provide successive power pulses every 90° across the interval. The **power stroke angles** ψ_i are the angles in the cycle at which the cylinders fire. They are defined by the crank-

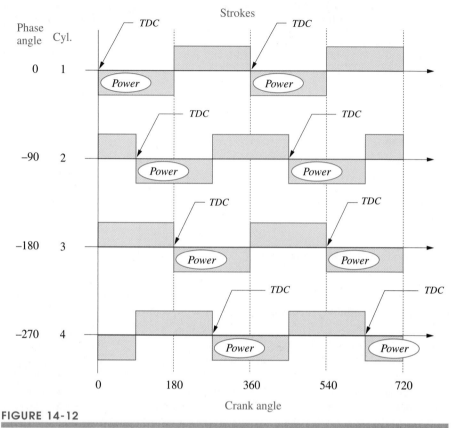

FIGURE 14-12

Two-stroke inline four-cylinder engine crank phase diagram with $\phi_i = 0, 90, 180, 270°$

shaft phase angles and the choice of firing order in combination and in this example are $\psi_i = 0, 90, 180,$ and $270°$. In general, ψ_i are not equal to ϕ_i. Their correspondence with the phase angles in this example results from choosing the consecutive firing order 1, 2, 3, 4.

For a **two-stroke engine**, the power stroke angles ψ_i must be *between 0 and 360°*. We always want them to be evenly spaced in that interval with a delta power stroke angle defined by equation 14.8c. For our four-cylinder, two-stroke engine, ideal power stroke angles are then $\psi_i = 0, 90, 180, 270°$, which we have achieved in this example.

We define the **delta power stroke angle** differently for each stroke cycle. For the two-stroke engine:

$$\Delta\psi_{two\ stroke} = \frac{360°}{n} \tag{14.8c}$$

For the four-stroke engine:

$$\Delta\psi_{four\ stroke} = \frac{720°}{n} \tag{14.8d}$$

The gas torque for a one-cylinder engine was defined in equation 13.8b (p. 675). The combined gas torque for all cylinders must sum the contributions of n cylinders, each phase-shifted by its power stroke angle ψ_i:

$$\mathbf{T}_{g21} \cong F_g\, r \sum_{i=1}^{n} \left\{ \sin\left(\omega t - \psi_i\right)\left[1 + \frac{r}{l}\cos\left(\omega t - \psi_i\right) \right] \right\} \hat{\mathbf{k}} \qquad (14.9)$$

Figure 14-13 (p. 722) shows the gas torque, inertia torque, and shaking force for this two-stroke four-cylinder engine plotted from program ENGINE. The shaking moment components are shown in Figure 14-11 (p. 719). Except for the unbalanced shaking moments, this design is otherwise acceptable. The inertia force and inertia torque are both zero which is ideal. The gas torque consists of uniformly shaped and spaced pulses across the interval, four per revolution. Note that program ENGINE plots two full revolutions to accommodate the four-stroke case; thus eight power pulses are seen. Open the file F14-13.eng in the program to exercise this example.

Four-Stroke Cycle Engine

Figure 14-14 (p. 723) shows a crank phase diagram for the *same crankshaft design* as in Figure 14-12 except that it is designed as a *four-stroke cycle engine*. There is now only one power stroke every 720° for each cylinder. The second negative block for each cylinder must be used for the intake stroke. Cylinder 1 is again fired first. An evenly spaced pattern of power pulses among the other cylinders is again desired but is now not possible with this crankshaft. Whether the firing order is 1, 3, 4, 2 or 1, 2, 4, 3 or 1, 4, 2, 3, or any other chosen, there will be both gaps and overlaps in the power pulses. The first firing order listed, 1, 3, 4, 2, has been chosen for this example. This results in the set of power stroke angles $\psi_i = 0, 180, 270, 450°$. These **power stroke angles** define the points in the **720° cycle** where each cylinder fires. Thus for a four-stroke engine, the power stroke angles ψ_i must be between 0 and 720°. We would like them to be evenly spaced in that interval with a delta angle defined by equation 14.8d. For our four-cylinder, four-stroke engine, the ideal power stroke angles would then be $\psi_i = 0, 180, 360, 540°$. We clearly have not achieved them in this example. Figure 14-15 (p. 723) shows the resulting gas torque. Open the file F14-15.eng in program ENGINE to exercise this example.

The uneven firing in Figure 14-15 is obvious. This uneven gas torque will be perceived by the operator of any vehicle containing this engine as rough running and vibration, especially at idle speed. At higher engine speeds the flywheel will tend to mask this roughness, but flywheels are ineffective at low speeds. It is this fact that causes most engine designers to *favor even firing over elimination of inertia effects* in their selection of crankshaft phase angles. The inertia force, torque, and moment are all functions of engine speed squared. But, as engine speed increases the magnitude of these factors, the same speed is also increasing the flywheel's ability to mask their effects. Not so with gas-torque roughness due to uneven firing. It is bad at all speeds and the flywheel won't hide it at low speed.

We therefore must reject this crankshaft design for our four-stroke, four-cylinder engine. Equation 14.8b (p. 719) indicates that we need a delta phase angle $\Delta\phi_i = 180°$ in our crankshaft to obtain even firing. We need four crank throws, and all crank phase angles must be less than 360°. So, we must repeat some angles if we use a delta phase

14

(a)

(b)

Inline 4-Cylinder
2-Stroke Cycle

Bore   = 3.00 in
Stroke = 3.54
B/S   = 0.85
L/R   = 3.50
m_A = 0.0284 bl
m_B = 0.0116 bl
P_{max} = 600 psi
RPM = 3400
Phase Angles:
0 90 180 270

Power Strokes:
0 90 180 270

(c)

FIGURE 14-13

Torque and shaking force in the two-stroke four-cylinder inline engine

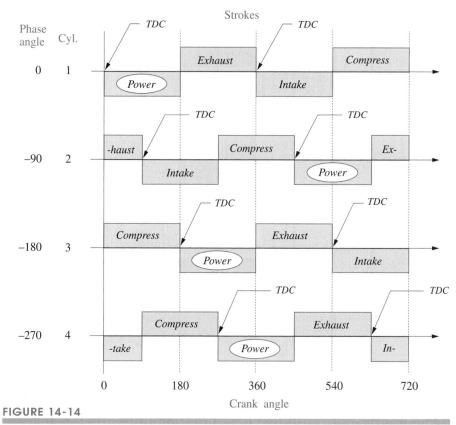

FIGURE 14-14

Four-stroke inline four-cylinder engine crank phase diagram with $\phi_i = 0, 90, 180, 270°$

Inline 4-Cylinder
4-Stroke Cycle

Bore = 3.00 in
Stroke = 3.54
B/S = 0.85
L/R = 3.50
m_A = 0.0284 bl
m_B = 0.0116 bl
P_{max} = 600 psi
RPM = 3400

Phase Angles:
0 90 180 270

Power Strokes:
0 90 180 270

FIGURE 14-15

Uneven firing four-stroke, four-cylinder inline engine with a 0, 90, 180, 270° crankshaft

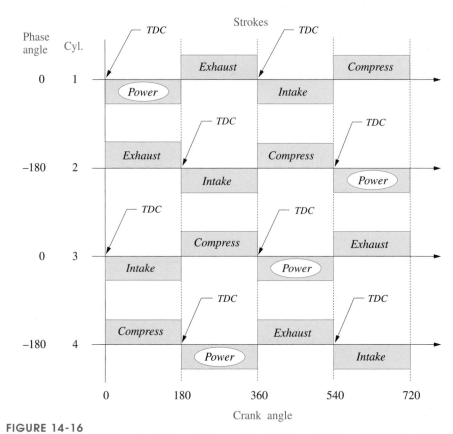

FIGURE 14-16

Even firing four-stroke, four-cylinder engine crank phase diagram with $\phi_i = 0, 180, 0, 180°$

angle of 180°. One possibility is $\phi_i = 0, 180, 0, 180°$ for the four crank throws.[*] The crank phase diagram for this design is shown in Figure 14-16. The power strokes can now be evenly spaced over 720°. A firing order of 1, 4, 3, 2 has been chosen which gives the desired sequence of power stroke angles, $\psi_i = 0, 180, 360, 540°$. (Note that a firing order of 1, 2, 3, 4 would also work with this engine.)[†]

The inertial balance condition of this design must now be checked with equations 14.3, 14.5 and 14.7 (pp. 714-717). These show that the primary inertia force is zero, but the primary moment, secondary force, secondary moment, and inertia torque are all non-zero as shown in Table 14-3. So, this even-firing design has compromised the very good state of inertia balance of the previous design in order to achieve even firing. The inertia torque variations can be masked by a flywheel. The secondary forces and moments are relatively small in a small engine and can be tolerated. The nonzero primary moment is a problem which needs to be addressed. To see the results of this engine configuration, run program ENGINE and select it from the *Example* pulldown menu. Then plot the results. See Appendix A for more detailed instructions on the use of program ENGINE.

We shall soon discuss ways to counter an unbalanced moment with the addition of balance shafts, but there is a more direct approach available in this example. Figure 14-17

[*] Note that 0, 180, 360, 540°, modulo 360 is the same as 0, 180, 0, 180.

[†] Note the pattern of acceptable firing orders (FO). Write two revolutions' worth of any acceptable FO, as in 1, 4, 3, 2, 1, 4, 3, 2. Any set of four successive numbers in this sequence, *either forward or backward*, is an acceptable FO. If we require the first to be cylinder 1, then the only other possibility here is the backward set 1, 2, 3, 4.

TABLE 14-3   Force and Moment Balance State of a 4-Cylinder, Inline Engine with a 0, 180, 0, 180° Crankshaft, and $z_1 = 0$, $z_2 = 1$, $z_3 = 2$, $z_4 = 3$

Primary forces:	$\sum_{i=1}^{n} \sin\phi_i = 0$	$\sum_{i=1}^{n} \cos\phi_i = 0$
Secondary forces:	$\sum_{i=1}^{n} \sin 2\phi_i = 0$	$\sum_{i=1}^{n} \cos 2\phi_i = 4$
Primary moments:	$\sum_{i=1}^{n} z_i \sin\phi_i = 0$	$\sum_{i=1}^{n} z_i \cos\phi_i = -2$
Secondary moments:	$\sum_{i=1}^{n} z_i \sin 2\phi_i = 0$	$\sum_{i=1}^{n} z_i \cos 2\phi_i = 6$

shows that the shaking moment is due to the action of the individual cylinders' inertial forces acting at moment arms about some center. If we consider that center to be point

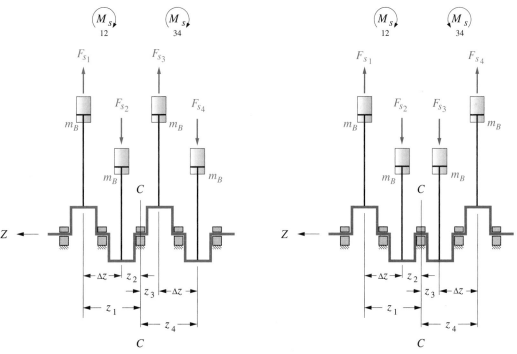

(a) Nonsymmetric 0, 180, 0, 180° crankshaft   (b) Symmetric 0, 180, 180, 0° crankshaft

FIGURE 14-17

Mirror-symmetric crankshafts cancel primary moments.

14

C in the middle of the engine, it should be apparent that any primary force-balanced crankshaft design which is mirror-symmetric about a transverse plane through point *C* would also have balanced primary moments as long as all cylinder spacings were uniform and all inertial forces equal. Figure 14-17a (p. 725) shows the 0, 180, 0, 180° crankshaft which is not mirror-symmetric. The couple $\mathbf{F}_{s_1}\Delta z$ due to cylinder pairs 1, 2 has the same magnitude and sense as the couple $\mathbf{F}_{s_3}\Delta z$ due to cylinders 3 and 4, so they add. Figure 14-17b shows the 0, 180, 180, 0° crankshaft which is *mirror-symmetric*. The couple $\mathbf{F}_{s_1}\Delta z$ due to cylinder pairs 1, 2 has the same magnitude but opposite sense to the couple $\mathbf{F}_{s_3}\Delta z$ due to cylinders 3, 4, so they cancel. We can then achieve both even firing and balanced primary moments by changing the sequence of crank throw phase angles to $\phi_i = 0$, 180, 180, 0° which is *mirror-symmetric*.

The crank phase diagram for this design is shown in Figure 14-18. The power strokes can still be evenly spaced over 720°. A firing order of 1, 3, 4, 2 has been chosen which gives the same desired sequence of power stroke angles, $\psi_i = 0$, 180, 360, 540°. (Note that a firing order of 1, 2, 4, 3 would also work with this engine.)[*] Equations 14.3, 14.5, and 14.7 (pp. 714-717) and Table 14-4 show that the primary inertia force and primary moment are both now zero, but the secondary force, secondary moment, and inertia torque are still nonzero.

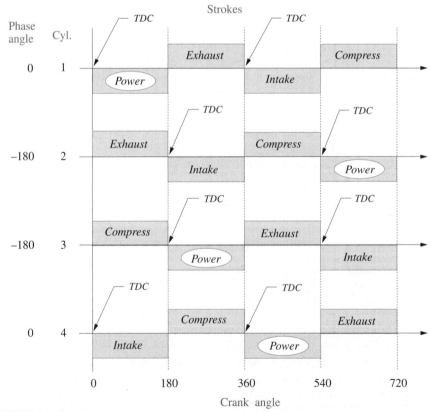

FIGURE 14-18

Even firing four-stroke, four-cylinder engine crank phase diagram with a mirror-symmetric 0, 180, 180, 0° crankshaft

[*] In inline engines, and within any one bank of a vee engine, a nonconsecutive firing order (i.e., not 1, 2, 3, 4) is usually preferred so that adjacent cylinders do not fire sequentially. This allows the intake manifold more time to recharge locally between intake strokes and the exhaust manifold to scavenge more efficiently.

14

TABLE 14-4   Force and Moment Balance State of a 4-Cylinder, Inline Engine
              with a 0, 180, 180, 0° Crankshaft, and $z_1 = 0$, $z_2 = 1$, $z_3 = 2$, $z_4 = 3$

Primary forces:	$\sum_{i=1}^{n} \sin\phi_i = 0$	$\sum_{i=1}^{n} \cos\phi_i = 0$
Secondary forces:	$\sum_{i=1}^{n} \sin 2\phi_i = 0$	$\sum_{i=1}^{n} \cos 2\phi_i = 4$
Primary moments:	$\sum_{i=1}^{n} z_i \sin\phi_i = 0$	$\sum_{i=1}^{n} z_i \cos\phi_i = 0$
Secondary moments:	$\sum_{i=1}^{n} z_i \sin 2\phi_i = 0$	$\sum_{i=1}^{n} z_i \cos 2\phi_i = 6$

This $\phi_i = 0$, 180, 180, 0° crankshaft is considered the best design trade-off and is the one universally used in these four-cylinder inline, four-stroke production engines. Figures 14-2 (p. 706) and 14-3 (p. 707) show such a four-cylinder design. Inertia balance is sacrificed to gain even firing for the reasons cited on p. 721. Figure 14-19 (p. 728) shows the gas torque, inertia torque, and total torque for this design. Figure 14-20 (p. 729) shows the secondary shaking moment, secondary shaking force component, and a polar plot of the total shaking force for this design. Note that Figures 14-20b and c are just different views of the same parameter. The polar plot of the shaking force in Figure 14-20c is a view of the shaking force looking at the end of the crankshaft axis with the piston motion horizontal. The cartesian plot in Figure 14-20b shows the same force on a time axis. Since the primary component is zero, this total force is due only to the secondary component. We will soon discuss ways to eliminate these secondary forces and moments.

To see the results of this engine configuration, run program ENGINE and select it from the *Examples* pulldown menu. Then *Plot* or *Print* the results. See Appendix A for more detailed instructions on the use of the program.

14.7   VEE ENGINE CONFIGURATIONS

The same design principles which apply to inline engines also apply to vee and opposed configurations. Even firing takes precedence over inertia balance and mirror symmetry of the crankshaft balances primary moments. In general, a vee engine will have similar inertia balance to that of the inline engines from which it is constructed. A vee-six is essentially two three-cylinder inline engines on a common crankshaft, a vee-eight is two four-cylinder inlines, etc. The larger number of cylinders allows more power pulses to be spaced out over the cycle for a smoother (and larger average) gas torque. The existence of a **vee angle** v between the two inline engines introduces an additional phase shift of the inertial and gas events which is analogous to, but independent of, the phase angle effects. This vee angle is the designer's choice, but there are good and poor choices. The same criteria of even firing and inertia balance apply to its selection.

14

(a)

(b)

Inline 4-Cylinder
4-Stroke Cycle

Bore   = 3.00 in
Stroke = 3.54
B/S   = 0.85
L/R   = 3.50
m_A = 0.0284 bl
m_B = 0.0116 bl
P_{max} = 600 psi
RPM = 3400

Phase Angles:
0 180 180 0

Power Strokes:
0 180 360 540

(c)

FIGURE 14-19

Torque in the four-stroke, four-cylinder inline engine with a 0, 180, 180, 0° crankshaft

(a)

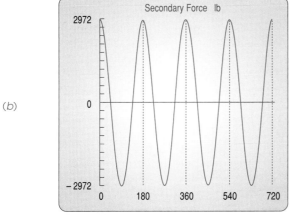

(b)

Inline 4-Cylinder
4-Stroke Cycle

Bore = 3.00 in
Stroke = 3.54
B/S   = 0.85
L/R   = 3.50
m_A = 0.0284 bl
m_B = 0.0116 bl
P_{max} = 600 psi
RPM = 3400

Phase Angles:
0 180 180 0

Power Strokes:
0 180 360 540

(c)

14

FIGURE 14-20

Shaking forces and moments in the four-stroke, four-cylinder 0, 180, 180, 0° engine

The **vee angle** $v = 2\gamma$ is defined as shown in Figure 14-21. Each bank is offset by its **bank angle** γ referenced to the central X axis of the engine. The crank angle ωt is measured from the X axis. Cylinder 1 in the right bank is the reference cylinder. Events in each bank will now be phase-shifted by its bank angle as well as by the crankshaft phase angles. These two phase shifts will superpose. Taking any one cylinder in either bank as an example, let its instantaneous crank angle be represented by:

$$\theta = \omega t - \phi_i \tag{14.10a}$$

Consider first a two-cylinder vee engine with one cylinder in each bank and with both sharing a common crank throw. The shaking force for a single cylinder in the direction of piston motion $\hat{u}$ with θ measured from the piston axis is:

$$\mathbf{F_s} \cong m_B r \omega^2 \left(\cos\theta + \frac{r}{l}\cos 2\theta \right) \hat{u} \tag{14.10b}$$

The total shaking force is the vector sum of the contributions from each bank.

$$\mathbf{F_s} = \mathbf{F_{s_L}} + \mathbf{F_{s_R}} \tag{14.10c}$$

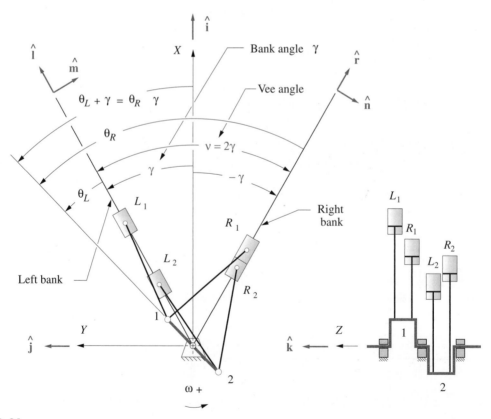

FIGURE 14-21

Vee-engine geometry

We now want to refer the crank angle to the central X axis. The shaking forces for the right (R) and left (L) banks, in the planes of the respective cylinder banks, can then be expressed as:

$$\mathbf{F}_{s_R} \cong m_B r \omega^2 \left[\cos(\theta+\gamma) + \frac{r}{l}\cos 2(\theta+\gamma) \right] \hat{\mathbf{r}}$$

$$\mathbf{F}_{s_L} \cong m_B r \omega^2 \left[\cos(\theta-\gamma) + \frac{r}{l}\cos 2(\theta-\gamma) \right] \hat{\mathbf{i}}$$

(14.10d)

Note that the bank angle γ is added to or subtracted from the crank angle for each cylinder bank to reference it to the central X axis. The forces are still directed along the planes of the cylinder banks. Substitute the identities:

$$\cos(\theta+\gamma) = \cos\theta\cos\gamma - \sin\theta\sin\gamma$$

$$\cos(\theta-\gamma) = \cos\theta\cos\gamma + \sin\theta\sin\gamma$$

(14.10e)

to get:

$$\mathbf{F}_{s_R} \cong m_B r \omega^2 \begin{bmatrix} \cos\theta\cos\gamma - \sin\theta\sin\gamma \\ + \frac{r}{l}(\cos 2\theta\cos 2\gamma - \sin 2\theta\sin 2\gamma) \end{bmatrix} \hat{\mathbf{r}}$$

$$\mathbf{F}_{s_L} \cong m_B r \omega^2 \begin{bmatrix} \cos\theta\cos\gamma + \sin\theta\sin\gamma \\ + \frac{r}{l}(\cos 2\theta\cos 2\gamma + \sin 2\theta\sin 2\gamma) \end{bmatrix} \hat{\mathbf{i}}$$

(14.10f)

Now, to account for the possibility of multiple cylinders, phase-shifted within each bank, substitute equation 14.10a for θ and replace the sums of angle terms with products from the identities:

$$\cos(\omega t - \phi_i) = \cos\omega t\cos\phi_i + \sin\omega t\sin\phi_i$$

$$\sin(\omega t - \phi_i) = \sin\omega t\cos\phi_i - \cos\omega t\sin\phi_i$$

(14.10g)

After much manipulation, the expressions for the contributions from right and left banks reduce to:

for the right bank:

$$\mathbf{F}_{s_R} \cong m_B r \omega^2 \begin{bmatrix} (\cos\omega t\cos\gamma - \sin\omega t\sin\gamma)\sum_{i=1}^{n/2}\cos\phi_i \\ + (\cos\omega t\sin\gamma + \sin\omega t\cos\gamma)\sum_{i=1}^{n/2}\sin\phi_i \\ + \frac{r}{l}(\cos 2\omega t\cos 2\gamma - \sin 2\omega t\sin 2\gamma)\sum_{i=1}^{n/2}\cos 2\phi_i \\ + \frac{r}{l}(\cos 2\omega t\sin 2\gamma + \sin 2\omega t\cos 2\gamma)\sum_{i=1}^{n/2}\sin 2\phi_i \end{bmatrix} \hat{\mathbf{r}}$$

(14.10h)

14

and for the left bank:

$$
\mathbf{F_{s_L}} \cong m_B r \omega^2
\begin{bmatrix}
(\cos\omega t \cos\gamma + \sin\omega t \sin\gamma) \displaystyle\sum_{i=n/2+1}^{n} \cos\phi_i \\[2em]
-(\cos\omega t \sin\gamma - \sin\omega t \cos\gamma) \displaystyle\sum_{i=n/2+1}^{n} \sin\phi_t \\[2em]
+\dfrac{r}{l}(\cos 2\omega t \cos 2\gamma + \sin 2\omega t \sin 2\gamma) \displaystyle\sum_{i=n/2+1}^{n} \cos 2\phi_i \\[2em]
-\dfrac{r}{l}(\cos 2\omega t \sin 2\gamma - \sin 2\omega t \cos 2\gamma) \displaystyle\sum_{i=n/2+1}^{n} \sin 2\phi_i
\end{bmatrix}
\hat{\mathbf{i}}
\qquad (14.10i)
$$

The summations in equations 14.10h and i (p. 731) give a set of sufficient criteria for **zero shaking force** through the second harmonic for each bank, similar to those for the inline engine in equation 14.3 (p. 714). We can resolve the shaking forces for each bank into components along and normal to the central X axis of the vee engine:[*]

$$
\begin{aligned}
F_{s_x} &= \left(F_{s_L} + F_{s_R}\right)\cos\gamma \,\hat{\mathbf{i}} \\
F_{s_y} &= \left(F_{s_L} - F_{s_R}\right)\sin\gamma \,\hat{\mathbf{j}} \\
\mathbf{F_s} &= F_{s_x}\,\hat{\mathbf{i}} + F_{s_y}\,\hat{\mathbf{j}}
\end{aligned}
\qquad (14.10j)
$$

Equations 14.10j provide additional opportunities for cancellation of shaking forces beyond the choice of phase angles; e.g., even with nonzero values of F_{s_L} and F_{s_R}, if γ is 90°, then the x component of the shaking force will be zero. Also, if $F_{s_L} = F_{s_R}$, then the y component of the shaking force will be zero for any γ. This situation obtains for the case of a horizontally opposed engine (see Section 14.8, p. 739). With some vee or opposed engines it is possible to get cancellation of shaking force components even when the summations in equation 14.10 are not all zero.

The **shaking moment** equations are easily formed from the shaking force equations by multiplying each term in the summations by the moment arm as was done in equations 14.6 (p. 717). The moments exist within each bank and their vectors will be orthogonal to the respective cylinder planes. For the right bank we define a moment unit vector $\hat{\mathbf{n}}$ perpendicular to the $\hat{\mathbf{r}}$-Z plane in Figure 14-21 (p. 730). For the left bank we define a moment unit vector $\hat{\mathbf{m}}$ perpendicular to the $\hat{\mathbf{I}}$-Z plane in Figure 14-21.

[*] The effect of overbalancing the crank throws is not included in equation 14.10j as shown. The crankshaft is assumed to be exactly balanced here. See Appendix G for the complete equations that include the effects of crank overbalance. Program ENGINE uses the equations from Appendix G to account for the effects of overbalancing in multicylinder engines.

$$
\mathbf{M_{s_R}} \cong m_B r \omega^2
\begin{bmatrix}
(\cos\omega t \cos\gamma - \sin\omega t \sin\gamma)\displaystyle\sum_{i=1}^{n/2} z_i \cos\phi_i \\[2em]
+(\cos\omega t \sin\gamma + \sin\omega t \cos\gamma)\displaystyle\sum_{i=1}^{n/2} z_i \sin\phi_i \\[2em]
+\dfrac{r}{l}(\cos 2\omega t \cos 2\gamma - \sin 2\omega t \sin 2\gamma)\displaystyle\sum_{i=1}^{n/2} z_i \cos 2\phi_i \\[2em]
+\dfrac{r}{l}(\cos 2\omega t \sin 2\gamma + \sin 2\omega t \cos 2\gamma)\displaystyle\sum_{i=1}^{n/2} z_i \sin 2\phi_i
\end{bmatrix}
\hat{\mathbf{n}}
\qquad (14.11a)
$$

14

$$\mathbf{M}_{s_L} \cong m_B r \omega^2 \begin{bmatrix} (\cos \omega t \cos \gamma + \sin \omega t \sin \gamma) \sum_{i=n/2+1}^{n} z_i \cos \phi_i \\[2ex] -(\cos \omega t \sin \gamma - \sin \omega t \cos \gamma) \sum_{i=n/2+1}^{n} z_i \sin \phi_t \\[2ex] +\frac{r}{l}(\cos 2\omega t \cos 2\gamma + \sin 2\omega t \sin 2\gamma) \sum_{i=n/2+1}^{n} z_i \cos 2\phi_i \\[2ex] -\frac{r}{l}(\cos 2\omega t \sin 2\gamma - \sin 2\omega t \cos 2\gamma) \sum_{i=n/2+1}^{n} z_i \sin 2\phi_i \end{bmatrix} \hat{\mathbf{m}} \qquad (14.11b)$$

The summations in equations 14.11a and b provide a set of sufficient criteria for **zero shaking moment** through the second harmonic for each bank, similar to those found for the inline engine in equations 14.7 (p. 717). Resolving the shaking moments for each bank into components along and normal to the central X axis of the vee engine gives:[*]

$$M_{s_x} = \left(M_{s_L} - M_{s_R}\right) \sin \gamma$$
$$M_{s_y} = \left(-M_{s_L} - M_{s_R}\right) \cos \gamma \qquad (14.11c)$$
$$\mathbf{M_s} = M_{s_x}\,\hat{\mathbf{i}} + M_{s_y}\,\hat{\mathbf{j}}$$

Equation 14.11c allows possible cancellation of shaking moment components for some vee or opposed configurations even when the summations in equations 14.11a and b are not all zero; e.g., if γ is 90°, then the y component of the shaking moment is zero.

The **inertia torques** from the right and left banks of a vee engine are:

$$\mathbf{T}_{i21_R} \cong \frac{1}{2} m_B r^2 \omega^2 \begin{bmatrix} \frac{r}{2l}\left(\sin(\omega t + \gamma)\sum_{i=1}^{n/2}\cos \phi_i - \cos(\omega t + \gamma)\sum_{i=1}^{n/2}\sin \phi_i\right) \\[2ex] -\left(\sin 2(\omega t + \gamma)\sum_{i=1}^{n/2}\cos 2\phi_i - \cos 2(\omega t + \gamma)\sum_{i=1}^{n/2}\sin 2\phi_i\right) \\[2ex] -\frac{3r}{2l}\left(\sin 3(\omega t + \gamma)\sum_{i=1}^{n/2}\cos 3\phi_i - \cos 3(\omega t + \gamma)\sum_{i=1}^{n/2}\sin 3\phi_i\right) \end{bmatrix} \hat{\mathbf{k}} \qquad (14.12a)$$

$$\mathbf{T}_{i21_L} \cong \frac{1}{2} m_B r^2 \omega^2 \begin{bmatrix} \frac{r}{2l}\left(\sin(\omega t - \gamma)\sum_{i=n/2+1}^{n}\cos \phi_i - \cos(\omega t - \gamma)\sum_{i=n/2+1}^{n}\sin \phi_i\right) \\[2ex] -\left(\sin 2(\omega t - \gamma)\sum_{i=n/2+1}^{n}\cos 2\phi_i - \cos 2(\omega t - \gamma)\sum_{i=n/2+1}^{n}\sin 2\phi_i\right) \\[2ex] -\frac{3r}{2l}\left(\sin 3(\omega t - \gamma)\sum_{i=n/2+1}^{n}\cos 3\phi_i - \cos 3(\omega t - \gamma)\sum_{i=n/2+1}^{n}\sin 3\phi_i\right) \end{bmatrix} \hat{\mathbf{k}} \qquad (14.12b)$$

[*] The effect of overbalancing the crank throws is not included in equation 14.11c as shown. The crankshaft is assumed to be exactly balanced here. See Appendix G for the complete equations that include the effects of crank overbalance. Program ENGINE uses the equations from Appendix G to account for the effects of overbalancing in multicylinder engines.

14

Add the contributions from each bank for the total. For **zero inertia torque** through the third harmonic in a vee engine it is sufficient (but not necessary) that:

$$\sum_{i=1}^{n/2} \sin\phi_i = 0 \qquad \sum_{i=1}^{n/2} \cos\phi_i = 0 \qquad \sum_{i=n/2+1}^{n} \sin\phi_i = 0 \qquad \sum_{i=n/2+1}^{n} \cos\phi_i = 0$$

$$\sum_{i=1}^{n/2} \sin 2\phi_i = 0 \qquad \sum_{i=1}^{n/2} \cos 2\phi_i = 0 \qquad \sum_{i=n/2+1}^{n} \sin 2\phi_i = 0 \qquad \sum_{i=n/2+1}^{n} \cos 2\phi_i = 0 \quad (14.12c)$$

$$\sum_{i=1}^{n/2} \sin 3\phi_i = 0 \qquad \sum_{i=1}^{n/2} \cos 3\phi_i = 0 \qquad \sum_{i=n/2+1}^{n} \sin 3\phi_i = 0 \qquad \sum_{i=n/2+1}^{n} \cos 3\phi_i = 0$$

Note that when equations 14.12a and b are added, particular combinations of ϕ_i and γ may cancel the inertia torque even when some terms of equation 14.12c are nonzero.

The **gas torque** is:

$$\mathbf{T}_{g21} \cong F_g\, r \sum_{i=1}^{n} \left(\sin\left[\omega t - (\psi_i + \gamma_k)\right]\right)\left\{1 + \frac{r}{l}\cos\left[\omega t - (\psi_i + \gamma_k)\right]\right\} \,\hat{\mathbf{k}} \qquad (14.13)$$

where the left bank has a *bank angle* $\gamma_k = +\gamma$ and the right bank a *bank angle* $\gamma_k = -\gamma$.

It is possible to design a vee engine that has as many crank throws as cylinders, but, for several reasons, this is not always done. The principal advantage of a vee engine over an inline of the same number of cylinders is its more compact size and greater stiffness. It can be about half the length of a comparable inline engine (at the expense of greater width), provided that the crankshaft is designed to accommodate two conrods per crank throw as shown in Figure 14-22. Cylinders in opposite banks then share a crank throw. One bank of cylinders is shifted along the crankshaft axis by the thickness of a conrod. The shorter, wider cylinder block and the shorter crankshaft are much stiffer in both torsion and bending than are those for an inline engine with the same number of cylinders. Figure 14-23 (p. 736) shows computer simulations of several bending and one torsional mode of vibration for a four throw crankshaft. The deflections are greatly exaggerated. The necessarily contorted shape of a crankshaft makes it difficult to control these deflections by design. If excessive in magnitude, they can lead to structural failure.

As an example we will now design the crankshaft for a four-stroke cycle, vee-eight engine. We could put two $\phi_i = 0, 180, 180, 0°$ four-cylinder engines together on one such crankshaft and have the same balance conditions as the four-stroke, four-cylinder engine designed in the previous section (primaries balanced, secondaries unbalanced). However, the motivation for choosing that crankshaft for the four-cylinder engine was the need to space the four available power pulses evenly across the cycle. Equation 14.8b (p. 719) then dictated a 180° delta phase angle $\Delta\phi_i$ for that engine. Now we have eight cylinders available and equation 14.8b defines a delta phase angle of 90° for optimum power pulse spacing. This means we could use the $\phi_i = 0, 90, 180, 270°$ crankshaft designed for the two-stroke four-cylinder engine shown in Figure 14-12 (p. 720), and take advantage of its better inertia balance condition as well as achieve even firing in the four-stroke eight-cylinder vee engine.

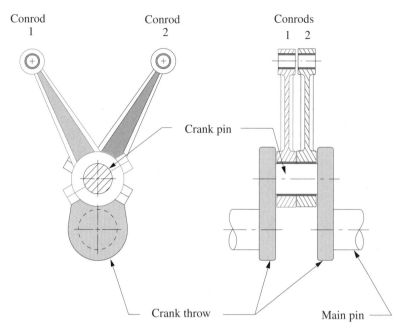

Conrod
1

Conrod
2

Conrods
1   2

Crank pin

Crank throw — Main pin

FIGURE 14-22

Two connecting rods on a common crank throw

The $\phi_i = 0, 90, 180, 270°$ four-cylinder crankshaft has all inertia factors equal to zero except for the primary and secondary moments. We learned that arranging the crank throws with mirror symmetry about the midplane would balance the primary moment. Some thought and/or sketches will reveal that it is not possible to obtain this mirror symmetry with any four-throw, 90° delta phase angle crankshaft arrangement. However, just as rearranging the order of the crank throws from $\phi_i = 0, 180, 0, 180°$ to $\phi_i = 0, 180, 180, 0°$ had an effect on the shaking moments, rearranging this crankshaft's throw order will as well. A crankshaft of $\phi_i = 0, 90, 270, 180°$ has all inertia factors equal to zero except for the primary moment. The secondary moment is now gone.[*] This is an advantage worth taking. We will use this crankshaft for the vee-eight and deal with the primary moment later.

Figure 14-24a shows the crank phase diagram for the right bank of a vee-eight engine with a $\phi_i = 0, 90, 270, 180°$ crankshaft. Figure 14-24b shows the crank phase diagram for the second (left) bank which is identical to that of the right bank (as it must be since they share crank throws), but it is *shifted to the right by the vee angle* 2γ. Note that in Figure 14-21 (p. 730), the two pistons are driven by conrods on a common crank throw with positive ω, and the piston in the right bank will reach TDC before the one in the left bank. Thus as we show it, the left bank's piston motions lag those of the right bank. Lagging events occur later in time, so we must shift the second (left) bank rightward by the vee angle on the crank phase diagram.

We would like to shift the second bank of cylinders such that its power pulses are evenly spaced among those of the first bank. A little thought (and reference to equation

[*] The explanation for this is quite simple. Equation 14.7b (p. 707) shows that second moments are a function of twice the phase angles and the cylinder moment arms. If you double the values of the original 0, 90, 180, 270° phase angle sequence and modulo them with 360, you get 0, 180, 0, 180° which is not mirror-symmetric. Doubling the new phase angle sequence of 0, 90, 270, 180°, modulo 360, gives 0, 180, 180, 0° which is mirror-symmetric. It is this symmetry of the doubled phase angles that causes cancellation of the second harmonic of the shaking moment.

14

FIGURE 14-23

Bending and torsional modes of vibration in a four-throw crankshaft
Courtesy of Chevrolet Division, General Motors Corp.

14.8b, p. 719) should reveal that, in this example, each four-cylinder bank has potentially 720/4 = 180° between power pulses. Our chosen crank throws are spaced at 90° increments. A 90° vee angle (bank angle $\gamma = 45°$) will be optimum in this case as the phase angles and bank angles will add to create an effective spacing of 180°. Every vee-engine design of four or more cylinders will have one or more optimum vee angles that will give approximately even firing with any particular set of crank phase angles.

Several firing orders are possible with this many cylinders. Vee engines are often arranged to fire cylinders in opposite banks successively to balance the fluid flow demands in the intake manifold. Our cylinders are numbered from front to back, first down the right bank and then down the left bank. The firing order shown in Figure 14-24b (p. 737)

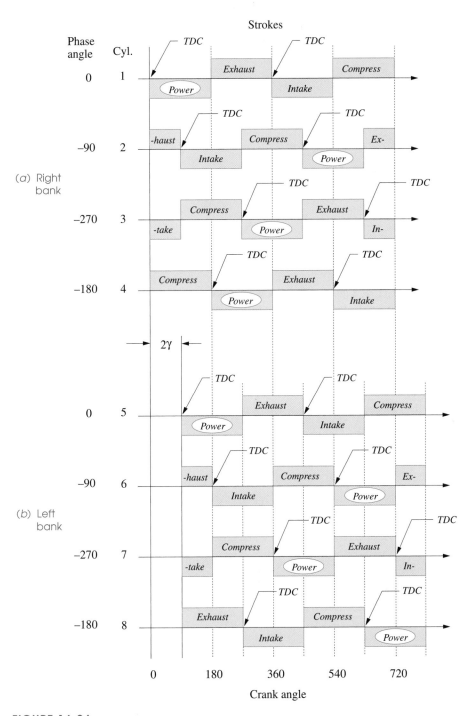

FIGURE 14-24

Four-stroke vee-eight crank phase diagram with 0, 90, 270, 180° crankshaft phase angles

14

90 deg Vee, 8-Cylinder
4-Stroke Cycle

Bore   = 2.50 in
Stroke = 2.55
B/S   = 0.98
L/R   = 3.50
m_A = 0.0418 bl
m_B = 0.0116 bl
RPM = 3400

Phase Angles:
0 90 270 180 0 90 270 180

Power Strokes:
0 90 180 270 360 450 540 630

FIGURE 14-25

Total torque in the 90° vee-eight engine with 0, 90, 270,180° crankshaft phase angles

is 1, 5, 4, 3, 7, 2, 6, 8 and results in power stroke angles ψ_i = 0, 90, 180, 270, 360, 450, 540, 630°. This will clearly give even firing with a power pulse every 90°.

Figure 14-25 shows the total torque for this engine design, which in this case is equal to the gas torque because the inertia torque is zero. Table 14-5 and Figure 14-26 show the only significant unbalanced inertial component in this engine to be the primary moment, which is quite large. The fourth harmonic terms have negligible coefficients in the Fourier series, and we have truncated them from the equations. We will address the balancing of this primary moment in a later section of this chapter.

Any vee-cylinder configuration may have one or more desirable vee angles which will give both even firing and acceptable inertia balance. However, vee engines of fewer than twelve cylinders will not be completely balanced by means of their crankshaft con-

TABLE 14-5   **Force and Moment Balance State of an 8-Cylinder, Vee Engine with a 0, 90, 270, 180° Crankshaft, and $z_1 = 0$, $z_2 = 1$, $z_3 = 2$, $z_4 = 3$**

Primary forces in each bank:	$\displaystyle\sum_{i=1}^{n} \sin\phi_i = 0$	$\displaystyle\sum_{i=1}^{n} \cos\phi_i = 0$
Secondary forces in each bank:	$\displaystyle\sum_{i=1}^{n} \sin 2\phi_i = 0$	$\displaystyle\sum_{i=1}^{n} \cos 2\phi_i = 0$
Primary moments in each bank:	$\displaystyle\sum_{i=1}^{n} z_i \sin\phi_i = -1$	$\displaystyle\sum_{i=1}^{n} z_i \cos\phi_i = -3$
Secondary moments in each bank:	$\displaystyle\sum_{i=1}^{n} z_i \sin 2\phi_i = 0$	$\displaystyle\sum_{i=1}^{n} z_i \cos 2\phi_i = 0$

14

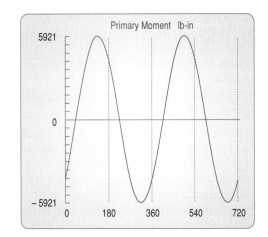

90 deg Vee, 8-Cylinder
4-Stroke Cycle

Bore   = 2.50 in
Stroke = 2.55
B/S   = 0.98
L/R   = 3.50
m_A = 0.0418 bl
m_B = 0.0116 bl
RPM   = 3400

Phase Angles:
0 90 270 180 0 90 270 180

Power Strokes:
0 90 180 270 360 450 540 630

FIGURE 14-26

Unbalanced primary moment in the 90° vee-eight engine with a 0, 90, 270, 180° crankshaft

figuration. The desirable vee angles will typically be an integer multiple (including one) or submultiple of the optimum delta phase angle as defined in equations 14.8 (pp. 718-720) for that engine. Ninety degrees is the optimum vee angle (2γ) for an eight-cylinder vee engine. To see the results for this vee-eight engine configuration, run program ENGINE and select the vee-eight from the *Examples* pulldown menu. See Appendix A for instructions on the use of the program.

14.8   OPPOSED ENGINE CONFIGURATIONS

An opposed engine is essentially a vee engine with a 180° vee angle. The advantage, particularly with a small number of cylinders such as two or four, is the relatively good balance condition possible. A four-stroke opposed (boxer) twin[*] with 0, 180° crank has even firing plus primary and secondary force balance, though all harmonics of its moment are nonzero. A four-stroke, opposed four-cylinder engine (flat four) with a 0, 180, 180, 0°, four-throw crank is even firing and has primary force and moment balance. Unlike its inline-four cousin with the same crankshaft, it also has secondary force balance from equation 14.10j. A four-stroke opposed six with a six-throw, 0, 120, 240, 60, 180, 300° crank has even firing and the same good balance condition as the inline six. Program ENGINE will calculate the parameters for opposed as well as vee and inline configurations.

14.9   BALANCING MULTICYLINDER ENGINES

With a sufficient number (m) of cylinders, properly arranged in banks of n cylinders in a multibank engine,[†] an engine can be inherently balanced. In a two-stroke engine with its crank throws arranged for even firing, all harmonics of shaking force will be balanced except those whose harmonic number is a multiple of n. In a four-stroke inline engine with its crank throws arranged for even firing, all harmonics of shaking force will be balanced except those whose harmonic number is a multiple of $n/2$. Primary shaking moment components will be balanced if the crankshaft is mirror-symmetric about the central transverse plane. A four-stroke inline configuration then requires at least six cylinders to

14

[*] As in the BMW R-series motorcycles.

[†] For an inline engine, $m = n$.

be inherently balanced through the second harmonic. We have seen that an inline four with a 0, 180, 180, 0° crankshaft has nonzero secondary forces and moments as well as nonzero inertia torque. The inline six with a mirror-symmetric crank of $\phi_i = 0, 240, 120, 120, 240, 0°$ will have zero shaking forces and moments through the second harmonic, though the inertia torque's third harmonic will still be present. To see the results of this six-cylinder inline engine configuration, run program ENGINE and select the inline six from the *Examples* pulldown menu.

A VEE-TWELVE is then the smallest vee engine with an inherent state of near perfect balance, as it is two inline sixes on a common crankshaft. We have seen that vee engines generally take on the balance characteristics of the inline banks from which they are made. Equations 14.10a to i and 14.11a and b (pp. 730-733) introduced no new criteria for balance in the vee engine over those already defined in equations 14.3 (p. 714) and 14.5 (p. 716) for shaking force and moment balance in the inline engine except for the possible cancellation available from the value of bank angle γ in equations 14.10j and 14.11c.[*] Open the file BMWV12.eng in program ENGINE to see the results for a vee-twelve engine. The common vee-eight engine with crankshaft phase angles of $\phi_i = 0, 90, 270, 180°$ has an unbalanced primary moment as does the inline four from which it is made. It is an example in program ENGINE.

UNBALANCED INERTIA TORQUES can be smoothed with a flywheel as was shown in Section 13.8 (p. 684) for the single-cylinder engine. Note that even an engine having zero inertial torque may require a flywheel to smooth its variations in gas torque. The total torque function should be used to determine the energy variations to be absorbed by a flywheel as it contains both gas torque and inertia torque (if any). The method of Section 11.11 (p. 605) also applies to calculation of the flywheel size needed in an engine, based on its variation in the total torque function. Program ENGINE will compute the areas under the total torque pulses needed for the calculation. See the referenced sections for the proper flywheel design procedure.

UNBALANCED SHAKING FORCES AND SHAKING MOMENTS can be cancelled by the addition of one or more rotating balance shafts within the engine. To cancel the primary components usually requires two balance shafts rotating at crank speed, one of which can be the crankshaft itself. To cancel the secondary components usually requires at least two balance shafts rotating at twice crank speed, gear or chain driven from the crankshaft. Figure 14-27a shows a pair of counterrotating shafts which are fitted with eccentric masses arranged 180° out of phase.[†] As shown, the unbalanced centrifugal forces from the equal, unbalanced masses will add to give a shaking force component in the vertical direction of twice the unbalanced force from each mass, while their horizontal components will exactly cancel. Pairs of counterrotating eccentrics can be arranged to provide a harmonically varying force in any one plane. The harmonic frequency will be determined by the rotational speed of the shafts.

If we arrange two pairs of eccentrics, with one pair displaced some distance along the shaft from the other, and also rotated 180° around the shaft from the first, as shown in Figure 14-27b, we will get a harmonically varying couple in one plane. There will be cancellation of the forces in one direction and summation in an orthogonal direction.

Thus to cancel the shaking moment in any plane, we can arrange a pair of shafts, each containing two eccentric masses displaced along those shafts, 180° out of phase, and gear the shafts together to rotate in opposite directions at any multiple of crankshaft speed.

---

[*] The angle γ is most effective when set to 90° for an opposed engine with 2γ = 180° since its cosine is zero, which cancels some components of the shaking force and moment.

[†] This is called a Lanchester balancer after its English inventor who developed it prior to World War I (c. 1913). It is still used in various kinds of machinery as well as in engines to cancel inertia forces.

14

To cancel the shaking force as well, it is only necessary to provide sufficient additional unbalanced mass in one of the pairs of eccentric masses to generate a shaking force opposite to that of the engine, over and above that needed to generate the forces of the couple.

In an inline or opposed engine, the unbalanced forces and moments are all confined to the single plane of the cylinders as they are due entirely to the reciprocating masses assumed concentrated at the wrist pin. (We are assuming that all crank throws are exactly balanced rotationally to cancel the effects of mass at the crank pin.) In a vee engine, however, the shaking forces and moments have both x and y components as shown in equations 14.10 and 14.11 (pp. 730-733) and in Figure 14-21 (p. 730). The shaking effects of each bank's pistons are acting within the plane of that bank's cylinders, and the bank angle γ is used to resolve them into x and y components using equations 14.10j and 14.11c (pp. 732-733).

VEE-TWIN   Primary force balance is possible in a 4-stroke vee-twin of any vee angle v if two crank pins are used. If $\phi_1 = 0$, the phase angle of the second pin ϕ_2 must be:[2]

$$\phi_2 = 180° - 2v \qquad\qquad (14.14a)$$

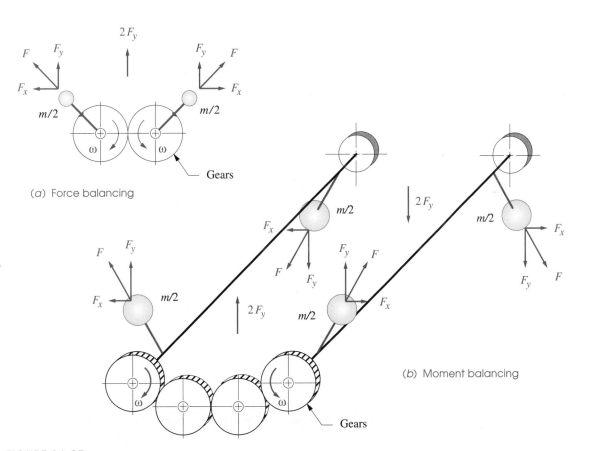

(a) Force balancing

(b) Moment balancing

FIGURE 14-27

Counterrotating eccentric masses can balance forces and moments.

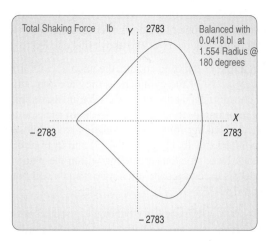

FIGURE 14-28

Shaking force in a 90° vee-twin engine (looking end-on to the crankshaft axis)

For even firing, the relationship must be:

$$\phi_2 = 360° - \nu \qquad (14.14b)$$

where ν is the vee angle as defined in Figure 14-21 (p. 730). The only value of ν that satisfies both criteria is 180° (opposed cylinders). All other vee-twin angles can have either even firing or primary balance, but not both.[*]

Figure 14-28 shows the two-dimensional shaking force present in a two-cylinder 90° vee, single-crank-pin engine, which satisfies equation 14.14a (p. 741) with $\phi_1 = \phi_2 = 0$, coalescing its "two crank pins" into one for this vee angle. The inertia force of each piston is confined to the reciprocating plane (bank) of that piston, but the vee angle between the cylinder banks creates the pattern shown when the primary and secondary components of each piston force are added vectorially. The shaking force of the 90° vee, single-crank pin twin has a rotating primary component of constant magnitude that can be cancelled with overbalanced counterweights on the crankshaft. However, its second harmonic is planar (in the YZ plane). To cancel it requires a pair of twice-speed balance shafts as shown in Figure 14-27a (p. 741).

VEE EIGHT The 90° vee-eight engine with a 0, 90, 270, 180° crankshaft, which has only an unbalanced primary moment, presents a special case. The 90° angle between the banks results in equal horizontal and vertical components of the primary shaking moment that reduces it to a couple of constant magnitude rotating about the crank axis at crankshaft speed in the same direction as the crank as shown in Figure 14-29. With this vee-eight engine, the primary moment can be balanced by merely adding two eccentric counterweights of proper size and opposite orientation to the crankshaft. No independent, second balance shaft is needed in the 90° vee-eight engine with this crankshaft. The 180° out-of-phase counterweights are typically placed near the ends of the crankshaft to obtain the largest moment arm possible and thus reduce their size.

VEE-SIX engines with a 3-throw, 2-conrod-per-throw, 0, 240, 120° crankshaft have unbalanced primary and secondary moments as does the three-cylinder inline from which

[*] Vee-twins for motorcycles have been made in a variety of vee angles: 45°, 48°, 50°, 52°, 60°, 75°, 80°, 90°, 180°, and possibly others. All but the 180° boxer twin have unbalanced secondary force and most have unbalanced primary force. Some have been fitted with balance shafts to reduce shaking. Many are also uneven firing, giving them a distinctive exhaust sound. The Harley Davidson single-crankpin, 45° vee-twin is one example whose sound (which the company unsuccessfully tried to patent) has been described as "potato-potato."

 14

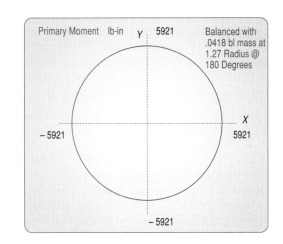

90 deg Vee 8 Cylinder
4 Stroke Cycle

Bore = 2.50 in
Stroke = 2.55
B/S = 0.98
L/R = 3.50
m_A = 0.0418 bl
m_B = 0.0116 bl
RPM = 3400

Phase Angles:
0 90 270 180 0 90 270 180

Power Strokes:
0 90 180 270 360 450 540 630

Primary Moment lb-in Y 5921   Balanced with
                                      .0418 bl mass at
                                      1.27 Radius @
                                      180 Degrees

– 5921   X   5921

– 5921

FIGURE 14-29

Primary moment in the 90° vee-eight engine (looking end-on to the crankshaft axis)

they are made and are not even firing. A 3-throw, 2-conrod-per-throw vee six needs a 120° vee angle for inherent balance. To reduce engine width, vee-sixes are most often made with a 60° vee angle which gives even firing with a 6-throw 0, 240, 120, 60, 300, 180° crank. This engine has unbalanced primary and secondary moments, each being a constant-magnitude rotating vector like that of the vee-eight shown in Figure 14-29. The primary component can be completely balanced by adding counterweights to the crank-shaft as done in the 90° vee-eight. Some manufacturers also add a single balance shaft in the valley of the-vee six, driven by gears at twice crankshaft speed to cancel the circular, constant-magnitude secondary shaking moment. Some vee-sixes use 90° vee angles to allow assembly on the same production line as 90° vee eights, but 3-throw 90° vee-sixes will run roughly due to uneven firing unless the crankshaft is redesigned to shift (or splay) the two conrods on each pin by 30°. This results in a 4-main bearing, 6-throw 0, 240, 120, 30, 270, 150° crankshaft that gives even firing but has nonconstant-magnitude primary and secondary shaking moments.

Calculation of the magnitude and location of the eccentric balance masses needed to cancel any shaking forces or moments is a straightforward exercise in **static balancing** (for forces) and **two-plane dynamic balancing** (for moments) as discussed in Sections 12.1 (p. 631) and 12.2 (p. 634), respectively. The unbalanced forces and moments for the particular engine configuration are calculated from the appropriate equations in this chapter. Two correction planes must be selected along the length of the balance shafts/crankshaft being designed. The magnitude and angular locations of the balance masses can then be calculated by the methods described in the noted sections of Chapter 12.

Secondary Harmonic Balancing of the **Four-Cylinder Inline Engine**

The four-cylinder inline engine with a 0, 180, 180, 0° crankshaft is one of the most widely used engines in the automobile industry. As described in a previous section, this engine suffers from unbalanced secondary force, moment, and torque. If the displacement of the engine is less than about 2.0 liters, then the magnitudes of the secondary forces may

14

be small enough to be ignored, especially if the engine mounts provide good vibration isolation of the engine from the passenger compartment. Above that displacement, objectionable noise, vibration, and harshness (NVH) may be heard and felt by the passengers at certain engine speeds where the frequency of the engine's second harmonic coincides with one of the natural frequencies of the body structure. Then some balancing is needed in the engine to avoid customer dissatisfaction.

Equation 14.2d (p. 714) defines the inline engine shaking force. Applying the relevant factors from Table 14-4 (p. 727) for this engine to the second harmonic term gives

$$\mathbf{F}_{s_2} \cong m_B r \omega^2 \frac{4r}{l} \cos 2\omega t \ \hat{\mathbf{i}} \tag{14.15}$$

The shaking torque for an inline engine is given by equation 14.4c (p. 716) in combination with equation 13.15f (p. 683). Taking only the second harmonic term and applying the relevant factors from Table 14-4 for this engine give

$$\mathbf{T}_{s_2} \cong 2 m_B r^2 \omega^2 \sin 2\omega t \ \hat{\mathbf{k}} \tag{14.16}$$

The principle of the Lanchester balancer, shown in Figure 14-27a (p. 741), can be used to counteract the secondary forces by driving its two counterrotating balance shafts at twice crankshaft speed with chains and/or gears. Figure 14-30 shows such an arrangement as applied to a Mitsubishi 2.6-liter, four-cylinder engine.[*]

H. Nakamura [3] improved on Lanchester's 1913 design by arranging the balance shafts within the engine so as to cancel the second harmonic of the inertia torque as well as the secondary inertia force. But, his arrangement does not affect the unbalanced secondary shaking moment. In fact, it is designed to impart zero net moment about a transverse axis to either balance shaft in order to minimize bending moments on the shafts, and so reduce bearing loads and friction losses. This feature is the subject and principal claim of Nakamura's patent on this design. [4]

Figure 14-31a shows a schematic of a conventional Lanchester balancer arranged with the two counterrotating balance shafts with their centers in a single horizontal plane transverse to the vertical plane of piston motion.[†]

The balance force from the two balance shafts combined is

$$\mathbf{F}_{bal} \cong -8 m_{bal} r_{bal} \omega^2 \cos 2\omega t \ \hat{\mathbf{i}} \tag{14.17}$$

where m_{bal} and r_{bal} are the mass and radius, respectively, of one balance weight.

Figure 14-31b shows Nakamura's arrangement of the balance shafts with one situated above the other in separate horizontal planes. The vertical offset $x_1 - x_2$ between the shafts, in combination with the oppositely directed but equal-magnitude horizontal components of the counterweights' centrifugal forces, creates a time-varying couple *about the crankshaft axis* defined as:

$$\mathbf{T}_{bal} \cong -4 m_{bal} r_{bal} \omega^2 \left(x_1 - x_2 \right) \sin 2\omega t - \left(y_1 + y_2 \right) \cos 2\omega t \ \hat{\mathbf{k}} \tag{14.18}$$

where x and y refer to the coordinates of the shaft centers referenced to the crankshaft center, and the subscripts 1 and 2 refer, respectively, to the balance shaft turning in directions the same as and opposite to that of the crankshaft.

[*] Also used at one time by Chrysler and Porsche under license from Mitsubishi.

[†] The arrangement shown in Figure 14-31a is only schematic of Lanchester's original design in which the crankshaft drove the balance shafts through right-angled helical gears with the balance shaft axes parallel to the y axis of Figure 14-31, i.e., transverse to, rather than parallel to, the crankshaft axis as shown here. See reference [4] for drawings of his original design.

FIGURE 14-30

Balance shafts used to eliminate the secondary unbalance in the four-cylinder inline engine
Courtesy of the Chrysler Corporation

The vertical components of the balance weights' centrifugal forces still add to provide force balance as in equation 14.17. The torque in equation 14.18 will have opposite sense to the shaking torque if the upper shaft turns in the same direction as, and the lower shaft turns in the opposite direction to the crankshaft.

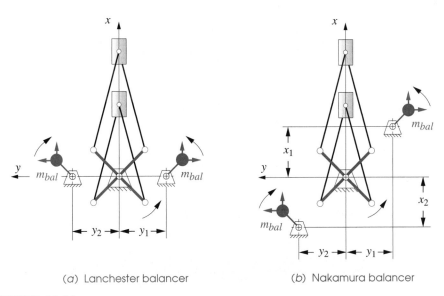

(a) Lanchester balancer   (b) Nakamura balancer

FIGURE 14-31

Two types of secondary balancer mechanisms for the four-cylinder inline engine

For force balance, equations 14.15 and 14.17 must sum to zero,

$$m_B r \omega^2 \frac{4r}{l} \cos 2\omega t - 8 m_{bal} r_{bal} \omega^2 \cos 2\omega t = 0$$

or
$$m_{bal} r_{bal} = \frac{r}{2l} m_B r \qquad\qquad (14.19)$$

which defines the mass-radius product needed for the balance mechanism.

For torque balance, equations 14.16 and 14.18 (p. 744) must sum to zero.

$$2 m_B r^2 \omega^2 \sin 2\omega t - 4 m_{bal} r_{bal} \omega^2 (x_1 - x_2) \sin 2\omega t - (y_1 + y_2) \cos 2\omega t = 0 \qquad (14.20a)$$

Substitute equation 14.19 in 14.20a.

$$2 m_B r^2 \sin 2\omega t - 4 \frac{r}{2l} m_B r (x_1 - x_2) \sin 2\omega t - (y_1 + y_2) \cos 2\omega t = 0 \qquad (14.20b)$$

For this equation to be zero for all ωt,

$$y_2 = -y_1$$
$$x_1 - x_2 = l$$
$$(14.20c)$$

So, if the balance shafts are arranged symmetrically with respect to the piston plane at any convenient locations y_1 and $-y_1$, and the distance $x_1 - x_2$ is made equal to the length of the connecting rod l, then the second harmonic of the inertia torque will be completely cancelled. Since the second harmonic is the only nonzero component of inertia torque in this engine as can be seen in Figure 14-20b (p. 729), it will now be completely balanced for shaking force and shaking torque (but not shaking moment).

There is also significant oscillation of the gas torque in a four-cylinder engine as shown in Figure 14-20a (p. 729). The gas torque is 180° out of phase with the inertia torque as can be seen in Figure 14-20b and so provides some natural cancellation as shown in the total torque curve of Figure 14-20c. The magnitude of the gas torque varies with engine load and so cannot itself be cancelled with any particular balance shaft geometry for all conditions. However, one engine speed and load condition can be selected as representative of the majority of typical driving conditions, and the balance system geometry altered to give an optimum reduction of total engine torque under those conditions. Nakamura estimates that gas torque magnitude is about 30% of the inertia torque under typical driving conditions and so suggests a value of $x_1 - x_2 = 0.7l$ for the best overall reduction of total torque oscillation in this engine. Note that the average value of the driving torque is not affected by balancing because the average torque of any rotating balance system is always zero.

A Perfectly Balanced Two-Cylinder Engine

Frederick Lanchester, in 1897, devised an extremely clever horizontally opposed engine arrangement[5] of Figure 14-32 that, with only two cylinders, completely cancelled all harmonics of inertia forces and moments. He recognized that the lateral motion of the connecting rods was a contributor and so provided two counterrotating crankshafts, driven by a total of six connecting rods, three per crank pin, with two upper rods straddling one lower rod for Z axis symmetry. The crank counterweights exactly balance the cranks.

(a) Right cylinder at TDC
and left nearing BDC

(b) Both cylinders
at mid-stroke

(c) Left cylinder at BDC
and right descending

FIGURE 14-32

Perfectly balanced Lanchester two-cylinder, horizontally opposed engine (1897)

The **colinear** opposed pistons exactly balance one another's linear accelerations and the scissors action of the multiple conrods exactly cancels all higher harmonics of motion. Clearly the work of genius. The genesis of his later (c. 1913) harmonic balancer of Figure 14-31a (p. 745) can also be seen here.

14.10 REFERENCES

1   **MacKay, R. F.** (1915). *The Theory of Machines*, Edward Arnold: London, p. 103.

2   **Jantz, J., and R. Jantz.** (2001). "Why They Shake and Why They Don't," Part 7, *Roadracing World*, Feb. 2001, pp. 32-35.

3   **Nakamura, H.** (1976). "A Low Vibration Engine with Unique Counter-Balance Shafts." SAE Paper: 760111.

4   **Nakamura, H., et al.** (1977). "Engine Balancer." Assignee: Mitsubishi Corp.: U.S. Patent 4,028,963.

5   **Bird, A., and F. Hutton-Stott.** (1965). *Lanchester Motor Cars*, Cassell: London, p. 137.

6   **Ibid.**, p. 26.

14.11 BIBLIOGRAPHY

Crouse, W. H. (1970). *Automotive Engine Design*, McGraw-Hill Inc., New York.

Jantz, J., and Jantz, R. (1994-2002). "Why They Shake and Why They Don't," Parts 1-9, *Roadracing World*, Sept. 1994, Feb. 1995, Mar. 1996, Mar. 1997, Mar. 1999, Apr.. 1999, Feb. 2001, Mar. 2001, Mar. 2002.

Jennings, G. (1979). "A Short History of Wonder Engines," *Cycle Magazine*, May 1979, p. 68ff.

Setright, L. J. K. (1975). *Some Unusual Engines*, Mechanical Engineering Publications Ltd., The Inst. of Mech. Engr., London.

Thomson, W. (1978). *Fundamentals of Automotive Balance*, Mechanical Engineering Publications Ltd., London.

14

14.12 PROBLEMS

14-1 Draw a crank phase diagram for a three-cylinder inline engine with a 0, 120, 240° crankshaft and determine all possible firing orders and select the best arrangement to give even firing for each stroke cycle for:

a. Four-stroke cycle   b. Two-stroke cycle

14-2 Repeat Problem 14-1 for an inline four-cylinder engine with 0, 90, 270, 180° crank.

14-3 Repeat Problem 14-1 for a 45° vee, four-cylinder engine with 0, 90, 270, 180° crank.

14-4 Repeat Problem 14-1 for a 45° vee, two-cylinder engine with 0, 90° crank.

14-5 Repeat Problem 14-1 for a 90° vee, two-cylinder engine with 0, 180° crank.

14-6 Repeat Problem 14-1 for a 180° opposed, two-cylinder engine with a 0, 180° crank.

14-7 Repeat Problem 14-1 for a 180° opposed, four-cylinder engine with 0, 180, 180, 0° crank.

†14-8 Calculate the shaking force, torque, and moment balance conditions through the second harmonic for the engine design in Problem 14-1.

†14-9 Calculate the shaking force, torque, and moment balance conditions through the second harmonic for the engine design in Problem 14-2.

†14-10 Calculate the shaking force, torque, and moment balance conditions through the second harmonic for the engine design in Problem 14-3.

†14-11 Calculate the shaking force, torque, and moment balance conditions through the second harmonic for the engine design in Problem 14-4.

†14-12 Calculate the shaking force, torque, and moment balance conditions through the second harmonic for the engine design in Problem 14-5.

†14-13 Calculate the shaking force, torque, and moment balance conditions through the second harmonic for the engine design in Problem 14-6.

†14-14 Calculate the shaking force, torque, and moment balance conditions through the second harmonic for the engine design in Problem 14-7.

14-15 Derive expressions, in general terms, for the magnitude and angle with respect to the first crank throw of the mass-radius products needed on the crankshaft to balance the shaking moment in a 90° vee-eight engine with a 0, 90, 270, 180° crankshaft.

14-16 Repeat Problem 14-15 for a 90° vee-six with a 0, 240, 120° crankshaft.

14-17 Repeat Problem 14-15 for a 90° vee-four with a 0, 180° crankshaft.

†14-18 Design a pair of Nakamura balance shafts to cancel the shaking force and reduce torque oscillations in the engine shown in Figure 14-19 (p. 728).*

14-19 Using program ENGINE, data in Table P14-1 and the crank phase diagram from Problem 14-1, determine the maximum force magnitudes on main pin, crank pin, wrist pin, and piston for a 2-stroke engine with even firing. Overbalance the crank, if necessary, to bring the shaking force down to at least half the unbalanced value.

14-20 Using program ENGINE, data in Table P14-1 and the crank phase diagram from Problem 14-2, determine the maximum force magnitudes on main pin, crank pin, wrist pin, and piston for a 4-stroke engine with even firing. Overbalance the crank, if necessary, to bring the shaking force down to at least half of the unbalanced value.

14

† These problems are suited to solution using *Mathcad, Matlab,* or *TKSolver* equation solver programs.

* More information on this engine design can be found in program ENGINE where it is one of the examples.

14-21 Using program ENGINE, data in Table P14-1 and the crank phase diagram from Problem 14-3, determine the maximum force magnitudes on main pin, crank pin, wrist pin, and piston for a 4-stroke engine with even firing. Overbalance the crank, if necessary, to bring the shaking force down to at least half of the unbalanced value.

14-22 Using program ENGINE, data in Table P14-1 and the crank phase diagram from Problem 14-4, determine the maximum force magnitudes on main pin, crank pin, wrist pin, and piston for a 2-stroke engine with even firing. Overbalance the crank, if necessary, to bring the shaking force down to at least half of the unbalanced value.

†*14-23 A four-cylinder inline engine with a 0, 180, 180, 0° crankshaft has a stroke of $S = 3.50$ in, a conrod length to crank radius ratio of $L/R = 3.75$, and an effective wrist pin mass of $m_B = 0.0215$ blob. Design a pair of Nakamura balance shafts to cancel the shaking force and reduce the oscillations in the engine.

†14-24 Repeat problem 14-23 with $S = 2.750$ in, $L/R = 3.00$, and $m_B = 0.0125$ blob.

14.13 PROJECTS

These are loosely structured design problems intended for solution using program ENGINE. All involve the design of one or more multicylinder engines and differ mainly in the specific data for the engine. The general problem statement is:

Design a multicylinder engine for a specified displacement and stroke cycle. Optimize the conrod/crank ratio and bore/stroke ratio to minimize shaking forces, shaking torque, and pin forces, also considering package size. Design your link shapes and calculate realistic dynamic parameters (mass, CG location, moment of inertia) for those links using the methods of Chapters 10-13. Dynamically model the links as described in those chapters. Balance or overbalance the linkage as needed to achieve the desired results. Choose crankshaft phase angles (and vee angles, if appropriate) to optimize the inertial balance of the engine. Choose a firing order and determine the power stroke angles to optimize even firing. Trade off inertia balance if necessary to achieve even firing. Design and size a minimum-weight flywheel by the method of Chapter 11 to smooth total torque. Write an engineering report on your design and analysis.

P14-1   Two-stroke cycle inline twin with a displacement of 1 liter.

P14-2   Four-stroke cycle inline twin with a displacement of 1 liter.

P14-3   Two-stroke cycle vee-twin with a displacement of 1 liter.

P14-4   Four-stroke cyclevee-twin with a displacement of 1 liter.

P14-5   Two-stroke cycle opposed twin with a displacement of 1 liter.

P14-6   Four-stroke cycle opposed twin with a displacement of 1 liter.

P14-7   Two-stroke cycle vee-four with a displacement of 2 liters.

P14-8   Four-stroke cycle vee-four with a displacement of 2 liters.

P14-9   Two-stroke cycle opposed four with a displacement of 2 liters.

P14-10 Four-stroke cycle opposed four with a displacement of 2 liters.

P14-11 Two-stroke inline five-cylinder with a displacement of 2.5 liters.

P14-12 Four-stroke inline five-cylinder with a displacement of 2.5 liters.

P14-13 Two-stroke cycle vee-six with a displacement of 3 liters.

TABLE P14-1
Data for Problems
14-19 to 14-22

Displacement	10.0
Bore	1.87
L/R ratio	3.00
r_{G2}/r	0.40
r_{G3}/l	0.36
Main pin dia	2.00
Crank pin dia	1.50
Idle rpm	600
Redline rpm	4000
Piston mass	0.015
Conrod mass	0.012
Crank mass	0.045
P_{gmax}	5500
Friction coeff	0.02
Flywheel coeff	0.10

14

---

* Answers in Appendix F.

† These problems are suited to solution using *Mathcad, Matlab,* or, *TKSolver* equation solver programs.

P14-14   Four-stroke cycle vee-six with a displacement of 3 liters.

P14-15   Two-stroke cycle opposed six with a displacement of 3 liters.

P14-16   Four-stroke cycle opposed six with a displacement of 3 liters.

P14-17   Two-stroke inline seven-cylinder with a displacement of 3.5 liters.

P14-18   Four-stroke inline seven-cylinder with a displacement of 3.5 liters.

P14-19   Two-stroke inline eight-cylinder with a displacement of 4 liters.

P14-20   Four-stroke inline eight-cylinder with a displacement of 4 liters.

P14-21   Two-stroke vee ten-cylinder with a displacement of 5 liters.

P14-22   Four-stroke vee ten-cylinder with a displacement of 5 liters.

P14-23   Four-stroke W-6 comprised of three banks of two with a displacement of 5 liters.*

P14-24   Four-stroke W-9 comprised of three banks of three with a displacement of 5 liters.*

P14-25   Four-stroke W-12 comprised of three banks of four with a displacement of 5 liters.*

P14-26   Four-stroke W-12 comprised of four banks of three with a displacement of 5 liters.*

P14-27   Design a family of vee engines to be built on the same assembly line. They must all have the same vee angle and use the same pistons, connecting rods, and strokes. Their crankshafts can each be different. Four configurations are needed: vee-four, vee-six, vee-eight, and vee-ten. All will have the same single-cylinder displacement of 0.5 liters. Optimize the single-cylinder configuration from which the multicylinder engines will be constructed for bore/stroke ratio and conrod/crank ratio. Then assemble this cylinder design into the above configurations. Find the best compromise of vee angle to provide a good mix of balance and even firing in all engines.

P14-28   Repeat Project P14-27 for a family of three-bank W engines: W-3, W-6, W-9, and W-12. The interbank angles must be the same for all models. See the built-in example W-12 engine in program ENGINE for more information on this unusual W configuration.

P14-29   In recent years some automobile manufacturers have made unusual vee configurations such as the VW-Audi VR15 which is a 15° vee-six. Obtain detailed information on this engine design and then analyze it with program ENGINE. Write a report that explains why the manufacturer chose this unusual arrangement and justify your conclusions with sound engineering analysis.

P14-30   Design an inline six- and an inline five-cylinder engine of the same displacement, say 2.5 liters. Analyze their dynamics with program ENGINE. Write an engineering report to explain why such manufacturers as Audi, Volvo, and Acura have chosen a five-cylinder inline over a six-cylinder of comparable torque and power output.

P14-31   Ferrari has produced vee-twelve engines in both 60° vee and horizontally opposed configurations. Design 3-liter versions of each and compare their dynamics. Write a report that explains why the manufacturer chose these arrangements and justify your conclusions with sound engineering analysis.

P14-32   Design and compare a 3-liter 90° vee-six, 60° vee-six, inline six, and 180° opposed six, examples of which are all in volume production. Explain their advantages and disadvantages and justify your conclusions with sound engineering analysis.

14

* A W engine has three banks of four cylinders or four banks of three cylinders on a common crankshaft. The VW-Audi W-12, as used in some Bentley models, is two VR15, 15-deg-vee sixes grafted together on a common crankshaft.

Chapter 15

CAM DYNAMICS

The universe is full of magical things
patiently waiting for our wits to grow sharper
EDEN PHILLPOTS

15.0 INTRODUCTION

Chapter 8 presented the kinematics of cams and followers and methods for their design. We will now extend the study of cam-follower systems to include considerations of the dynamic forces and torques developed. While the discussion in this chapter is limited to examples of cams and followers, the principles and approaches presented are applicable to most dynamic systems. The cam-follower system can be considered a useful and convenient example for the presentation of topics such as creating lumped parameter dynamic models and defining equivalent systems as described in Chapter 10. These techniques as well as the discussion of natural frequencies, effects of damping, and analogies between physical systems will be found useful in the analysis of all dynamic systems regardless of type.

In Chapter 10 we discussed the two approaches to dynamic analysis, commonly called the forward and the inverse dynamics problems. The forward problem assumes that all the forces acting on the system are known and seeks to solve for the resulting displacements, velocities, and accelerations. The inverse problem is, as its name says, the inverse of the other. The displacements, velocities, and accelerations are known, and we solve for the dynamic forces that result. In this chapter we will explore the application of both methods to cam-follower dynamics. Section 15.1 (p. 752) explores the forward solution. Section 15.3 (p. 763) will present the inverse solution. Both are instructive in this application of a force-closed (spring-loaded) cam-follower system and will each be discussed in this chapter.

15

TABLE 15-1   Notation Used in This Chapter

c = damping coefficient
c_c = critical damping coefficient
k = spring constant
F_c = force of cam on follower
F_s = force of spring on follower
F_d = force of damper on follower
m = mass of moving elements
t = time in seconds
T_c = torque on camshaft
θ = camshaft angle, in degrees or radians
ω = camshaft angular velocity, rad / sec
ω_d = damped circular natural frequency, rad / sec
ω_f = forcing frequency, rad / sec
ω_n = undamped circular natural frequency, rad / sec
x = follower displacement, length units
$\dot{x} = v$ = follower velocity, length / sec
$\ddot{x} = a$ = follower acceleration, length / sec^2
ζ = damping ratio

15.1   DYNAMIC FORCE ANALYSIS OF THE FORCE-CLOSED CAM-FOLLOWER

Figure 15-1a shows a simple plate or disk cam driving a spring-loaded, roller follower. This is a force-closed system which depends on the spring force to keep the cam and follower in contact at all times. Figure 15-1b shows a lumped parameter model of this system in which all the **mass** which moves with the follower train is lumped together as m, all the springiness in the system is lumped within the **spring constant** k, and all the **damping** or resistance to movement is lumped together as a damper with coefficient c. The sources of mass which contribute to m are fairly obvious. The masses of the follower stem, the roller, its pivot pin, and any other hardware attached to the moving assembly all add together to create m. Figure 15-1c shows the free-body diagram of the system acted upon by the cam force F_c, the spring force F_s, and the damping force F_d. There will of course also be the effects of mass times acceleration on the system.

Undamped Response

Figure 15-2 shows an even simpler lumped parameter model of the same system as in Figure 15-1 but which omits the damping altogether. This is referred to as a *conservative model* since it conserves energy with no losses. This is not a realistic or safe assumption in this case but will serve a purpose in the path to a better model which will include the damping. The free-body diagram for this mass-spring model is shown in Figure 15-2c. We can write Newton's equation for this one-*DOF* system:

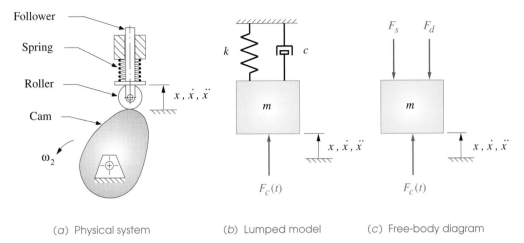

(a) Physical system   (b) Lumped model   (c) Free-body diagram

FIGURE 15-1

One-*DOF* lumped parameter model of a cam-follower system including damping

$$\sum F = ma = m\ddot{x}$$

$$F_c(t) - F_s = m\ddot{x}$$

From equation 10.16 (p. 557):

$$F_s = k x$$

then:

$$m\ddot{x} + k x = F_c(t) \qquad\qquad (15.1a)$$

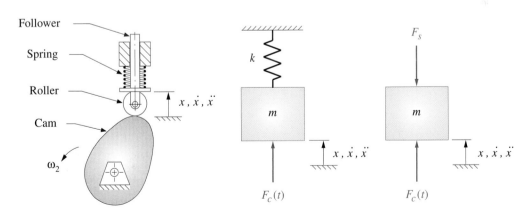

(a) Physical system   (b) Lumped model   (c) Free-body diagram

FIGURE 15-2

One-*DOF* lumped parameter model of a cam-follower system without damping

15

This is a second-order ordinary differential equation (ODE) with constant coefficients. The complete solution will consist of the sum of two parts, the transient (homogeneous) and the steady state (particular). The homogeneous ODE is

$$m\ddot{x} + k\,x = 0$$

$$\ddot{x} = -\frac{k}{m}x$$

(15.1b)

which has the well-known solution

$$x = A\cos\omega t + B\sin\omega t$$

(15.1c)

where A and B are the constants of integration to be determined by the initial conditions. To check the solution, differentiate it twice, assuming constant ω, and substitute in the homogeneous ODE.

$$-\omega^2\left(A\cos\omega t + B\sin\omega t\right) = -\frac{k}{m}\left(A\cos\omega t + B\sin\omega t\right)$$

This is a solution provided that:

$$\omega^2 = \frac{k}{m} \qquad\qquad \omega_n = \sqrt{\frac{k}{m}}$$

(15.1d)

The quantity ω_n (rad/sec) is called the *circular natural frequency* of the system and is the frequency at which the system wants to vibrate if left to its own devices. This represents the *undamped natural frequency* since we ignored damping. The *damped natural frequency* will be slightly lower than this value. Note that ω_n is a function only of the physical parameters of the system m and k; thus it is completely determined and unchanging with time once the system is built. By creating a one-*DOF* model of the system, we have limited ourselves to one natural frequency which is an "average" natural frequency usually close to the lowest, or fundamental, frequency.

Any real physical system will also have higher natural frequencies which in general will not be integer multiples of the fundamental. In order to find them we need to create a multi-degree-of-freedom model of the system. The fundamental tone at which a bell rings when struck is its natural frequency defined by this expression. The bell also has overtones which are its other, higher, natural frequencies. The fundamental frequency tends to dominate the transient response of the system.[1]

The circular natural frequency ω_n (rad/sec) can be converted to cycles per second (hertz) by noting that there are 2π radians per revolution and one revolution per cycle:

$$f_n = \frac{1}{2\pi}\omega_n \quad \text{hertz}$$

(15.1e)

The constants of integration, A and B in equation 15.1c, depend on the initial conditions. A general case can be stated as

When $t = 0$,   let $x = x_0$ and $v = v_0$,   where x_0 and v_0 are constants

which gives a general solution to the homogeneous ODE 15.1b of:

$$x = x_0\cos\omega_n t + \frac{v_0}{\omega_n}\sin\omega_n t$$

(15.1f)

15

Equation 15.1f can be put into polar form by computing the magnitude and phase angle:

$$X_0 = \sqrt{x_0^2 + \left(\frac{v_0}{\omega_n}\right)^2} \qquad\qquad \phi = \arctan\left(\frac{v_0}{x_0\omega_n}\right)$$

then:

$$x = X_0 \cos\left(\omega_n t - \phi\right) \qquad\qquad (15.1g)$$

Note that this is a pure harmonic function whose amplitude X_0 and phase angle ϕ are a function of the initial conditions and the natural frequency of the system. It will oscillate forever in response to a single, transitory input if there is truly no damping present.

Damped Response

If we now reintroduce the damping of the model in Figure 15-1b and draw the free-body diagram as shown in Figure 15-1c, the summation of forces becomes:

$$F_c\left(t\right) - F_d - F_s = m\ddot{x} \qquad\qquad (15.2a)$$

Substituting equations 10.16 (p. 557) and 10.17c (p. 559):

$$m\ddot{x} + c\dot{x} + kx = F_c\left(t\right) \qquad\qquad (15.2b)$$

HOMOGENEOUS SOLUTION   We again separate this differential equation into its homogeneous and particular components. The homogeneous part is:

$$\ddot{x} + \frac{c}{m}\dot{x} + \frac{k}{m}x = 0 \qquad\qquad (15.2c)$$

The solution to this ODE is of the form:

$$x = Re^{st} \qquad\qquad (15.2d)$$

where R and s are constants. Differentiating versus time:

$$\dot{x} = Rse^{st}$$
$$\ddot{x} = Rs^2 e^{st}$$

and substituting in equation 15.2c:

$$Rs^2 e^{st} + \frac{c}{m}Rse^{st} + \frac{k}{m}Re^{st} = 0$$

$$\left(s^2 + \frac{c}{m}s + \frac{k}{m}\right)Re^{st} = 0 \qquad\qquad (15.2e)$$

For this solution to be valid either R or the expression in parentheses must be zero as e^{st} is never zero. If R were zero, then the assumed solution, in equation 15.2d, would also be zero and thus not be a solution. Therefore, the quadratic equation in parentheses must be zero.

$$\left(s^2 + \frac{c}{m}s + \frac{k}{m}\right) = 0 \qquad\qquad (15.2f\,)$$

15

This is called the characteristic equation of the ODE and its solution is:

$$s = \frac{-\dfrac{c}{m} \pm \sqrt{\left(\dfrac{c}{m}\right)^2 - 4\dfrac{k}{m}}}{2}$$

which has the two roots:

$$s_1 = -\frac{c}{2m} + \sqrt{\left(\frac{c}{2m}\right)^2 - \frac{k}{m}}$$

(15.2g)

$$s_2 = -\frac{c}{2m} - \sqrt{\left(\frac{c}{2m}\right)^2 - \frac{k}{m}}$$

These two roots of the characteristic equation provide two independent terms of the homogeneous solution:

$$x = R_1 e^{s_1 t} + R_2 e^{s_2 t} \qquad \text{for } s_1 \neq s_2 \qquad (15.2h)$$

If $s_1 = s_2$, then another form of solution is needed. The quantity s_1 will equal s_2 when:

$$\sqrt{\left(\frac{c}{2m}\right)^2 - \frac{k}{m}} = 0 \qquad \text{or:} \qquad \frac{c}{2m} = \sqrt{\frac{k}{m}}$$

and:

$$c = 2m\sqrt{\frac{k}{m}} = 2m\omega_n = c_c \qquad (15.2i)$$

This particular value of c is called the **critical damping** and is labeled c_c. The system will behave in a unique way when critically damped, and the solution must be of the form:

$$x = R_1 e^{s_1 t} + R_2 t e^{s_2 t} \qquad \text{for } s_1 = s_2 = -\frac{c}{2m} \qquad (15.2j)$$

It will be useful to define a dimensionless ratio called the **damping ratio** ζ which is the actual damping divided by the critical damping.

$$\zeta = \frac{c}{c_c}$$

(15.3a)

$$\zeta = \frac{c}{2m\omega_n}$$

and then:

$$\zeta\omega_n = \frac{c}{2m} \qquad (15.3b)$$

The damped natural frequency ω_d is slightly less than the undamped natural frequency ω_n and is:

$$\omega_d = \sqrt{\frac{k}{m} - \left(\frac{c}{2m}\right)^2} \qquad (15.3c)$$

We can substitute equations 15.1d (p. 754) and 15.3b into equations 15.2g to get an expression for the characteristic equation in terms of dimensionless ratios:

$$s_{1,2} = -\omega_n \zeta \pm \sqrt{\left(\omega_n \zeta\right)^2 - \omega_n^{\,2}}$$

(15.4a)

$$s_{1,2} = \omega_n \left(-\zeta \pm \sqrt{\zeta^2 - 1}\right)$$

This shows that the system response is determined by the damping ratio ζ which dictates the value of the discriminant. There are three possible cases:

CASE 1:	$\zeta > 1$	Roots real and unequal
CASE 2:	$\zeta = 1$	Roots real and equal
CASE 3:	$\zeta < 1$	Roots complex conjugate

(15.4b)

Let's consider the response of each of these cases separately.

CASE 1: $\zeta > 1$ *overdamped*

The solution is of the form in equation 15.2h and is:

$$x = R_1 e^{\left(-\zeta + \sqrt{\zeta^2 - 1}\right)\omega_n t} + R_2 e^{\left(-\zeta - \sqrt{\zeta^2 - 1}\right)\omega_n t}$$

(15.5a)

Note that since $\zeta > 1$, both exponents will be negative making x the sum of two decaying exponentials as shown in Figure 15-3. This is the transient response of the system to a disturbance and dies out after a time. There is no oscillation in the output motion. An example of an overdamped system which you have probably encountered is the tone arm on a good-quality record turntable with a "cueing" feature. The tone arm can be lifted up, then released, and it will slowly "float" down to the record. This is achieved by putting a large amount of damping in the system, at the arm pivot. The arm's motion follows an exponential decay curve such as in Figure 15-3.

CASE 2: $\zeta = 1$ *critically damped*

The solution is of the form in equation 15.2j and is:

$$x = R_1 e^{-\omega_n t} + R_2 t e^{-\omega_n t} = \left(R_1 + R_2 t\right) e^{-\omega_n t}$$

(15.5b)

This is the product of a linear function of time and a decaying exponential function and can take several forms depending on the values of the constants of integration, R_1 and R_2, which in turn depend on initial conditions. A typical transient response might look like Figure 15-4 (p. 758). This is the transient response of the system to a disturbance, which response dies out after a time. There is fast response but no oscillation in the output motion. An example of a critically damped system is the suspension system of a new sports car in which the damping is usually made close to critical in order to provide crisp handling response without either oscillating or being slow to respond. A critically damped system will, when disturbed, return to its original position within one bounce. It may overshoot but will not oscillate and will not be sluggish.

(a) Term 1

+

(b) Term 2

=

(c) Total response

FIGURE 15-3

Transient response of an overdamped system

15

(a) Term

X

(b) Term 2

=

(c) Total response

FIGURE 15-4

Transient response of a critically damped system

CASE 3: $\zeta < 1$ *underdamped*

The solution is of the form in equation 15.2h (p. 756) and s_1, s_2 are complex conjugate. Equation 15.4a (p. 757) can be rewritten in a more convenient form as:

$$s_{1,2} = \omega_n\left(-\zeta \pm j\sqrt{1-\zeta^2}\right) \qquad\qquad j = \sqrt{-1} \qquad (15.5c)$$

Substituting in equation 15.2h:

$$x = R_1 e^{\left(-\zeta+j\sqrt{1-\zeta^2}\right)\omega_n t} + R_2 e^{\left(-\zeta-j\sqrt{1-\zeta^2}\right)\omega_n t}$$

and noting that:

$$y^{a+b} = y^a y^b$$

$$x = R_1\left[e^{-\zeta\omega_n t} e^{\left(j\sqrt{1-\zeta^2}\right)\omega_n t}\right] + R_2\left[e^{-\zeta\omega_n t} e^{\left(-j\sqrt{1-\zeta^2}\right)\omega_n t}\right]$$

factor:

$$x = e^{-\zeta\omega_n t}\left[R_1 e^{\left(j\sqrt{1-\zeta^2}\right)\omega_n t} + R_2 e^{\left(-j\sqrt{1-\zeta^2}\right)\omega_n t}\right] \qquad (15.5d)$$

Substitute the Euler identity from equation 4.4a (p. 185):

$$x = e^{-\zeta\omega_n t}\left\{\begin{array}{l} R_1\left[\cos\left(\sqrt{1-\zeta^2}\,\omega_n t\right)+j\sin\left(\sqrt{1-\zeta^2}\,\omega_n t\right)\right] \\ \\ + R_2\left[\cos\left(\sqrt{1-\zeta^2}\,\omega_n t\right)-j\sin\left(\sqrt{1-\zeta^2}\,\omega_n t\right)\right] \end{array}\right\}$$

and simplify: $\qquad\qquad\qquad\qquad\qquad\qquad\qquad\qquad\qquad (15.5e)$

$$x = e^{-\zeta\omega_n t}\left\{(R_1 + R_2)\left[\cos\left(\sqrt{1-\zeta^2}\,\omega_n t\right)+(R_1 - R_2)j\sin\left(\sqrt{1-\zeta^2}\,\omega_n t\right)\right]\right\}$$

Note that R_1 and R_2 are just constants yet to be determined from the initial conditions, so their sum and difference can be denoted as some other constants:

$$x = e^{-\zeta\omega_n t}\left\{A\left[\cos\left(\sqrt{1-\zeta^2}\,\omega_n t\right)+B\sin\left(\sqrt{1-\zeta^2}\,\omega_n t\right)\right]\right\} \qquad (15.5f)$$

We can put this in polar form by defining the magnitude and phase angle as:

$$X_0 = \sqrt{A^2 + B^2} \qquad\qquad \phi = \arctan\frac{B}{A} \qquad (15.5g)$$

then:

$$x = X_0 e^{-\zeta\omega_n t}\cos\left[\left(\sqrt{1-\zeta^2}\,\omega_n t\right)-\phi\right] \qquad (15.5h)$$

This is the product of a harmonic function of time and a decaying exponential function where X_0 and ϕ are the constants of integration determined by the initial conditions.

15

Figure 15-5 shows the transient response for this **underdamped case**. The response overshoots and oscillates before finally settling down to its final position. Note that if the damping ratio ζ is zero, equation 15.5g reduces to equation 15.1g (p. 755) which is a pure harmonic.

An example of an underdamped system is a diving board which continues to oscillate after the diver has jumped off, finally settling back to zero position. Many real systems in machinery are underdamped, including the typical cam-follower system. This often leads to vibration problems. It is not usually a good solution simply to add damping to the system as this causes heating and is very energy inefficient. It is better to design the system to avoid the vibration problems.

PARTICULAR SOLUTION Unlike the homogeneous solution which is always the same regardless of the input, the particular solution to equation 15.2b (p. 755) will depend on the forcing function $F_c(t)$ which is applied to the cam-follower from the cam. In general the output displacement x of the follower will be a function of similar shape to the input function but will lag the input function by some phase angle. It is quite reasonable to use a sinusoidal function as an example since any periodic function can be represented as a Fourier series of sine and cosine terms of different frequencies (see equation 13.2, p. 670, equation 13.3, p. 671, and the footnote on p. 671).

Assume the forcing function to be:

$$F_c(t) = F_0 \sin \omega_f t \tag{15.6a}$$

where F_0 is the amplitude of the force and ω_f is its circular frequency. Note that ω_f is unrelated to ω_n or ω_d and may be any value. The system equation then becomes:

$$m\ddot{x} + c\dot{x} + kx = F_0 \sin \omega_f t \tag{15.6b}$$

The solution must be of harmonic form to match this forcing function and we can try the same form of solution as used for the homogeneous solution.

$$x_f(t) = X_f \sin(\omega_f t - \psi) \tag{15.6c}$$

where:

X_f = amplitude

ψ = phase angle between applied force and displacement

ω_f = angular velocity of forcing function

The factors X_f and ψ are not constants of integration here. They are constants determined by the physical characteristics of the system and the forcing function's frequency and magnitude. They have nothing to do with the initial conditions. To find their values, differentiate the assumed solution twice, substitute in the ODE, and get:

$$X_f = \frac{F_0}{\sqrt{\left(k - m\omega_f^2\right)^2 + \left(c\omega_f\right)^2}}$$

$$\psi = \arctan\left[\frac{c\omega_f}{\left(k - m\omega_f^2\right)^2}\right] \tag{15.6d}$$

(a) Term 1

X

(b) Term 2

=

(c) Total response

FIGURE 15-5

Transient response of an underdamped system

15

Substitute equations 15.1d (p. 754), 15.2i (p. 756), and 15.3a (p. 756) and put in dimensionless form:

$$\frac{X_f}{F_0/k} = \frac{1}{\sqrt{\left[1-\left(\frac{\omega_f}{\omega_n}\right)^2\right]^2 + \left(2\zeta\frac{\omega_f}{\omega_n}\right)^2}}$$

(15.6e)

$$\psi = \arctan\left[\frac{2\zeta\frac{\omega_f}{\omega_n}}{1-\left(\frac{\omega_f}{\omega_n}\right)^2}\right]$$

The ratio ω_f/ω_n is called the **frequency ratio**. Dividing X_f by the static deflection F_0/k creates the **amplitude ratio** which defines the relative dynamic displacement compared to the static.

COMPLETE RESPONSE   The complete solution to our system differential equation for a sinusoidal forcing function is the sum of the homogeneous and particular solutions:

$$x = X_0 e^{-\zeta\omega_n t}\cos\left[\left(\sqrt{1-\zeta^2}\,\omega_n t\right)-\phi\right] + X_f \sin(\omega_f t - \psi)$$

(15.7)

The homogeneous term represents the **transient response** of the system which will die out in time but is reintroduced any time the system is again disturbed. The **particular term** represents the **forced response** or **steady-state response** to a sinusoidal forcing function which will continue as long as the forcing function is present.

Note that the solution to this equation, shown in equations 15.5 (pp. 757-758) and 15.6, depends only on two ratios, the damping ratio ζ which relates the actual damping relative to the critical damping, and the *frequency ratio* ω_f/ω_n which relates the forcing frequency to the natural frequency of the system. Koster[1] found that a typical value for the damping ratio in cam-follower systems is $\zeta = 0.06$, so they are underdamped and can **resonate** if operated at frequency ratios close to 1.

The initial conditions for the specific problem are applied to equation 15.7 to determine the values of X_0 and ϕ. Note that these constants of integration are contained within the homogeneous part of the solution.

15.2   RESONANCE

The natural frequency (and its overtones) are of great interest to the designer as they define the frequencies at which the system will **resonate**. The single-*DOF* lumped parameter systems shown in Figures 15-1 and 15-2 (p. 753) are the simplest possible to describe a dynamic system, yet they contain all the basic dynamic elements. Masses and springs are energy storage elements. A mass stores kinetic energy, and a spring stores potential energy. The damper is a dissipative element. It uses energy and converts it to heat. Thus all the losses in the model of Figure 15-1 occur through the damper.

These are "pure" idealized elements which possess only their own characteristics. That is, the spring has no damping and the damper no springiness, etc. Any system that contains more than one energy storage device such as a mass and a spring will possess at least one natural frequency. If we excite the system at its natural frequency, we will set up the condition called resonance in which the energy stored in the system's elements will oscillate from one element to the other at that frequency. The result can be violent oscillations in the displacements of the movable elements in the system as the energy moves from potential to kinetic form and vice versa.

Figures 15-6a and b show the amplitude and phase angle, respectively, of the displacement response X of the system to a sinusoidal input forcing function at various frequencies ω_f. The forcing frequency ω_f is the angular velocity of the cam. These plots normalize the forcing frequency as a frequency ratio ω_f / ω_n. The amplitude X is normalized by dividing the dynamic deflection x by the static deflection F_0 / k that the same force amplitude would create on the system. Thus at a frequency of zero, the output is

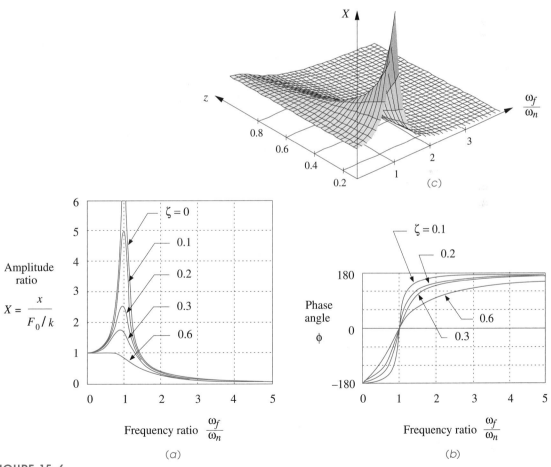

FIGURE 15-6

Amplitude ratio and phase angle of system response

one, equal to the static deflection of the spring at the amplitude of the input force. As the forcing frequency increases toward the natural frequency ω_n, the amplitude of the output motion, for zero damping, increases rapidly and becomes theoretically infinite when $\omega_f = \omega_n$. Beyond this point the amplitude decreases rapidly and asymptotically toward zero at high frequency ratios.

The effects of damping ratio ζ can best be seen in Figure 15-6c (p. 761), which shows a 3-D plot of forced vibration amplitude as a function of both frequency ratio ω_f / ω_n and damping ratio ζ. The addition of damping reduces the amplitude of vibration at the natural frequency, but very large damping ratios are needed to keep the output amplitude less than or equal to the input amplitude. About 50 to 60% of critical damping will eliminate the resonance peak. Unfortunately, most cam-follower systems have damping ratios of less than about 10% of critical. At those damping levels, the response at resonance is about five times the static response. This will create unsustainable stresses in most systems if allowed to occur.

It is obvious that we must avoid driving this system at or near its natural frequency. One result of operation of an underdamped cam-follower system near ω_n can be **follower jump**. The system of follower mass and spring can oscillate violently at its natural frequency and leave contact with the cam. When it does reestablish contact, it may do so with severe impact loads that can quickly fail the materials.

The designer has a degree of control over resonance in that the system's mass m and stiffness k can be tailored to move its natural frequency away from any required operating frequencies. A common rule of thumb is to design the system to have a fundamental natural frequency ω_n at least ten times the highest forcing frequency expected in service, thus keeping all operation well below the resonance point. This is often difficult to achieve in mechanical systems. One tries to achieve the largest ratio ω_n / ω_f possible nevertheless. It is important to adhere to the fundamental law of cam design and use cam programs with finite jerk in order to minimize vibrations in the follower system.

Some thought and observation of equation 15.1d (p. 754) will show that we would like our system members to be both light (low m) and stiff (high k) to get high values for ω_n. Unfortunately, the lightest materials are seldom also the stiffest. Aluminum is one-third the weight of steel but is also about one-third as stiff. Titanium is about half the weight of steel but also about half as stiff. Some of the exotic composite materials such as carbon fiber/epoxy offer better stiffness-to-weight ratios but their cost is high and processing is difficult. Another job for *Unobtainium 208!*

Note in Figure 15-6 that the amplitude of vibration at large frequency ratios approaches zero. So, if the system can be brought up to speed through the resonance point without damage and then kept operating at a large frequency ratio, the vibration will be minimal. An example of systems designed to be run this way are large devices that must run at higher speed such as electrical power generators. Their large mass creates a lower natural frequency than their required operating speeds. They are "run up" as quickly as possible through the resonance region to avoid damage from their vibrations and "run down" quickly through resonance when stopping them. They also have the advantage of long duty cycles of constant-speed operation in the safe frequency region between infrequent starts and stops.

15

15.3 KINETOSTATIC FORCE ANALYSIS OF THE FORCE-CLOSED CAM-FOLLOWER

The previous sections introduced **forward dynamic analysis** and the solution to the system differential equation of motion (equation 15.2b, p. 755). The applied force $F_c(t)$ is presumed to be known, and the system equation is solved for the resulting displacement x from which its derivatives can also be determined. The **inverse dynamics**, or **kinetostatics**, approach provides a quick way to determine how much spring force is needed to keep the follower in contact with the cam at a chosen design speed. The displacement and its derivatives are defined from the kinematic design of the cam based on an assumed constant angular velocity ω of the cam. Equation 15.2b can be solved algebraically for the force $F_c(t)$ in a spring-loaded cam-follower system provided that values for mass m, spring constant k, preload F_{pl}, and damping factor c are known in addition to the displacement, velocity, and acceleration functions.

Figure 15-1a (p. 753) shows a simple plate or disk cam driving a spring-loaded, roller follower. This is a force-closed system which depends on the spring force to keep the cam and follower in contact at all times. Figure 15-1b shows a lumped parameter model of this system in which all the **mass** that moves with the follower train is lumped together as m, all the springiness in the system is lumped within the **spring constant** k, and all the **damping** or resistance to movement is lumped together as a damper with coefficient c.

The designer has a large degree of control over the system spring constant k_{eff} as it tends to be dominated by the k_s of the physical return spring. The elasticities of the follower parts also contribute to the overall system k_{eff} but are usually much stiffer than the physical spring. If the follower stiffness is in series with the return spring, as it often is, equations 10.19 (p. 562) show that the softest spring in series will dominate the effective spring constant. Thus the return spring will virtually determine the overall k unless some parts of the follower train have similarly low stiffness.

The designer will choose or design the return spring and thus can specify both its k and the amount of preload deflection x_0 to be introduced at assembly. Preload of a spring occurs when it is compressed (or extended if an extension spring) from its *free length* to its initial assembled length. This is a necessary and desirable situation as we want some residual force on the follower even when the cam is at its lowest displacement. This will help maintain good contact between the cam and follower at all times. This spring preload $F_{pl} = kx_0$ adds a constant term to equation 15.2b which becomes:

$$F_c(t) = m\ddot{x} + c\dot{x} + kx + F_{pl} \tag{15.8a}$$

or:

$$F_c(t) = m\ddot{x} + c\dot{x} + k(x + x_0) \tag{15.8b}$$

The value of m is determined from the effective mass of the system as lumped in the single-*DOF* model of Figure 15-1 (p. 753). The value of c for most cam-follower systems can be estimated for a first approximation to be about 0.05 to 0.10 of the critical damping c_c as defined in equation 15.2i (p. 756). Koster[1] found that a typical value for the damping ratio in cam-follower systems is $\zeta = 0.06$.

Calculating the damping c based on an assumed value of ζ requires specifying a value for the overall system k and for its effective mass. The choice of k will affect both the natural frequency of the system for a given mass and the available force to keep the joint

15

closed. Some iteration will probably be needed to find a good compromise. A selection of data for commercially available helical coil springs is provided in Appendix D. Note in equation 15.8 (p. 763) that the terms involving acceleration and velocity can be either positive or negative. The terms involving the spring parameters k and F_{pl} are the only ones that are always positive. So, to keep the overall function always positive requires that the spring force terms be large enough to counteract any negative values in the other terms. Typically, the acceleration is larger numerically than the velocity, so the negative acceleration usually is the principal cause of a negative force F_c.

The principal concern in this analysis is to keep the cam force always positive in sign as its direction is defined in Figure 15-1. The cam force is shown as positive in that figure. In a force-closed system the cam can only push on the follower. It cannot pull. The follower spring is responsible for providing the force needed to keep the joint closed during the negative acceleration portions of the follower motion. The damping force also can contribute, but the spring must supply the bulk of the force to maintain contact between the cam and follower. If the force F_c goes negative at any time in the cycle, the follower and cam will part company, a condition called **follower jump**. When they meet again, it will be with large and potentially damaging impact forces. The follower jump, if any, will occur near the point of maximum negative acceleration. Thus we must select the spring constant and preload to guarantee a positive force at all points in the cycle. In automotive engine valve cam applications, follower jump is also called *valve float*, because the valve (follower) "floats" above the cam, also periodically impacting the cam surface. This will occur if the cam rpm is increased to the point that the larger negative acceleration makes the follower force negative. The "redline" maximum engine rpm often indicated on its tachometer is to warn of impending valve float above that speed which will damage the cam and follower.

Program DYNACAM allows the iteration of equation 15.8 to be done quickly for any cam whose kinematics have been defined in that program. The program's *Dynamics* button will solve equation 15.8 for all values of camshaft angle, using the displacement, velocity, and acceleration functions previously calculated for that cam design in the program. The program requires values for the effective system mass m, effective spring constant k, preload F_{pl}, and the assumed value of the damping ratio ζ. These values need to be determined for the model by the designer using the methods described in Sections 10.11 (p. 556) and 10.12 (p. 559). The calculated force at the cam-follower interface can then be plotted or its values printed in tabular form. The system's natural frequency is also reported when the tabular force data are printed.

✎ EXAMPLE 15-1

Kinetostatic Force Analysis of a Force-Closed (Spring-Loaded) Cam-Follower System.

Given: A translating roller follower as shown in Figure 15-1 (p. 753) is driven by a force-closed radial plate cam which has the following program:

                Segment 1:   Rise 1 inch in 50° with modified sine acceleration
                Segment 2:   Dwell for 40°
                Segment 3:   Fall 1 inch in 50° with cycloidal displacement
                Segment 4:   Dwell for 40°

Segment 5: Rise 1 inch in 50° with 3-4-5 polynomial displacement
Segment 6: Dwell for 40°
Segment 7: Fall 1 inch in 50° with 4-5-6-7 polynomial displacement
Segment 8: Dwell for 40°
Camshaft angular velocity is 18.85 rad/sec.
Follower effective mass is 0.0738 in-lb-sec^2 (blob).
Damping is 15% of critical ($\zeta = 0.15$).

Problem: Find the spring constant and spring preload to maintain contact between cam and follower and calculate the dynamic force function for the cam. Find the system natural frequency with the selected spring. Keep the pressure angle under 30°.

Solution:

1   Calculate the kinematic data (follower displacement, velocity, acceleration, and jerk) for the specified cam functions. The acceleration for this cam is shown in Figure 15-7 and has a maximum value of 3504 in/sec^2. See Chapter 8 to review this procedure.

2   Calculate the pressure angle and radius of curvature for trial values of prime circle radius, and size the cam to control these values. Figure 15-8 (p. 766) shows the pressure angle function and Figure 15-9 (p. 766) the radii of curvature for this cam with a prime circle radius of 4 in and zero eccentricity. The maximum pressure angle is 29.2° and the minimum radius of curvature is 1.7 in. Figure 8-51 (p. 463) shows the finished cam profile. See Chapter 8 to review these calculations.

3   With the kinematics of the cam defined, we can address its dynamics. To solve equation 15.8 (p. 763) for cam force, we must assume values for the spring constant k and the preload F_{pl}. The value of c can be calculated from equation 15.3a (p. 756) using the given mass m, the damping factor ζ, and assumed k. The kinematic parameters are known.

Segment Number	Function Used	Start Angle	End Angle	Delta Angle
1	ModSine rise	0	50	50
2	Dwell	50	90	40
3	Cycloid fall	90	140	50
4	Dwell	140	180	40
5	345 poly rise	180	230	50
6	Dwell	230	270	40
7	4567 poly fall	270	320	50
8	Dwell	320	360	40

(a) Cam program specifications

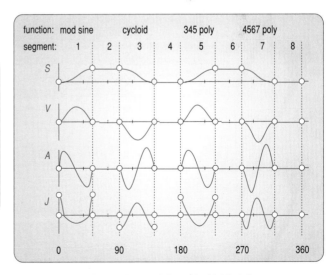

(b) Plots of cam-follower's *S V A J* diagrams

FIGURE 15-7

S V A J diagrams for Examples 15-1 and 15-2 (p. 769)

15

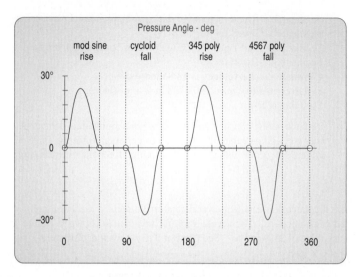

FIGURE 15-8

Pressure angle plot for Examples 15-1 and 15-2 (p. 769)

4 Program DYNACAM does this computation for you. The dynamic force that results from an assumed k of 150 lb/in and a preload of 75 lb is shown in Figure 15-10a. The damping coefficient $c = 0.998$. Note that the force dips below the zero axis in two places during negative acceleration. These are locations of follower jump. The follower has left the cam during the fall because the spring does not have enough available force to keep the follower in contact with the rapidly falling cam. Open the file E15-01.cam in DYNACAM and provide the specified k and F_{pl} to see this example. Another iteration is needed to improve the design.

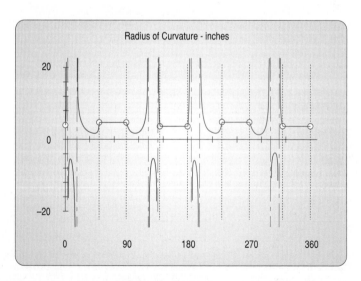

FIGURE 15-9

Radius of curvature of a four-dwell cam for Examples 15-1 and 15-2

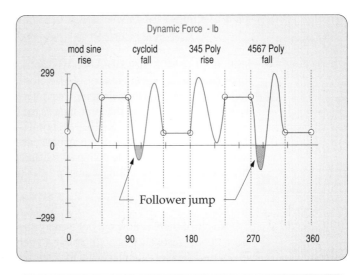

(a) Insufficient spring
force allows
follower jump

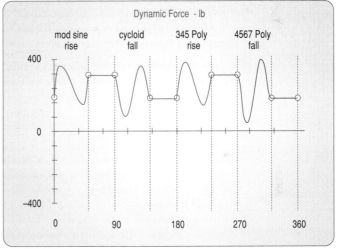

(b) Sufficient spring
force keeps the
dynamic force positive

FIGURE 15-10

Dynamic forces in a force-closed cam-follower system

5   Figure 15-10b shows the dynamic force for the same cam with a spring constant of $k = 200$ lb/
in and a preload of 150 lb. The damping coefficient $c = 1.153$. This additional force has lifted
the function up sufficiently to keep it positive everywhere. There is no follower jump in this
case. The maximum force during the cycle is 400.4 lb. A margin of safety has been provided
by keeping the minimum force comfortably above the zero line at 36.9 lb. Run example #5 in
the program and provide the specified spring constant and preload values to see this example.

6   The undamped and damped fundamental natural frequencies can be calculated for the system
from equations 15.1d (p. 754) and 15.3c (p. 756), respectively, and are:

$$\omega_n = 52.06 \text{ rad/sec;} \qquad\qquad \omega_d = 51.98 \text{ rad/sec}$$

15

15.4   KINETOSTATIC FORCE ANALYSIS OF THE FORM-CLOSED CAM-FOLLOWER

Section 8.1 (p. 402) described two types of joint closure used in cam-follower systems, **force closure** and **form closure**. Force closure uses an open joint and requires a spring or other force source to maintain contact between the elements. Form closure provides a geometric constraint at the joint such as the cam groove shown in Figure 15-11a or the conjugate cams of Figure 15-11b. No spring is needed to keep the follower in contact with these cams. The follower will run against one side or the other of the groove or conjugate pair as necessary to provide both positive and negative forces. Since there is no spring in this system, its dynamic force equation 15.8 (p. 763) simplifies to:

$$F_c(t) = m\ddot{x} + c\dot{x} \qquad\qquad (15.9)$$

Note that there is now only one energy storage element in the system (the mass), so, theoretically, resonance is not possible. There is no natural frequency for it to resonate at. This is the chief advantage of a form-closed system over a force-closed one. Follower jump will not occur, short of complete failure of the parts, no matter how fast the system is run. This arrangement is sometimes used in high-performance or racing engine valve trains to allow higher redline engine speeds without valve float. In engine valve trains, a form-closed cam-follower valve train is called a *desmodromic* system.

As with any design, there are trade-offs. While the form-closed system typically allows higher operating speeds than a comparable force-closed system, it is not free of all vibration problems. Even though there is no physical return spring in the system, the follower train, the camshaft, and all other parts still have their own spring constants which abruptly shift from one side of the cam groove to the other. There cannot be zero clearance between the roller follower and the groove and still have it operate. Even if the clearance is very small, there will still be an opportunity for the follower to develop some velocity in its short trip across the groove, and it will impact the other side. Track cams of the type shown in Figure 15-11a typically fail at the points where the acceleration reverses sign, due to many cycles of crossover shock. Note also that the roller follower has to reverse direction every time it crosses over to the other side of the groove. This causes significant follower slip and high wear on the follower compared to an open, force-closed cam where the follower will have less than 1% slip.

Because there are two cam surfaces to machine and because the cam track, or groove, must be cut and ground to high precision to control the clearance, form-closed cams tend to be more expensive to manufacture than force-closed cams. Track cams usually must be ground after heat treatment to correct the distortion of the groove resulting from the high temperatures. Grinding significantly increases cost. Many force-closed cams are not ground after heat treatment and are used as-milled. Though the conjugate cam approach avoids the groove tolerance and heat treat distortion problems, there are still two matched cam surfaces to be made per cam. Thus, the desmodromic cam's dynamic advantages come at a significant cost premium.

We will now repeat the cam design of Example 15-1 (p. 764), modified for desmodromic operation. This is simple to do with program DYNACAM by setting the spring constant and preload values to zero, which assumes that the follower train is a rigid body. A more accurate result can be obtained by calculating and using the effective spring constant

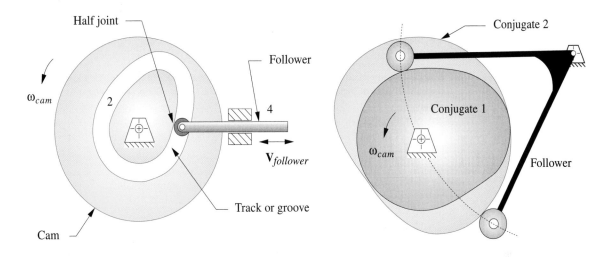

(a) Form-closed cam with translating follower   (b) Conjugate cams on common shaft

FIGURE 15-11

Form-closed cam-follower systems

of the combination of parts in the follower train, once their geometries and materials are defined. The dynamic forces will now be negative as well as positive, but a form-closed cam can both push and pull.

EXAMPLE 15-2

Dynamic Force Analysis of a Form-Closed (Desmodromic) Cam-Follower System.

Given: A translating roller follower as shown in Figure 15-11a is driven by a form-closed radial plate cam which has the following program:

Segment 1: Rise 1 inch in 50° with modified sine acceleration
Segment 2: Dwell for 40°
Segment 3: Fall 1 inch in 50° with cycloidal displacement
Segment 4: Dwell for 40°
Segment 5: Rise 1 inch in 50° with 3-4-5 polynomial displacement
Segment 6: Dwell for 40°
Segment 7: Fall 1 inch in 50° with 4-5-6-7 polynomial displacement
Segment 8: Dwell for 40°
Camshaft angular velocity is 18.85 rad/sec.
Follower effective mass is 0.0738 in-lb-sec^2 (blob).
Damping is 15% of critical ($\zeta = 0.15$).

Problem: Compute the dynamic force function for the cam. Keep the pressure angle < 30°.

15

Solution:

1 Calculate the kinematic data (follower displacement, velocity, acceleration, and jerk) for the specified cam functions. The acceleration for this cam is shown in Figure 15-7 (p. 765) and has a maximum value of 3504 in/sec^2. See Chapter 8 to review this procedure.

2 Calculate radius of curvature and pressure angle for trial values of prime circle radius, and size the cam to control these values. Figure 15-8 (p. 766) shows the pressure angle function and Figure 15-9 (p. 766) the radii of curvature for this cam with a prime circle radius of 4 in and zero eccentricity. The maximum pressure angle is 29.2° and the minimum radius of curvature is 1.7 in. Figure 8-51 (p. 463) shows the finished cam profile. See Chapter 8 to review these calculations.

3 With the kinematics of the cam defined, we can address its dynamics. To solve equation 15.9 (p. 768) for the cam force, we assume zero values for the spring constant k and the preload F_{pl}. The value of c is assumed to be the same as in Example 15-1 (p. 764), 1.153. The kinematic parameters are known.

4 Program DYNACAM does this computation for you. The dynamic force that results is shown in Figure 15-12. Note that the force is now more nearly symmetric about the axis and its peak absolute value is 289 lb. Crossover shock will occur each time the follower force changes sign. Open the file E15-02.cam in DYNACAM to see this example.

Compare the dynamic force plots for the force-closed system (Figure 15-10b, p. 767) and the form-closed system (Figure 15-12). The absolute peak force magnitude on either side of the track in the form-closed cam is less than that on the spring-loaded one. This shows the penalty that the spring imposes on the system in order to keep the joint closed. Thus, either side of the cam groove will experience lower stresses than will the open cam, except for the areas of crossover shock mentioned on p. 768.

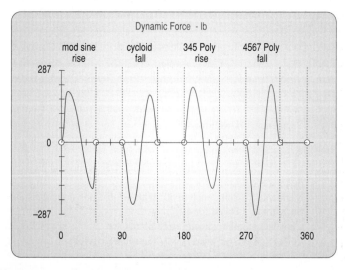

FIGURE 15-12

Dynamic force in a form-closed cam-follower system

15.5   KINETOSTATIC CAMSHAFT TORQUE

The kinetostatic analysis assumes that the camshaft will operate at some constant speed ω. As we saw in the case of the fourbar linkage in Chapter 11 and with the slider-crank mechanism in Chapter 13, the input torque must vary over the cycle if the shaft velocity remains constant. The torque is calculated from the power relationship, ignoring losses.

$$\text{Power in} = \text{Power out}$$

$$T_c\omega = F_f V$$

$$T_c = \frac{F_f V}{\omega} = \frac{\left(F_c \cos\phi\right) V}{\omega} \tag{15.10}$$

where T_c is camshaft torque, ω is camshaft angular velocity, F_c is force between cam and follower along the common normal, and F_f is the component of F_c in the direction of follower velocity V as defined by the pressure angle ϕ.

Once the cam force F_c has been calculated from either equation 15.8 (p. 763) or 15.9 (p. 768), T_c is easily found since V, ϕ, and ω are known for all values of cam angle θ. Figure 15-13a shows the camshaft input torque needed to drive the force-closed cam designed in Example 15-1 (p. 764). Figure 15-13b shows the camshaft input torque needed to drive the form-closed cam designed in Example 15-2 (p. 769). Note that the torque required to drive the force-closed (spring-loaded) system is significantly higher than that needed to drive the form-closed (track) cam. The spring force is also extracting a penalty here as energy must be stored in the spring during the rise portions which will tend to slow the camshaft. This stored energy is then returned to the camshaft during the fall portions, tending to speed it up. The spring loading causes larger oscillations in the torque.

A flywheel can be sized and fitted to the camshaft to smooth these variations in torque just as was done for the fourbar linkage in Section 11.11 (p. 605). See that section for the design procedure. Program DYNACAM integrates the camshaft torque function pulse by pulse and prints those areas to the screen. These energy data can be used to calculate the required flywheel size for any selected coefficient of fluctuation.

One useful way to compare alternate cam designs is to look at the torque function as well as at the dynamic force. A smaller torque variation will require a smaller motor and/or flywheel and will run more smoothly. Three different designs for a single-dwell cam were explored in Chapter 8. (See Examples 8-6, p. 435, 8-7, p. 436, and 8-8, p. 438.) All had the same lift and duration but used different cam functions. One was a double harmonic, one cycloidal, and one a sixth-degree polynomial. On the basis of their kinematic results, principally acceleration magnitude, we found that the polynomial design was superior. We will now revisit this cam as an example and compare its dynamic force and torque among the same three programs.

EXAMPLE 15-3

15

Comparison of Dynamic Torques and Forces Among Three Alternate Designs of the Same Cam.

Given: A translating roller follower as shown in Figure 15-1 (p. 753) is driven by a force-closed radial plate cam which has the following program:

Design 1

Segment 1:   Rise 1 inch in 90° double harmonic displacement
Segment 2:   Fall 1 inch in 90° double harmonic displacement
Segment 3:   Dwell for 180°

Design 2

Segment 1:   Rise 1 inch in 90° cycloidal displacement
Segment 2:   Fall 1 inch in 90° cycloidal displacement
Segment 3:   Dwell for 180°

Design 3

Segment 1:   Rise 1 inch in 90° and fall 1 inch in 90° with polynomial displacement. (A single polynomial can create both rise and fall.)
Segment 2:   Dwell for 180°

(a) Force-closed (spring-loaded) cam-follower

(b) Form-closed (desmodromic) cam-follower

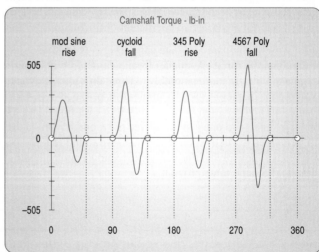

FIGURE 15-13

Input torque in force- and form-closed cam-follower systems

Camshaft angular velocity is 15 rad/sec. Follower effective mass is 0.0738 in-lb-sec^2 (blob). Damping is 15% of critical ($\zeta = 0.15$).

Find: The dynamic force and torque functions for the cam. Compare their peak magnitudes for the same prime circle radius.

Solution: Note that these are the same kinematic cam designs as are shown in Figures 8-27 (p. 437), 8-28 (p. 438), and 8-30 (p. 440).

1   Calculate the kinematic data (follower displacement, velocity, acceleration, and jerk) for each of the specified cam designs. See Chapter 8 to review this procedure.

2   Calculate the radius of curvature and pressure angle for trial values of prime circle radius, and size the cam to control these values. A prime circle radius of 3 in gives acceptable pressure angles and radii of curvature. See Chapter 8 to review these calculations.

3   With the kinematics of the cam defined, we can address its dynamics. To solve equation 15.1a (p. 753) for the cam force, we will assume a value of 50 lb/in for the spring constant k and adjust the preload F_{pl} for each design to obtain a minimum dynamic force of about 10 lb. For design 1 this requires a spring preload of 28 lb; for design 2, 15 lb; and for design 3, 10 lb.

4   The value of damping c is calculated from equation 15.2i (p. 756). The kinematic parameters x, v, and a are known from the prior analysis.

5   Program DYNACAM will do these computations for you. The dynamic forces that result from each design are shown in Figure 15-14 (p. 774) and the torques in Figure 15-15 (p. 775). Note that the force is largest for design 1 at 82-lb peak and least for design 3 at 53-lb peak. The same ranking holds for the torques which range from 96 lb-in for design 1 to 52 lb-in for design 3. These represent reductions of 35% and 46% in the dynamic loading due to a change in the kinematic design. Not surprisingly, the sixth-degree polynomial design which had the lowest acceleration also has the lowest forces and torques and is the clear winner. Open the files E08-06.cam, E08-07.cam, and E08-08.cam in program DYNACAM to see these results.

15.6   MEASURING DYNAMIC FORCES AND ACCELERATIONS

As described in previous sections, cam-follower systems tend to be underdamped. This allows significant oscillations and vibrations to occur in the follower train. Dynamic forces and accelerations can be measured fairly easily in operating machinery. Compact, piezoelectric force and acceleration transducers are available that have frequency response ranges in the high thousands of hertz. Strain gages provide strain measurements that are proportional to force and have bandwidths of a kilohertz or better.

Figure 15-16 (p. 776) shows acceleration and force curves as measured on the follower train of a single overhead camshaft (SOHC) valve train in a 1.8-liter four-cylinder inline engine. [2] The nonfiring engine was driven by an electric motor on a dynamometer. The camshaft is turning at 500, 2000, and 3000 rpm (1000, 4000, and 6000 crankshaft rpm), respectively, in the three plots of Figure 15-16a, b, and c. Acceleration was measured with a piezoelectric accelerometer attached to the head of one intake valve, and the force was calculated from the output of strain gages placed on the rocker arm for that intake valve. The theoretical follower acceleration curve (as designed) is superposed on

15

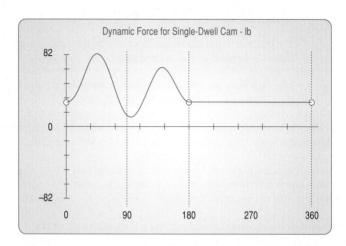

(a) Double harmonic rise -
      double harmonic fall

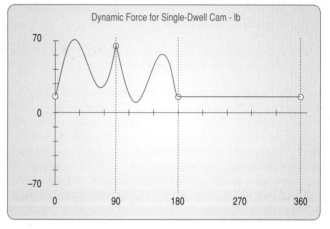

(b) Cycloidal rise -
      cycloidal fall

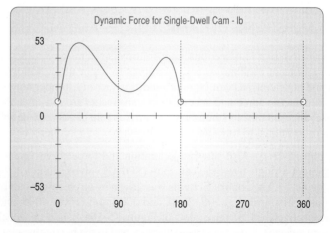

(c) Sixth-degree
      polynomial

FIGURE 15-14

Dynamic forces in three different designs of a single-dwell cam

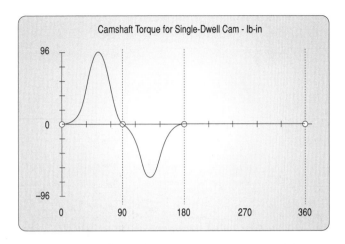

(a) Double harmonic rise -
double harmonic fall

(b) Cycloidal rise -
cycloidal fall

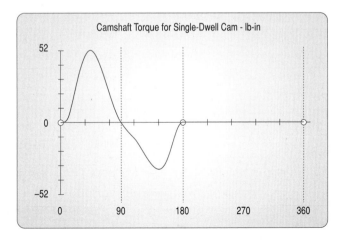

(c) Sixth-degree
polynomial

15

FIGURE 15-15

Dynamic input torque in three different designs of a single-dwell cam

(a) 500 rpm

(b) 2000 rpm

(c) 3000 rpm

········· Measured acceleration of valve   ——— Theoretical acceleration of valve   — – – Force on rocker arm

FIGURE 15-16

Valve acceleration and rocker arm force in a single overhead cam valve train the measured acceleration curve. All acceleration measurements are converted to units of mm/deg^2 (i.e., normalized to camshaft speed) to allow comparison with one another and with the theoretical acceleration curve.

At 500 camshaft rpm, the measured acceleration closely matches the theoretical acceleration curve with some minor oscillations due to spring vibration. At 2000 camshaft rpm, significant oscillation in the measured acceleration is seen during the first positive and in the negative acceleration phase. This is due to the valve spring vibrating at its natural frequency in response to excitation by the cam. This is called "spring surge" and is a significant factor in valve spring fatigue failure. At 3000 camshaft rpm, the spring surge is still present but is less prominent as a percentage of total acceleration. The frequency content of the cam's forcing function passed through the first natural frequency of the valve spring at about 2000 camshaft rpm, causing the spring to resonate. The same effects can be seen in the rocker arm force. Everything in a machine tends to sympathetically vibrate at its own natural frequency when excited by any input forcing function. Sensitive transducers such as accelerometers will pick up these vibrations as they are transmitted through the structure.

15.7 PRACTICAL CONSIDERATIONS

Koster[1] proposes some general rules for the design of cam-follower systems for high-speed operation based on his extensive dynamic modeling and experimentation.

To minimize the positional error and residual acceleration:

1   Keep the total lift of the follower to a minimum.

2   If possible, arrange the follower spring to preload all pivots in a consistent direction to control backlash in the joints.

3   Keep the duration of rises and falls as long as possible.

4   Keep follower train mass low and follower train stiffness high to increase natural frequency.

5   Any lever ratios present will change the effective stiffness of the system by the square of the ratio. Try to keep lever ratios close to 1.

6   Make the camshaft as stiff as possible **both in torsion and in bending**. This is perhaps the most important factor in controlling follower vibration.

7   Reduce pressure angle by increasing the cam pitch circle diameter.

8   Use low backlash or antibacklash gears in the camshaft drive train.

15.8 REFERENCES

1   **Koster, M. P.** (1974). *Vibrations of Cam Mechanisms.* Phillips Technical Library Series, Macmillan Press Ltd.: London.

2   **Norton, R. L., et al.** (1998). "Analyzing Vibrations in an IC Engine Valve Train." SAE Paper: 980570.

15

15.9 BIBLIOGRAPHY

Barkan, P., and R. V. McGarrity. (1965). "A Spring Actuated, Cam-Follower System: Design Theory and Experimental Results." *ASME J. Engineering for Industry,* August, pp. 279-286.

Chen, F. Y. (1975). "A Survey of the State of the Art of Cam System Dynamics." *Mechanism and Machine Theory*, **12**, pp. 210-224.

Chen, F. Y. (1982). *Mechanics and Design of Cam Mechanisms*. Pergamon Press: New York, p. 520.

Freudenstein, F. (1959). "On the Dynamics of High-Speed Cam Profiles." *Int. J. Mech. Sci.*, **1**, pp. 342-349.

Freudenstein, F., et al. (1969). "Dynamic Response of Mechanical Systems." IBM: New York Scientific Center, Report No. 320-2967.

Hrones, J. A. (1948). "An Analysis of the Dynamic Forces in a Cam System." *Trans ASME*, **70**, pp. 473-482.

Johnson, A. R. (1965). "Motion Control for a Series System of *N* Degrees of Freedom Using Numerically Derived and Evaluated Equations." *ASME J. Eng. Industry*, pp. 191-204.

Knight, B. A., and H. L. Johnson. (1966). "Motion Analysis of Flexible Cam-Follower Systems." ASME Paper: 66-Mech-3.

Matthew, G. K., and D. Tesar. (1975). "Cam System Design: The Dynamic Synthesis and Analysis of the One Degree of Freedom Model." *Mechanisms and Machine Theory*, **11**, pp. 247-257.

Matthew, G. K., and D. Tesar. (1975). "The Design of Modeled Cam Systems Part I: Dynamic Synthesis and Design Chart for the Two-Degree-of-Freedom Model." *Journal of Engineering for Industry,* November, pp. 1175-1180.

Midha, A., and D. A. Turic. (1980). "On the Periodic Response of Cam Mechanisms with Flexible Follower and Camshaft." *J. Dyn. Sys. Meas. Control*, **102**, December, pp. 225-264.

Norton, R. L. (2002). *Cam Design and Manufacturing Handbook*. Industrial Press: New York.

15.10 PROBLEMS

Program DYNACAM *may be used to solve these problems where applicable. Where units are unspecified, work in any consistent units system you wish. Appendix D contains some pages from a catalog of commercially available helical coil springs to aid in designing realistic solutions to these problems. Other spring information can be found on the Web.*

*‡15-1   Design a double-dwell cam to move a 2-in-dia roller follower of mass = 2.2 bl from 0 to 2.5 inches in 60° with modified sine acceleration, dwell for 120°, fall 2.5 inches in 30° with cycloidal motion, and dwell for the remainder. The total cycle must take 4 sec. Size a return spring and specify its preload to maintain contact between cam and follower. Calculate and plot the dynamic force and torque. Assume damping of 0.2 times critical. Repeat for a form-closed cam. Compare the dynamic force, torque, and natural frequency for the form-closed design and the force-closed design.

*‡15-2   Design a double-dwell cam to move a 2-in-dia roller follower of mass = 1.4 bl from 0 to 1.5 inches in 45° with 3-4-5 polynomial motion, dwell for 150°, fall 1.5 inches in 90° with 4-5-6-7 polynomial motion, and dwell for the remainder. The total cycle must take 6 sec. Size a return spring and specify its preload to maintain contact between cam and follower. Calculate and plot the dynamic force and torque. Assume damping of 0.1 times critical. Repeat for a form-closed cam. Compare the dynamic force, torque, and natural frequency for the form-closed design and the force-closed design.

* Answers in Appendix F.

† These problems are suited to solution using *Mathcad, Matlab,* or *TKSolver* equation solver programs.

‡ These problems are suited to solution using program DYNACAM which is on the attached DVD.

15

TABLE P15-1
Problem 15-6

	m	k	c
a	1.2	14	1.1
b	2.1	46	2.4
c	30.0	2	0.9
d	4.5	25	3.0
e	2.8	75	7.0
f	12.0	50	14.0

*†‡15-3 Design a single-dwell cam to move a 2-in-dia roller follower of mass = 3.2 bl from 0 to 2 inches in 60°, fall 2 inches in 90°, and dwell for the remainder. The total cycle must take 5 sec. Use a seventh-degree polynomial. Size a return spring and specify its preload to maintain contact between cam and follower. Calculate and plot the dynamic force and torque. Assume damping of 0.15 times critical. Repeat for a form-closed cam. Compare the dynamic force, torque, and natural frequency for the form-closed design and the force-closed design.

*†‡15-4 Design a three-dwell cam to move a 2-in-dia roller follower of mass = 0.4 bl from 0 to 2.5 inches in 40°, dwell for 100°, fall 1.5 inches in 90°, dwell for 20°, fall 1 inch in 30°, and dwell for the remainder. The total cycle must take 10 sec. Choose suitable programs for rise and fall to minimize dynamic forces and torques. Size a return spring and specify its preload to maintain contact between cam and follower. Calculate and plot the dynamic force and torque. Assume damping of 0.12 times critical. Repeat for a form-closed cam. Compare the dynamic force, torque, and natural frequency for the form-closed design and the force-closed design.

*†‡15-5 Design a four-dwell cam to move a 2-in-dia, 1.25 bl mass roller follower from 0 to 2.5 inches in 40°, dwell for 100°, fall 1.5 inches in 90°, dwell for 20°, fall 0.5 inch in 30°, dwell for 40°, fall 0.5 inch in 30°, and dwell for the remainder. The total cycle is 15 sec. Choose suitable programs for rise and fall to minimize dynamic forces and torques. Size a return spring and specify its preload to maintain contact between cam and follower. Calculate and plot the dynamic force and torque. Assume damping of 0.18 times critical. Repeat for a form-closed cam. Compare the dynamic force, torque, and natural frequency for the form-closed design and the force-closed design.

*†‡15-6 A mass-spring damper system as shown in Figure 15-1b (p. 753) has the values shown in Table P15-1. Find the undamped and damped natural frequencies and the value of critical damping for the system(s) assigned.

†15-7 Figure P15-1 shows a cam-follower system. The dimensions of the solid, rectangular 2 x 2.5 in cross-section aluminum arm are given. The cutout for the 2-in-diameter, 1.5-in-wide steel roller follower is 3 in long. Find the arm's mass, center of gravity location, and mass moment of inertia about both its CG and the arm pivot. Create a linear, one-DOF lumped mass model of the dynamic system referenced to the cam-follower and calculate the cam-follower force for one revolution. The cam is a pure eccentric with eccentricity = 0.5 in and turns at 500 rpm. The spring has a rate of 123 lb/in and a preload of 173 lb. Ignore damping.

†‡15-8 Repeat Problem 15-7 for a double-dwell cam to move the roller follower from 0 to 2.5 inches in 60° with modified sine acceleration, dwell for 120°, fall 2.5 inches in 30° with cycloidal motion, and dwell for the remainder. Cam speed is 100 rpm. Choose a suitable spring rate and preload to maintain follower contact. Select a spring from Appendix D. Assume a damping ratio of 0.10.

†‡15-9 Repeat Problem 15-7 for a double-dwell cam to move the roller follower from 0 to 1.5 inches in 45° with 3-4-5 polynomial motion, dwell for 150°, fall 1.5 inches in 90° with 4-5-6-7 polynomial motion, and dwell for the remainder. Cam speed is 250 rpm. Choose a suitable spring rate and preload to maintain follower contact. Select a spring from Appendix D. Assume a damping ratio of 0.15.

†‡15-10 Repeat Problem 15-7 for a single-dwell cam to move the follower from 0 to 2 inches in 60°, fall 2 inches in 90°, and dwell for the remainder. Use a seventh-degree polynomial. Cam speed is 250 rpm. Choose a suitable spring rate and preload to maintain follower contact. Select a spring from Appendix D. Assume a damping ratio of 0.15.

* Answers in Appendix F.

† These problems are suited to solution using *Mathcad, Matlab,* or *TKSolver* equation solver programs.

‡ These problems are suited to solution using program DYNACAM which is on the attached DVD.

15

FIGURE P15-1

Problems 15-7 (p. 779) to 15-11

$s = e \cos \omega t$

FIGURE P15-2

Problems 15-12 to 15-14

---

†‡15-11   Repeat Problem 15-7 for a double-dwell cam to move the roller follower from 0 to 2 inches in 45° with cycloidal motion, dwell for 150°, fall 2 inches in 90° with modified sine motion, and dwell for the remainder. Cam speed is 200 rpm. Choose a suitable spring rate and preload to maintain follower contact. Select a spring from Appendix D. Assume a damping ratio of 0.15.

†‡15-12   The cam in Figure P15-2 is a pure eccentric with eccentricity $e = 20$ mm and turns at 200 rpm. Follower mass = 1 kg. The spring has a rate of 10 N/m and a preload of 0.2 N. Find the follower force over one revolution. Assume a damping ratio of 0.10. If there is follower jump, redefine the spring rate and preload to eliminate it.

†‡15-13   Repeat Problem 15-12 using a cam with a 20-mm symmetric double harmonic rise and fall (180° rise -180° fall). See Chapter 8 for cam formulas.

†‡15-14   Repeat Problem 15-12 using a cam with a 20-mm 3-4-5-6 polynomial rise and fall (180° rise -180° fall). See Chapter 8 for cam formulas.

†‡15-15   Design a double-dwell cam to move a 50-mm-dia roller follower of mass = 2 kg from 0 to 45 mm in 60° with modified sine acceleration, dwell for 120°, fall 45 mm in 90° with 3-4-5 polynomial motion, and dwell for the remainder. The total cycle must take 1 sec. Size a return spring and specify its preload to maintain contact between cam and follower. Select a spring from Appendix D. Calculate and plot the dynamic force and torque. Assume damping of 0.25 times critical. Repeat for a form-closed cam. Compare the dynamic force, torque, and natural frequency for the form-closed design and the force-closed design.

---

† These problems are suited to solution using *Mathcad, Matlab,* or *TKSolver* equation solver programs.

---

‡ These problems are suited to solution using program DYNACAM which is on the attached DVD.

†‡15-16   Design a single-dwell cam using polynomials to move a 50-mm-dia roller follower of mass = 10 kg from 0 to 25 mm in 60°, fall 25 mm in 90°, and dwell for the remainder. The total cycle must take 2 sec. Size a return spring and specify its preload to maintain contact between cam and follower. Select a spring from Appendix D. Calculate and plot the dynamic force and torque. Assume damping of 0.15 times critical. Repeat for a form-closed cam. Compare the dynamic force, torque, and natural frequency for the form-closed design and the force-closed design.

15

†‡15-17 Design a four-dwell cam to move a 50-mm-dia roller follower of mass = 3 kg from 0 to 40 mm in 40°, dwell for 100°, fall 20 mm in 90°, dwell for 20°, fall 10 mm in 30°, dwell for 40°, fall 10 mm in 30°, and dwell for the remainder. The total cycle must take 10 sec. Choose suitable programs for rise and fall to minimize dynamic forces and torques. Size a return spring and specify its preload to maintain contact between cam and follower. Calculate and plot the dynamic force and torque. Assume damping of 0.25 times critical. Repeat for a form-closed cam. Compare the dynamic force, torque, and natural frequency for the form-closed design and the force-closed design.

†‡15-18 Design a cam to drive an automotive valve train whose effective mass is 0.2 kg. ζ = 0.3. Valve stroke is 12 mm. Roller follower is 10 mm diameter. The open-close event occupies 160° of camshaft revolution; dwell for remainder. Use one or two polynomials for the rise-fall event. Select a spring constant and preload to avoid jump to 3500 rpm. Fast opening and closing and maximum open time are desired.

†‡15-19 Figure P15-3 shows a cam-follower system that drives slider 6 through an external output arm 3. Arms 2 and 3 are both rigidly attached to the 0.75-in-dia shaft X-X, which rotates in bearings that are supported by the housing. The pin-to-pin dimensions of the links are shown. The cross sections of arms 2, 3, and 5 are solid, rectangular 1.5 × 0.75 in steel. The ends of these links have a full radius equal to one-half of the link width. Link 4 is 1-in-dia × 0.125 wall round steel tubing. Link 6 is a 2-in-dia × 6-in-long solid steel cylinder. Find the effective mass and effective spring constant of the follower train referenced to the cam-follower roller if the spring at A has a rate of 150 lb/in with a preload of 60 lb. Then determine and plot the kinetostatic follower force and camshaft torque over one cycle if the cam provides a 3-4-5 polynomial double-dwell angular motion to roller arm 2 with a rise of 10° in 90 camshaft degrees, dwell for 90°, fall 10° in 90°, and dwell for the remainder. The camshaft turns 100 rpm.

†‡15-20 Repeat problem 15-19 for a cam that provides a double-dwell cycloidal displacement rather than a 3-4-5 polynomial displacement.

† These problems are suited to solution using *Mathcad*, *Matlab*, or *TKSolver* equation solver programs.

‡ These problems are suited to solution using program DYNACAM which is on the attached DVD.

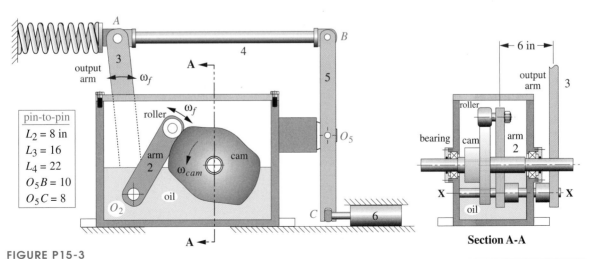

pin-to-pin
$L_2 = 8$ in
$L_3 = 16$
$L_4 = 22$
$O_5B = 10$
$O_5C = 8$

FIGURE P15-3

Problem 15-19 to 15-20

15

†‡15-21   A single-dwell cam-follower system similar to that shown in Figure 15-1a (p. 753)
provides a two-segment polynomial for a rise of 35 mm in 75°, a fall of 35 mm in 120°,
and a dwell for the remainder of the cycle. Using equations 15.8 (p. 763) and 15.10
(p. 771), calculate and plot the dynamic force and torque for one cycle if the roller follower train weighs 2.34 N, the system has a damping ratio of $\zeta = 0.06$, and the spring
has a rate of 1.5 N/mm with a preload of 10 N. The cam turns at 20 rpm.

†‡15-22   A single-dwell cam-follower system similar to that shown in Figure 15-1a (p. 753)
provides a two-segment polynomial for a rise of 35 mm in 75°, a fall of 35 mm in 120°,
and a dwell for the remainder of the cycle. Using equation 15.8, size a return spring
and specify its preload to maintain contact between cam and follower. Then calculate
and plot the dynamic force for one cycle if the roller follower train weighs 3.55 N, the
system has a damping ratio of $\zeta = 0.06$, and the cam turns at 100 rpm.

†‡15-23   A single-dwell cam-follower system similar to that shown in Figure 15-1a (p. 753)
provides a constant velocity to the follower of 100 mm/sec for 2 sec then returns to
its starting position with a total cycle time of 3 sec. Using equations 15.8 and 15.10,
calculate and plot the dynamic force and torque for one cycle if the roller follower train
weighs 4.5 N, the system has a damping ratio of $\zeta = 0.06$, and the spring has a rate of
2.5 N/mm with a preload of 50 N.

†‡15-24   A single-dwell cam-follower system similar to that shown in Figure 15-1a (p. 743)
provides a constant velocity to the follower of 100 mm/sec for 2 sec then returns to
its starting position with a total cycle time of 3 sec. Using equation 15.8, size a return
spring and specify its preload to maintain contact between cam and follower. Then
calculate and plot the dynamic force for one cycle of the roller.

15-25   As stated in Section 15.2 (p. 762), "A common rule of thumb is to design the [cam-
follower] system to have a fundamental frequency ω_n at least ten times the highest
forcing frequency expected in service..." Since this is not always possible, calculate
and plot the amplitude ratio resulting from $5 \le \omega_n / \omega_f \le 10$ for damping ratios of 0,
0.02, 0.04, 0.06, 0.08, and 0.10.

15-26   A cam-follower system similar to that shown in Figure P15-2 (p. 780) has a lumped
mass of 0.1 kg, lumped stiffness of 0.10 N/mm, and a damping coefficient of 0.40 kg/s.
If the forcing frequency is 50 rpm, determine the resulting amplitude ratio.

15-27   A cam-follower system similar to that shown in Figure P15-1 (p. 780) has a lumped
mass of 2.5 kg, lumped stiffness of 4 N/mm, and a damping coefficient of 12.0 kg/s. If
the forcing frequency is 80 rpm, determine the resulting amplitude ratio.

---
† These problems are suited
to solution using *Mathcad,
Matlab,* or *TKSolver* equation solver programs.

---
‡ These problems are suited
to solution using program
DYNACAM which is on the
attached DVD.

15.11 VIRTUAL LABORATORY

L15-1   View the video *Cam Machine Virtual Laboratory* that is on the DVD. Open the file
Virtual Cam Machine Lab.doc from the DVD and follow the instructions as directed by
your professor. Focus on the dynamic force measurements.

15

Chapter 16

CAM- AND SERVO-DRIVEN MECHANISMS

A professor must have a theory just as a dog must have fleas

H. L. MENCKEN

16.0 INTRODUCTION

A servomechanism uses feedback to control the system's output, such as its position and velocity. A desired control signal is applied to the motor and a transducer measures the output and feeds back information on its actual position. Any difference between the control signal and the fed-back output is an error that is amplified and used to force the system in a direction to reduce or eliminate the error. This is an application of control theory. An example is the cruise-control system in an automobile which takes as input a set speed and feeds back the vehicle's actual speed. Any difference is used to adjust throttle position until the error is brought as close to zero as possible.

Servomotors were introduced briefly in Chapter 2. They now are being used frequently in modern machinery, in part because they have become less costly than in the past. They offer many advantages over conventional motors because they can provide constant speed against dynamic variations in load torque due to their closed-loop operation. An encoder is built into or attached to the motor that provides a large number of equispaced pulses per revolution. This train of pulses is fed back to the computer-driven motor controller, which varies the current to the motor to maintain constant speed or provide any programmed variation in speed. The motor can also be made to hold an angular position against a load, thus creating a dwell on the output motion of any mechanism

16

driven by the motor. Many motion functions can be programmed into the controller to accelerate the motor to a set speed and decelerate it to a different speed or to zero. It can be made to follow any motion function, similar to a cam-driven follower.

Cams were discussed in Chapter 8 and are used in all kinds of machinery, most often to drive a linkage mechanism. A **function generator** was defined in Chapter 3 as *the correlation of an output motion with an input motion in a mechanism*. The cam-follower mechanism was also noted to be a *flexible and useful function generator*. A **servomotor** can generate any output motion that can be programmed into it and that can be dynamically achieved within the limits of the motor's torque capability. So, we can now add the servomotor to linkage and cam-follower mechanisms as a potential means for function generation. More importantly, the combination of a servomotor and a linkage can create an even more useful device for the solution of some motion generation problems. In fact, in many cases cam-driven-linkage mechanisms are being replaced by servomotor-driven linkages in machinery.

16.1   SERVOMOTORS

Both electric and hydraulic servomotors are widely available. We will limit this discussion to electric servomotors, both rotary and linear. The first electric servomotors were direct current (DC) and used an analog tachometer to provide velocity feedback. This allows the motor to run at close to constant speed in the face of dynamically varying torque loads. As the motor slows down or speeds up under changing load, the tachometer signal that is being fed back to the controller provides a measurement of the actual velocity. The difference between the command or setpoint velocity and the actual velocity constitutes an error. The current to the motor is automatically and continuously adjusted to minimize the error between the setpoint velocity and actual velocity.

Modern servomotors come in many varieties, among them DC servomotors based on conventional DC motor designs, alternating current (AC) servomotors based on induction motors, and the so-called brushless servomotor, which uses a synchronous AC motor that locks onto the frequency of the AC supplied. For this, the mains' AC is rectified to DC in the brushless motor's controller then inverted back to AC at different frequencies as dictated by feedback from the digital encoder on the motor. The frequency of the amplified current sent to the motor is varied by the digital motor controller and amplified to control motor speed. Because the motor's angular position is fed back to the motor controller, this type of servomotor can be made to move to a specified angular location, hold that position, and return to zero, all with controlled velocity and acceleration. The train of pulses from the position encoder or resolver is differentiated to obtain information on motor angular velocity in addition to angular position information. The velocity feedback controls motor speed. A block diagram of this system is shown in Figure 16-1.

Linear servomotors are essentially a rotary motor that has been "unwrapped" into a straight line. They are made as both moving coil and moving magnet types. A moving coil linear motor has a coil moving within an assembly of powerful magnets arranged to form a "track." There is an air gap between coil and magnets. An applied DC voltage causes the coil to move with constant force. A moving magnet type swaps the roles of coil and magnets with a magnet being pulled through the coil by its magnetic field. With either type, reversing polarity reverses motion and position feedback is provided by a

FIGURE 16-1

Typical servomotor system

linear optical or inductive encoder along the track. Linear servomotors, also called linear actuators, are capable of very high force, high speed, and high accelerations. Most linear servomotors require that the moving element be guided in bearings to maintain the air gap between the elements. Modern "Maglev" high speed trains[*] work on the same principle with the linear motor extending the length of the track and the magnetic flux doing double duty by lifting (levitating) the mass of the train off the track to reduce friction to the very low levels of an air bearing.

16.2   SERVO MOTION CONTROL

We will not address the control of servos beyond the mathematics of their motion control functions. The topic of programming and controlling servos lies well beyond the scope of this text and many books exist that cover this topic in detail. Some are noted in the bibliography of this chapter.

Servo Motion Functions

The same rules surrounding dynamic motion described for cam-follower systems in Chapter 8 also apply to servo motion. Newton's second law is universal and applies regardless of the means used to provide the motion. Thus the *fundamental law of cam design* defined in Chapter 8 can be restated and applied to servomotor motions as:

The servomechanism function must be piecewise continuous through the first and second derivatives of displacement across the entire interval (360 degrees or one cycle).

Corollary:

The jerk function must be finite across the entire interval (360 degrees or one cycle).

We will call this the *fundamental law of servomechanism design.*

The same set of acceptable functions defined for cams in Chapter 8 can be used for servo motions, and the ones listed as unacceptable in Chapter 8 should be avoided. Unfortunately, some suppliers of servo controllers provide these dynamically inferior functions as choices in their servo programming software. One of the largest suppliers (manufacturer A) offers only two types of acceleration functions in its basic controller, one of which, called trapezoidal velocity, is the same as the constant acceleration (parabolic displacement) function shown in Figure 8-13 on p. 418. This function has infinite jerk at its ends because the acceleration rises instantly from zero to its maximum value and for that reason is rejected as a usable motion function since it violates the fundamental law. The other offering from this supplier is called an "S-curve."

S-CURVES   This term is used by most servo controller manufacturers but is defined differently depending on the manufacturer. The name comes from the fact that any displacement function between two dwells resembles a laid-down letter S as can be seen in Figure 8-23 on p. 430. The same manufacturer A mentioned above defines its S-curve as having constant jerk, linear acceleration, parabolic velocity, and cubic displacement. While this function does not "break the fundamental law," it is a poor choice dynamically when compared to other possible functions. In fact, their S-curve is actually a trapezoidal acceleration function as shown in Figure 8-14 on p. 419.

* Japan Railways operates *Maglev* (short for Magnetic Levitation) vehicles at speeds up to 350 mph on their 20-kilometer guideway in Yamanashi Prefecture. *http://www.21stcentury sciencetech.com/articles/ Summer03/maglev2.html*

TABLE 16-1   Servo Motion Functions Offered by Manufacturer A

Manufacturer's Term	Cam Function Term	Figure	Eqns.	Fund. Law	Comments
Trapezoidal velocity	Constant acceleration	8-13	8.13–8.21	No	Not an acceptable choice
S-curve	Trapezoidal acceleration	8-14	—	Yes	Marginally acceptable choice

TABLE 16-2   Some of the Servo Motion Functions Offered by Manufacturer B

Manufacturer's Term	Cam Function Term	Figure	Eqns.	Fund. Law	Comments
Constant position	Dwell	—	—	Yes	Match to adjacent segments
Constant velocity	Constant velocity	8-35	—	No	Match to adjacent segments
Constant acceleration	Constant acceleration	8-13	8.13–8.21	No	Not an acceptable choice
Simple harmonic	Simple harmonic motion	8-9	8.13–8.21	No	Not an acceptable choice
Triple harmonic	Fourier-3 function	—	—	Yes	Low harmonic content
Cycloidal (their S-curve)	Cycloidal	8-12	8.12	Yes	Very smooth, lowest vibration
Modified trapezoidal	Modified trapezoidal	8-15	8.13–8.21	Yes	High vibration
Modified sine	Modified sine	8-16	8.13–8.21	Yes	Low peak velocity, low vibration
Sine-constant-cosine	SCCA	8-18	8.13–8.21	Yes	Depends on b, c, d values
Polynomial	Polynomial	8-25, 8-26	8.24	Yes	Depends on boundary conds.
Quadratic spline	Quadratic spline	—	—	No	Discontinuous acceleration
Cubic spline	Cubic spline	—	—	Yes	Discontinuous jerk

A second major manufacturer B of servo controllers that offers many good choices of functions also has an S-curve but defines it to be sine acceleration or cycloidal displacement, a dynamically superior motion when used between dwells. So the designer needs to be careful when selecting both the controller and the functions offered within that controller's software. Table 16-1 shows the servo functions offered by manufacturer A, and Table 16-2 shows a subset of those offered by manufacturer B.' Both tables show the manufacturer's name for the function, the standard cam design name for the same function, and references a figure and equation(s) for that function as defined elsewhere in this text. Each function's compliance with the fundamental law is also noted and comments on the function are provided.

16.3   CAM-DRIVEN LINKAGES

Chapter 8 showed how to design cams to obtain suitable s, v, a, j motions and Chapter 15 showed how to determine dynamic forces and torques in simple cam-follower systems. Both of these chapters were limited to systems using a single translating link as the follower. This is sufficient to explore the basics of cam design, but the fact is that most real cam-follower systems have the cam driving one link of a multibar linkage, referred to as a cam-driven linkage. There are a number of advantages to this arrangement. The cam provides a controlled set of s, v, a, j functions to the input link, and the linkage geometry modifies that motion, perhaps to deliver the motion to a location remote from the cam as shown in Figure 10-11 on p. 566. The linkage can be of any suitable configuration but the most common cam-driven combinations are a fourbar triple-rocker linkage, a fourbar slider with either the crank or the slider driven by the cam, or a Watt II sixbar with either a rocker or slider output link.

Depending on the particular linkage geometry, it may only slightly modify the s, v, a, j functions as input by the cam, or it could radically alter them. The designer can choose to define the s, v, a, j motion at the cam roller follower and let the linkage geometry modify those functions at the linkage output link (end effector). Or, the designer can choose to apply the desired s, v, a, j motion to the end effector and recalculate the cam shape to account for the linkage's distortion of the motion. In the latter case, the cam will have a different function cut into its shape.

The pushrod overhead valve mechanism of Figure 16-2a is an example of the first type mentioned above. Even if the rocker has unequal-length arms, the motion of the valve will be essentially the same shape in s, v, a, j as that of the tappet because the tappet motion is parallel to the valve motion and the rocker's mid-position is orthogonal to both motions, giving transmission angles of 90° at midstroke. Any ratio in the rocker arms will serve to amplify the magnitudes of s, v, a, j but will essentially preserve their shapes with small distortion. The overhead camshaft mechanism of Figure 16-2b has significant geometry change between the cam-roller motion and valve motion due to the differences in angle between the vectors of valve displacement and roller displacement. In this case, the designer applied the desired s, v, a, j functions to the valve-stem end of the rocker and calculated what the corresponding functions needed to be at the cam to get that result given the linkage geometry. The cam was then contoured to generate the modified displacement function that gave the proper motion at the valve.

(a) Overhead-valve train   (b) Overhead-camshaft valve train

FIGURE 16-2

Cam-driven linkages

16

In either case the task becomes that of combining the kinematics of the chosen s, v, a, j functions, applied at the desired end of the mechanism, with the kinematics of the linkage to obtain the motion of the other end. All the mathematical tools we need for this task have been developed in previous chapters of this book. All that remains is to apply them, which we will now do with some examples.

✎EXAMPLE 16-1

Cam-Driven Fourbar Slider With Motion Functions Applied to the Input Link

Given: A cam-driven fourbar crank-slider with the geometry shown in Figure 16-3 is driven by a cam with a constant velocity motion program similar to that developed in Example 8-12 (p. 450) and Figure 8-42 (p. 453). The slider is the end effector and must chase a constant velocity conveyor for at least 6 in at 10 in/sec with minimal velocity error and return to the start position to repeat the cycle.[*] The motion functions are applied to the cam-follower arm. The linkage was designed in a CAD system, which determined the geometry of Figure 16-3 and that 30° of crank rotation gives 6.435 in of constant velocity motion at 1 Hz.

Problem: Compare the displacement and velocity functions applied to the cam follower (link 2) and the resulting functions at the end effector of the linkage (link 4).

Solution:

1   The equations for the position of the crank-slider linkage were developed in Section 4.6 (p. 191), its velocity in Section 6.7 (p. 315), and acceleration in Section 7.3 (p. 358) and are repeated here for your convenience.

$$\theta_{3_1} = \arcsin\left(\frac{a\sin\theta_2 - c}{b}\right) \tag{4.16a}$$

$$d = a\cos\theta_2 - b\cos\theta_3 \tag{4.16b}$$

$$\omega_3 = \frac{a}{b}\frac{\cos\theta_2}{\cos\theta_3}\omega_2 \tag{6.22a}$$

$$\dot{d} = -a\omega_2\sin\theta_2 + b\omega_3\sin\theta_3 \tag{6.22b}$$

$$\alpha_3 = \frac{a\alpha_2\cos\theta_2 - a\omega_2^2\sin\theta_2 + b\omega_3^2\sin\theta_3}{b\cos\theta_3} \tag{7.16d}$$

$$\ddot{d} = -a\alpha_2\sin\theta_2 - a\omega_2^2\cos\theta_2 + b\alpha_3\sin\theta_3 + b\omega_3^2\cos\theta_3 \tag{7.16e}$$

The terminology used in the equations is shown in Figure 16-3.

2   The cam program has two segments: 220° of constant velocity at 49.06605°/s for 30° of follower arm (crank) motion, and a fifth-degree polynomial return motion over 140°. Total angular displacement of the crank is 33.2° in segment 2.

3   The boundary conditions for the polynomial in segment 1 are shown in Table 16-3. The boundary conditions for the polynomial in segment 2 are shown in Table 16-4. The normalized displacement equations for the cam functions as derived with program DYNACAM with $x = \theta/\beta$ are as follows:[†]

---

[*] The reader might reasonably wonder why anyone would bother to add the complexity of a linkage to the system when the same output motion could be obtained from a cam directly driving a translating follower. One reason would be that it was not possible to mount and power the cam adjacent to the end effector. Another might be that the stroke would require a large cam to get acceptable pressure angles. One can multiply the stroke of the cam follower by adding a linkage.

[†] See Chap 8, p. 411 for an explanation of normalized variables.

16

linear dimensions in inches

$\theta_{2_{min}} = 46.95°$
$\theta_{2_{max}} = 80.16°$
$\Delta\theta = 33.20°$
$\Delta d = 6.767$
$R_{prime} = 6$

$a = 12$
$b = 36$
$c = 0$
$r = 8$

offset $c = 0$

FIGURE 16-3

Cam-driven slider-crank linkage

TABLE 16-3
Boundary Conditions

Example 16-1, Segment 1

When $\theta = 0$:
$\quad s = 0°$,
$\quad v = 49.066\ 05\ °/sec$

TABLE 16-4
Boundary Conditions

Example 16-1, Segment 2

When $\theta = 0$:
$\quad s = 30.000\ 05°$,
$\quad v = 49.066\ 05\ °/sec$,
$\quad a = 0$

When $\theta = 140°$
$\quad s = 0°$,
$\quad v = 49.066\ 05\ °/sec$
$\quad a = 0$

Segment 1:

$$S = 30.00\,x \quad \text{deg}$$

$$V = \frac{\omega}{\beta}(30) = \frac{2\pi}{3.8397}(30) = 49.066 \quad \text{deg/sec} \tag{16.1}$$

$$A = \frac{\omega^2}{\beta^2}(0) \quad \text{deg/sec}^2$$

Segment 2:

$$S = -294.546x^5 + 736.364x^4 - 490.909x^3 + 19.091x + 30.00 \quad \text{deg}$$

$$V = \frac{\omega}{\beta}\left(-1472.73x^4 + 2495.46x^3 - 1472.73x^2 + 19.091\right) \quad \text{deg/sec} \tag{16.2}$$

$$A = \frac{\omega^2}{\beta^2}\left(-5890.91x^3 + 7486.38x^2 - 2945.46x\right) \quad \text{deg/sec}^2$$

4 These S, V, A equations are the driving functions for the input crank of the linkage where $\omega = 2\pi$ rad/sec for a 1-Hz cycle, and β is the angular duration of the motion in radians—3.8397 rad for segment 1 and 2.4435 rad for segment 2.

5 The equations for the kinematic behavior of the linkage from step 1 must be solved at a set of discrete points (e.g., every degree of the cam cycle) to determine the behavior of the slider d. Equation 16.1 applies to the first 220° (constant velocity) and equation 16.2 to the last 140°.

6 This is obviously a job for a computer and one was used to solve these equations. A good approach is to use an equation solver such as *Matlab, Mathcad,* or *TKSolver*. Plots of the polynomial motions from program DYNACAM are shown in Figure 16-4 (p. 790).

16

FIGURE 16-4

S, V, A functions as applied to cam-follower arm in Example 16-1

7 Figure 16-5a shows the displacement of the follower arm and the slider superposed and nor-malized to percent for comparison. The nonlinear behavior of the linkage distorts the entire function. Figure 16-5b shows the normalized velocities of follower arm and slider superposed. Note that the follower arm has constant angular velocity during the first segment but the slider does not have constant linear velocity.

8 Figure 16-6 shows the percent error between the crank arm and slider in both displacement and velocity. The differences in this example are not extreme, mainly because the crank only rotates through $30°$ during the constant velocity motion. The larger the angle of crank rotation, the larger the distortion in the output functions.

9 If the magnitude of the deviation from constant velocity is small enough to still allow proper function, then this solution will be adequate as is. If the error is not tolerable, then the solution is to apply the desired motion function to the slider and calculate the linkage's deviation in reverse to redefine the cam function to correct it. The next example will do this.

(a) Displacement   (b) Velocity

16

FIGURE 16-5

Normalized input and output displacement and velocity for Example 16-1

FIGURE 16-6

Normalized error in displacement and velocity for Example 16-1

✎ EXAMPLE 16-2

Cam-Driven Fourbar Slider With Motion Functions Applied to the Output Link

Given: A cam-driven fourbar crank-slider with the geometry shown in Figure 16-3 (p. 789) is driven by a cam with a constant velocity motion program similar to that developed in Example 8-12 and Figure 8-42 (pp. 450-453). The slider is the end effector and must chase a constant velocity conveyor for at least 6 inches at 10 in/sec with minimal velocity error and then return to the start position to repeat the cycle. The motion functions are applied to the slider and converted to modified functions to be applied to the cam-follower arm with a modified contour cam running at 1 Hz.

Problem: Compare the displacement, velocity, and acceleration functions between the cam follower (link 2) and the end effector of the linkage (link 4) and compare the cam profile for this solution to that of Example 16-1.

Solution:

1   The linkage geometry is the same as shown in Figure 16-3.

2   The equations that define the motion of the slider-crank when driven at the slider rather than at the crank take on a different form than those shown in Example 16-1. The vector loop equation for both cases is the same, so equations 4.14a, b, c, and 4.15a, b (p. 192) apply to both cases. The derivation starts with the above noted equations for either mode of operation. The simultaneous solution of equations 14.15a, b for the crank-driven mode is done by simple substitution as shown in equations 4-16a, b (p. 193). However, solving for the slider-driven mode with slider position d as the input is more complex because both variables now sought, θ_2 and θ_3, are contained in transcendental expressions in both equations 4.15a and 4.14b. This requires the derivation done in equations 4.18 to 4.21 (pp. 194-195) to find θ_2 and θ_3 as functions of slider position d.

3   The derivation for crank velocity is done in equations 6.24 on p. 321 and for crank acceleration in equations 7.17, p. 364. The results of these derivations for θ_2, ω_2, and α_2, as a function of slider position d, velocity $\dot{d}$, acceleration $\ddot{d}$, and linkage geometry a, b, c (shown in Figure 16-3, p. 789) are used in the following steps.

16

$$K_1 = a^2 - b^2 + c^2 + d^2$$
$$K_2 = -2ac$$
$$K_3 = -2ad$$
$$A = K_1 - K_3$$
$$B = 2K_2$$
$$C = K_1 + K_3$$

$$\theta_{2_{1,2}} = 2\arctan\left(\frac{-B \pm \sqrt{B^2 - 4AC}}{2A}\right) \tag{16.3}$$

4　The angular velocities ω_2 and ω_3 as a function of slider velocity and link geometry are:

$$\omega_2 = \frac{\dot{d}\cos\theta_3}{a(\cos\theta_2\sin\theta_3 - \sin\theta_2\cos\theta_3)}$$

$$\omega_3 = \frac{a\omega_2\cos\theta_2}{b\cos\theta_3} \tag{16.4}$$

5　The angular accelerations α_2 and α_3 as a function of slider acceleration and link geometry are:

$$\alpha_2 = \frac{b\omega_3^2 - a\omega_2^2\cos(\theta_2 - \theta_3) - \ddot{d}\cos\theta_3}{a\sin(\theta_2 - \theta_3)}$$

$$\alpha_3 = \frac{a\alpha_2\cos\theta_2 - a\omega_2^2\sin\theta_2 + b\omega_3^2\sin\theta_3}{b\cos\theta_3} \tag{16.5}$$

TABLE 16-5
Boundary
Conditions

Example 16-2, Segment 1

When $\theta = 0$:
　$S = 0$,
　$V = 6.114$ in/sec

6　The cam program has two segments: 220° of constant velocity at 10 in/s over 30° of follower arm (crank) motion, and a fifth-degree polynomial return motion over 140°. Total displacement is 6.767 in.

7　The boundary conditions for the polynomial in segment 1 are shown in Table 16-5. The boundary conditions for the polynomial in segment 2 are shown in Table 16-6. The s, v, a functions as derived with program DYNACAM are:

TABLE 16-6
Boundary
Conditions

Example 16-2, Segment 2

When $\theta = 0$:
　$s = 6.114\,210$ in,
　$v = 10$ in/sec,
　$a = 0$

When $\theta = 140°$
　$s = 0$,
　$v = 10$ in/sec,
　$a = 0$

Segment 1:
$$S = 6.114x \text{ in}$$

$$V = \frac{\omega}{\beta}(6.114) = \frac{2\pi}{3.8397}(6.114) = 10 \text{ in/sec} \tag{16.6}$$

$$A = 0$$

Segment 2:
$$S = -60.029x^5 + 150.073x^4 - 100.049x^3 + 3.891x + 6.114 \text{ in}$$

$$V = \frac{\omega}{\beta}\left(-300.145x^4 + 600.292x^3 - 300.147x^2 + 3.891\right) \text{ in/sec} \tag{16.7}$$

$$A = \frac{\omega^2}{\beta^2}\left(-1200.58x^3 + 1800.88x^2 - 600.294x\right) \text{ in/sec}^2$$

16

FIGURE 16-7

S, V, A functions as applied to slider in Example 16-2

8 Equation 16.6 applies to the first 220° and equation 16.7 to the last 140°. The normalized variable $x = \theta/\beta$ runs from 0 to 1. The S, V, A diagrams for these polynomial motions as applied to the slider are shown in Figure 16-7. Note that the displacement is now in length units rather than degrees as was the case in the previous example (Figure 16-4, p. 790).

9 The equations for the kinematic behavior of the linkage from equations 16.3 through 16.5 must be solved at a set of discrete points (e.g., every degree of the cam cycle) using the functions of equations 16.7 and 16.8 as input to the slider to determine the behavior of the crank arm. The crank arm displacement is then used to calculate the required cam profile.

10 This is obviously a job for a computer and one was used to solve these equations. A good approach is to use an equation solver such as *Matlab*, *Mathcad*, or *TkSolver*. The problem can be solved numerically using discrete representations of the motion functions calculated at every degree or fraction thereof.

11 Figure 16-8a shows the displacement of the slider and the crank arm superposed and normalized to percent for comparison. The nonlinear behavior of the linkage distorts the entire function. Figure 16-8b shows the normalized velocities of follower arm and slider superposed.

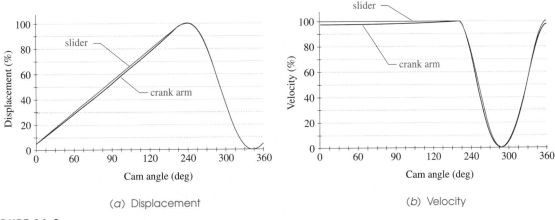

(a) Displacement   (b) Velocity

FIGURE 16-8

Normalized input and output displacement and velocity for Example 16-2

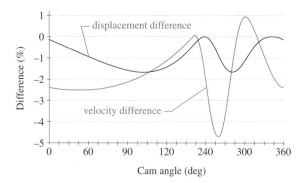

FIGURE 16-9

Difference in displacement and velocity between slider and cam follower in Example 16-2

Note that the slider has constant linear velocity during the first segment but the crank arm does not have constant angular velocity.

12   Figure 16-9 shows the percent difference between the crank arm and slider in both displacement and velocity. The distorted crank arm functions serve to correct the error introduced by linkage geometry and give the designed motions at the slider.

13   Figure 16-10 shows the difference in radius to the roller center for the cam of Example 16-1 and that of Example 16-2. The first has the desired functions applied to its follower. The second has the functions applied to the slider with the cam profile recalculated to remove the linkage geometry error. The largest radial difference between the cams is nearly 0.2 in.

Whatever the linkage configuration, the exercise outlined in the above two examples will apply. The cam-motion functions and the equations that define the linkage motion transfer function must be combined, working from input to output, or vice versa, depending on whether you want to apply the theoretical cam functions to the input or output ends of the mechanism.

16.4   SERVO-DRIVEN LINKAGES

Applications that have traditionally used cam-driven linkages for pick-and-place operations, conveyor chasing motions, and other assembly tasks are increasingly using linkage mechanisms directly driven by a servomotor through a gear reducer. The pros and cons of this approach versus the traditional cam-driven linkage are addressed in a later section of this chapter.

If the need is simply to move the end effector of the linkage from a start position (perhaps in a dwell) to a final position and hold it there in dwell before returning, then appropriate mathematical functions from the collection of acceptable double-dwell cam-follower functions can be applied directly to the servo axis. This situation is analogous to the *Critical End Position* (CEP) cam design case described in Chapter 8 where the path motion between the end positions is not critical. Nevertheless, when selecting motion functions to drive the servo axis, be sure to observe the *fundamental law of servomecha-*

16

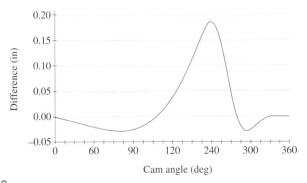

FIGURE 16-10

Difference in cam profile radius with functions applied to cam follower versus end effector

nism design described in an earlier section of this chapter. The distortion of motion between the endpoints introduced by the linkage may not be detrimental to function in the CEP case except to the degree that it increases accelerations on the members. Links must be designed to withstand dynamic loads from their accelerations and motors selected to provide the needed torque. Most mechanical failures of this type of mechanism occur during acceleration when starting or stopping, especially during an emergency stop when the servomotors must dynamically brake the system to a stop in one or a few machine cycles.

In cases where critical intermediate positions or velocities are specified between the endpoints, distortion of the motion functions can be a problem. This is analogous to the *critical path motion* (CPM) case described for cam mechanisms in Chapter 8. Also, if the input link driven by the servomotor (i.e., the crank of a crank-slider, fourbar, or sixbar mechanism) needs to rotate through a significant angle, it will give greater distortion than was the case in the two previous examples. In such cases, it can become critical to modify the servomotor's control function to account for the linkage geometry's distortion and create the desired precision motion at the linkage end effector. The best way to illustrate this process is with an example.

EXAMPLE 16-3

Slider-Crank Linkage Driven by a Servomotor to Perform a Forming Operation

Given: A slider-crank linkage mechanism (Figure 16-13, p. 797) has been designed to provide the geometry for a suitable motion of a die in a forming press. The required motion begins from a dwell and moves the slider through a 2-in stroke ending at TDC with crank angle $\theta_2 = 0$. The die needs to be brought to zero velocity for an instant at 1.588-in stroke then advanced into the material. It is then pulled back to dwell at 1.400 in before returning to zero. The slider motion timing diagram is shown in Figure 16-11 (p. 796). Rather than design a custom cam to provide the motion and dwells to the linkage crank, it is desired to use a servomotor directly driving the input link of the crank-slider mechanism through a gearbox to obtain the necessary motions and dwells. Machine speed is 120 cycles per min or 2 Hz.

Find: The motion function needed to program a servomotor to drive a crank-slider linkage that provides the required linkage output motion program as defined at the slider.

16

FIGURE 16-11

Timing diagram for slider in Example 16-3

Assume: The position and velocity accuracy required at intermediate points in the stroke requires that the motion function be applied to the slider and mathematically transformed to a function at the servo axis that accounts for linkage geometry.

Solution:

1   The functions chosen to provide the motions defined in Figure 16-11 are:

        Segment 1:   Polynomial over 60° with the boundary conditions of Table 16-7.
        Segment 2:   Dwell at 2 in for 20°.
        Segment 3:   Cycloidal fall to 1.4 in over 20°.
        Segment 4:   Dwell at 1.4 in for 20°.
        Segment 5:   Cycloidal fall to zero over 60°
        Segment 6:   Dwell at zero in for 180°.

    The S, V, A functions for the slider motion as calculated with program DYNACAM are shown in Figure 16-12.

2   The geometry of the slider-crank linkage to be driven by the servomotor was chosen based on packaging requirements and is shown in Figure 16-13. The maximum displacement at TDC when $\theta_2 = 0$ is $d = a + b = 9.8$ in. A 2-in stroke requires that the starting displacement $d_{start} = 9.8 - 2 = 7.8$ in. Equations 16.4 are used to find the starting crank angle at d_{start}.

$$K_1 = a^2 - b^2 + c^2 + d^2 = 2.4^2 - 7.4^2 + 0 + 7.8^2 = 11.840$$
$$K_2 = -2ac = -2(2.4)(0) = 0$$
$$K_3 = -2ad = -2(2.4)(7.8) = -37.440$$
$$A = K_1 - K_3 = 11.840 - (-37.440) = 49.280 \qquad (a)$$
$$B = 2K_2 = 2(0) = 0$$
$$C = K_1 + K_3 = 11.840 + (-37.440) = -25.600$$

$$\theta_{start_{1,2}} = 2\arctan\left(\frac{-B \pm \sqrt{B^2 - 4AC}}{2A}\right)$$

$$= 2\arctan\left(\frac{0 \pm \sqrt{0 - 4(49.280)(-25.600)}}{2(49.280)}\right) \qquad (b)$$

$$\theta_{start_1} = 71.564° \qquad \theta_{start_2} = -71.564°$$

TABLE 16-7
Boundary
Conditions

Example 16-3, Segment 1

When $\theta = 0°$:
  $s = 0, v = 0, a = 0$

When $\theta = 35°$:
  $s = 1.588, v = 0$

When $\theta = 60°$
  $s = 2, v = 0, a = 0$

FIGURE 16-12

S, V, A functions as applied to slider in Example 16-3

Because of symmetry, either value could be used depending on the desired direction of crank rotation. We will choose clockwise crank rotation with $\theta_{start} = 71.564°$, $\theta_{end} = 0°$.

3   It is only necessary to convert the displacement function with the linkage geometry in order to create a usable servo control function because the typical servomotor controller needs only that function for control.[*] Note that the effects of the chosen velocity, acceleration, and jerk functions are all contained within the displacement function because it comes from the integration of those functions.[†] Most servo controllers are capable of importing a table of displacement data, which in this case will be generated by the displacement function of Figure 16-12 as modified by the linkage geometry. This data table of crank angles can be generated at any resolution desired from every degree to any fraction of a degree.

4   The values of $d_{start} + d$ as defined by the linear displacement function of Figure 16-12 for each increment of machine angle over the cycle from 0 to 360° are used as input to equation 16.4 (p.

[*] Note that this is the same situation when a cam profile is generated. Only the displacement function is used to create the cam contour, but it contains all the mathematics of the higher derivatives used in the cam design.

[†] The reason the velocity function was converted in Examples 16-1 and 16-2 was to show the effects of linkage geometry on the constant velocity portion of the functions. Only the displacement function would be used to program the servo controller. The slope of the displacement function encodes the velocity.

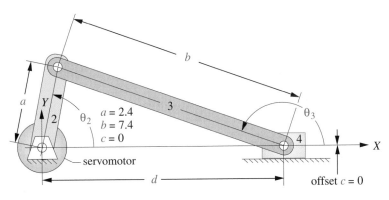

FIGURE 16-13

Servomotor-driven slider-crank linkage

FIGURE 16-14

Normalized input and output displacements for crank and slider of Example 16-3

792) to calculate their corresponding angular displacements θ_2 to be applied to the servo axis. Do not confuse the linkage crank angles θ_2 with the timing angles shown as the independent variable in Figures 16-12 (p. 797) and 16-14. Those angles represent fractions of one machine cycle, which is always one revolution of the main timing shaft of the machine. In a fully servomotor-controlled machine with no physical timing shaft, this is often an "imaginary" or virtual "master" axis that serves as a timing reference for all "slave" servomotors to follow.

5 Figure 16-14 shows the displacement functions for both the slider linear displacement and the linkage crankshaft angular displacement over the first five segments. The 180° dwell of segment 6 is omitted since the crank is then stationary at the start position. Each function is normalized to a range of 0 to 1 to allow their superposition. Note the large differences between the two functions as compared to those of the previous two examples. In those cases, the total rotation of the crank was less and avoided the regions near BDC or TDC where geometric distortion is greatest. This example takes the crank to TDC and the differences between the functions is seen to increase as it approaches TDC.

6 While this design does what the client requested, a close inspection of the crank arm displacement function of Figure 16-14 at 60 and 80 degrees of timing angle shows this to be a dynamically poor solution. There is such a rapid change of slope at those points that it will have very large acceleration. It comes close to violating the fundamental law described earlier in this chapter. This results from the large amount of distortion introduced by the nonlinear slider-crank function. An improvement will be sought in the next example.

✏️ **EXAMPLE 16-4**

Improvement to the Design of Example 16-3 (p. 795)

Given: The design of Example 16-3 proved to be dynamically poor. In that example the required position and velocity constraints were applied to the slider motion and the resulting function was transformed by equations relating slider motion to crank motion. The constraints require that the slider move through a 2-in stroke ending at TDC with crank angle $\theta_2 = 0$. The slider needs to be brought to zero velocity for an instant at 1.588-in stroke then advanced into the material. It is then pulled back to dwell at 1.400 in before returning to zero. Rather than apply those constraints directly to the slider motion as was done in Example 16-3, we will instead transform the constraints from linear parameters to angular parameters first and then apply those angular parameters to the motion of the servo-driven crank arm. Machine speed is 120 cycles per min or 2 Hz.

Find: The motion function needed to program a servomotor to drive a crank-slider linkage that provides the required linkage output motion parameters as defined for specific positions of the slider.

Assume: The position and velocity accuracy required at intermediate points in the stroke only requires that those specific parameters be transformed to corresponding parameters at the servo axis. The shape of the functions between those points is not critical.

1   The functions chosen to provide the motions will be similar to those defined in step 1 of Example 16-3 but will instead be applied directly to the crank arm motion.

2   The start and end angles $\theta_{start} = 71.564°$, $\theta_{end} = 0°$ required to achieve the 2-in stroke were calculated in Step 2 of Example 16-3 and will be the same here.

3   The crank angles corresponding to the intermediate positions can be calculated by the same method used for the start angle in step 2 of Example 16-3. There must be a point d_1 having instantaneous zero velocity when the slider has moved through 1.588 in of its 2-in stroke. The start position of the slider as referenced to the crank pivot was calculated in step 2 of Example 16-3 to be 7.8 in. The point d_1 will be at $7.8 + 1.588 = 9.388$ in. The corresponding crank angle for that position is found from equations 16.4 to be 29.721°.

4   The dwell at a slider stroke of 1.4 in is $7.8 + 1.4 = 9.6$ in from the crank pivot. Equations 16.4 show this slider position to correspond to a crank angle of 36.10°.

5   The boundary conditions for crank motion then become:

Start at crank angle of 71.564°.
Segment 1:   B-spline over 60° with the boundary conditions of Table 16-8.[*]
Segment 2:   Dwell at zero crank degrees for 20°.
Segment 3:   Cycloidal rise to 36.19° over 20°.
Segment 4:   Dwell at 36.19° for 20°.
Segment 5:   Cycloidal rise to 71.564° over 60°.
Segment 6:   Dwell at 71.564° for 180°.

6   The resulting function superposed on the solution for Example 16-3 is shown in Figure 16-15 (p. 800). Note the smoother contour approaching 60 and leaving 80 timing degrees. The slope discontinuity is gone but the critical points are still being maintained. This is a dynamically superior design to that of Example 16-3.

7   It should now be apparent that if any intermediate positions or velocities during the stroke are critical, it is necessary to modify the servo function by the linkage geometry to ensure proper function. If only the end positions are critical (the CEP case), then one may be able to get away with applying the chosen motion functions directly to the input crank. This could nevertheless cause unwanted distortion to the acceleration function and negatively affect the torque function as well. The engineer would be wise to check these effects in each case.

TABLE 16-8
Boundary
Conditions
_____
Example 16-4, Segment 1
_____

When $\theta = 0$:
$s = 71.546°, v = 0, a = 0$

When $\theta = 35°$:
$s = 29.721°, v = 0$

When $\theta = 60°$:
$s = 0°, v = 0, a = 0$
_____

* This set of boundary conditions did not allow an acceptable polynomial solution, and a B-spline was needed to solve it. B-splines give better control over motions and their derivatives than polynomials. They are not described in this book, but information on their mathematics and use can be found in reference 2.

16

FIGURE 16-15

Normalized input displacements for the cranks of Examples 16-3 (p. 795) and 16-4 (p. 798)

16.5   OTHER LINKAGES

The approach used for a servo-driven slider-crank linkage in the previous example also can be used for any other linkage mechanism. For a fourbar linkage, the desired motion functions will be applied to the output rocker, link 4, and modified by the linkage displacement transfer function as define by equations 4.8a, 4.10a, and 4.10b. Note that the variables θ_2 and θ_4 must be temporarily interchanged since we want to solve for θ_2 as a function of θ_4 to which the servo motion is applied. Simply consider link 4 to be link 2 and vice versa for the purpose of this calculation. A Watt II sixbar can be solved in the same manner, considering the output link to which the servo functions are to be applied as the temporary "input" link in order to modify the applied servo function to one that can be applied at the true input crank of the mechanism to generate the proper motion.

16.6   CAM-DRIVEN VERSUS SERVO-DRIVEN MECHANISMS

One can sometimes hear the comment among engineers "Cams are obsolete. I can do anything with a servo that you can with a cam." The author heard this said 40 years ago and occasionally still does today. Is this really true? If so, why are cams still used? Cams have been around a long time. Leonardo DaVinci's sketches show many examples of cams. Even older illustrations from ancient civilizations show cam-like devices used for various purposes. Servomechanisms are more recent, developed in the second half of the twentieth century. Originally they were only rotary devices (motors) but now are also available as linear actuators. Both systems are extremely useful and valuable, but each has its strong points and limitations.

Flexibility

The greatest strength of servo systems is their inherent programmability. With moderate effort by skilled personnel, a servo can be reprogrammed in a short time to provide a different motion within the limits of its mechanical package. The position, velocity, and acceleration profiles of its motion can be changed in software and more than one version can be stored in memory, allowing the operator to change the motion with a button press.

16

A cam can provide the same motions as a servo, but its motion program is encoded "in hardware" in its shape. To change the motion program requires remachining its contour. Since its motion program and contour are typically designed with software, and modern manufacturing techniques allow a cam to be made in a day or two, the cam motion is capable of change as well. But it requires new hardware.

Cost

Servomotors and their controllers are expensive, typically costing anywhere from $1000 for a 1/4 hp example to thousands of dollars for large examples. Cams can cost anywhere from a few hundred dollars for an open radial cam to several times as much for a more complicated barrel cam, and even more for a three-dimensional cam. A less expensive (non-servo) motor can be used to drive the camshaft.

Reliability

Once a properly designed cam is properly manufactured, it can last a very long time. If it is properly lubricated, one should expect many millions of cycles from the cam and follower before it fails. Modern automobile engines are capable of going 200,000 miles without camshaft replacement. Conservatively assuming an average speed of 35 mph and an average engine speed of 1500 rpm, this amounts to about a quarter-billion camshaft revolutions without failure. Cams of the sort used in production machinery do not lead such a happy life since they are not bathed in filtered oil, but are often run dry, or with minimal lubrication, in less than pristine environments. Nevertheless, some production machine cams last several years on machines that accumulate as many as 100 million cycles per year. Servomotors and controllers have high reliability as well. Electronic components can fail and bearings can wear out, but these systems have long lives if properly sized.

Complexity

Cams and servo systems both require engineering-level personnel to be correctly designed. Specialized training in cam design or servo programming and tuning is needed for these tasks. The mathematical requirements for their motion functions are essentially the same for both systems. Dynamics is dynamics whether achieved by servo or cam. F always equals mA. The ratio of load inertia to servomotor inertia must be kept in proper range by the designer to achieve the desired dynamics. A cam does not have this requirement. The desired servo motion program can only be achieved by proper tuning of the system controller. If no changes are made to the driven system, the tuning can remain untouched for the life of the machine. Once a properly designed cam is manufactured, it needs no further tuning. If servo systems require some reprogramming during their lifetimes, it is typically done by highly trained personnel. Cams can be serviced by mechanics and technicians.

Robustness

Cam-driven machines maintain synchrony and phasing between mechanisms mechanically. If the power fails, the machine comes to a stop with no change in phasing. If power is lost to a servo-driven system, all control over relative position is lost and crashes can occur with expensive consequences. Servo controller programs also have been known to lose their phasing even with no power loss and then require a machine stop to re-home the drives, meaning to get them all back to their zero position.

Packaging

Cam-follower systems can be quite compact (think of an overhead camshaft valve train) or large if the motion must be transferred over some distance, as in many assembly machines with camshafts below and tooling above the base plate. In some cases, a small servomotor mounted at the tooling can provide the needed motions without the complexity of the linkage system. Linear servos lend themselves well to this situation if the desired output motion is rectilinear translation, as they can be mounted to act directly on the end effector. They can achieve large displacements also. The resultant reduction of mass and increased stiffness of the follower train can have significant dynamic advantages.

Load Capacity

Cams have inherently high mechanical advantage. This makes them very suitable for applications that require the generation of high forces or torques or ones that have large inertia loads. One example is the compaction of the contents in a dry-cell battery. The required forces are thousands of pounds and cams are typically used to drive a compacting ram into the battery to reduce its volume. Another is the insertion/bonding of lid liners in food jars which requires around 7 000 lb force, also done with cam mechanisms.

A servomotor driving the input crank to a linkage or other mechanism through a gearbox has the gearbox ratio as a constant mechanical advantage. The linkage, of course, introduces its own additional, variable mechanical advantage. The same linkage, driven by a cam acting on a roller attached to the input crank, substitutes a variable mechanical advantage for that of the servo gearbox's constant one. A linear servomotor acting directly on the end effector has no mechanical advantage. Though these devices can generate large forces, a cam with its infinite mechanical advantage will still generate much more force.

As an example, consider the same slider-crank linkage to be driven through the same crank angle by either a cam acting on a roller placed at the joint between crank and connecting rod or a servomotor driving the crank directly. Each can provide the identical angular displacement, velocity, and acceleration profiles to the crank. The mass of the linkage is the same in both cases. Let the motion profile be a 3-4-5-6 polynomial rise-fall-dwell function applied to the crank. Calculate the torque required by the motor driving the cam and the motor directly driving the crank and you will find that the servomotor directly driving the crank must apply a torque about 20 times that required on the camshaft. Now, a 20:1 reduction gearbox will even things out, after a fashion. But, if the purpose of the mechanism is to provide large external forces at some point in the stroke, the cam's variable (and potentially infinite) mechanical advantage may be preferable to the gearbox's constant mechanical advantage.

When dealing with high speeds and high loads, the motor's so-called velocity-torque envelope can become a limitation. The product of angular velocity and torque is power, and any motor has a particular angular velocity-torque curve as shown in Figures 2-40 through 2-42, pp. 76-78. When a servomotor's size is increased to obtain the required torque for a high-load application, its rotor's polar moment of inertia also increases as the square of its radius. The result is a disproportionate amount of torque being required to accelerate the rotor and thus not being available to drive the load. This results in an oversized motor compared to one driving a cam that runs at constant speed and amplifies the force or torque through its variable mechanical advantage. A known example is an application that uses 47% of the servomotor's torque capacity just to accelerate the motor rotor each cycle.

16.7 REFERENCES

1   **Norton, R. L.** (2009). *Cam Design and Manufacturing Handbook* 2ed. Industrial Press Inc.: New York, pp. 417-420.

2   **Ibid**, Chapter 5.

16.8 BIBLIOGRAPHY

Dote, Y., and S. Kinoshita (1990). *Brushless Servomotors*. Clarendon Press: Oxford.

Humphrey, W. M. (1973). *Introduction to Servomechanism System Design*. Prentice-Hall: Englewood Cliffs, NJ.

Younkin, G. W.. (2003). *Industrial Servo Control Systems*. Marcel Dekker: New York.

16.9 PROBLEMS

Program DYNACAM *may be used to solve these problems where applicable. Where units are unspecified, work in any consistent units system you wish.*

16-1   Design a double-dwell motion to translate a slider from 0 to 75 mm in 60° with modified sine acceleration, dwell for 120°, fall 75 mm in 30° with cycloidal motion, and dwell for the remainder. Calculate and plot the displacement function of the slider using DYNACAM and modify that function to drive the crank of a nonoffset crank-slider mechanism with dimensions: crank = 100 mm, coupler = 300 mm such that at maximum stroke of the slider, the crank is at zero degrees. The slider must follow the specified motion. Plot the resulting crank input function.

16-2   Repeat Problem 16-1 for a slider function for a 65-mm rise with 3-4-5 polynomial motion over 80°, dwell for 100°, fall over 40° with 3-4-5 polynomial motion, and dwell for the remainder.

16-3   Design a single-dwell polynomial motion to translate a slider from 0 to 75 mm and return to 0 in 120° and dwell for the remainder. Total cycle time is 6 sec. Calculate and plot the displacement function of the slider using DYNACAM and modify that function to drive the crank of a non-offset crank-slider mechanism with the dimensions crank = 90 mm, coupler = 270 mm such that at maximum stroke of the slider, the crank is at 180°. The slider must follow the specified motion. Plot the resulting crank input function.

16-4   Using the constant velocity motion program of Example 16-1 (p. 788) applied to the slider of a nonoffset fourbar crank-slider linkage with a constant velocity stroke of 150 mm at 40 mm/s, calculate and plot the *s, v, a, j* functions for this motion applied directly to the slider using program DYNACAM. Total cycle time is 6 sec. Calculate the modified program needed to drive the crank of the linkage to obtain the specified motion at the slider. Crank = 200 mm, coupler = 600 mm, crank start angle = 110°.

16

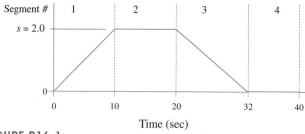

FIGURE P16-1

Timing diagram for Problems 16-6 and 16-7

16-5   Repeat Problem 16-4 for slider stroke of 5.5 in at 0.75 in/sec, a crank length of 7.5 in, a coupler length of 18 in, and a crank start angle of 120°. Total cycle time is 12 sec.

16-6   A nonoffset crank-slider with crank and coupler lengths of 3.25 in and 10.875 in, respectively, is to be driven by a servo through a gear reducer. The required slider timing diagram is shown in Figure P16-1. The slider moves through the distance s in. Choose a set of servo functions from Table 16-2 (p. 786) to drive the crank with the lowest resulting vibration and calculate the resulting crank displacement for one cycle of 40-sec duration. Based on this crank input, calculate the resulting slider motion and compare the two by plotting their normalized displacements. The crank angle at $t = 0$ is 70 deg, and the crank turns clockwise to accomplish the first motion of the slider.

16-7   Repeat Problem 16-6 using the timing diagram of Figure P16-1 but with a total stroke of 75 mm. The crank length is 110 mm, the coupler length is 275 mm, and the initial crank angle is 90°.

Appendix A

COMPUTER PROGRAMS

I really hate this damned machine;
I wish that they would sell it.
It never does quite what I want
But only what I tell it.
FROM THE FORTUNE DATABASE, BERKELEY UNIX

A.0   INTRODUCTION

In addition to the downloadable version of the commercial simulation program Working Model, there are four custom computer programs, written by the author, provided with this text: programs LINKAGES, MATRIX, DYNACAM, and ENGINE. These are student editions of the programs for educational use only. For commercial applications, professional versions with extended capabilities are available at **http://www.designofmachinery.com/**. Program LINKAGES is based on the mathematics derived in Chapters 4 to 7 and 10 to 11 and use the equations presented therein to solve for position, velocity, acceleration, forces, and torques in fourbar, fivebar, sixbar, and slider linkages. Program DYNACAM is a cam design program based on the mathematics derived in Chapters 8 and 15. Program ENGINE is based on the mathematics derived in Chapters 13 and 14. Program MATRIX is a general linear simultaneous equation solver. All have similar choices for display of output data in the form of tables and plots. All the programs are designed to be user friendly and reasonably "crashproof." The author encourages users to email reports of any "bugs" or problems encountered in their use to him at norton@wpi.edu.

Learning Tools

All the custom programs provided with this text are designed to be learning tools to aid in the understanding of the relevant subject matter and *are specifically not intended to be used for commercial purposes in the design of hardware* **and must not be so used**. It is quite possible to obtain inappropriate (but mathematically correct) results to any problem solved with these programs, due to incorrect or inappropriate input of data. The user is expected to understand the kinematic and dynamic theory underlying the program's structure and to also understand the mathematics on which the program's algorithms are based. This information on the underlying theory and mathematics is derived and described in the noted chapters of this text. Most equations used in the programs are derived or presented in this textbook.

Disclaimer and Limitations on Use

Student editions of these programs are made available with this book and carry a limited-term license restricted to educational use in course work for up to 1 year. If you wish to use the program for the benefit of a company or for any commercial purpose, then you must obtain the professional edition of the same program. **The student editions may not be used commercially!** The professional editions typically offer more features and better accuracy than the student editions. Commercial software for use in design or analysis needs to have built-in safeguards against the possibility of the user providing incorrect, inappropriate, or ridiculous values for input variables, in order to guard against erroneous results due to user ignorance or inexperience. **The student editions of the custom programs provided with this text are not commercial software and deliberately do not contain such safeguards against improper input data**, on the premise that to do so would "short-circuit" the student's learning process. We learn most from our failures. These programs provide an educational environment to explore failure of your designs "on paper" and in the process to come to a more thorough and complete understanding of the subject matter. **The author and publisher are not responsible for any damages which may result from the use or misuse of these programs.**

A.1   GENERAL INFORMATION

Hardware/System Requirements

These programs will run in Windows 2000/NT/XP/Vista/Windows7, but settings changes are needed in Vista as described in the installation instructions. DYNACAM and LINKAGES will operate properly in both 32 and 64-bit operating systems. MATRIX and ENGINE will run in any 32-bit system but may display some anomalies when run in the newer 64-bit systems such as Windows 7-64 bit. Replacements are in development. A DVD drive is needed, as is a hard disk drive. A Pentium III (or faster) equivalent processor is recommended with at least 128 MB of RAM. More RAM is preferred.

Installing the Software

The install.exe files contain the executable program files plus all necessary Dynamic Link Library (DLL) and other ancillary files needed to run the programs. Run the SETUP or Install file for each program to install all of its files on your hard drive. The program name will appear in the list under the *Start/Program/DesignofMachinery* menu after installation and can be run from there. DYNACAM and LINKAGES can be updated from that menu also.

User Manual

The user manual is provided as a PDF file on the book's DVD. It is intended to be used while running the programs. To see a screen referred to, bring it up within the program to follow its discussion. An instructional video and the user manual are also accessible from the help menus within the programs if the computer is connected to the Internet.

A

Appendix **B**

MATERIAL PROPERTIES

For selected engineering materials. Many other alloys are available.

The following tables contain approximate values for strengths and other specifications of a variety of engineering materials compiled from various sources. In some cases, the data are minimum recommended values, and in other cases data are from a single test specimen. These data are suitable for use in the engineering exercises contained in this text but should not be considered as statistically valid representations of specifications for any particular alloy or material. The designer should consult the materials' manufacturers for more accurate and up-to-date strength information on materials used in engineering applications or conduct independent tests of the selected materials to determine their ultimate suitability to any application.

Table B-1   Physical Properties of Some Engineering Materials

Data from Various Sources.[*] These Properties are Essentially Similar for All Alloys of the Particular Material

Material	Modulus of Elasticity E		Modulus of Rigidity G		Poisson's Ratio ν	Weight Density γ	Mass Density ρ	Specific Gravity
	Mpsi	GPa	Mpsi	GPa		lb/in³	Mg/m³	
Aluminum alloys	10.4	71.7	3.9	26.8	0.34	0.10	2.8	2.8
Beryllium copper	18.5	127.6	7.2	49.4	0.29	0.30	8.3	8.3
Brass, bronze	16.0	110.3	6.0	41.5	0.33	0.31	8.6	8.6
Copper	17.5	120.7	6.5	44.7	0.35	0.32	8.9	8.9
Iron, cast, gray	15.0	103.4	5.9	40.4	0.28	0.26	7.2	7.2
Iron, cast, ductile	24.5	168.9	9.4	65.0	0.30	0.25	6.9	6.9
Iron, cast, malleable	25.0	172.4	9.6	66.3	0.30	0.26	7.3	7.3
Magnesium alloys	6.5	44.8	2.4	16.8	0.33	0.07	1.8	1.8
Nickel alloys	30.0	206.8	11.5	79.6	0.30	0.30	8.3	8.3
Steel, carbon	30.0	206.8	11.7	80.8	0.28	0.28	7.8	7.8
Steel, alloys	30.0	206.8	11.7	80.8	0.28	0.28	7.8	7.8
Steel, stainless	27.5	189.6	10.7	74.1	0.28	0.28	7.8	7.8
Titanium alloys	16.5	113.8	6.2	42.4	0.34	0.16	4.4	4.4
Zinc alloys	12.0	82.7	4.5	31.1	0.33	0.24	6.6	6.6

[*] Properties of Some Metals and Alloys, International Nickel Co., Inc., NY; Metals Handbook, American Society for Metals, Materials Park, OH.

Table B-2   Mechanical Properties of Some Wrought-Aluminum Alloys

Data from Various Sources.[*] Approximate Values. Consult Manufacturers for More Accurate Information

Wrought-Aluminum Alloy	Condition	Tensile Yield Strength (2% Offset)		Ultimate Tensile Strength		Fatigue Strength at 5E8 Cycles		Elongation over 2 in	Brinell Hardness
		kpsi	MPa	kpsi	MPa	kpsi	MPa	%	HB
1100	Sheet annealed	5	34	13	90			35	23
	Cold rolled	22	152	24	165			5	44
2024	Sheet annealed	11	76	26	179			20	-
	Heat treated	42	290	64	441	20	138	19	-
3003	Sheet annealed	6	41	16	110			30	28
	Cold rolled	27	186	29	200			4	55
5052	Sheet annealed	13	90	28	193			25	47
	Cold rolled	37	255	42	290			7	77
6061	Sheet annealed	8	55	18	124			25	30
	Heat treated	40	276	45	310	14	97	12	95
7075	Bar annealed	15	103	33	228			16	60
	Heat treated	73	503	83	572	14	97	11	150

[*] Properties of Some Metals and Alloys, International Nickel Co., Inc., NY; Metals Handbook, American Society for Metals, Materials Park, OH.

B

Table B-3   Mechanical Properties of Some Carbon Steels

Data from Various Sources. * Approximate Values. Consult Manufacturers for More Accurate Information

SAE / AISI Number	Condition	Tensile Yield Strength (2% Offset)		Ultimate Tensile Strength		Elongation over 2 in	Brinell Hardness
		kpsi	MPa	kpsi	MPa	%	HB
1010	Hot rolled	26	179	47	324	28	95
	Cold rolled	44	303	53	365	20	105
1020	Hot rolled	30	207	55	379	25	111
	Cold rolled	57	393	68	469	15	131
1030	Hot rolled	38	259	68	469	20	137
	Normalized @ 1650°F	50	345	75	517	32	149
	Cold rolled	64	441	76	524	12	149
	Q&T @ 1000°F	75	517	97	669	28	255
	Q&T @ 800°F	84	579	106	731	23	302
	Q&T @ 400°F	94	648	123	848	17	495
1035	Hot rolled	40	276	72	496	18	143
	Cold rolled	67	462	80	552	12	163
1040	Hot rolled	42	290	76	524	18	149
	Normalized @ 1650°F	54	372	86	593	28	170
	Cold rolled	71	490	85	586	12	170
	Q&T @ 1200°F	63	434	92	634	29	192
	Q&T @ 800°F	80	552	110	758	21	241
	Q&T @ 400°F	86	593	113	779	19	262
1045	Hot rolled	45	310	82	565	16	163
	Cold rolled	77	531	91	627	12	179
1050	Hot rolled	50	345	90	621	15	179
	Normalized @ 1650°F	62	427	108	745	20	217
	Cold rolled	84	579	100	689	10	197
	Q&T @ 1200°F	78	538	104	717	28	235
	Q&T @ 800°F	115	793	158	1 089	13	444
	Q&T @ 400°F	117	807	163	1 124	9	514
1060	Hot rolled	54	372	98	676	12	200
	Normalized @ 1650°F	61	421	112	772	18	229
	Q&T @ 1200°F	76	524	116	800	23	229
	Q&T @ 1000°F	97	669	140	965	17	277
	Q&T @ 800°F	111	765	156	1 076	14	311
1095	Hot rolled	66	455	120	827	10	248
	Normalized @ 1650°F	72	496	147	1 014	9	13
	Q&T @ 1200°F	80	552	130	896	21	269
	Q&T @ 800°F	112	772	176	1 213	12	363
	Q&T @ 600°F	118	814	183	1 262	10	375

* *SAE Handbook*, Society of Automotive Engineers, Warrendale, PA; *Metals Handbook*, American Society for Metals, Materials Park, OH.

B

Table B-4   Mechanical Properties of Some Cast-Iron Alloys

Data from Various Sources.* Approximate Values. Consult Manufacturers for More Accurate Information

Cast-Iron Alloy	Condition	Tensile Yield Strength (2% Offset)		Ultimate Tensile Strength		Compressive Strength		Brinell Hardness
		kpsi	MPa	kpsi	MPa	kpsi	MPa	HB
Gray cast iron—Class 20	As cast	–	–	22	152	83	572	156
Gray cast iron—Class 30	As cast	–	–	32	221	109	752	210
Gray cast iron—Class 40	As cast	–	–	42	290	140	965	235
Gray cast iron—Class 50	As cast	–	–	52	359	164	1 131	262
Gray cast iron—Class 60	As cast	–	–	62	427	187	1 289	302
Ductile iron 60-40-18	Annealed	47	324	65	448	52	359	160
Ductile iron 65-45-12	Annealed	48	331	67	462	53	365	174
Ductile iron 80-55-06	Annealed	53	365	82	565	56	386	228
Ductile iron 120-90-02	Q & T	120	827	140	965	134	924	325

* *Properties of Some Metals and Alloys*, International Nickel Co., Inc., NY; *Metals Handbook*, American Society for Metals, Materials Park, OH.

Table B-5   Properties of Some Engineering Plastics

Data from Various Sources.* Approximate Values. Consult Manufacturers for More Accurate Information

Material	Approximate Modulus of Elasticity E †		Ultimate Tensile Strength		Ultimate Compressive Strength		Elongation over 2 in	Max Temp	Specific Gravity
	Mpsi	GPa	kpsi	MPa	kpsi	MPa	%	°F	
ABS	0.3	2.1	6.0	41.4	10.0	68.9	5-25	160-200	1.05
20-40% glass filled	0.6	4.1	10.0	68.9	12.0	82.7	3	200-230	1.30
Acetal	0.5	3.4	8.8	60.7	18.0	124.1	60	220	1.41
20-30% glass filled	1.0	6.9	10.0	68.9	18.0	124.1	7	185-220	1.56
Acrylic	0.4	2.8	10.0	68.9	15.0	103.4	5	140-190	1.18
Fluoroplastic (PTFE)	0.2	1.4	5.0	34.5	6.0	41.4	100	350-330	2.10
Nylon 6/6	0.2	1.4	10.0	68.9	10.0	68.9	60	180-300	1.14
Nylon 11	0.2	1.3	8.0	55.2	8.0	55.2	300	180-300	1.04
20-30% glass filled	0.4	2.5	12.8	88.3	12.8	88.3	4	250-340	1.26
Polycarbonate	0.4	2.4	9.0	62.1	12.0	82.7	100	250	1.20
10-40% glass filled	1.0	6.9	17.0	117.2	17.0	117.2	2	275	1.35
HMW polyethylene	0.1	0.7	2.5	17.2	–	–	525	–	0.94
Polyphenylene oxide	0.4	2.4	9.6	66.2	16.4	113.1	20	212	1.06
20-30% glass filled	1.1	7.8	15.5	106.9	17.5	120.7	5	260	1.23
Polypropylene	0.2	1.4	5.0	34.5	7.0	48.3	500	250-320	0.90
20-30% glass filled	0.7	4.8	7.5	51.7	6.2	42.7	2	300-320	1.10
Impact polystyrene	0.3	2.1	4.0	27.6	6.0	41.4	2-80	140-175	1.07
20-30% glass filled	0.1	0.7	12.0	82.7	16.0	110.3	1	180-200	1.25
Polysulfone	0.4	2.5	10.2	70.3	13.9	95.8	50	300-345	1.24

* *Modern Plastics Encyclopedia*, McGraw-Hill, New York; *Machine Design Materials Reference Issue*, Penton Publishing, Cleveland, OH.
† Most plastics do not obey Hooke's law. These apparent moduli of elasticity vary with time and temperature.

Appendix C

GEOMETRIC PROPERTIES

**DIAGRAMS AND FORMULAS TO CALCULATE THE FOLLOWING
PARAMETERS FOR SEVERAL COMMON GEOMETRIC SOLIDS**

V = volume

m = mass

C_g = location of center of mass

I_x = second moment of mass about x axis = $\int \left(y^2 + z^2 \right) dm$

I_y = second moment of mass about y axis = $\int \left(x^2 + z^2 \right) dm$

I_z = second moment of mass about z axis = $\int \left(x^2 + y^2 \right) dm$

k_x = radius of gyration about x axis

k_y = radius of gyration about y axis

k_z = radius of gyration about z axis

(a) Rectangular prism

$V = abc$

$x_{CG} @ \dfrac{c}{2}$

$I_x = \dfrac{m\left(a^2 + b^2\right)}{12}$

$k_x = \sqrt{\dfrac{I_x}{m}}$

$m = V \cdot \text{mass density}$

$y_{CG} @ \dfrac{b}{2}$

$I_y = \dfrac{m\left(a^2 + c^2\right)}{12}$

$k_y = \sqrt{\dfrac{I_y}{m}}$

$z_{CG} @ \dfrac{a}{2}$

$I_z = \dfrac{m\left(b^2 + c^2\right)}{12}$

$k_z = \sqrt{\dfrac{I_z}{m}}$

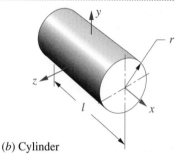

(b) Cylinder

$V = \pi r^2 l$

$x_{CG} @ \dfrac{l}{2}$

$I_x = \dfrac{mr^2}{2}$

$k_x = \sqrt{\dfrac{I_x}{m}}$

$m = V \cdot \text{mass density}$

y_{CG} on axis

$I_y = I_z = \dfrac{m\left(3r^2 + l^2\right)}{12}$

$k_y = k_z = \sqrt{\dfrac{I_y}{m}}$

z_{CG} on axis

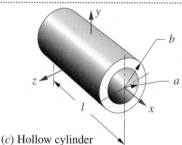

(c) Hollow cylinder

$V = \pi\left(b^2 - a^2\right)l$

$x_{CG} @ \dfrac{l}{2}$

$I_x = \dfrac{m\left(a^2 + b^2\right)}{2}$

$k_x = \sqrt{\dfrac{I_x}{m}}$

$m = V \cdot \text{mass density}$

y_{CG} on axis

$I_y = I_z = \dfrac{m\left(3a^2 + 3b^2 + l^2\right)}{12}$

$k_y = k_z = \sqrt{\dfrac{I_y}{m}}$

z_{CG} on axis

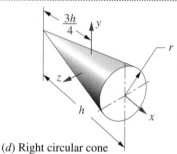

(d) Right circular cone

$V = \pi\dfrac{r^2 h}{3}$

$x_{CG} @ \dfrac{3h}{4}$

$I_x = \dfrac{3}{10}mr^2$

$k_x = \sqrt{\dfrac{I_x}{m}}$

$m = V \cdot \text{mass density}$

y_{CG} on axis

$I_y = I_z = \dfrac{m\left(12r^2 + 3h^2\right)}{80}$

$k_y = k_z = \sqrt{\dfrac{I_y}{m}}$

z_{CG} on axis

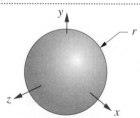

(e) Sphere

$V = \dfrac{4}{3}\pi r^3$

x_{CG} at center

$I_x = I_y = I_z = \dfrac{2}{5}mr^2$

$k_x = k_y = k_z = \sqrt{\dfrac{I_x}{m}}$

$m = V \cdot \text{mass density}$

y_{CG} at center

z_{CG} at center

Appendix D

SPRING DATA

The following catalog pages of helical compression and extension spring data were provided courtesy of the *Hardware Products Co., Chelsea, Massachusetts*
http://www.hardwareproducts.com/

Other spring information can be found on the Web at:

http://www.leespring.com/

http://www.cookspring.com/

http://www.allrite.com/

http://www.springsfast.com/

http://www.asbg.com/

http://www.centuryspring.com/

COMPRESSION SPRINGS

Will go in hole In.		7/16			1/2				5/8				3/4				7/8			
Wire Dia. In.		.031	.047	.062	.047	.062	.078	.094	.047	.062	.078	.094	.062	.078	.094	.125	.062	.078	.094	.125
7/16 Catalog No.		247	248	249																
Price Code		HB	HB	HC																
lbs./in.		12	55	180																
Max. Defl.		.32	.23	.16																
1/2 Catalog No.		250	251	252	283	284	285	286												
Price Code		HB	HB	HC	HB	HB	HE	HE												
lbs./in.		10	47	150	37	110	320	840												
Max. Defl.		.37	.27	.19	.29	.22	.15	.10												
5/8 Catalog No.		253	254	255	287	288	289	290	331	332	333	334								
Price Code		HB	HB	HC	HB	HB	HE	HE	HB	HB	HD	HE								
lbs./in.		7.9	36	175	29	85	240	610	20	54	140	320								
Max. Defl.		.47	.35	.25	.38	.29	.20	.14	.43	.35	.27	.20								
3/4 Catalog No.		256	257	258	291	292	293	294	335	336	337	338	375	376	377	378				
Price Code		HB	HB	HC	HB	HB	HE	HE	HB	HB	HD	HE	HD	HD	HF	HG				
lbs./in.		6.4	29	90	23	68	185	470	16	43	105	250	32	78	170	650				
Max. Defl.		.58	.44	.32	.48	.37	.26	.18	.53	.44	.34	.26	.59	.40	.32	.19				
7/8 Catalog No.		259	260	261	295	296	297	298	339	340	341	342	379	380	381	382	419	420	421	422
Price Code		HB	HB	HC	HB	HB	HE	HE	HB	HB	HD	HE	HD	HD	HF	HG	HF	HF	HJ	HK
lbs./in.		5.4	24	75	19	56	155	384	13	36	90	204	27	65	140	520	21	49	100	350
Max. Defl.		.68	.52	.39	.57	.44	.32	.23	.64	.53	.42	.32	.58	.48	.39	.24	.62	.53	.44	.30
1 Catalog No.		262	263	264	299	300	301	302	343	344	345	346	383	384	385	386	423	424	425	426
Price Code		HB	HB	HC	HB	HB	HE	HE	HB	HC	HD	HE	HD	HD	HF	HG	HF	HF	HJ	HK
lbs./in.		4.7	21	65	17	48	130	320	11	31	77	170	23	55	115	430	18	42	86	290
Max. Defl.		.79	.60	.45	.66	.51	.37	.27	.74	.62	.49	.38	.68	.57	.46	.29	.73	.63	.53	.36
1¼ Catalog No.		265	266	267	303	304	305	306	347	348	349	350	387	388	389	390	427	428	429	430
Price Code		HB	HC	HC	HC	HC	HF	HF	HC	HC	HD	HE	HE	HE	HF	HG	HG	HG	HJ	HK
lbs./in.		3.7	16	50	13	38	100	245	9.0	24	59	130	18	42	89	320	14	32	66	220
Max. Defl.		1.0	.77	.58	.84	.66	.49	.35	.94	.80	.64	.49	.88	.74	.60	.39	.94	.81	.69	.47
1½ Catalog No.		268	269	270	307	308	309	310	351	352	353	354	391	392	393	394	431	432	433	434
Price Code		HB	HC	HC	HC	HC	HF	HF	HD	HD	HE	HF	HE	HE	HF	HG	HG	HG	HJ	HK
lbs./in.		3.1	14	41	11	31	83	200	7.4	20	48	105	15	34	72	260	12	26	53	175
Max. Defl.		1.2	.94	.70	1.0	.81	.60	.43	1.1	.98	.78	.61	1.1	.91	.74	.48	1.1	1.0	.85	.59
1¾ Catalog No.		271	272	273	311	312	313	314	355	356	357	358	395	396	397	398	435	436	437	438
Price Code		HC	HD	HE	HC	HD	HF	HF	HD	HD	HC	HF	HE	HE	HF	HG	HG	HG	HJ	HK
lbs./in.		2.6	11	35	9.1	26	70	170	6.2	17	41	90	12.4	29	61	216	9.9	22	45	147
Max. Defl.		1.4	1.1	.84	1.2	.96	.71	.52	1.3	1.1	.93	.73	1.3	1.1	.89	.58	1.35	1.2	1.0	.71
2 Catalog No.		274	275	276	315	316	317	318	359	360	361	362	399	400	401	402	439	440	441	442
Price Code		HD	HD	HE	HC	HD	HF	HF	HD	HD	HE	HF	HE	HE	HF	HG	HG	HG	HJ	HL
lbs./in.		2.3	10	30	7.9	23	60	145	5.4	14	35	77	11	25	52	185	8.6	19	38	125
Max. Defl.		1.6	1.3	.96	1.4	1.1	.82	.60	1.5	1.3	1.1	.85	1.4	1.2	1.0	.68	1.5	1.4	1.2	.83
3 Catalog No.		277	278	279	319	320	321	322	363	364	365	366	403	404	405	406	443	444	445	446
Price Code		HD	HE	HE	HF	HF	HG	HG	HD	HD	HE	HF	HF	HF	HG	HK	HJ	HJ	HK	HL
lbs./in.		1.5	6.6	20	5.2	15	39	94	3.6	9.4	23	50	7	16	34	115	5.6	12	25	80
Max. Defl.		2.4	1.9	1.4	2.1	1.7	1.2	.93	2.4	2.0	1.6	1.3	2.2	1.9	1.6	1.0	2.4	2.1	1.8	1.3
4 Catalog No.		280	281	282	323	324	325	326	367	368	369	370	407	408	409	410	447	448	449	450
Price Code		HE	HE	HE	HE	HF	HG	HG	HE	HE	HF	HG	HF	HF	HG	HL	HJ	HJ	HL	HM
lbs./in.		1.1	4.9	15	3.9	11	29	69	2.6	6.9	17	37	5.2	12	25	86	4.2	9.2	18	59
Max. Defl.		3.3	2.6	2.0	2.8	2.3	1.7	1.2	3.2	2.7	2.2	1.8	3.0	2.6	2.1	1.4	3.2	2.8	2.5	1.8
6 Catalog No.					327	328	329	330	371	372	373	374	411	412	413	414	451	452	453	454
Price Code					HF	HF	HG	HG	HF	HF	HG	HG	HG	HG	HJ	HN	HK	HK	HL	HO
lbs./in.					2.5	7.0	17	45	1.8	4.6	11	24	3.4	7.9	16	56	2.7	6.0	12	38
Max. Defl.					4.4	3.5	2.5	2.	4.8	4.2	3.4	2.7	4.6	3.9	3.3	2.2	4.9	4.3	3.8	2.7
8 Catalog No.													415	416	417	418	455	456	457	458
Price Code													HJ	HJ	HK	HO	HL	HL	HM	HP
lbs./in.													2.6	6	11	40	2.0	4.5	8.9	28
Max. Defl.													6.1	5.2	4.5	3.0	6.5	5.8	5.1	3.7
Maximum Load		3.7	12.7	29	11	25	45	88	8.3	19	38	66	15.8	31.2	54	125	13.4	26.3	45	105
Will work free over		.347	.315	.285	.375	.345	.313	.281	.505	.475	.443	.411	.585	.554	.522	.460	.700	.670	.638	.576
Pitch		.195	.141	.128	.173	.151	.141	.141	.259	.214	.188	.177	.284	.240	.217	.204	.371	.306	.268	.239
Solid Stress (000 omitted)		125	118	113	118	113	109	105	118	113	109	105	113	109	105	99	113	109	105	99

(FREE LENGTHS — left margin label)

Pounds per inch figure is a constant for each spring, and represents the number of pounds required to compress the spring 1". To compress the spring ½" or ¼" requires ½ or ¼ of this value.

Maximum Deflection is the amount spring deflects to give the maximum load. This value subtracted from the free length gives the solid or compressed length.

NOTE: Stock springs can be ordered in stainless steel or plated. Prices quoted upon request.

Wire Diameter — Outside Diameter — Free Length

FREE LENGTHS

	Will go in hole In.	1				1 1/4			1 1/2			2			3			4			6	
	Wire Dia. In.	.078	.094	.125	.187	.094	.125	.187	.125	.187	.250	.187	.250	.375	.250	.375	.500	.375	.500	.750	.750	1.000
1	Catalog No.	459	460	461	462																	
	Price Code	HK	HL	HL	HR																	
	lbs./in.	34	67	210	1500																	
	Max. Defl.	.67	.58	.41	.19																	
1 1/4	Catalog No.	463	464	465	466	499	500	501														
	Price Code	HL	HM	HM	HS	HN	HN	HM														
	lbs./in.	26	52	160	1100	35	100	600														
	Max. Defl.	.87	.76	.55	.26	.85	.67	.37														
1 1/2	Catalog No.	467	468	469	470	502	503	504	526	527	528											
	Price Code	HL	HM	HN	HS	HN	HO	HT	HR	HX	HAC											
	lbs./in.	21	42	130	870	29	82	460	60	300	1200											
	Max. Defl.	1.0	.93	.69	.34	1.1	.84	.48	.95	.60	.35											
1 3/4	Catalog No.	471	472	473	474	505	506	507	529	530	531											
	Price Code	HL	HM	HN	HS	HN	HO	HT	HR	HX	HAC											
	lbs./in.	18	35	108	712	24	68	379	50	244	960											
	Max. Defl.	1.3	1.1	.83	.41	1.3	1.0	.59	1.1	.74	.44											
2	Catalog No.	475	476	477	478	508	509	510	532	533	534	553	554	555								
	Price Code	HM	HN	HO	HT	HO	HP	HU	HS	HZ	HAE	HAA	HAG	HZZ								
	lbs./in.	16	30	93	600	21	59	320	43	200	800	115	390	3000								
	Max. Defl.	1.4	1.3	.97	.49	1.4	1.2	.70	1.3	.87	.53	1.1	.77	.34								
3	Catalog No.	479	480	481	482	511	512	513	535	536	537	556	557	558	577	578	579					
	Price Code	HN	HO	HP	HZ	HP	HR	HAA	HT	HAD	HAL	HAE	HAN	HZZ	HAR	HZZ	HZZ					
	lbs./in.	10	19	59	370	13	37	200	27	130	480	73	230	1650	105	560	2300					
	Max. Defl.	2.2	2.0	1.5	.79	2.2	1.8	1.1	2.1	1.4	.89	1.8	1.3	.61	1.8	1.1	.64					
4	Catalog No.	483	484	485	486	514	515	516	538	539	540	559	560	561	580	581	582	598	599	610		
	Price Code	HP	HR	HS	HAC	HS	HT	HAD	HW	HAG	HAQ	HAJ	HAR	HZZ	HAT	HZZ	HZZ	HZZ	HZZ	HZZ		
	lbs./in.	7.4	14	43	270	9.9	27	144	20	93	340	53	170	1150	76	390	1500	210	720	4600		
	Max. Defl.	3.0	2.7	2.1	1.1	3.0	2.5	1.5	2.8	1.9	1.2	2.5	1.8	.88	2.5	1.6	.96	2.1	1.4	.8		
6	Catalog No.	487	488	489	490	517	518	519	541	542	543	562	563	564	583	584	585	600	601	611	616	621
	Price Code	HR	HT	HU	HAD	HU	HW	HAE	HX	HAJ	HAT	HAM	HAW	HZZ	HAZ	HZZ	HZZ	HZZ	HZZ	HZZ	HZZ	HZZ
	lbs./in.	4.9	9.4	28	175	6.5	18	93	13	60	220	34	105	710	49	240	920	130	430	2840	850	3500
	Max. Defl.	4.6	4.1	3.2	1.7	4.7	3.9	2.4	4.3	3.0	1.9	3.8	2.8	1.4	4.0	2.6	1.6	3.4	2.4	1.3	1.9	1.4
8	Catalog No.	491	492	493	494	520	521	522	544	545	546	565	566	567	586	587	588	602	603	612	617	622
	Price Code	HS	HU	HW	HAL	HW	HX	HAM	HAA	HAP	HAW	HAR	HAZ	HZZ	HBD	HZZ	HZZ	HZZ	HZZ	HZZ	HZZ	HZZ
	lbs./in.	3.6	7.0	21	125	4.8	13	68	9.6	44	160	25	79	510	36	175	660	95	310	2050	630	2500
	Max. Defl.	6.2	5.6	4.3	2.3	6.3	5.2	3.2	5.9	4.1	2.7	5.2	3.9	1.9	5.4	3.6	2.2	4.5	3.4	1.8	2.7	2.0
12	Catalog No.	495	496	497	498	523	524	525	547	548	549	568	569	570	589	590	591	604	605	613	618	623
	Price Code	HT	HW	HZ	HAP	HX	HAA	HAR	HAC	HAU	HBA	HAZ	HBE	HZZ	HBK	HZZ	HZZ	HZZ	HZZ	HZZ	HZZ	HZZ
	lbs./in.	2.4	4.6	14	84	3.2	8.7	45	6.3	29	105	16	52	330	23	110	420	61	195	1325	400	1580
	Max. Defl.	9.4	8.4	6.5	3.5	9.5	7.9	5.0	8.9	6.2	4.1	8.0	5.9	3.0	8.3	5.5	3.5	7.3	5.3	2.6	4.3	3.1
16	Catalog No.								550	551	552	571	572	573	592	593	594	606	607	614	619	624
	Price Code								HAE	HAW	HBD	HAZ	HBG	HZZ	HBL	HZZ	HZZ	HZZ	HZZ	HZZ	HZZ	HZZ
	lbs./in.								4.7	21	74	12	38	240	17	83	310	45	145	975	300	1170
	Max. Defl.								11.9	8.5	5.6	10.7	8.0	4.1	11.3	7.5	4.0	10	7.3	3.8	6.1	4.3
24	Catalog No.											574	575	576	595	596	597	608	609	615	620	625
	Price Code											HBA	HBL	HZZ	HBP	HZZ	HZZ	HZZ	HZZ	HZZ	HZZ	HZZ
	lbs./in.											7.8	23.4	150	11.4	54	200	29	94	640	175	760
	Max. Defl.											16.3	12.1	7.0	17.1	11.5	7.3	15.2	11.1	5.8	9.4	6.5
	Maximum Load	23	39	90	295	30	69	224	57	180	428	131	307	1000	195	624	1470	449	1040	3700	2000	4800
	Will work free over	784	752	690	565	1.00	.940	.815	1.19	1.06	.940	1.52	1.39	1.14	2.33	2.08	1.83	3.08	2.83	2.25	4.25	3.75
	Pitch	.382	.328	.279	.268	.481	.384	.327	.516	.403	.388	.596	.518	.514	.917	.741	.736	1.08	.969	1.00	1.3	1.4
	Solid Stress (000 omitted)	109	105	99	90	105	99	90	99	90	85	90	85	77	85	77	73	77	73	70	70	65

Pounds per inch figure is a constant for each spring, and represents the number of pounds required to compress the spring 1". To compress the spring 1/2" or 1/4" requires 1/2 or 1/4 of this value.

Maximum Deflection is the amount spring deflects to give the maximum load. This value subtracted from the free length gives the solid or compressed length.

NOTE: Stock springs can be ordered in stainless steel or plated. Prices quoted upon request.

Hardware Products Company, Inc.

D

EXTENSION SPRINGS

Regular hook   Regular loop

← Length →

ORDER BY:
SE LENGTH x O.D. x WIRE DIA.

SPECIFY HOOKS OR LOOPS

The figures given for "Maximum Extension" and "lbs. per inch" are for a spring 1" long. For other lengths multiply the "Maximum Extension" and divide the "lbs. per inch" by the length in inches. The "Maximum Load" and "Initial Tension" remain constant for any length.

Example: A spring ½" diam. .062" wire and 4" long will have a safe maximum extension of 3.2" and it will require 4 lbs. to deflect it 1 in. The spring will hold approximately 3.3 lbs. before it starts to extend, and will hold a maximum of 16.1 lbs. without permanent stretch. If 8.5 lbs. is hung on the spring it will deflect 1.3". 8.5 lbs. minus 3.3 lbs. divided by 4 lbs. per inch equals 1.3".

NOTE: Stock springs can be ordered in stainless steel or plated. Prices quoted upon request.

Outside dia.	Wire dia.	Catalog No.	Price code	Safe maximum load in pounds	Safe maximum extension - in.	Approx. initial tension in pounds	Pound per inch extension	Stress at max. load (000 omitted)	Weight per foot (lbs.)
1/8	.012	01	EHE	.6	1.9	.07	.27	100	.012
	.016	02	EHD	1.3	.9	.2	1.2	93	.015
	.023	03	EHD	4.2	.35	.9	9.0	90	.02
5/32	.012	04	EHE	47	3.5	.01	.12	100	.015
	.016	05	EHD	1.1	1.7	.15	.55	93	.019
	.023	06	EHD	3.2	.7	.5	3.9	90	.027
3/16	.016	07	EHD	.87	2.5	1	3	93	.024
	.023	08	EHD	2.6	1.0	.4	2.2	90	.032
	.031	09	EHD	6.5	.45	1.5	10.7	88	.04
7/32	.016	10	EHD	.75	4.0	.01	.18	93	.028
	.023	11	EHD	2.3	1.6	.32	1.2	90	.039
	.031	12	EHD	5.5	.7	1.0	6.5	88	.048
1/4	.023	13	EHD	1.8	1.9	.26	8	90	.044
	.031	14	EHD	4.7	1.0	.75	3.8	88	.055
	.047	15	EHE	16.0	.3	3.5	40.0	83	.082
5/16	.023	16	EHE	1.5	3.5	.16	.38	90	.058
	.031	17	EHE	3.6	1.6	.55	1.9	88	.072
	.047	18	EHE	12.5	.9	2.2	10.8	83	.108
3/8	.031	19	EHE	2.9	2.5	.37	1.0	88	.084
	.047	20	EHE	10.5	9	1.7	9.5	83	.13
	.062	21	EHF	23.0	.39	5.3	45.0	79	.16
7/16	.031	22	EHF	2.5	3.5	.26	63	88	.105
	.047	23	EHF	8.5	1.2	1.4	5.7	83	.163
	.062	24	EHF	20.0	.6	4.3	26.0	79	.2
1/2	.047	25	EHF	7.3	1.6	1.1	3.7	83	.18
	.062	26	EHF	17.0	.8	3.3	16.0	79	.23
	.078	27	EHG	34.0	.45	8.0	57.0	77	.28
	.094	28	EHJ	57.0	.25	16.0	160.0	74	.32
5/8	.047	29	EHG	6.0	3.0	.7	1.7	83	.24
	.062	30	EHG	13.3	1.4	2.1	7.6	79	.3
	.078	31	EHG	27.0	.9	5.2	23.0	77	.37
	.094	32	EHJ	45.0	.4	11.0	73.0	74	.44
3/4	.062	33	EHG	10.5	2.2	1.5	4.1	79	.36
	.078	34	EHJ	22.0	1.3	3.5	14.0	77	.46
	.094	35	EHJ	36.0	7	8.0	38.0	74	.51
	.125	36	EHK	85.0	.3	22.0	180.0	69	.64
7/8	.062	37	EHK	9.2	3.3	1.1	2.4	79	.4
	.078	38	EHK	18.0	1.7	2.6	8.7	77	.59
	.094	39	EHL	31.0	1.0	6.0	25.0	74	.64
	.125	40	EHM	72.0	.5	17.0	107.0	69	8
1	.078	41	EHL	16.0	2.5	2.0	5.5	77	.67
	.094	42	EHL	26.0	1.5	4.5	13.7	74	.70
	.125	43	EHN	65.0	.75	14.0	68.0	69	.90
	.187	44	EHW	200.0	.23	60.0	600.0	63	1.4
1¼	.094	45	EHM	71.0	2.6	2.8	6.8	74	.94
	.125	46	EHO	47.0	1.2	9.0	31.0	69	1.3
	.187	47	EHZ	148.0	.3	40.0	290.0	63	1.8
1½	.125	48	EHS	39.0	1.9	6.0	17.0	69	1.4
	.187	49	EHAA	122.0	.6	33.0	150.0	63	2.2
	.250	50	EHAC	290.0	.27	90.0	720.0	60	2.6
2	.187	51	EHAD	90.0	1.3	20.0	54.0	63	3.1
	.250	52	EHAG	210.0	.6	55.0	260.0	60	3.7

Carried in stock in 3-foot lengths · cut to length and looped to order.

Hardware Products Company, Inc.

191 WILLIAMS STREET • CHELSEA, MA 02150

D

Appendix E

COUPLER CURVE ATLASES

E.1 HRONES AND NELSON ATLAS OF FOURBAR LINKAGES

The entire Hrones and Nelson coupler curve atlas is on the DVD as PDF files. Figure 3-17 (p. 126) in Section 3.6 shows one page from this atlas and describes how to use it. Read the first chapter within the Hrones and Nelson atlas for more information on how it is arranged and how to use it. The video *Coupler Curves and Linkage Atlases* on the DVD gives detailed instructions on its use and shows an example. Once you extract a trial linkage geometry from the atlas, use program LINKAGES on the DVD to investigate its behavior and to vary the linkage geometry.

E.2 ZHANG, NORTON, HAMMOND ATLAS OF GEARED FIVEBAR LINKAGES

The entire Zhang atlas is on the DVD as PDF files. A sample page is shown in Figure 3-23 (p. 131). Read the first chapter within the Zhang atlas for information on how it is arranged and how to use it. See Sections 3.6 (p. 122), 4.8 (p. 197), 6.8 (p. 324), and 7.4 (p. 371) for more information on the geared fivebar linkages. The video *Coupler Curves and Linkage Atlases* on the DVD gives a brief overview of this atlas. Once you extract a trial linkage geometry from the atlas, use program LINKAGES on the DVD to investigate its behavior and to vary the linkage geometry.

A summary of the parameters in the Zhang atlas is:

Alpha = Coupler Link 3 / Link 2

Beta = Ground Link 1 / Link 2

Lambda = Gear Ratio = Gear 5 / Gear 2.

Phase angle is noted on each plot of a coupler curve.

The dots along curves are at every 10 degrees of Link 2's rotation.

Linkage is symmetrical: Link 2 = Link 5 and Link 3 = Link 4

Note that the *Lambda* in the atlas is the inverse of the λ that is defined in Sections 4.8 (p. 197), 6.8 (p. 324), and 7.4 (p. 371). See also Figures P4-4 (p. 218), P6-4 (p. 331), and P7-4 (p. 383). For example, a gear ratio *Lambda* of 2 in the Zhang atlas corresponds to a λ of 0.5 in the text and in program LINKAGES. (The difference merely corresponds to a mirroring of the linkage from left to right.)

E

Appendix **F**

ANSWERS TO SELECTED PROBLEMS

CHAPTER 2   **KINEMATICS FUNDAMENTALS**

2-1

a. 1	b. 1	c. 2	d. 1	e. 7	f. 1	g. 4
h. 4	i. 4	j. 2	k. 1	l. 1	m. 2	n. 2
o. 4	p. as many as it has sections less one		q. 3			

2-3   a. 1   b. 3   c. 3   d. 3   e. 2

2-4   a. 6   b. 6   c. 5   d. 4, but 2 are dynamically coupled*   e. 4   f. 3

2-5   force-closed

2-6

    a.   pure rotation
    b.   complex planar motion
    c.   pure translation
    d.   pure translation
    e.   pure rotation
    f.   complex planar motion
    g.   complex planar motion

2-7   a. 0   b. 1   c. 1   d. 3

2-8

    a. structure - $DOF = 0$
    b. mechanism - $DOF = 1$
    c. mechanism - $DOF = 1$
    d. mechanism - $DOF = 3$

2-15   a. Grashof   b. non-Grashof   c. special-case Grashof

* Dynamically coupled means that, at speed, leaning the bike to the side results in its turning to the side to which it is leaning. So the angular freedom of this machine in the plane of the road is coupled with its ability to rotate about its long axis (lean). Except at very low speed, you steer a motorcycle by pushing down (toward the ground) on the handlebar on the inside of the turn, rather than by actually turning the handlebar in the direction of the turn. If you are moving the bike with your feet to park it, then you turn the handlebar. But at any significant speed, the gyroscopic effect takes over and leaning the bike makes it turn. This is true of a pedal bike as well if it has sufficient forward speed.

(a) One possible solution to Problem 3-3   (b) One possible solution to Problem 3-5

FIGURE S3-1

Solutions to Problems 3-3 and 3-5

2-21

a. $M = 1$	b. $M = 1$	c. $M = 1$
d. $M = 1$	e. $M = -1$ (a paradox)	f. $M = 1$
g. $M = 1$	h. $M = 0$ (a paradox)	

2-24   a. $M = 1$ b. $M = 1$

2-26 $M = 1$

2-27 $M = 0$

2-35 $M = 1$, fourbar slider-crank

2-61   a. $M = 3$ b. $M = 2$ c. $M = 1$

2-62   a. $M = 1$ b. $M = 2$ c. $M = 4$

CHAPTER 3   GRAPHICAL LINKAGE SYNTHESIS

3-1

a.   path generation
b.   motion generation
c.   function generation
d.   path generation
e.   path generation

Note that synthesis problems have many valid solutions. We cannot provide a "right answer" to all of these design problems. Check your solution with a cardboard model and/or by putting it into one of the programs supplied with the text.

3-3   See Figure S3-1.

3-5   See Figure S3-1.

3-6   See Figure S3-2.

3-8   See Figure S3-3.

FIGURE S3-2

Unique solution to
Problem 3-6

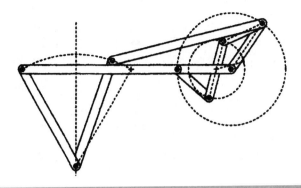

FIGURE S3-3

One possible solution to Problem 3-8

3-10   The solution using Figure 3-17 is shown in Figure S3-4. (Use program LINKAGES to check your solution.)

3-22   The transmission angle ranges from 31.5° to 89.9°.

3-23   Grashof crank-rocker. Transmission angle ranges from 58.1° to 89.8°.

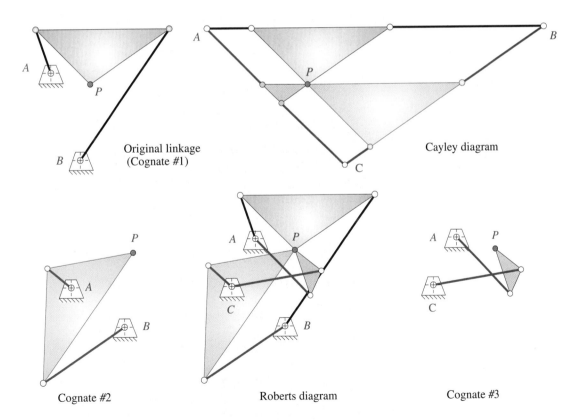

Original linkage (Cognate #1)

Cayley diagram

Cognate #2

Roberts diagram

Cognate #3

FIGURE S3-4

Solution to Problem 3-10. Finding the cognates of the fourbar linkage shown in Figure 3-17 (p. 126)

F

3-31 $L_1 = 160.6, L_2 = 81.3, L_3 = 200.2, L_4 = 200.2$ mm.

3-36   Grashof double-rocker. Works from 56° to 158° and from 202° to 310°. Transmission angle ranges from 0° to 90°.

3-39   Non-Grashof triple-rocker. Toggles at ±116°. Transmission angle ranges from 0° to 88°.

3-42   Non-Grashof triple-rocker. Toggles at ±55.4°. Transmission angle ranges from 0° to 88.8°.

3-79   Link 2 = 1, link 3 = link 4 = link 1 = 1.5. Coupler point is at 1.414 @ 135° versus link 3. Put these data into program LINKAGES to see the coupler curve.

CHAPTER 4   POSITION ANALYSIS

4-6 and **4-7**   See Table S4-1 and the file P07-04*row*.4br.

4-9 and **4-10**   See Table S4-2.

4-11 and **4-12**   See Table S4-3.

4-13   See Table S4-1.

4-14   Open the file P07-04*row*.4br[†] in program LINKAGES to see this solution.[*]

4-15   Open the file P07-04*row*.4br[†] in program LINKAGES to see this solution.[*]

4-16   See Table S4-4.

4-17   See Table S4-4.

4-21   Open the file P04-21.4br in program LINKAGES to see this solution.[*]

4-23   Open the file P04-23.4br in program LINKAGES to see this solution.[*]

4-25   Open the file P04-25.4br in program LINKAGES to see this solution.[*]

4-26   Open the file P04-26.4br in program LINKAGES to see this solution.[*]

4-29   Open the file P04-29.4br in program LINKAGES to see this solution.[*]

4-30   Open the file P04-30.4br in program LINKAGES to see this solution.[*]

4-31 $r_1 = -6.265, \ r_2 = -0.709.$

CHAPTER 5   ANALYTICAL LINKAGE SYNTHESIS

These files can be found in the PROBLEM SOLUTIONS folder on the DVD included with this text.

† *The letter x in the filename represents the row number from the table of problem data.*

5-8   Given: $\alpha_2 = -62.5°,$ $P_{21} = 2.47,$ $\delta_2 = 120°$

For left dyad: Assume: $z = 1.075, \phi = 204°$ $\beta_2 = -27°$

Calculate: $\mathbf{W} = 3.67$ @ $-113.5°$

For right dyad: Assume: $s = 1.24,$ $\psi = 74°$ $\gamma_2 = -40°$

Calculate: $\mathbf{U} = 5.46$ @ $-125.6°$

TABLE S4-1   Solutions for Problems 4-6, 4-7, and 4-13

Row	θ_3 Open	θ_4 Open	Trans Ang	θ_3 Crossed	θ_4 Crossed	Trans Ang
a	88.8	117.3	28.4	−115.2	−143.6	28.4
c	−53.1	16.5	69.6	173.3	103.6	69.6
e	7.5	78.2	70.7	−79.0	−149.7	70.7
g	−16.3	7.2	23.5	155.7	132.2	23.5
i	−1.5	103.1	75.4	−113.5	141.8	75.4
k	−13.2	31.9	45.2	−102.1	−147.3	45.2
m	−3.5	35.9	39.4	−96.5	−135.9	39.4

TABLE S4-2   Solutions for Problems 4-9 to 4-10

Row	θ_3 Open	Slider Open	θ_3 Crossed	Slider Crossed
a	180.1	5.0	−0.14	−3.0
c	205.9	9.8	−25.90	−4.6
e	175.0	16.4	4.20	−23.5
g	212.7	27.1	−32.70	−14.9

TABLE S4-3   Solutions for Problems 4-11 to 4-12

Row	θ_3 Open	θ_4 Open	R_B Open	θ_3 Crossed	θ_4 Crossed	R_B Crossed
a	232.7	142.7	1.79	−79.0	−169.0	1.79
c	91.4	46.4	2.72	208.7	163.7	11.20
e	158.2	128.2	6.17	−36.2	−66.2	9.63

TABLE S4-4   Solutions for Problems 4-16 to 4-17

Row	θ_3 Open	θ_4 Open	θ_3 Crossed	θ_4 Crossed
a	173.6	−177.7	−115.2	−124.0
c	17.6	64.0	−133.7	180.0
e	−164.0	−94.4	111.2	41.6
g	44.2	124.4	−69.1	−149.3
i	37.1	120.2	−67.4	−150.5

F

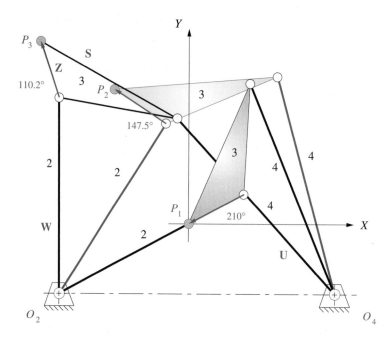

FIGURE S5-1

Solution to Problem 5-11. Open the file P05-11 in program LINKAGES for more information

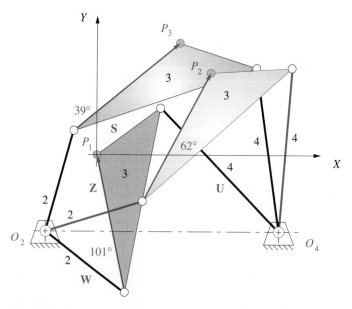

FIGURE S5-2

Solution to Problem 5-15. Open the file P05-15 in program LINKAGES for more information

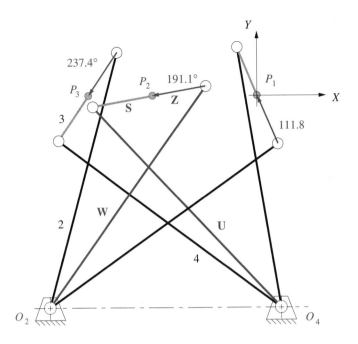

FIGURE S5-3

Solution to Problem 5-19. Open the file P05-19 in program LINKAGES for more information

5-11 See Figure S5-1 for the solution. The link lengths are:

Link 1 = 4.35,   Link 2 = 3.39,   Link 3 = 1.94,   Link 4 = 3.87

5-15 See Figure S5-2 for the solution. The link lengths are:

Link 1 = 3.95,   Link 2 = 1.68,   Link 3 = 3.05,   Link 4 = 0.89

5-19 See Figure S5-3 for the solution. The link lengths are:

Link 1 = 2,   Link 2 = 2.5,   Link 3 = 1,   Link 4 = 2.5

5-26 Given:

$\alpha_2 = -45°$, $P_{21} = 184.78$ mm, $\delta_2 = -5.28°$

$\alpha_3 = -90°$, $P_{31} = 277.35$ mm, $\delta_3 = -40.47°$

$O_{2x} = 86$ mm $O_{2y} = -132$ mm

$O_{4x} = 104$ mm $O_{4y} = -155$ mm

For left dyad: Calculate: $\beta_2 = -85.24°$ $\beta_3 = -164.47°$

Calculate: $W = 110.88$ mm $\theta = 124.24°$

Calculate: $Z = 46.74$ mm $\phi = 120.34°$

For right dyad: Calculate: $\gamma_2 = -75.25°$ $\gamma_3 = -159.53°$

Calculate: $U = 120.70$ mm $\sigma = 104.35°$

Calculate: $S = 83.29$ mm $\psi = 152.80°$

F

5-33 Given: $\alpha_2 = -25°,$ $P_{21} = 133.20$ mm, $\delta_2 = -12.58°$

$\alpha_3 = -101°,$ $P_{31} = 238.48$ mm, $\delta_3 = -51.64°$

$O_{2x} = -6.2$ mm $O_{2y} = -164.0$ mm

$O_{4x} = 28.0$ mm $O_{4y} = -121.0$ mm

For left dyad: Calculate: $\beta_2 = -53.07°$ $\beta_3 = -94.11°$

Calculate: $W = 128.34$ mm $\theta = 118.85°$

Calculate: $Z = 85.45$ mm $\phi = 37.14°$

For right dyad: Calculate: $\gamma_2 = -77.26°$ $\gamma_3 = -145.66°$

Calculate: $U = 92.80$ mm $\sigma = 119.98°$

Calculate: $S = 83.29$ mm $\psi = 65.66°$

5-35 Given: $\alpha_2 = -29.4°,$ $P_{21} = 99.85$ mm, $\delta_2 = 7.48°$

$\alpha_3 = -2.3°,$ $P_{31} = 188.23$ mm, $\delta_3 = -53.75°$

$O_{2x} = -111.5$ mm $O_{2y} = 183.2$ mm

$O_{4x} = -111.5$ mm $O_{4y} = -38.8$ mm

For left dyad: Calculate: $\beta_2 = 69.98°$ $\beta_3 = 139.91°$

Calculate: $W = 100.06$ mm $\theta = 150.03°$

Calculate: $Z = 306.82$ mm $\phi = -49.64°$

For right dyad: Calculate: $\gamma_2 = -4.95°$ $\gamma_3 = -48.81°$

Calculate: $U = 232.66$ mm $\sigma = 62.27°$

Calculate: $S = 167.17$ mm $\psi = -88.89°$

CHAPTER 6   VELOCITY ANALYSIS

6-4 and **6-5**   See Table S6-1 and the file P07-04*row*.4br.

6-6 and **6-7**   See Table S6-2.

6-8 and **6-9**   See Table S6-3.

6-10 and **6-11**   See Table S6-4.

6-16 $V_A = 12$ in/sec @ 124.3°, $V_B = 11.5$ in/sec @ 180°, $V_C = 5.65$ in/sec @ 153.3°, $\omega_3 = -5.69$ rad/sec.

6-47   Open the file P06-47.4br in program LINKAGES to see this solution.[*]

6-48   Open the file P06-48.4br in program LINKAGES to see this solution.[*]

6-49   Open the file P06-49.4br in program LINKAGES to see this solution.[*]

6-51   Open the file P06-51.4br in program LINKAGES to see this solution.[*]

6-62   Open the file P06-62.4br in program LINKAGES to see this solution.[*]

6-65 $V_A = 94.5$ in/sec, $V_B = 115.2$, $V_{slip} = 162.8$, $V_{trans} = 65.9$, $\omega_3 = -70$ rad/sec.

[*] These files can be found in the PROBLEM SOLUTIONS folder on the DVD included with this text.

F

TABLE S6-1   Solutions for Problems 6-4 to 6-5

Row	ω_3 Open	ω_4 Open	V_P Mag	V_P Ang	ω_3 Crossed	ω_4 Crossed	V_P Mag	V_P Ang
a	−6.0	−4.0	40.8	58.2	−0.66	−2.66	22.0	129.4
c	−12.7	−19.8	273.8	−53.3	−22.7	−15.7	119.1	199.9
e	1.85	−40.8	260.5	−12.1	−23.3	19.3	139.9	42.0
g	76.4	146.8	798.4	92.9	239.0	168.6	1 435.3	153.9
i	−25.3	25.6	103.1	−13.4	56.9	6.0	476.5	70.4
k	−56.2	−94.8	436.0	−77.4	−55.6	−16.9	362.7	79.3
m	18.3	83.0	680.8	149.2	7.73	−57.0	571.3	133.5

TABLE S6-2   Solutions for Problems 6-6 to 6-7

Row	V_A Mag	V_A Ang	ω_3 Open	V_B Mag Open	ω_3 Crossed	V_B Mag Crossed
a	14	135	−2.47	−9.9	2.47	−9.92
c	45	−120	5.42	−41.5	−5.42	−3.54
e	250	135	−8.86	−189.7	8.86	−163.80
g	700	60	−28.80	738.9	28.80	−38.90

TABLE S6-3   Solutions for Problems 6-8 to 6-9

Row	V_A Mag	V_A Ang	ω_3 Open	V_{slip} Open	V_B mag Open	ω_3 Crossed	V_{slip} Crossed	V_B Mag Crossed
a	20.0	120.0	−10.3	33.5	41.2	−3.6	−4.25	14.6
c	240.0	135.0	23.7	73.0	142.5	−14.9	130.5	89.4
e	180.0	−15.0	−2.7	−176.0	5.4	5.7	162.0	11.5

TABLE S6-4   Solutions for Problems 6-10 to 6-11

Row	ω_3 Open	ω_4 Open	ω_3 Crossed	ω_4 Crossed
a	32.6	16.9	−75.2	−59.6
c	10.7	−2.6	−8.2	5.1
e	−158.3	−81.3	−116.8	−193.9
g	−8.9	−40.9	−48.5	−16.5
i	−40.1	47.9	59.6	−28.4

F

TABLE S7-1   Solutions for Problems 7-3 to 7-4

Row	α_3 Open	α_4 Open	A_P Mag	A_P Ang	α_3 Crossed	α_4 Crossed	A_P Mag	A_P Ang
a	26.1	53.3	419	240.4	77.9	50.7	298	−11.3
c	−154.4	−71.6	4 400	238.9	−65.2	−148.0	3 554	100.6
e	331.9	275.6	10 260	264.8	1 287.7	1 344.1	19 340	−65.5
g	−23 510.0	−19 783.0	172 688	191.0	−43 709.0	−47 436.0	273 634	−63.0
i	−344.6	505.3	9 492	−81.1	121.9	−728.0	27 871	150.0
k	−2 693.0	−4 054.0	56 271	220.2	311.0	1 672.1	27 759	−39.1
m	680.8	149.2	35 149	261.5	9 266.1	10 303.0	63 831	103.9

TABLE S7-2   Solutions for Problems 7-5 to 7-6

Row	A_A Mag	A_A Ang	α_3 Open	A_B Mag Open	A_B Ang Open	α_3 Crossed	A_B Mag Crossed	A_B Ang Crossed
a	140	−135	25	124	180	−25	74	180
c	676	153	−29	709	180	29	490	180
e	12 500	45	−447	6 653	0	447	11 095	0
g	70 000	150	−1 136	62 688	180	1 136	58 429	180

TABLE S7-3   Solutions for Problems 7-7 to 7-8

Row	α_3 Open	α_4 Open	A_{slip} Open	α_3 Crossed	α_4 Crossed	A_{slip} Crossed
a	130.5	130.5	−128.5	−9.9	−9.9	19.0
c	−212.9	−212.9	1 078.8	−217.8	−217.8	−728.2
e	896.3	896.3	−1 818.6	595.6	595.6	1 822.6

TABLE S7-4   Solutions for Problem 7-9

Row	α_3 Open	α_4 Open	α_3 Crossed	α_4 Crossed
a	3 191	2 492	−6 648	−5 949
c	314	228	87	147
e	2 171	−6 524	7 781	5 414
g	−22 064	−23 717	−5 529	−29 133
i	−5 697	−3 380	−2 593	−7184

F

CHAPTER 7   ACCELERATION ANALYSIS

7-3 and **7-4**   See Table S7-1 and the file P07-04*row*.4br.

7-5 and **7-6**   See Table S7-2.

7-7 and **7-8**   See Table S7-3.

7-9   See Table S7-4.

7-12   176.9 in/sec^2.

7-21 A_A = 26.26 m/sec^2 @ 211.1°, A_B = 8.328 m/sec^2 @ –13.9°.

7-24 A_A = 16 m/sec^2 @ 237.6°, A_B = 12.01 m/sec^2 @ 207.4°, α_4 = 92 rad/sec^2.

7-28 A_A = 39.38 m/sec^2 @ –129°, A_B = 39.7 m/sec^2 @ –90°.

7-39   Open the file P07-39.4br in program LINKAGES to see this solution.[*]

7-40   Open the file P07-40.4br in program LINKAGES to see this solution.[*]

7-41   Open the file P07-41.4br in program LINKAGES to see this solution.[*]

7-42   Open the file P07-42.4br in program LINKAGES to see this solution.[*]

7-44   Open the file P07-44.4br in program LINKAGES to see this solution.[*]

7-56   Tipover at 19.0 to 20.3 mph; load slides at 16.2 to 19.5 mph.

7-76 A_D = 7 554.1 in/sec^2 @ 150.8°, α_6 = 692.98 rad/sec^2.

7-78 A_A = 677.1 in/sec^2 @ –119.7°, A_B = 1 337.5 in/sec^2 @ –26.09°, A_P = 730.37 in/sec^2
@ –53.65°, α_4 = 431.175 rad/sec^2

7-87 A_C = 37.5 in/sec^2 @ 90°

CHAPTER 8   CAM DESIGN

Most of the problems in this cam chapter are design problems with more than one correct
solution. Use program DYNACAM to check your solution obtained with *Mathcad, Matlab,
Excel,* or *TKSolver* and also to explore various solutions and compare them to find the
best one for the constraints given in each problem.

8-1   See Figure S8-1a.

8-2   See Figure S8-1b.

8-4 ϕ = 4.9°.

8-6 ϕ = 13.8°.

CHAPTER 9   GEAR TRAINS

9-1   Pitch diameter = 4.8, circular pitch = 0.628, addendum = 0.20, dedendum = 0.25, tooth
thickness = 0.314, and clearance = 0.050.

9-5   a. p_d = 10,   b. p_d = 6

9-6   Assume a minimum no. of teeth = 16, then: pinion = 16t and 1.600-in pitch dia. Gear =
112t and 11.240-in pitch dia. Contact ratio = 1.68.

[*] These files can be found
in the PROBLEM SOLUTIONS
folder on the DVD included
with this text.

F

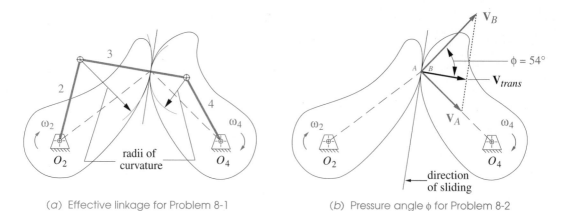

(a) Effective linkage for Problem 8-1   (b) Pressure angle ϕ for Problem 8-2

FIGURE S8-1

Solutions to Problems 8-1 and 8-2

9-7 Assume a minimum no. of teeth = 16, then: pinion = 16t and 3.20-in pitch dia. Gear = 96t and 19.20-in pitch dia. An idler gear of any dia. is needed to get the positive ratio. Contact ratio = 1.67.

9-10 Three stages of 4:1, 4:1, and 5:1 give –80:1. Stage 1 = 20t (d = 1.67 in) to 80t (d = 6.67in). Stage 2 = 20t (d = 1.67 in) to 80t (d = 6.67in). Stage 3 = 18t (d = 1.5 in) to 90t (d = 7.5 in).

9-12 The square root of 120 is > 10 so will need three stages. 5 × 4 × 6 = 150. Using a mini-mum no. of teeth = 18 gives 18:90, 18:72, and 18:108 teeth. Pitch dias. are 3.6, 18 and 21.6 in. An idler (18t) is needed to make the overall ratio positive.

9-14 The factors 4 × 7 = 28. The ratios 24:96 and 15:105 revert to same center distance of 7.5 in. Pitch dias. are 1.875, 3, 12, and 13.125 in.

9-16 The factors 6.5 × 10 = 65. The ratios 22:143 and 15:150 revert to same center distance of 10.3125 in. Pitch dias. are 2.75, 17.875, 1.875, and 18.75.

9-19 The factors 2 × 1.5 = 3. The ratios 15:30 and 18:27 revert to the same center distance of 3.75. Pitch dias. are 2.5, 5, 3, and 4.5. The reverse train uses the same 1:2 first stage as the forward train, so it needs a second stage of 1:2.25 which is obtained with a 12:27 gearset. The center distance of the 12:27 reverse stage is 3.25 which is less than that of the forward stage. This allows the reverse gears to engage through an idler of any suitable diameter to reverse output direction.

9-21 For the low speed of 6:1, the factors 2.333 × 2.571 = 6. The ratios 15:35 and 14:36 revert to the same center distance of 3.125. Pitch dias. are 1.875, 4.375, 1.75, and 4.5. The second speed train uses the same 1:2.333 first stage as the low-speed train, so it needs a second stage of 1:1.5 which is obtained with a 20:30 gearset which reverts to the same center distance of 3.125. The additional pitch dias. are 2.5 and 3.75. The reverse train also uses the same 1:2.333 first stage as both forward trains, so it needs a second stage of 1:1.714 which is obtained with a 14:24 gearset. The center distance of the 14:24 reverse stage is 2.375 which is less than that of the forward stages. This allows the reverse gears to engage through an idler of any suitable diameter to reverse output direction.

9-25 a. ω_2 = 790,   c. ω_{arm} = –4.544,   e. ω_6 = –61.98

TABLE S9-1   Solution to Problem 9-29
Possible Ratios for Two-Stage Compound Gear Train to Give the Ratio 2.718 28

Pinion 1	Gear 1	Ratio 1	Pinion 2	Gear 2	Ratio 2	Train Ratio	Abs Error
25	67	2.68	70	71	1.014	2.718 285 71	5.71E-06
29	57	1.966	47	65	1.383	2.718 268 53	1.15E-05
30	32	1.067	31	79	2.548	2.718 279 57	4.30E-07
30	64	2.133	62	79	1.274	2.718 279 57	4.30E-07
31	48	1.548	45	79	1.756	2.718 279 57	4.30E-07
31	64	2.065	60	79	1.317	2.718 279 57	4.30E-07
31	79	2.548	75	80	1.067	2.718 279 57	4.30E-07
35	67	1.914	50	71	1.420	2.718 285 71	5.71E-06

9-26   a. $\omega_2 = -59$,   c. $\omega_{arm} = 61.54$,   e. $\omega_6 = -63.33$

9-27   a. 577.7 rpm and 4.33 to 1,   b. $x = 577.7 \times 2 - 800 = 355.5$ rpm

9-29   See Table S9-1 for solution. The third row has the smallest error and smallest gears.

9-35 $\eta = 0.963$.

9-37 $\eta = 0.996$.

9-39 $\omega_1 = 979.6$ rpm, $\omega_2 = 2742.9$ rpm.

9-41 $\omega_1 = -293.9$ rpm, $\omega_3 = -587.8$ rpm.

9-43 $\omega_G = -18.6$ rpm, $\omega_F = -187.7$ rpm.

9-67 $\phi = 26.23°$.

9-69   Gear ratio = 2.4 and contact ratio = 1.698. Circular pitch = 0.785, base pitch = 0.738,
pitch dia.= 6.25 and 15, outside dia. = 6.75 and 15.5, center dist. = 10.625, addendum =
0.250, dedendum = 0.313, whole depth = 0.562 5, clearance = 0.063 (all in inches).

9-71   Four stages with factors 6 x 5 x 5 x 5 x 5 = 750: Stage 1 = 14t to 84t. Stages 2, 3, 4 = 14t
to 70t. Output in same direction as input due to even number of stages.

CHAPTER 10   DYNAMICS FUNDAMENTALS

10-1 CG @ 8.77 in from handle end, $I_{zz} = 0.394$ in-lb-sec², $k = 9.35$ in.

10-2 CG @ 8.08 in from handle end, $I_{zz} = 0.221$ in-lb-sec², $k = 8.95$ in.

10-4

a. $x =$	3.547,	$y = 4.8835$,	$z = 1.4308$,	$w = -1.3341$
b. $x = -62.029$,		$y = 0.2353$,	$z = 17.897$,	$w = 24.397$

10-6

a.	In series:	$k_{eff} = 3.09$,	Softer spring dominates
b.	In parallel:	$k_{eff} = 37.4$,	Stiffer spring dominates

F

10-9

    a.   In series: $c_{eff} = 1.09$,   Softer damper dominates
    b.   In parallel: $c_{eff} = 13.7$,   Stiffer damper dominates

10-12 $k_{eff} = 12$ N/mm, $m_{eff} = 0.688$ kg

10-14 $k_{eff} = 225$ N/mm, $m_{eff} = 58.5$ kg

10-20   Effective mass in 1st gear = 0.054 bl, 2nd gear = 0.096 bl, 3rd gear = 0.216 bl, 4th gear = 0.863 bl.

10-21   Effective spring constant at follower = 308.35 lb/in.

10-25   Effective spring constant = 111.1 N/mm, effective mass = 27 kg.

10-26 $x = 5.775$ in.

10-34 I_{crank} about pivot = 1 652 kg-mm^2, I_{rocker} about pivot = 18 420 kg-mm^2, $I_{coupler}$ about CG = 2106 kg-mm^2 (both couplers are the same).

10-35 $x = 774$ mm to strike point of ball.

CHAPTER 11   DYNAMIC FORCE ANALYSIS

11-3   Open file P11-03row.sld in program LINKAGES to check your solution.[*]

11-4   Open file P11-03row.sld in program LINKAGES to check your solution.[*]

11-5   Open file P11-05row.4br in program LINKAGES to check your solution.[*]

11-6   Open file P11-05row.4br in program LINKAGES to check your solution.[*]

11-7   Open file P11-07row.4br in program LINKAGES to check your solution.[*]

11-12 $F_{12x} = -1851$ N, $F_{12y} = 1315$ N; $F_{14x} = 1047$ N, $F_{14y} = -3156$ N; $F_{32x} = 479$ N, $F_{32y} = -275$ N; $F_{43x} = 53.7$ N, $F_{43y} = -1087$ N; $T_{12} = -45.3$ N-m

11-13   Open file P11-13.4br in program LINKAGES to check your solution.[*]

11-14 $F_{12} = 1\ 308$ lb, $F_{32} = 1\ 290$ lb, $F_{43} = 1\ 290$ lb, $F_{14} = 710$ lb, $F_{hand} = 63.2$ lb, $JFI = 0.645$.

11-25 $T_{12} = 463$ lb-in

11-40   Mass moment of inertia needed in flywheel = 11.8 bl-in^2. Many flywheel geometries are possible. Assuming a steel cylinder with a radius of 9.0 in, thickness = 1.474 in.

CHAPTER 12   BALANCING

12-1

    a. $m_b r_b = 0.934$, $\theta_b = -75.5°$

    c. $m_b r_b = 5.932$, $\theta_b = 152.3°$

    e. $m_b r_b = 7.448$, $\theta_b = -80.76°$

[*] These files can be found in the PROBLEM SOLUTIONS folder on the DVD included with this text.

12-5

a. $m_a r_a = 0.814$, $\theta_a = -175.2°$, $m_b r_b = 5.50$, $\theta_b = 152.1°$

c. $m_a r_a = 7.482$, $\theta_a = -154.4°$, $m_b r_b = 7.993$, $\theta_b = 176.3°$

e. $m_a r_a = 6.254$, $\theta_a = -84.5°$, $m_b r_b = 3.671$, $\theta_b = -73.9°$

12-6 $W_a = 3.56$ lb, $\theta_a = 44.44°$, $W_b = 2.13$ lb, $\theta_b = -129.4°$

12-7 $W_a = 4.2$ lb, $\theta_a = -61.8°$, $W_b = 3.11$ lb, $\theta_b = 135°$

12-8 These are the same linkages as in Problem 11-5. Open the file P11-05*row*.4br in program LINKAGES to check your solution.[*] Then use the program to calculate the flywheel data.

12-9 Open the file P12-09.4br in program LINKAGES to check your solution.[*]

12-14 $R_3 = 5.85$ in, $\theta_3 = s-142.11°$, $R_4 = 1.13$ in, $\theta_4 = 120°$

12-16 $W_4 = 14.48$ lb, $\theta_4 = 89.15°$, $W_5 = 5.04$ lb, $\theta_5 = 83.90°$

12-18 $d_3 = 18.95$ mm, $\theta_3 = -147.46°$, $d_4 = 20.8$ mm, $\theta_4 = 28.94°$

12-38 Plane 2: $e = 0.113$, $\theta = -152.15°$. Plane 3: $e = 0.184$, $\theta = 19.36°$.

CHAPTER 13 ENGINE DYNAMICS

13-1 Exact solution $= -42\,679.272$ in/sec @ 299.156° and 200 rad/sec

Fourier series approximation $= -42\,703.631$ in/sec @ 299.156° and 200 rad/sec

Error $= -0.0571\%$ $(-0.000\,571)$

13-3 Gas torque = 2040 (approx.), Gas force = 3142

13-5 Gas torque = 2039.53 (approx.), Gas torque = 2039.91 (exact)

Error $= 0.0186\%$ $(0.000\,186)$

13-7

a. $m_b = 0.007\,48$ at $l_b = 7.2$, $m_p = 0.012\,51$ at $l_p = 4.31$
b. $m_b = 0.008\,00$ at $l_b = 7.2$, $m_a = 0.012\,00$ at $l_a = 4.80$
c. $I_{model} = 0.691\,2$, Error $= 11.48\%$ $(0.114\,8)$

13-9 $m_{2a} = 0.018$ at $r_a = 3.5$, $I_{model} = 0.220\,5$, Error $= -26.5\%$ (-0.265)

13-11 Open the file P13-11.eng in program ENGINE to check your solution.[*]

13-14 Open the file P13-14.eng in program ENGINE to check your solution.[*]

13-19 Open the file P13-19.eng in program ENGINE to check your solution.[*]

CHAPTER 14 MULTICYLINDER ENGINES

Use program ENGINE to check your solutions.[*]

14-23 *mr product* on the balance shafts = 5.017E–3 bl-in or 1.937 lb-in.

[*] These files can be found in the PROBLEM SOLUTIONS folder on the DVD included with this text.

F

TABLE S15-1
Solutions to Problem 15-6

	ω_n	ω_d	c_c
a	3.42	3.38	8.2
b	4.68	4.65	19.7
c	0.26	0.26	15.5
d	2.36	2.33	21.2
e	5.18	5.02	29.0
f	2.04	1.96	49.0

CHAPTER 15   CAM DYNAMICS

15-1 to **15-5** Use program DYNACAM to solve these problems. There is not any *one right answer* to these design problems.

15-6 See Table S15-1.

15-7 to **15-19** Use program DYNACAM to solve these problems. There is not any *one right answer* to these design problems.

G

EQUATIONS FOR UNDER- OR OVERBALANCED MULTICYLINDER ENGINES

G.1 INTRODUCTION

Chapter 14 developed the equations for shaking forces, moments, and torques in multicylinder engines of inline and vee configurations. In Chapter 14, it is assumed that the crank throws are all exactly balanced, an assumption that greatly simplifies the equations. However, some multicylinder engines overbalance the crank throws to reduce main bearing forces. This also can have an effect on shaking forces and moments.

This appendix provides replacement equations for the simplified versions in Chapter 14, and these equations do not assume exactly balanced crank throws.[*] The equation numbers used here correspond to those in Chapter 14 and can be substituted for the simplified ones if desired. In the equations that follow, m_A is the effective crank pin mass and m_B the effective wrist pin mass as defined in Chapter 13. The parameters m_c and r_c represent, respectively, the counterweight mass of any one crank throw and the radius to the counterweight's CG. All other parameters are the same as defined in Chapters 13 and 14.

[*] These complete equations are used in program ENGINE.

For an inline engine (Section 14.3, p. 711) the shaking forces for an engine with an under- or overbalanced crankshaft are:

$$
\begin{aligned}
F_{s_x} \cong \Bigg[& (m_A + m_B) r\omega^2 \left[\cos\omega t \sum_{i=1}^{n} \cos\phi_i + \sin\omega t \sum_{i=1}^{n} \sin\phi_i \right] \\
& + m_c r_c \omega^2 \left[\cos(\omega t + \pi) \sum_{i=1}^{n} \cos\phi_i + \sin(\omega t + \pi) \sum_{i=1}^{n} \sin\phi_i \right] \\
& + \frac{m_B r^2 \omega^2}{l} \left[\cos 2\omega t \sum_{i=1}^{n} \cos 2\phi_i + \sin 2\omega t \sum_{i=1}^{n} \sin 2\phi_i \right] \Bigg] \hat{\mathbf{i}}
\end{aligned}
$$

$$(14.2d)$$

$$
\begin{aligned}
F_{s_y} \cong \Bigg[& m_A r\omega^2 \left[\sin\omega t \sum_{i=1}^{n} \cos\phi_i - \cos\omega t \sum_{i=1}^{n} \sin\phi_i \right] \\
& + m_c r_c \omega^2 \left[\sin(\omega t + \pi) \sum_{i=1}^{n} \cos\phi_i - \cos(\omega t + \pi) \sum_{i=1}^{n} \sin\phi_i \right] \Bigg] \hat{\mathbf{j}}
\end{aligned}
$$

For an inline engine (Section 14.3, p. 711) the shaking moments for an engine with an under- or overbalanced crankshaft are:

$$
\begin{aligned}
M_{s_x} \cong \Bigg[& (m_A + m_B) r\omega^2 \left[\cos\omega t \sum_{i=1}^{n} z_i \cos\phi_i + \sin\omega t \sum_{i=1}^{n} z_i \sin\phi_i \right] \\
& + m_c r_c \omega^2 \left[\cos(\omega t + \pi) \sum_{i=1}^{n} z_i \cos\phi_i + \sin(\omega t + \pi) \sum_{i=1}^{n} z_i \sin\phi_i \right] \\
& + \frac{m_B r^2 \omega^2}{l} \left[\cos 2\omega t \sum_{i=1}^{n} z_i \cos 2\phi_i + \sin 2\omega t \sum_{i=1}^{n} z_i \sin 2\phi_i \right] \Bigg] \hat{\mathbf{i}}
\end{aligned}
$$

$$(14.6b)$$

$$
\begin{aligned}
M_{s_y} \cong \Bigg[& m_A r\omega^2 \left[\sin\omega t \sum_{i=1}^{n} z_i \cos\phi_i - \cos\omega t \sum_{i=1}^{n} z_i \sin\phi_i \right] \\
& + m_c r_c \omega^2 \left[\sin(\omega t + \pi) \sum_{i=1}^{n} z_i \cos\phi_i - \cos(\omega t + \pi) \sum_{i=1}^{n} z_i \sin\phi_i \right] \Bigg] \hat{\mathbf{j}}
\end{aligned}
$$

For a vee or opposed engine (Sections 14.7, p. 727, and 14.8, p. 739) the shaking forces for an engine with an under- or overbalanced crankshaft are:

$$F_{s_x} \cong \left(F_{s_L} + F_{s_R} \right) \cos\gamma + m_A r \omega^2 \left[\cos\omega t \sum_{i=1}^{n} \cos\phi_i + \sin\omega t \sum_{i=1}^{n} \sin\phi_i \right]$$

$$+ m_c r_c \omega^2 \left[\cos(\omega t + \pi) \sum_{i=1}^{n} \cos\phi_i + \sin(\omega t + \pi) \sum_{i=1}^{n} \sin\phi_i \right] \hat{\mathbf{i}}$$

(14.10j)

$$F_{s_y} \cong \left(F_{s_L} - F_{s_R} \right) \sin\gamma + m_A r \omega^2 \left[\sin\omega t \sum_{i=1}^{n} \cos\phi_i - \cos\omega t \sum_{i=1}^{n} \sin\phi_i \right]$$

$$+ m_c r_c \omega^2 \left[\sin(\omega t + \pi) \sum_{i=1}^{n} \cos\phi_i - \cos(\omega t + \pi) \sum_{i=1}^{n} \sin\phi_i \right] \hat{\mathbf{j}}$$

$$\mathbf{F_s} = F_{s_x}\hat{\mathbf{i}} + F_{s_y}\hat{\mathbf{j}}$$

For a vee or opposed engine (Sections 14.7, p. 727, and 14.8, p. 739) the shaking moments for an engine with an under- or overbalanced crankshaft are:

$$M_{s_x} \cong \left(M_{s_L} + M_{s_R} \right) \cos\gamma + m_A r \omega^2 \left[\cos\omega t \sum_{i=1}^{n} z_i \cos\phi_i + \sin\omega t \sum_{i=1}^{n} z_i \sin\phi_i \right]$$

$$+ m_c r_c \omega^2 \left[\cos(\omega t + \pi) \sum_{i=1}^{n} z_i \cos\phi_i + \sin(\omega t + \pi) \sum_{i=1}^{n} z_i \sin\phi_i \right] \hat{\mathbf{i}}$$

(14.11c)

$$M_{s_y} \cong \left(M_{s_L} - M_{s_R} \right) \sin\gamma + m_A r \omega^2 \left[\sin\omega t \sum_{i=1}^{n} z_i \cos\phi_i - \cos\omega t \sum_{i=1}^{n} z_i \sin\phi_i \right]$$

$$+ m_c r_c \omega^2 \left[\sin(\omega t + \pi) \sum_{i=1}^{n} z_i \cos\phi_i - \cos(\omega t + \pi) \sum_{i=1}^{n} z_i \sin\phi_i \right] \hat{\mathbf{j}}$$

$$\mathbf{M_s} = M_{s_x}\hat{\mathbf{i}} + M_{s_y}\hat{\mathbf{j}}$$

Note that inertia torque is unaffected by crankshaft balance condition because, at constant angular velocity, the acceleration vector of the crank pin mass is centripetal and has no moment arm. The moment of inertia added to the crankshaft by any overbalance mass will increase the flywheel effect of the crankshaft and thus reduce its willingness to change rotational speed in transient angular acceleration. But, the size of the engine's physical flywheel can be reduced to compensate for the more massive crankshaft.

G

INDEX

I

I

I

I

I

I

I

DVD CATALOG

KEY TO FILENAME SUFFIXES	
DYNACAM	.CAM
ENGINE	.ENG
FIVEBAR*	.5BR
FOURBAR*	.4BR
LINKAGES	.BAR
MATLAB	.M
MATRIX	.MTX
SIXBAR*	.6BR
SLIDER*	.SLD
TKSOLVER	.TKW
WORKING MODEL	.WM2D, WM3

---

* These files can be opened by program LINKAGES

LINKAGE ATLASES Folder

Contains PDF file of atlases of coupler curves for fourbar and geared fivebar linkages.

Hrones and Nelson Fourbar Atlas

Zhang et al. Geared Fivebar Atlas

PDF PROBLEM WORKBOOK Folder

Contains PDF files of all the figures needed to solve the text's end-of-chapter problems. Each PDF file contains one problem figure and all of the problem statements associated with it. They are grouped in subfolders by chapter and their filenames are the same as the figure number or problem number involved. These files provide the student with a printable workbook of illustrated problems in which graphical problem solutions can be directly worked out or analytical solution results recorded.

PROBLEM SOLUTIONS Folder

Data files that solve selected problems in the text.

Chapter 3 Subfolder
P03-14.4br
P03-22.4br
P03-23.4br
P03-36.4br
P03-42.4br

Chapter 4 Subfolder
P04-21.4br
P04-23.4br
P04-25.4br
P04-26.4br
P04-29.4br
P04-30.4br

Chapter 5 Subfolder
P05-08.4br
P05-11.4br
P05-15.4br
P05-19.4br
P05-26.4br

Chapter 6 Subfolder
P06-47.4br
P06-48.4br
P06-49.4br
P06-51.4br
P06-62.4br

Chapter 7 Subfolder
P07-04a.4br
P07-04c.4br
P07-04e.4br

P07-04g.4br
P07-04i.4br
P07-04k.4br
P07-04m.4br
P07-39.4br
P07-40.4br
P07-41.4br
P07-42.4br
P07-44.4br

Chapter 10 Subfolder
P10-04a.mtr
P10-04b.mtr

Chapter 11 Subfolder
P11-03a.sld
P11-03c.sld
P11-03e.sld
P11-03g.sld
P11-04a.tkw
P11-05a.tkw
P11-05a.4br
P11-05c.4br
P11-05e.4br
P11-05g.4br
P11-06a.tkw
P11-06c.tkw
P11-06e.tkw
P11-06g.tkw
P11-07a.4br
P11-07c.4br
P11-07e.4br
P11-12.4br
P11-13.4br

Chapter 12 Subfolder
P12-09.4br

Chapter 13 Subfolder
P13-11.eng
P13-14.eng
P13-19a.eng
P13-19b.eng

PROGRAM MANUAL Folder

Contains a PDF file of the user manual for programs LINKAGES, DYNACAM, ENGINE, *and* MATRIX.

TKSOLVER FILES Folder

TKSolver model files.

The TKsolver program is needed to run these files and is not included with this text. See www.uts.com.

Gears.tk Subfolder
Compound.tkw
Revert.tkw
Triple.tkw

Linkages.tk Subfolder
3 position FixPivots.tkw
3 position.tkw
Cognate.tkw
Coupler.tkw
DragSlider.tkw
Eq04-02.tkw
Ex11-04.tkw
Figure P05-05.tkw
Fivebar.tkw
Fourbar.tkw
Inverted slider-crank.tkw
SCCA.tkw
Slider_Cmpr.tkw
Slider.tkw
Soni Cognate.tkw
Symmetric.tkw
Transport.tkw
Virtual Work.tkw

Misc.tk Subfolder
CamCalc.tkw
Constrnt.tkw
Cubic.tkw
Cycloid.tkw
F04-18.tkw
Pressang.tkw
SCCA.tkw
Student.tkw

VIRTUAL LABS Folder

Contains data taken from the fourbar and cam machines shown in the Virtual Lab videos. These data can be used to do the lab exercises, also detailed in this folder.

Cam Machine Lab & Data Files

Fourbar Linkage Lab & Data Files

WORKING MODEL FILES Folder

Chapter 2 Subfolder

Working Model 2D Files
02-10b.wm2d - Scotch Yoke
02-12a.wm2d - Geneva
02-12b.wm2d - Ratchet and Pawl
02-12c.wm2d - Linear Geneva
02-13.wm2d - Slider-Crank
02-14abc.wm2d - Stephenson Inversion
02-14de.wm2d - Watt Inversions
02-15.wm2d - Grashof Inversions
02-16.wm2d - Non-Grashof Inversions
02-19b.wm2d - Geared Fivebar
02-20.wm2d - Desk Lamp
P2-01f.wm2d - Overhead Valve
P2-03.wm2d - Front End Loader
P2-04c.wm2d - Radial Engine
P2-04d.wm2d - Walking Beam
P2-04e.wm2d - Drafting Arm
P2-04g.wm2d - Drum Brake

P2-04h.wm2d - Compression Chamber
P2-05a.wm2d - Chebyshev Mechanism
P2-05b.wm2d - Kempe SL Mechanism
P2-07.wm2d - Throttle Mechanism
P2-08.wm2d - Scissors Jack
P2-10.wm2d - Watt's Engine
P2-13.wm2d - Crimping Tool
P2-14.wm2d - Pick and Place
P2-15.wm2d - Power Hacksaw
P2-16.wm2d - Powder Press
P2-18.wm2d - Oil Field Pump

Working Model 3D Files

P2-01h.wm3 - Cylindrical Cam

Chapter 3 Subfolder

Working Model 2D Files

03-04.wm2d - Example 3-1
03-05.wm2d - Example 3-2
03-07b.wm2d - Example 3-4
03-09c.wm2d - Example 3-6
03-11.wm2d - 3-Position Synthesis
03-12b.wm2d - 4br Quick Return
03-13a.wm2d - 6br Quick Return
03-14.wm2d - Quick-Return Shaper
03-14.-*wm2d - Quick-Return Shaper
03-15.wm2d - Coupler Curves
03-17.wm2d - Coupler Curve Atlas
03-17a.wm2d - Coupler Curve Atlas
03-18.wm2d - Camera Film Advance
03-18-*.wm2d - Camera Film Advance
03-19a.wm2d - Auto Suspensions
03-19a-*.wm2d - Auto Suspensions
03-24a.wm2d - Roberts Diagram
03-25a.wm2d - Roberts Diagram
03-25b.wm2d - Roberts Diagram
03-26.wm2d - Chebyschev Cognates
03-26a.wm2d - Roberts Diagram
03-26b.wm2d - Chebyschev Cognates
03-26b-*.wm2d - Chebyschev Cognates
03-27c.wm2d - Curvilinear Trans.
03-27d.wm2d - Curvilinear Trans.
03-28.wm2d - GFBM 4br Cognate
03-28-*.wm2d - GFBM Cognates (alt.)
03-29.wm2d - Straight-Line Linkages
03-29a.wm2d - Watt Straight-Line
03-29b.wm2d - Watt's Engine
03-29c.wm2d - Roberts Straight-Line
03-29d.wm2d - Chebyschev SL
03-29e.wm2d - Hoeken Straight-Line
03-29f.wm2d - Evans Straight-Line
03-29g.wm2d - Peaucellier Strt-Line
03-31c.wm2d - Single-Dwell—Rocker
03-31d.wm2d - Single-Dwell—Slider
03-32.wm2d - Double-Dwell Linkage
03-34.wm2d - 180° Rocker Output
03-35.wm2d - Washing Machine
03-36.wm2d - 360° Rocker Output
P3-03.wm2d - Treadle Wheel
P3-07.wm2d - Walking Beam
P3-08.wm2d - Loom Laybar Drive

Chapter 4 Subfolder

Working Model 2D Files

04-16.wm2d - Double Rocker Toggle
P4-01.wm2d - Fourbar Analysis
P4-02.wm2d - Slider-Crank Analysis
P4-03.wm2d - Inverted Slider-Crank
P4-05c.wm2d - Radial Engine
P4-05d.wm2d - Walking Beam
P4-05e.wm2d - Drafting Machine
P4-05g.wm2d - Drum Brake
P4-05h.wm2d - Compression Chamber
P4-06.wm2d - Pick and Place
P4-07.wm2d - Power Hacksaw
P4-09.wm2d - Walking Beam Conveyor
P4-11.wm2d - Loom Laybar Drive
P4-14.wm2d - Treadle Wheel
P4-18.wm2d - Elliptical Trammel

Chapter 6 Subfolder

Working Model 2D Files

06-05c.wm2d - Instant Centers
06-10b.wm2d - Instant Centers
06-11.wm2d - Rock Crusher
06-12.wm2d - Suspension
06-14a.wm2d - Centrodes 1
06-14b.wm2d - Centrodes 2
06-14c.wm2d - Centrodes 3
06-14d.wm2d - Centrodes 4
06-15a.wm2d - Centrodes 5
06-15b.wm2d - Centrodes 6
06-17a.wm2d - Cycloidal Motion
P6-01.wm2d - Fourbar Analysis
P6-02.wm2d - Slider-Crank Analysis
P6-03.wm2d - Inverted Slider-Crank
P6-08c.wm2d - Radial Engine
P6-08d.wm2d - Walking Beam
P6-08e.wm2d - Drafting Machine
P6-08g.wm2d - Drum Brake
P6-08h.wm2d - Compression Chamber
P6-15.wm2d - Power Hacksaw
P6-16.wm2d - Pick and Place
P6-18.wm2d - Powder Press
P6-19.wm2d - Walking Beam Conveyor
P6-21.wm2d - Toggle Pliers
P6-23.wm2d - Surface Grinder
P6-29.wm2d - Drum Pedal
P6-30.wm2d - Oil Field Pump
P6-32.wm2d - Elliptical Trammel

Working Model 3D Files

06-12.wm3 - Bump Steering

Chapter 7 Subfolder

Working Model 2D Files

P7-01.wm2d - Fourbar Analysis
P7-02.wm2d - Slider-Crank Analysis
P7-03.wm2d - Inverted Slider-Crank
P7-08c.wm2d - Radial Engine
P7-08d.wm2d - Walking Beam

P7-08e.wm2d - Drafting Machine
P7-08g.wm2d - Drum Brake
P7-08h.wm2d - Compress Chamber
P7-15.wm2d - Power Hacksaw
P7-16.wm2d - Pick and Place
P7-19.wm2d - Walking Beam
P7-20.wm2d - Surface Grinder
P7-24.wm2d - Drum Pedal

Chapter 8 Subfolder

Working Model 2D Files

08-02a.wm2d - Translating Follower
08-02b.wm2d - Oscillating Follower
08-03a.wm2d - Roller Follower
08-03c.wm2d - Flat-Faced Follower
08-39.wm2d - Cam and Follower
08-48.wm2d - Radii of Curvature
E8-02.wm2d - Example 8-2
E8-03.wm2d - Example 8-3
E8-04.wm2d - Example 8-4
E8-07.wm2d - Example 8-7

Working Model 3D Files

08-03a.wm3 - Roller Follower
08-04.wm3 - Cylindrical Cam

Chapter 9 Subfolder

Working Model 2D Files

09-01b.wm2d - Internal Gearset
09-04.wm2d - External Gearset
09-05.wm2d - Involute Curves
09-06.wm2d - Tooth Engagement
09-19.wm2d - Rack and Pinion
09-28.wm2d - Compound Gear Train
09-33.wm2d - Planetary Gearset

Working Model 3D Files

09-16.wm3 - Helical-Parallel Gears
09-17.wm3 - Helical-Crossed Gears
09-18.wm3 - Worm and Worm Gear
09-21.wm3 - Bevel Gears
09-30.wm3 - Gear Trains
09-34.wm3 - Planetary Gearset
09-44a.wm3 - Transmission - High
09-44b.wm3 - Transmission - Low
09-44c.wm3 - Transmission - Reverse
09-51.wm3 - Drive Train
P9-02.wm3 - Compound Epicyclic
P9-03_open.wm3 - Differential
P9-03_locked.wm3 - Differential

Chapter 10 Subfolder

Working Model 2D Files

10-11a.wm2d - Valve Train

Chapter 13 Subfolder

Working Model 2D Files

13-01.wm2d - Vee-Eight Engine

THE
END

SYSTEM REQUIREMENTS

Windows® operating systems

- Pentium-based or better processors
- Windows® 2000/NT/XP/Vista/Windows7
- 128 megabytes of RAM
- 60 megabytes of free hard disk space
- DVD drive

INSTALLATION INSTRUCTIONS FOR DESIGN OF MACHINERY PROGRAMS AND FILES

1. In Windows 2000/NT/XP/Vista/Windows7, you MUST be LOGGED ON as ADMINISTRATOR with FULL PRIVILEGES to install these programs! If you do not have full operating system privileges, the installation will fail.
2. Quit all applications currently running on your Windows computer.
3. Insert the DVD into your computer's DVD drive.
4. Open the Custom Programs folder on the DVD and then open the folder for the program you want to install.
5. Double-click on the Programname_Install.exe file within that program's folder.
6. Follow the instructions on the screen, taking all defaults.
7. It will create a Program Group called "Design of Machinery" and install each program in that group. This will place all of the programs conveniently under one item on your Windows Start/Programs menu.
8. As the installation proceeds, you may see messages asking if you want to replace a newer file in your system with an older version from the DVD. Always answer YES to KEEP any NEWER files that are already on your computer.
9. When installation is complete, the program name will appear on the Start Menu under "Design of Machinery."
10. The programs can be uninstalled with the Uninstall link in the Start Menu or with the Windows "Add/Remove Programs" utility in the Control Panel. Some programs can be updated from the Start Menu as well.
11. If running Vista, read the "Instructions for Running Design of Machinery Programs in Vista" on the DVD.

A list of the contents of the DVD is provided on p. 855 and on the DVD. Specific instructions and video tutorials on the operation of each program are on the DVD and in help files within the programs. The data files for each program can be copied from the DVD to your hard drive with standard Windows copying techniques. See the Windows platforms system hardware requirements above.

INSTALLATION INSTRUCTIONS FOR THE WORKING MODEL PROGRAM

Windows® platforms only

1. The first owner of this textbook may purchase a student version of Working Model at a substantial discount from the normal price. The student version is fully functional and will run for six months from the date of purchase.
2. Using a web browser, connect to the web site: http://www.design-simulation.com/4871970E2B0671F43D1A
3. On the web page, enter the requested word from the specified book page and the serial number found on the DVD label. Then click the Validate button.
4. If the word and the serial number are valid, you will be taken to the Working Model on-line store where you can complete the purchase. Visa, Mastercard, American Express, and Paypal are accepted. You will need to create an account using a valid email address and provide your billing address.
5. After the purchase is complete, you will receive an email message with a hyperlink that will download the Working Model 2D installer. Click on that link to download the installer.
6. Once the installer download has completed, run it and follow the prompts.
7. The first time you start Working Model it must be activated. This is a 1-2 minute process that takes place over the internet. Once the product is activated, it is ready to be used.
8. You can deactivate Working Model from the computer it is on and install and activate it on another computer.